TEXT BOOK
OF
INORGANIC CHEMISTRY

TEXT BOOK
OF
INORGANIC CHEMISTRY

By
Amit Arora

D P H

DISCOVERY PUBLISHING HOUSE
NEW DELHI-110002

First Published – 2005

Reprinted – 2017

ISBN: 978-81-8356-013-9

Text Book of Inorganic Chemistry

Published by:

DISCOVERY PUBLISHING HOUSE PVT. LTD.
4383/4B, Ansari Road Darya Ganj
New Delhi - 110 002 (India)
Phone: +91-11-23279245, 43596064-65
Fax: +91-11-23253475
E-mail: discoverypublishinghouse@gmail.com
sales@discoverypublishinggroup.com
web: www.discoverypublishinggroup.com

Printed at:
Infinity Imaging Systems
Delhi

Preface

The present title "Inorganic Chemistry" has been designed for undergraduate and postgraduate of all Indian Universities. The aim of this book is to provide a concise modern text of inorganic chemistry which is large enough to cover the essentials, yet short enough to be interesting. It provides a simple and logical theoretical framework into which the reader should be able to fit his factus knowledge.

There has been considerable interest in organo-metallic compounds, some of which are manufactured on a large scale. There has also been great interest in the role of inorganic materials in biological systems (chlorophyll, hemoglobin, vitamin B_{12} and nitrogen fixation) and a public awareness of the toxicity of various materials, most notably lead and mercury.

All material is presented at a solid pedagogical level, with close attention given to achieving a clear presentation that students can understand and hopefully enjoy.

The author tried hard to be accurate and upto date in statement and realises the impossibility of completely avoiding errors therefore, the author will greatly appreciate having his attention called to any questionable statement.

The author is deeply grateful to or his expertise and advice during the preparation of the present title.

The author expresses his gratitude Mr. Wasan and staff of M/s Discovery Publishing House for their whole hearted co-operation in the publication of this book.

—Author

Preface

Contents

1

Structure of the Atom

Atoms Today

Atoms are the fundamental building blocks that make up all matter. All atoms of the same element are essentially the same. Atoms of different elements are different.

Scientists now know that there are three major kinds of particles that make up atoms: protons, neutrons, and electrons. Protons and neutrons are located at the center of an atom, the part of the atom called the **nucleus**. The electrons are in motion at some distance from the nucleus. Protons, neutrons, electrons, and still other particles that are smaller than atoms are called **subatomic particles.** See Figure.

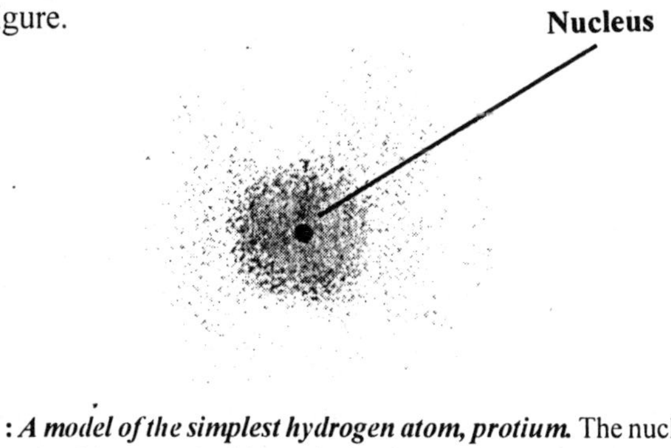

Fig 1.1 : ***A model of the simplest hydrogen atom, protium.*** The nucleus is located at the center of the atom. In this particular atom, a single electron moves about the nucleus in the space shown by the colour blue. Where the blue colour is densest there is the greatest likelihood of finding the electron at any one instant. More will be said this model later in the chapter

Mass and electrical charge are characteristics that differentiate one subatomic particle from another. A **proton** has one unit of positive charge, and an **electron** one unit of negative charge. **Neutrons** have no charge—they are electrically neutral. A unit of positive charge neutralizes a unit of negative charge. So a particle consisting of one proton and one electron; would have no net charge.

The masses of a proton and a neutron are very nearly equal (they are identical to three significant figures). But the mass of each particle is about 1840 times greater than the mass of an electron. Because the mass of an electron is so small in comparison, the total mass of the electrons in an atom is considered to be a negligible part of the atom's mass. That means that just about all the mass of an atom is in its nucleus. See Figure

Characteristics of Subatomic Particles

	Approximate mass	**Charge**	**Location in atom**
Electron	$\frac{1}{1837}$ u *	1–	Normally at relatively large distances from the nucleus
Proton	1u	1+	Part of the nucleus
Neutron	1u	No	Part of the nucleus

* The letter *u* is an abbreviation for *atomic mass unit.*

Fig. The three particles making up atoms, their approximate masses, their charges, and their locations within the atom.

Historical Background

Since the beginning of history, people have held differing views on the nature of matter. One view is that a solid body can be divided and subdivided into smaller and smaller pieces without limit. This view is known as the **continuous theory of matter.** But the Greek philosophers Democritus and Leucippus, who lived more than 2000 years ago, had another idea. They suggested that matter is made up of particles so small and indestructible that they cannot be divided into anything smaller. This view is known as the **discontinuous theory of matter.** The philosophers called such indestructible particles *atomos,* the Greek word for "indivisible."

Calculation of Net Charge

Particle	Number of particles		Charge per particle		Total charge
proton	15	×	1+	=	15+
neutron	16	×	0	=	0
electron	18	×	1–	=	18–
				Net Charge	3–

Fig. *The net charge on the particle described in Sample Problem 1 is 3–.*

Dalton's Atomic Theory

John Dalton, an English chemist, first stated the law of multiple proportions in 1803. He had observed this law at work in compounds of carbon and hydrogen, and in the oxides of nitrogen (compounds of nitrogen and oxygen).

Dalton saw that this law, together with the law of conservation of mass and the law of definite proportions, gave strong support to the idea of atoms. He reasoned that only matter composed of indivisible, fundamental particles could form compounds that illustrate the law of multiple proportions. In order to account for these laws, Dalton, proposed a theory of matter based on the existence of atoms.

Dalton's atomic theory includes four basic ideas:

1. All elements are composed of atoms, which are indivisible and indestructible particles.
2. All atoms of the same element are exactly alike; in particular, they all have the same mass.
3. Atoms of different elements are different; in particular, they have different masses.
4. Compounds are formed by the joining of atoms of two or more elements. In any compound, the atoms of the different elements in the compound are joined in a definite whole-number ratio, such as 1 to 1, 2, to 1, 3, to 2, etc.

Dalton's, theory accounted very well for the three laws that have been mentioned.

Law of conservation of mass. If atoms have definite masses and cannot be divided or destroyed, then a chemical change is simply a rearrangement of atoms. Their total mass should be the same before and after the chemical reaction.

Law of definite proportions. If atoms of different elements combine in a definite ratio whenever they form a certain compound the ratio of the masses in the compound should also be fixed. It will depend only on the masses of the different atoms and the ratio in which they combine.

Two Compounds of Elements A and B	
Compound 1	**Compound 2**
(a) 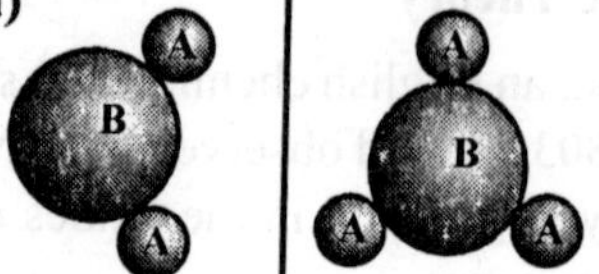	
(b) 2000 atoms A combine with 1000 atoms B	3000 atoms A combine with 1000 atoms B
(c) $\frac{2000\ A}{1000\ B} = \frac{2\ A}{1B}$	$\frac{3000\ A}{1000\ B}$ $\frac{3\ A}{1B}$
(d) $\dfrac{\frac{2\ A}{1B} \text{ from Compound 1}}{\frac{3\ A}{1B} \text{ from Compound 2}} = \frac{2}{3}$ a simple whole-number ratio	

Fig. 1.2 : A comparison of two compounds of the same elements shows that the laws of multiple proportions is consistent with Dalton's atomic theory.

Law of multiple proportions. According to Dalton's theory, atoms of different elements always combine in whole-number ratios. For example, in one compound of elements A and B, the ratio of A atoms to B atoms may be 2:1. In another compound, the ratio may be 3:1 See Figure 1.2. In the first compound, there will be 2000 A atoms combined with 1000 B atoms. In the second compound, there will be 3000 A atoms combined with 1000 B atoms. See Figure 6-6(b). The ratio of A atoms that combine with a fixed number of B atoms in these two compounds is therefore 2:3. See Figure 6-6(c) and (d). Because all A atoms have the same mass, the ratio of *masses* of A that combine with a fixed *mass* of B will also be 2:3. Thus the theory accounts for the law of multiple proportions.

Updating the Atomic Theory

Since Dalton's time, scientists have learned a great deal about the structure of matter. They no longer think of atoms as solid, indestructible, particles. Atoms have many parts and a complex organization. However, the essential ideas of Dalton's theory are still useful today. Elements *are* made of atoms. Compounds *do* form by the joining of atoms in fixed whole-number ratios. And atoms are *not* permanently changed by chemical reactions. They are only rearranged into different combinations.

There are three major differences between the modern atomic theory and Dalton's theory:

1. In the modern theory, atoms are not indivisible. They are made up of smaller particles—electrons, protons, and neutrons. The arrangement of electrons within an atom is temporarily altered when a chemical change is taking place.
2. Atoms can be changed from one element to another, but not by chemical reactions. Such changes occur only during certain nuclear reactions. There are the subject of Chapter 26.
3. Atoms of the same element are not all exactly alike. They are alike in those characteristics that determine the chemical properties of an element. But atoms of the same element can and do have different masses.

You may wonder how the laws of definite proportion and multiple proportions can hold true if the atoms of an element can have different masses. Wouldn't the mass ratios into the combination? The answer lies in the fact that in every natural sample of an element, the atoms of various masses are always present in virtually identical proportions. As a result, the atoms of the element have a certain *average* mass, which is always very nearly the same for that element. In terms of mass ratios in compounds, each element seems to consist of atoms that all have the same mass—the average mass for that element. We will refer again to this fact when we come to discuss the subject to atomic mass later in this chapter.

Subatomic particles of matter. The idea that atoms are indivisible persisted for many years after the postulations of Dalton, and it has been only in the last fifty years or so that evidence has accumulated which indicates that atoms may actually be decomposed and resolved into subatomic particles much smaller than any known atom. Atoms of all of the elements are now regarded as composed of three subatomic particles of matter:

1. *Electrons:* units of matter carrying negative charges of electricity.
2. *Protons:* units of matter carrying positive charges.
3. *Neutrons:* neutral units of matter of about the same mass as the proton.

The electron. The electron was the first of the subatomic particles to be discovered. Knowledge of the electron began with a study of the discharge of electricity through gases at reduced pressures. For this study, a so-called cathode ray tube (Figure 1.3) may be used. It consists of a tube of glass in which are sealed two metal discs, and a glass side arm through which the tube may be evacuated by means of a pump. The metal discs acts as electrodes between which a discharge of electricity will take place when the electrodes are connected to a source of high potential such as an induction coil. The nature of the discharge, then, is studied as the pressure is reduced. At atmospheric pressure an irregular violet coloured discharge takes place. As the pressure is reduced, the discharge becomes pink, and finally the whole tube glows. At very low pressures, streaks of white light seem to emerge at right angles from the negative pole or cathode and to cause the appearance of a greenish fluorescence on the glass walls where the rays strike. These rays were originally referred to as *cathode rays* and at first were thought to be light rays. An obstacle placed between the cathode and the glass wall casts a sharp shadow on the wall opposite the cathode, thus indicating that the rays travel in straight lines. A pinwheel placed in the path of the rays revolves, or if the rays are allowed to strike a piece of metal foil, the foil becomes hot. These observations suggested that the rays were material in nature. In 1879 Sir William Crookes showed that the rays were bent in an are by an electromagnetic field. Subsequently, Sir J. J. Thomson showed that the rays were deflected toward a positively charged plate in an electric field and therefore must possess a negative charge. These negatively charged particles issuing from the cathode came to be called *electrons.* By observing the extent or magnitude of the electron's deflection in magnetic and electric fields. Thomson determined that the mass of a single electron is about $\frac{1}{1837}$ that of the hydrogen atom. Since electrons can be generated by using any cathode or any gas in the tube it would seem that electrons are constituents of atoms of all elements".

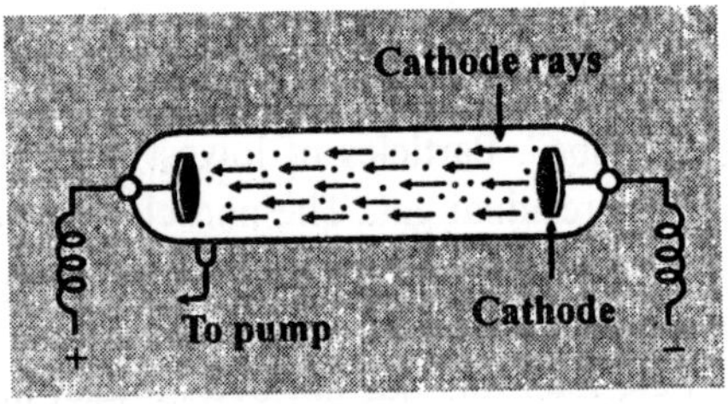

Fig. 1.3 : Cathode ray tube.

The proton. After electrons were found to be generated in a cathode ray tube, it was natural to ask whether or not positively charged particles were emanating from the region of the positive pole or anode of the tube at the same time that electrons were being generated at the cathode. In 1886 Goldstein found that if the cathode is perforated with a cylindrical hole, a beam of rays passes through the cathode into a region behind the cathode and away from the anode. These rays were called *positive rays or canal rays.* "When these rays in magnetic and electric fields were studied in much the same manner as the cathode rays were investigated, it was shown that they possess a unit positive charge or an integral multiple of this charge, and are composed of particles of the residual gas used in the tube". Particles of lowest mass were found when hydrogen was the residual gas. The particles consist of hydrogen atoms with a unit positive charge; in other words, each particle is a hydrogen atom which has lost one electron. Since a smaller positively charged particle than this has not been found, there is strong evidence that these particles are fundamental building units in the structure of atoms of all of the elements. These subatomic particles are called *protons.*

The neutron. In 1932 Chadwick observed that neutral particles of a mass approximately that of the proton were produced when beryllium was exposed to the radiations from a radioactive substance. It was at first thought that each particle consisted of one proton and one electron; however, these neutral units of matter, which are called *neutrons,* are now regarded as a new and fundamentally distinct kind of particle.

The Rutherford Model of the Atom

The next major advance in understanding the nature of the atom resulted from the work of the distinguished British scientist, Ernest Rutherford. In 1909, Rutherford and his assistants did a famous experiment that proved that atoms are not like the solid spheres proposed by Dalton. The experiment had developed from Rutherford's interest in *alpha particles.* Rutherford had earlier identified **alpha particles** as being positively charged helium atoms. Alpha particles are one of the types of particles released when uranium and some other radioactive elements disintegrate.

In his experiment, Rutherford used a very thin sheet of gold as a target. On one side of the foil was a lead box containing a radioactive substance. A small hole in the box permitted a narrow stream of alpha particles to shoot out. These particles were directed at right angles to the surface of the foil. Surrounding the foil was a screen coated with zinc sulfide. Each time an alpha particle hit this zinc sulfide coating, a flash of light was produced at the point of contact. By observing these flashes of light, it was possible to see whether the alpha particles that passed through the foil had been deflected from their straight-line path.

Nearly all the alpha particles passed straight through as though there were no foil there. Rutherford interpreted this observation to mean that the atoms of the foil consisted mostly of *empty space!* In addition to the many particles that passed straight through the foil, Rutherford also noticed that a few particles were deflected at large angles of 90° or more. See Figure.

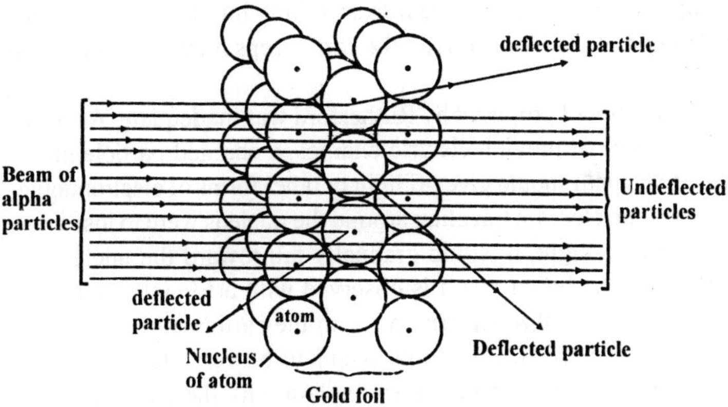

Fig. 1.4 : In Rutherford's experiment, most of the alpha particles passed straight through the gold foil as though it were not there. A few particles, were deflected from their straight-line path, some by 90° or more.

Rutherford interpreted these observations to mean that each atom contained a small, dense, positively charged central portion, or nucleus. The positively charged gold nuclei repelled the positively charged alpha particles deflecting the alpha particles that came near, much as a steel helmet, would deflect, a rifle bullet. However, because the nuclei were so small, not many "hits" were scored. Most of the alpha particles passed straight through the "empty" portion of the atoms.

Rutherford repeated this experiment using other metal foils. Each time, he got similar results. He concluded that atoms in general consists of a small, massive, positively charged central portion (the nucleus) surrounded mainly by " empty" space. In this model, electrons moved about in the empty space that made up most of the atom's volume. The negative charge of the electrons offset the positive charge of the nucleus, thereby accounting for the atom as a whole being electrically neutral.

Shortcomings of Rutherford's Model

As you have been seen, Rutherford's model depicted the atom as having a positively charged nucleus were one or more negatively charged electrons of very small mass. Because opposite charges attract, the electrons should be strongly attracted toward the nucleus by electric forces. Rutherford assumed that the motion of the electrons around the nucleus keeps them from falling into the nucleus, just as the motion of the moon around the earth keeps it from falling into the earth.

The model proposed by Rutherford was inadequate. According to the theory of moving electric charges, a charged particle moving in a curved path should give off light or other forms of electromagnetic energy. If an electron traveling around the nucleus were to lose energy by giving off light, it would gradually fall toward the nucleus. See Figure 1.5. A similar result is observed with artificial satellites. As they lose energy through friction with the earth's atmosphere, they drop to lower orbits. But atoms seems to be very stable structures. An adequate atomic model must explain why the electrons do not give off energy and collapse into the nucleus.

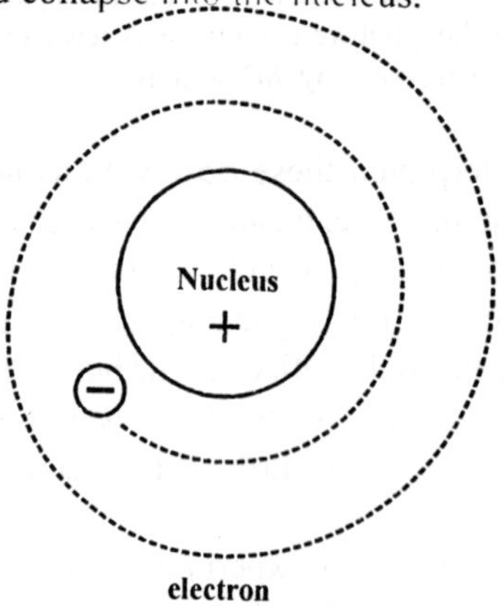

Fig. 1.5 : A charged particle traveling in a curved path radiates energy. If applied to electrons orbiting an atom, then the electrons would continually give off energy. Eventually they would slow down and spiral in toward the nucleus, causing the atom of collapse.

The Bohr atom

In 1913 Neil Bohr, a Danish physicist, considered the dilemma of the noncollapsing atom. He said that there must be some as yet unknown physical principles which describe electrons in atoms. Although Bohr's theory was shown to have serious flaws, Bohr was brave enough to question classical physics, and his work encouraged others to find out why classical physics fails for small particles and to develop the new physics, quantum mechanics.

Bohr started out by assuming that when a substance is heated and emits light, it must do so because its atoms have absorbed energy from the flame or electric discharge. Bohr then suggested that it is actually the electrons which absorb energy and then re-emit it as light. But, Bohr asked, why is the radiation limited to certain specific wavelengths, each of which produces a line in the element's spectrum? With what must have been a flash of inspiration he saw that there could be only one explanation for the discrete wavelengths. He reasoned that these would result if *the energy of an electron in an atom is quantized.* This means that an electrons in an atom is not free to have just any amount of energy, but only *certain, specific, discrete amounts of energy.*

What is the connection between the energy of an electron and the wavelength of the light emitted from an atom? Why did Bohr see that quantized electronic energy levels account for line spectra? At the beginning of the twentieth century Max Planck and Albert Einstein, using entirely different approaches, demonstrated that all electromagnetic radiation behaves as if it were composed of little energy packets, now called *photons.* They showed that each photon has an *energy* which is proportional to the *frequency* of the light:

$$E_{photon} = h\nu \qquad (1)$$

where the proportionality constant, h, is called *Planck's constant* (and has the value 6.63×10^{-34} J s). We have already seen that frequency, wavelength, and speed of light are related by the equation

$$\nu\lambda = c \qquad \text{or} \qquad \nu = \frac{c}{\lambda}$$

so we can write Eq. (1) s

$$E_{photon} = \frac{hc}{\lambda} \tag{2}$$

This equation shows that if electromagnetic radiation has a single, discrete wavelength, then the energy of its photons must have a single value (since h and c are both constants). Bohr pointed out that if an electron in an atom drops from one quantized energy level to another, it should be expected to emit light of a single energy and therefore of a single wavelength.

Bohr described the origin of line spectra as follows: Each atom of an element has a set of available quantized energies, or *energy levels,* for its electrons. Normally the atom is in its *ground state;* that is, all the electrons are in the lowest available levels. When the atom absorbs energy from a flame or electric discharge, its electrons are raised into higher energy levels. The atom is now said to be in an *excited state.* But now there are unoccupied lower energy levels and so each electron drops down from a higher quantized level E_2 to a lower one, E_1. For a given electron the *difference* between these energies is the amount of energy lost by the electron and equals the energy of the photon of electromagnetic energy (light) which is emitted. In other words,

$$E_2 - E_1 = E_{photon} = \frac{hc}{\lambda} \tag{3}$$

Thus a discrete spectral line of wavelength λ is accounted for by the quantized electronic energies E_2 and E_1. In a given atom many such transitions from higher to lower level are possible. Each contributes toward a single discrete line in the spectrum of that element. (Of course, many such transitions in many atoms are necessary to produce a line which is visible.)

Bohr Theory of Hydrogen Atom

The Rutherford nuclear atom provides the theoretical basis for understanding much of chemistry. Moseley's atomic number, which is characteristic of atomic species, represents (indirectly) the number of electrons in the extranuclear region of a neutral atom. The number and distribution of extranuclear electrons are of interest because the electron configurations dictate the physical and chemical characteristics of the atoms. Our present description of extranuclear electronic distributions arises from considering the source of the lines observed in the emission spectra of the elements. The original interpretation of emission spectra using quantum theory was given by Bohr. Although the Bohr theory of atomic structure is no longer significant in modern chemistry, it does represent an important point in the development of chemical theory. As early as 1811, Helmholtz suggested that electrical forces play an important role in the formation of molecules. The Bohr theory focused attention on the extranuclear region of atoms with far-reaching consequences.

The key experimental result on the which the Bohr theory focuses in the fact that the emission spectrum of each element consists of distinctive groups of lines characteristic of the element. The lines within a group usually exhibit a relatively simple relationship. For example, the general regularity observed for the hydrogen spectrum in the visible— ultraviolet region (the Balmer series, shown in Fig. 1.6) is simply expressed by the empirical relationship

$$\bar{\nu} = R\left(\frac{1}{n'^2} - \frac{1}{n^2}\right) \quad (4)$$

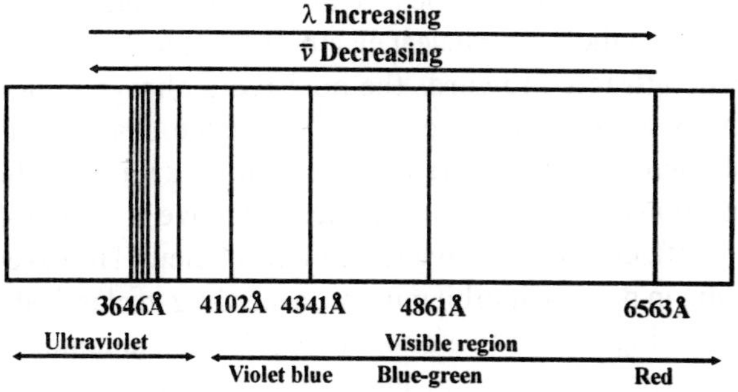

Fig. 1.6 : The Balmer series of lines in the spectrum of atomic hydrogen. The series limit occurs at 3646 Å.

where $\bar{\nu}$ is the position of a line in wave numbers ($\bar{\nu} = 1/\lambda$), R is an empirical constant determined to be 109737,309 cm, n' is an integer, and n an integer dependent upon n' as follows :

$$n = (n' + 1), (n' + 2), (n' + 3),.... \quad (5)$$

Four series of lines exist in the hydrogen spectrum (Fig. 1.7), each with characteristics values of n' and n.

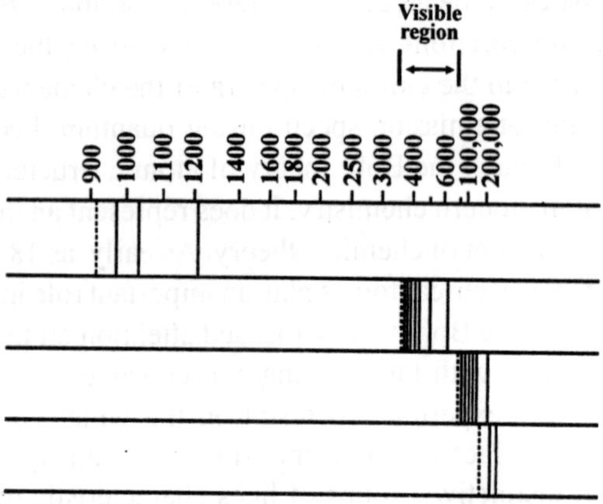

Fig. 1.7 : The four series of spectral lines for hydrogen atoms. The series have been separated clarity, and the dotted lines represent the series limit.

The fact that discrete lines are observed in the emission spectra of the elements indicates that processes occur within the atom which involve discrete amounts of energy. Moreover, these processes involve different amounts of energy for different atoms. It should be clear from the picture of the nuclear atom which is developed in the previous section that the extra-nuclear electrons (one, in the case of hydrogen) will be attracted towards the positively charged nucleus. If the electrons are stationary the system is unstable, since the electrons will be drawn into the nucleus. This simple argument leads to the suggestion that the extranuclear electron move in an orbit-like motion which offsets the attractive force of the nucleus. However, a particle moving in nonlinear orbit must be, according to classical mechanics, accelerating. Under these conditions the fundamentals of classical electrodynamics lead to an impossible situation, since an accelerating electric charge emits electromagnetic radiation. If the motion of the electron about the nucleus is contained to a circular path, electrodynamic principles dictate that energy must be radiated from the system. Thus, the electron should eventually lose all its energy in the form of radiation and fall into the nucleus. If the constituent particles which comprise atoms obey classical laws, all atoms should be in various states of losing energy and the radiation emitted from a collection of atoms should contain all possible frequencies. Under such conditions atomic spectra should exhibit continua, a conclusion not reflected in experimental fact (Fig. 1.8) Bohr suggested an atomic model which incorporated quantum arguments to account for the line spectra of the elements.

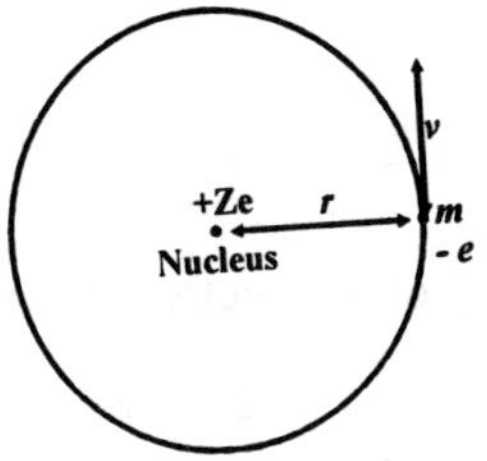

Fig. 1.8 : In the Bohr model of the hydrogen atom, an electron with charge $-e$ and mass m is constrained to move about a nucleus with charge $+Ze$. At any moment the electron has a velocity v, vector of which is tangential to the orbital path.

The Bohr model of the hydrogen atom assumes that an electron of mass m and charge—e moves with a velocity v about a nucleus with a charge of $+ Ze$ and is constrained by the charge on the nucleus to a circular orbit (Fig.).

In the formalim of the Bohr theory the following postulates are made:

1. Atomic systems possess a number of states (called stationary states) for which no emission of radiant energy occurs even if the classical laws of electrodynamics predict such to be the case because of the motion of charged particles relative to each other.
2. Emission or absorption of radiant energy corresponds a transition between stationary states.
3. The equilibrium of the system in a stationary state is governed by Newton's laws of motion.
4. The angular momentum of an electron in the lowest stationary state is equal to $h/2\pi$; in any higher state the angular momentum is an integral number of times this basic unit $(nh/2\pi)$.

The first postulate eliminates the embarrassing conclusions which would follow if *some* of the classical laws were obeyed, the second postulate relates the model of spectroscopic observations; the third postulate provides a means of calculating certain parameters of the system; and the fourth postulate explicitly introduces, the principal quantum number n which takes on integral values. The direct application (I) of these postulates to the model of the hydrogen atom shown in Fig. leads to expressions for the radii of the allowed or Bohr orbits:

$$r = \frac{n^2}{z}\frac{h^2}{4\pi^2 e^2 m} \quad (6)$$

and for the energy of an electron in a stationary state :

$$E_n = -\frac{1}{n^2}\frac{z^2 2\pi^2 e^4 m}{h^2} \quad (7)$$

Equations (6) and (7) indicate that the size of the Bohr orbits, as well as the energies of the stationary states, are quantized.

In the Bohr view of atomic structure, the spectrum of hydrogen arises because hydrogen atoms are excited to different stationary states. When each atom returns to the ground state a discrete amount of energy is emitted which is dependent upon the difference of the two states. According to Planck the frequency of electromagnetic radiation is related to its energy by the relationship

$$E = h\nu. \quad (8)$$

The energy difference $h\nu$ between two states, which corresponds to a line in the spectrum, is given by

$$h\nu = E_n - E_n'. \quad (9)$$

Equations (7) and (9) lead to Eq. (10), which predicts the frequency dependence of the series of lines found in the spectrum of hydrogen shown in Fig.

$$\nu = \frac{z^2 2\pi^2 m e^4}{h^2}\left(\frac{1}{n'^2} - \frac{1}{n^2}\right). \quad (10)$$

The energies of the stationary states and the assignment of the transitions for some of the lines in the hydrogen spectrum are given in Fig. The Lyman, Balmer, Paschen and Brackett series of lines shown in Fig. arise from transitions from higher energy levels with principal quantum numbers 1, 2, 3, and 4, respectively.

The Bohr theory not only successfully predicts the gross features of the hydrogen spectrum shown in Fig. but it also provides the value of the Rydberg constant *R* in Eq. (9). The relationship between the frequency v and wave number $\bar{v}$ is given by

$$\bar{v} = \frac{1}{\lambda} = \frac{v}{c},$$

which permits Eq. (16) to be converted into

$$\bar{v} = \frac{z^2 2\pi^2 me^4}{h^3 c}\left(\frac{1}{n'^2} - \frac{1}{n^2}\right). \tag{11}$$

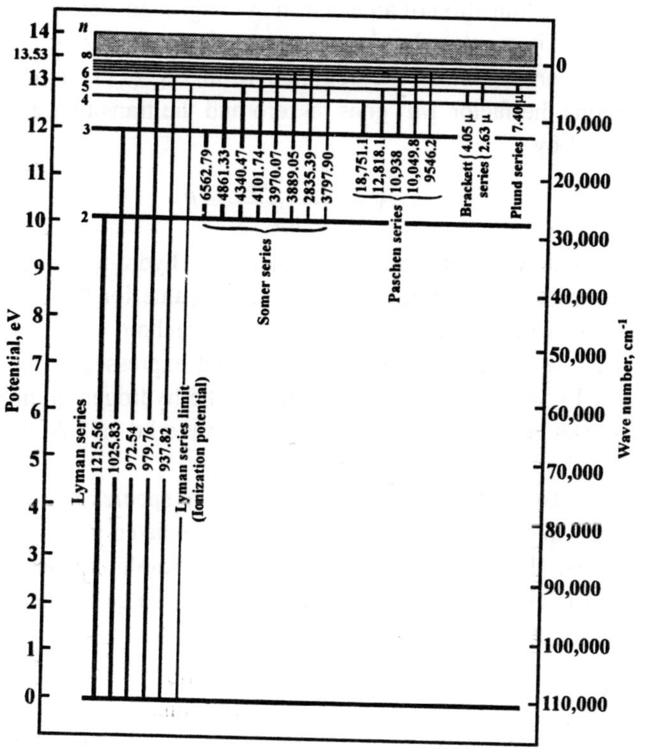

Fig. 1.9 : The energy levels or the hydrogen atom as described by the Bohr theory. The wavelengths of individual lines in each series are given in angstrom units, except in the cases of the Brackett and Pfund series, where they are expressed in microns.

A comparison of Eq. (11) with Eq. (4) shows that they are of the same form and that the Rydberg constant R can be identified with the factor ($z^2 2\pi 2me^4/h^3c$); evaluation of this factor using the best values of fundamental constants gives 109,700 cm^{-1}, which is in remarkable agreement with the experimental value.

The Bohr theory also successfully predicts other characteristics of the hydrogen atom. For example, the energy required to remove an electron from a hydrogen atom, as in Eq. (12):

$$H \rightarrow H^+ + e^- \qquad (12)$$

(i.e., the atom's ionization potential) corresponds in the Bohr theory to the transition of an electron in the ground state ($n = 1$) to a state where it is not under the influence of the nucleus ($n = \infty$).This transition energy can be calculated from Eq. (11); under these conditions the factor $1/n^2$ goes to zero and the transition energy is expressed by

$$\bar{\nu} = \frac{z^2 2\pi^2 me^4}{h^3 c} = R. \qquad (13)$$

In other words the ionization potential of hydrogen corresponds to the Rydberg constant; the experimental value for the ionization potential as determined from a variety of techniques is 13.6 eV, whereas the value of the Rydberg constant expressed in the same units is 13.54 eV. Finally, the radius of the hydrogen atom in the ground state can be estimated from Eq. (16) to be 0.53 Å, which is of the order expected for atomic species.

The Bohr theory is applicable to any species which has a hydrogen-like structure, i.e., one electron associated with a nucleus. Thus, Eqs. (16), (17), and (10)-(11) are applicable to species such as He^+, Li^{2+}, Be^{3+}, etc. by using the appropriate value for the nuclear charge in the these systems.

Bohr Sommerfeld Model of Atom

(Modification of the Bohr Theory). The Bohr theory was outstandingly successful in describing hydrogen-like atoms, but it failed to account for the spectra of atomic systems containing more than one electron. In addition, other spectroscopic observations could not be accounted for by the simple Bohr theory. For example, certain lines in the hydrogen spectrum observed under high resolution consists of groups of closely spaced lines, and some lines in the spectrum split to several components when the emitting source is placed in a strong magnetic field. The appearance of several lines under certain conditions implies the existence of several energy levels with very similar energies.

In 1916 Sommerfeld suggested a modification of the Bohr theory to account for the multiplicity of lines observed under high resolution. Sommerfeld's extension provided for elliptical orbits, with the atomic nucleus at a focus, as well as for circular Bohr orbits. The shape of the elliptical orbit is governed by another quantum number, designated by l. The angular momentum of the electron in an elliptical orbit is related to l in much the same way as its angular momentum in a circular orbit is defined by Bohr's fourth postulate:

$$mvr = \frac{h}{2\pi}(l+1)^{1/2}. \tag{14}$$

The quantum condition described by Eq. (14) restricts the electron to certain allowed ellipses, and the quantum number l defines the ellipticity of the orbit. In the Sommerfeld modification, the values of l, sometimes called the angular or secondary quantum number, are determined by the value of the principal quantum number n:

$$l = 0, 1, 2, 3, 4, ..., (n-1). \tag{15}$$

In effect, the number of orbits of nearly equal energy is specified by the value of n; that is for $n = 1$, one Bohr-Sommerfeld orbit is possible with $l = 0$; for $n = 2$, two Bohr Sommerfeld orbits are possible ($l = 0, 1$), etc.

The splitting of certain spectral lines by a magnetic field, called the Zeeman effect (after its discoverer), suggests that energy levels are split when the emitting atoms are placed in a magnetic field. This phenomenon was interpreted in terms of a magnetic quantum number m which defined the orientation of the angular-momentum vector of an electron in a Bohr-Sommerfeld orbit with the external magnetic field. The fact that spectral lines are split into a finite number of *discrete* lines indicates that only a finite number of orientations are possible; i.e., the interaction of the angular momentum with the magnetic field is *quantized*. The values of the magnetic quantum number are integral and depend upon the value of l:

$$m = l, (l-1), (l-2), \ldots, 0, -1, -2, \ldots, -(l-1), -l. \qquad (16)$$

There are a total of $2l + 1$ possible values of m.

Finally, a fourth quantum number was suggested to account for a doubling of certain spectral lines, a phenomenon which could not be interpreted in terms of the previous quantum numbers. In 1925 Uhlenbeck and Goudsmit proposed that an electron spinning on an axis as it moved in a Bohr-Sommerfeld orbit would have two possible orientations as shown in Fig. The electron moving in an orbit exhibits an orbital angular momentum which is represented by a vector perpendicular to the plane of the orbit (l in Fig.). A spinning electron **generates** spin angular momentum, and the total energy of the system depends upon the orientation of the electron spin with respect to its orbital motion. There are two possible orientations of the electron spin which lead to two values of the spin quantum number s. *viz.* $+\frac{1}{2}$ or $-\frac{1}{2}$ in terms of $(h/2\pi)$- units of angular momentum.

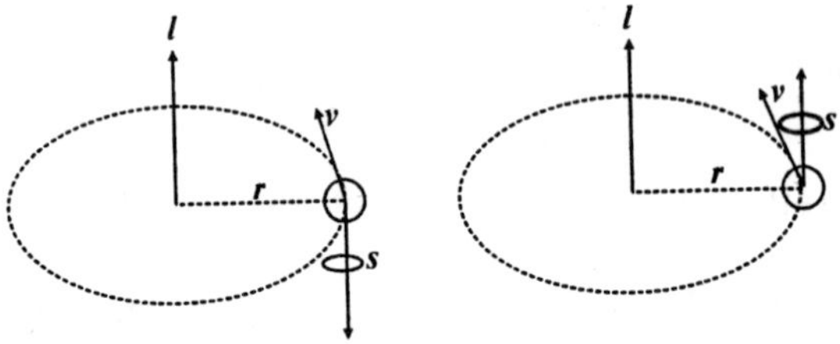

Fig. 1.10 : The electron-spin momentum s can be oriented in one of two ways with respect to the orbital momentum l.

The Bohr postulates lead to a resolution of the difficulties created by Rutherford's nuclear atom by breaking with classical mechanics and permitting atomic systems to possess only certain allowed—or quantized—energies. The quantum numbers in the Bohr-Sommerfeld model were devised to describe the energy of such systems. Unfortunately other irreconcilable problems existed with the Bohr-Sommerfeld model for atomic structure.

Table : The four main quantum numbers

	Symbol	Values
Principal quantum number	n	1, 2, 3 ...
Azimuthal or subsidiary quantum number	l	0, 1, ... (n – 1)
Magnetic quantum number	m	–1, ... 0, ... +1
Spin quantum number	s	$\pm \frac{1}{2}$

The Dual Nature of Electrons-Particles or Waves

The planetary theory of atomic structure put forward by Rutherford and Bohr describes the atom as a central nucleus surrounded by electrons in certain orbits. The electrons is thus considered as a particle. In the 1920s it was shown that moving particles such as electrons behaved in some ways as waves. This is an important concept in explaining the electronic structure of atoms.

For some time light has been considered as either particles or waves. Certain materials such as potassium emit electrons when irradiated with visible light, or in some cases with ultraviolet light. This is called the photoelectric effect. It is explained by light travelling as particles called photons. If a photon collides with an electron, it can transfer its energy to the electron. If the energy of the photon is sufficiently large it can remove the electron from the surface of the metal. However, the phenomena of diffraction and interference of light can only be explained by assuming that light behaves as waves. In 1924, de Brogie postulated that the same dual character existed with electrons—sometimes they are considered as particles, and at other times it is more convenient to consider them as waves. Experimental evidence for the wave nature of electrons was obtained when diffrac-tion rings were observed photographically when a stream of electrons was passed through a thin metal foil. Electron diffraction has now become a useful tool in determining molecular structure, particularly of gases. Wave mechanics is a means of studying the build-up of electron shell in atoms, and the shape of orbitals occupied by the electrons.

Heisenberg Uncertainty Principle

The Bohr theory of the hydrogen atom presupposed that the electron could be regarded as a discrete particle whose position and velocity could be measured simultaneously and with high accuracy. However, for something as small as an electron this is impossible; suppose we could devise a method of determining the position of an electron by directing short wavelength radiation onto it, then the effect on the radiation would allow us to determine the electron's position, but the energy associated with a photon of radiation is comparable with the energy of an electron and the velocity of the electron would change, i.e. the velocity or momentum (product of the mass and velocity) of the electron would now be 'uncertain'. Heisenberg (1927) was the first to realize this difficulty, which he expressed in his famous 'Uncertainty Principle'.

If the position of an electron is known with great precision its velocity (and hence its momentum) is known with far less accuracy, and likewise if the velocity (and hence the momentum) of an electron is known with great precision its position becomes uncertain.

This principle is often expressed as:

$$(\Delta x)(\Delta mv) \simeq 2h$$

where Δx is the uncertainty in position and Δmv is the uncertainty in momentum and h is Planck's constant (6.625×10^{-34} J-sec). The product of these two uncertainties is incredibly small and need not concern us in our dealings with macroscopic systems, but for minute objects like electrons this relationship is of great importance. A direct consequence of the Uncertainty Principle is that precise statements about the position and velocity of an electron must be replaced by less precise ones. The best we can do is to state the probable position and velocity of a particular electron.

Three years before Heisenberg put forward his Uncertainty Principle, the French physicist de Broglie had predicted that waves should be associated with moving electrons, and furthermore, that the wavelength associated with electrons could be calculated from the equation :

$$\lambda = h/mv$$

where h is Planck's constant, m is the mass of the electron, and v is the velocity of the electron. This prediction was verified by Davisson and Germer (1927) when they demonstrated that an electron beam was diffracted by a nickel crystal. Since diffraction phenomena can only be explained on a wave theory, we must conclude that electrons have wavelike characteristics associated with them under certain conditions. Are electrons particulate in nature or is their behaviour best interpreted in terms of waves? Certainly the experiments carried out to determine the mass and the charge carried by an electron are explicable in terms of a particulate nature but diffraction experiments need a wave theory. Little is gained by arguing for or against either of these theories and it must be accepted that electrons, like light ,show a dual behaviour. Some experiments are, therefore, best interpreted by assuming electrons to be discrete particles, while others are more in accordance with a wave theory.

Quantum numbers. The electrons in an atom group themselves into successive shells (or energy levels) in which there are 2 electrons in the first shell, 8 in the second, 18 in the third, 32 in the fourth, and so on. However, spectral studies indicate that the electrons in a given shell do not move in exactly the same orbits or energy levels. The electrons in the different orbits are best described by sets of numbers known as *quantum numbers,* of which there are four:

1. The *principal quantum number, **designated by*** **n,** corresponds to the shell in which the electron moves. Instead of describing an electron by the numerical value of its principal quantum number—for example, 1, 2, 3, etc.—it is sometimes the practice to employ the capital letters *K, L, M.* Thus a *K* electron is an electron having a principal quantum number of unity (1) for an *L* electron the value of n is 2, and so on.

2. The *azimuthal quantum number,* designated by l, gives a measure of the angular momentum of an electron in its orbital motion about the nucleus. The introduction of this term indicates that the main shells of electrons may be considered as being made up of one or more *subshells.* These subshells of electrons possess *nearly* the same energy as one another, but not exactly so. The term l may have values of 0 to $(n - 1)$. Hence for the first main shell where $n = 1$, l can have only a value of 0. For the first shell, then, the subshells and main shell would coincide. When $n = 2$, l can have values of 0 and 1. This would correspond to two subshells within the second main shell. When $n = 3$, l may have the values 0, 1, 2, corresponding to 3 subshells.

Table 1.11 : Distribution of Electrons of the Elements

Element	Symbol	Atomic Number	Number of Electrons in various energy levels K	L	M	N	O	P	Q
Hydrogen	H	1	1						
Helium	He	2	2						
Lithium	Li	3	2	1					
Beryllium	Be	4	2	2					
Boron	B	5	2	3					
Carbon	C	6	2	4					
Nitrogen	N	7	2	5					
Oxygen	O	8	2	6					
Fluorine	F	9	2	7					
Neon	Ne	10	2	8					
Sodium	Na	11	2	8	1				
Magnesium	Mg	12	2	8	2				
Aluminum	Al	13	2	8	3				
Silicon	Si	14	2	8	4				
Phosphorus	P	15	2	8	5				
Sulfur	S	16	2	8	6				
Chlorine	Cl	17	2	8	7				
Argon	Ar	18	2	8	8				
Potassium	K	19	2	8	8	1			
Calcium	Ca	20	2	8	8	2			
Scandium	Sc	21	2	8	9	2			
Titanium	Ti	22	2	8	10	2			
Vanadium	V	23	2	8	11	2			
Chromium	Cr	24	2	8	13	1			
Manganese	Mn	25	2	8	13	2			
Iron	Fe	26	2	8	14	2			
Cobalt	Co	27	2	8	15	2			
Nickel	Ni	28	2	8	16	2			
Copper	Cu	29	2	8	18	1			
Zinc	Zn	30	2	8	18	2			
Gallium	Ga	31	2	8	18	3			
Germanium	Ge	32	2	8	18	4			
Arsenic	As	33	2	8	18	5			
Selenium	Se	34	2	8	18	6			
Bromine	Br	35	2	8	18	7			
Krypton	Kr	36	2	8	18	8			
Rubidium	Rb	37	2	8	18	8	1		
Strontium	Sr	38	2	8	18	8	2		

Yttrium	Y	39	2	8	18	9	2	
Zirconium	Zr	40	2	8	18	10	2	
Niobium	Nb	41	2	8	18	12	1	
Molybdenum	Mo	42	2	8	18	13	1	
Technetium	Tc	43	2	8	18	14	1	
Ruthenium	Ru	44	2	8	18	15	1	
Rhodium	Rh	45	2	8	18	16	1	
Palladium	Pd	46	2	8	18	18	0	
Silver	Ag	47	2	8	18	18	1	
Cadmium	Cd	48	2	8	18	18	2	
Indium	In	49	2	8	18	18	3	
Tin	Sn	50	2	8	18	18	4	
Antimony	Sb	51	2	8	18	18	5	
Tellurium	Te	52	2	8	18	18	6	
Iodine	I	53	2	8	18	18		
Xenon	Xe	54	2	8	18	18	8	
Cesium	Cs	55	2	8	18	18	8	1
Barium	Ba	56	2	8	18	18	8	2
Lanthanum	La	57	2	2	18	18	9	2
Cerium	Ce	58	2	8	18	20	8	2
Praseodynium	Pr	59	2	8	18	21	8	2
Neodymium	Nd	60	2	8	18	22	8	2
Promethium	Pm	61	2	8	18	23	8	2
Samarium	Sm	62	2	8	18	24	8	2
Europium	Eu	63	2	8	18	25	8	2
Gadolinium	Gd	64	2	8	18	25	9	2
Terbium	Tb	65	2	8	18	27	8	2
Dysprosium	Dy	66	2	8	18	28	8	2
Holmium	Ho	67	2	8	18	29	8	2
Erbium	Er	68	2	8	18	30	8	2
Thulium	Tm	69	2	8	18	31	8	2
Ytterbium	Yb	70	2	8	18	32	8	2
Lutetium	Hf	71	2	8	18	32	9	2
Hafmium	Hf	72	2	8	18	32	10	2
Tantalum	Ta	73	2	8	18	32	11	2
Tungsten	Ways	74	2	8	18	32	12	2
Rhenium	Re	75	2	8	18	32	13	2
Osmium	Os	76	2	8	18	32	14	2
Iridium	Ir	77	2	8	18	32	15	2
Platinum	Pt	78	2	8	18	32	17	1
Gold	Au	79	2	8	18	32	18	1
Mercury	Hg	80	2	8	18	32	18	2
Thallium	Tl	81	2	8	18	32	18	3

Lead	Pb	82	2	8	18	32	18	4	
Bismuth	Bi	83	2	8	18	32	18	5	
Polonium	Po	84	2	8	18	32	18	6	
Astatine	At	85	2	8	18	32	18	7	
Radon	Rn	86	2	8	18	32	18		
Francium	Fr	87	2	8	18	32	18	8	1
Radium	Ra	88	2	8	18	32	18	8	2
Actinium	Ac	89	2	8	18	32	18	9	2
Thorium	Th	90	2	8	18	32	18	10	2
Proactinium	Pa	91	2	8	18	32	20	9	2
Uranium	U	92	2	8	18	32	21	9	2
Neptunium	Np	93	2	8	18	32	22	9	2
Plutonium	Pu	94	2	8	18	32	23	9	2
Americium	Am	95	2	8	18	32	24	9	2
Curium	Cm	96	2	8	18	32	25	9	2
Berkelium	Bk	97	2	8	18	32	26	9	2
Californium	Cf	98	2	8	18	32	27	9	2
Einsteinium	E	99	2	8	18	32	28	9	2
Fermium	Fm	100	2	8	18	32	29	9	2
Mendelevium	Mv	101	2	8	18	32	30	9	2
Nobelium	No	102	2	8	18	32	31	9	2

3. *Magnetic quantum number,* designated by m, accounts for the behaviour of the electrons in a magnetic field, m may have values from $-l$ to $+l$, including 0.

Hence for the first main shell, where $n = 1$, $l = 0$, and m can only a value of 0. In the second main shell, where $n = 2$, l can have the values 0 and 1. When $l = 0$, $m = 0$, and when $l = 1$, m may be -1, 0, or $+1$.

4. *Spin quantum number,* designated by s, takes into account the spin of an electron about its own axis in a clockwise or counterclockwise direction. The direction of the spin is designated by + or –.

If we follow the above set of rules for quantum numbers, we may easily calculate the number of electrons in the various subshells of the atoms:

For the first shell where $n = 1$, $l = 0$, and $m = 0$, $s = +$ or $-$. Since, according to Pauli's exclusion principle, no more than one electron can have the same values from the four quantum numbers, the first shell can have only two electrons.

$$n = 1, l = 0, m = 0, s = +$$
$$n = 1, l = 0, m = 0, s = -$$

For the second main shell:

$$n = 2, l = 0, m = 0, s = +$$
$$s = -$$

2 electrons

$$n = 2, l = 1, m = -1, s = +$$
$$s = -$$
$$m = 0, s = +$$
$$s = -$$
$$m = +1, s = +$$
$$s = -$$

6 electrons

The second shell, then, consists of two subshells, the first of which holds two electrons and the second six electrons, to make a total of eight. Following the same computation, it can shown that the third main shell consists of three subshells, the first with 2 electrons, the second with 6 electrons, and the third with 10 electrons to make a total of 18. The fourth main shell would have 4 subshells holding 2, 6, 10, and 14 electrons respectively, to make a total of 32.

The various subshell are frequently designated by letters, *s* for the 2 electron subshell, *p* for the 6, *d* for the 10, and *f* for the 14. Thus the maximum electrons in the various subshells could be described as

2 *s* electrons
6 *p* electrons
10 *d* electrons
14 *f* electrons

The maximum number of electrons in the various subshells of the main energy levels may be shown in diagrammatic fashion:

Nucleus

The above notations are convenient in describing the electronic configuration of an atom. For example, neon, with 10 electrons, may be designated as $1s^2$, $2s^2$, $2p^6$, which indicates 2 *s* electrons only in the first shell, 2 *s* electrons and 6 *p* electrons in the second shell. The electron orbits of iron, which has 26 electrons, are designated:

$$1s^2,\ 2s^2,\ 2p^6,\ 3s^2,\ 3p^6,\ 3d^6,\ 4s^2$$

Atomic numbers. It may be noted that, as we proceed from one element to another, the number of protons in the atom increases by one in each instance. Thus the element lithium has three protons, beryllium has four protons, boron has five protons, and so on. If we arrange the elements according to their increasing atomic weights, we may assign numbers to the chemical elements, starting with hydrogen as one. These numbers will represent the number of protons present in the atom. The number, called the *atomic number*. Since the number of electrons in a neutral atom must equal the number of protons, the atomic number is equal to the number of electrons in the atom.

Mass numbers. The mass of an tom is made up primarily of the protons and neutrons present in the nucleus; the contribution of electrons to the mass is so slight that it may ordinarily be neglected. If the mass of each proton and neutron is regarded as unity (this is approximately so), the mass of the atom will be an integral or whole number equal to the sum of the numbers of protons and neutrons present. This integral number is called the *mass number*. The constituents of the nucleus, both protons and neutrons, are often referred to as *nucleons*. Hence the mass number of an atom is the number of its nucleons.

The Schrodinger Wave Equation

For a standing wave (such as a vibrating string) of wavelength λ, whose amplitude at any point along x may be described by a function f(x), it can be shown that :

$$\frac{d^2 f(x)}{dx^2} = -\frac{4\pi^2}{\lambda_2} f(x)$$

If an electron is considered as a wave which moves in only one dimension then:

$$\frac{d^2\psi}{dx^2} = -\frac{4\pi^2}{\lambda^2}\psi$$

An electron may move in three directions *x, y* and *z* so this becomes

$$\frac{\partial^2\psi}{\partial x^2} + \frac{\partial^2\psi}{\partial y^2} + \frac{\partial^2\psi}{\partial z^2} = -\frac{4\pi^2}{\lambda_2}\psi$$

Using the symbol Δ instead of the three partial differentials, this is shortened to

$$\nabla^2\psi = -\frac{4\pi^2}{\lambda^2}\psi$$

The de Broglie relationship states that

$$\lambda = \frac{h}{mv}$$

(where h is Planck's constant, *m* is the mass of an electron and *v* its velocity); hence:

$$\nabla^2\psi = -\frac{4\pi^2 m^2 v^2}{h^2}\psi$$

or

$$\nabla^2\psi + \frac{4\pi^2 m^2 v^2}{h^2}\psi = 0 \qquad \text{(i)}$$

However, the total energy of the system *E* is made up of the kinetic energy *K* plus the potential energy *V*

$$E = K + V$$

so

$$K = E - V$$

But the kinetic energy $= \frac{1}{2}mv^2$ so

$$\frac{1}{2}mv^2 = E - V$$

and

$$v^2 = \frac{2}{m}(E - V)$$

Substituting for v^2 in equation (i) gives the well-known form of the Schrodinger equation

$$\nabla^2\psi + \frac{8\pi^2 m}{h^2}(E - V)\psi = 0$$

Acceptable solutions to the wave equation, that is solutions which are physically possible, must have certain properties:

1. ψ must be continuous.
2. ψ must be finite.
3. ψ must be single valued.
4. The probability of finding the electron over all the space from plus infinity to minus infinity must be equal to one.

The probability of finding an electron at a point *x, y, z,* is ψ^2, so

$$\int_{-\infty}^{+\infty} \psi^2 \mathrm{dx\ dy\ dz} = 1$$

Several wave functions called ψ_1, ψ_2, ψ_3, will satisfy these conditions to the wave equation, and each of these has a corresponding energy E_1, E_2, E_3.....Each of these wave functions ψ_1, ψ_2, etc. is called an *orbital,* by analogy with the *orbits* in the Bohr theory. In a hydrogen atom, the single electron normally occupies the lowest of the energy levels E_1. This is called the ground state. The corresponding wave function ψ_1 describes the orbital, that is the volume in space where there is a high probability of finding the electron.

For a given type of atom, there are a number of solutions to the wave equation which are acceptable, and each orbital may be described uniquely by a set of three quantum numbers, *n, l* and *m.* (These are the same quantum numbers—principal, subsidiary and magnetic—as were used in the Bohr theory).

The subsidiary quantum number *l* describes the shape of the orbital occupied by the electron. *l* may have values 0, 1, 2, or 3. When $l = 0$, the orbital is spherical and is called an *s* orbital; when $l = 1$, the orbital is dumb-bell shaped and is called a *p* orbital; when $l = 2$, the orbital is double dumb-bell shaped and is called a *d* orbital; when $l = 3$ a more complicated *f* orbital is formed. The letters *s, p, d* and *f* come from the spectroscopic terms *s*harp, *p*rincipal, *d*iffuse and *f*undamental, which were used to describe the lines in the atomic spectra.

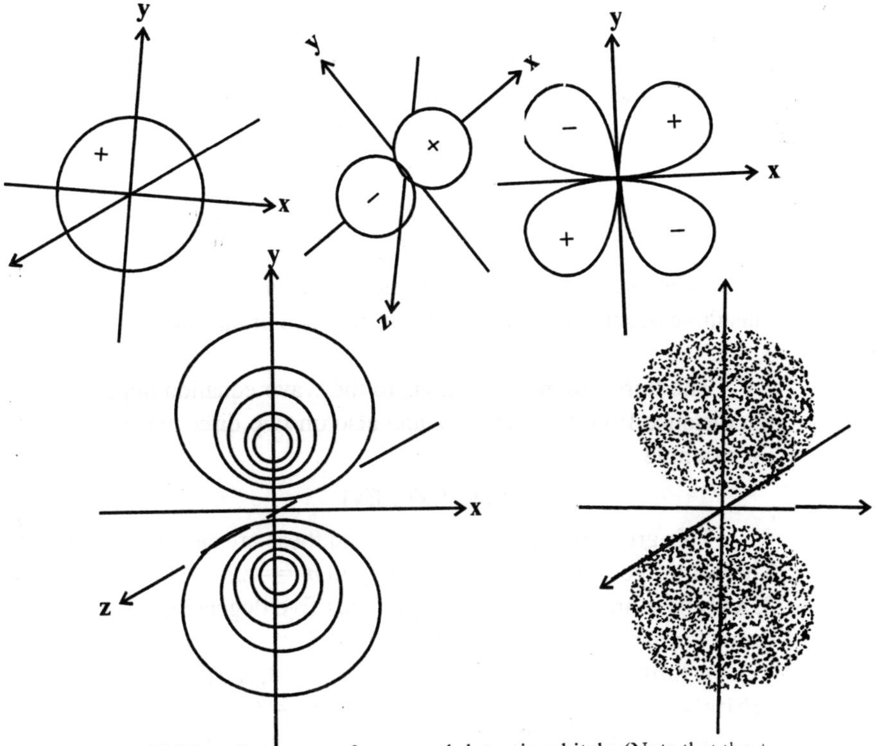

Fig. 1.11 : (a) Wave functions ψ for *s*, *p* and *d* atomic orbitals. (Note that the + and – signs refer to symmetry, not charge.) (b) Different way of representing ψ^2 for a 2*p* orbital (as a contour diagram or as a 90% boundary surface).

Examination of a list of all the allowed solutions to the wave equation shows that the orbitals fall into groups.

In the first group of solutions the value of the wave function ψ, and hence the probability of finding the electron ψ^2 depends only on the distance *r* from the nucleus, and is the same in all directions.

$$\psi = f(r)$$

This leads to a spherical orbital, and occurs when the subsidiary quantum number *l* is zero. These are called *s* orbitals. When $l = 0$, the magnetic quantum number $m = 0$, so there is only one such orbital for each value of *n*.

In the second group of solutions to the wave equation, ψ depends both on the distance from the nucleus, and on the direction in space (*x*, *y* or *z*). Orbitals of this kind occur when the subsidiary quantum number $l = 1$. These are called *p* orbitals and there are three possible values of the magnetic quantum number ($m = -1, 0$ and $+1$). There are therefore three orbitals which are identical in energy, shape and size, which differ only in their direction in space. These three solutions to the wave equation may be written

$$\psi_x = f(r) \, . \, f(x)$$
$$\psi_y = f(r) \, . \, f(y)$$
$$\psi_z = f(r) \, . \, f(z)$$

Orbitals that are identical in energy are termed degenerate, and thus three degenerate *p* orbitals occur for each of the values of $n = 2, 3, 4...$

The third group of solutions to the wave equation depend on the distance from the nucleus *r* and also on two direction in space, for example

$$\psi = f(r) \, . \, f(x) \, . \, f(y)$$

This group of orbitals has $l = 2$, and these are called *d* orbitals. There are five solutions corresponding to $m = -2, -1, 0, +1$ and $+2$, and these are all equal in energy. Thus five degenerate *d* orbitals occur for each of the values of $n = 3, 4, 5....$

A further set of solutions occurs when $l = 3$, and these are called *f* orbitals. There are seven values of *m:* – 3, – 2, – 1, 0, + 1, +2, and +3, and seven degenerate *f* orbitals are formed when $n = 4, 5, 6...$

Table 1.12 : Atomic orbitals

Principal quantum number *n*	Subsidiary quantum number *l*	Magnetic quantum numbers m	Symbol
1	0	0	1*s* (one orbital)
2	0	0	2*s* (one orbital)
2	1	– 1, 0, + 1	2*p* (three orbitals)
3	0	0	3*s* (one orbital)
3	1	– 1, 0, + 1,	3*p* (three orbitals)
3	2	– 2, – 1, 0, + 1, +2,	3*d* (five orbitals)
4	0	0	4*s* (one orbital)
4	1	– 1, 0, + 1,	4*p* (three orbitals)
4	2	– 2, – 1, 0, + 1, +2,	4*d* (five orbitals)
4	3	– 3, – 2, – 1, 0, + 1, +2, – 3	Of (seven orbitals)

Radial and Angular Functions

The Schrodinger equation can be solved completely for the hydrogen atom and for related ions which have only one electron such as He^+ and Li^{2+}. For other atoms only approximate solutions can be obtained. For most calculations, it is simpler to solve the wave equation if the cartesian coordinates *x, y* and *z* are converted into polar coordinates *r, ϕ and θ.* The coordinates of the point A measured from the origin are x, y, and z in cartesian coordinates, and *r*, θ, and ϕ in polar coordinates. It can be seen from Figure 1.12 that the two sets of coordinates are related by the following expressions:

$$z = r \cos \theta$$

$$y = r \sin \theta \sin \phi$$

$$x = r \sin \theta \cos \phi$$

The Schrodinger equation is usually written:

$$\nabla^2 \psi + \frac{8\pi^2 m}{h^2}(E - V)\,\psi = 0$$

Where

$$\nabla^2 \psi = \frac{\partial^2 \psi}{\partial x^2} + \frac{\partial^2 \psi}{\partial y^2} + \frac{\partial^2 \psi}{\partial z^2}$$

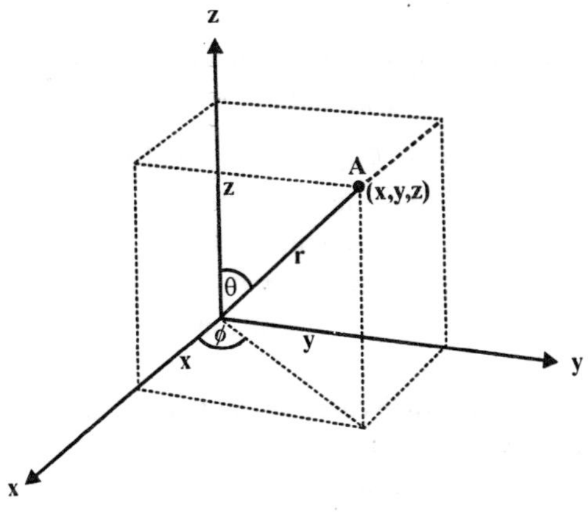

Fig. 1.12 : The relationship between Cartesian and polar coordinates.

Changing to polar coordinates, $\nabla^2\psi$ becomes

$$\frac{1}{r^2}\frac{\partial}{\partial r}\left(r^2\frac{\partial\psi}{\partial r}\right)+\frac{1}{r^2\sin^2\theta}\cdot\frac{\partial^2\psi}{\partial\phi^2}+\frac{1}{r^2\sin\theta}\cdot\frac{\partial}{\partial\theta}\left(\sin\theta\frac{\partial\psi}{\partial\theta}\right)$$

The solution of this is of the form

$$\psi = R(r).\Theta(\theta).\Phi(\phi) \tag{1.6}$$

$R(r)$ is a function that depends on the distance from the nucleus, which in turn depends on the quantum numbers n and l

Θ (θ) is a function of θ, which depends on the quantum numbers l and m

$\Phi(\phi)$ is a function of ϕ, which depends only on the quantum number m

Equation (1.6) may be rewritten

$$\psi = R\,(r)_{nl}.\,A_{ml}$$

This splits the wave function into two parts which can be solved separately:

1. R (r) the radial function, which depends on the quantum numbers n and l.
2. A_{ml} the total angular wave function, which depends on the quantum numbers m and l.

The radial function R has no physical meaning, but R^2 gives the probability of finding the electron in a small volume dv near the point at which R is measured. For a given value of r the number of small volumes is $4\pi r^2$, so the probability of the electron being at a distance r from the nucleus is $4pr^2R^2$. This is called the radial distribution function. Graphs of the radial distribution function for hydrogen plotted against r are shown in Figure 1.13.

These diagrams show that the probability is zero at the nucleus (as $r = 0$), and by examining the plots for 1s, 2s and 3s that the most probable distance increases markedly as the principal quantum number increases. Furthermore, by comparing the plots for 2*s* and 2*p*, or 3*s*, 3*p* and 3*d* it can be seen that the most probable radius decreases slightly as the subsidiary quantum number increases. All the *s* orbitals except the first one (1*s*) have a shell-like structure, rather like an onion or a hailstone, consisting of concentric layers of electron density. Similarly, all but the first *p* orbitals (2*p*) and the first *d* orbitals (3*d*) have a shell structure.

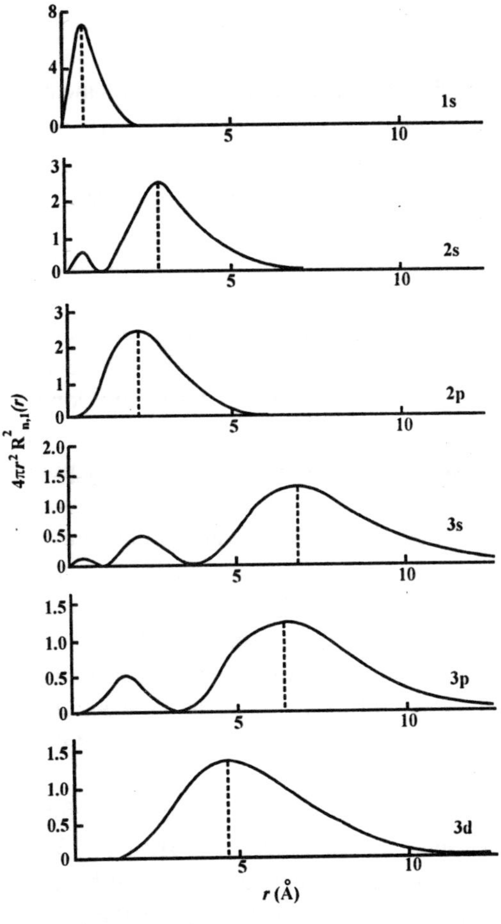

Fig. 1.13 : Radial distribution functions for various orbitals in the hydrogen atom.

The angular function A depends only on the direction, and is independent of the distance from the nucleus (r). Thus A^2 is probability of finding an electron at a given direction θ, ϕ at any distance from the nucleus to infinity. The angular functions A are plotted as polar diagrams in Figure 1.14. It must be emphasized that *these polar diagrams do* **not** *represent the total wave function* ψ, but only the angular part of the wave function. (The total wave function is made up from contributions from both the radial and the angular functions.)

$$y = R(r).\ A$$

Thus the probability of finding an ȇlectron simultaneously at a distance r and in a given direction θ, ϕ is $\psi^2 r.\ \theta,\phi$.

$$\psi^2_{r,\theta,\phi} = R^2(r).\, A^2(\theta,\phi)$$

Polar diagrams that is drawings of the angular part of the wave function, are commonly used to illustrate the overlap of orbitals giving bonding between atoms. Polar diagrams are quite good for this purpose, as they show the signs + and – relating to the symmetry of the angular function. For bonding like signs must overlap. These shapes are slightly different from the shapes of the total wave function. There are several points about such diagrams:

1. It is difficult to picture an angular wave function as a mathematical equation. It is much easier to visualize boundary surface, that is a solid shape, which for example contains 90% of the electron density. To emphasize that ψ is a continuous function, the boundary surfaces have been extended up to the nucleus in Figure 1.14. For p orbitals the electron density is zero at the nucleus, and some texts show a p orbital as two spheres which do not touch.

2. These drawings show the symmetry for the $1s$, $2p$ $3d$ orbitals. However, in the others, $2s$, $3s$, $4s$...., $3p$, $4p$, $5p$..., $4d$, $5d$....the sign (symmetry) changes inside the boundary surface of the orbital. This is readily seen as nodes in the graphs of the radial functions (Figure 1.13).

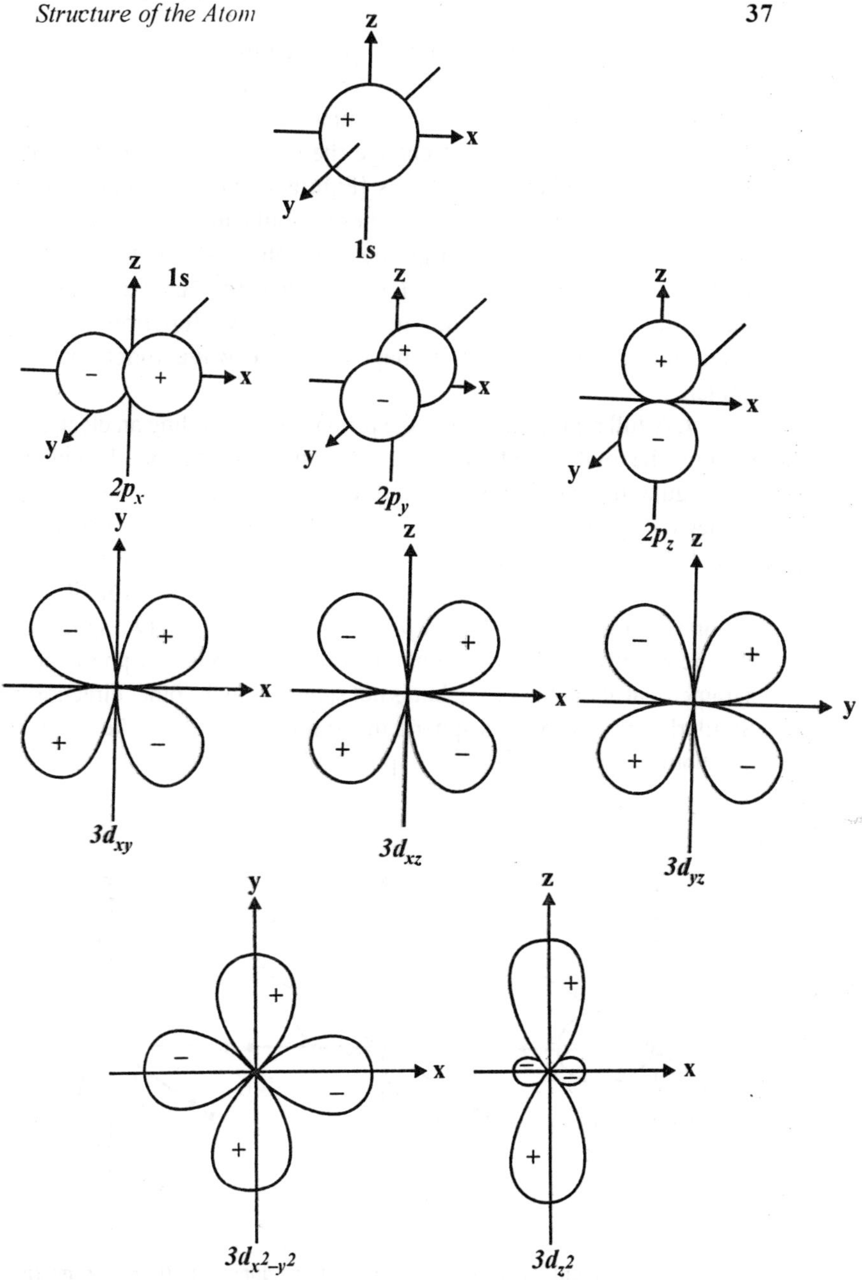

Fig. 1.14 : Boundary surface for the angular part of the wave function $A(\theta, \phi)$ for the 1s, 2p and 3d orbitals for a hydrogen atom shown as polar diagrams.

3. The probability of finding an electron at a direction θ, ϕ is the wave function squared. A^2, or more precisely $\psi_\theta^2\psi_\phi^2$. The diagrams in Figure 1.9 are of the angular part of the wave function. A, not A^2. Squaring does not change the shape of an *s* orbital, but it elongates the lobes of *p* orbitals (Figure 1.15). Some books use elongated *p* orbitals, but strictly these should **not** have signs, as squaring removes any sign from the symmetry. Despite this, many authors draw shape approximating to the probabilities, i.e. squared wave functions, and put the signs of the wave function on the lobes, and refer to both the shapes and the wave functions as orbitals.

4. A full representation of the probability of finding an electron requires the total wave function squared and includes both the radial and angular probabilities squared. It really needs a three-dimensional model to display this probability, and show the shapes of the orbitals. It is difficult to do this adequately on a two-dimensional piece of paper, but a representation is shown in Figure 1.16. The orbitals are not drawn to scale. Note that the *p* orbitals are not simply two spheres, but are ellipsoids of revolution. Thus the $2p_x$ orbital is spherically symmetrical about the *x* axis, but is not spherical in the other direction. Similarly the p_x orbital is spherically symmetrical about the *y* axis, and both the p_z and the $3d_{z2}$ are spherically symmetrical about the *z* axis.

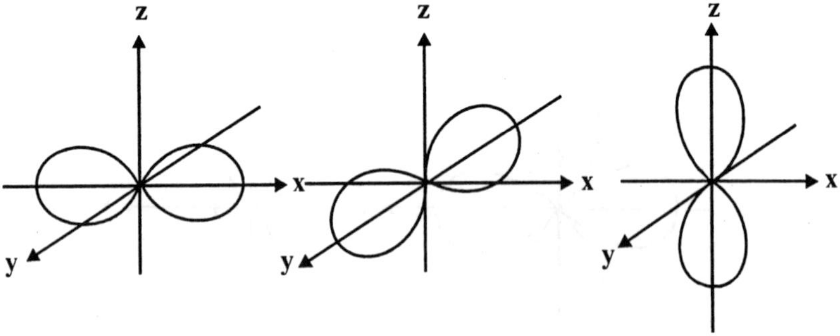

Fig. 1.15 : The angular part of the wave function squared A^2 (θ, ϕ) for the 2p orbitals for a hydrogen atom.

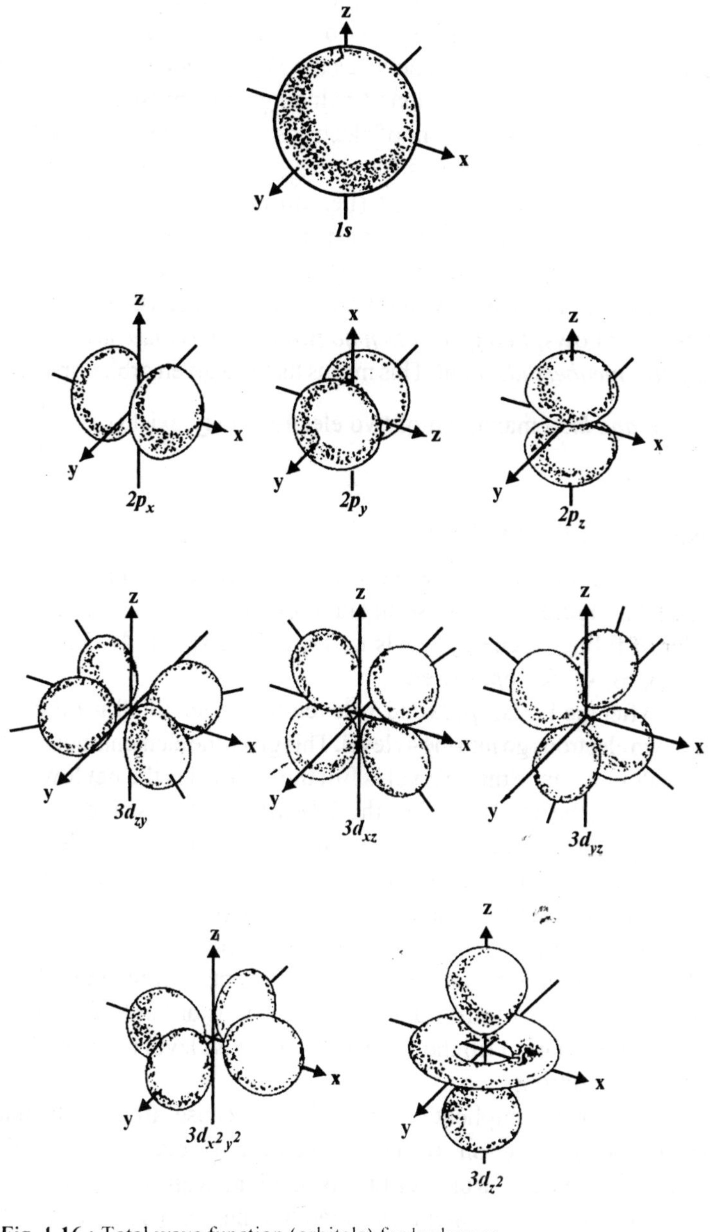

Fig. 1.16 : Total wave function (orbitals) for hydrogen.

Electron having the same spin strongly repel each other and tend to occupy different regions of space. This is a result of a fundamental law of nature known as the *Pauli exclusion principle.* It states that total wave functions (including spin) must change their signs on exchange of any pair of electrons in the system. Briefly, this means that if two electrons have the same spin they must have different spatial wave functions (i.e., different orbitals) and if they occupy the same orbital they must have paired spins. The Pauli principle and the so-called Pauli repulsive forces have far-reaching consequences in chemistry, but for our present discussion it may be stated as follows: *In a given atom no two electrons may have all four quantum numbers identical.* This means that in a given orbital specified by *n, l, and m*, a maximum of two electrons may exist ($s = +\frac{1}{2}$ and $s = -\frac{1}{2}$).

Sequence of Energy Levels

It is important to know the sequence in which the energy levels are filled. Figure 1.13 is a useful aid. From this it can be seen that the order of filling of energy levels is; 1*s*, 2*s*, 2*p*, 3*s*, 3*p*, 4*s*, 3*d*, 4*s*, 5*s*, 4*d*, 5*p*, 6*s*, 4*f*, 5*d*, 6*p*, 7*s*, etc.

After the 1*s*, 2*s*, 2*p*, 3*s* and 3*p* levels have been filled at argon, the next two electrons go into the 4*s* level. This gives the elements potassium and calcium. Once the 4*s* level is full the 3*d* level is the net lowest in energy, not the 3*p* level. Thus the 3*d* starts to fill to scandium. The elements from scandium to copper have two electrons in the 4*s* level and an incomplete 3*d* level, and all behave in a similar manner chemically. Such a series of atoms is known as a transition series.

A second transition series starts after the 5*s* orbitals has been filled, at strontium, because in the next element, yttrium, the 4*d* level begins to fill up. A third transition series starts at lanthanum where the electrons start to fill the 5*d* level after the 6*d* level has been filled with two electrons.

A further complication arises here because after lanthanum, which has one electron in the 5*d* level, the 4*f* level begins to fill, giving the elements from cerium to lutetium with from one to 14*f* electrons. These are sometimes called the inner transition elements, but are usually known as the lanthanides or rare earth metals.

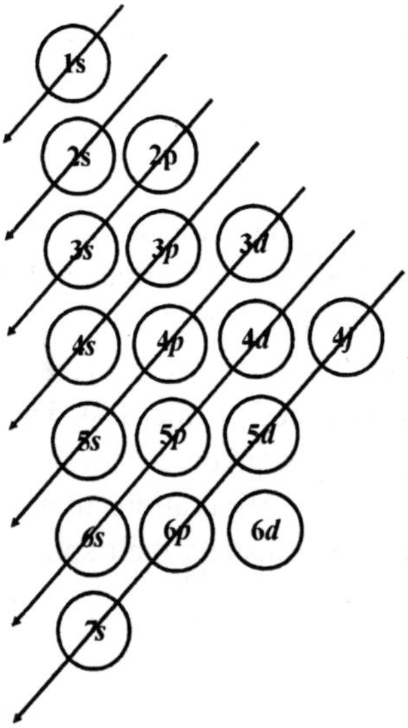

Fig. 1.17 : Sequence of filling energy levels

Atomic states, term symbols, and Hund's rule

It is convenient to be able to specify the energy angular momentum and spin multiplicity of an atom by a symbolic representation. For example, for the hydrogen atom we may define *S, P, D,* and *F* states, depending upon whether the single electron occupies an *s, p, d,* or *f* orbitals. The ground state of hydrogen $1s^1$, is a $1S$ state; a hydrogen atom excited to a $2p^1$ configuration is in a $2P$ state; etc. For polyelectronic atoms, an atom in a *P* state has the same total angular momentum (for all electrons) as a hydrogen atom in a *P* state. Corresponding to states *S, P, D, F,* .... are quantum numbers $L = 0, 1, 2, 3, 4, ...$, which parallel the l values for *s, p, d, f,* ... orbitals. Likewise, there is quantum number S (not to be confused with the *S* state just mentioned) that is the summation of all the electronic spins. For a closed shell or subshell, obviously $S = 0$, since all electrons are paired. Somewhat less obviously, under these conditions $L = 0$, since all of the orbital momenta cancel. This greatly simplifies working with states and term symbols.

The chemist frequently uses a concept known as *multiplicity,* originally derived from the number of lines shown in a spectrum. It is related to the number of unpaired electrons and, in general, is given by the expressions $2S + 1$. Thus, if $S = 0$, the multiplicity is one and the state is called a single; if $S = \frac{1}{2}$ the multiplicity is two and the state is a doublet; $S = 1$ is a triplet; etc. Hund's *rule of maximum multiplicity* (i.e., the greatest value of S). Consider a carbon atom (= $1s^2\ 2s^2\ 2p^2$). We may ignore the closed $1s^2$ and $2s^2$. The two $2p$ electrons may be paired in the same orbitals ($S = 0$) or have parallel spins in different orbitals ($S = 1$). Hund's rules predict that the latter will be the ground state, i.e., a triplet state. It happens that in this state $L = 1$, so we may say that the ground state of carbon is ^{3}P (pronounced "triplet-P"). The ^3P" is said to be the *term symbol.*

It is convenient for many purposes to draw "box diagrams" of electron configurations in which boxes represent individual orbitals, and electrons and their spins are indicated by arrows:

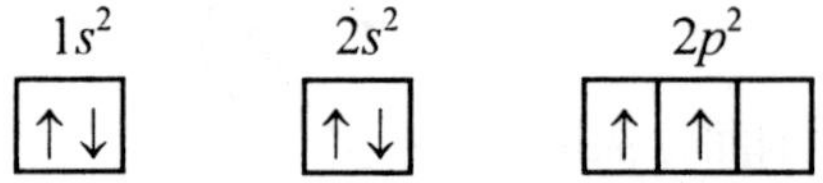

Such devices can be very useful for bookkeeping, providing pigeonholes in which to place electrons. However, the reader is warned that they can be misleading if improperly used, especially with respect to term symbols.

2

The Periodic Table

Origin

There are more than 100 elements. If you were to pick a few elements at random and compare them, you probably would find great differences in their properties. However, there are certain elements that are similar to one another in many ways. During the nineteenth century when most of the elements were being discovered, many chemists tried to classify the elements according to their similarities. Johann Dobereiner (der-buh-RINE-ner) (1780-1849), a German chemist, and the Englishman John Newlands (1837-1898) were among those who noted that the atomic masses of elements seemed to be related to chemical properties. Lothar Meyer (1835 1895), a German chemist, developed this idea further. Meyer stated that there was a pattern of repetition in the properties of the known elements when they were arranged in order of their atomic masses.

The Russian chemist Dimitri Mendeleev (duh-MEE-tree menduh-LAY-ef) (1834-1907) was the first to publish the classification of the elements that became the basis for the system used today. In 1869, he developed a table of the elements that was so carefully planned and contained such detail that he is credited as the major contributor to a successful classification of the elements.

ПЕРИОДИЧЕСКАЯ СИСТЕМА ЭЛЕМЕНТОВ

[illegible]	[illegible]	I	II	III	IV	V	VI	VII	VIII			0
1	I	H 1 1,008										He 2 4,003
2	II	Li 3 6,940	Be 4 9,02	5 B 10,82	6 C 12,010	7 N 14,008	8 O 16,000	9 F 19,00				Ne 10 20,183
3	III	Na 11 22,997	Mg 12 24,32	13 Al 26,97	14 Si 28,06	15 P 30,98	16 S 32,06	17 Cl 35,457				Ar 18 39,944
4	IV	K 19 39,096	Ca 20 40,08	Sc 21 45,10	Ti 22 47,90	V 23 50,95	Cr 24 52,01	Mn 25 54,93	Fe 26 55,85	Co 27 58,94	Ni 28 58,69	
	V	29 Cu 63,57	30 Zn 65,38	31 Ga 69,72	32 Ge 72,60	33 As 74,91	34 Se 78,96	35 Br 79,916				Kr 36 83,7
5	VI	Rb 37 85,48	Sr 38 87,63	Y 39 88,92	Zr 40 91,22	Nb 41 92,91	Mo 42 95,95	Ma 43 —	Ru 44 101,7	Rh 45 102,91	Pd 46 106,7	
	VII	47 Ag 107,88	48 Cd 112,41	49 In 114,76	50 Sn 118,70	51 Sb 121,76	52 Te 127,61	53 J 126,92				Xe 54 131,3
6	VIII	Cs 55 132,91	Ba 56 137,36	La 57 * 138,92	Hf 72 178,6	Ta 73 180,88	W 74 183,92	Re 75 186,31	Os 76 190,2	Ir 77 193,1	Pt 78 195,23	
	IX	79 Au 197,2	80 Hg 200,61	81 Tl 204,39	82 Pb 207,21	83 Bi 209,00	84 Po 210	85 —				Rn 86 222
7	X	87 —	Ra 88 226,05	Ac 89 227	Th 90 232,12	Pa 91 231	U 92 238,07					

* ЛАНТАНИДЫ 58–71

Ce 58 140,13	Pr 59 140,92	Nd 60 144,27	61 —	Sm 62 150,43	Eu 63 152,0	Gd 64 156,9
Tb 65 159,2	Dy 66 162,46	Ho 67 164,94	Er 68 167,2	Tu 69 169,4	Yb 70 173,04	Cp 71 174,99

Fig. 2.1 : ***Mendeleev's periodic table.*** This is one of several versions of the periodic table that Mendeleev prepared.

Mendeleev arranged the elements in order of their atomic masses. In an early version of his work, he placed elements with similar properties in horizontal rows. If no known element had the expected properties to fit a particular place in the table, he left that place empty. See Figure 2.1. Mendeleev assumed that elements not yet discovered would fill the empty places. He even predicted some of the properties of these undiscovered elements. The value of mendeleev's table was confirmed when elements were discovered that had the expected properties.

Mendeleev's table was called the periodic table. It was based on the original *periodic law,* which stated that the physical and chemical properties of elements were periodic functions of their atomic masses. This meant that if elements were arranged in order of increasing atomic masses, those with similar properties would show up at regular intervals, or periods. These similar elements were called families. For example, the halogen family contains the chemically similar elements fluorine, chlorine, bromine, and iodine. Lithium, sodium, potassium, rubidium, and cesium make up another family called the alkali metals.

Mendeleev realized that there were a few irregularities in his chart. For example, iodine has a *smaller* atomic mass than tellurium. However, iodine is chemically similar to chlorine and bromine. If placed in the column for those elements, iodine *follows* tellurium, as if its atomic mass were *greater*. See Figure 2.2. Mendeleev placed all the elements in groups based on similar properties, disregarding the fact that occasionally an element seemed to be out of order because its atomic mass was less than, not greater than, the atomic mass of the elements that preceded it. Mendeleev believed further study would show that his placement of the elements, according to their properties was the correct ordering.

Moseley and the modern periodic law. Henry Moseley (1887-1915) was a brilliant English physicist. He performed experiments that revealed the atomic numbers of several elements. In 1914, Moseley suggested that the elements in the periodic table be arranged in order of increasing atomic number instead of atomic mass. Today, scientists know that electron structure is the major factor responsible for the chemical properties of an element. Electron structure is related to atomic number, which does in fact determine the position of an element in the periodic table. See Figure 2.2.

The **modern periodic law,** as proposed by Moseley, states: The chemical and physical properties of the elements are periodic functions of their atomic numbers. Put another way, the law states that when the elements are arranged in order of increasing atomic number, there is a periodic repetition of their properties.

The Modern Periodic Table

The repetition expressed in the periodic law is the basis for the modern *periodic table,* in which families of elements, such as halogens, noble gases, and alkali metals, are placed in vertical columns. Such a periodic table is shown in Fig. 2.3 and also inside the front cover of this book. The horizontal rows in the table are called *periods*. They are numbered from 1 through 7 using arabic numerals. Note that they vary greatly in length from the first period (two elements) to the sixth (32 elements) and seventh (potentially 32 elements) periods.

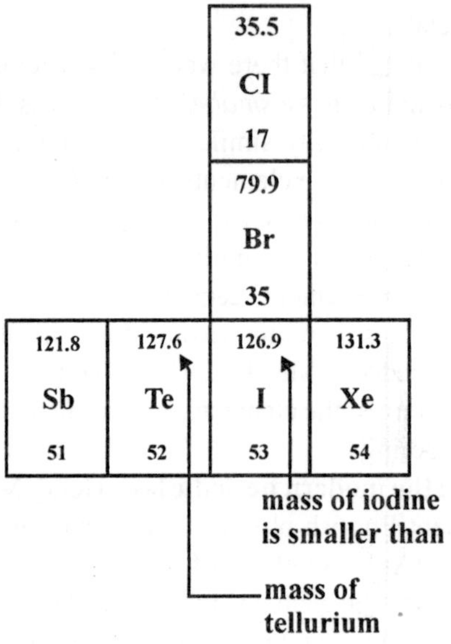

Fig. 2.2 : A small section of the periodic table, showing an irregularity in the arrangement of atomic masses (top numerals).

The vertical columns in the periodic table are called *groups*. Each of these is a family of elements. Some groups consist of five or six elements each. These are the so-called *main, representative,* or *A groups*. They are numbered from one through seven, using roman numerals and the letter A. Thus the halogens, for instance, are the elements of group VIIA (the A designation is sometimes omitted, and they are thus simply referred to as group VII). Also included with the main-group elements is group 0, the noble gases. The shorter groups are called *subgroups* or *B groups*. Thus, for example, copper, silver, and gold together constitute group IB, also known as the *copper subgroup*.

Notice these special characteristics of the periodic table (Fig. 2.3):

Groups	I A	II A	III B	IV B	V B	VI B	VIIB	VIIIB	VIIIB	VIIIB	I B	II B	III A	IV A	V A	VI A	VII A	0
1									1 H									2 He
Periods	3 Li	4 Be											5 B	6 C	7 N	8 O	9 F	10 Ne
	11 Na	12 Mg											13 Al	14 Si	15 P	16 S	17 Cl	18 Ar
	19 K	20 Ca	21 Sc	22 Ti	23 V	24 Cr	25 Mn	26 Fe	27 Co	28 Ni	29 Cu	30 Zn	31 Ga	32 Ge	33 As	34 Se	35 Br	36 Kr
	37 Rb	38 Sr	39 Y	40 Zr	41 Nb	42 Mo	43 Tc	44 Ru	45 Rh	46 Pd	47 Ag	48 Cd	49 In	50 Sn	51 Sb	52 Te	53 I	54 Xe
	55 Cs	56 Ba	57 La	72 Hf	73 Ta	74 W	75 Re	76 Os	77 Ir	78 Pt	89 Au	80 Hg	81 Ti	82 Pb	83 Bi	84 Po	85 At	86 Rn
	87 Fr	88 Ra	89 Ac	104 Rf	105 Ha	106												

*Lanthanoids	58 Ce	59 Pr	60 Nd	61 Pm	62 Sm	63 Eu	64 Gd	65 Tb	66 Dy	67 Ho	68 Es	69 Tm	70 Yb	71 Lu
+Actionoids	90 Th	91 Pa	92 U	93 Np	94 Pu	95 Am	96 Cm	97 Bk	98 Cf	99 Es	100 Fm	101 Md	102 No	103 Lr

Fig. 2.3 : The periodic table to the elements.

1. Hydrogen is placed all by itself. It is one of the two members of the first period, and its properties set it aside from all other elements. (Some versions of the periodic table place hydrogen above lithium in group IA and sometimes above fluorine in group VIIA, but it is clearly neither an alkali metal nor a halogen.)
2. Group VIIIB (also sometimes called, simply, group VIII) consists of a total of nine elements.
3. The sixth period includes the 14 elements from $Z = 58$, cerium through $Z = 71$, lutetium. These elements follow lanthanum ($Z = 57$) and are called the *lanthanoids*. They would be shown in the appropriate location, except that this would make the table too wide. This sequence is shown separately and at the bottom of the main portion of the table.
4. Similarly, the seventh period includes the 14 *actinoids* following actinium ($Z = 89$). Both the lanthanoids and actionids are considered to be part of group IIIB, the scandium subgroup.
5. The most important thing to note about the periodic table as a whole is that *a new period, or cycle, starts when a repetition in properties begins in the sequence of elements*. Thus lithium ($Z = 3$) is an alkali metal, and when the next alkali metal, sodium ($Z = 11$), rolls around, it is placed under lithium, and so on. In this way each group is a family of elements having similar properties.

Chemistry of The Elements in Relation to Their Position in The Periodic Table

We can now proceed to more detailed commentary on the chemical reactivity and types of compounds formed by the elements. The periodic table forms the basis for the discussion, starting with the simplest chemistry, namely that of hydrogen, and proceeding to the heaviest elements.

Hydrogen $1s^1$

The chemistry of hydrogen depends on three electronic processes:

1. *Is valence electron loss.* It forms the proton, H^+, its small charge result in a unique ability to distort the electron clouds

surrounding other atoms. The proton *never* exists as such except in gaseous ion beams. It is invariably associated with other atoms or molecules. Although the hydrogen ion in water is commonly written as H^+, it is actually H_3O^+.

2. *Acquisition of an electron.* The H atom can acquire an electron forming the *hydride ion,* H– with the He $1s^2$ structure. This ion exists only in crystalline hydrides of the most electropositive metals for example, NaH, CaH_2.
3. *An electron pair bond formation.* Nonmetals and even many metals can form covalent bonds to hydrogen.

The chemistry of hydrogen compounds is highly dependent on the nature of the element and on the nature of other groups or ligands also present. The extent to which compounds dissociate in polar solvents and act as acids.

$$HX \rightleftarrows H^+ + X^-$$

is particularly dependent on the nature of X.

All H-X bonds necessarily have some polar character with the dipole oriented $\overset{\delta+}{H}-\overset{\delta-}{X}$ or $\overset{\delta-}{H}-\overset{\delta+}{X}$. The term "hydride" is usually given to those compounds in which the negative end of the dipole is on hydrogen, for example, in SiH_4, $\overset{\delta+}{Si}-\overset{\delta-}{H}$. However, although HCl as $\overset{\delta+}{H}-\overset{\delta-}{Cl}$. is a strong acid in aqueous solution, nevertheless, it is a gas and is properly termed a covalent hydride.

Helium, $1s^2$ and the Noble Gases, $ns^2\ np^6$

The second element He, Z = 2 has the closed $1s$ shell; its very small size leads to some physical properties that are unique to liquid helium. The physical properties of the other noble gases vary systematically with size. Although the first ionization energies are high, consistent with their chemical inertness, the values fall steadily as the size of the atom increases. The ability to enter into chemical combination with other atoms should increase with decreasing ionization potential and decreasing energy of promotion to states with unpaired electrons—that is, $ns^2np^6 \rightarrow ns^2np^5\ (n + 1)s$.

Elements of the First Period

The third element, Li, Z = 3 has the structure $1s^2 2s$. With increasing Z, electrons enter the 2*s* and 2*p* levels until the closed shell configuration, $1s^2 2s^2 2p^6$, is reached at neon. The seven elements Li to F constitute the first members of the *groups* of elements.

Although these elements have many properties in common with the heavier elements of their respective groups—which is to be expected in view of the similarity in the outer electronic structures of the *gaseous* atoms—they nevertheless show highly individual behaviour in many important respects. We have already seen that O_2 and N_2 form diatomic molecules whereas their congeners, S and P, form polyatomic molecules or chains. Indeed, the differences between the chemistries of B,C, N, and O and Al, Si, P, and S and the heavier members of these groups are sufficiently striking that in many ways it is not useful to regard the first period elements as prototypes for their congeners. The closest analogies between the first row and the heavier elements are for Li and F, followed by Be.

The increase in nuclear charge and consequent changes in the extranuclear structure result in extremes of physical and chemical properties. In Fig. 2.4 are given the first ionization enthalpies. The low ionization enthalpy for *lithium* is in accord with facile loss of an electron to form the Li^+ ion, which occurs both in solids and in solution. It accords also with the high reactivity of lithium to oxygen, nitrogen, water, and many other elements.

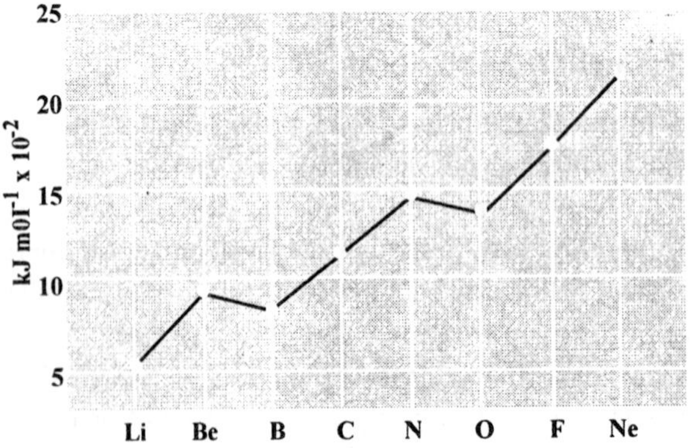

Fig. 2.4 : *First ionization enthalpies of elements Li-Ne.*

For *beryllium* the first (899 kJ mol^{-1}) and especially the second (1757 kJ mol^{-1}) ionization enthalpies are sufficiently high that total loss of both electrons to give Be^{2+} does not occur even with the most electronegative elements. Even in BeF_2 the Be-F bonds have appreciable covalent character. The ion in aqueous solution, $[Be(H_2O)_4]^{2+}$, is very strongly aquated and undergoes, hydrolysis quite readily to give species with Be(OH) bonds.

For the succeeding elements, the absence of any simple cations under any conditions is to be expected from the high ionization enthalpies. Note that the values (Fig. 2.4) for B,C, and N increase regularly but that they are lower than the values which would be predicted by extrapolation from Li and Be. This arises because *p* electrons are less penetrating than *s* electrons—they are therefore shielded by the *s* electrons and are removed more easily. Another discontinuity occurs between N and O. This one is due to the fact that the 2*p* shell is half-full, that is $p_xp_yp_z$ at N. The *p* electrons added in O, F, and Ne thus enter *p* orbitals that are already single occupied. Hence, they are partly repelled by the *p* electron already present in the same orbital and are thus less tightly bound.

The electron attachment energies rise from Li to F and the electro-negativity also rises.

Boron, $2s^22p$ has no cationic chemistry and is bound covalently in all its compounds, such as oxoanions, organoboron compounds sand hydrides.

Amino *formation* first appears for carbon, which forms C_2^{2-} and some other polyatomic ions, although the existence of C^{4-} is uncertain N^{3-} ions are stable in nitrides of highly electropositive elements. Oxide, O^{2-}, and fluoride, F^-, are common in solids but observe that O^{2-} ions cannot exist in aqueous solutions, compare,

$$O^{2-} + H_2O = 2OH^- \qquad K > 10^{22}$$
$$F^- + H_2O = HF + OH^- \qquad K = 10^{-7}$$

Carbon is a true nonmetal and its chemistry is dominated by single, double and triple bonds to itself or to nitrogen, oxygen, and a few other elements. What distinguishes carbon from other elements is its unique ability to form chains of carbon-carbon bonds (called catenation) in compounds—as distinct from the element itself.

Nitrogen. Nitrogen gas, N_2, is relatively unreactive because of the great strength of the $N \equiv N$ bond and its electronic structure (page 67). Nitrogen compounds are covalent, usually involving three single bonds, although multiple bonds such as $C \equiv N$ or $Os \equiv N$ can exist. With electropositive elements, ionic nitrides containing N^{3-} may be formed.

Oxygen. The diatomic molecule has two unpaired electrons (page 66) and consequently is very reactive. There is an extensive chemistry with covalent bonds as in $(CH_3)_2$ C = O, $(C_2H_5)_2O$, CO, SO_3, and the like. However, well-defined oxide ions O^{2-}, O_2^-, and O_2^{2-} exist in crystalline solids. Hydroxide ions, OH^-, exist both in solids and in solutions, although in hydroxylic solvents the OH^- ion is doubtless hydrated via hydrogen bonds.

Fluorine is extremely reactive due largely to the low bond energy in F_2. This is a result in part, of repulsions by nonbonding, electrons. Ionic compounds containing F^- ions and covalent compounds containing X-F bonds are well established. Owing to the high electronegativity of fluorine, such covalent bonds are generally quite polar in the sense X^+ - F^-.

The Second Period

The elements Na, Mg, Al, Si, P, S, and Cl, constitute the second short period. Although their outer electronic structures are similar to those of the first shell period, the chemistries differ considerably. In particular, the chemistries of P, S and, to a large extent, chlorine are completely different from their corresponding first-row element. The second-row elements do, however, give a be guide to the chemistries of the heavier elements in their respective groups that the first-row members. It is the nonmetallic elements in particular that substantially. The main reasons for this are as follows.

1. It is not possible to form *pπ-pπ* multiple bonds such as Si = Si, Si = O P = P. Most likely, this is because, in order to approach close enough to get overlap of *pπ* atomic orbitals, the heavier atoms would encounter large repulsive forces due to overlapping of their filled inner shells, whereas the small *compound* inner shell of the first-row elements, that is, just $1s^2$, does not produce this repulsion.

The results is that, as we have seen, the nature of the elements themselves strikingly different. As a dramatic example, silicon has no chemistry comparable to the carbon chemistry of C=C, C ≡ C, >C=O, and >C=N bonds; that no analogs of alkenes, alkynes, ketones, nitriles, and the like. Whereas CO_2 gas, SiO_2 forms an infinite polymer.

2. The possibility of using the 3*d* orbitals then allows promotion to valence states leading to formation of five or six bonds. Hence, we find compounds such as PCl_5 or SF_6 and silicon can form species with 5 and 6 coordination also as in SF_6^{2-}. For silicon, even where there is some analogy with carbon chemistry, as in compounds with single bonds, the reactions and mechanisms operating in silicon chemistry may be vastly different. A simple example is the unreactivity of CCl_4 toward water, whereas $SiCl_4$ is instantly hydrolyzed.

Nontransition Elements

The first and especially the second-row elements just discussed give some guide of the chemistry of the elements in their respective groups. The main features are as follows.

Group I. All the elements (Table 2.1) are highly electropositive giving +1 ions. Of all the groups in periodic table, these metals show most clearly the effect of increasing size and mass on chemical properties. Thus, as example the following *decrease* from Li to Cs : (a) melting points and heat of sublimation of the metals; (b) lattice energies of salts except those with very small anion (because of irregular radius ratio effects); (c) effective hydrated radii and hydration energies, and (d) strength of covalent bonds in M_2 molecules.

Table 2.1 : *Some Properties of Group I Elements*

Element	Electronic configuration	mp°C	Ionic radius Å	$E°V^a$
Li	[He]2*s*	180	0.60	−3.0
Na	[Ne]3*s*	98	0.96	−2.7
K	[Ar]4*s*	64	1.33	−2.9
Rb	[Kr]5*s*	39	1.48	−3.0
Cs	[Xe]6*s*	29	1.69	−3.0
Fr[b]	[Rn]7*s*	–	–	–

a For M+ (Aq) + *e* = M(*s*).

b All isotopes are radioactive with short half-lives.

Group II. Some properties of the elements are given in Table 2.2. Calcium Sr, Ba, and Ra are also highly electropositive forming +2 ions. Systematic grow trends are again shown, for example, by increasing (a) hydration tendencies, (b) is solubilities of sulfates, and (c) thermal stabilities of carbonates or nitrates.

Table 2.2 : *Some Properties of Group II Elements*

Element	Electronic configuration	mp °C	Ionic radius M^{2+}, Å	E°V[a]
Be	$[He]2s^2$	1280	0.31	– 1.85
Mg	$[Ne]3s^2$	650	0.65	– 2.37
Ca	$[Ar]4s^2$	840	0.99	– 2.87
Sr	$[Kr]5s^2$	770	1.13	– 2.89
Ba	$[Xe]6s^2$	725	1.35	– 2.90
Ra	$[Rn]7s^2$	700	1.40	– 2.92
Zn	$[Ar]3d^{10}4s^2$	420	0.74	– 0.76
Cd	$[Kr]4d^{10}5s^2$	320	0.97	– 0.40
Hg	$[Xe]4s^{14}5d^{10}6s^2$	– 39	1.10	+ 0.85

a For M^{2+} (aq) + 2e = M(s).

Zinc, Cd, and Hg are also classed in Group II, as IIB. They have two *s* electrons outside filled *d* shell since they follow Cu, Ag, and Au, respectively, as the first, second, and third transition series elements. The chemistries of Zn and Cd are quite similar, but the polarizing power of the M^{2+} ions is larger then would be predicted by comparing the radii with those of the Mg-Ra group. This can be associated with the greater ease of distortion of the filled *d* shell compare with the noble gas shell of the Mg-Ra ions. Zinc and Cd are quite electropositive resembling Mg in their chemistry, although there is a greater tendency to form complexes with NH_3, halide ions, and CN^-.

Mercury is unique. It has a high *positive* potential, and the Hg^{2+} ion does resemble Zn^{2+} or Cd^{2+}. For example, the formation constants for, say, half ions are orders of magnitude greater than for Cd^{2+}. Mercury also readily forms mercurous ion, which has a metal-metal bond, $^+Hg\text{-}Hg^+$.

Group III. Some properties of the elements are given in Table 2.3. This group is quite a numerous one, since it contains the Group IIIA elements, Sc, Y, and Ac, and the Group IIIB elements. Al, Ga, In, and Tl. In addition, all of lanthanide elements could be included, since their chemistry is similar to that of IIIA elements.

Table 2.3 : *Some Properties of Group III Elements*

Element	Electronic configuration	mp °C	Ionic radius Å[c]	$E^{\circ}V$[a]
Sc	$[Ar]3d^14s^2$	1540	0.81	– 1.88
Y	$[Kr]4d^15s^2$	1500	0.93	– 2.37
La	$[Xe]5d^16s^2$	920	1.15	– 2.52
Acb	$[Rn]6d^17s^2$	1050	1.11	~ – 2.6
Al	$[Ne]3s^23p$	660	0.050	– 1.66
Ga	$[Ar]3d^{10}4s^24p$	30	0.62	– 0.53
In	$[Kr]\ 4d^{10}5s^25p$	160	0.81	– 0.34
Ti	$[Xe]\ 4f^{14}5d^{10}6s^26p$	300	0.95	+ 0.72

[a] For M^{3+} (aq) + 3 e = M(s).
[b] Isotopes are all radioactive.
[c] For M^{3+}

However, we consider the lanthanides separately because of their special position in the periodic table. Notice that in the Sc-Ac group the 3-valence electrons are d^1s^2 compared with s^2p^1 for the Al–TI group. Despite this occupancy of the *d* levels the elements show no transition metal-like chemistry. They are highly electropositive metals, and their chemistry is primarily one of the + 3 ions which have the noble gas configuration.

Scandium with the smallest ionic radius has chemical behaviour intermediate between that of Al, which has a considerable tendency to covalent bond formation, and the mainly ionic natures of the heavier elements.

Gallium, In and TI, like Al are borderline between ionic and covalent in compounds, even though the metals are quite electropositive and they form M^{3+} ions.

The + 1 state becomes progressively more stable as the group is descended, and for Tl the Tl^{I}-Tl^{III} relationship is a dominant factor of the chemistry. The occurrence of an oxidation state two below the group valence is sometimes contributed to the *inert pair effect,* which first makes itself evident here. It could be considered to apply in the low reactivity of mercury, but it is more pronounced still in Groupes IV and V. The term refers to the resistance of a pair of *s* electrons to be lost or to participate in covalent bond formation. Thus Hg is difficult to oxidize allegedly because it contains only an inert pair ($6s^2$), TI forms TI^{I} rather than TI^{III} because of the inert pair in the valence shell ($6s^2 6p$), and so on. The concept of the inert pair tells us little, if anything, about the ultimate reasons for the stability of lower oxidation states. It is a useful label. The true cause of the phenomenon is not intrinsic inertness—that is, unusually high ionization energy of the s^2 pair, but rather the decreasing strength of bonds as a group is descended. Thus the sum of the second and third ionization enthalpies is lower for In (4506 kJ Mol^{-1}) than for Ga (4920 kJ mol^{-1}) with TI (4825 kJ mol^{-1}) intermediate. There is, however, a steady decrease in the mean thermochemical bond energies, for example, in the chlorides, GA 242, In 206, and TI 153 kJ mol^{-1}.

Group IV. Some properties of the elements are given in Table 8-4. Note that we are now considering the Group IVB in the periodic table, since Group IVA comprises the transition metals Ti, Zr, and Hf, whose chemistry we consider separately. This pattern holds true for the remaining Groups V to VII.

There is no more striking example of the enormous discontinuity in proper between the first-and second-row elements, followed by a relatively smooth character toward metallic character, than in Group IVB. Carbon is nonmetallic; silicon also nonmetallic but little of the chemistry of Si can be inferred from that of Germanium is much like silicon, although it shows more metallic-type behaviour in its chemistry. Tin and lead are metals and both have some metal-like chemistry especially in the divalent state.

The main chemistry in the IV oxidation state for all the elements is essential one that involves covalent and molecular compounds. Typical examples $GeCl_4$ and $PbEt_4$. There is a decrease in the tendency to *catenation,* which is a feature of carbon chemistry in the order $C > Si > Ge \simeq Sn \simeq Pb$. This partly due to the diminishing strength of the C-C, Si-Si, and the like, both. The strengths of covalent bonds to other atoms also generally decrease in going from C to Pb.

The divalent state. Although in CO the *oxidation* state of C is *forms* taken to be 2, this *is* only a formalism and carbon uses more than two valence electrons in bonding. True divalence is found only in *carbenes* such as : CF_2, these species are very reactive due to accessibility of the sp^2 hybridized lone pair. The divalent compounds of the other Group IV elements can be regarded carbenelike in the sense that they are angular with a lone pair and can ready to undergo an oxidative addition reaction to give two new born to the element, for example,

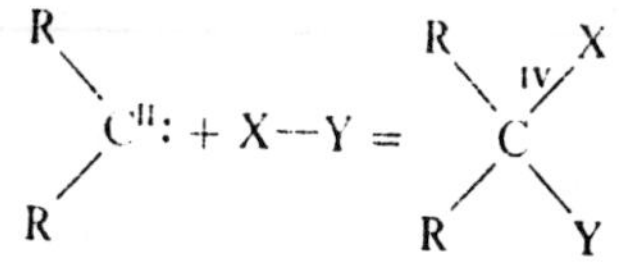

The "inert pair" concept is not very useful, especially since the bonding electrons are known *not* to be inert in a stereochemical sense. Index the two electrons behave much as a lone pair in every respect.

The increase in stability of the divalent state cannot be attributed to ionization energies as they are very similar in all cases. Factors that doubtless govern to relative stabilities are (i) promotion energies, (ii) bond strengths in covalent compounds and (iii) lattice energies in ionic compounds.

For CH_4, the factor that stabilizes CH_4 relative to $CH_2 + H_2$ despite the much higher promotional energy required in forming CH_4 is the great strength of the C—H bonds. If we now have a series of reactions

$$MX_2 + X_2 = MX_4$$

in which the M—X bond energies are decreasing, as they do from Si → Pb, then it is possible that bond energy may become too small to compensate for the M^{II}—M^{iv} promotion energy and MX_2 then becomes the more stable.

The change in this group is shown by the reactions:

$GeCl_2 + Cl_2 = GeCl_4$	(very rapid at 25°)
$SnCl_2 + Cl_2 = SnCl_4$	(slow at 25°)
$PbCl_2 + Cl_2 = PbCl_4$	(only under forcing condition)

Also, even $PbCl_4$ decomposes readily while $PbBr_4$ and PbI_4 do not exist, probably because of the reducing power of Br^- and I^-.

It is difficult to give any rigorous argument on lattice energy effects, since there is no evidence for the existence of M^{4+} ions and in only a few compounds are there even Pb^{2+} ions.

Group V. Some properties of these elements are given in Table 2.4.

Like nitrogen, phosphorous is essentially covalent in all its chemistry but arsenic, antimony, and bismuth show increasing tendencies to cationic behaviour.

Table 2.4 : *Some Properties of Group VB Elements*

Element	Electronic configuration	mp °C	Covalent radius Å	Ionic radius Å
P	$[Ne]3s^23p^3$	44	1.10	2.12 (P^{3-})
As	$[Ar]3d^{10}4s^24p^3$	814 (36 an)	1.21	
Sb	$[Kr]4d^{10}5s^25p^3$	603	1.14	0.92 (Sb^{3+})
Bi	$[Xe]4f^{14}5d^{10}6s^26p^3$	271	1.52	1.08 (Bi^{3+})

Although electron gain to achieve the electronic structure of the next noble gas is conceivable (as in N^{3-}), considerable energies are involved so that anionic compounds are rare. Similarly, loss of valence electrons is difficult because of high ionization energies. There are no + 5 ions and even the + 3 ions are not simple, being SbO^+ and BiO^+. BiF_3 seems predominantly ionic.

The increasing metallic character is shown by the oxides that change from acidic phosphorus to basic for bismuth, and by halides that have increasing ionic character.

Group VI. Table 2.5 gives some properties.

Table 2.5 : *Some Properties of Group VIB Elements*

Element	Electronic configuration	mp °C	Covalent radius Å	Ionic (X^{2-}) radius Å
S	$[Ne]3s^23p^4$	119	1.03	1.90
Se	$[Ar]3d^{10}4s^24p^4$	217	1.17	2.02
Te	$[Kr]4d^{10}5s^25p^4$	450	1.37	2.22
Po	$[Xe]4f^{14}5d^{10}6s^26p^4$	254	-	2.30

These atoms can achieve the configuration of the noble gas by forming :

1. The *chalconide* ions, M^{2-}, in salts of highly electropositive elements.
2. Two electron-pair bonds as in H_2S or $SeCl_2$.
3. Anionic species with one bond as in Hs^-.
4. Three bonds and one positive charge as in sulfonium ions R_3S^+.

There are also compounds in formal oxidations states IV and VI with 4, 5, covalent bonds, for example, $SeCl_4$, SeF_5^-, and TeF_6.

Only for polonium is there any evidence for cationic behaviour, but there gradual changes in properties with increasing size and decreasing electronegative such as:

(a) Decreasing stability of the hydrides H_2X.
(b) Increasing metallic character of the elements themselves.
(c) Increasing tendency to form anionic complexes such as SeBr $TeBr_6^{2-}$, PoI_6^{2-}.

Group VII. Some properties are given in Table 2.6.

The halogens atoms are only one electron short of the noble configuration and the elements form the anion X^- or a single covalent bond. Their chemistry completely nonmetallic. The changes in behaviour with increasing size are aggressive, and, with the exception of the Li-Cs group, there are closer similarity within the group than in any other in the periodic table.

The halogens can form compounds in higher formal oxidations states, mainly halogen fluorides such as ClF_3, ClF_5, BrF_5, and IF_7 and oxo compounds.

Table 2.6 : Some Properties of Group VIIB Elements.

Element	Electronic configuration	mp °C	Radius bp °C	Covalent X^- A	radius Å
F	$[He]2s^2 2p^5$	– 233	– 118	1.19	0.71
Cl	$[Ne]3S^2 3p^5$	– 103	– 34.6	1.70	0.97
Br	$[Ar]3d^{10} 4s^2 4p^5$	– 7.2	58.8	1.87	1.14
At[a]	$[Xe]4f^{14} 5d^{10} 6s^2 6p^5$	–	–	–	–

[a] All isotopes are radioactive with short half-lives.

No evidence exists for cationic behaviour with ions of the type X^+. However, F^+ I_2+, Cl_3^+, and Br_3^+ and several iodine cations are known. When a halogen forms a bond to another atom more electrónegative than itself, for example, ICI, bond will be polar with a positive charge on the heavier halogen.

Transition Elements of the *d* and *f* Blocks

The transition elements may be strictly defined as those that *as elements,* have partly filled *d* or *f* shells. We adopt a broader definition and the include also elements that have partly filled *d* or *f* shells in *compounds*. This means that we treat the *change metals,* Cu, Ag, and Au, as transition metals, since Cu^{II} has a $3d^9$ configuration, Ag^{II} has a $4d^9$ configuration, and Au^{II} has a $5d^8$ configuration. Appropriately also a consider these elements as transition elements because their chemical behaviour is quite similar to that of other transition elements.

There are thus 56 transition elements, counting the heaviest elements through one of atomic number 104. All these elements have certain common properties :

1. They are all metals.
2. They are practically all hard, strong, high-melting, high-boiling metals that conduct heat and electricity well.
3. They form alloys with one another and with other metallic elements.
4. Many of them are sufficiently electropositive to dissolve in mineral acids, though a few are "noble"—that is, they have such low electrode potentials that they are unaffected by simple acids.
5. With very few exceptions, they exhibit variable valence, and their ions and compounds are coloured in one if not all oxidation states.
6. Because of partially filled shells they form at least some paragmagnetic compounds.

This large number of transition elements is subdivided into three main groups : (a) the main transition elements or *d*-block elements, (b) the lanthanide elements, and "(c) the actinide elements.

The main transition group of *d* block includes those elements that have partially filled *d* shells only. Thus, the elements scandium, with the outer electron configuration $4s^2 3d$ is, the lightest member. The eight succeeding elements, the *first transition series,* Ti, V, Cr, Mn, Fe, Co, Ni, and Cu, all have partly filled 3*d* shells either in the ground state of the free atom (all except Cu) or in one or more of their chemically important ions (all excepts Sc). At zinc the configuration is $3d^{10}4s^2$; and this element forms no compound in which the $3d$ shells is ionized, nor does this ionization occur in any of the next nine elements. It is not until we come to yttrium, with ground-state outer electron configuration $5s^2 4d$, that we meet the next transition element. The following eight elements, Zr, Nb, Mo, Tc, Ru, Rh, Pd, and Ag, all have partially filled $4d$ shells whether in the free element (all but Ag) or in one or more of the chemically important ions (all but Y). This group of nine elements constitutes the *second transition series.*

Again there follows a sequence of elements in which there are never *d*-shell vacancies under chemically significant conditions until we reach the element lanthanum, with an outer electron configuration in the ground state of $6s^2 5d$. Now, if the pattern we have observed twice before were to be repeated, there would follow 8 elements with enlarged, but not complete, sets of $5d$ electrons. This does not happen however. The of shell now becomes slightly more stable than the $5d$ shell and, through the next 14 elements enter the $4f$ shell until at lutetium it becomes filled. Lutetium thus has the outer electron configuration $4f^{14}5d6s^2$. Since both La and Lu have partially filled d shells and no other partially filled shells, it might be argued that both of them should be considered as *d*-block elements. However, for chemical reasons, is would be unwise to classify them in this way, since all of the 15 elements La ($Z = 57$) through Lu ($Z = 71$) have very similar chemical and physical properties, those of lanthanum being in a sense prototypal; hence, these elements are called the *lanthanides*.

The shielding of one f electron by another from the effects of the nuclear change is quite weak on account of the shapes of the f orbitals. Hence, with increasing atomic number and nuclear charge, the effective nuclear charge experienced by each $4f$ electron increases. This causes a shrinkage in the radii of the atoms or ions as one proceeds from La to Lu (see Table 26-1). This accumulation of successive shrinkages is called the *lanthanide contraction.* It has a profound effect on the radii of subsequent elements, which are smaller than might have been anticipated from the increased mass. Thus Zr^{4+} and Hf^{4+} have almost identical radii despite the atomic numbers of 40 and 72, respectively.

For practical purposes, the *third transition series* begins with hafnium, having the ground-state outer electron configuration $6s^2 5d^2$, and embraces the elements Ta, W, Re, Os, Ir, Pt, and Au, all of which have partially filled $5d$ shells in one or more chemically important oxidation states as well as (excepting Au) in the neutral atom.

Continuing on from mercury, which follows gold, we come via the noble-gas radon and the radioelements. Fr and Ra to actinium, with the outer electron configuration $7s^2 6d$. Here we might expect, by analogy to what happened at lanthanum that in the following elements electrons would enter the $5f$ orbitals, producing a lanthanide-like series of 15 elements. What actually occurs is, unfortunately, not as simple. Although immediately following lanthanum, the $4f$ orbitals become decisively more favourable than the $5d$ orbitals for the electrons entering in the succeeding elements, there is apparently not so great a difference between the $5f$ and $6d$ orbitals until later. Thus, for the elements immediately following Ac, and their ions, there may be electrons in the $5f$ or $6d$ orbitals, or both. Since it appears that later on, after 4 or 5 more electrons have been added to the Ac configuration, the $5f$ orbitals do become definitely the more stable, and since the elements from about americium on do show moderately homologous chemical behaviour, it has become accepted practice to call the 15 elements beginning with Ac the *actinide elements.*

There is an important distinction, based on electronic structures, between the three classes of transition elements. For the *d*-block elements the partially filled shells are *d* shells, $3d$, $4d$, or $5d$. These *d* orbitals project well out to the periphery of the atoms and ions so that the electrons occupying them are strongly influenced by the surroundings of the ion and, in turn, are able to influence the environments very significantly. Thus, many of the properties of an ion with a partly filled *d* shell are quite sensitive to the number and arrangement of the *d* electrons present. In marked contrast to this, the $4f$ orbitals in the lanthanide elements are rather deeply buried in the atoms and ions. The electrons that occupy them are largely screened from the surroundings by the overlying shells ($5s$, $5p$) of electrons, and therefore reciprocal interactions of the $4f$ electrons and the surroundings of the atom or the ion are of relatively little chemical significance. This is why the chemistry of all the lanthanides is so homologous, whereas there are seemingly erratic and irregular variations in chemical properties as one passes through a series of *d*-block elements. The behaviour of the actinide elements lies between those of the two types described above because the 5*f* orbitals are not so well shielded as are the 4*f* orbitals, although they are not so exposed as are the *d* orbitals in the *d*-block elements.

Periodicity in Electronic Configurations

At this point we will return to a further consideration of the building up (*Aufbau*) of the electronic configurations of the elements.

When we compare electronic configurations with the periodic table we find that each period is started by adding an electron to a *new* (previously unoccupied) shell. Thus hydrogen and the group IA elements have an ns^1 configuration, where n is the principal quantum number of the "last," or outermost shell. This shell is almost always referred to as the *valence shell*, and so from now on we will use this term. (Valence means *combining capacity*, and the electrons in the outermost shell are largely responsible for the way in which most atoms form bonds, or combine, with other atoms.) *Note :* The valence shell of an atom is the occupied shell with the highest principal quantum number n.

The main-group elements have valence-shell populations ranging from one through eight electrons, the respective configurations being ns^1 and ns^2 for groups IA and IIA and ns^2np^1 through ns^2ns^6 for groups IIIA through 0. These elements are often called the *representative elements.* The valence-shell configuration ns^2np^6 is an especially stable one, as can be seen by the low reactivities of the noble gases. This configuration is often called an *octet*, and the generalization that it is especially stable is known as the *octet rule.*

The B-group elements in the periodic table are most commonly called the *transition elements.* Each horizontal series of transition elements corresponds to the belated filling of the *d* subshell of the $(n - 1)$ shell of these atoms. Since any *d* subshell can accommodate a total of 10 electrons, this filling builds up 10 transition elements in periods 4 and 5. Across the lanthanoid and actinoid series the subshell being filled in each case is the *f* subshell of the $(n - 2)$ shell. The *f* subshells have a maximum population of 14 electrons; so there are 14 lanthanoids and 14 actinoids.

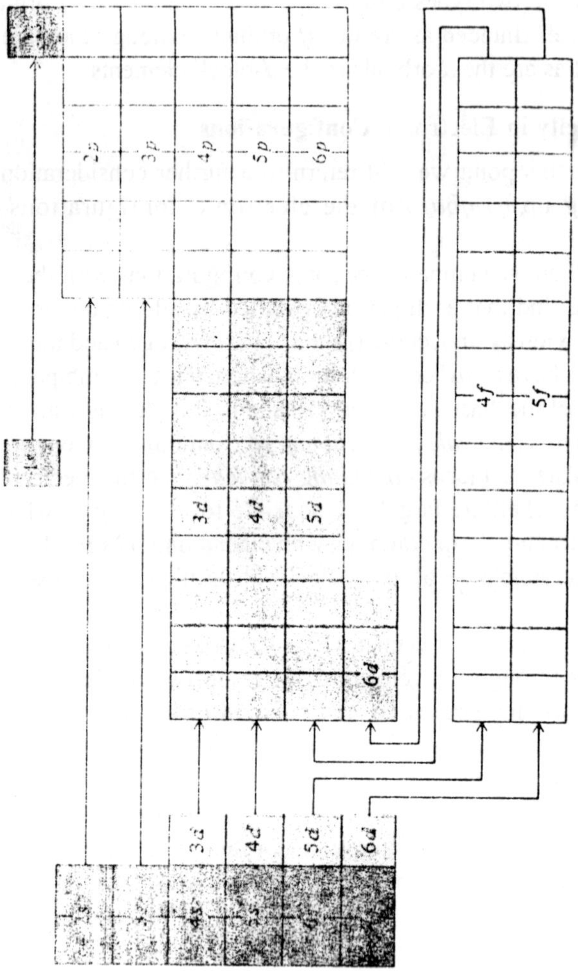

Fig. 2.5 : The periodic table and electronic configurations.

The periodic table can be used to predict the electronic configurations of most atoms. Figure 2.5 summarizes the relationship between orbital filling and the periodic table. In each indicated region of the table the "last" electrons are added to the designated subshell; in groups IA and IIA to the *ns* subshell, in groups IIIA through 0 to the *np* subshell, in the transition elements to the $(n-1)d$ subshell, and in the lanthanoids and actinoids to the $(n-2)f$ subshell.

Valence and the periodic table. Elements within any group exhibit a characteristics valence. For the *A* series of elements the number of the group corresponds to the number of electrons in the outermost orbit, and hence we may predict the valence of a particular element from its position in the table. The alkali metals of Group I*A* exhibit a valence of +1, since the atoms easily lose the one electron in the outer orbit. The halogens of Group VII*A* exhibit a characteristic valence of – 1, since one electron is readily taken up, and this, added to the seven already present in the outer orbit, gives a stable configuration of eight. On the same basis, elements of Group III*A* would be expected to exhibit a characteristics valence of +3, those of group VI*A* a valence of –2, and so on.

The rare gases of Group VIII*A*, each of which has a stable outer configuration of electrons, would not be expected to give up or take up electrons and hence would have a zero valence. The inert or inactive characteristics of these elements are explained on this basis.

Although valences as measured by the loss or gain electrons are quite clear cut for the *A* series of elements, they are much less well defined for the transition elements (*B* series). In the *A* series the outer orbit alone determines valence characteristics, while with the *B* series (in which there is at least one incomplete inner orbit) the incomplete orbit contributes also to valence characteristics. Most of the transition elements have one or two electrons in the outer shell. It appears that one or two electrons from an incomplete inner orbit may be lost in chemical change and these, added to the one or two in the outer orbit, give rise to several possibilities of valence among the transition elements. For example, the iron atom with the normal configuration 2, 8, 14, 2, may exhibit a valence of + 2 by loss of the two outer electrons or a valence + 3 when an additional electron is lost from the incomplete third orbit.

Changes in properties across a period. As the atomic number increases across a period, from Group 1 through Group VII, the elements become progressively less electropositive (donor of electrons) and more electronegative (acceptor of electrons). Lithium readily loses electrons to form positive ions, whereas oxygen and fluorine readily accept electrons to form negative ions; beryllium and nitrogen are usually non-ionic, forming covalent compounds; boron and carbon are always covalent in their bonds with other atoms. Similar relations apply to the elements of period 3.

As the atomic number increases across a period, the atomic radius decreases. A fluorine atom is considerably smaller than a lithium atom, and a chlorine atom is smaller than a sodium atom. This decrease in size, which brings the outer electrons closer to the positive nucleus, is a contributing factor in the decrease in electron-donating power across a period.

The strongly electropositive elements are called **metals.** The electronegative elements are called **nonmetals.** The elements in the middle groups, which can act either as electron donors or electron acceptors, are called **amphoteric elements**. Therefore, *as the atomic number increases in a period, the elements become less metallic,* or in other words more nonmetallic.

The elements of Group 0 are very different from the other elements. Reference has already been made to the fact that helium and neon are inert elements. All the members of this group have complete outer shells of electrons (a stable structure) and do not undergo any ordinary reactions with the other elements. Since all the elements of this group are gases, they are called the **inert gases.**

Position of hydrogen in the classification. The position of hydrogen in the periodic classification is a problem in some respects. Since it has one electron in its outer shell, it is usually placed in Group I with lithium, sodium, and potassium. However, it is very different from these strong metals. Hydrogen is a gas at ordinary temperatures, and the metals are solids. Also, hydrogen does not form definite ionic compounds but rather forms covalent compounds. Since it falls between boron and silicon in the scale of electronegativity, it is apparently a weak electron donor compared to the metals. However, in its more common compounds it does have a covalence of 1+, and its position in Group I is therefore somewhat logical.

Nevertheless, there are some reasons for placing hydrogen in Group VII. Here it would be present with the gases fluorine and chlorine. Also, hydrogen needs one more electron to attain the stable structure of helium. It can accept one electron from strong electron donors like lithium and sodium to form the negative hydride ion, H^-. In this respect it acts like fluorine when the fluorine atom accepts an electron to form fluoride ion, F^-. However, these compounds of hydrogen are much less common than those in which hydrogen has a covalence of 1+, and the placement of hydrogen in Group VII is not favoured by many chemists.

Changes in properties within a group. In addition to the fact that the elements having similar chemical properties appear in a group, there are certain regularities in the variation of the properties of the elements within a group. This fact is helpful in summarizing the chemical characteristics of the elements.

Consider the elements in Group I. It has been determined experimentally that potassium is more active than sodium, which is in turn more active than lithium. This change in activity of the metals as the atomic number increases can readily be explained in terms of the sizes of the atoms.

The atomic structures are

Li	Na	K
(2,1)	(2,8,1)	(2,8,8,1)
3+	11+	19+

The numbers 3+, 11+, and 19+ represent the positive charge on the nucleus. The atom of potassium is larger than the atom of sodium by one extra shell of electrons. Therefore, the valence electron of potassium is farther removed from the nucleus than the valence electron of sodium. Even though the nuclear charge (number of protons in the nucleus) is eight greater than that of the sodium atom, the valence electron of potassium is less tenaciously held than that of sodium because of the increased atomic radius. Potassium therefore assumes its positive valence of 1 more readily than sodium.

The same effect is noted in Group II. Calcium is more active (loses its two valence electrons more readily) than magnesium or beryllium because of the larger atomic radius of the calcium atom. For the same reason, magnesium is more active than beryllium.

This effect of increased nuclear charge and atomic radius on chemical activity is to be noted also in the groups of electronegative elements. Experiments shows that fluorine is more active than chlorine as an acceptor of electrons. The structures of fluorine and chlorine are

F	Cl
(2,7)	(2,8,7)
9+	17+

Each has seven electrons in its outer shell but the nuclear charge of chlorine is eight greater than that of fluorine. This increased charge tends to draw the electrons of the outer shell towards the nucleus; nevertheless, the external shell of chlorine, the third shell, is farther from the nucleus than the external shell of fluorine. Since the extra electron forming the negative ion will go into the external shell, the attractive force of the positive nucleus of the smaller fluorine atom will be greater than that of the larger chlorine atom. Even though the positive charge is eight greater in the atom of chlorine, the greater distance more than counteracts the increased charges on the nucleus. The result is that fluorine accepts an electron more readily than chlorine and is therefore more active in forming the negative ion.

The tendency to share electrons with a more negative element, such as oxygen, is in the reverse order. Chlorine forms oxides in which the chlorine atom is positive with respect to oxygen. Since the seven valence electrons of chlorine are farther from the positive nucleus than those of fluorine, chlorine can more readily donate these electrons in covalent bonds. The atom of fluorine cannot donate electrons to another element because its small size in relation to its nuclear charge makes it the most electronegative (least electropositive) of all elements. The same effect is noted in Group VI. Oxygen is more active as an electron acceptor than sulfur. Sulfur, being more electropositive, has a greater tendency to show positive valences in covalent compounds.

Atomic Radius and Periodicity

Definitions of atomic radius. The radius of an atom is the closest distance to which one atom will approach another atom of any size under certain specified conditions. The first, **covalent atomic**

radius, refers to the effective distance between the nucleus of the atom and its valence shell when that atom has formed a covalent bond by the sharing of electrons. If the two bonded atoms are of the same element—as, for example, the two hydrogen atoms in H_2—the covalent atomic radius can be measured as half the distance between the nuclei of the two bonded atoms.

The second, the **van der Waals radius,** is half the distance between the nuclei of identical atoms at their point of closest approach when no bond is formed. An example is half the distance between the nuclei of two neighbouring helium atoms, which do not bond with one another. A third type of radius is the atomic radius in metals, defined as half the distance between nuclei of atoms when arranged in a metal-like crystalline structure.

All three types of atomic radius are shown in Figure 2.6. Atomic radii are measured in nanometers (nm; 1 nm = 10^{-9} m.) A table of values for the three types of atomic radius appears in Appendix C.

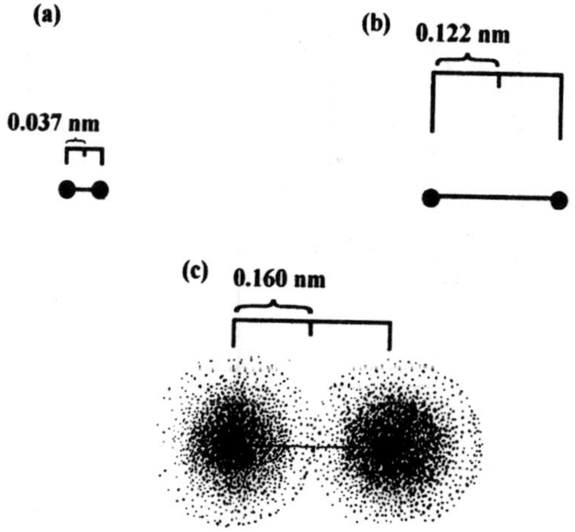

Fig. 2.6 : Three types of atomic radius.

(a) *Covalent atomic radius.* In the H_2 molecule the covalent atomic radius of hydrogen is half distance between the nuclei of the bonded hydrogen atoms.

(b) *Van der Waals radius.* The van der Waals radius is half the distance between the nuclei of identical atoms that formed no bond. Helium atoms are shown here.

(c) *Atomic radius in metals* The atomic radius in metals is half the distance between two nuclei in a crystalline metallic solid form of the element. Magnesium atoms are shown here.

In general, the van der Waals radius is greater than the corresponding covalent atomic radius. However, a few nonmetals show the opposite, relationship. For most elements, the atomic radius in metals, for a metal-like crystalline arrangement of atoms, is greater than the covalent atomic radius. Covalent atomic radii are expected to have the smallest values because the bonds that are present tend to hold the atoms more closely together.

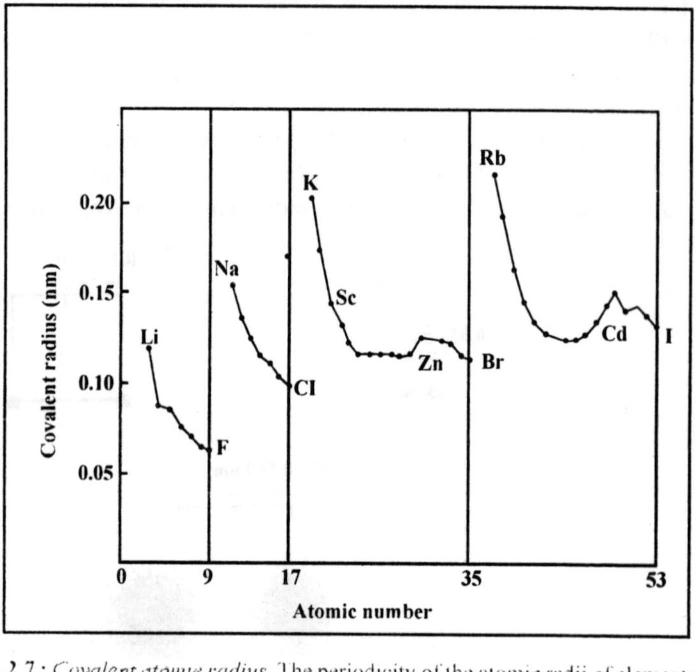

Fig. 2.7 : *Covalent atomic radius.* The periodicity of the atomic radii of elements is shown in this graph. Note how the high values reoccur for the Group I (Li, Na, K, and Rb) at the beginning of each period and how the low values reoccur for the Group 17 elements (F, Cl, Br, and I) at the end of each period.

Atomic radius as a periodic property. Atomic radius is another periodic property of the elements. Note the trends in size shown graphically in Figure 2.7 and with spheres in Figure 2.8. The major factor determining atomic radius is the number of electrons in the atom.

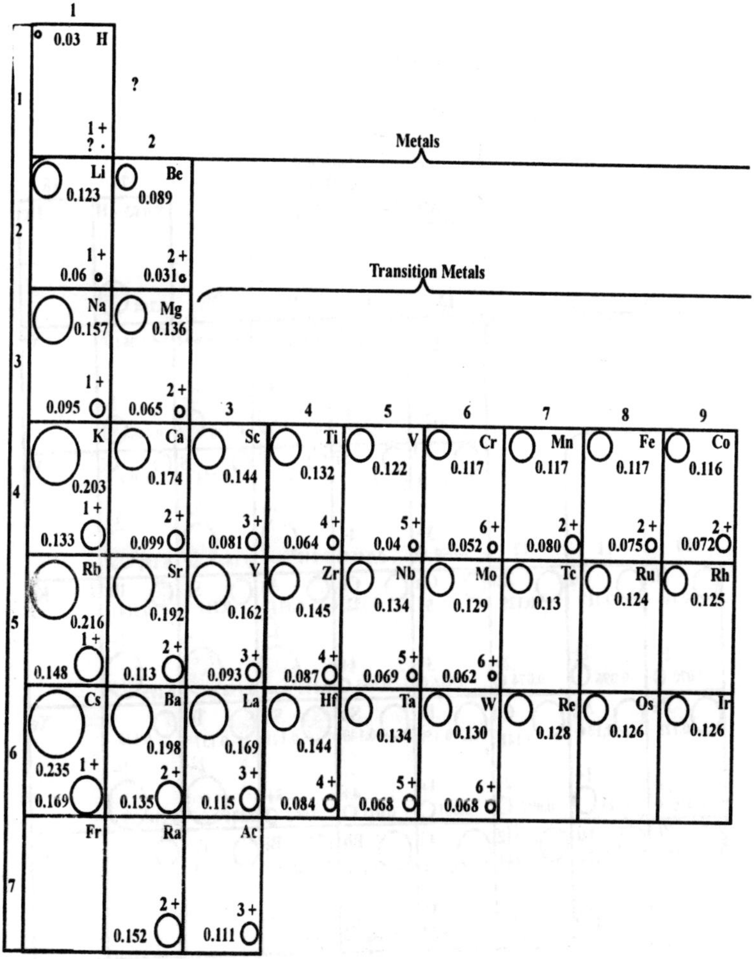

Fig. 2.8 : *Covalent atomic radii and ionic radii* Within each period, as atomic number increases, atomic radius decreases. Within each group, as atomic number increases. Ionic radii, shown in the lower right-hand corner of each box

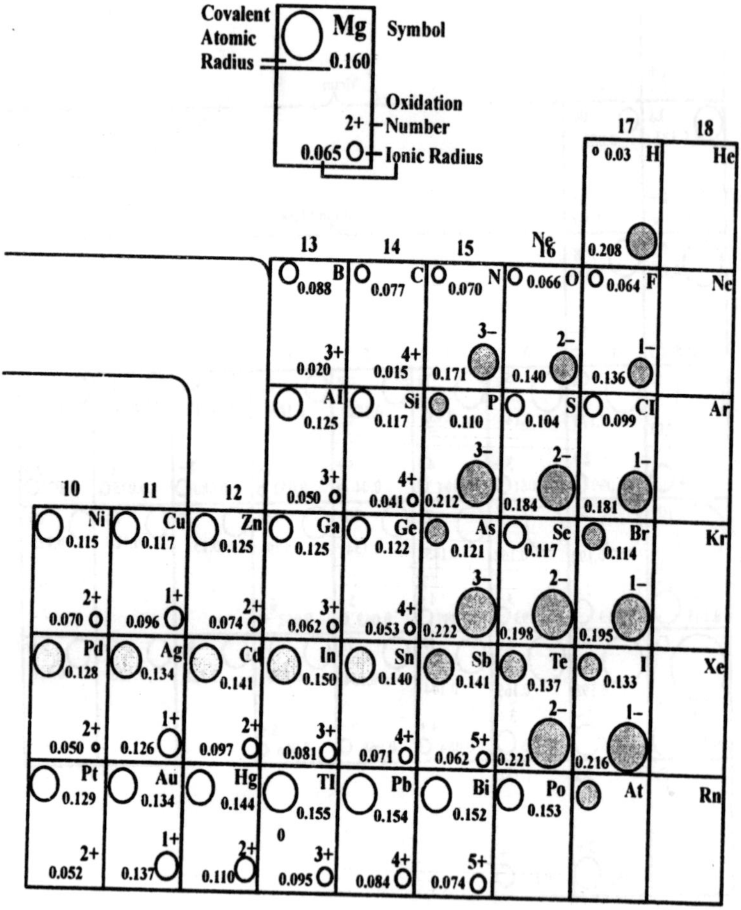
Covalent Atomic Radius
Mg 0.160
Symbol
Oxidation Number
2+
0.065
Ionic Radius
10
11
12
13
14
15
16
17
18
0.03 H
0.208
He
B 0.088 3+ 0.020
C 0.077 4+ 0.015
N 0.070 3– 0.171
O 0.066 2– 0.140
F 0.064 1– 0.136
Ne
Al 0.125 3+ 0.050
Si 0.117 4+ 0.041
P 0.110 3– 0.212
S 0.104 2– 0.184
Cl 0.099 1– 0.181
Ar
Ni 0.115 2+ 0.070
Cu 0.117 1+ 0.096
Zn 0.125 2+ 0.074
Ga 0.125 3+ 0.062
Ge 0.122 4+ 0.053
As 0.121 3– 0.222
Se 0.117 2– 0.198
Br 0.114 1– 0.195
Kr
Pd 0.128 2+ 0.050
Ag 0.134 1+ 0.126
Cd 0.141 2+ 0.097
In 0.150 3+ 0.081
Sn 0.140 4+ 0.071
Sb 0.141 5+ 0.062
Te 0.137 2– 0.221
I 0.133 1– 0.216
Xe
Pt 0.129 2+ 0.052
Au 0.134 1+ 0.137
Hg 0.144 2+ 0.110
Tl 0.155 0 3+ 0.095
Pb 0.154 4+ 0.084
Bi 0.152 5+ 0.074
Po 0.153
At
Rn

Within a period of elements (reading across the periodic table from left to right), each atom has the same number of occupied energy levels in its kernel. As the atomic number increases, the increasing number of protons attracts the valence electrons more closely to the nucleus. Therefore, within a period, atomic radius generally decreases as atomic number increases. (Note that there are no atomic radii for the noble gases because they usually do not form covalent bonds.)

Within a group of elements (reading down a column of the periodic table), each element has the same electron arrangement in the valence shell. As the atomic number increases, however, each succes-sive element has a larger kernel with more occupied energy levels, placing the valence shell farther form the nucleus. In addition, the shielding effect, which decreases the attraction between the nucleus and the valence electrons, allows the valence electrons to be located farther from the nucleus. As a result, within a group, as the atomic number of the elements increases, atomic radius generally increases.

Figure 2.9 is a summary of several trends in the properties of elements.

Ionic Radius

In many chemical changes, atoms are converted into ions by taking on or losing electrons. Atoms of metals are more likely to give off electrons to become positively charged ions. Atoms of nonmetals tend to take on electrons to become negatively charged ions. For a monatomic ion, which is formed from a single atom, the **ionic radius** is the effective distance from the nucleus of the ion to its outer shell of electrons. For many monatomic ions, this outer shell consists of eight electrons, the filled *s* and *p* sublevels of the highest principal energy level. "Effective" in the definition is a reminder that ions, like atoms, have electrons in the form of a diffuse cloud of negative charge. Therefore an ion has no identifiable edge.

Comparing atomic and ionic radii. When an atom takes on one or more electrons to become a negative ion, its radius increases. Most atoms of nonmetals take on enough electrons so that there are eight electrons in the valence shell. These added electrons increase the forces of repulsion between electrons, pushing them farther apart. As a result, the radius of the ion is greater than the radius of its parent atom. See Figure 2.10.

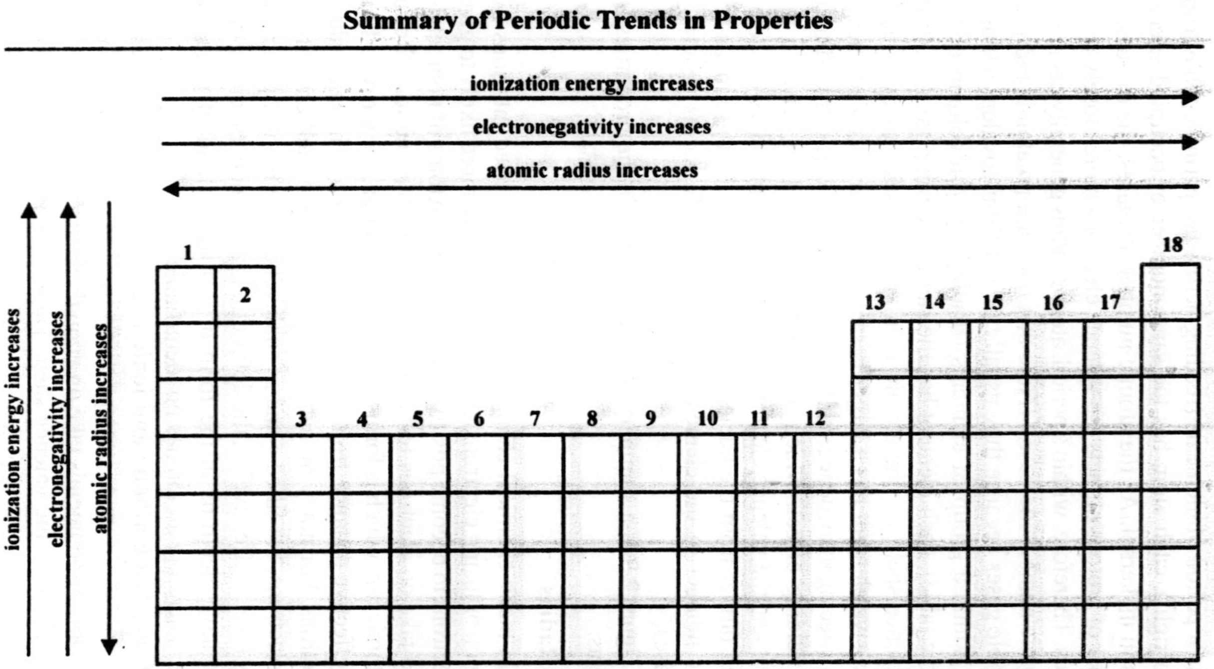

Fig. 2.9. ***Trends within periods and groups in some properties of the elements.*** How do the trends for atomic radius differ from those for the other properties shown here?

When one ore more electrons are removed from an atom, the atom's radius decreases. A typical metal forms a positive ion when it loses some or all of its valence electrons. Usually, the entire outer energy level is lost, causing a large decrease in the radius of the atom as it is converted to an ion. See Figure 2.11. Within a group of elements, as the atomic number increases, ionic radius increases. With each successive member of the group, a new occupied energy level is found in the kernel, placing the valence shell farther from the nucleus. This causes an increase in the radius of the ion.

Trends in ionic radius within a period do not follow a clear pattern. It is not easy to compare the ions of metals with the ions of nonmetals within a period because both nuclear charge and electron configuration change as the ions from the atoms.

Fig. 2.10 : ***Atomic and ionic radii of a non-metal.*** When a nitrogen atom adds three electrons to form a nitride, ion N^3, the radius, increase because of the added electrons.

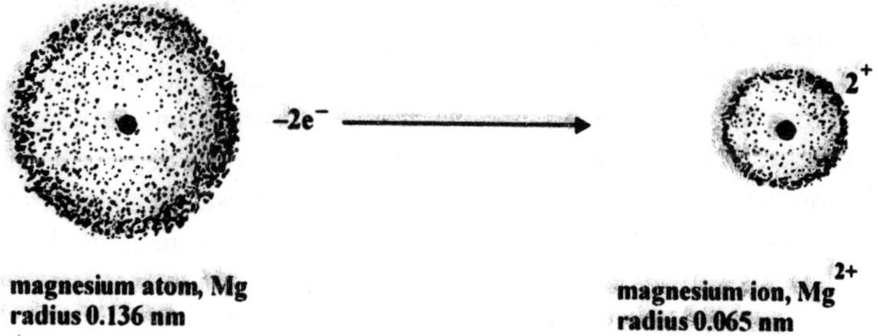

Fig. 2.11 : ***Atomic and ionic radii of a metal.*** When a magnesium atom loses two electrons to form a magnesium ion, Mg^{2+}, the radius decrease because of this loss of electrons.

Ionization Energy

When energy is absorbed by an isolated atom in its ground state, an electron may be raised from one quantized energy level to another. If enough energy is supplied, the electron may be completely removed for the atom, leaving behind a *positive ion*. The electron most easily removed from an atom is the one least strongly bound to the atom. The process of forming a positive ion from an isolated, neutral, ground-state atom by removing an electron from it is called *ionization,* and the minimum energy required to accomplish this is called the *ionization energy*. Since an isolated atom is essentially free from the perturbing influence of any nearby atom, the term implies an atom in the gaseous state. Thus ionization energy is the energy necessary to cause the following process to occur :

$$M(g) \rightarrow M^{+}(g) + e^{-}$$

Since more than one electron may be removed from an atom, the energy required to cause the above process to occur is often called the *first* ionization energy. The *second* ionization energy is the energy required to remove a second electron that is, to cause the following process to occur :

$$M^{+}(g) \rightarrow M2^{+}(g) + e^{-}$$

Ionization energies are usually expressed in electronvolts per atom, kilocalories per mole, or kilojoules per mole. We will use the SI unit, kJ mol^{-1}. (Conversion factors are given in App. B.)

The way in which ionization energies vary with atomic number provides another good illustration of the periodic law. Figures 2.12 shows this variation for the first six periods. In general the ionization energy starts low at the beginning of each periods and then increases across the period. How can we account for this trend? As nuclear charge increases across a period the outer-shell electrons become more strongly bound to the nucleus. The decreases in atomic radius across a period also tends to make these electrons more tightly bound. These two effects combine to produce an increase in ionization energy across a period.

Notice the high ionization energies for the noble gases. For all of these except helium the removal of an electron breaks up the *octet* of eight electrons in the outer shell. As we have previously mentioned, a valence-shell octet is an especially stable arrangement, and the high ionization energies of neon, argon, krypton, xenon, and radon are evidence of this stability. (Helium has only a K shell and can therefore accommodate only two electrons in its outer shell. These two electrons serve the same stability-producing function for helium that an octet does for the rest of the noble gases.

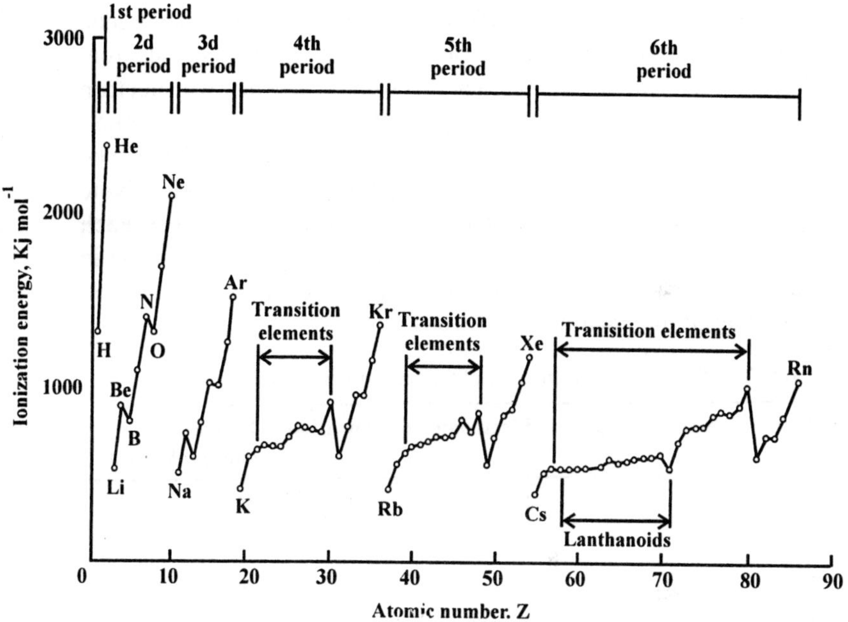

Figure 2.12 : First ionization energies.

Notice, however, that the trend across a period is not without some irregularities, even in periods 2 and 3, where there are no transition elements. Look at the values of the ionization energies for the second period, as shown in Table 7-5. The table and the graph of Fig. 2.12 both show that the ionization energy for boron ($Z = 5$) is actually *lower* than that for beryllium ($Z = 6$). The same is true for oxygen ($Z = 8$) as compared to nitrogen ($Z = 7$). These irregularities in the general trend for increasing ionization energy across a period may be accounted for as follows : In B the electron to be removed is in a $2p$ orbital, while in Be it is in a $2s$ orbital. Electrons in $2s$ orbitals have maxima in their probability-density curves which are closer to the nucleus. (Each *s* electron has a maximum which is next to the nucleus, in fact.) Thus a 2*s* electron (1) is more strongly bound to the nucleus and (2) can partially screen a 2*p* electron from the nucleus. This makes it easier to remove the 2*p* electron from B than the 2*s* electron from Be. The case of oxygen is slightly different. The ionization energy of O is lower than would otherwise be expected, because in this case the electron comes from a 2*p* orbital which has another electron it it. The two electron occupying the same orbital (region of space) repel each other, making it easier to remove one. Similar irregularities (which can be similarly explained) can be found in periods 3, 4, 5, and 6. Note also the very slow increase in ionization energy across each transition-element series. This is caused by the gradual increase in nuclear charge which is almost compensated by the increased screening of the inner-shell electrons.

Table 2.7 : First ionization energies of the second-period elements

Atom	Nuclear charge	Electronic configuration	Ionization energy, kJ mol^{-1}
Li	3 +	$1s^2 2s^1$	520
Be	4 +	$1s^2 2s^2$	899
B	5 +	$1s^2 2s^2 2p_x^{\ 1}$	801
C	6 +	$1s^2 2s^2 2p_x^{\ 1} 2p_y^{\ 1}$	1086
N	7 +	$1s^2 2s^2 2p_x^{\ 1} 2p_y^{\ 1} 2p_z^{\ 1}$	1402
O	8 +	$1s^2 2s^2 2p_x^{\ 2} 2p_y^{\ 1} 2p_z^{\ 1}$	1314
F	9 +	$1s^2 2s^2 2p_x^{\ 2} 2p_y^{\ 2} 2p_z^{\ 1}$	1681
Ne	10 +	$1s^2 2s^2 2p_x^{\ 2} 2p_y^{\ 2} 2p_z^{\ 2}$	2081

How does ionization energy vary down a group in the periodic table? Figure 7-5 shows that it decreases. The decrease is additionally shown for the alkali metals (group IA) in Table 2.8. The increased screening effect of the inner-shell electrons (recall that the number of shells increases going down any group) largely compensates for the increased nuclear charge, so that the main consideration is the increase in atomic radius. Thus ionization energy decreases; it becomes easier to remove an electron.

Table 2.8 : First ionization energies of the alkali metals (group IA)

Atom	Nuclear charge	Electronic configuration	Ionization energy, kJ mol^{-1}
Li	3 +	$[He]2s^1$	520
Na	11 +	$[Ne]3s^1$	496
K	19 +	$[Ar]4s^1$	419
Rb	37 +	$[Kr]5s^1$	403
Cs	55 +	$[Xe]6s^1$	376

The second, third, fourth, and so on, ionization energies are the energies needed to remove a second, third, fourth, and so on, electron, respectively. In Fig. 2.13 are shown the first through the sixteenth ionization energies for sulfur (Z = 16). The pronounced increase in successive ionization energies is a result of the fact that each electron must leave an ion which is more highly charged than the ion left by the previous electron. Since the attractive force between the departing electron and the ion increases accordingly, ionization energy increases. Note, however the large jump at the seventh ionization energy and again at the fifteenth. These can not be explained on the basis of nuclear charge increase alone. The seventh electron must come from a completed octet in the L shell, and this shell is closer to the nucleus than the M shell, which has supplied the first through the sixth electrons. Again at the fifteenth a sharp increase occurs, this time because the previously unused K shell must be broken into. Removal of an electron from an outer-shell noble-gas configuration (two electrons in the case of the K shell; eight electrons for all others) requires an extra amount of energy.

Electron Affinity

An atom can pick up an electron, too. The resulting particle is a *negative* ion, and the process can be written

$$X(g) + e^- \rightarrow X^-(g)$$

This process is normally accompanied by the *release* of energy, the amount of which measures how tightly the electron becomes bound to the atom. The *electron affinity* of an atom is accordingly defined as the *amount of energy released when a gaseous (isolated) ground-state atom gains an electron.*

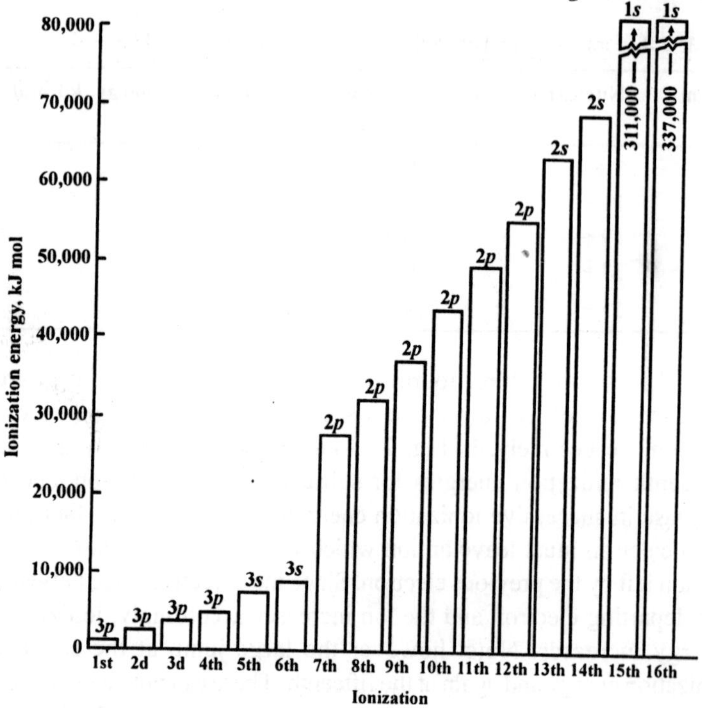

Fig. 2.13 : Ionization energies for sulphur. (The "origin" of each electron is shown.)

Electron affinities are difficult to measure, and accurate values are not known for all elements. Values for the elements of the first three periods are shown plotted against atomic number in Fig. 2.14. (Not all of these have been experimentally obtained; a few are taken from theoretical calculations.) Some of the electron affinities are negative; this means that energy is *absorbed* when the electron is added.

The periodicity in electron affinity shows clearly in Fig. 2.14. Across a period, the values generally increase as nuclear charge increases. Each alkali metal (Li, Na) has a small (positive) electron affinity. The next elements, the alkaline-earth metals (Be and Mg), have very negative values, however. This can be explained by the fact that these atoms must accept the added electron in the 2*p* and 3*p* subshell, respectively. Screening by the 2*s* and 3*s* subshells lowers the effective nuclear charge "felt" by these 2*p* and 3*p* electrons; so electron affinity is low.

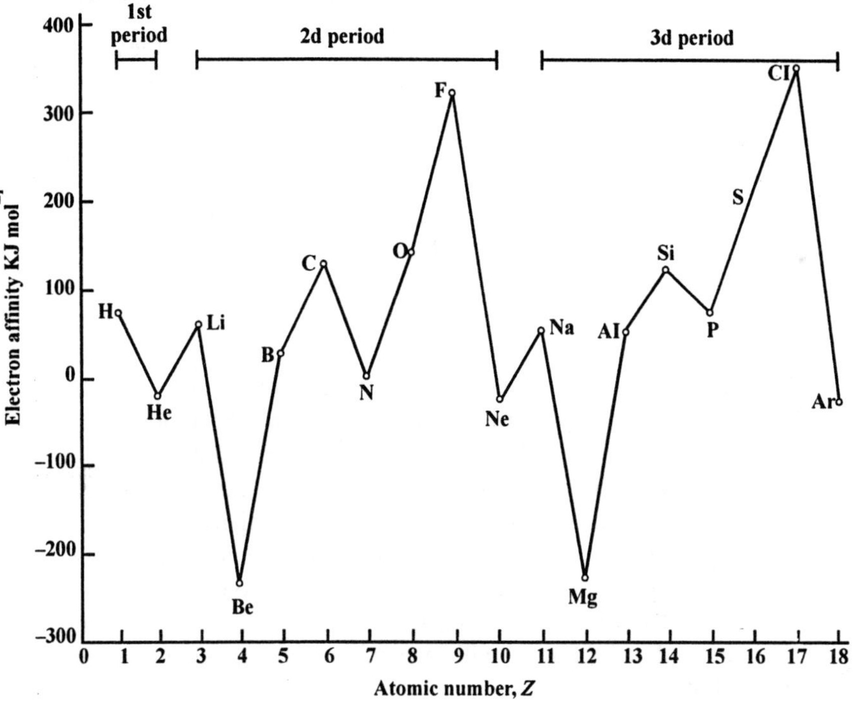

Fig. 2.14 : Electron affinities of the elements of the first three periods. (A few values have been calculated theoretically.)

Table 2.9 : Electron affinities of the halogens (group VIIA)

Atom,	Nuclear charge	Electronic configuration	Electron affinity, kJ mol^{-1}
F	9 +	$[He]2s^22p^5$	333
Cl	17 +	$[Ne]3s^23p^5$	348
Br	35 +	$[Ar]4s^24p^5$	324
I	53 +	$[Kr]5s^25p^5$	296

After these elements the electron affinity increases across the period until it drops down suddenly at group VA. This sudden decrease is accounted for by the fact that the added electron must enter an already half-full $2p$ (in the case of N) or $3p$ (in P) subshell. Repulsion between the two electrons in the same orbital lowers the electron affinity. From here on the increasing nuclear charge increases the electron affinity until it reaches a maximum in each period at a halogen (group VII). Addition of an electron to a halogen atom completes the octet in the valence shell. Finally, the electron affinities of the noble gases are low; each of these atoms shows little tendency to start a new shell.

Going down a group electron affinities undergo a general decrease. This is a result of the fact that the valence shell is progressively farther from the nucleus with progressively more inner shells screening the nuclear charge. Values for the halogens are shown in Table 2.9. Each of these elements has a comparatively high electron affinity; adding an electron completes the outer-shell octet and creates a noble-gas configuration. The value for fluorine, is however, out of line, being smaller than that of chlorine. As a matter of fact, all of the elements of the second period have what appear to be anomalously low electron affinities. (See Fig. 2.14.) This is apparently an indirect result of the small sizes of the atoms of these elements. Electron repulsion within the compact L shell seems to partially compensate for the high attraction from the nucleus, thus reducing electron affinity.

As is the case with ionization energies, electron affinities may be expressed in any convenient energy units. Electronvolts (per atom) are commonly used but we will be consistent and use kilojoules per mole of atoms.

Electronegativity and Periodicity

Another property of elements that is related to the forces of attraction between the positive nucleus and negative electrons in atoms is electronegativity. **Electronegativity** is a measure of the ability of an atom of an element that is chemically combined with another element to attract electrons to itself. The forces that hold atoms of chemically combined elements together are called chemical bonds. Put another way, electronegativity is a measure of the force of attraction that exists between an atom and a shared pair of electrons in a covalent bond. *Covalent bonds* are formed when pairs of electrons are shared between atoms.

The principles that determine the electronegativity of an element are similar to those that apply to ionization energy. The scale of electronegativity proposed by Linus Pauling (1901-) is used most often. Values range from a low of 0.7 for several metals in Group 1 to a high of 4.0 for fluorine in Group VIIA. Figure 2.15 gives electronegativity values for a number of elements.

Period	1	2	13	14	15	16	17	18
1	H 2.2							He
2	Li 1.0	Be 1.5	B 2.0	C 2.6	N 3.1	O 3.5	F 4.0	Ne
3	Na 0.9	Mg 1.2	Al 1.5	Si 1.9	P 2.2	S 2.6	Cl 3.2	Ar
4	K 0.8	Ca 1.0	Ga 1.6	Ge 1.9	As 2.0	Se 2.6	Br 2.9	Kr
5	Rb 0.8	Sr 1.0	In 1.7	Sn 1.8	Sb 2.1	Sb 2.3	I 2.7	Xe
6	Cs 0.7	Ba 0.9	Tl 1.8	Pb 1.8	Bi 1.9	Po 2.0	At 2.2	Rn
7	Fr 0.7	Ra 0.9						

Fig. 2.15 : *Electronegativity.* Electronegativity values are given on the Pauling scale, using a value of 4.0 for fluorine. No electronegativity values are assigned to Group 18 elements because these elements do not commonly share electrons.

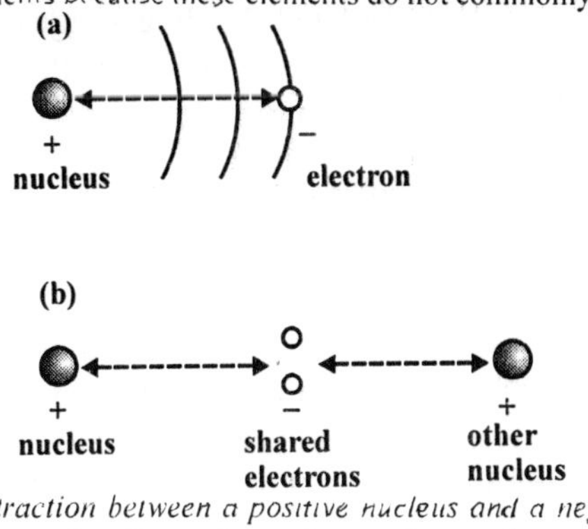

Fig. 2.16 : *Attraction between a positive nucleus and a negative electron.* (a) ionization of the nucleus for a valence electron that is to be removed from an atom. (b) Electronegativity is a measure of the attraction of the nucleus for a shared pair of electrons.

Pauling

Pauling pointed out that since reactions of the type:

$$A_2 + B_2 \rightarrow 2AB$$

are almost always exothermic, the bond formed between the two atoms A and B must be stronger than the average of the single bond energies of A-A and B-B molecules. For example:

$H_{2\,(gas)} + F_{2(gas)} \rightarrow 2HF_{(gas)}$ $\Delta H = -\,5396$ kJ mol^{-1}

$H_{2\,(gas)} + Cl_{2(gas)} \rightarrow 2HCl_{(gas)}$ $\Delta H = -\,1852$ kJ mol^{-1}

$H_{2\,(gas)} + Br_{2(gas)} \rightarrow 2HBr_{(gas)}$ $\Delta H = -\,727$ kJ mol^{-1}

The bonding molecular orbital for AB (ϕ_{AB}) is made up from contributions from the wave functions for the appropriate atomic orbitals (ψ_A and ψ_B).

$$\phi_{AB} = (\psi_A) + \text{constant}\ (\psi_B)$$

If the constant is greater than 1, the molecular orbital is concentrated on the B atom, which therefore acquires a partial negative charge, and the bond is partly polar.

$$\overset{\delta^+}{A}\text{—}\overset{\delta^-}{B}$$

Conversely, if the constant is less than 1, atom A gains a partial negative charge. Because of this partial ionic character, the A-B bond is stronger than would be expected for a pure covalent bond. The extra bond energy is called delta Δ.

Δ = (actual bond energy) – (energy for 100% covalent bond)

The bond energy can be measured, but the energy of a 100% covalent bond energy be calculated. Pauling suggested the 100% covalent bond energy be calculated as the geometric mean of the covalent energies of A-A and B-B molecules.

$$E_{100\%\ \text{covalent A–B}} = \sqrt{(E_{A-A} \cdot E_{B-B})}$$

The bond energy in A-A and B-B molecules can be measured and so :

$$\Delta = (\text{actual bond energy}) - \sqrt{(E_{A-A} \cdot E_{B-B})}$$

Pauling states that the electronegativity difference between two atoms is equal to $0.208\sqrt{\Delta}$, where Δ is the extra bond energy in kcal mol^{-1} (converting the equation to SI units gives $0.017\sqrt{\Delta}$, where Δ is measured in kJ mol^{-1}.)

Pauling evaluated $0.208\sqrt{\Delta}$ for a number of bonds and called this the electronegativity difference between A and B. Repeating Pauling's calculations with SI units for energy, we can evaluate $0.17\sqrt{\Delta}$:

Bond	d(kJ mol^{-1})	$0.10187\sqrt{\delta}$	
C-H	24.3	0.50	i.e. $\chi C - \chi H = 0.50$
H-Cl	102.3	1.02	i.e. $\chi Cl - \chi H = 1.02$
N-H	105.9	1.04	i.e. $\chi N - \chi H = 0.04$

(χ(chi)) = electronegativity of atom)

If $\chi H = 0$ then the electronegativity values for C, Cl and N would be 0.50, 1.02 and 1.04 respectively. Pauling changed the origin of the scale from $\chi H = 0$ to $\chi H = 2.05$ to avoid having any negative values in the table of values, and this made the value for C become 2.5 and the value for F become 4.0. At the same time the values for a number of other elements approximated to whole numbers : Li = 1.0, B = 2.0, N = 3.0. Thus by adding 2.05 to the values calculated in this way we can obtain the usually accepted electronegativity values (Table 2.10).

If two atoms have similar electronegativity, that is a similar tendency to attract electrons, the bond between them will be predominantly covalent. Conversely a large difference in electronegativity leads to a bond with a high degree of polar character, that is a bond that is predominantly ionic.

Table 2.10 : Pauling's electronegativity coefficients (for the most common oxidation states of the elements)

						H 2.1
Li 1.0	Be 1.5	B 2.0	C 2.5	N 3.0	O 3.5	F 4.0
Na 0.9						Cl 3.0
K 0.8						Br 2.8
Rb 0.8						I 2.5
Cs 0.7						

Rather than have two extreme forms of bond (ionic and covalent), Pauling introduced the idea that the ionic character of a bond varies with the difference in electronegativity as shown in Figure 6.3. This graph is based on the ionic characters HI 4% ionic, HBr 11%, HCl 19% and Hf 45%, which are known from dipole measurements. Fifty per cent ionic character occurs when the electronegativity difference between the atoms is about 1.7, so for a

Table 2.11 : Pauling electronegativity values

Period \ Group	1	2	3	4	5	6	7	8	9	10	11	12	13	14	15	16	17	18
1	H 2.1																H 2.1	He
2	Li 1.0	Be 1.5											B 2.0	C 2.5	N 3.0	O 3.5	F 4.0	Ne
3	Na 0.9	Mg 1.2											Al 1.5	Si 1.8	P 2.1	S 2.5	Cl 3.0	Ar
4	K 0.8	Ca 1.0	Sc 1.3	Ti 1.5	V 1.6	Cr 1.6	Mn 1.5	Fe 1.8	Co 1.8	Ni 1.8	Cu 1.9	Zn 1.6	Ga 1.6	Ge 1.8	As 2.0	Se 2.4	Br 2.8	Kr
5	Rb 0.8	Sr 1.0	Y 1.2	Zr 1.4	Nb 1.6	Mo 1.8	Tc 1.9	Ru 2.2	Rh 2.2	Pd 2.2	Ag 1.9	Cd 1.7	In 1.7	Sn 1.8	Sb 1.9	Te 2.1	I 2.5	Xe
6	Cs 0.7	Ba 0.9	La 1.1	Hf 1.3	Ta 1.5	W 1.7	Re 1.9	Os 2.2	Ir 2.2	Pt 2.2	Au 2.4	Hg 1.9	Tl 1.8	Pb 1.8	Bi 1.9	Po 2.0	At 2.2	Rn
7	Fr 0.7	Ra 0.9	Ac 1.1															

larger difference than this a bond is more ionic than covalent. Similarly, if the electronegativity difference is less than 1.7, the bond is more covalent than ionic. It is better to describe a bond such as one of those in BF_3 as 63% ionic, rather than just ionic.

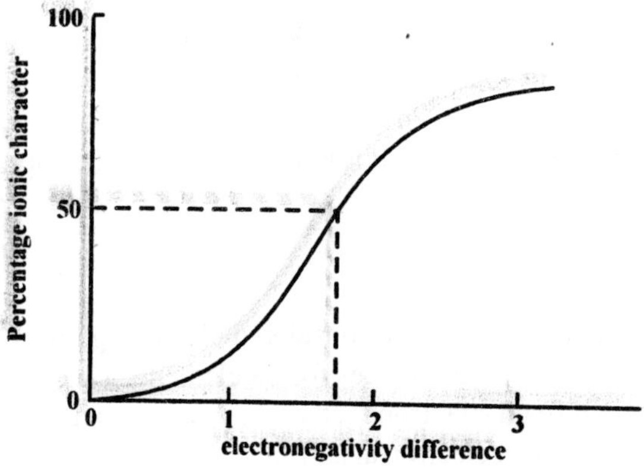

Figure 2.17 : Electronegativity difference.

Mulliken

In 1934, Mulliken suggested an alternative approach to electronegativity based on the ionization energy and electron affinity of an atom. Consider two atoms A and B. If an electron is transferred from A to B, forming ions A^+ and B^-, then the energy change is the ionization energy of atom A (I_A) minus the electron affinity of atom B (E_B), that is $I_A - E_B$. Alternatively, if the electron was transferred the other way to give B^+ and A^- ions, then the energy change would be $I_B - E_A$. If A^+ and B^- are actually formed, then this process required less energy, and

$$(I_A - E_B) < (I_B - E_A)$$

Rearranging

$$(I_A + E_A) < (I_B + E_B)$$

Thus Mulliken suggested that electronegativity could be regarded as the average of the ionization energy and the electron affinity of an atom.

$$Electronegativity = \frac{(I + E)}{2}$$

Mulliken used I and E values measured in electron volts, and the values were about 2.8 times larger than the Pauling values. We now measure I and E in kJ mol^{-1}. The energy 1 eV/molecule = 96.48 kJ mol^{-1}, so the commonly accepted Pauling values are more nearly obtained by performing this calculation $(I + E)$ $(2 \times 2.8 \times 96.48)$ or $(I + E)/540$.

This method has a simple theoretical basis, and also has the advantage that different values can be obtained for different oxidation states of the same element. It suffers from the limitation that only a few electron affinities are known. It is more usual to use the approach based on bond energies.

Allred and Rochow

In 1958 Allred and Rochow considered electronegativity in a different way, and worked out values for 69 elements. (See Further Reading.) They defined electronegativity as the attractive force between a nucleus and an electron at a distance equal to the covalent radius. This force F *is* electrostatic, and is given by :

$$F = \frac{e^2 . Z_{effective}}{r^2}$$

where e is the charge on an electron, r is the covalent radius, and $Z_{effective}$ is the effective nuclear charge. The latter is the nuclear charge modified by screening factors for the orbital electrons. The screening factors vary depending on the principal quantum number (the shell that the electron occupies), and the type of electron, *s, p, d* or *f.* Screening factors have been worked out by Slater, so this provides a convenient method of calculating electronegativity values. These F values may be converted to electronegativity values on the Pauling scale of values using an empirical relationship :

$$\chi = 0.744 + \frac{0.359 Z_{effective}}{r^2}$$

The electronegativity values so obtained agree quite closely with those obtained by Pauling and Mulliken.

As the oxidation number of an atom increases, the attraction for the electron increases, so the electronegativity should also increase.

Allred and Rochowl's method gives slightly different values :

Mo(II)	2.18	Fe(II)	1.83	Tl(I)	1.62	Sn(II)	1.80
Mo(III)	2.19	Fe(III)	1.96	Tl(III)	2.04	Sn(IV)	1.96
Mo(IV)	2.24						
Mo(V)	2.27						
Mo(VI)	2.35						

Allred and Rochow's method depends on measuring covalent radii (and these are obtained with great accuracy by X-ray crystallography) so it might be expected to yield very accurate electronegativity values.

Metallic properties

If we proceed across a period in the periodic table, we observe a gradual transition in character of the elements from *metallic* to *nonmetallic*. What is a metal? A metal is a substance whose physical properties include high electrical and thermal conductivity and a remarkable luster or shine. One chemical property of a metal is its ability to form a compound with hydrogen and oxygen, a *hydroxo* compound, which is known as a *base*. Nonmetals are just the opposite. They are poor or nonconductors of heat and electricity, have a dull luster, and form hydroxo compounds which are known as *acids*. For now it is enough to know that they are classes of compounds.)

In the second period there is a gradual change from the typically metallic properties of lithium ($Z = 3$) to the typically nonmetallic properties of fluorine ($Z = 9$). The noble gas neon then intervenes between fluorine and the next metal, sodium ($Z = 11$), where the sequence from metal to nonmetal starts again. Elements between the metals on the left and the nonmetals on the right are intermediate in properties and are called *metalloids*, or *semimetals*.

In the long periods 4 through 7, a much longer interval is needed for the gradation from metal to nonmetal to occur. In the fourth period, for instance, potassium ($Z = 19$) and calcium ($Z = 20$) are, excellent metals, but the appearance of clearly nonmetallic properties does not occur until the last part of the period. The transition elements scandium ($Z = 21$) through zinc ($Z = 30$) are more metallic than nonmetallic, and are thus often called *transition metals*.

Uses of the periodic table. For the chemist, the periodic table is a very useful classification of the elements. Instead of learning the properties of more than 100 elements, the chemist may master the general properties of each of the groups. If an element is known to be in a certain group, then by association with other members of the same group, the properties of which are known, its properties may be predicted. As a matter of fact, before the discovery of many of the chemical elements, the properties of the unknown elements were predicted with a high degree of accuracy. Thus Mendeleev predicted the properties of eka-silicon, later named germanium, many years before it was discovered. How accurate his predictions were is evident from Table 2.11.

Table 2.11 : Comparison of Predicted Properties of Eka-silicon with Known Properties of Germanium.

PROPERTY	EKA-SILICON	GERMANIUM
Atomic weight	72	72.6
Characteristic valence	4	4
Density	5.5	5.47
Colour	Gray	Grayish-white
Reaction with water	Will decompose steam with difficulty	No reaction with steam
Action of acids	Slight	No reaction with HCl, oxidized by HNO_3
Action of bases	Slight	No reaction
Properties of the chloride	Boiling point below 100°, density 1.9	Boiling point 86°, density 1.9
Properties of the oxide	White, refractory, density 4.7	White refractory, density 4,703

Imperfections of the classification. Despite its great usefulness, the periodic classification is not without its faults.

Hydrogen perhaps does not fit into the table entirely satisfactorily. It is usually placed in Group I*A*, since it does possess one valence electron (like the alkali metals) and combines with nonmetals such as the halogens and sulfur. However, the latter compounds are covalent in character and not like the electrovalent combination of alkali metals with the halogens. In some respects hydrogen resembles the members of Group VII*A*, since it appears to be nonmetallic in character and furthermore, like the halogens, may combine with the alkali metals to form ionic compounds. For example, sodium hydride (NaH) is ionic in character, like NaCl.

In the short form table, in which the *A* and *B* series fall within the same group, there is valid criticism that the two series are not similar in many respects. For example, Group I contains the most active metals—potassium, sodium, and so on—and also the very inactive metals—silver, copper, and gold. This objection is removed in the long form table, in which each series of elements is in a separate group.

In either table, a group of fifteen elements with very similar chemical and physical properties-the *rare earth* or *lanthanide* series, must be fitted into a single space. Of course, the long form table could be further elongated to take care of these elements, but this would make the table unwieldy without adding to its usefulness. In any event, these elements should perhaps be considered separately, since they constitute a unique group as far as structural characteristics are concerned. In these elements, as successive electrons are added, the electrons enter a deep inner orbit (the fourth) until the orbit has its capacity of thirty-two electrons, while the two outer orbits remain unchanged. Consequently, the chemical characteristics of these elements, which are governed mainly by the valence electrons in the two outer orbits, are very nearly the same for this entire group of fifteen elements. A second rare earth series, called the *actinide* series, starts with element number 89; its members, like those of the lanthanide series, are very similar in chemical behaviour.

However, the advantages of making possible the study of the properties of a group of similar elements rather than the more tedious study of each element far outweigh any disadvantages of the periodic classification.

Polarizing Power and Polarizability—Fajans's Rules

Consider making a bond theoretically by bringing two ions A^+ and B^- together to their equilibrium distance. Will the bond remain ionic, or will it become covalent? Ionic and covalent bonding are two extreme types of bonding, and almost always the bonds formed are intermediate in type, and this is explained in terms of polarizing (that is deforming) the shape of the ions.

The type of bond between A^+ and B^- depends on the effect one ion has on the other. The positive ion attracts the electrons on the negative ion and at the same time it repels the nucleus, thus distorting or polarizing the negative ion. The negative ion will also polarize the positive ion, but since anjons are usually large, and cations small, the effect of a large ion on a small one will be much less pronounced. If the degree of polarization is quite small, then the bond remains largely ionic. If the degree of polarization is large, electrons are drawn from the negative ion towards the positive ion, resulting in a high concentration of electrons between the two nuclei, and a large degree of covalent character results.

The extent to which ion distortion occurs depends on the power of an ion to distort the other ion (that is on its polarizing power) and also on how susceptible the ion is to distortion (that is on its polarizability). Generally the polarizing power increases as ions become smaller and more highly charged. The polarizability of a negative ion is greater than that of a positive ion since the electrons are less firmly bound because of the differences in effective nuclear charge. Large negative ions are more polarizable than small ones.

Fajans put forward four rules which summarize the factors favouring polarization and hence covalency

1. *A small positive ion favours covalency.*
   In small ions the positive charge is concentrated over a small area. This makes the ion highly polarizing, and very good at distorting the negative ion.
2. *A large negative ion favours covalency.*
   Large ions are highly polarizable, that is easily distorted by the positive ion, because the outermost electrons are shielded from the charge on the nucleus by filled shells of electrons.
3. *Large charges on either ion, or on both ions, favour covalency.*
   This is because a high charge increases the amount of polarization.
4. *Polarization, and hence covalency, is favoured if the positive ion does* **not** *have a noble gas configuration.*
   Examples of ions which do not have a noble gas configuration include a few main group elements such as Ti^{+}, Pb^{2+} and Bi^{3+} many transition metal ions such as Ti^{3+}, V^{3+}, Cr^{2+}, Mn^{2+} and Cu^{+}, and some lanthanide metal ions such as Ce^{3+} and Eu^{2+}. A noble gas configuration is the most effective at shielding the nuclear charge, so ions without the noble gas configuration will have high charges at their surfaces, and thus be highly polarizing.

3

Nuclear Chemistry

The discovery of radioactivity. In 1895, Wilhelm Roentgen (1845-1923) discovered X rays, a type of radiation later shown to be highly energetic electromagnetic waves with very short wavelengths. A year later, Henri Becquerel (1852-1909) found that uranium ores emit radiation that resembles X rays in some of its properties. Both kinds of radiation are invisible and can pass through materials opaque to visible light. Both were able to affect a photographic plate wrapped in black paper.

After working with Becquerel, Marie Curie (1867-1934) and her husband Pierre Curie (1859-1906) isolated from uranium ores two new elements, polonium and radium. They found that radium gives off radiation at a rate 2 million times that of uranium. The intensity of this radiation made it possible for the Curies to observe additional properties of it. While passing through air, the radiation made the air a conductor of electricity. That is, gas molecules in the air became ionized. It caused phosphorescent substances, such as zinc sulfide, to glow brightly. When exposed to this radiation bacteria and other small organisms were killed. The temperature near the radium was several degrees higher than that of the surroundings. This showed that the radium was releasing energy.

It later was shown that this radiation was produced when the nuclei of atoms changed, producing atoms of other elements. Such a change of one element into an entirely different element (or elements) is called **transmutation.** The radium atoms (atomic number 881) broke down to produce radon (atomic number 86) and helium (atomic number 2). This type of change, in which radiation is produced while the nuclei of certain elements spontaneously disintegrate to produce other elements, is called radioactivity. It also is known as **radioactive decay.**

Radioactive rays. By means of an electric field, Rutherford was able to identify three types of radiation from radioactive substances. A bit of radioactive substance was placed in a depression in a block of lead (Figure 3.1).

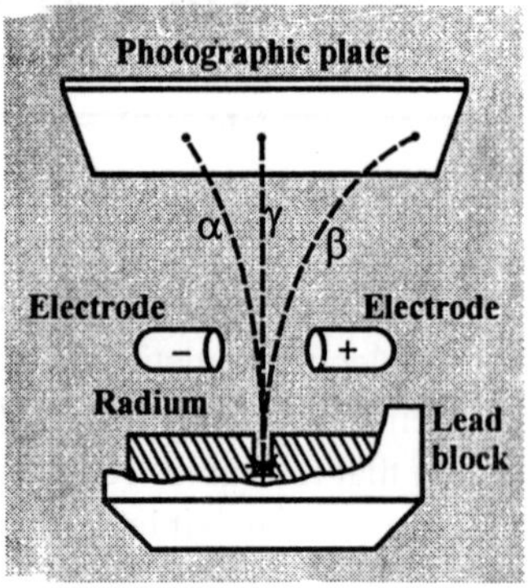

Fig. 3.1 : Three kinds of rays produced from the radioactive disintegration of radium.

All rays except those moving directly upward were absorbed by the lead. Electrodes and a photographic plate to record the emitted radiation were placed above the block of lead. This experiment shows that the radiation may be resolved into three kinds: one kind unaffected by the electrodes, the other two kinds showing opposite charges as evidenced by the bending of the rays in opposite directions. From the extent and the direction of bending, it is possible to calculate the masses and charge of the radiation. The three types of radiation were named *alpha* (α), *beta* (β), and *gamma* (γ) rays.

Alpha rays, positively charged particles with a mass four times that of the hydrogen atom, have been shown by spectral studies to be nuclei of helium atoms, each particle consisting of 2 neutrons and 2 protons. These particles are expelled from radioactive substances at moderate velocities (10,000 to 12,000 miles per second) and have a low penetrating power since they may be stopped by thin sheets of paper. Air through which alpha particles pass becomes ionized due to electrons knocked out of molecules in the air on impact with these relatively heavy particles. If this ionized air is saturated with water vapour and suddenly cooled, the ionized particles in the air act as nuclei for the condensation of small droplets of water, and these particles are rendered visible as a fog. C. T R. Wilson was able to follow the path of an alpha particle by photographing the fog tracks thus produced.

The expulsion of alpha particles from radioactive material may also be observed with the aid of a *spinthariscope*, a device consisting of a narrow tube which is fitted at one end with a lens and at the other end with a screen coated with zinc sulfide (Figure 3.2). If a piece of radium is placed in front of the screen and the screen is observed through the lens, flashes of light may be seen as a result of alpha particles colliding with zinc sulfide particles.

Beta rays are negatively charged particles and are identical with electrons. They are sometimes referred to as "high speed electrons," since their speed sometimes approaches that of light (186,000 miles per second). These rays are somewhat more penetrating than alpha particles, passing through thin layers of metals.

Gamma rays are of high frequency and are identical with X rays of very short wave length. They possess relatively high penetrating power and like alpha and beta rays affect a photographic plate. They carry no charge. Gamma rays constitute the portion of radioactive emission which affected the photographic plate of Becquerel even though the plate had been wrapped in several layers of black paper.

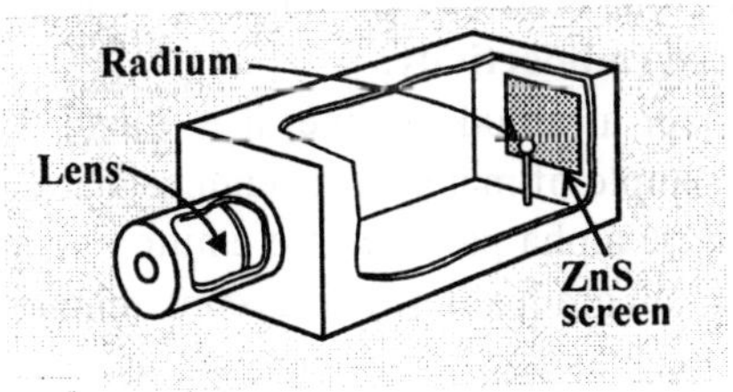

Fig. 3.2 : Spinthariscope.

The properties of alpha particles, beta particles, and gamma rays are contrasted in Figure 3.3.

Types of Radiation

	Alpha rays or particles	Beta rays or particles	Gamma rays
nature	sometimes behave like particles; sometimes like waves	sometimes behave like particles; sometimes like waves	electromagnetic waves of extremely short wavelength
speed	about 1/10 the speed of light	approaching the speed of light	speed of light
mass	4 atomic mass units	0.00055 atomic mass unit	0
penetrating power	relatively weak (can be stopped by a single sheet of paper)	greater than alpha (can be stopped by a thin sheet of aluminum)	very penetrating (several centimeters of led needed to stop them)
ionizing ability	will ionize gas molecules	will ionize gas molecules	will ionize the atoms in flesh, causing severe damage to the

Fig. 3.3 : *Comparison of alpha and beta particles and gamma rays.*

Natural Radioactivity

Isotopes of elements that ate identified by the number of their protons and neutrons are called **nuclides.** Most nuclides found in nature are stable, but some are not. These unstable nuclides are called radioactive isotopes or radioisotopes. There are no stable isotopes of elements with atomic numbers above 83. All such elements as they occur on the earth today are radioactive. The radioactive decay of naturally occurring elements is called **natural radioactivity.**

All but two of the lighter elements occur on the earth as stable isotopes. Radioactive isotopes of the lighter, stable elements can be made artificially by bombarding the nuclei of stable isotopes with high-energy particles. The radioactive decay of nuclei made unstable artificially is called **induced radioactivity.**

Alpha emission. Over a period of time, it is found that the amount of uranium in a sample of material containing that element decreases as more and more alpha particles are emitted. A sample that starts out as pure uranium gradually becomes contaminated with atoms of other elements such as thorium, radon, radium, and lead. These changes are nuclear changes because elements not originally present are being produced. They can be represented by nuclear equations. The first change that occurs in the transmutation of uranium is the emission of an alpha particle by the uranium atom. This is known as **alpha** emission. It can be represented by the nuclear word equation :

Uranium atom $\rightarrow$ thorium atom + alpha particle (Eq. 1)

This word equation represents a nuclear reaction with the production of thorium, an element not originally present. A **nuclear reaction** is a reaction that involves changes in the nuclei of atoms. In these changes both the charge and the mass of a nucleus may change. Therefore, when writing a symbol equation for a nuclear reaction, the charge and the mass of each particle involved in the reaction must be expressed.

Experiments show that the isotope of uranium that emits an alpha particle to produce thorium is uranium-238. The equation for the nuclear reaction expressed in symbols is:

$$^{238}_{92}U \rightarrow ^{234}_{90}TH + ^{4}_{2}He \qquad \textbf{(Eq. 2)}$$

The subscript represents the number of positive or negative charges on the particle. The charge is assumed to be positive unless indicated otherwise. Because each uranium isotope has 92 protons in its nucleus (atomic number 92), the subscript 92 is written at the lower left of the symbol U. Likewise, the subscript 90 is used for thorium and the subscript 2 is used for the helium nucleus. Each alpha particle emitted contains two protons, and each thorium nucleus contains 90 protons (atomic number 90). The nuclear equation is balanced because the sums of the superscripts on each side of the equation are equal, and the sums of the subscripts are equal. The sums of the superscripts indicate conservation of mass, and the sums of the subscripts indicate conservation of charge.

The periodic table shows that the element with the atomic number 91 is protactinium, Pa. The mass number of the protactinium isotope when Beta particle is ejected from the nucleus, a process in the nucleus takes place which is as through a neutron decompose to give a proton and electron

$$^{1}_{0}n \rightarrow ^{1}_{1}P + ^{0}_{-1}e$$

The proton remain in the nucleus, but electron is ejected at high speed. The formation of proton in the nucleus increased. The net positive change on the nucleus by one unit, thus the atomic number of the atom is increased by one unit, reejection of an electron from the nucleus of an atom cause no appreciable change in the mass of the nucleus, because the mass of the electron is relatively very small. For c.g.,

$$^{232}_{90}Tn \rightarrow ^{2332}_{91} Pa + ^{0}_{-1}e$$

The emission of gamma rays does not after either the atomic number or the mass number of the atom, because gamma-rays possess neither mass nor change. The emission is associated with energy charge within the nucleus.

Radioactive Displacement laws

1. Emission of an α particle produces an element which is four mass units lighter and the atomic number decreases by two. The daughter element is therefore two places to the left of the parent in the periodic table.
2. When a β particle is emitted the mass number remains the same. However, the atomic number increases by one and the new element is one place to the right of the parent in the periodic system. These changes are shown in the following series:

$$^{223}_{88}Ra \xrightarrow{\alpha} ^{219}_{86}Rn \xrightarrow{\alpha} ^{215}_{84}Po \xrightarrow{\alpha} ^{211}_{82}Pb \xrightarrow{\beta} ^{211}_{83}Bi \xrightarrow{\alpha} ^{207}_{81}Tl \xrightarrow{\beta} ^{207}_{82}Pb$$

Radioactive Decay Series

The heavy radioactive elements may be grouped into four decay series. The common radioactive elements thorium, uranium and actinium occur naturally and belong to three different series named after them. They are the parent members of their respective series and have the longest half life periods. They decay by-a series of α and β emissions, and produce radioactive elements which are successively more stable until finally a stable isotope is reached. All three series terminate with lead $\left(^{206}_{82}Pb,\ ^{207}_{82}Pb \text{ and } ^{208}_{82}Pb\right)$. Following the discovery of the artificial post-uranium elements, the neptunium series has been added, which ends with bismuth, $^{209}_{93}Bi$.

Thorium ($4n$) series
Neptunium ($4n + 1$) series
Uranium ($4n + 2$) series
Actinium ($4n + 3$) series

The numbers in brackets indicate that the parent and all the members of a particular series have mass numbers exactly divisible by four, or divisible by four with a remainder of one, two or three.

There is no natural cross-linking between the four series, although this can be performed artificially.

Measuring Radioactivity

Many methods have been used for detecting radioactivity and observing its properties. One of the oldest is the *Geiger-Muller counter*, shown in Fig. 3.4. Radiation enters the counter tube through the thin window at the end. It strikes argon atoms inside the tube, ionizes them, and causes an electrical discharge between the central wire and the outer shell. This discharge is detected and each such event is counted electronically.

Even older than the Geiger-Muller counter is the use of phosphorescence for detecting high-energy radiation. Today *scintillation counters* can be regarded as descendants of the early phosphorescent screen. In these, a specially doped crystal of an alkali-metal halide gives off a flash of light when high-energy radiation hits it. This light is detected by a photomultiplier tube, an ultrasensitive photocell, which is in turn connected to an amplifier and counter.

Photographic emulsions have been used for many years for observing the tracks of high-energy particles. Used similarly are *cloud chambers* in which a particle's path is made visible by a trail of droplets of water or other liquid condensed from a supersaturated vapour. *Bubble chambers* reveal a particle's path by a trail of tiny bubbles in liquid hydrogen, and *spark chambers* by a trail of spark discharges between thin, oppositely charged electrodes. In the cloud, bubble, and spark chambers the tracks are usually photographed to provide a permanent record.

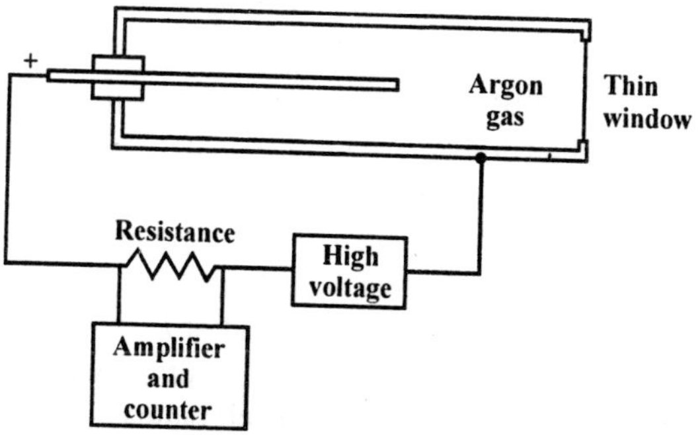

Fig. 3.4 : The Geiger-Muller Counter

Radioactive Decay

The decay of a radioactive element is a random process and is not influenced by external factors such as temperature changes. The rate of decay is directly proportional to the number of atoms present, following an exponential law, the rate of decay decreasing with time. Kinetically the process is a first order reaction and can be expressed by the equation :

$$-\frac{dN}{dt} = kN \qquad (1)$$

where N is the number of radioactive atoms present, t is the time, and k is the velocity constant (p. 121). Rearrangement of equation (1) followed by integration gives:

$$-\int \frac{dN}{N} = k\,dt$$

or $$-\log_e N = kt + \text{const.}$$

If the number of radioactive atoms present at time t = 0 is No, then

$$-\log_e N_o = \text{const.}$$

and $-\log_e N_o = kt - \log_e N_o$ or $\log_e(N_o/N) = kt$

Conversion of Napierian logarithms to ordinary logarithms gives

$$2.303 \log_{10} (N_o/N) = kt \qquad (2)$$

The Half life

Let $t\frac{1}{2}$ represent the time needed for the original activity to decrease to a half, i.e. $N_o = 2N$, then

$$2.303 \log_{10} (No/\tfrac{1}{2}N_o) = 2.303 \log_{10}2 = kt\tfrac{1}{2}$$

or $$0.6932 = kt\tfrac{1}{2}$$

or $$t\tfrac{1}{2} = \frac{0.6932}{k}$$

As equation (3) shows, the time taken for half the activity to disappear ($t\frac{1}{2}$), or the half-life as it is often called, is independent of the number of radioactive atoms. The half-life of a particular radioactive element is a characteristic constant of that element, and so too is the value of k which is termed the radioactive decay constant; it is often represented by the symbol λ instead of k.

If radioactive decay proceeds by a number of individual steps, i.e. if the first radioactive decay product is itself radioactive and thus decays further, then each stage in the process is represented by its own characteristic rate equation and thus by its own half-life and radioactive decay constant.

Artificial Radioactivity *(Induced Radioactivity)*

The alpha emission and beta emission discussed earlier occur naturally and spontaneously. Nothing needs to be done to get unstable elements to decay, and the rate at which they decay cannot be changed. Scientists have learned, however, how to transmute elements that normally are stable. In 1919, while working with an alpha emitter, Ernest Rutherford discovered that stable nitrogen could be transmuted into oxygen when the alpha particles were allowed to shoot into a sample of nitrogen. The equation for this change is:

$$^{14}_{7}N + ^{4}_{2}He \rightarrow ^{17}_{8}O + ^{1}_{1}H \qquad \text{(Eq. 9)}$$

alpha particle — symbol for proton, the nucleus of the most common isotope of hydrogen

This was the first instance of artificially produced nuclear transmutation. This event marked the beginning of many experiments in which various elements were made by bombarding the nuclei of other elements with fast-moving atomic or subatomic particles. Among the particles used for this purpose are electrons, protons, neutrons, alpha particles (helium nuclei), and deuterons (nuclei of deuterium atoms). It was with a transmutation experiment that James Chadwick discovered the neutron in 1932. See Figure 3.5.

$^{4}_{2}He$	+	$^{9}_{4}Be$	$\longrightarrow$	$^{12}_{6}C$	+	$^{1}_{0}n$
alpha particle		**beryllium-9**		**carbon-12**		**neutron**
2p		**4p**		**6p**		**0 p**
2n		**5n**		**6n**		**1 n**

Fig. 3.5 : ***Discovery of the neutron.*** When James Chadwick bombarded beryllium-9 with alpha particles, carbon-12 atoms were formed, and neutrons were emitted. Chadwick is credited with the discovery of the neutron as a result of this transmutation experiment.

Although the oxygen produced in Rutherford's experiment was a rare isotope, it was stable. It was not until 1934 that an unstable, or radioactive, nuclide was produced in the laboratory by bombarding another element. Radioactivity produced in this manner, known as artificial radioactivity or induced radioactivity, first occurred when Irene Joliot-Curie (1897-1956) and her husband Frederic Joliot-Curie (1900-1958) bombarded aluminum with alpha particles. When the alpha bombardment was stopped, the aluminum sample continued to give off radiation. In other words, a radioactive isotope had been produced. Investigation revealed that the initial bombardment had transformed the aluminum nuclei to nuclei of phosphorus-30. This isotope of phosphorus is radioactive. It decays by beta emission to an isotope of silicon. The two steps of the reaction can be represented as follows:

$${}^{4}_{2}He + {}^{27}_{13}Al \rightarrow {}^{30}_{15}P + {}^{1}_{0}N \qquad \text{(Eq. 10)}$$

$${}^{15}_{30}P \rightarrow {}^{30}_{14}Su + {}^{0}_{1}e \qquad \text{(Eq, 11)}$$

Radioactive phosphorus-30 (Equation 11) was the first radioactive isotope to be made in a laboratory. The particle in Equation 11 with the symbol $\left({}^{0}_{1}e\right)$ has the same mass as an electron but carries one unit of positive charge rather than one unit of negative charge. This particle is called a **positron,** or positive electron. Particles that are opposite in charge but have the same mass are called **antiparticles.** The positron is the antiparticle of the electron. The name given to the collection of antiparticles is **antimatter.**

Table 3.1 : Fundamental Particles of Matter

Subatomic Particles	Electric Charge	Relative Mass
Nucleon proton	+ 1	1836
neutron	0	1839
Electron. positron	– 1, + 1	1
Leptons (of diverse energy; such as photons and neutrinos)	0	0
π-mesons (pions)	– 1, 0, + 1	264-273
μ-mesons (muons)	– 1, + 1	207
K-mesons (several are known)		Masses between pions and nucleons
Hyperons (several are known)		Masses greater than nucleons

A positron is almost equal in mass with, but opposite in charge to, an electron. This knowledge of the existence of particles having like mass and other like physical characteristics but with opposite electrical charge seems to hold for many of the minute particles that make up matter. There are now known to be positive and negative *μ-mesons* (muons) and positive and negative *π-mesons* (pions) that weigh more than two hundred times as much as electrons. A variety of *K-mesons* have been detected in some radioactive decay processes, and these have a weight of about 966 electrons. *Hyperons* are particles heavier than protons. *Leptons* is the generic name given to a group of zero-mass units which are questionably called particles; these include *photons* and *neutrinos.* A photon is at times referred to as a *light quantum.* In summation, is should be honestly stated that scientists do not know in any detailed certainty what part all these particles play in the matter-energy aspects of the universe or even of a small part of the earth.

Nuclear Stability

Although an extensive discussion of the factors which contribute to stability and instability in nuclei is beyond the scope of this book, a few observations are in order. First, with the exception of ${}^{1}_{1}H$, all stable nuclei contain at least one neutron. Second, as the number of protons in the nucleus increases, the number of neutrons *per proton* increases in the stable nuclei. Apparently neutrons are necessary to keep a nucleus from flying apart as a result of proton-proton repulsions, and the more protons which are present in the nucleus, the greater the neutron/proton ratio must be for the nucleus to be stable. Third, when there are more than 83 protons in a nucleus, no number of neutrons will stabilize it. In the periodic table bismuth ($Z = 83$) is the last element having a stable isotope.

The Belt of Stability

Figure 24-4 shows a graph of the numbers of neutrons and of protons for all stable nuclei. These nuclei are found in a *belt of stability,* a region on the graph in which the neutron, proton ratio is close to 1 for the lighter nuclei, but in which the ratio increases as the number of protons increases. Thus in ${}^{6}_{3}Li$ the ratio is 1:1, in ${}^{48}_{110}Cd$ it is 1.29:1. and in ${}^{80}_{202}Hg$ it is 1.53:1.

NUCLEAR PROCESSES

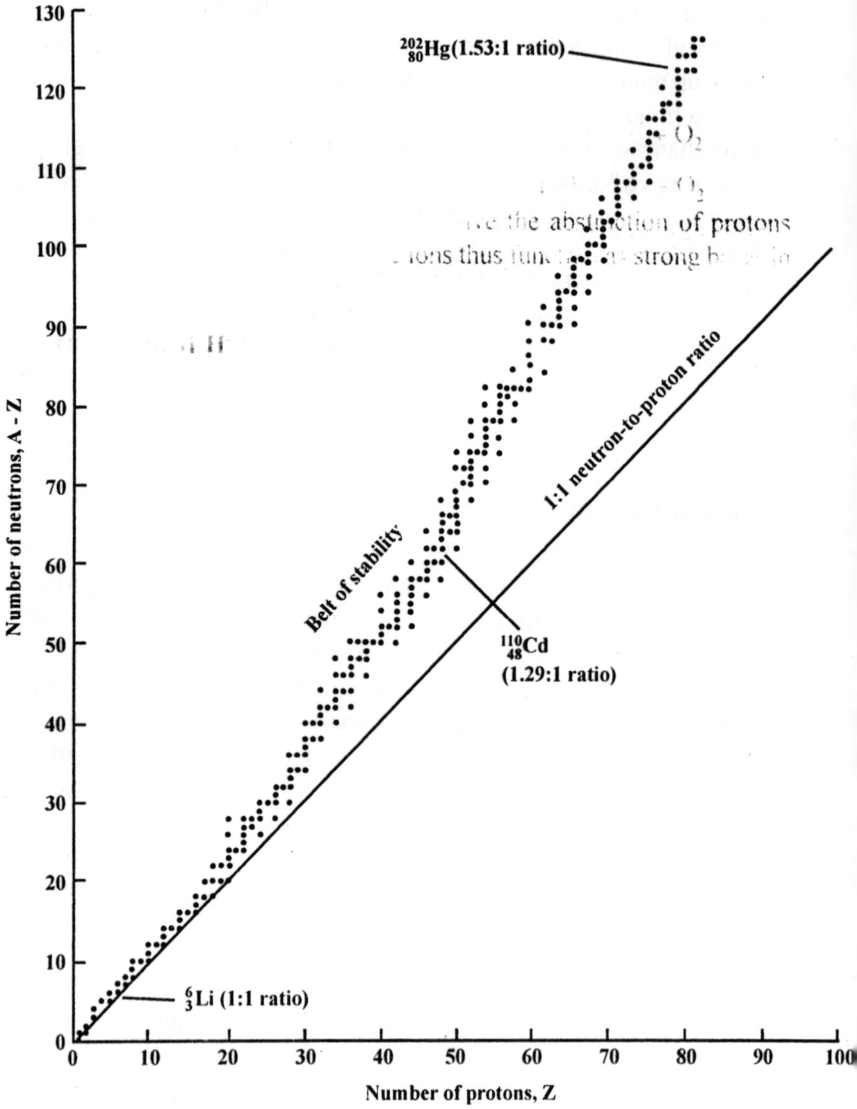

Fig. 3.6 : The belt of stability

On this kind of neutron-proton plot unstable nuclei lie either above, below, or beyond the end of the belt of stability. Those lying above the belt have too high a neutron/proton ratio, those below, too low a ratio, and those beyond the belt simply have too many nucleons to be stable. Nuclei in these regions tend to undergo transformation into nuclei within the belt, or, at least, closer to it.

Nuclei lying *above* the belt of stability lower their neutron/proton ratios by β decay or less commonly, neutron emission. In the β *decay* process

$$^{133}_{54}Xe \rightarrow {}^{133}_{55}Cs + {}^{0}_{-1}e$$

the daughter cesium nuclide is stable, but in some cases several successive decays are necessary before the nucleus gets into the belt of stability. For example, antimony 131 must undergo three consecutive β decays to form a stable nucleus:

$$^{131}_{51}Sb \rightarrow {}^{131}_{52}Te + {}^{0}_{-1}e$$

$$^{131}_{52}Te \rightarrow {}^{131}_{53}I + {}^{0}_{-1}e$$

$$^{131}_{53}I \rightarrow {}^{131}_{54}Xe + {}^{0}_{-1}e$$

An occasionally observed process for lowering the neutron/proton ratio is *neutron emission.* An example is

$$^{90}_{36}Kr \rightarrow {}^{89}_{36}Kr + {}^{1}_{0}n$$

Nuclei lying *below* the belt of stability raise their neutron/proton ratios either by β^+ (positron) emission or by electron capture. An example of β^+ *emission* is

$$^{105}_{48}Cd \rightarrow {}^{105}_{47}Ag + {}^{1}_{0}e$$

In *electron capture* (EC) a low-energy electron is captured by the nucleus

$$^{127}_{54}Xe + {}^{0}_{-1}e \xrightarrow{EC} {}^{127}_{53}I$$

and the energy lost in the process is often emitted as an x ray.

Sometimes it is not an unfavourable neutron/proton ratio which produces decay. Rather, the total number of nucleons may just be so large that the nuclear binding force is not strong enough to hold them together. In other words, nuclide is *beyond* the end of the belt of stability. This situation often leads to alpha emission, because in this way the nucleus can get rid of two protons and two neutrons at the same time:

$$^{211}_{84}Po \rightarrow ^{207}_{82}Pb + ^{4}_{2}He$$

If the daughter nuclide is far enough beyond the belt of stability, several successive emissions may occur, as in the uranium 238 decay series.

Binding energy, mass defect, packing fraction. Variations in atomic weights from whole numbers may be explained by the following factors :

1. Isotopes;
2. Weight of electrons (can become a second decimal point factor);
3. Neither protons or neutrons have unit mass, and atoms have varied neutron-proton ratios;
4. A small amount of mass is converted to what is called *binding energy* as atoms are formed from their nucleons.

Recent and exacting work with the mass spectrograph shows that *atomic masses are always less than the additive weights of their particle content.* For example, we may compute the mass loss as helium is formed from two protons and two neutrons :

$$2\,_1H^1 + 2\,_0n^1 \rightarrow \,_2He^4 + \text{binding energy}$$

$$2 \times 1.00812 + 2 \times 1.00893 \rightarrow 4.0039. + 0.03020 \text{ mass units}$$

Similarly, for all atoms a fractionally small mass loss exists in atomic mass as compared to the total of nuclear particle masses. There is variation in the amount of binding energy per nuclear particle as diverse-sized atoms are built up. To further describe this extremely important relationship, the terms *mass defect* and *packing fraction* have been carefully designated as follows:

For helium:

$$\text{Mass defect} = \text{atomic mass} - \text{mass number}$$

$$\text{Mass defect} = 4.00390 - 4 = .00390$$

$$\text{Packing fraction} = \frac{\text{mass defect}}{\text{mass number}}$$

For helium : Packing fraction $= \frac{.00390}{4} = .00975$

Consider the packing fraction curve. Note that a low in the curve exists near iron of mass number 56. This means the mass defect for iron is small as related to its mass number. *The low mass defect is an index of high binding energy, since much of the excess mass of the neutron and proton nuclear content has been dissipated to energy.* Iron would seem to be about the most stable atom from this concept. Some evidence indicates that the earth's core consists largely of iron and nickel, and meteorites have a high iron and nickel content. Perhaps this is related to the nuclear stability of these elements.

Note that the packing-fraction curve gradually rises from iron to uranium and the transuranium elements (elements above uranium in atomic number). These larger-mass-number elements of higher-packing fraction and large mass defect have mass to be converted into energy in any fission of them into more stable elements in the middle range of mass numbers. This accounts for the very large amounts of energy accompanying atomic fission.

Consider the high packing fraction of the lighter elements. These elements, which have comparatively large mass defect compared to atomic number, present distinct possibilities for conversion of mass to energy. The nuclear reactions—hydrogen plus neutrons to helium with energy emission, and lithium plus hydrogen to two helium atoms—have been developed.

Nuclear Fission

Sometimes a nucleus which is far beyond the end of the belt of stability will split into two major pieces, rather than emitting a succession of alpha particles. This process, *nuclear fission*, is one of the modes, though not a common one, by which uranium 235 can spontaneously decay:

$$^{235}_{92}U \rightarrow ^{140}_{56}Ba + ^{92}_{36}Kr + 3\ ^{1}_{0}n$$

As can be seen from the equation, the uranium nucleus divides itself into fragments with three neutrons left over.

The fission of uranium 235 described above is a natural, *spontaneous* process. Fission can be *induced*, however, when a *uranium* 235 nucleus captures a slow, or *thermal*, neutron:

$$^{235}_{92}U + ^{1}_{0}n \rightarrow \left(^{236}_{92}U\right) \rightarrow ^{90}_{38}Sr + ^{143}_{54}Xe + 3\ ^{1}_{0}n$$

This equation represents only one of many ways the nucleus can split. Other ways uranium 235 can undergo fission include

$$^{235}_{92}U + ^{1}_{0}n \rightarrow \left(^{236}_{92}U\right) \rightarrow ^{94}_{36}Kr + ^{139}_{53}Ba + 3\ ^{1}_{0}n$$

and $$^{235}_{92}U + ^{1}_{0}n \rightarrow \left(^{236}_{92}U\right) \rightarrow ^{90}_{38}Sr + ^{144}_{54}Xe + 2\ ^{1}_{0}n$$

Thus the fission fragments are variable. When many uranium 235 nuclides undergo fission, a collection of many different fragments distributed over a wide range of masses is produced. Figure 24-5 shows the yields of various products of the induced fission of uranium 235. Nuclear fission is the energy-producing process used in nuclear weapons and nuclear reactors.

The Atomic Bomb. With the discovery of Pu, two substances, U^{235} and Pu^{239}, were available for the manufacture of atomic bombs. Plutonium manufacture possesses certain definite advantages over the recovery of U^{235} from ordinary uranium. The separation of U^{235} from U^{238} is exceedingly difficult, since *physical means*, which take advantage of the line relatively small differences in the mass of the two isotopes, must be employed. Thermal effusion, gaseous effusion, and electromagnetic methods have been used to effect the separation, but many stages of purification are necessary. Plutonium, when produced in a uranium pile, can be separated from unconverted uranium by *chemical means,* since Pu is a different element and as such exhibits a different set of chemical properties.

Relatively little information has been released regarding the construction of the atom bomb itself. It was known early in the studies of this project, however, that a certain "critical size" of material was necessary for a chain reaction to be self-perpetuating and for an explosion to take place. In a small piece of U^{235} or Pu^{239}, in which the ratio of surface area to volume is large, too many neutrons escape through the surface, and consequently a chain reaction is not maintained. As the size of the piece is increased, the ratio of surface area to volume decreases, and there is more chance that neutrons will be captured before escaping through the surface. In producing an atomic bomb, it seems logical that pieces of subcritical size might be brought together rapidly, possibly by using one or more pieces as projectiles and another as a target in a firing mechanism. Just how pieces of subcritical size are brought together to form an effective fissionable mass in a bomb is still a secret.

In the explosion of an atomic bomb, a very high temperature is attained; desert sand was sintered at some distance from actual bomb tests at Los Alamos, New Mexico. Written accounts tell of vapourization of the metal of the tower holding the test bomb. Just how high a temperature is reached is difficult to estimate, as man has not heretofore attained temperatures above about 4000° C. Temperatures in the order of millions of degrees are known to be reached. This excessively high temperature consideration leads us to some understanding as to the nature of possible temperature effects in an atomic bomb explosion. Probably no large mass of material in an atomic bomb blast is heated as high as the internal part of the sun, though certainly some very high temperature is reached locally. The atomic bomb explodes and causes such air expansion—and later contraction on cooling—that cyclonic wind or blast effects arc produced. The radioactive fragments, clinging to dust, are very dangerous.

The elements that are products of fission are commonly isotopes of quite high atomic weight for the given elements. In reverting to stable isotopes, intense beta and gamma ray activity is involved. This is the marked after-effect so dangerous to biological tissue. Radioactive strontium and cobalt isotopes have been publicized as particularly dangerous products of a fission bomb.

Fission Reactors

A nuclear reactor is a device in which a nuclear chain reaction is carried out at a controlled rate. A fission reactor is one in which the controlled chain reaction is a fission reaction. Such a reactor is usually a nuclear furnace, in which nuclear energy is converted to heat. A typical fission reactor uses blocks or cylinders of U-235 or Pu-239 (a plutonium isotope) as fuel.

The nature of the fuel in a reactor depends on the purpose, design, and power requirements. The blocks of fuel are surrounded by graphite, which slows down the neutrons produced by the fission.

Graphite and other substances used to slow down neutrons are called **moderators.** Slow neutrons not only have a greater chance of being captured by the nuclei of the atoms of fuel but also give up energy in the form of heat when their velocity is reduced. As the products of fission scatter through the reactor, their energy is dissipated as heat. The number of neutrons available in the reactor is regulated by means of **control rods made of cadmium** steel, which can absorb neutrons. The number of new neutrons available for fission is kept equal to the number used up by fission. In this manner the reaction rate is controlled, and the reactor becomes a furnace in which heat is generated at a constant rate. See Figure 3.7.

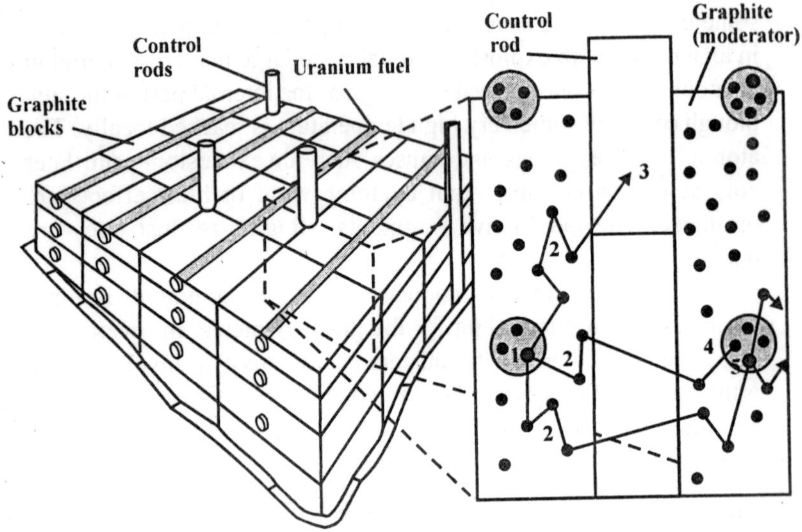

Fig. 3.7 : ***The principle of the nuclear reactor.*** One of the U-235 nuclei in the uranium fuel, like that at (1), undergoes fission, releasing energy and several high-speed neutrons. Several things can happen to the neutrons. Most of them are slowed by colliding with the atoms of the moderator (2). Some of them are then absorbed by the atoms of the control rods (3). Still others, like that at (4), are captured by atoms of U-238, producing U-239, which decays by beta emission to Pu-239, a fissionable plutonium isotope also usable as a reactor fuel. Finally, in the most essential step for continuing the reaction, some neutrons, like that at (5), re captured by other atoms of U-235, resulting in fissions and a repetition of the steps described.

Nuclear fuels. Less than 1% of natural uranium is fissionable U-235, which can be used as a nuclear fuel. To make pure U-235 or enriched uranium (containing more than 1% U-235), the isotopes must be separated. Because isotopes of an element are chemically alike, separation by chemical means is difficult. Separation makes use of the fact that the gaseous hexafluorides of the two isotopes diffuse through a porous wall at different rates.

Because making U-235 is slow and expensive, many reactors are fueled with enriched uranium. The U-238 changes to plutonium-239, a fissionable isotope that in turn can be used as a nuclear fuel. A fission reactor using enriched uranium that not only produces energy but manufactures another fuel is a *breeder reactor.*

The production of plutonium in a breeder reactor occurs as a result of three nuclear reactions. The capture of a neutron changes U-238 to the radioactive isotope U-239. The radioisotope undergoes beta decay, resulting in a new element, *neptunium,* atomic number 98. Like U-239, neptunium is radioactive, and through beta decay it forms plutonium, element 94. The equations for the nuclear reactions leading to the formation of plutonium are:

$$^{238}_{92}\text{U} + {}^{1}_{0}\text{n} \rightarrow {}^{239}_{92}\text{U} \qquad \text{(Eq. 16)}$$

$$^{239}_{92}\text{U} \rightarrow {}^{239}_{93}\text{Np} + {}^{0}_{-1}\text{e} \qquad \text{(Eq. 17)}$$

$$^{239}_{93}\text{NP} \rightarrow {}^{239}_{94}\text{Pu} + {}^{0}_{-1}\text{e} \qquad \text{(Eq. 18)}$$

Plutonium-239 has a half-life of 24,000 years. It is fissionable and can be used as a source of atomic energy. Uranium is the element of highest atomic number (92) found naturally on the earth. Neptunium (atomic number 98) and plutonium (atomic number 94) were the first elements with atomic numbers higher than that of uranium to be synthesized. Elements with atomic numbers greater than that of uranium are called **transuranium elements**. Additional transuranium elements have been synthesized in the laboratory. As of 1988, the list included elements 98 to 109.

Uses of fission reactors. The neutrons produced in fission reactors have been used to synthesize new elements and to make useful radioisotopes. The large amounts of heat released by fission have been used to make steam for the production of electricity. Nuclear electric power plants are in operation in countries around the world. Submarines and surface vessels also are powered by atomic energy. Nuclear reactors require heavy and costly shielding, making them impractical for powering trains and aircraft.

A number of environmental problems have been created by the use of nuclear reactors. Their fission products are highly radioactive and cannot be discarded as ordinary wastes. They must be stored for a long time or disposed of in special ways. One process is to remove the long-lived radioactive nuclides from the waste and reuse them in new fuel rods. The remaining wastes then are mixed with molten glass and converted into cylinders that can be packaged and buried.

Another problem is the heating of water used as a coolant in some reactors. When this water enters streams and lakes, it can be hot enough to kill marine life. Water rendered too hot for marine life is said to be thermally polluted.

Nuclear Fusion

Conversion of very light nuclei to heavier ones also results in an increase in binding energy per nucleon and so should release large amounts of energy, even larger than in fission. Such reactions are called *fusion reactions* because in them smaller nuclei fuse to form larger ones.

The source of energy in the sun is a series of reactions, the net result of which is the fusion of four protons to form a single $^{4}_{2}He$ nucleus. The reaction undoubtedly takes place in steps. One possibility is

$$^{1}_{1}H + ^{1}_{1}H \rightarrow ^{2}_{1}H + ^{0}_{1}e$$

$$^{2}_{1}H + ^{1}_{1}H \rightarrow ^{3}_{2}He$$

$$^{3}_{2}He + ^{1}_{1}H \rightarrow ^{4}_{2}He + ^{0}_{1}e$$

Net: $$4\,^{1}_{1}H \rightarrow ^{4}_{2}He + 2\,^{0}_{1}e$$

The only successful "practical" application of fusion reactions has been in so-called hydrogen, or thermonuclear, bombs. One problem with fusion reactions is getting them to start. In order for two light nuclei to fuse they must have extremely high energies so that the electron clouds in the atoms' extranuclear regions do not prevent nuclei from getting close to each other. What is needed is an ultrahigh temperature, about 10^{8}°C. In a thermonuclear weapon a fission reaction is used to provide the high energies needed to initiate fusion. In one device an ordinary fission bomb is surrounded by a shell of lithium deuteride. Neutrons from the fission reaction are captured by lithium nuclei,

$$^{6}_{3}\text{Li} + ^{1}_{0}\text{n} \rightarrow ^{4}_{2}\text{He} + ^{3}_{1}\text{H}$$

and under the high-energy conditions supplied by the fission reaction the tritium undergoes fusion with deuterium :

$$^{3}_{1}\text{H} + ^{2}_{1}\text{H} \rightarrow ^{4}_{2}\text{He} + ^{1}_{0}\text{n}$$

In a so-called fission-fusion- fission bomb a shell of $^{238}_{92}$U surrounds the fusion device. Normally this isotope of uranium will not undergo fission, but under energy-rich conditions it will. The "advantage" of this weapon is that $^{238}_{92}$U is relatively inexpensive because it is by far the most abundant isotope of this element.

Controlling nuclear fusion so it can be used for the production of usable power is a problem which has challenged scientists and engineers for over 20 years. Presumably reactions employing $^{1}_{1}$H and $^{2}_{1}$H would be employed, and although only about 0.0 15 percent of the hydrogen on earth is deuterium, there is so much water in the oceans that the total potentially available amount of this isotope of hydrogen is very large. But in addition to the problem of "igniting" a fusion reaction there is the difficulty of containing a reaction mixture at a temperature high enough to vapourize all matter Currently experiments are being performed using a battery of high-power lasers all aimed at a very small sample of hydrogen. It is hoped that the lasers will concentrate enough energy to start the fusion reaction. Other experiments are being run in which ultrahigh-temperature mixtures are contained in a "magnetic bottle," an arrangement of intense magnetic fields which prevents escape of the reaction mixture.

Hydrogen Bomb

Fusion of light elements requires temperature of the order of millions of degrees. Such high temperatures may be obtained in "fission bombs", and it is likely that the latter are used as detonates for fusion bombs. One possible arrangement of the ingredients in a hydrogen or fusion bomb may be such as shown in Fig. 3.8.

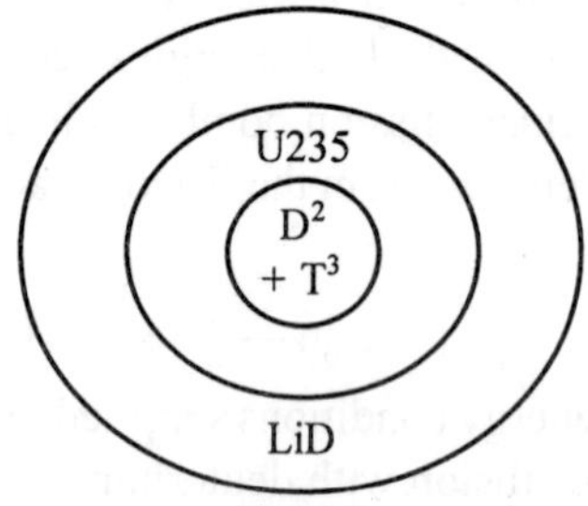

Fig. 3.8

The core of the device contains the hydrogen isotopes, deuterium and tritium. Around the core is a layer of fissionable matter such as U^{235} which, when set off, detonates the interior. By implosion, the following reach on between the hydrogen isotopes in the core is initiated :

$$^{2}_{1}D + ^{3}_{1}T \rightarrow ^{4}_{2}He + ^{1}_{0}n$$

This reaction in relates a flood of energy and further raises the temperature which is then sufficiently high to explode the outer layer of the bomb, which consists of Lithium deuteride, $^{6}_{3}Li$, $^{2}_{1}D$.The following sequences of reaction would give a self maintaining chain process :

$$^{6}_{3}Li + ^{1}_{0}n \rightarrow ^{4}_{2}He + ^{3}_{1}T$$

$$^{2}_{1}D + ^{3}_{1}T \rightarrow ^{4}_{2}He + ^{1}_{0}n \text{ and so on.}$$

In this arrangement the mixture of the isotopes of hydrogen acts as on booster to increase the temperature.

Use of radioactive elements; tracer elements. Radioactive isotopes, prepared either in the cyclotron or in the uranium pile, open a new field in medicine. Because radioactive isotopes retain the chemical properties of the ordinary variety of these elements, they can be substituted in compounds used by the body. If one of these compounds containing a radioactive isotope is introduced into the body, its progress throughout the body can be traced by use of a detector of the radiations. Such an instrument, the *Geiger counter*, is highly sensitive and can be used to determine the areas which the *tagged* element enters. For instance, when radioactive iodine salts are introduced into the body, they are taken up by the thyroid gland to form the substance thyroxine. The body needs only a little of this compound but cannot function without it. Ordinary analytical methods are not adaptable to the detection of the small amounts which are present in the body. However, if the thyroxine contains radioactive iodine, the amount can be accurately determined with the Geiger counter. The length of time between the introduction of radioactive iodine and its accumulation in the thyroid gland serves as a measure of the activity of the gland.

Similar use is made of radioactive sodium. If salt solution containing radioactive sodium is introduced into the circulation, the rate of its distribution can serve as an indicator of the efficiency of the circulatory system. If diseased areas have restricted circulation, tests with the Geiger counter will indicate the areas so affected. In the past, if an amputation were necessary because of gangrene, the surgeon could only make an intelligent guess as to the exact point at which healthy tissue existed. The tagged atoms of sodium show quite accurately the point at which normal circulation ends. The radioactive elements in these cases are called **tracer elements.**

Various radioactive elements such as phosphorus, strontium, gold, and numerous others have been suggested or tried as treatments for cancer. Many organs of the body selectively store up certain elements. If the element supplied is radioactive, it provides radiations within the tissues which may be efficacious in curing the cancer. As yet the field of cancer research is too new for us to expect in the immediate future a universal cancer cure by radioactivity. It will remain necessary in the future as in the past for the person afflicted to get early and competent medical attention for any condition which might be, or might become, cancerous.

Use is being made of radioactive tracer elements in the study of plant metabolism. Studies are going forward on the nature of photosynthesis by the use of radioactive carbon and radioactive metals.

Uses of Radioisotopes

Carbon-14 Dating

The carbon dioxide in the atmosphere contains small amounts of the radioactive isotope ${}^{14}_{6}C$ which is produced by the action of neutrons on the nitrogen (the neutrons are produced by the cosmic rays which are present in the atmosphere):

$$^{14}_{7}N + ^{1}_{0}n \rightarrow ^{14}_{6}C + ^{1}_{1}H$$

Since plants and trees take up carbon dioxide and convert it into tissue material and animals, including humans, eat plants, all living matter contains the same small proportion of carbon-14 as occurs in the atmosphere. After death the uptake of carbon dioxide ceases and the level of carbon-14 in the body and in plant tissues falls ($^{14}_{6}C$ has a half-life of 5600 years). For instance, a tree felled 5600 years ago will only contain half the activity of a growing tree. Paper and papyrus have a vegetable origin and can thus be dated by the carbon-14 method, e.g. the Dead Sea scrolls have been dated by this method.

The detection of wear in an engine

The incorporation of a radioactive isotope of a metal in the piston rings of engines allows the accurate measurement of wear by friction. The engine is run under test for a specified period and the amount of radioactivity in the lubricating oil determined. Tests such as this have been directly responsible for the development of more efficient engine oils.

The non-equivalence of the sulphur atoms in the thiosulphate anion

The thiosulphate anion is formed when a solution of a sulphite is boiled with sulphur. If the sulphur (but not the sulphite) contains radioactive sulphur, $^{16}_{35}S$, then, of course, the radioactivity will be present in the final thiosulphate:

$$^{16}_{35}S + SO^{2-}_{3} \rightarrow \left[^{35}_{16}SSO_3\right]^{2-}$$

When the thiosulphate is broken down it is found that the radioactivity appears in the precipitated sulphur and not in the sulphur dioxide:

$$\left[^{35}_{16}SSO_3\right]^{2-} + 2H^+ \rightarrow ^{35}_{16}S + H_2O + SO_2$$

This experiment clearly shows that the two sulphur atoms in the thiosulphate anion do not occupy equivalent positions.

4

Oxidation and Reduction

Oxidation and Reduction

In the early days of chemistry, a substance was said to be oxidized after it had reacted with oxygen. The reaction itself was called **oxidation.** A familiar example of oxidation in this sense is the rusting of iron. In this reaction, iron in the presence of moisture reacts with oxygen from the air to produce ordinary rust, iron oxide.

Another example of oxidation is the burning of fuels. When wood, natural gas, or coal burns, the oxygen in the air combines with the carbon and hydrogen in the fuel to produce water and the oxides of carbon.

Today, the term *oxidation* still is used in this sense, but it has a broader meaning, as well. Oxidation refers today to a category of reactions with some characteristics similar to those of the reactions of oxygen just described.

Originally, **reduction** referred to a chemical reaction in which a compound lost oxygen:

$$2Fe_2O_3\,(s) + 3C\,(s) \rightarrow 4Fe\,(l) + 3CO_2\,(g) \qquad \text{(Eq. 1)}$$

In Equation 1, the iron (III) oxide is being reduced. This is because, during the reaction, oxygen is removed from the compound, leaving elemental iron. Today, the term *reduction* has been given a second, broader meaning.

To describe reactions in which oxidation and reduction occur in the broader sense of those terms, chemists use oxidation numbers. The next three sections explain their determination and use.

Oxidation Numbers

The **oxidation number,** or oxidation state, of an atom is the *apparent* charge assigned to an atom of an element. The charge is called *apparent* because, while in some cases an oxidation number is the actual charge on an ion, in other cases there is no evidence for the presence of a true electric charge on an atom that has been assigned an oxidation number. Oxidation numbers, therefore, are merely a convenient device for analyzing types of reactions discussed in this chapter.

The Rules for Determining Oxidation Numbers

1. The oxidation number of an atom in the uncombined state is zero. For example, the oxidation numbers of the following atoms are all 0 :
   - An atom of phosphorus in the P_4 molecule.
   - An atom of chlorine in the Cl_2 molecule.
   - An atom of sodium. Na, when *not* part of a compound.
2. The oxidation number of a monatomic ion is equal to the charge on the ion.

ion:	Mg^{2-} in $MgBr_2$;	$Cl-$ ion in NH_4Cl;	O^{2-} ion in Na_2O
oxidation number :	+2	−1	−2

3. In most compounds, the oxidation number of hydrogen is +1, except in the case of the hydride compounds of the metals from Groups 1 and 2. In hydrides, the oxidation number of hydrogen is −1. For example.

compound:	H_2SO_4	NH_3	H_2O	NaH (sodium hydride)
oxidation number of hydrogen:	+1	+1	+1	−1

4. In the vast majority of compounds, the oxidation number of oxygen is −2. In peroxides the oxidation number of oxygen is −1. In OF_2, its oxidation number is +2. For example,

compound:	H_2O	Fe_2O_3	$KClO_3$	H_2O_2	OF_2
oxidation number of oxygen:	−2	−2	−2	−1	+2

5. In binary compounds of nonmetals, the more electronegative element is assigned the negative oxidation number. For example. in PCl_3, chlorine is more electronegative than phosphorus. Each chlorine atom partially gains one electron, giving each chlorine atom an oxidation number of −1 and the phosphorus atom an oxidation number of −3.
6. The sum of the oxidation numbers for all the atoms in a neutral molecule or formula is zero. The sum of the oxidation numbers for all atoms represented by a polyatomic ion is equal to the charge on the ion. For example.

H_2O: $(2 \times + 1) + (-2) = 0$

oxidation number of each of 2 H atoms

HCO_3^- : $(+1) + (+4) - (3 \times -2) = -1$

oxidation number of each of :3 oxygen atom — charge on ion

Figure 4.1 : How oxidation numbers should be determined.

Oxidation numbers are assigned to atoms in chemical formulas. Rules have been developed for assigning oxidation numbers to atoms. These rules are given in Figure 4.1.

Sample Problem 1

Using the rules for assigning oxidation numbers given in Figure 4.1, assign oxidation numbers to each element in the compound calcium chloride $CaCl_2$.

Solution

Calcium fluoride is an ionic compound composed of the calcium ion, Ca^{2+}, and the chloride ion, Cl^-. The calcium ion is a monatomic ion with a charge of 2+. Hence, by Rule 2, the oxidation number of the calcium atom in calcium chloride is +2. The chloride ion is a monatomic ion with a charge of 1–. Hence by Rule 2, the oxidation number of an atom of chlorine in calcium chloride is – 1.

An example of oxidation-reduction. *Oxidation* is defined simply as a change in which the oxidation number of an element increases. Such a change can take place only at the expense of a decrease in the oxidation number of some other element. The latter is termed *reduction.* For example, consider the following reaction

$$\overset{0}{2\,Al} + \overset{0}{3\,S} \rightarrow \overset{+3}{Al_2}\overset{-2}{S_3}$$

in which the oxidation number of each element is written above the symbol. Aluminum and sulfur in the free state have, in accordance with our rules, an oxidation number of 0; whereas in the compound, aluminum sulfide, the oxidation number of aluminum is +3 and that of sulfur is –2. In this reaction, aluminum gains in oxidation number from 0 to +3 and hence is *oxidized;* whereas sulfur loses in oxidation number from 0 to –2 and is *reduced.*

Let us examine this reaction in another way. As it proceeds, each atom of aluminium gives up 3 electrons and each atom of sulfur takes up 2 electrons. In the reaction as a whole, 6 electrons given up by the 2 aluminum atoms are taken up by the 3 sulfur atoms. A transfer of electrons has taken place, and the resultant compound Al_2S_3 is electrovalent; it is composed of Al^{+++} and S^{--} ions. In this particular case, Al gains in oxidation number or loses electrons;

likewise, sulfur loses in oxidation number or gains electrons. It is often convenient when dealing with electrovalent substances to discuss these changes in terms of a gain or loss in electrons as well as in terms of changes in oxidation number. To summarize: for reactions involving a transfer of electrons,

Oxidation is *gain* in oxidation number or *loss* of electrons
Reduction is *loss* in oxidation number or *gain* of electrons.

Oxidation and reduction are mutually dependent processes; if electrons are taken up by one substance, they must be given up by another. There can be no oxidation without reduction, nor reduction without oxidation.

The union of aluminum and sulfur may be written as taking place in two steps with *e* representing an electron:

$$2\,Al \rightarrow 2\,Al^{+++} + 6\,e \text{ (oxidation)} \quad (1)$$

$$6\,e + 3\,S \rightarrow 3\,S^{--} \text{ (reduction)} \quad (2)$$

By adding the two steps and cancelling the electrons, we obtain the net equation which was written above.

Oxidizing and reducing agents. Oxygen, sulfur, or other nonmetals, as they combine with metals, are recognized as *oxidizing agents*; therefore, the metals must be the *reducing agents*. In all oxidation-reduction chemical changes, the following relationship exists:

Oxidizing agent = electron receiver = loser in oxidation number
Reducing agent = electron giver = gainer in oxidation number

The receiver-loser and giver-gainer relationship is somewhat confusing, but it is helpful to consider that a particle giving away negative charge must itself become more positive. If it receives a negative charge, it becomes less positive. In the combination

$$2\,Fe + 3\,Cl_2 \rightarrow 2\,FeCl_3\ (2\,Fe^{+++},\ 6\,Cl^{-})$$

Cl_2 is the oxidizing agent; each Cl receives 1 electron and loses 1 in oxidation number. Fe is the reducing agent: each Fe gives 3 electrons and gains 3 in oxidation number. In the replacement

$$Zn + CuSO_4 \text{ (solution)} \rightarrow ZnSO_4 \text{ (solution)} + Cu$$

or $$Zn + Cu^{++} + SO_4^{--} \rightarrow Zn^{++} + SO_4^{--} + Cu$$

Cu^{++} is the oxidizing agent: it receives 2 electrons and loses 2 in oxidation number. Zn is the reducing agent: it gives 2 electrons and gains 2 in oxidation number.

Balancing oxidation-reduction equations. To balance these equations, we make use of the rule that the total *gain in oxidation numbers must equal the total loss in oxidation numbers of the elements.* Consider the reaction

+2

$$\overset{+3\ -1}{FeCl_3} + \overset{+1\ -2}{H_2S} \rightarrow \overset{+2\ -1}{FeCl_3} + \overset{+1\ -1}{HCl} + \overset{0}{S}$$

1

in which the oxidation number of each element is placed above the symbol. Two of the elements undergo a change in oxidation number; each iron atom loses 1 (from +3 to +2) and each sulfur atom gains 2 (from –2 to 0) in oxidation number. The loss of 1 in oxidation number of Fe is indicated by –1, and the gain of 2 in oxidation number of S is indicated by +2. To balance the gain and loss, 2 iron atoms will be required for each sulfur atom, and this will call for a ratio of 2 $FeCl_3$ to 1 H_2S. This must be the ratio in which these two substances react:

$$2\ FeCl_3 + 1\ H_2S \rightarrow 2\ FeCl_2 + 2\ HCl + S$$

Having obtained the ratio of the oxidizing agent, $FeCl_3$, to the reducing agent H_2S, it is easy to complete the balancing of the equation on the right side. In this equation, the change in oxidation number of sulphur (2) is made the coefficient of $FeCl_3$, while the change in oxidation number of iron (1) is the coefficient of H_2S.

In general, we may employ the following steps in balancing oxidation-reduction equations:

1. Write the oxidation number of each of the elements in each compound above the symbol of the element in the equation.
2. Determine which elements have changed in oxidation number and the amount of change. Usually two elements only will have changed; one gains in oxidation number while the other loses.

3. Balance the gain and loss in oxidation numbers. This may be accomplished automatically by placing the number representing the change in the oxidizing agent in front of the formula for the reducing agent, and vice versa.
4. Complete the balancing of the two elements which have changed oxidation number on the right side of the equation, keeping in mind that the ratio of oxidizing agent to reducing agent as determined from the change in oxidation numbers must be maintained. Balance hydrogen and oxygen atoms after all other elements have been balanced.

Let us illustrate the outlined steps by balancing an equation:

(1) 0 1+5–2 +1+5–2 +2–2 +1–2

$$P + HNO_3 \rightarrow HPO_3 + NO + H_2O$$

–3

0 +5 +5 +2

(2) $P + HNO_3 \rightarrow HPO_3 + NO + H_3O$

5

(3) $3\,P + 5\,HNO_3 \rightarrow HPO_3 + NO + H_2O$

(4) $3\,P + 5\,HNO_3 \rightarrow 3\,HPO_3 + 5\,NO + H_2O$ (balanced)

Oxidation-reduction equations are more difficult to balance by ordinary trial and error methods than those which involve no change in oxidation number. The "change in oxidation number" method for balancing oxidation-reduction equations, however, is rapid and accurate.

BALANCING EQUATIONS FOR REDOX REACTIONS—NO SOLVENT PRESENT. In general when you are faced with balancing an equation for a nonaqueous redox reaction, you will be provided with formulas for all reactants and products. The method used, the *oxidation-number method* is outlined below :

BALANCING REDOX EQUATIONS : OXIDATION-NUMBER METHOD

Step 1 Assign oxidation number to all atoms.

Step 2 Note which atoms appear to lose and which appear to gain electrons, and determine how many electrons are lost and gained.

Step 3 If there is more than one atom losing or gaining electrons in a formula unit, determine the total loss or gain of electrons per formula unit.

Step 4 Make the gain of electrons by the oxidizing agent equal to the loss by the reducing agent by inserting the appropriate coefficient before the formula of each on the left-hand side of the equation. (See the following examples.)

Step 5 Complete the balancing of the equation by inspection. First balance atoms which have gained or lost electrons; second, atoms other than O or H; third, O atoms; and last, H atoms.

Example 1

Problem : Balance the following equation

$$MnO_2 + KClO_3 + KOH \rightarrow K_2MnO_4 + KCl + H_2O$$

Solution :

Step 1: $\underset{+4-2}{MnO_2} + \underset{+1+5-2}{KClO_3} + \underset{+1-2+1}{KOH} \rightarrow \underset{+1+6-2}{K_2MnO_4} + \underset{+1-1}{KCl} + \underset{+1-2}{H_2O}$

Step 2: Manganese, Mn, changes oxidation number from +4 to +6 and so appears to lose two electrons. Cl changes from +5 to –1 and so appears to gain six electrons.

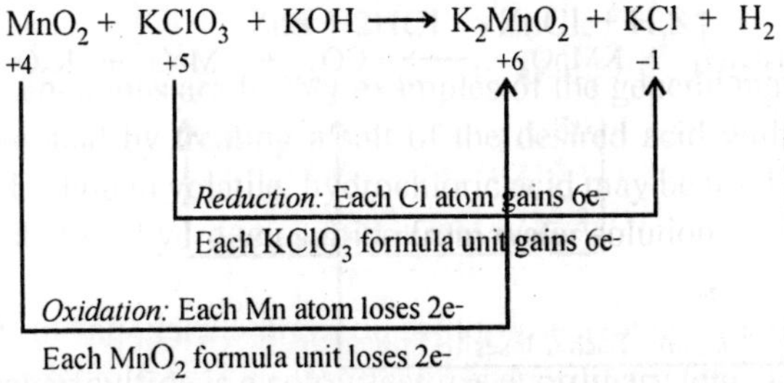

Step 3: As is shown above, since there is only one Mn atom in each MnO_2 formula unit, the electron loss per formula unit is two. Also, since there is but one Cl atom in each $KClO_3$ formula unit, the electron gain per formula unit is six.

Step 4: $3MnO_2 + KClO_3 + KOH \rightarrow K_2MnO_4 + KCl + H_2O$

Just look at the left-hand side for now. By placing a 3 before MnO_2, we are saying that if one MnO2 formula unit loses two electrons, then three will lose 3 × 2, or six, electrons. This will balance the electrons lost (six) against the electrons gained (six) by one $KClO_3$ formula unit. Now that this 3 : 1 ratio (MnO_2/KCl_3) has been established, *it must not be changed.*

Step 5 : Add coefficient to the right-hand side to balance the Mn and Cl atoms, because they were involved in the electron transfer.

$$3MnO_2 + KClO_3 + KOH \rightarrow 3K_2MnO_4 + KCl + H_2O$$

Then, balance the K atoms (note : the coefficient before $KClO_3$ must be changed)

$$3MnO_2 + KClO_3 + 6KOH \rightarrow 3K_2MnO_4 + KCl + 3H_2O$$

then, O atoms

$$3MnO_2 + KClO_3 + 6KOH \rightarrow 3K_2MnO_4 + KCl + 3H_2O$$

and last, H atoms. But note that they are already balanced.

Example 2

Problem: Balance the following equation

$$H_2C_2O_4 + KMnO_4 \rightarrow CO_2 + MnO + K_2O + H_2O$$

Solution :

Step 1 : $H_2C_2O_4 + K\ MnO_4 \rightarrow CO_2 + MnO + K_2O + H_2O$

+1 +3 –2 +1 +7 –2 +4 –2 +2 –2 +1 –2 +1 –2

Step 2 :

$H_2C_2O_4 + KMnO_4 \rightarrow CO_2 + MnO + K_2O + H_2O$

–3 +7 +4

$H_2C_2O_4 + KMnO_4 \longrightarrow CO_2 + MnO + K_2O + H_2O$

+3 +7 +4 +2

Reduction: Each Mn gains 5e-

Oxidation: Each C loses 1e-

Step 3:

$H_2C_2O_4 + KMnO_4 \rightarrow CO_2 + MnO + K_2O + H_2O$

–3 +7 +4 +2

$H_2C_2O_4 + KMnO_4 \longrightarrow CO_2 + MnO + K_2O +$

+3 +7 +4 +2

Each $KMnO_4$ gains 5e-

Since there are two C atoms per formula unit, each $H_2C_2O_4$ loses 2 X 1, or 2e-

Step 4 : $5H_2C_2O_4 + 2KMnO_4 \rightarrow CO_2 + MnO + K_2O + H_2O$

Step 5 : $5H_2C_2O_4 + 2KMnO_4 \rightarrow 10CO_2 + 2MnO + K_2O + 5H_2O$

BALANCING EQUATIONS FOR REDOX REACTIONS—AQUEOUS SOLUTIONS. For reactions which occur in aqueous solution you may be given only a skeletal equation, one in which only the principal reactants and products are given. You must complete the equation, as well as balance it. There are two methods for doing this. These methods, the *oxidation-number method* and the *half-reaction method,* are different and should not be confused with each other. Each has its advantages and disadvantages.

BALANCING REDOX EQUATIONS (AQUEOUS SOLUTIONS) : OXIDATION-NUMBER METHOD

Step 1 Assign oxidation numbers to all atoms.

Step 2 Note which atoms appear to lose and which appear to gain electrons, and determine how many electrons are lost and gained.

Step 3 If more than one atom in a formula unit loses or gains electrons, determine the total loss or gain per formula unit.

Step 4 Make the gain of electrons by the oxidizing agent equal to the loss by the reducing agent by inserting an appropriate coefficient before the formula of each on the left-hand side of the equation. (See the following examples.)

Step 5 Balance the atoms which have gained or lost electrons by adding appropriate coefficients on the right.

Step 6 Balance all other atoms except for O and H.

Step 7 Balance the charge (the sum of all the ionic charges) so that it is the same on both sides, by adding either H^+ or OH^-.

(a) If the reaction takes place in *acidic* solution, add H^+ ions to the side deficient in positive charges.

(b) If the reaction takes place in *basic* solution, add OH^- ions to the side deficient in negative charges.

Step 8 Balance O atoms by adding H_2O to the appropriate side. Check to see that the H atoms are now balanced. (They will be, if you have made no mistakes.)

Example 3

Problem : Complete and balance the following equation for a reaction which takes place in acidic solution :

$$Cr_2O_7^{2-} + Fe^{2+} \rightarrow Cr^{3+} + Fe^{3+} \text{ (acidic)}$$

Solution : Step 1:

$$Cr_2O_7^{2-} + Fe^{2+} \rightarrow Cr^{3+} + Fe^{3+}$$
$$+6 \; -2 \quad +2 \quad +3 \quad +3$$

Step 2 :

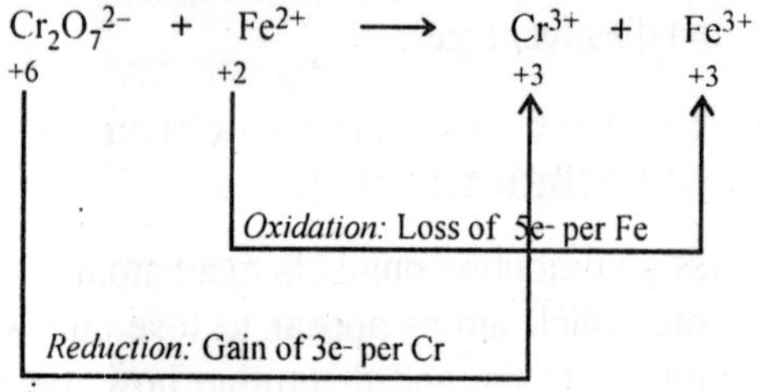

Step 3 :

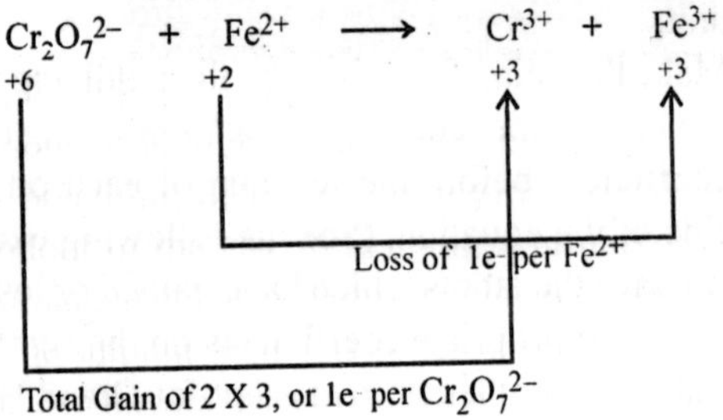

Step 4: $Cr_2O_7^{2-} + 6Fe^{2+} \rightarrow Cr^{3+} + Fe^{3+}$

Step 5: $Cr_2O_7^{2-} + 6Fe^{2+} \rightarrow 2Cr^{3+} + 6Fe^{3+}$

Step 6: Done!

Step 7:

Total charge on left = –2 + 6 (+2) = +10.

Total charge on right = 2 (+3) + 6 (+3) = +24.

Added positive charge needed on *left* = +14.

$$14H^+ + Cr_2O_7^{2-} + 6Fe^{2+} \rightarrow 2Cr^{3+} + 6Fe^{3+}$$

Step 8 : $14\,H^+ + Cr_2O_7^{2-} + 6Fe^{2+} \rightarrow 2Cr^{3+} + 6Fe^{3+} + 7H_2O$.

Example 4

Problem: Complete and balance the following equation for a reaction in *basic* solution :

$$CrO_4^{2-} + Fe(OH)_2 \rightarrow CrO_2^- + Fe(OH)_3 \text{ (basic)}$$

Solution :

Step 1:

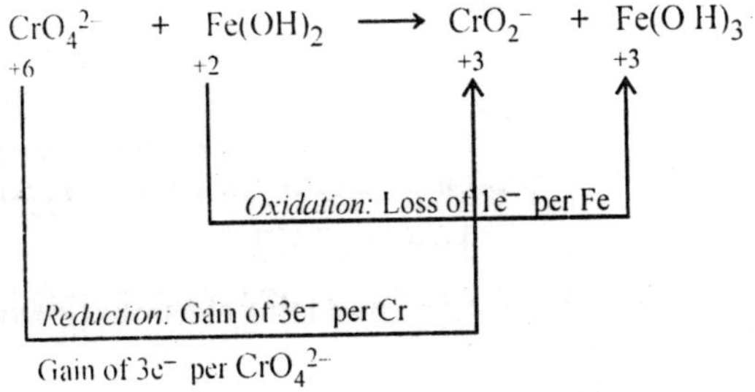

Step 4:

$$CrO_4^{2-} + 3Fe(OH)_2 \rightarrow CrO_2^- + Fe(OH)_3$$

Step 5:

$$CrO_4^{2-} + 3Fe(OH)_2 \rightarrow CrO_2^- + 3Fe(OH)_3$$

Step 6 : Done!

Step 7 :

Total charge on left = –2 + 3 (0) = –2.

Total charge on right = –1 + 3 (0) = –1.

Added negative charge needed on *right* = –1.

$$CrO_4^{2-} + 3Fe(OH)_2 \rightarrow CrO_2^- + 3Fe(OH)_3\ OH^-$$

Step 8 :

$$2H_2O + CrO_4^{2-} + 3Fe(OH)_2 \rightarrow CrO_2^- + Fe(OH)_3 + OH^-$$

The second method for balancing redox equations for reactions which occur in aqueous solution is the *half-reaction method,* also known as the *ion-electron method.* Note that in it you do *not* assign any oxidation numbers.

BALANCING REDOX EQUATIONS (AQUEOUS SOLUTIONS) :

HALF-REACTION METHOD

Step 1 Separate the skeletal equation into two half-reactions, one an oxidation and the other a reduction.

Step 2 Balance each half-reaction separately according to this sequence :

a Balance all atoms other than H and O by inspection.

b Balance O atoms by adding H_2O to the appropriate side.

c Balance the H atoms. The way this is done depends on whether the solution is *acidic or basic.*

i In *acidic* solution, add the appropriate number of H^+ to the side deficient in H.

ii In *basic* solution, instead of adding H^+ ions, *add one H_2O molecule to the side deficient in H plus one OH^- ion to the opposite side for each needed H atom.* You may wish to cancel out H_2O molecules duplicated on each side at this point.

d Balance the charge by adding electrons (e^-) to the side deficient in *negative* charge.

Step 3 Multiply each balanced half-reaction by an appropriate number in order to balance the electron loss against the electron gain Then add the two half-reactions.

Step 4 Subtract out (cancel) anything which appears on both sides. (The electrons should all disappear in this step.)

Example 5

Problem : Using the half-reaction method complete and balance the following equation for a reaction occurring in *acidic* solution.

$Cr_2O_7^{2-} + Fe^{2+} \rightarrow Cr^{3+} + Fe^{3+}$ (acidic)

Solution : Step 1 : $Cr_2O_7^{2-} \rightarrow Cr^{3+}$	$Fe^{2+} \rightarrow Fe^{3+}$
Step 2 : a $Cr_2O_7^{2-} \rightarrow 2Cr^{3+}$	(done!)
b $Cr_2O_7^{2-} \rightarrow 2Cr^{3+} + 7H_2O$	(done!)
c $14H^+ + Cr_2O_7^{2-} \rightarrow 2Cr^{3+} + 7H_2O$	(done!)
d $6e- + 14H^+ \ Cr_2O_7^{2-} \rightarrow 2Cr^{3+} + 7H_2O$	$Fe^{2+} \rightarrow Fe^{3+} + e^-$
Step 3 : $6e^- \ 14H^+ + Cr_2O_7^{2-} \rightarrow 2Cr^{3+} + 7H_2O$	$[Fe^{2+} \rightarrow Fe^{3+} \ e^-] \times 6$

$6e^- \ 14H^+ + Cr_2O_7^{2-} \rightarrow 2Cr^{3+} + 7H_2O$

$6 \ Fe^{2+} \rightarrow 6 \ Fe^{3+} + 6e^-$

$6e^- \ 14H^+ + Cr_2O_7^{2-} + 6Fe^{2+} \rightarrow 2Cr^{3+} \ 6 \ Fe^{3+} + 7H_2O + 6e^-$

Step 4 : $14H^+ + Cr_2O_7^{2-} + 6Fe^{2+} \rightarrow 2Cr^{3+} \ 6 \ Fe^{3+} + 7H_2O$

Example 6

Problem : Complete and balance the following equation for a reaction taking place in *basic* solution using the half-reaction method :

$$CrO_4^{2-} + Fe(OH)_2 \rightarrow CrO_2^- + Fe(OH)_3 \text{ (basic)}$$

Solution:

Step 1 : $CrO_4^{2-} \rightarrow CrO_2^-$	$Fe(OH)_2 \rightarrow Fe(OH)_3$
Step 2 : a (done)	(done!)
b $CrO_4^{2-} \rightarrow CrO_2^- + 2H_2O$	$H_2O + Fe(OH)_2 \rightarrow Fe(OH)_3$
c $4H_2O + CrO_4^{2-} \rightarrow CrO_2 + 2H_2O + 4OH^-$	$OH + H_2O + Fe(OH)_2 \rightarrow Fe(OH)_3 + H_2O$
$2H_2O + CrO_4^{2-} \rightarrow CrO_2^- + 4OH^-$	$OH^- + Fe(OH)_2 \rightarrow Fe(OH)_3$
Step 3 : $3e^- + 2H_2O + CrO_4^{2-} \rightarrow CrO_2^- + 4OH^-$	$[OH^- + Fe(OH)_2 \rightarrow Fe(OH)_3 + e^-] \times 3$

$$3e^- + 2H_2O + CrO_4^{2-} \rightarrow CrO_2^- + 4OH^-$$
$$3OH^- + 3Fe(OH)_2 \rightarrow 3Fe(OH)_3 + 3e^-$$

$$3e^- + 2H_2O + 3OH^- + CrO_4^{2-} + 3Fe(OH)_2 \rightarrow CrO_2^- + 3Fe(OH)_3 + 4OH^- + 3e^-$$

Step 4 : $2H_2O + CrO_4^{2-} + 3Fe(OH)_2 \rightarrow CrO_2^- + 3Fe(OH)_3 + OH^-$.

At this point a comparison of the oxidation-number and half-reaction methods is in order. For *nonsolution reactions* the oxidation-number is usually superior to the half-reaction method. Fro *reactions in aqueous solution* the half-reaction method is the slower of the two, but for most people the surer. A major advantage of the half-reaction method is the lack of necessity of assigning oxidation numbers. Individual half-reactions are used in describing electrode reactions in electrochemical cells. The biggest advantage of the oxidation-number method is its potential speed. With practice you will find that you can do much of it in your head.

Ion-electron method of balancing equations. This method is applicable only to ionic reactions; however, since most reactions in solution are between ions, the method proves convenient and useful. The method takes into account only those ions in solution which actually enter into reaction; in other words, only the net reaction in shown. Those ions which remain unaltered in the solution do not appear in the equation. The oxidation-reduction equation is broken down into two partial equations or steps called *half-reactions;* one equation represents the oxidation step, the other the reduction step. The number of electrons transferred in each step is indicated, and each equation is balanced both atomically and electrically. The gain and loss of electrons is then balanced by multiplying step by appropriate number.

Example 1 :

$$FeCl_3 + SnCl_2 \rightarrow FeCl_2 + SnCl_4 \text{ (not balanced)}$$

From an ionic standpoint, we should write down only the ions which enter into the chemical change—namely Fe^{+++}, which is reduced to Fe^{++}, and Sn^{++}, which is oxidized to Sn^{++++}. The two half-reactions would be

$$Sn{+}{+} \rightarrow Sn^{++++} + 2\,e \text{ (oxidation)}$$

$$Fe^{+++} + e \rightarrow Fe^{++} \text{ (reduction)}$$

Each step is now balanced electrically as well as atomically. Since 2 electrons appear in the first step, we may balance electrons by multiplying the second equation by 2

$$2\,Fe^{+++} + 2\,e \rightarrow 2Fe^{++}$$

Now the two steps may be added (electrons are cancelled), and we obtain the net ionic reaction

$$Sn^{++} \rightarrow Sn^{++++} + 2e$$

$$2\,Fe^{+++} + 2e \rightarrow 2\,Fe^{++}$$

$$\overline{Sn^{++} + 2\,Fe^{+++} \rightarrow Sn^{++++} + 2\,Fe^{++}}$$

Note that the final step is also balanced electrically; the net charge on each side of the equation is +8.

Example 2 : Copper reacts with dilute nitric acid to form cupric ions, water, and nitric oxide. Applying the ion-electron method to this reaction, copper is oxidized from the free state to cupric ion :

$$Cu \rightarrow Cu{+}{+} + 2\,e \text{ (oxidation) (1)}$$

Meanwhile, nitric acid is reduced to NO and H_2O, the latter two substances being un-ionized :

$$H^+ + NO_3^- \rightarrow NO + H_2O$$

The latter equation is balanced atomically as follows :

$$4\ H^+ + NO_3^- \rightarrow NO + 2\ H_2O$$

This equation is now balanced atomically but not electrically, since the net charge on the left is +3, and the net charge on the right side is 0. The addition of 3 *e* to the left side will balance the step electrically :

$$3\ e + 4\ H^+ + NO_3^- \rightarrow NO + 2\ H_2O \qquad (2)$$

To balance the gain and loss of electrons, equation (1) may be multiplied by 3 and equation (2) by 2, then :

$$3Cu \rightarrow 3\ Cu{+}{+} + 6e$$

$$6e + 8H^+ + 2\ NO_3^- \rightarrow 2\ NO + 4\ H_2O$$

$$3\ Cu + 8\ H^+ + 2\ NO_3^- \rightarrow 3\ Cu^{++} + 2NO + 4\ H_2O$$

Example 3 : It is further assumed in the ion-electron method that H^+ ions and H_2O molecules may be used to balance each half-reaction atomically if the reaction is being carried out in acid solution, or OH^- ions and H_2O molecules may be used if the reaction is in basic solution. To illustrate, permanganate ion will oxidize sulfide ion to free sulfur in acid solution

$$S^{--} \rightarrow S + 2\ e \qquad (1)$$

and the permanganate ion is reduced to manganous ion

$$MnO_4^- \rightarrow Mn^{++}$$

To balance the latter step atomically, we may add 8 hydrogen ions to the left side, in which case 4 molecules of water will appear on the right side :

$$8\ H^+ + MnO_4^- \rightarrow Mn^{++} + 4\ H_2O$$

5 electrons are then necessary on the left side to balance electrically:

$$5\ e + 8\ H^+\ MnO_4^- \rightarrow Mn^{++} + 4\ H_2O \qquad (2)$$

Combining steps (1) and (2) and balancing gain and loss of electrons,

$$5\ S^{--} \rightarrow 5\ S + 10\ e$$

$$10\ e + 16\ H^+ + 2\ MnO_4^- \rightarrow 2\ Mn^{++} + 8\ H_2O$$

$$5\ S^{--} + 16\ H^+ + 2\ MnO_4^- \rightarrow 5\ S + 2\ Mn^{++} + 8\ H_2O$$

5

Chemical Bonding: Covalent Bonding

The hydrogen molecule—the simplest case of covalent bonding. In ionic bonding, electrons leave the valence shell of metallic atoms to become part of the valence shells of nonmetallic atoms. However, many elements do not form bonds this way. Very often, an atom of a nonmetal combines with an atom of the same or another nonmetal in such a manner that no complete transfer of electrons takes place. Instead, a valence orbital containing an electron from one atom overlaps a valence orbital containing an electron from another atom. Figure ?? is a diagram of the overlapping electron clouds from two hydrogen atoms bonded covalently to form a diatomic hydrogen. H_2, molecule. Because the nuclei of both atoms have a positive charge. they are attracted to the shared concentration of negative charge that exists between them. This attraction of the nuclei of both atoms for the concentration of negative charge is the **covalent bond** that binds the atoms together.

When a covalent bond has formed between two hydrogen atoms, two *s* electron clouds overlap. These two electrons are attracted to both hydrogen nuclei at the same time. In this arrangement, the two electrons and the two nuclei are in a lower energy state than when arranged as separate atoms. Because of this lowering of energy, a stable covalent bond forms between the atoms.

The Lewis structure

This method of thinking about bonding, learned in high school and too often forgotten in graduate school or before, is a most useful *first step* in thinking about molecules. Before delving into whether valence bond theory or molecular orbital theory is likely to be more helpful, whether valence shell electron pair repulsion theory should be used, or even whether a compound is likely to be composed of molecules, simple ions, or molecular ions, a Lewis structure should be sketched. The following is a brief review of the rules for Lewis structures:

1. *Normally two electrons pair up to form each bond.* This is a consequence of the Pauli exclusion principle—two electrons must have paired spins if they are both to occupy the same region of space between the nuclei and attract both. The definition of a bond as a shared *pair* of electrons, however, is overly restrictive, and we shall see that the early emphasis on electron pairing in bond formation is unnecessary and even misleading.

2. *For most molecules there will be a maximum of eight electrons in the valence shell (=Lewis octet structure).* This is absolutely necessary for atoms of the elements lithium through fluorine since they have only four orbitals (an *s* and three *p* orbitals) in the valence shell. It is quite common, as well, for atoms of other elements that often utilize only their *s* and *p* orbitals. Under these conditions the sum of shared pairs (bonds) and unshared pairs (lone pairs) must equal the number of orbitals—four. This is the *maximum*, and for elements having fewer than four valence electrons, the octet will usually not be filled. The following compounds illustrate these possibilities:

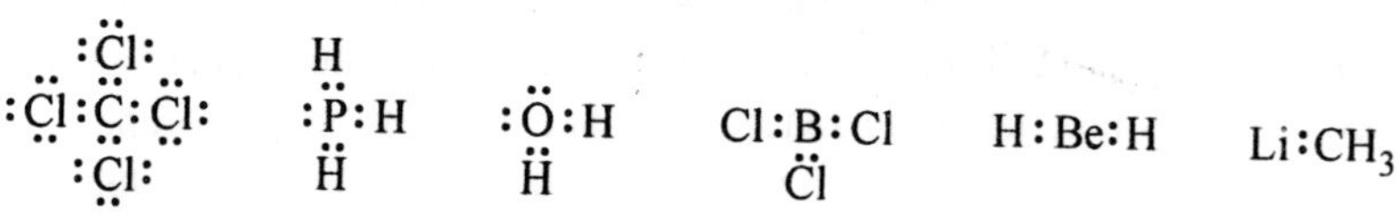

3. *For elements with avaialable d orbitals, the valence shell can be expanded beyond an octet.* Since *d* orbitals first appear in the third energy level, these elements are all in period 3 or beyond and are non-metals in their higher valence compounds and transition metals in complexes. In the non-metals, where the number of valence

electrons is usually the limiting factor, we have maximum covalencies of 5, 6, 7, and 8 in Groups VA, VIA, VIIA, and VIIIA. Factors determining coordination number in complexes are of several kinds, and discussion of them will be deferred. Examples of molecules or ions containing more than eight electrons in the valence shell are :

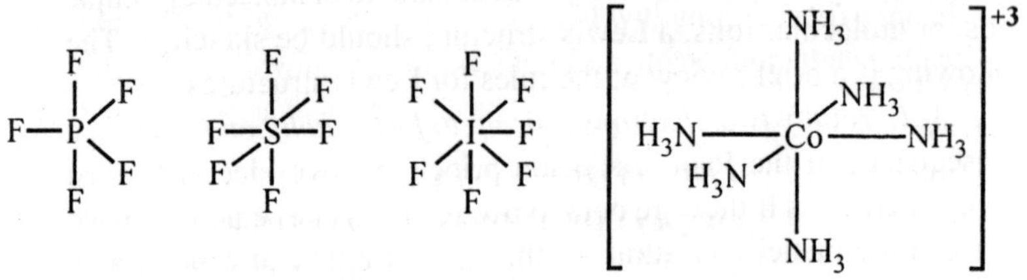

4. *It has been assumed implicity in all of these rules that the molecule will seek the lowest overall energy.* This means that, in general, the maximum number of bonds will form, that the strongest possible bonds will form, and that the arrangement of the atoms in the molecule will be such as to minimize adverse repulsion energies.

Sidgwick—Powell Theory

In 1940 Sidgwick and Powell reviewed the structures of molecules then known. They suggested that for molecules and ions that only contain single bonds, the approximate shape can be predicted from the number of electron pairs in the outer or valence shell of the central atom. The outer shell contains one or more bond pairs of electrons, but it may also contain unshared pairs of electrons (lone pair). Bond pairs and lone pairs were taken as equivalent, since all electron pairs take up some space, the since all electron pairs repel each other. Repulsion is minimized if the electron pairs are orientated in space as far apart as possible.

1. If there are two pairs of electrons in the valence shell of the central atom, the orbitals containing them will be oriented at 180° to each other. It follows that if these orbitals overlap with orbitals from other atoms to form bonds, then the molecule formed will be linear.
2. If there are three electron pairs on the central atom, they will be at 120° to each other. giving a plane triangular structure.

3. For four electron pairs the angle is 109°28', and the shape is tetrahedral.
4. For five pairs, the shape is a trigonal bipyramid.
5. For six pairs the angles are 90° and the shape is octahedral.

Bonding Theory

The two "contenders for the throne" of bonding theory are *valence bond theory* (VBT) and *molecular orbital theory* (MOT). The allusion is an apt one since it seems that much of the history of these two theories consisted of contention between their respective proponents as to which was "best." This is indeed unfortunate inasmuch as they are but opposite sides of the same coin. Sometimes overzealous proponents of one theory give the impression that the other is "wrong." Granted that any theory can be used unwisely, it remains nonetheless true that neither theory can be "true" to the *exclusion* of the other since both reduce to the same common and fundamental quantum-mechanical ideas. As used in inorganic chemistry, however, they do have distinctive aspects. Given a specific question one theory may prove distinctly superior in insight, ease of calculation, or simplicity and clarity of results, but a different question may reverse the picture completely. Surely the inorganic chemist who does not become thoroughly familiar with *both* theories is like the carpenter who refuses to carry a saw because he already has a hammer! He is severely limiting his skills by limiting his tools.

The Shapes of Molecule—the VSEPR Model

The valence-shell electron-pair repulsion model—VSEPR (VES-purr) for short—provides a way to predict the shapes of molecules.

Figure 5.1 : In the ammonia molecule, six of the electrons about the central nitrogen atom are shared pairs (electron pairs shared with the three hydrogen atoms). The remaining two electrons in the octet are an unshared pair.

In its simplest version, the shape of molecule is determined by first drawing a dot diagram for the molecule under consideration, starting with central atom in the molecule. The electron pairs of the central atom are of two types. They are either shared pair (a pair of electrons in a covalent bond) or unshared pair. See Figure 5.1)

According to the VSEPR model, each pair of electrons surrounding the central atom is considered to repel all the other electron pairs around that atom. This repulsion causes each electron pair to take a position about the central atom as far away as possible from the other electron pairs. In the case of a carbon atom bonded to four hydrogen atoms (a molecule of methane), the mutual repulsion of electron pairs produces a molecule with the shape of a regular tetrahderon. (A *regular* tetrahedron is one in which the four faces all are equilateral triangles having the same area.) The nucleus of the carbon atom is at the centre of the tetrahedron. The four pairs of electrons and the four hydrogen atoms bonded by these electrons are spaced symmetrically around the carbon nucleus so that they face the four corners of the tetrahedron. See Figure 5.2.

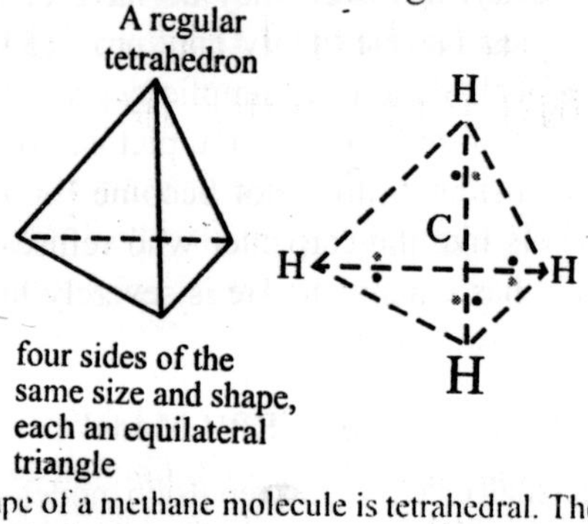

Figure 5.2 : The shape of a methane molecule is tetrahedral. The carbon atom is at the centre of the tetrahedron. The four pairs of shared electrons and the hydrogen atoms bonded by these electron pairs are directed towards the corners of the tetrahdron.

The angles formed by any two hydrogen atoms and the central carbon atom are all 109.47°. This bond angle often is called the **tetrahedral angle**. See Figure 5.3. In many instances in which a central atom is bonded to four other atoms, a molecule with a tetrahedral shape results. In carbon, it is the four sp^3 hybrid orbitals that account for this structure.

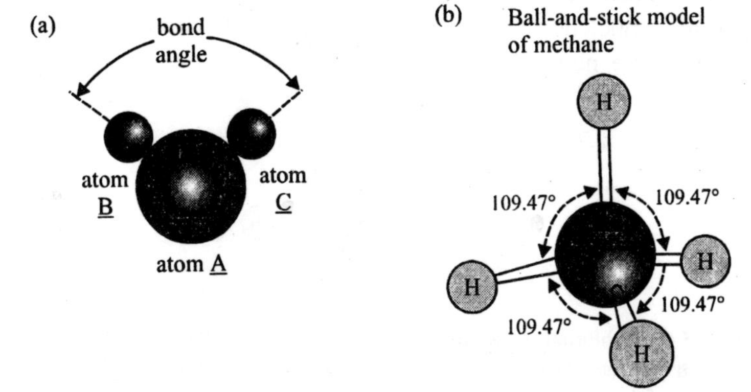

Figure 5.3 : ***Bond angles.*** (a) In this three-atom molecule, the bond angle beween atoms *B* and *C* is formed by the straight lines connecting the nucleus of atom *A* to the nuclei of atoms B and *C*. This angle would be referred to as the angle *BAC*. (b) The bond angle formed by any two hydrogen atoms and the central carbon atom in the tetrahedral methane molecule is 109.47°, often is called the tetrahedral angle.

In some molecules, not all the electron pairs about the central atom are shared by atoms of the same element. An example is chloromethane, CH_3Cl. See Figure 15.26. In chloromethane, only three of the atoms bonded to the carbon are hydrogen atoms. The fourth is a chlorine atom. Chlorine is a highly electronegative element. The shared electrons in the bond between the chlorine and carbon atoms are held much closer to the chlorine atom than to the carbon atom. This is in marked contrast with the positions of the shared electrons between the hydrogen atoms and the carbon atom. As a result, the shape of the chloromethane molecule differs slightly from that of the methane molecule. The molecule has a tetrahedral shape, but it is not that of a regular tetrahedron. Not all four faces of the tetrahedron have the same shape. Likewise, the bond angles vary from the tetrahedral angle.

The shape of the ammonia molecule. In ammonia. NH_3, the nitrogen atom is the central atom. As was true of the carbon atom in methane, the nitrogen atom has four pairs of valence electrons. However, ammonia differs from methane in that only three of the four pairs are shared pairs. See Figure 5.5. The unshared pair has a greater repulsive effect than the shared pairs. This is because there is no positive nucleus on one side of the unshared electron pair to help dissipate the negative charge of the electrons. According to

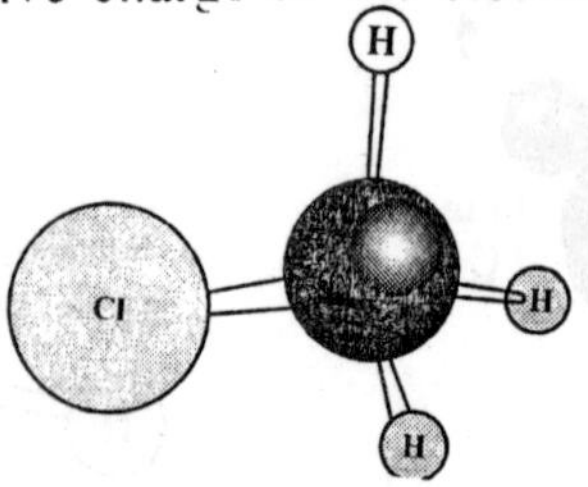

Figure 5.4 : In chloromethane, the chlorine atom has a greater attractionfor its pair of shared electrons than to the hydrogen atoms for their shared pairs. As a result, the shape of the molecule is a distorted tetrahderon.

the VSEPR model, the greater repulsive effect of the unshared pair tends to push the shared electron pairs closer together. By experiment, it has been found that each of the H—N—H bonds forms an angle of about 107°. This is in contrast with the 109.47° angle found for the H—C—H bonds in methane. The resulting shape of an ammonia molecule is that of a pyramid. The three hydrogen atoms form the base, and the nitrogen atom is at the apex. See Figure 5.5(b).

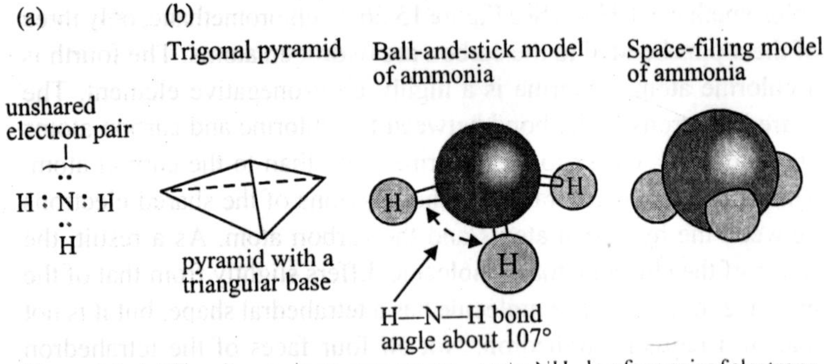

Figure 5.5 : (a) The nitrogen in an ammonia molecule, NH_3, has four pair of electrons about it. However, only three of these are shared with hydrogen atoms. The fourth pair is unshared. (b) An ammonia molecule has the shape of a pyramid with the nitrogen atom at the apex and the three hydrogen atoms forming a triangular base.

The shape of the water molecule. The oxygen atom in water is the central atom. Like the carbon in methane, it also has four pairs of valence electrons surrounding it. However, only two of these are shared pairs, those that form covalent bonds with the hydrogen atoms. The other two pairs are unshared. See Figure 5.6(a). As is true in the case of ammonia, these unshared pairs have a greater repulsive effect than the shared pairs. The combined repulsive effect of the two unshared electron pairs is to produce an H—O—H bond angle smaller than the H—C—H bond angle (109°.47°) in methane and the N—H—N bond angle (107°) in ammonia. In water, the H—O—H bond angle is only 105°. The shape of the water molecule is referred to as "bent" or angular. See Figure 5.6(b).

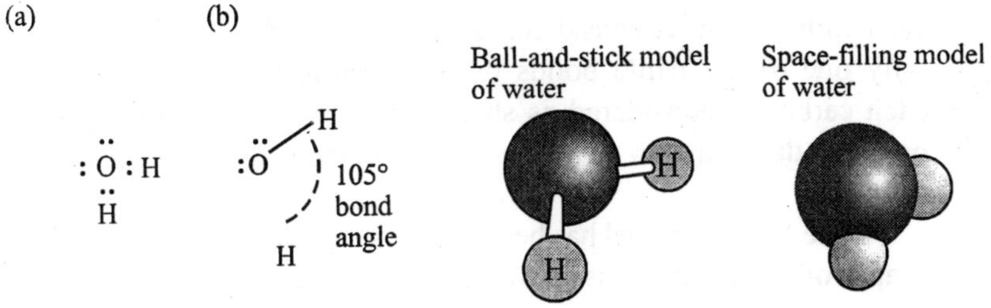

Figure 5.6 : (a) The oxygen atom in a water molecule has two pairs of shared electrons and two pairs of unshared electrons. (b) The four electrons pairs are directed at the corner of a tetrahedron, with oxygen at the center. However, because only two of the electron pairs in the oxygen atom are shared pairs, the water molecule has a bent or angular shape.

Molecular Shapes Predicted by the VSEPR Theory

Total pairs	Electrons shared pairs	Unshared pairs	General shape	Approximate bond angle (degrees)	Examples
4	4	0	tetrahedral	109.47	CH_4,$CHCl_3$
	3	1	pyramidal	<109.47	NH_3,PCl_3
	2	2	bent	<109.47	H_2O, H_2S
	1	3	linear	(any 2-atom molecular is linear)	
3	3	0	trigonal planar	120	BF_3, SO_3
	2	1	bent	<120	SO_2
	1	2	linear	(any 2-atom molecule is linear)	
2	2	0	linear	180	BeH_2, C_2H_2
	1	1	linear	(any 2-atom molecular is linear)	
1	1	0	linear	(any 2-atom molecular is linear)	

Figure 5.7 : Based on the number of shared and unshared electron pairs about the central atom in a molecule, the valence-shell electron-pair repulsion model (VSEPR model) predicts the shapes of a larger number of molecules with a high degree of success.

The geometry associated with various combinations of shared and unshared electron pairs is given in Figure 5.7. The first column in Figure 5.7 lists the total number of pairs of electrons surrounding the central atom. This total is simply the sum of the shared pairs and the unshared pairs. In the case of a central atom that is bonded to another atom by a double or triple bond, the second and third shared pairs of electrons in the bond are not counted because they do not affect the shape of the molecule. For example, in a molecule of acetylene.

$$H-C\equiv C-H$$

each carbon atom is considered a central atom. However, because only one of the triple bonds between these atoms is counted, each carbon is considered as sharing two pairs of electrons, one pair with the hydrogen atom, the other with the neighbouring carbon atom.

The VSEPR model has been highly successful in predicting the shapes of polyatomic ions as well as the shapes of molecules.

2 The Valence-Bond Method

Imagine that the two hydrogen atoms which will eventually constitute an H_2 Molecule are sufficiently separated so that the electrons on each hydrogen atom feel the presence of only one nucleus: i.e., there is no inter-action between the atoms. The wave describing this system is given by

$$\psi_A = \psi_A(1)\psi_B(2) \tag{1}$$

where $\psi_A(1)$ and $\psi_B(2)$ are the 1*s* functions for hydrogen, as described in Chapter 1:

$$\psi_{1s} = \frac{1}{\sqrt{\pi}}\left(\frac{1}{a_0}\right)^{3/2} e^{-r/a_0}. \tag{2}$$

If the hydrogen atoms are brought together, as they exist in the hydrogen molecule, it would be expected that the wave function described by Eq. (1) is a poor approximation to the solution of the Schrödinger equation for the system. However, in a hydrogen molecule it is not possible to distinguish electrons as being associated with a specific nucleus. Thus, for a hydrogen molecule there are two acceptable approximate wave functions which arise because of the indistinguishability of electrons:

$$H_A^{\cdot(1)(2)\cdot}H_B;\ \psi_I = \psi_A(1)\psi_B(2) \tag{3}$$

$$H_A^{\cdot(2)(1)\cdot}H_B;\ \psi_{II} = \psi_A(2)\psi_B(1) \tag{4}$$

Neither Eq. (3) nor Eq. (4) is a good solution to the Schrödinger equation for the system, but a linear combination* of the two approximations is a better solution :

$$\psi = C_1\psi_I + C_2\psi_{II}. \tag{5}$$

The coefficient C_1 and C_2 reflect the extent to which ψ_I and ψ_{II} contribute to the combined function ψ, which is normalized to ensure that the total probability of finding each electron in all space is unity. Since ψ_I and ψ_{II} represent states of equal energy the relationship between C_1 and C_2 is simply

$$C_1 = \pm C_2, \tag{6}$$

giving two possible combinations of ψ_I and ψ_{II}:

$$\psi_+ = C_1(\psi_I + \psi_{II}) \tag{7}$$

$$\psi_- = C_1(\psi_I - \psi_{II}) \tag{8}$$

The best value for the constant C_1 is that for which the energy associated with the trial function is a minimum. Practically, C_1 can be valued evaluated by substituting Eq. (7) setting $\partial E/\partial C_1 = 0$, and solving for C_1.

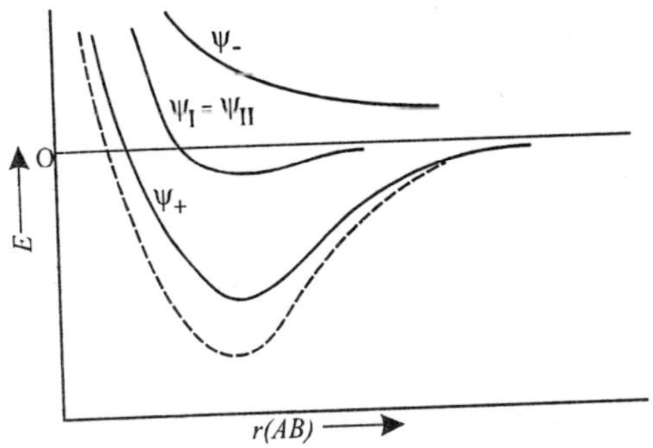

Fig. 5.8 : The energy for the H_2 system as a function of internuclear separation for various trial functions The experimental curve is given as a dotted line.

It should be recalled at this point that any trial function for ψ leads to a value for the energy of the system and that the variation principle guides us to the best trial function. Thus, it should be possible to calculate the energy of the system at a given internuclear separation using ψ_I, ψ_{II}, ψ, or indeed any trial function. The energy of the system as a function of internuclear separation for the various trial functions is shown in Fig. 5.8.

The numerical results, i.e., the depth of the potential well and the internuclear distance at the energy minimum, are shown in Table 4.1. Taking the reference state of H_2 as the energy associated with two unbound hydrogen atoms, the dissociation energy of H_2.

$$H_2 + 104 \text{ kcal mole}^{-1} \rightarrow 2H \quad (9)$$

becomes one of the fiducial for how well the trial function approximates the true wave function. The second experimental quantity in this respect is the equlibrium internuclear distance (Chapter 3), which for H_2 is 0.74 Å.

Table 5.1 : Calculated bond energy and internuclear separation for the hydrogen molecule using several trial functions[a]

Wave function	*E*. kcal mole^{-1}	*r*(*AB*) at min., Å
ψ_{II} [Eq. (11) or (12)]	6	0.90
ψ_+ [Eq. (15]	72.4	0.87
ψ_+' [Eq. (20)]	80.3	0.77

[a] The experimental values for these quantities are $E = 108$ kcal mole^{-1} and $r = 0.74$ Å.

The results shown in Fig. 5.8 and Table 5.1 indicate that the trial function ψ_I [Eq. (3)] is a very poor approximation to the true wave function because the binding energy and the internuclear distance calculated with it are considerably different from the experimental values. The linear combination ψ_- [Eq. (8)] represents a nonbonded state (Fig. 5.8), but ψ_+ [Eq. (7)] gives values for the binding energy and the internuclear distance considerably closer than ψ_I. Apparently ψ_+ is a much better approximation than ψ_I.

A closer approximation to the ture wave function can be obtained by considering the possibility of placing both electrons on one of the hydrogen atoms:

$$H_A \overset{+(1)\cdot}{\underset{(2)\cdot}{}} H_B^{\,-}; \quad \psi_{\text{III}} = \psi_B(1)\psi_B(2) \tag{9}$$

$$H_A \overset{\cdot(1)-}{\underset{\cdot(2)}{}} H_B^{\,+}; \quad \psi_{\text{IV}} = \psi_A(1)\psi_A(2). \tag{10}$$

There are of course two arrangements for this possibility. If hydrogen molecules were to exist with either of these arrangements of electrons we would expect it to exhibit properties characteristic of ionic compounds. This, of course, is not the case, and it should not be surprising that neither ψ_{III} nor ψ_{IV} are good approximations of the true wave function for H_2. However, just as the linear combination of ψ_{I} and ψ_{II} gave a better approximation than either function by itself, it might be tempting to add ψ_{III} and ψ_{IV} to ψ_+ [Eq. (7)] in the hope of obtaining an even better approximation:

$$\psi_+' = C_1(\psi_{\text{I}} + \psi_{\text{II}}) + C_3(\psi_{\text{III}} + \psi_{\text{IV}}). \tag{11}$$

Since ψ_{III} and ψ_{IV} are also states of equal energy there is only one constant, C_3, to be evaluated [cf. the discussion of Eq. (5)]. The new trial function ψ_+', approximating the solution to the wave equation, give rise to the energy of the system at any internuclear separation. The best value of C_3 Can be obtained by application of the variation principle. The results using ψ_+' as a trial solution to the wave equation (Fig. 5.8 and Table 5.1) show that it is indeed a better approximation to the ture wave function since both the binding energy (80.3 kcal) and the internuclear distance (0.77 Å) are closer to the experimentally observed values for these quantities.

Further adjustments to the trial function can be made; these are interpreted in terms of the shielding effect of electrons on the nuclei and electron-electron interaction. The effect of nuclear charge screening is usually introduced into this calculation by considering an effective nuclear charge Z^* rather than the true nuclear charge. This procedure provides another adjustable parameter in the trial function which can be minimized with respect to the energy of the system. Finally, it has been possible to create a trial function in an essentially empirical manner to obtain virtually complete agreement between the theoretical and the experimental values for the binding energy and the equilibrium internuclear separation. The best results obtained arise from using a 13-term polynomial as the solution of the wave equation; this solution gives a discrepancy of only 0.6% and 0.1% in the calculated and experimental values of the bond energy and bond distance, respectively.

The valence-bond description presented here for the structure of H_2 serves to illustrate explicity serveral general points about the method. First, a trial function which is a reasonably good solution to the wave equation for a system can be obtained by the linear combination of a series of approximations which themselves may be much poorer approximations. The latter usually incorporate atomic orbitals and a consideration of various possible distributions of the valence electrons in the system. There is, of course, a very practical reason for using atomic wave functions in this process; explicit mathematical forms for hydrogenic atomic functions are known. Thus, in the case of H_2 a good approximation to the solution of the wave equation arises from combining the wave functions corresponding to the distribution of electrons shown in Fig. 5.9. The pictorial representations shown in Fig. 5.9 are similar to the structural formulas to which chemists are accustomed, but their meaning is not the same. Accordingly, the representations shown in Fig. 5.9 are called valence-bond structures, canonical forms, or resonance forms. They are the individual representations of the wave functions which are linearly combined to approximate the solution to the wave equation for the system. Canonical forms do not exist as such in the chemical sense. The true structure is said to be resonance hybrid of the canonical forms; i.e., the true wave function can be approximated by a mixture of the wave functions represented by the canonical forms. This mathematical relationship is indicated by the double-headed arrows in Fig. 5.9.

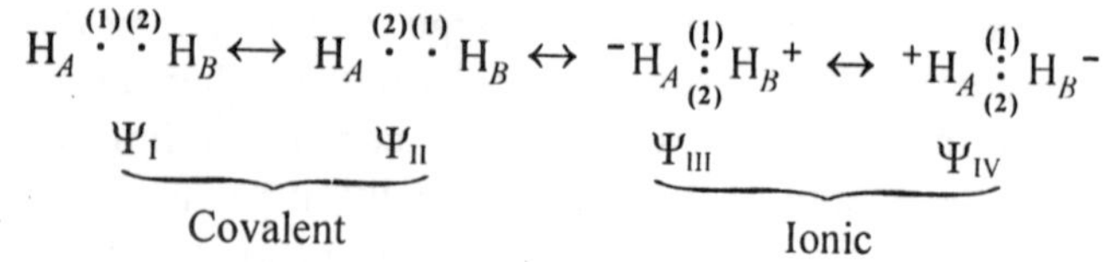

Fig. 5.9 : Representations of the four principal canonical forms which contribute to the best trial function for H_2.

A second point can be made concerning the valence-bond method. The canonical forms are broadly divided into a group which contains forms with no charge distribution, called covalent forms, and a group containing forms with a charge distribution, called ionic forms (Fig. 5.9). The addition of wave functions associated with ionic forms to the final approximation [cf. Eq. (11) invariably gives a better solution to the wave equation; that is the calculated binding energy and internuclear distance are closer to the experimental values. Thus, in general, the valence-bond method gives an approximation [Eq. (12)] to the true wave function for a molecule which contains contributions from covalent and ionic canonical forms:

$$\psi = \psi_{\text{covalent}} + \psi_{\text{ionic}}. \qquad (12)$$

A third point should be stressed in reviewing the strategy of the valence-bond method. The indication whether any trial function is acceptable as a solution to the wave equation for a given system comes from a comparison of experimental parameters such as bond distance and bond energy with the calculated values. Indeed, such experimentally determined parameters can often be used as a guide to the types of canonical forms which might be important contributors to the resonance hybrid. For example, the Lewis formulation for carbon dioxide (1) suggests that the molecule contains a pair of carbon-oxygen double bonds. This structure suggests that the C—O bond

:Ö::C::Ö:

distance should be similar to that observed in ketones (R_2CO), which is 1.22 Å. The bond distance experimentally observed for CO_2 is 1.16 Å, considerably shorter than that expected on the basis of (1); this distance is not, however, as short as that in carbon monoxide (1.13 Å), the Lewis formulation of which corresponds to a triple bond (: C : : :O:). Thus, if ψ_I is the wave function for the canonical form which corresponds to the best Lewis structure [i.e., (1)], we might expect that ψ_I is a completely satisfactory trial function for the solution to the Schrödinger equation for this system. However, it is possible to obtain a better approximation to the solution by considering canonical forms that incorporate triple bond; two such forms exist [(2) and (3)] and they happen to be ionic structures. In valence-bond terms, the true structure of CO_2 can be expressed as the resonance hybrid of the canonical forms where the best approximation to the wave function is given by

$$\psi = C_1\psi_I + C_2\,\psi_{II} + C_3\,\psi_{III} \tag{13}$$

A comparison of the energy expected for a given canonical form with the experimentally determined energy content of the molecule can also indicate how well that canonical form represents the structure of the molecule. For example, if the Lewis formulation of CO_2 (1) is used as the canonical form leading to the wave function, the energy content of the molecule might be expected to be twice the C=O bond energy (2 × 179 = 358 kcal mole^{-1}). The energy of formation of CO_2 (383 kcal mol^{-1}) is markedly different than represented by the Lewis structure (1). In other words the wave function corresponding to this canonical form has to be modified to give a solution yielding a lower energy. As was pointed out earlier in this discussion, the addition of ionic canonical forms to the resonance hybrid leads to lower-energy wave functions. The arguments based on bond distances for CO_2 also result in the same general conclusion.

Finally, it must be emphasized that the process by which acceptable solutions to the wave equation are constructed need not necessarily be related to chemical considerations. Presumably need not necessarily be related to chemical considerations. Presumably any wave function is acceptable which satisfies the Schrödinger equation using the criteria outlined in this section. The valence-bond method of constructing suitable wave functions reflects; on additive principle which has long been used in other areas of chemistry; that is, the characteristics of a system composed of a number of components can be expressed in terms of the individual characteristics of the components. This type of additivity of characteristics is reflected, for example, in Raoult's law and the establishment of a table of half-cell potentials. The valence-bond method appears to be an extension of these general ideas, in the sense that the electronic distributions of molecules is expressed as a linear combination of the electronic distributions of collections of atomic systems.

Hybrid orbitals Carbon forms countless compounds in which its atoms bond covalently to four other atoms. The simplest of these is methane, CH_4. How can we describe the four covalent bonds in this molecule in terms of orbital overlap? The ground-state electronic configuration of C is

C (Z = 6):	⇅	↑	↑	↑	↑
	1s	2s	2p		

Carbon thus appears to be able to form only two covalent bonds by contributing each of its two unpaired electrons to a shared pair. But the short-lived methylene (CH_2) molecule is much less stable than CH_4.

In the methane molecule each H atom is located at the corner of a regular tetrahedron, shown inscribed in a cube in the drawing, so that the relationship between these two regular solids can be seen. In CH_4 all bond lengths are the same and the angle between each C—H bond and any of the other three is the *tetrahedral angle*, 109.5°. The observed tetrahedral structure of methane is what we expect after applying VSEPR theory to this molecule.

According to the Lewis structure for methane the carbon

```
   H
   ··
H : C : H
   ··
   H
```

evidently uses all four of its valence electrons so that four C—H bonds can be formed. It is not too difficult to see how carbon can form four bonds. Suppose that one of the 2*s* electrons is *promoted* to the vacant, but higher energy, 2*p* orbital.

Now the C atom apears to be ready to form four σ bonds by overlap of its 2*s* and 2*p* orbitals with the 1*s* orbitals of four H atoms. The difficulty here is that if the bonding occurred this way, the CH_4 molecule would *not* be tetrahedral. Instead, its shape would be like that shown in Fig. 5.10 (*a*) through *c* are shown the three C—H bonds which would result from overlap of the three 2*p* orbitals of C with the 1*s* orbitals of three H atoms. The fourth bond might go almost anywhere, because an *s* orbital is spherically symmetrical and good overlap is possible from any direction. If the last H is located as far away as possible from the other H atoms in order to minimize interelectronic repulsion, then it goes in the position indicated in *d*. The entire proposed CH_4 structure is shown in Fig. 5.10 (*e*). If these orbitals were used in bonding, methane would evidently have the

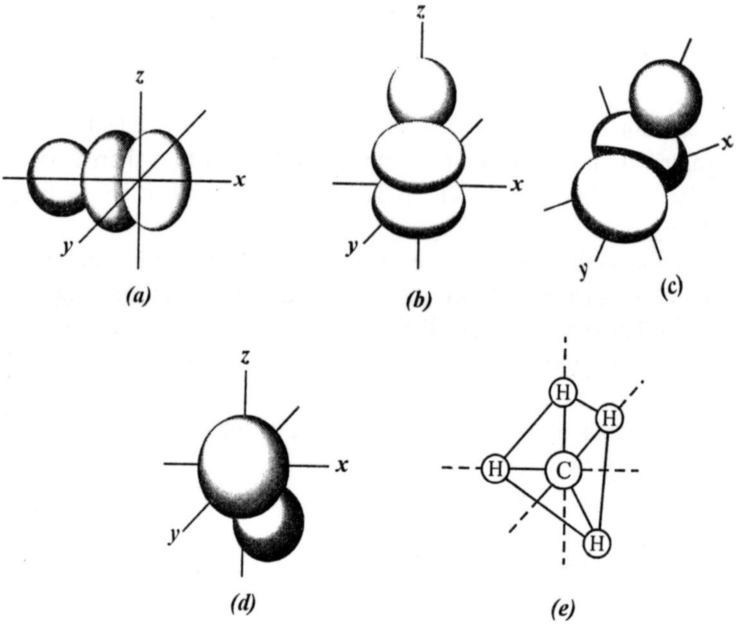

Fig. 5.10 : Bonding in methane: an *incorrect* model. (a) First bond: $2p_z - 1s$ overlap. (b) Second bond: $2p_z - 1s$ overlap. (*c*) Third bond : $2p_z - 1s$ overlap. (d) Fourth bond: $2s - 1s$ overlap. (*e*) Complete CH_4 molecule: a trigonal pyramid (*incorrect*).

shape of a trigonal pyramid *but it does not*. In CH_4 all bond angles are equal and all H atoms are equivalent.

The experimentally determined structure of methane is tetrahedral. How we account for it using the *s* and *p* orbitals of carbon? The answer is that the ground-state set of *s* and *p* orbitals of carbon is replaced by a new set which is suitable for forming four *equivalent* bonds, each at the tetrahedral angle from each of the others. This may sound like a kind of orbital sleight of hand, and so in order to aid understanding of this replacement, we will pause of consider first two simpler cases, the bonding of beryllium and boron.

sp Hybrid orbitals

Beryllium ($Z = 4$) forms a hydrogen compound which at high temperatures exists as discrete BeH_2 molecules. The ground-state electronic configuration of a Be atom is

$$\text{Be } (Z = 4): \quad \underset{1s}{\underline{\uparrow\downarrow}} \quad \underset{2s}{\underline{\uparrow\downarrow}} \quad \underbrace{\_\_\ \_\_\ \_\_}_{2p}$$

The two bonds in BeH_2 are found to be oriented at 180° from each other; that is, bonds, its 2*s* and one of its 2*p* orbitals are replaced by a pair of new orbitals, and these new orbitals, *hybrid orbitals*, are used for bonding. (Recall that each orbital corresponds to a solution, a *wave function*, to the Schödinger wave equation. Because the wave equation is a differential equation, any set of its solutions can be combined mathematically to form a new set of wave functions which are also solutions. These new wave functions are said to be hybrids of the original ones and correspond to a set of hybrid orbitals.

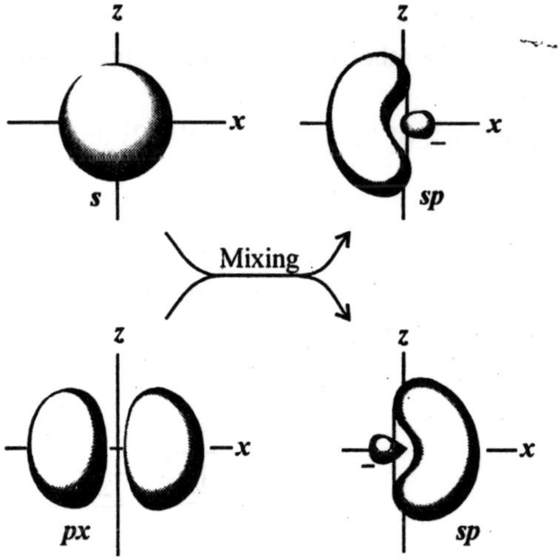

Fig. 5.11 : Formation of *sp* hybrid orbitals.

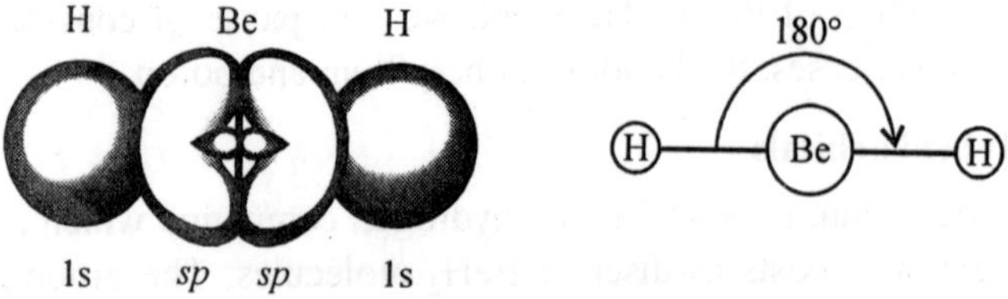

Fig. 5.12 : *The BeH$_2$ Molecule*

Perhaps some pictures will help. At the left of Fig. 5.11 are shown an *s* and a *p* orbital. In the illustration the plus and minus signs are *not charges*. Each is the algebraic sign of the wave function in the designated lobe of the orbital. Now we will combine or *mix* the orbitals, first (upper-right drawing) by *adding* the *p* to the *s*. The result is a hybrid orbital, in which the density of electronic charge has increased where the original wave functions had the same sign and has decreased where they had opposite signs. This hybrid orbital, called an *sp* orbital, is highly directional; overlap is favoured in the direction of its large major lobe. *Subtraction* of the original *s* and *p* orbitals (lower-right drawing) yields the second hybrid orbital. It is eqivalent to the first, but points 180° away. Thus by combining or mixing two *nonequivalent* orbital (one is an s and the other, a *p*) we have obtained two *equivalent* *sp* hybrid orbitals.

We can keep track of the mixing of the 2*s* and 2*p* orbitals in beryllium as follows:

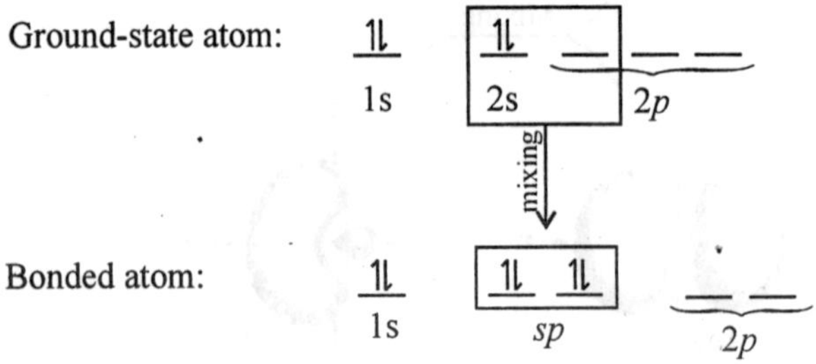

The beryllium *sp* orbitals overlap with hydrogen 1*s* orbitals (the hydrogens' electrons are shown in the above orbital diagram as coloured arrows) to produce two σ bonds and a resulting linear molecule, as is shown in Fig. 5.12.

*sp*² Hybrid orbitals

Other combinations of orbitals may be involved in hybridization. Boron ($Z = 5$) uses its 2*s* orbital and *two* of its 2*p* orbitals to form a set of *three* equivalent sp^2 hybrid orbitals, as shown below:

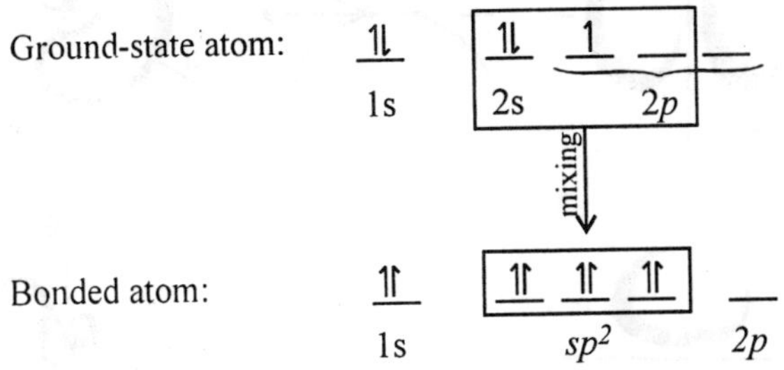

Each of the three sp^2 hybrids has much the same shape as an *sp* hybrid orbital, but the three are oriented at 120° from each other, as is shown in Fig. 5.13.

If each of the sp^2 hybrid orbitals overlaps with a 1*s* orbital of an H atom, the result is a boron hydride, or borane (BH_3), molecule. This molecule, shown in Fig. 9.23 is predicted to have a planar structure in which each H occupies the corner of an equilateral triangle. BH_3 is however, an almost hypothetical molecule, having been observed only as a short-lived intermediate in certain rapid reactions. We have used it as an example because of the analogy with BeH_2 (just discussed) and with CH_4 (to be discussed next). Boron does form similar trigonal planar molecules of the type BX_3, where X is a halogen atom, F, Cl, Br, or I.

***sp*³ Hybrid orbitals** Finally, we return to carbon and its hydrogen compound, methane (CH_4). Here the carbon atom uses its 2s and all three of its 2*p* orbitals to form four equivalent sp^3 hybrid orbitals.

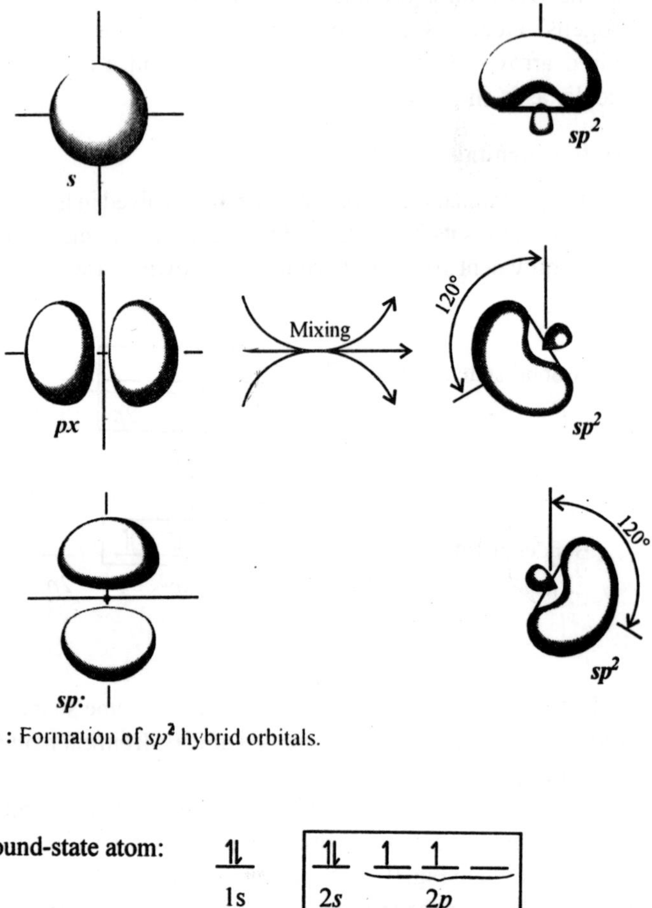

Fig. 5.13 : Formation of sp^2 hybrid orbitals.

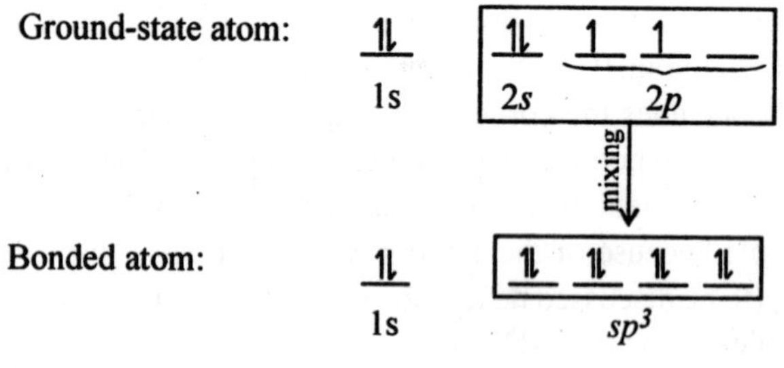

These orbitals look much like the sp and sp^2 hybrids but point towards the corners of a regular tetrahedron (Fig. 5.15). This accounts for the tetrahedral shape of the CH_4 molecule.

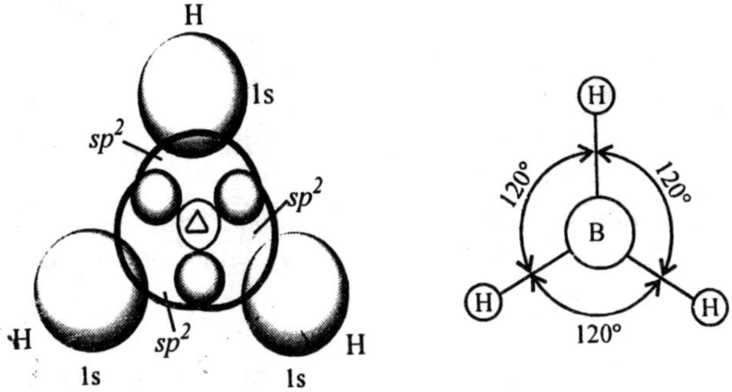

Fig. 5.14 : BH_3: probable structure.

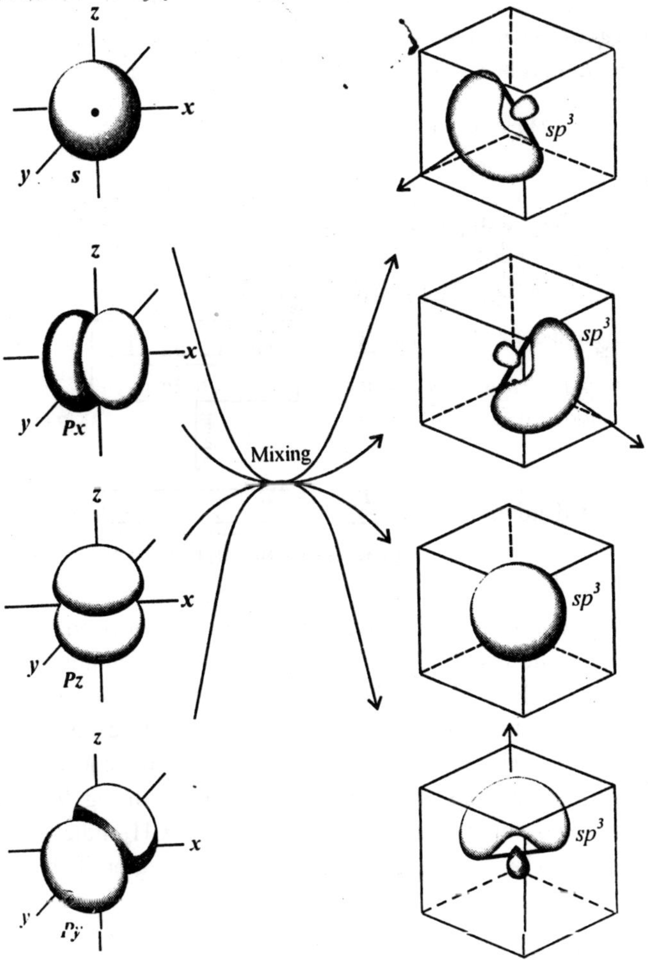

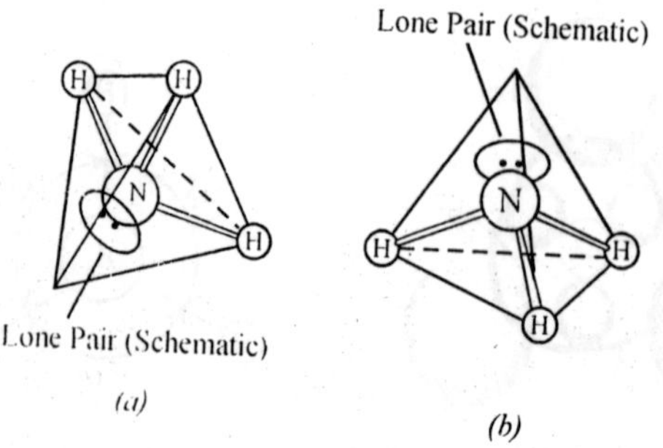

Fig. 5.16 : The NH_3 molecule In (a) the orientation is similar to that of CH_4 in (b) the there H atoms from a base.

The ammonia molecule There are many small molecules in which the bonding is best described in terms of sp^3 hybrid orbitals. In

$$\begin{array}{c} \cdot\cdot \\ H : N : H \\ \cdot\cdot \\ H \end{array}$$

ammonia, NH_3, the mixing of the *s* and *p* orbitals of N (Z = 7) takes place essentially as it does with C in CH_4:

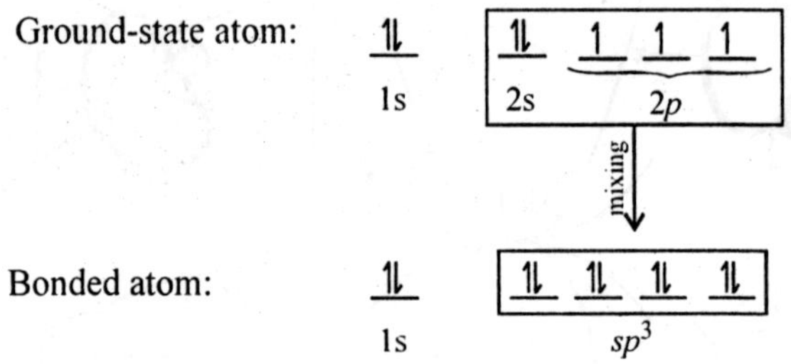

In this case, however, because N has one more electron than C, only three of the resulting sp^3 hybrid orbitals are available for forming bonds with H atoms. The fourth contains a *lone pair*. Therefore, NH_3 has a geometrical shape like that of CH_4, except for the missing H atom. This is shown in Fig. 9.16, in which the location of the lone pair of electrons is shown as a colour oval. In Fig. 9.16a the NH_3 is drawn so that you can compare it with the CH_4 molecule. In Fig. 5.16b it has been flipped over so that it is sitting on its base of three H atoms, the lone pair pointing upward. The shape of the molecule is that of a trigonal pyramid with the three H's defining the base.

In ammonia the H—N—H bond angles are 107.3°, which is a little *less* than the tetrahedral angle 109.5°. This is a result of repulsion between the lone pair of electrons and the three bonding pairs. The charge cloud of a lone pair of electrons is more diffuse, spread out over more space than that of a bonding pair, which is more tightly compressed into the region betweeen the bonded atoms. The spreading out of the lone-pair charge cloud means that lone pair-bonding pair repulsions (Fig. 5.17) are stronger than bonding pair-bonding pair repulsions in NH_3. The result is the pushing together of the bonding pairs, and so the H atoms adjust their positions for best overlap.

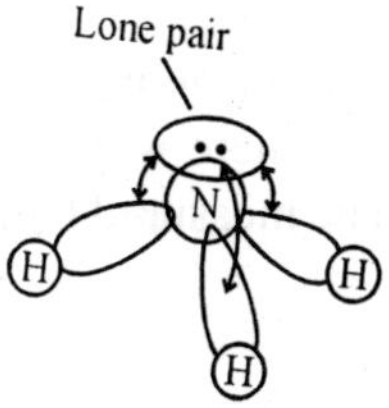

Fig. 5.17 : Lone pair-bonding pair repulsion in NH_2.

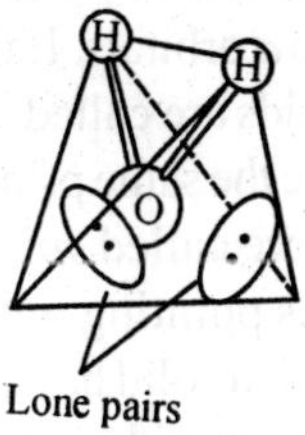

Fig. 5.18 : The H_2O molecule.

The water molecule The bonding in H_2O is similar to that in NH_3. The Lewis structure of water is

$$\begin{array}{c} H:\ddot{\underset{\cdot\cdot}{O}}: \\ H \end{array}$$

The four orbitals in the valence shell of O (Z = S) hybridize

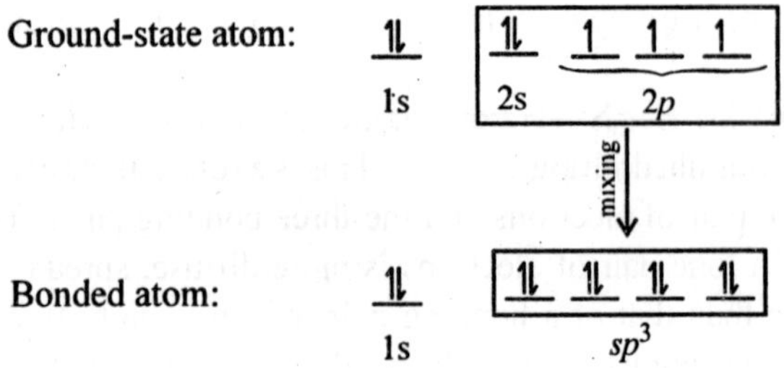

and the four sp^3 hybrid orbitals are occupied by two lone pairs and two bonding pairs of electrons. The resulting structure is described as *bent*, or *angular* (Fig. 5.18).

In H_2O there is an even greater shrinking of the bond angle from the tetrahedral angle 109.5° than there is in NH_3, this being a result of the *two* lone pairs in H_2O. The measured bond angle in H_2O is only 104.5°.

Other hybrid orbital sets

Other possibilities exist for the mixing of pure atomic orbitals to form sets of hybrid oritals. The most important of these is the hybridization of one *s*, three *p*, and two *d* orbitals. If the *d* orbitals are from the $n-1$ shell of the atom, the hybrids are called d^2sp^3. If they are from the valence shell, that is, if they have the same principal quantum number as the *s* and *p* orbitals, then they are called sp^3d^2 orbitals. In either case these orbitals have major lobes pointing out toward the corners of a regular *octahedron*, an eight-sided solid having faces that are identical equilateral triangles. It is evident that with such hybridization the valence shell of the central atom no longer contains just an octet but has been expanded to 12 electrons. Octahedral hybrid orbitals are used to account for the structure of sulphur hexafluoride, SF_6.

Table 5.2 shows a summary of the more important sets of hybrid orbitals, their geometries, and some examples.

Sigma and Pi Bonds

All the bonds formed in these examples result from end to end overlap of orbitals and are called sigma σ bonds. In σ bonds the

Table 5.2 : Some impotant sets of hybrid orbitals

Hybrid orbital set	Geometry*		Examples
sp	Linear		BeF_2 $CdBr_2$ $HgCl_2$
sp^2	Trigonal planar	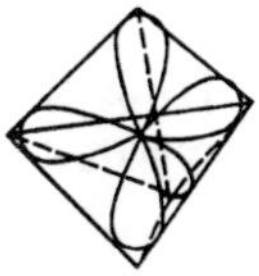	BF_3 $B(CH_3)_3$ GaI_2
sp^3	Tetrahedral	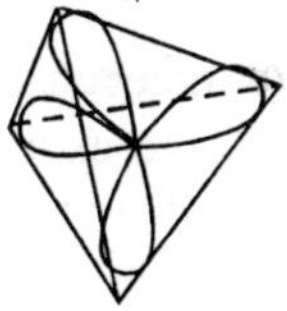	$TiCl_4$ CCl_4 SiF_4
dsp^3 or sp^3d	Trional bipyramidal	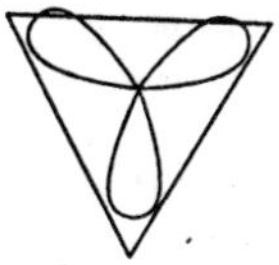	PCl_5 $MoCl_5$ $TaCl_5$
d^2sp^3 or sp^3d^2	Octahedral		SF_6 SbF_6^- $CrCl_6^{3-}$

* In these drawing only the major jobes are shown, and these are considerably slimmed down in order to emphasize their directional charactoristics.

electron density is concentrated in between the two atoms, and on a line joining the two atoms. Double or triple bonds occur by the sideways overlap of orbitals., giving pi π bonds. In π bonds the electron density also concentrates between the atoms, but on either side of the line joining the atoms. The shape of the molecule is determined by the σ bonds (and lone pairs) but not by the π bonds. Pi bonds merely shorten the bond lengths.

Consider the structure of the carbon dioxide molecule. Since C is typically four-valent and O is typically two-valent, the bonding can be simply represented.

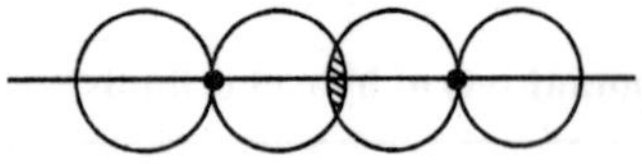

(a)

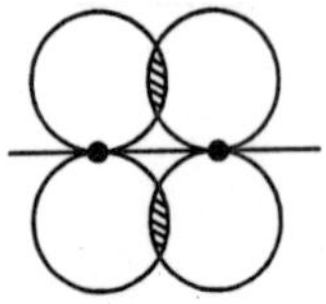

(b)

Fig. 5.19 : Sigma and pi overlap: (a) sigma overlap (lobes point along the nuclei); (b) pi overlap (lobes are at right angles to the line joining the nuclei).

O=C=O

Triatomic molecules must be either linear or angular. In CO_2, the C atom must be excited to provide four unpaired electrons to form the four bonds required.

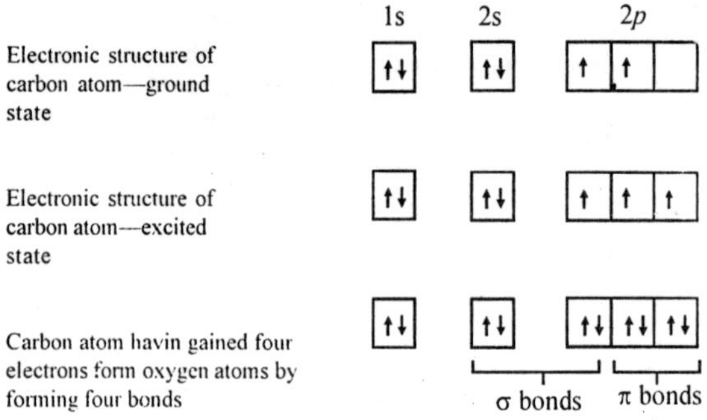

There are two σ bonds and two π bonds in the molecule. Pi orbitals are ignored in determining the shape of the molecule. The remaining *s* and *p* orbitals are used to form the σ bonds. (These could be hybridized and the two sp^2 orbitals will point in opposite directions. Alternatively VSEPR theory suggests that these two orbitals will be oriented as far apart as possible.) These two orbitals overlap with *p* orbitals from two O atoms, forming a linear molecule with a bond angle of 180°. The $2p_y$ and $2p_z$ orbitals on C used for π bonding are at right angles to the bond, and overlap sideways with *p* orbitals on the O atoms at either side. This π overlap shortens the C—O distances, but does not affect the shape.

The sulphur dixoide molecule SO_2 may be considered in a similar way. S shows oxidation states of (+II), (+IV) and (+VI), whilst O is two-valent. The structure may be represented:

$$O=S=O$$

Triatomic molecules are either linear or bent. The S atom must be excited to provide four unpaired electrons

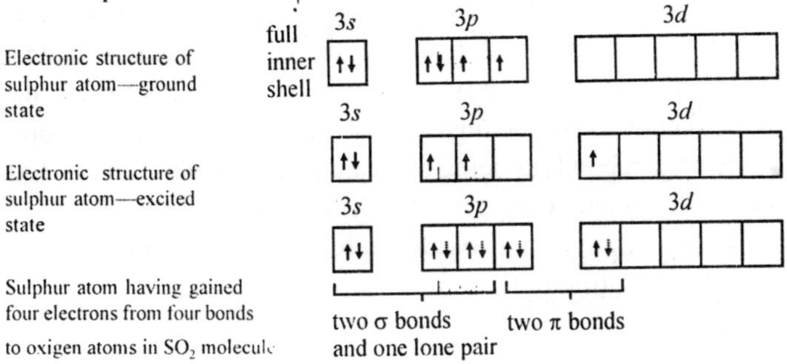

The two electrons pairs which forms the π bonds do not affect the shape of the molecule. The remaining three orbitals point to the corners of a triangle, and result in a planar triangular structure for the molecule with two corners occupied by O atoms and one corner occupied by a lone pair. The SO_2 molecule is thus angular or V shaped (Figure 5.20).

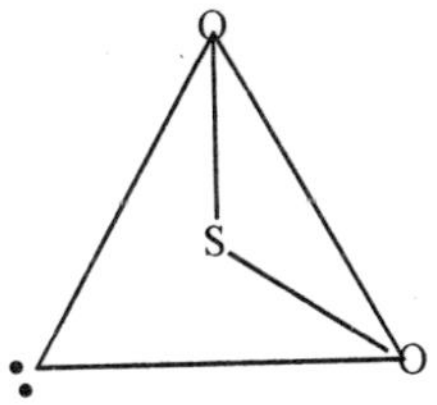

Fig 5.20 : Sulphur dioxide molecule.

The π bonds do not alter the shape, but merely shorten the bond lengths. The bond angle is reduced from the ideal value of 120° to 119°30' because of the repulsion by the lone pair of electrons. Problems arise when we examine exactly which AOs are involved in π overlap. If the σ bonding occurs in the *xy* plane then π overlap can occur between the $3p_z$ orbital on S and the $2p_z$ orbital on one O atom to give one π bond. The second π bond involves a *d* orbital. Though the $3d_{z^2}$ orbital on S is in the correct orientation for π overlap with the $2p_z$ orbital on the other O atom, the symmetry of the $3d_{z^2}$ orbital is wrong (both lobes have a + sign) whilst for a *p* orbital one lobe is + and the other –. Thus overlap of these orbitals does not result in bonding. The $3d_{xz}$ orbital on S is in the correct orientation, and has the correct symmetry to overlap with the $2p_z$ orbital on the second O atom, and could give the second π bond. It is surprising that π bonds involving *p* and *d* orbitals both have the same energy (and bond length). This calls into question whether it is correct to treat molecules with two π bonds as containing two discrete π bonds. A better approach is to treat the π bonds as being delocalized over several atoms. Examples of this treatment are given near the end of this chapter.

In the sulphur trioxide molecule SO_3 valency requirements suggest the structure

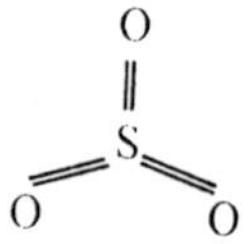

The central S atom must be excited to provide six unpaired electrons to form six bonds.

Electronic structure of sulphur atom—excited state

Sulphur atom having gained six electrons from six bonds to oxygen atoms in SO_3 molecule

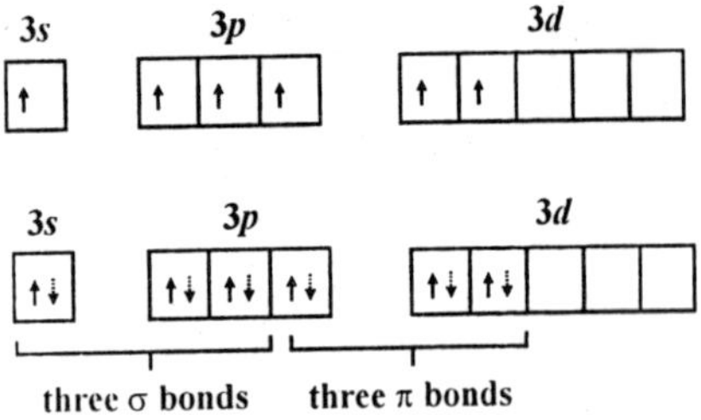

The three π bonds are ignored in determining the shape of the molecule. The three σ orbitals are directed towards the corners of an equilateral triangle, and the SO_3 molecule is completely regular plane triangle (Figure 5.21). The π bonds shorten the bond lengths, but do not affect the shape. This approach explains the σ bonding and shape of the molecule, but the explanation of π bonding is unsatisfactory. It presumes:

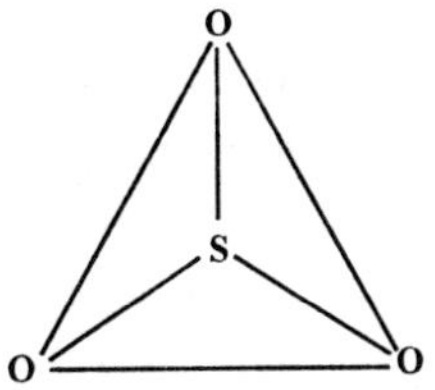

Fig. 5.21 : Sulphure trioxide moleeule

1. That one $3p$ and two $3d$ orbitals on S are in the correct orientation to overlap sideways with the $2p_y$ or $2p_z$ orbitals on three different O atoms, and
2. That the π bonds formed are all of equal strength.

This calls into question the treatment of π bonds. In molecules with more than one π bond, or molecules where the π bond could equally well exist in more than one position, it is better to treat the π bonding as being delocalized over several atoms rather than localized between two atoms. This approach is developed near the end of this chapter.

The molecular orbital theory

The molecular orbital theory of covalent bonding considers the ways in which electrons are influenced by the presence of two or more nuclei, the electrons is bonded atoms being said to occupy molecular orbitals. In the case of the hyrogen molecule there will be repulsive forces operating between the two nuclei and between the two electrons; in addition there will be attractive forces operating between the nuclei and the two electrons. These complications mean that precise solutions of the energies and wave functions corresponding to molecular orbitals are out of the question, but approximate solutions are possible. We shall do no more than consider the molecular orbital theory in a purely qualitative sense.

Let us first consider the hydrogen molecule. If two hydrogen atoms are first considered to be widely separated then each electron is only influenced by its own nucleus, each electron occupying a $1s$ atomic orbital. As the hydrogen atoms approach one another both electrons come under the influence of the two hydrogen nuclei and eventually a chemical bond is established; under these conditions the separate atomic orbitals can be said to merge and constitute molecu!ar orbitals. An important point here is that the number of possible molecular orbitals is equal to the number of separate atomic orbitals in the separated atoms. For the hydrogen molecule there are two possible molecular orbitals. The two molecular orbitals are obtained by adding and subtracting the two atomic orbitals (the linear comibination of atomic orbitals, abbreviated as LCAO method). The two possible molecular orbitals are represented thus :

$$\psi_{\text{bonding}} = \psi_{A.1s} + \psi_{B.1s}$$
$$\psi_{\text{antibonding}} = \psi_{A.1s} - \psi_{B.1s}$$

where the letters A and B are used to distinguish the two hydrogen atoms. A bonding molecular orbital is obtained by adding together the separate atomic orbitals and this results in the build-up of negative charge between the two nuclei; an antibonding molecular orbital is obtained by subtracting the two atomic orbitals and this leads to a draining away of negative charge between the two nuclei. A visual picture of what is involved can be gained by considering the overlap of the two atomic orbitals (Fig. 5.22).

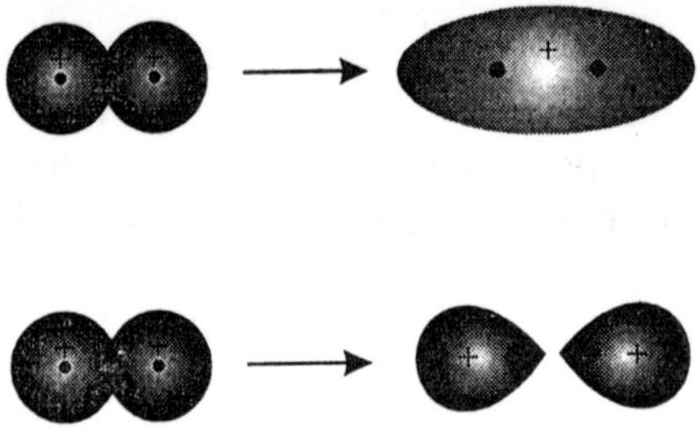

Bonding molecular orbital (σls)

Antibonding molecular orbital (σ*

Fig. 5.22. The combination of hydrogen 1*s* atomic orbitals to give a bonding and an antibonding molecular orbital.

In the case of hydrogen the wave function corresponding to the two 1*s* orbitals is positive and the wave function corresponding to the bonding molecular orbital is also positive (abbreviated to σ 1*s*). Subtraction of the two atomic orbitals gives a molecular orbital which has positive and negative regions (abbreviated π^*1s).

In general, a bonding molecular orbital is of lower energy than the two separate atomic orbitals, while the reverse is true for an antibonding orbital. Furthermore, two electrons with opposite spins can occupy each molecular orbital (two electrons with opposite spins can occupy each atomic orbital by the Pauli exclusion principle. Thus, in the hydrogen molecule, both electrons occupy the bonding molecular orbital, a chemical bond being established because this bonding molecular orbital is of lower energy than the two singly occupied atomic orbitals of the hydrogen atoms (Fig. 5.23).

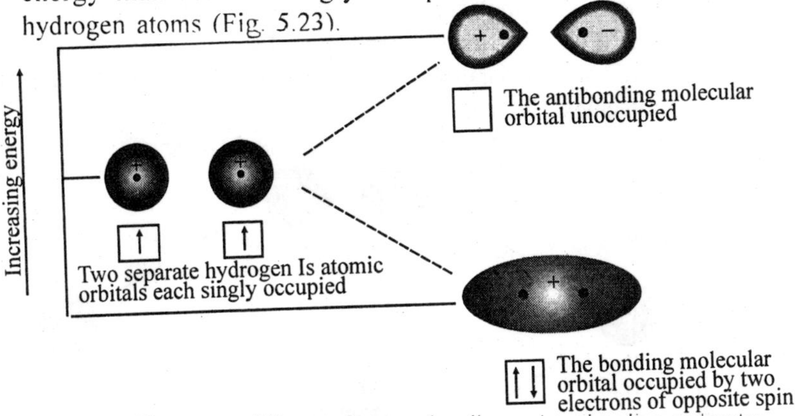

Fig. 5.23 : The energy difference between bonding and antibonding molecular orbitals

Before proceeding to discuss the application of molecular orbital theory to more complex molecules it is necessary to state that atomic orbitals will only combine to give molecular orbitals under certain conditions :

(a) The atomic orbitals must overlap to an appreciable extent, otherwise there will be no significant build-up of charge between the nuclei and therefore no chemical bond will be established.

(b) In the region of overlap the wave functions of the separate atomic orbital must be of the same sign, otherwise there will be no build-up of charge between the two nuclei.

(c) The energies of the separate atomic orbitals must be of comparable magnitude.

It will be seen that these conditions are fulfilled for the simple example we have been discussing.

The helium molecule and helium molecule-ion

Helium atoms contain two electrons occupying the 1*s* atomic orbitals, and the electronic configuration of the helium atom is $1s^2$. If these atomic orbitals are combined they will give one bonding and one antibonding molecular orbital and, since four electrons are involved, both of these orbitals will be doubly occupied. However, the lower energy of the bonding molecular orbital (as compared with the separate atomic orbitals) is effectively cancelled out by the higher energy of the antibonding molecular orbital and no chemical bond is established; helium molecules, He_2, cannot be formed (Fig. 5.24).

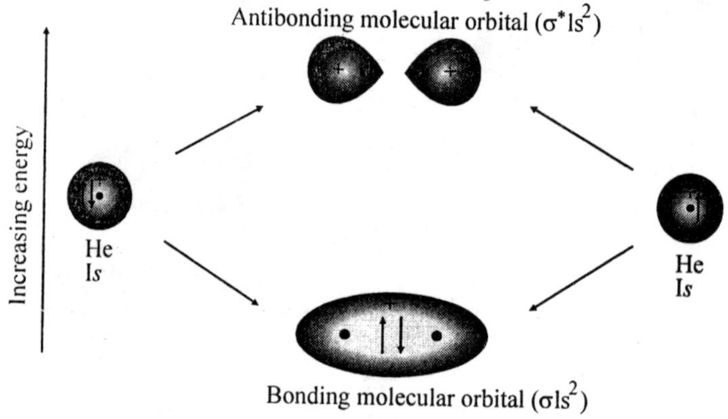

Fig. 5.24 : The higher energy of the antibonding molecular orbital cancels out the lower energy of the bonding molecular orbital and no He_2 molecule is formed

The entity He_2^+ molecule ion can be obtained by subjecting helium gas at a low pressure to a high energy discharge. One of the successes of molecular orbital theory is that it explains the existence of this species. The He_{2+} species contains three elctrons and two can be accommodated in a bonding molecular orbital, while the third can occupy the antibonding molecular orbital; there is some overall bonding and this species can exist.

Molecular orbitals formed by the overlap of *p* atomic orbitals

The criterion that molecular orbitals can be formed by the overlapping of atomic orbitals, proSvided their wave functions are of the same sign in the region of overlap, implies that *p*-type atomic orbitals can overlap in two distinct ways. Thus two $2p_x$ orbitals can overlap along the x-axis and addition and substraction of their wave functions gives rise to a bonding and an antibonding molecular orbital. The two $2p_x$ atomic orbitals can also overlap laterally and again two molecular orbitals (one bonding and the other one antibonding) can be obtained. These methods of overlap are shown in Fig. 5.25 and in Fig. 5.26.

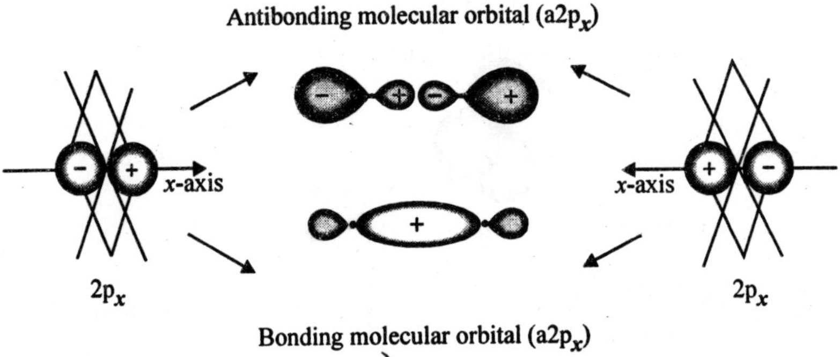

Fig. 5.25 : Molecular orbitals formed by the overlap of $2p_x$ atomic orbitals along the x-axis.

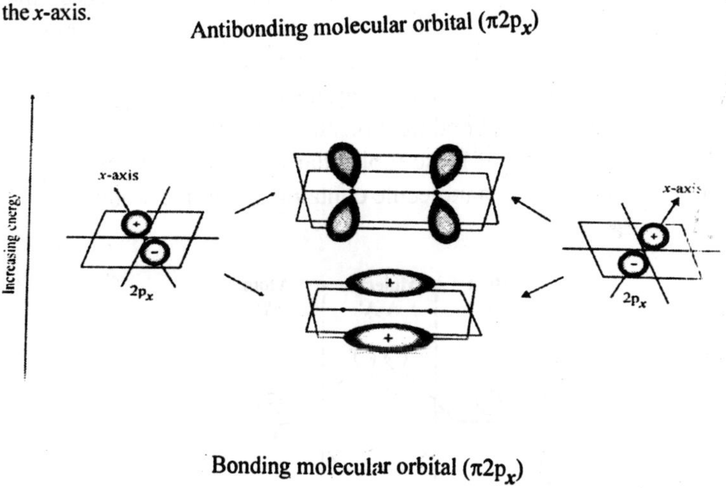

Figure 5.2 : Molecular orbitals formed by the lateral overlap of $2p_x$ atomic orbitals

In general the energy of the $\sigma 2p_x$ molecular orbital is of lower energy than the $\pi 2p_x$ molecular orbital. The energy of the $\pi^* 2p_x$ molecular orbital is of lower energy than the $a^* 2px$ molecular orbital. Thus the energy increases in the order :

$$\sigma 2p_x < \pi 2p_x < \pi^* 2p_x < \sigma^* 2p_x$$

It is impotant to realise that a $2p_x$ atomic orbital cannot be combined with say, a $2p_z$ atomic orbital of another atom since the wave functions are not of the same sign in the region of overlap; the negative portion cancels out the positive portion.

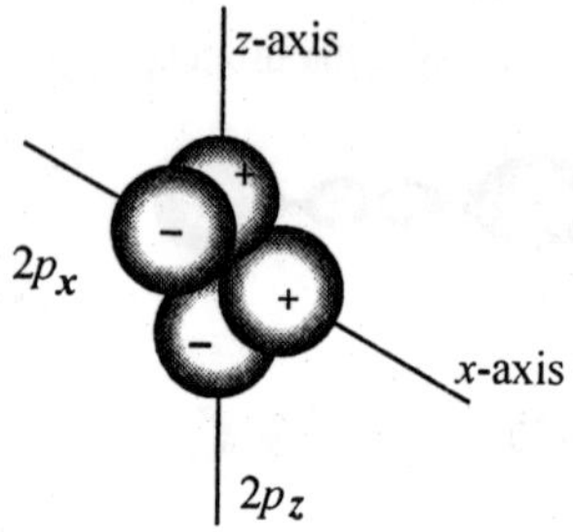

Fig. 5.27 : A $2p_x$ and a $2p_z$ atomic orbital overlap to form a molecular orbital.

We have considered the formation of molecular orbitals by the overlapping of $2p$ atomic orbitals and it only remains to see how $2s$ atomic orbitals overlap. The type of overlapping here is similar to that for the overlapping of two 1s atomic orbitals, thus a bonding molecular orbital ($\sigma 2s$) and an antibonding molecular orbital ($\sigma^* 2s$) are formed. The energy sequence of these molecular orbitals in a homonuclear diatomic molecule, i.e. a molecule containing two like atoms,

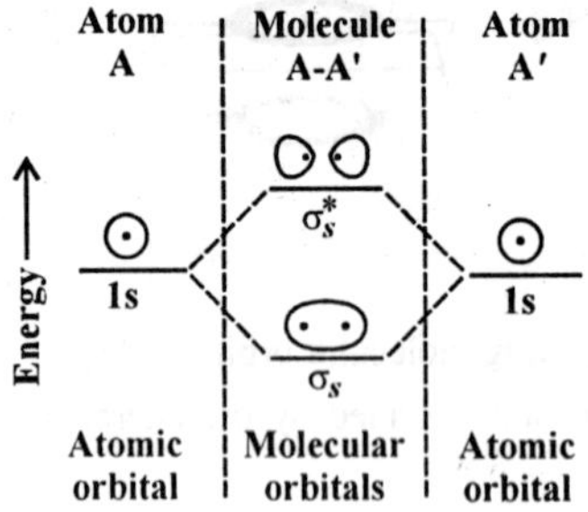

Fig. 5.28 : Relative energies of σ_s orbitals in homonuclear diatomic molecules.

The filling of molecular orbitals we described the *Aufbau* procedure in which electrons are added one by one to an AO filling diagram in order to build up the electronic configurations of atoms.

We will now use a similar technique, but will add electrons to a filling diagram composed of MO energy levels. We wish to build up the *ground-state* configurations of homonuclear diatomic molecules, and so we will add electrons starting at the bottom of the diagram and work upward toward higher-energy MOs. The MO filling diagram consists of Figs. 5.28 through 5.30. At the start, however, we need consider only the low-energy end of the diagram, that is, those MOs derived from the K shells of the bonded atoms (Fig. 5.28).

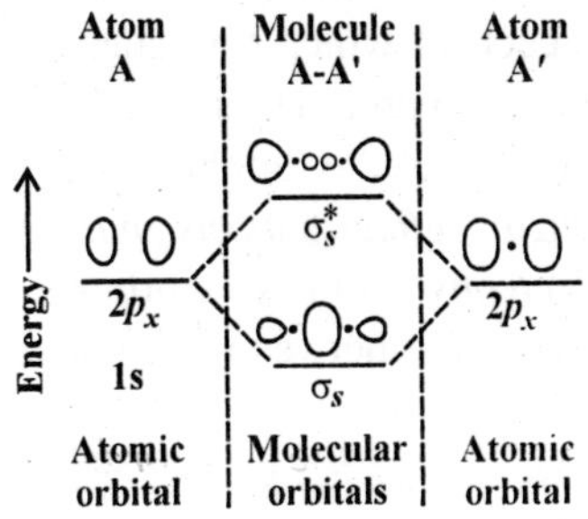

Fig. 5.29 : Relative energies of σ_z orbitals in homonuclear diatomic molecules.

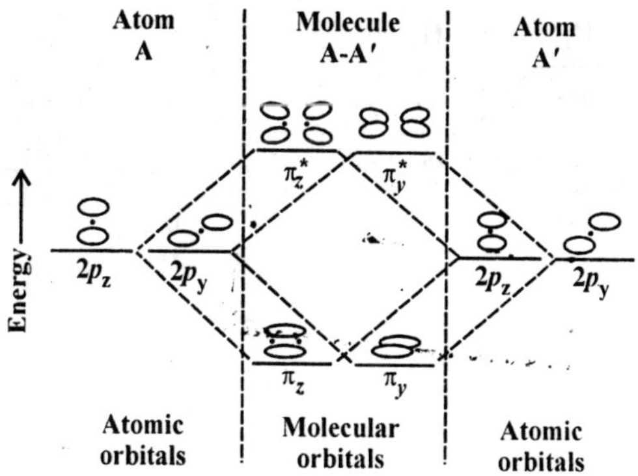

Fig. 5.30 : Relative energies of π_y and π_z orbitals in homonuclear diatomic molecules.

H_2. The simplest molecule is hydrogen. Figure 5.31 shows orbital populations for two unbonded ground-state H atoms at the left and right, together with that of the ground-state H_2 molecule, in the middle of the diagram. The two 1*s* electrons end up as a pair (antiparallel spins) in the bonding σ1s orbital of H_2 and constitute a single bond. The electronic configuration of H_2 can be written

$$H_2: (\sigma 1s)^2$$

He_2. Next consider the molecule which might be formed from two atoms of helium, each of which furnishes two electrons to the molecule. This is two more than in H_2, so the MO population would look like that in Fig. 5.32. But the (antibonding) σ1*s* orbital is now filled, and *its destabilizing effect cancels out the stabilizing effect of the two bonding electrons* (in the σ1*s* orbital). The result is that there is no net attractive force between He atoms, and so He_2 does not exist.

In molecular-orbital theory the *bond order* is defined as

$$\text{Bond order} = \frac{\text{bonding electrons} - \text{antibonding electrons}}{2}$$

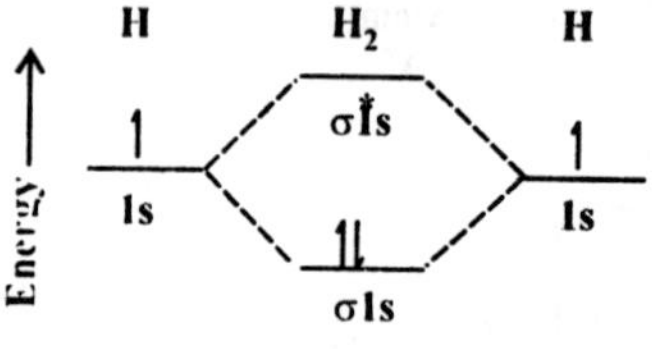

Fig. 5.31 : MO population diagram for H_2.

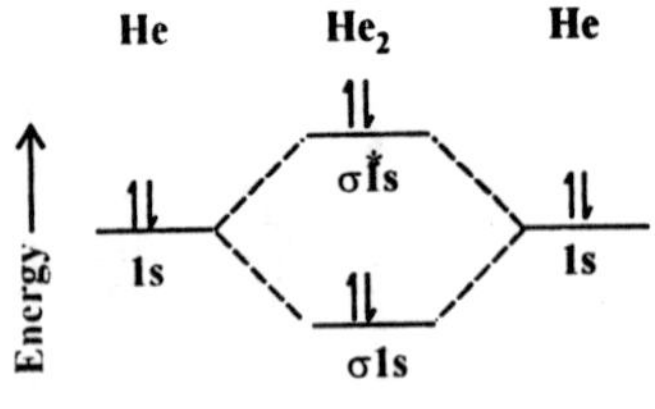

Fig. 5.32 : MO population diagram for the hypothetical He_2 molecule.

Thus the bond order in the H_2 molecule is

$$.\frac{2-0}{2} =$$

while that in He_2 is

$$\frac{2-2}{2} = 0$$

The H_2^+ ion should have an orbital occupancy like that of H_2, but with one less electron. Therefore, its electronic configuration is

$$H_2^+: (\sigma 1s)^1$$

The bond order in H_2^+ is $\frac{1}{2}(1-0)$, or $\frac{1}{2}$. This means that the H_2^+ particle should exist, its atoms held together by a *half bond*. The H_2^+ ion does indeed exist.

Li_2. Now consider the Li_2 molecule.The valence electrons of the two Li atoms are used to populate a new $\sigma 1s$ MO as is shown in Fig. 9.36. The 1*s* atomic orbitals are essentially unperturbed and are not shown in the diagram. The configuration is much like that of H_2, and the bond order, which can be determined from the valence electrons only, is equal to $\frac{1}{2}(2-0)$, or 1. 1*s* the electronic configuration of Li_2 can be shown as

$$Li_2: (\sigma 1s)^2$$

With a bond order of 1 the Li_2 molecule is predicted to exist. Neither liquid nor solid Li consists of Li_2 molecules, but diatomic molecules are indeed found in gaseous lithium.

Be_2. Moving on to the hypothetical Be_2 molecule, we find a situation like that in He_2. The atomic number of beryllium is 4, and the "seventh" and "eighth" electrons in the Be_2 molecule add to the 2*s* orbital. The destabilizing effect of the filled $\sigma 2s$ orbital cancels out the stabilizing effect of the filled σs orbital, the bond order is zero, and therefore the Be_2 molecule should not be stable. Indeed, Be_2 has not been observed.

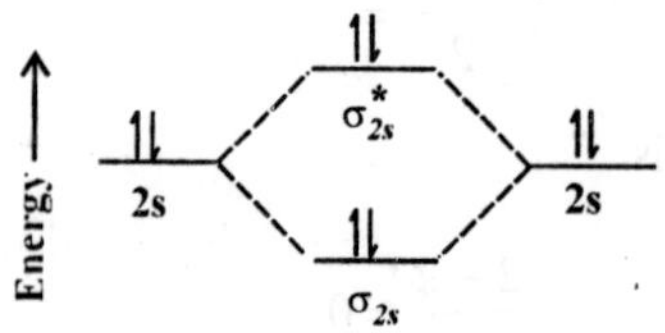

Fig. 5.33 : MO population diagram for Li_2.

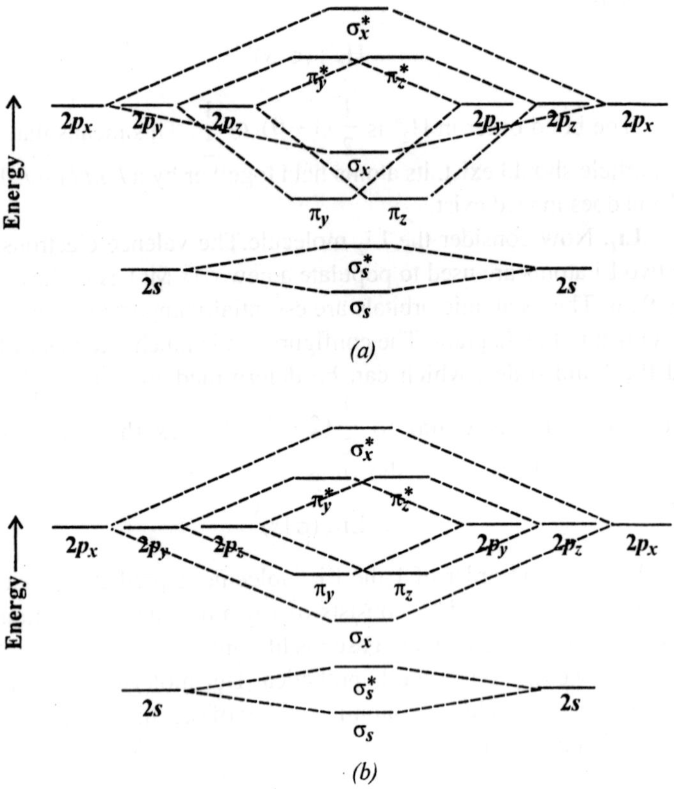

Fig. 5.34 : MO energies: B_2 through Ne_2: (a) B_2, C_2, and N_2; (b) O_2, F_2, and Ne_2

Next consider sequence B_2, C_2, N_2, O_2, F_2 and Ne_2, as we work across the rest of the second period constructing homonuclear diatomic molecules. But since we now need some more molecular orbitals, we go to the σ and π orbitals of Figs. 5.29 and 5.30. When we put these two diagrams together, however, we run into a small difficulty. The relative energy of the π_y and π_x orbitals is less than that of the σ_x orbital for B_2 through N_2, but *greater* for the remainder of the sequence. Thus the orbital energies for B_2, C_2, and N_2 are as shown in Fig. 5.34a and for O_2, F_2 and Ne_2, as shown in Fig. 5.34b. The main difference between Fig. 5.34 a and b is the relative energy of the σ_z compared to the π_y and π_x orbitals.

The change in sequence of MO energies between N_2 and O_2 occurs because the σ_z and σ^*_z MOs actually have some *s* character, a fact which we had to ignore, when we decided to use the "one AO plus one AO yields two MOs" simplification. (The amount of *s* character in these orbitals decreases as the nuclear charge increases across the period. Because of this the σ_z energy drops below the $\pi_y - \pi_x$ energy at O_2.)

B_2. the MO populations in B_2, C_2 and N_2. In the first molecule. B_2, there is a single electron in each of the π_y and π_x orbitals. Since these are bonding orbitals, and since in all lower energy levels the antibonding electrons have exactly compensated for the bonding electrons, the bond order is 1. (We could call the bond a single bond but it is perhaps better described as two half bonds.) Note that the π_y and π_z orbitals are equal in energy, and so the two electrons do not pair up in one orbital. By spreading out, the electrons can occupy different regions in space, thus avoiding each other and reducing interelectronic repulsion. The electronic configuration of B_2 is written as

$$B_2 : (\sigma 1s)^2\ (\sigma^* 1s)^2\ (\sigma 2s)^2\ (\sigma^* 2s)^2\ [(\pi 2p_y)^1\ (\pi 2p_z)^1\ ]$$

Strong experimental support for this configuration comes from magnetic measurements. B_2 is paramagnetic, and measurements indicate that two unpaired electrons are present per molecule.

Like Li_2, B_2 is not a molecule you can find in a bottle on the shelf of the chemistry stockroom. Elemental boron is usually encountered as a solid in which the arrangment of B atoms is complex. At very high temperatures, however, B_2 molecules can be detected in the gaseous state.

C_2. Adding two more electrons (one from each atom) yields the configuration for C_2. These electrons are added to the π_y and π_z MOs, filling each. All electrons are now paired, and so C_2 is not paramagnetic. The bond order in C_2 can be seen to be 2, because there are four more bonding than antibonding electrons in the molecule. The electronic configuration of C_2 is

$$C_2 : (\sigma 1s)^2\ (\sigma^* 1s)^2\ (\sigma 2s)^2\ (\sigma^* 2s)^2\ [(\pi 2p_y)^2\ (\pi 2p_z)^2]$$

Because of its nonzero bond order, C_2 should, and does, exist, having been detected at high temperatures.

N_2. The last MO population is that of the nitrogen molecule N_2. It has a *net* excess of six bonding electrons, which corresponds to a bond order of 3. These electrons occupy the π_y, π_z, and σ_z orbitals, as shown, giving N_2 the electronic configuration

$$N_2 : (\sigma 1s)^2\ (\sigma^* 1s)^2\ (\sigma 2s)^2\ (\sigma^* 2s)^2\ [(\pi 2p_y)^2\ (\pi 2p_z)^2]\ (\sigma 2p_z)^2$$

N_2 is, of course, very stable and is encountered every day in every breath we take. Magnetic measurements confirm that all electrons are paired in N_2; it is not paramagnetic.

The MO model of the N_2 molecule correlates especially well with the VB picture. The six electrons in the π_y, π_z, and σ_z orbitals correspond to the six electrons shown in the VB Lewis structure.

$$:N:::N:$$

O_2. The addition of two more electrons to the N_2 configuration gives the orbital population of O_2. Note that these two electrons must go into antibonding orbitals resulting in a bond order decrease (from 3 in N_2) to 2. The lower bond order is consistent with the fact that O_2 has a smaller bond energy and a longer bond length than does N_2. The electronic configuration of O_2 is

F_2. The addition of two more electrons give us the population shown in the center of Fig. 5.35. Since the (antibonding) π^* orbitals are both filled, the bond order in F_2 is only 1. This agrees with the experimentally determined bond energy and length, both of which are about what would be expected for a single bond. Also, F_2 is found not to be paramagnetic, which is consistent with the lack of unpaired electrons. The configuration of F_2 is

F_2: $(\sigma 1s)^2$, $(\sigma^* 1s)^2$, $(\sigma 2s)^2$, $(\sigma^* 2s^2)$, $(\sigma 2p^2_x)$, $[(\pi 2p^2_y)$, $(\pi 2p^2_z)]$ $[(\pi^* 2p^2_y)$, $(\pi^* 2p^2_z)]$

Ne_2. Adding two more electrons fills the σ^*_z orbital, reducing the bond order to zero. Ground-state Ne_2 has never been observed.

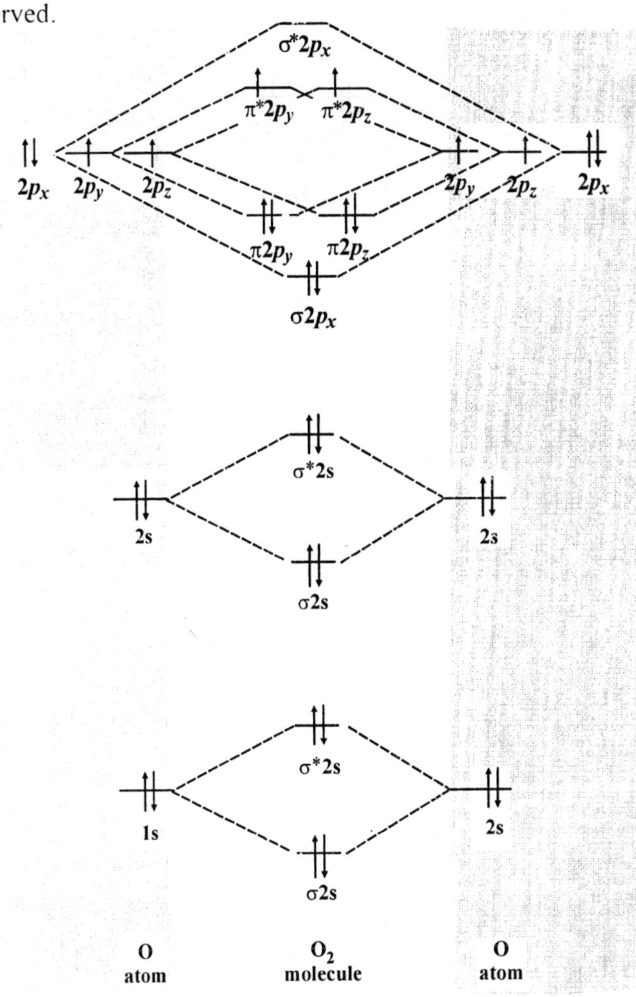

Fig. 5.35 : Electronic configuration, atomic and molecular orbitals for oxygen.

O_2^{2-} ion

In a similar way sodium peroxide Na_2O_2 contains the peroxide ion O_2^{2-}. This ion has 18 electrons, arranged:

$$\sigma 1s^2, \sigma^* 1s^2, \sigma 2s^2, \sigma^* 2s^2, \sigma 2p_x^2, [\pi 2p_y^2, \pi 2p_z^2], [\pi^* 2p_y^2, \pi^* 2p_z^2]$$

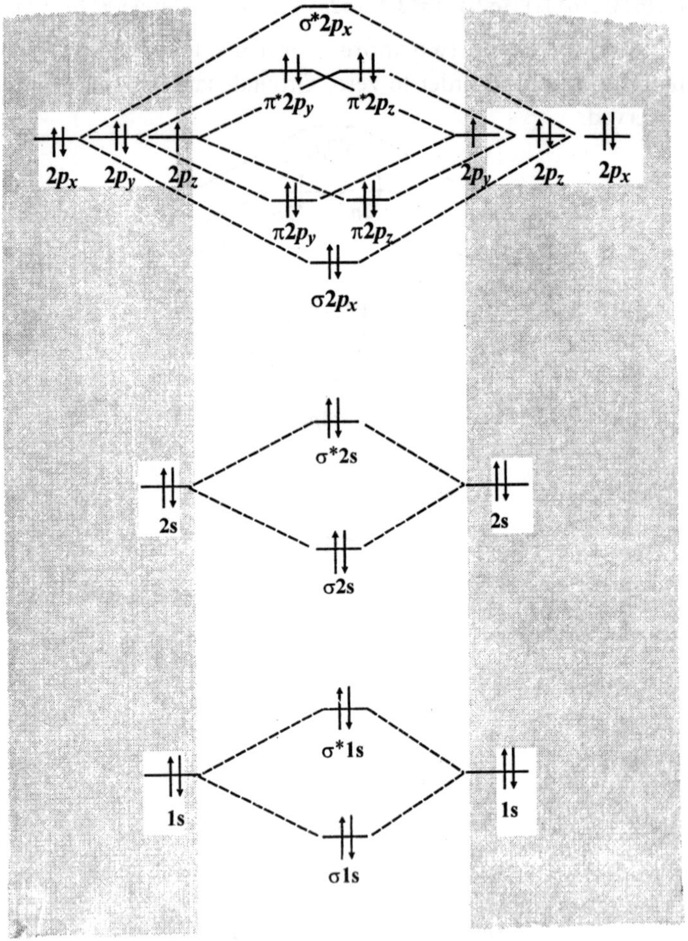

Fig. 5.36 : Electronic configuration, atomic molecular orbitals from fluorine

Once again the inner shell takes no part in bonding. The bonding and antibonding 2*s* orbitals completely cancel each other. One σ bond forms from the filled $2p_x$ orbital. Both the bonding $2p_y$ and $2p_z$ orbitals are cancelled out by their corresponding antibonding orbitals. Thus the bond order is one, that is a single bond. Alternatively the bond order may be calculated as (bonding—antibonding)/2, that is (10 – 8)/2 = 1.

Examples of Molecular Orbital Treatment for Heteronuclear Diatomic Molecules

The same principles apply when combining atomic orbitals from two different atoms as applied when the atoms were identical, that is:

1. Only atomic orbitals of about the same energy can combine effectively.
2. They should have the maximum overlap.
3. They must have the same symmetry.

Since the two atoms are different, the energies of their atomic orbitals are slightly different. A diagram showing how they combine to form molecular orbitals is given in figure 5.37.

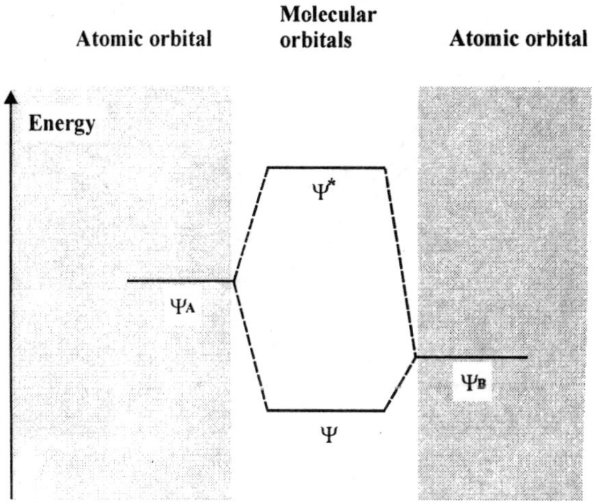

Fig. 5.37 : The realtive energy levels of atomic orbitals and molecular orbitals for a heteronuclear diatomic molecule AB.

The problem is that in many cases the order of MO energy levels in not known with certainty. Thus we will consider first some examples where the two different atoms are close to each other in the periodic table, and consequently it is reasonable to assume that the order of energies for the MOs are the same as for homonuclear molecules.

NO molecule

The nitrogen atom has 2 + 5 = 7 electrons, and the oxygen atom has 2+6=8 electrons, making 15 electrons in the molecule. The order of energy levels of the various MOs are the same as for homonuclear diatomic molecules heavier than C_2, so the arrangement is:

$$\sigma 1s^2, \sigma^* 1s^2, \sigma 2s^2, \sigma^* 2s^2, \sigma 2p^2_x, [\pi 2p^2_y, \pi 2p^2_z], [\pi^* 2p^1_y, \pi^* 2p^1_z]$$

This is shown in Figure 5.38.

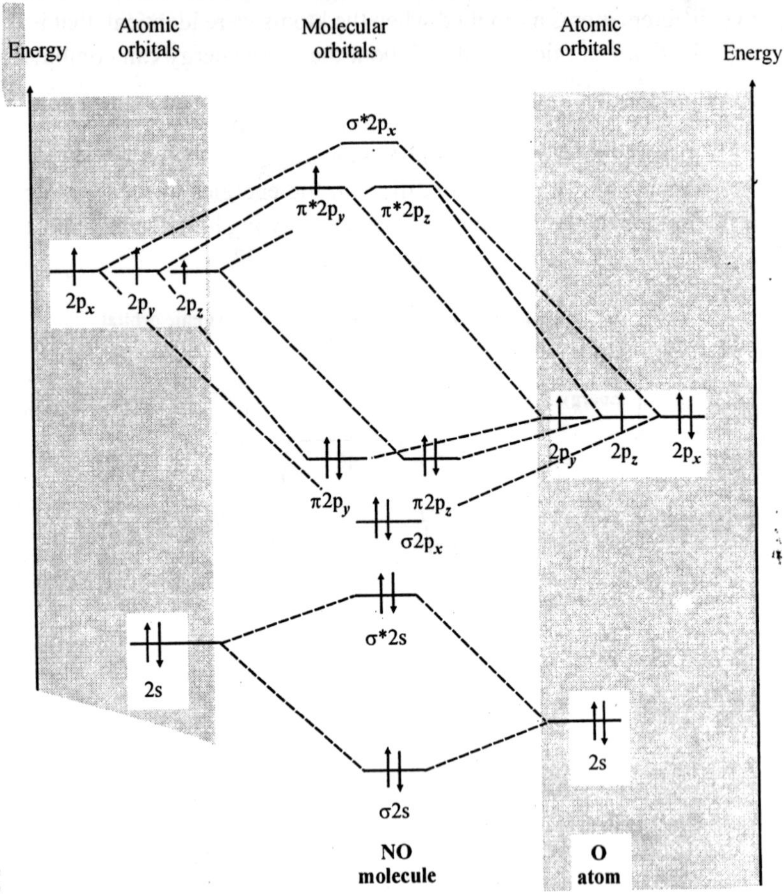

Fig. 5.38 : Electronic configuration, atomic orbitals and molecular orbitals for nitric oxide. (This diagram is essentially the same as that for homonuclear diatomic molecules such as N_2, O_2 of F_2. The difference is that the atomic energy levels of N and O are not the same. The $\sigma 1s$ and $\sigma^* 1s$ MOs are omitted for simplicity).

The inner shell is non-bonding. The bonding and antibonding $2s$ orbitals cancel, and a σ bond is formed by the filled $\sigma 2p^2_x$ orbital. A π bond is formed by the filled $\pi 2p^2_z$ orbital. The half-filled $\pi^* 2p^1_y$ half cancels the filled $\pi 2p^2_y$ orbital, thus giving half a bond. The bond order is thus $2\frac{1}{2}$, that is in between a double and a triple bond. Alternatively the bond order may be worked out as (bonding—antibonding)/2, that is $(10-5)/2\frac{1}{2}$. The molecule is paramagnetic since it contains an unpaired electron.In NO there is a significant difference of about 250 kJ mol^{-1} in the energy of the AOs involved, so that combination of AOs to give MOs is less effective than in O_2 or N_2. The bonds are therefore weaker than might be expected. Apart from this the molecular orbital pattern (Figure 5.38) is similar to that for homonuclear diatomic molecules. Removal of one electron to make NO^+ results in a shorter and stronger bond because the electron is removed from an antibonding orbital, thus increasing the bond order to 3.

CO molecule

The carbon atom has 2 + 4 = 6 electrons, and the O atom has 2 + 6 = 8 electrons, so the CO molecule contains 14 electrons. In this case we are rather less certain of the order of energies of the MOs, since they are different for C and O. Assume the order is the same as for light atoms like C:

$$\sigma 1s^2,\ \sigma^* 1s^2,\ \sigma 2s^2,\ \sigma^* 2s^2,\ [\pi 2p^2_y,\ \pi 2p^2_z],\ \sigma 2p^2_y.$$

This is shown in Figure 5.39.

The inner shell is non-bonding, and the bonding and antibonding $2s$ orbitals cancel, leaving one σ and two π bonds—and thus a bond order of 3. Alternatively the bond order may be calculated using the formula (bonding—antibonding)/2, that is $(10-4)/2 = 3$. This simple picture is not adequate, since if CO is ionized to give CO^+ by removal of the electron from the $\sigma 2p_x$ orbital then the bond order should be reduced to $2\frac{1}{2}$ and the bond length increased. In fact the bond length in CO is 1.128 Å and in CO^+ it is 1.115 Å. Thus the bond length decreases when we expected it to increase, and it indicates that the electron must have been removed from an antibonding orbital. The problem remains if we assume the order of energy for the MOs is the same as for atoms heavier than C, since this only reverses the position of the $\sigma 2p_x$ and the ($\pi 2p_y$ and $\pi 2p_z$) MOs. The most likely explanation of the bond shortening when CO is changed to CO^+ is that the $\sigma 2s$ and $\sigma^* 2s$ molecular orbitals differ in energy more than is shown in the figure. This means that they are wider apart, and the $\sigma^* 2s$ MO is higher in energy than the $\sigma 2p_x$, $\pi 2p_y$ and $\pi 2p_z$ MOs. This illustrates very plainly that the order of MO energy levels for simple homonuclear diatomic molecules used above is not automatically applicable when two different types of atoms are bonded together, and it is certainly incorrect in this particular heteronuclear case.

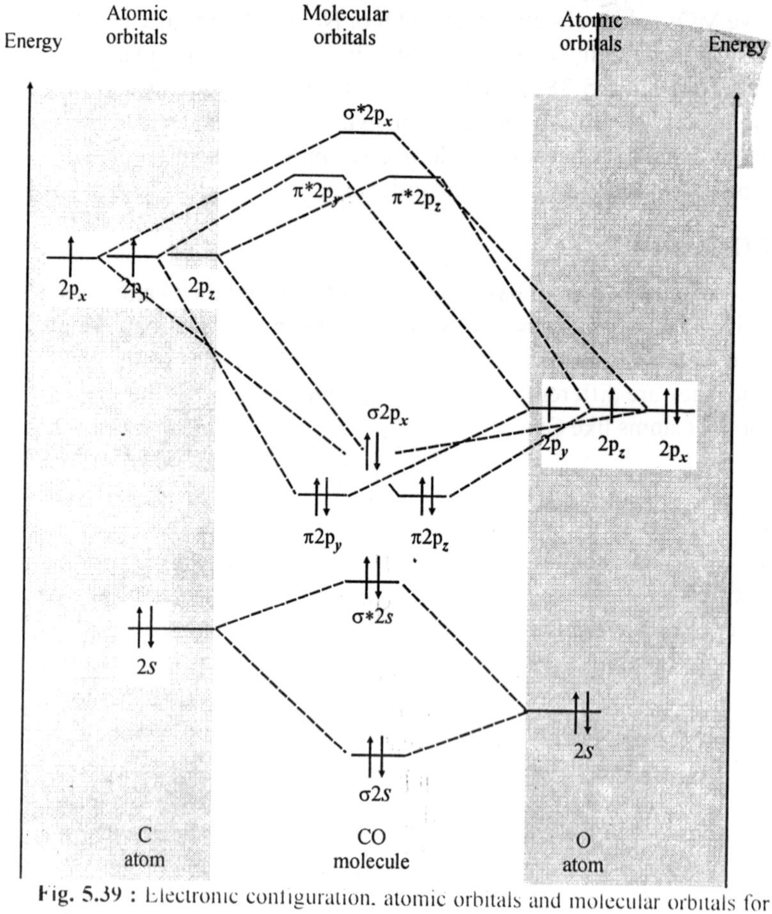

Fig. 5.39 : Electronic configuration, atomic orbitals and molecular orbitals for carbon monoxide. (The $\sigma^* 1s$ MOs are omitted for simplicity.)

HCl molecule

With heteronuclear atoms it is not obvious which AOs should be combined by the LCAO method to form MOs. In addition because the energy levels of the AOs on the atoms are not identical, some MOs will contain a bigger contribution from one AO than the other. This is equivalent to saying that the MO 'bulges' more towards one atom, or the electrons in the MO spend moe time round one atom than the other. Thus some degree of change separation $\delta+$ and $\delta-$ occurs, resulting in a dipole. Thus partial ionic contributions may play a significant part in the bonding.

Consider the HCl molecule. Combination between the hydrogen $1s$ AO and the chlorine $1s$, $2s$, $2p$ and $3s$ orbitals can be ruled out because their energies are too low. If overlap occurred between the chlorine $3p_y$ and $3p_z$ orbitals it would be non-bonding because the positive lobe of hydrogen will overlap equally with the positive and negative lobes of the chlorine orbitals. Thus the only effective overlap is with the chlorine $3p_x$ orbital. The combination of H $1s^1$ and Cl p^1_x gives both bonding and antibonding orbitals, and the two electrons occupy the bonding MO, leaving the antibonding MO empty. It is assumed that all the chlorine AOs except $3p_x$ are localized on the chlorine atom and retain their original AO status, and the $3s$, $3p_y$ and $3p_z$ orbitals are regarded as non-bonding lone pairs.

This over-simplification ignores any ionic contribution such as can be shown with the valence-bond resonance structures

$$H^+Cl^- \text{ and } H^-Cl^+$$

The former would be expected to contribute significantly, resulting in a stronger bond.

4. Molecular-Orbital and Valence-Bond Methods : A Comparison

Both the molecular-orbital and valence-bond methods of describing covalently bound species are attempts to obtain approximate solutions of the equations which govern the characteristics properties of molecules. Thus, in a very real sense, neither is the right or the wrong method. Rather, each method has a different utility under a given set of circumstances. The valence-bond method has found great favour with some chemists since the process of constructing an approximate solution involves many components that are familiar to chemists (e.g., canonical structures than can be conveniently represented by the usual chemical symbolism). Such an easy correspondence of ideas may often lead to misconceptions because these parallelisms are not complete. For example, although several canonical forms can be written as a guide to the form of the approximate solution to the wave equation, it does not follow that these canonical forms have a distinct existence and form a kind of system at equilibrium. In general, the valence-bond method is very useful in describing ground-state phenomena.

On the other hand, the molecular-orbital method gives a useful description of covalently bonded systems if more detailed knowledge of electronic structure is necessary. For example, phenomena involving considerations of the excited states of molecules, as in molecular spectra and photo-chemistry, are generally more easily understood using molecular-orbital arguments. Further refinements to obtain better wave functions using either method require significantly more extensive calculations than have been indicated here.

6

Ionic Bonding

The Ionic Bond

Ionic solids are held together by the electrostatic attraction between the negative and positive ions. There will be replusion if ions of the same charge are adjacent, and attraction will occurs when positive ions are surrounded by negative ions and vice-versa. The attractive forces will be maximum when each ions is surrounded by the greatest possible no. of oppositely charged ions.

Properties

Several properties distinguish ionic compounds from covalent compounds. These may be related rather simple to the crystal structure of ionic compounds, namely, a lattice composed of positive and negative ions in such a way that the attractive forces between oppositely charged ions are maximized and the repulsive forces between ions of the same charge are minimized. Before discussing some of the possible geometries, a few simple properties of ionic compounds may be mentioned :

1. Ionic compounds tend to have very low electrical conductivities as solids[1] but conduct electricity quite well when molten. This conductivity is attributed to the presence of *ions*, atoms charged either positively or negatively, which are free to move under the influence of an electric field. In the solid the ions are bound tightly in the lattice and are not free to migrate and carry electrical current.

It should be noted that we have no absolute *proof* of the existence of ions in solid sodium chloride, for example, though our best evidence will be discussed later in this chapter (pp. 71-72). The fact that ions are found when sodium chloride is melted or dissolved in water does not prove that they existed in the solid crystal. However, their existence in the solid is usually assumed, since the properties of these materials may readily be interpreted in terms of electrostatic attractions.

2. Ionic compounds tend to have high melting points. Ionic bonds usually are quite *strong* and they are *omnidirectional.* The second point is quite important, since ignoring it could lead one to conclude that ionic bonding was much stronger than covalent bonding—which is *not* the case. We shall see that substances containing strong, multidirectional covalent bonds, such as diamond, also have very high melting points. The high melting point of sodium chloride, for example, results from the strong electrostatic attractions between the sodium cations and the chloride anions, and from the lattice structure, in which each sodium ion attracts six chloride ions, each of which in turns atracts six sodium ions, etc., throughout the crystal.

3. Ionic compounds usually are very hard but brittle substances. The hardness of ionic substances follows naturally from the argument presented above, expected in this case we are relating the multivalent attractions between the ions with *mechanical* separation rather than separation through thermal energy. The tendency towards brittleness results from the nature of ionic bonding. If one can apply suficient force to displace the ions slightly (e.g., the length of one-half of the unit cell in NaCl), the formerly attractive forces become repulsive as anion-anion and cation-cation contacts occur; hence the crystal flies apart. This accounts for the well-known cleavage properties of many minerals.

4. Ionic compounds are often soluble in polar solvents with high permittivities (dielectric constants). The energy of interaction of two charged particles is given by

$$E = \frac{q^+ q^-}{4\pi r \varepsilon} \tag{3.1}$$

where q^+ and q^- are the charges, r is the distance of separation, and ε is the permittivity of the medium. The permittivity of a vacuum, ε_0, is $8.85 \times 10^{-12} C^2 m^{-1} J^{-1}$. For common polar solvents, however, the permittivity values are considerably higher. For example, the permittivity is $7.25 \times 10^{-10} C^2 m^{-1} J^{-1}$. for water, $2.9 \times 10^{-10} C^2 m^{-1} J^{-1}$ for acetonitrile, and $2.2 \times 10^{-10} C^2 m^{-1} J^{-1}$ for ammonia, giving relative permittivities of 82 ε_0 (H_2O), 33 ε_0 (CH_3CN), and 25 ε_0 (NH_3). Since the permittivity of ammonia is 25 times that of a vacuum, the attraction between ions dissolved in ammonia, for example, is only 4% as great as in the absence of solvent. For solvents with higher permittivities the effect is even more pronounced.

Another way of looking at this phenomenon is to consider the interaction between the dipole moments of the polar solvent and the ions. Such solvation will provide considerable energy to offset the otherwise unfavourable energetics of breaking up the crystal lattice.

Occurrence of ionic bonding

Simple ionic compounds form only between very active metallic elements and very active nonmetals.[2] Two important requisites are that the ionization energy to form the cation and the electron affinity to form the anion must be energetically favourable. This does not mean that these two reactions must be exothermic (an impossibility—see Problem 3.8), but means, rather, that they must not cost too much energy. Thus the requirements for ionic bonding are : (1) The atoms of one element must be able to lose one or two (rarely three) electrons without undue energy input and (2) the atoms of the other element must be able to accept one or two electrons (almost never three) without undue energy input. This restricts ionic bonding to compounds between the most active metals—Groups IA, IIA, part of IIIA, and some lower oxidation states of the transition metals (forming cations)—and the most active nonmetals—Groups VIIA, VIA, and nitrogen (forming anions).[3] All ionization energies are endothermic, but for the metals named above they are not prohibitively so. Electron affinities are exothermic only for the halogens but are not excessively endothermic for the chalcogens and nitrogen.

The formation of ions

An ion is formed when an atom gains or loses one or more electrons. Loss of an electron by an atom yields a psotive ion, or *cation*,

$$M(g) \rightarrow M^+(g) + e^-$$

while gain of an electron yields a negative ion, or *anion*,

$$X(g) + e^- \rightarrow X^-(g)$$

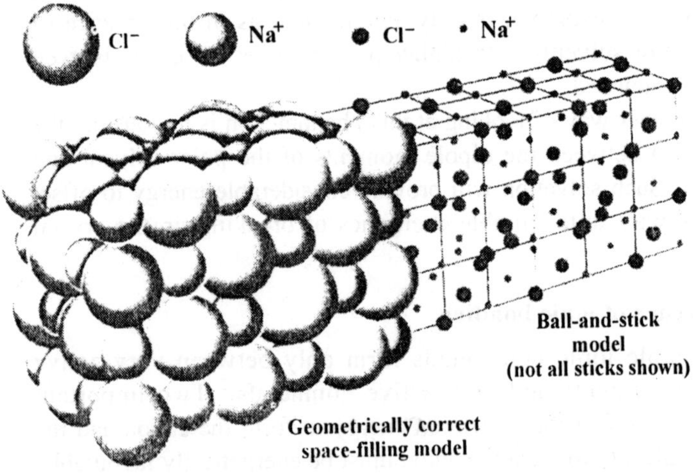

Fig. 6.1 : The sodium chloride structure.

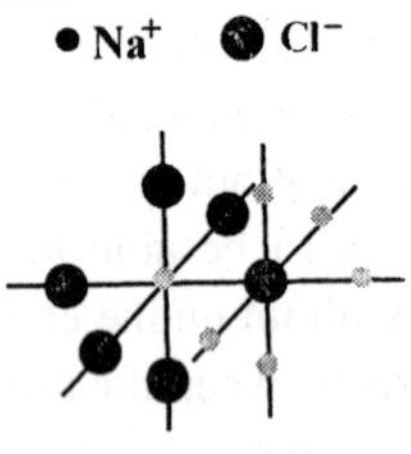

Fig. 6.2 : A small portion of the NaCl structure. This drawing shows that an Na^+ ions froms six ionic bonds with its six nearest-neighbour Cl^- ions. Similarly, a Cl^- ions forms six bonds with its six nearest-neighbour Na^+ ions.

If the electron lost by M is gained by X, then the overall process can be represented by

$$M(g) + X(g) \rightarrow M^+(g) + X^-(g)$$

The M^+ and X^- ions thus formed are then attracted to each other because they have opposite electrical charges :

$$M^+(g) + X^-(g) \rightarrow M^+X^-(g)$$

This attraction is called an *ionic* bond, or, sometimes, an *electrovalent* bond. It is the electrostatic attraction between ions of opposite charge. In the sequence of events just described, the end result is a gaseous M^+X^- unit which is called an *ion pair.*

Gaseous ion pairs are found at very high temperatures. Ionic bonds are more commonly encountered in *ionic solids*, interlocking structures containing huge numbers of positive and negative ions, not just one of each. In solid NaCl (sodium chloride, table salt) the Na^+ and Cl^- ions are not arranged in pairs; rather, they are ordered in three-dimensions, as shown in Fig. 8.1. On the left is a picture of a space-filling, geometrically accurate model of the NaCl structure. The smaller sodium ions are shown as spheres in contact with the larger chloride ions. On the right the ions have all been "shrunk," and the resulting ball-and-stick model affords a view into the interior of the structure.

In solid NaCl each Na^+ has six Cl^- ions as its nearest neighbours. Similarly, the immediate environment of each Cl^- ion consists of six Na^+ ions. (See Fig. 6.2) Thus each Na^+ forms not just one but six ionic bonds, as does each Cl^-. (Each of the two kinds of ions forms six bonds; so the formula must be NaCl. Do you see why this is so?)

Lewis structures As early as 1906 J.J. Thomson expressed his belief that electrons would ultimately be shown to provide the key to chemical bonding. In 1916 the American chemist G.N. Lewis developed a method for keeping track of electrons in atoms and molecules. This method, still useful today, makes use of a diagram, now called a *Lewis structure*, in which the chemical symbol for an atom is surrounded by a number of dots (or, sometimes, litle o's or x's, etc.) corresponding to the number of electrons in the valence shell of the atom.[1] A sodium atom has one valence-shell electron; so its Lewis structure is

$$Na\cdot$$

where the dot can be located at the left, right, top, or bottom of the symbol. A chlorine atom, on the other hand, has seven valence-

1. Lewis structures are also sometimes called electronic, or dot, symbols and formulas.

shell electrons; so its Lewis structure is

$$:\overset{\cdot\cdot}{\underset{\cdot\cdot}{cl}}:$$

It makes no difference on which of the four sides of Cl the single dot is shown.

The symbol in a Lewis structure represents the nucleus plus all inner-shell electrons, this combination often being called the *core*, or *kernel*, of the atom. Thus we have for a magnesium (Z = 12) atom.

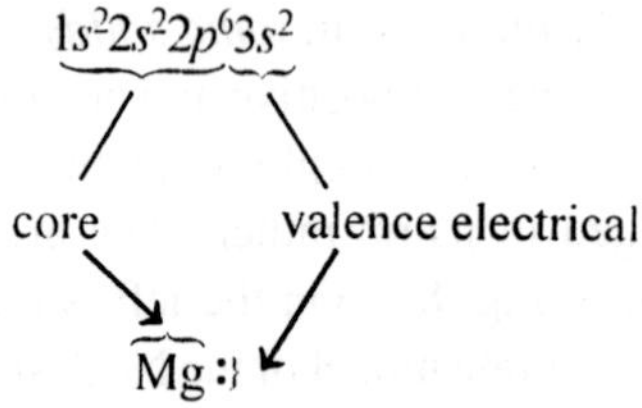

The dots in a Lewis structure are best grouped to show whether or not the electrons are paired. For an aluminum atom (Z = 13) we have

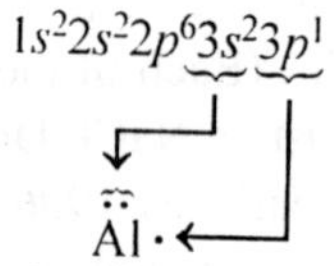

Here the valence electrons are shown as a pair (the two 3*s* electrons) and a single electron (the 3*p*). The Lewis structures for the atoms of the third period are shown in Table 6.1, together with their electronic configurations.

Octet rule

As we have mentioned, noble-gas atoms (except for He) have an especially stable valence-shell electronic configuration: ns^2np^6, where n is the principal quantum number of the valence shell. These elements have high ionization energies and low electrons affinities, and they show little tendency to undergo chemical reaction. The *octet rule* is a statement of the stability of the ns^2np^6 valence-shell configuration. Atoms which can achieve this configuration by the addition of only a few electrons tend to do so, that is, tend to *complete the octet*. In adding electrons the atom becomes a negative ion. Thus the *chloride ion* (the *-ide* suffix tell us that it is a *negative* ion) is formed when one electron adds to a chlorine atom :

Table 6.1 : Lewis structures for atoms of the third period

Atom	Electronic configuration	Lewis structure
Na (Z = 11)	$1s^22s^22p^63s^1$	$\text{Na}\cdot$
Mg (Z = 12)	$1s^22s^22p^63s^2$	$\text{Mg}:$
Al (Z = 13)	$1s^22s^22p^63s^23p^1$	$\dot{\text{Al}}:$
Si (Z = 14)	$1s^22s^22p^63s^23p^2$	$\cdot\dot{\text{Si}}:$
P (Z = 15)	$1s^22s^22p^63s^23p^3$	$\cdot\underset{\cdot}{\dot{\text{P}}}:$
S (Z = 16)	$1s^22s^22p^63s^23p^4$	$\cdot\underset{\cdot}{\ddot{\text{S}}}:$
Cl (Z = 17)	$1s^22s^22p^63s^23p^5$	$:\underset{\cdot}{\ddot{\text{Cl}}}:$
Ar (Z = 18)	$1s^22s^22p^63s^23p^6$	$:\underset{\cdot\cdot}{\ddot{\text{Ar}}}:$

$$\text{Cl}\,(1s^22s^22p^63s^23p^5) + e^- \rightarrow \text{Cl}^-\ (1s^22s^22p^63s^23p^2)$$

or $$:\underset{\cdot\cdot}{\ddot{\text{Cl}}}\cdot\ + e^- \rightarrow \left[\,:\underset{\cdot\cdot}{\ddot{\text{Cl}}}:\,\right]^-$$

Here the negative sign is written to remind us that the resulting particle is an ion, and brackets are used to keep the negative sign from geting lost among the dots.

What about positive ions and the octet rule? When an atom has but a few valence electrons and has an octet in the second shell the outside, it tends to lose is valence electrons, thereby exposing the octet. In this way the resulting positive ion ends up with an octet in what is now its outer shell. Thus the sodium atom tends to lose its vaience electrons to form a sodium ion:

$$\text{Na}\,(1s^22s^22p^63s^1) \rightarrow \text{Na}^+\ (1s^22s^22p^6) + e^-$$

or $$\text{Na}\cdot \rightarrow \text{Na}^+ + e^-$$

Here the eight electrons in the L shell of Na^+ are not indicated; in the Lewis structure dots are used to count electrons in the valence shell of the *original sodium atom only.*

Lewis structures and ionic compunds To write the Lewis structure for an ionic compound, we write structures for the individual ions. Thus the Lewis structure for NaCl is

$$Na^+ \left[:\ddot{\underset{\cdot\cdot}{Cl}}: \right]^-$$

Note that this does not mean that each Na^+ is bonded to only one Cl^-. It means merely that equal number of Na^+ and Cl^- ions are present in the coumpound.

The octet rule helps us to predict stoichiometry, that is, atomic combining ratios, in ionic compounds. In the NaCl example one electron was transferred from one Na atom to one Cl atom :

In sodium oxide, however, the situation is different. Oxygen has only *six* valence electrons and thus needs *two* to complete its octet:

$$:\ddot{\underset{\cdot}{O}}\cdot + 2e^- \rightarrow \left[:\ddot{\underset{\cdot\cdot}{O}}: \right]^{2-}$$

Since a sodium atom has only one valence electron to lose (and its *second* ionization energy is very large), *two* sodium atoms are required to furnish two electrons to a single oxygen atom :

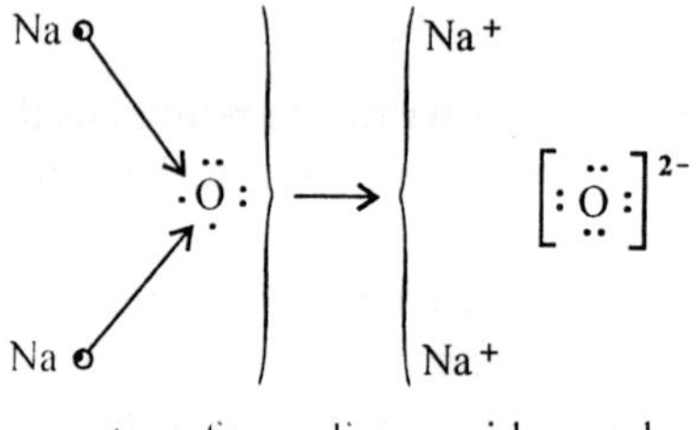

The Lewis sructure for sodium oxide can be written as

$$2Na^+ \left[:\ddot{\underset{\cdot\cdot}{O}}: \right]^{2-}$$

The empirical formula of sodium oxide is, therefore, Na_2O.

Lewis structure for aluminum oxide

Aluminium (Al) (group IIIA) has three valence electrons; O (group VIA) has six. We expect each Al to lose three electrons and O to gain two. It must be that *two* Al atoms *each* give up three electrons and that this total of six electrons is added to *three* O atoms to form three O^{2-} ions :

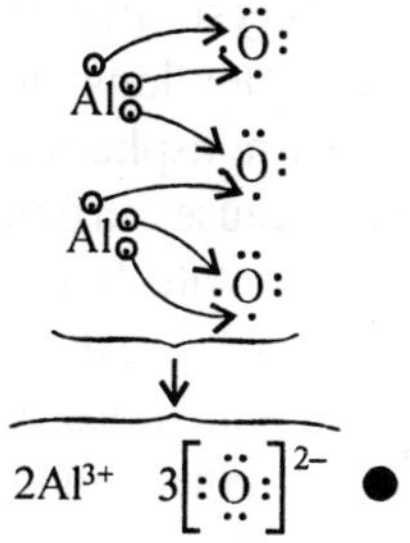

It is not necessary to work out the Lewis structure for an ionic compound in order to determine the compound's empirical formula *if* the charges of the ions are known. Note that in a correct empirical formula the sum of all the positive charges due to the cations equals to the sum of the negative charges due to the anions.

Sodium-Chloride-Type Structure

Sodium chloride and a variety of other ionic uni-univalent ionic substances crystallize in a cubic structure. The sodium-chloride-type structure consists of two interpenetrating face-centred cubic lattices; the unit cell for sodium chloride is shown in Fig. 6.3. An important descriptive feature of crystal structures is the number of nearest neighbours about each ion. In the case of ionic crystals, the nearest neighbours of an ion of a given charge are ions with the opposite charge. Thus, in the case of sodium chloride (Fig. 6.3) each sodium ion is surrounded by six chloride ions which can be imagined as occupying the corners of an octahedron; a similar arrangement of sodium ions exists about each chloride ion. The number of nearest neighbours for a given ion in a crystal is called the coordination number; we can speak of the coordination number of either the cation or the anion. In the case of the sodium-chloride-type structure, both the cation and the anion exhibit octahedral coordination, which is a natural consequence of the lattice type. We shall see that the cation and anion need not have the same coordination number in all structural types.

All of the alkali-metal halides (except CsCl, CsBr, and CsI), the oxides and sulfides of the alkaline earths, and AgF, AgCl, and AgBr possess the sodium-chloride structure. Many other compounds with the general stoichiometry MX in which either the cation, the anion, or both are complex ions rather than simple monoatomic ions also exhibit the sodium-chloride-type structure. Thus, the ionic compounds

$[Co(NH_3)_6]$ $[TiCl_6]$ and NH_4I crystallize in the sodium-chloride structure; in these cases the complex ions are, in effect, spherically symmetrical. In some instances nonspherical ions can achieve an effective spherical symmetry because sufficient kinetic energy can be attained to permit the ion to rotate freely in the crystal lattice. Common examples of such behaviour are the high-temperature forms of KOH, KSH, and KCN, all of which possess the sodium-chloride structure. At lower temperatures, where rotation is restricted, these substances attain different structures.

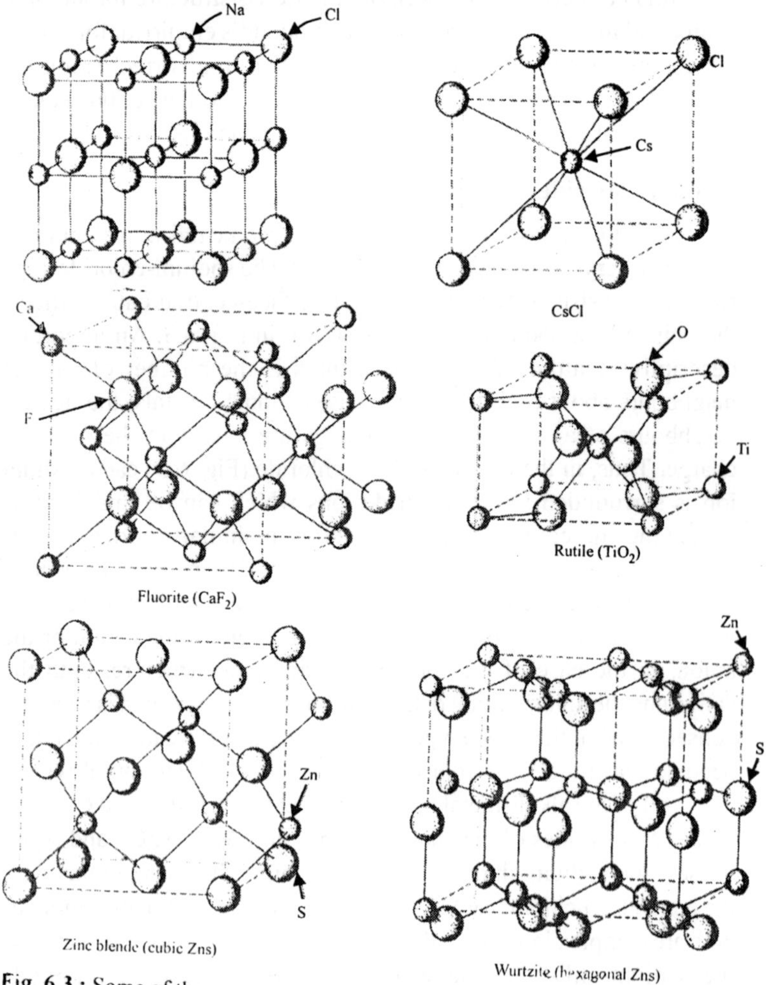

Fig. 6.3 : Some of the more common structure types found ionic substances.

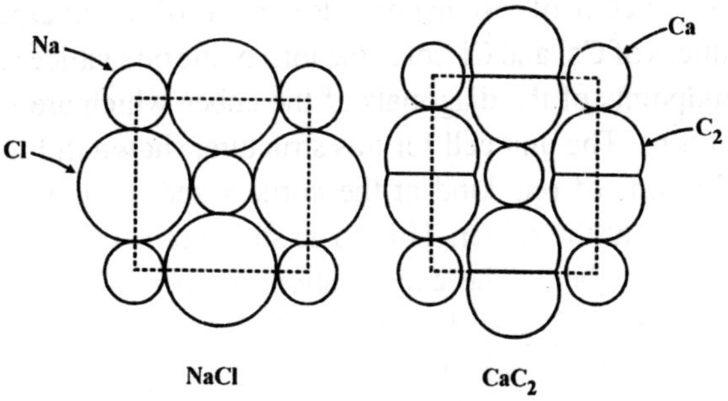

Fig. 6.4 : The unit cell for CaC_2 is a sodium-chloride-type structure in which latice has been distorted to accommodate the linear C_2^{2-} ions. Only one layer of the structure is shown here.

Many ionic substances of the form MX which contain complex ions that cannot attain spherical symmetry by rotation within the lattice crystallize in a distorted sodium-chloride-type structure. For example, the structure of CaC_2 is essentially that of sodium chloride except that the lattice is distorted in one direction because of the shape of the C_2^{2-} ions which are all arranged so that their long axes are parallel (Fig. 6.4). Similarly, $CaCO_3$ and $NaNO_3$ crystallize in a distorted cubic structure that can be related to the NaCl structure by replacing the spherical anions in the latter by non-planar CO_3^{2-} or NO_3^- ions. Since these complex ions are flat, the basic NaCl-type sstructure distorts to accommodate them (Fig. 6.5).

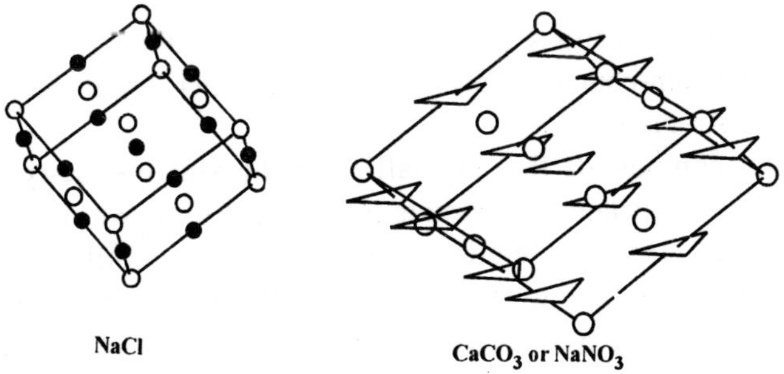

Fig. 6.5 : Both $CaCO_3$ and $NaNO_3$ crystallize in a sodium-chloride-type structure which is distorted to accommodate the flat anionic units (CO_3^{2-} or NO_3^-). This process, of course, leads to a noncubic lattice.

Cesium-Chloride-Type Structure

The structure of cesium chloride consists of two interpenetrating cubic lattices of Cs^+ and Cl^- ions; the ions of the one lattice are located at the midpoints of the diagonals of the cubes which are formed by the other ions. The unit cell for this structure, shown in Fig. 6.3, is a cube with ions of one kind at the corners and a single ion of the opposite kind at the center. This basic structure is often called body centered cubic (bcc). The coordination number of both the cation and anion is 8 (cubic coordination).

The cesium-chloride structure is exhibited by the cesium halides (except CsF), TlCl, the rubidium halides (except RbF) at pressures above 5000 kg cm^{-2}, and RbCl at low temperatures. As is the case in the sodium-chloride-type structure, certain complex anions are nearly spherically symmetrical in the crystal. Thus, $K[SbF_6]$, $Ag[NbF_6]$, $[Be(H_2O)_4]SO_4$, and $[Ni(H_2O)_6]$ $[SnCl_6]$ contain nearly spherical complex ions and crystallize in the CsCl-type structure. The anions in CsCN, TICN, and CsSH attain spherical symmetry by rotation within the crystal lattice, and these substances exhibit the CsCl structure; the anions in NH_4CN and the low-temperature form of $K[HF_2]$ cannot, so these substances crystallize in a distorted CsCl-type structure.

Zinc-Blende-Type Structures The other common structure observed for binary compounds is that exhibited by ZnS (zinc blende). As shown in Fig. 6.3, the cations are surrounded by a tetrahedral array of anions. The zinc-blende-type structure arises from a cubic closet-packed arrangement of anions in which one-half of the tetrahedral holes are filled with cations. In addition to ZnS, the sulfur, selenium, and tellurium com-pounds of zinc, cadmium, and mercury attain the zinc-blende structure.

Many compounds that crystallize in the zinc-blende-type structure also can attain the wurtzite-type structure, shown in Fig. 2.4, which is closely related. Compounds which exhibit two crystalline forms are said to be dimorphic. In the wurtzite structure, as in the zinc-blende-type structure, the anions form a tetrahedral array about the cations; however, the anions occupy a hexagonal closest-packed structure rather than a cubic closest-packed structure.

Calcium fluoride (fluorite) structure

In fluorite, each Ca^{2+} ion is surrounded by eight F^- ions, giving a body-centred cubic arrangement of F^- round Ca^{2+}. Since there are twice as many F^- ions as Ca^{2+} ions, the coordination number of both ions is different, and four Ca^{2+} ions are tetrahedrally arranged around each F^- ion. The coordination numbers are therefore 8 and 4, so this is called an 8 : 4 arrangement. The fluorite structure is found when the radius ratio is 0.73 or above.

An alternative description of the structure is that the Ca^{2+} ions form a face-centred cubic arrangement. The Ca^{2+} ions are too small to touch each other, so the structure is not close packed. However, the structure is related to a close-packed arrangement, since the Ca^{2+} occupy the same relative positions as for a cubic close-packed structure, and the F^- ions occupy all the tetrahedral holes.

Rutile structure

TiO_2 exists in three forms called anatase, brookite and rutile. The rutile structure is found in many crystals where the radius ratio is between 0.41 and 0.73. This suggests a coordination number of 6 for one ion, and from the formula it follows that the coordination number of the other ion must be 3. This is a 6 : 3 structure. Each Ti^{4+} is octahedrally surrounded by six O^{2-} ions and each O^{2-} ion has three Ti^{4+} ions round it is a plane tri-angular arrangement.

The rutile structure is not close packed. The unit cell, i.e. the repeating unit of this structure, is not a cube, since one of the axes is 30% shorter than the other two. It is convenient to describe it as a considerably distorted cube (though the distortion is rather large). The structure may then be described as a considerably distorted body-centred cubic lattice of Ti^{4+} ions. Each Ti^{4+} ion is surrounded octahedrally by six O^{2-} ions, and the O^{2-} are in positions of threefold coordination, that is each O^{2-} is surrounded by three Ti^{4+} ions at the corners of an equilateral triangle. Three-coordination is not common in solids. There are no examples of three-coordination in compounds of the type AX, but there is another example in the compounds of type AX_2, that is CdI_2, though in this case the shape is not an equilateral triangle. The structure of $CaCl_2$ is also a 6 : 3 structure, and is similar to CdI_2. These are described later.

There are only a few cases where the radius ratio is below 0.41. Examples include silica SiO_2 and beryllium fluoride BeF_2. These have coordination numbers of 4 and 2, but radius ratio predictions are uncertain since they are appreciably covalent.

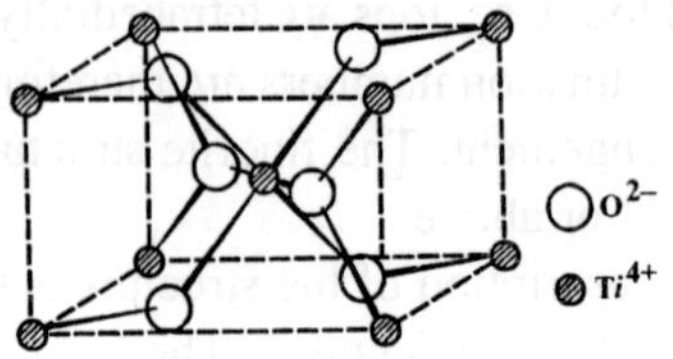

β-cristobalite (silica) structure

Silica SiO_2 exists in six different crystalline forms as quartz, cristobalite and tridymite, each with an α and β-cristobalite is related to zinc blende, with two interpenetrating close-packed lattices, one lattice arising from Si occupying the S^{2-} positions (i.e. the tetrahedral holes in the first lattice). The oxygen atoms lie midway between the Si atoms, but are shifted slightly of the line joining the Si atoms, so the bond angle Si—O—Si is not 180°. The radius ratio predicts a coordination number of 4, and this is a 4 : 2 structure.

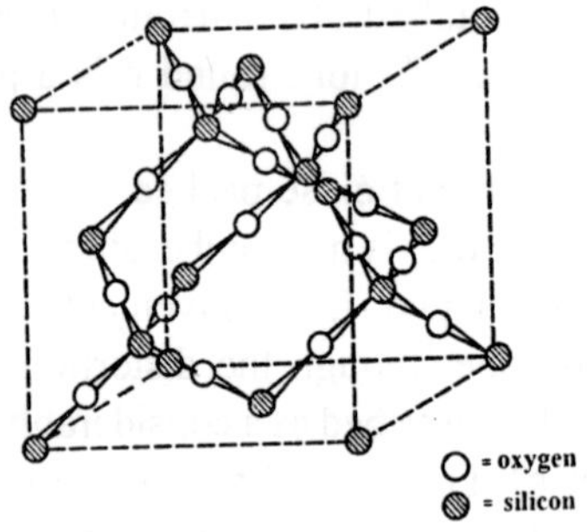

Radius Ratio

It is not difficult to calculate the size of the octahedral hole in a lattice of closest-packed anions. Fig. 6.6 illustrates the geometric arrangement resulting from six anions in contact with each other and with a cation in the octahedral hole. Simple geometry allows us to fix the diagonal of the square as $2r_- + 2r_+$. The angle formed by the diagonal in the corner must be 45°, so we can say :

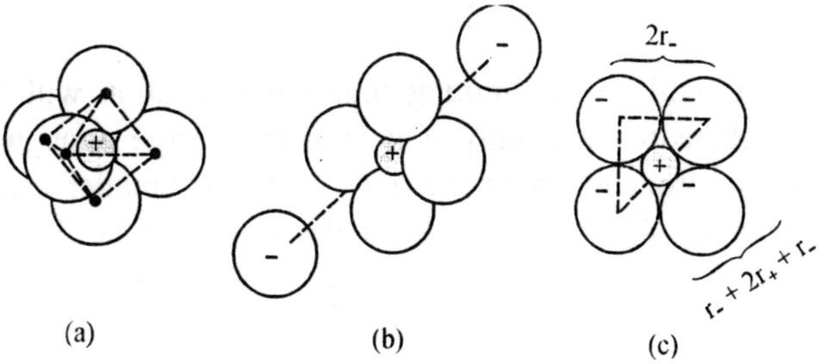

Fig. 6.6 : (a) Small cation (dashed line) in octahedral hole formed by six anions. (b) Dissection of octahedron to illustrate geometric relationships shown in (c).

Table 6.2 : Radius ratio and coordination number

Coordination number	Geometry	Limiting radius ratio[a]	Possible lattice structures
4	Tetrahedral		Wurtzite, zinc blende
		0.414; 2.42	
6	Octhaderal		NaCl, rutile
		0.732; 1.37	
8	Cubic		CsCl, fluorite
		1.000	
12	Dodecahedral		*b*

a The second ratio is merely the reciprocal of the first. It is often convenient to have both values.

b Coordination number 12 is not found in simple ionic crystals. It occurs in complex metal and in closest-packed lattices of atoms.

$$\frac{2r_-}{2r_- + 2r_+} = \cos 45° = 0.707$$

$$r_- = 0.707r_- + 0.707r_+$$

$$0.293r_- = 0.707r_+$$

$$\frac{r_+}{r_-} = \frac{0.293}{0.707} = 0.414$$

This will be the limiting ratio since a cation will be stable in an octahedral hole only if it is at least large enough to keep the anions from touching, i.e. $r_+/r_- > 0.414$. Smaller cations will preferentially fit into tetrahedral holes in the lattice. By a similar geometric calculation it is possible to determine that the lower limit for tetrahderal coordination is $r_+/r_- > 0.225$. For radius ratios ranging from 0.225 to 0.414, tetrahedral sites will be preferred. Above 0.414, octahedral coordination is favoured. By similar calculations it is possible to find the ratio when one cation can accommodate eight anions (0.732) or twelve anions (1.000). A partial list of limiting radius ratio values is given in Table 6.2.

The use of radius ratios to rationalize structures and to predict coodination numbers may be illustrated as follows.[24] Consider beryllium sulfide, in which $r_{Be^{+2}}/r_{S^{-2}} = 59$ pm/170 pm = 0.35. We should thus expect a coordination number of 4 as the Be^{+2} ion fits most readily into the *tetrahedral* holes of the closest-packed lattice, and indeed this is found experimentally: BeS adopts a wurtzite structure.

In the same way we can predict that sodium ions will prefer *octahedral* holes in a closest-packed lattice of chloride ions r_{Na^+}/r_{Cl^-} = 116 pm/167 pm = 0.69), forming the well-known sodium chloride lattice with a coordination number of 6.

With larger cations, such as cesium, the radius ratio (r_{Cs^+}/r_{Cl^-} = 181 pm/167 pm = 1.08) increases beyond the acceptable limit for a coordination number of 6; the coordination number of the cations (and anions) increases to 8, and the cesium chloride lattice results. As we have seen, although this is an efficient structure for cations and anions of about the same size, it cannot be directly related to a closest-packed structure of anions.

1. Since crystal radii vary slightly with coordination number, values from Table 3.4 were taken for C.N. = 6 as an "average" value.

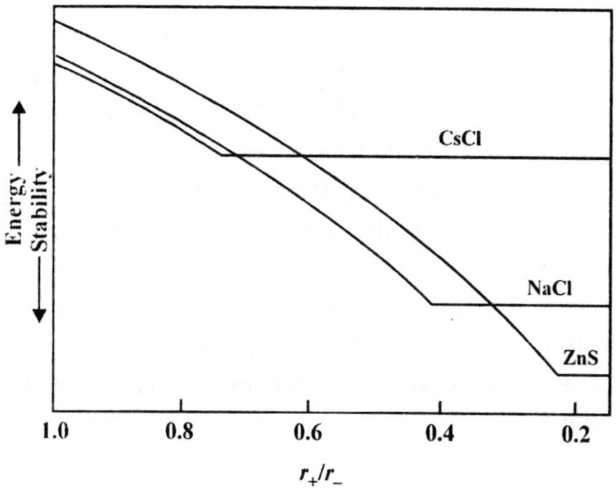

Fig. 6.7 : The total energy of a cubic lattice of rigid anions as a function of r_+ with r_- fixed, for different coordination configurations. When the anions come into mutual contact as a result of decreasing r_+ their repulsion determines the lattice constant and the cohesive energy becomes constant when expressed in terms of r_-. Thus near the values of r_+/r_- at which anion-anion contact takes place the radius ratio model predicts phase transitions to structures of successively lower coordination numbers.

Table 6.2 indicates that a coordination number of 12 should be possible when the radius ratio is 1.00. Geometrically it is possible to fit 12 atoms about a central atom (see the discussion of closest packing in metals), but it is impossible to obtain mutual 12-coordination of cations and anions because of the limitations of geometry. Twelve-coordination does occur in complex crystal structures of mixed metal oxides in which one metal acts as one of the closest-packing atoms and others fit into octahedral holes, but a complete discussion of such structures is more appropriate in a book devoted to the structures of solids.

The change in coordination number as a result of the ratio of ionic radii is shown graphically in Fig. 6.7. In general, as the cation decreases in size the lattice is stabilized (lattice energy becomes more negative) until anion-anion contact occurs. Further shrinkage of the lattice is impossible without a reduction in coodination number; therefore, zinc sulfide adopts the wurtzite or zinc blende structure, gaining additional energy over what would be possible in a structure with a higher coordination number. Note that although there is a significant difference between coordination numbers 4 and 6, there is little difference between 6 and 8. We might therefore expect small factors to tip the balance for or against 6- and 8-coordination and the radius rules to be obeyed inexactly.

In a 1 : 1 or 2 : 2 salt the appropriate radius ratio is obviously the ratio of the smaller ion (usually the cation) to the larger to determine how many of the latter will fit around the smaller ion. In compounds containing different numbers of cations and anions (e.g., SrF_2, TiO_2, Li_2O, Rb_2S) it may not be immediately obvious how to apply the ratio. In such cases it is usually best to perform two calculations. For example, consider SrF_2:

$$\frac{r_{Sr^{+2}}}{r_{F^-}} = \frac{132}{119} = 1.11 \qquad \text{maximum C.N. of } Sr^{+2} = 8$$

$$\frac{r_{F^-}}{r_{Sr^{+2}}} = \frac{119}{132} = 0.90 \qquad \text{maximum C.N. of } F^- = 8$$

Now there must be twice as many fluoride ions as strontium ions, so the coordination number of the strontium ion must be twice as large as that of fluoride. Coordination numbers of 8 (Sr^{+2}) and 4 (F^-) are compatible with the maximum allowable coordination numbers and with the stoichiometry of the crystal. Strontium fluoride crystallizes in the fluorite lattice.

A second example is SnO_2:

$$\frac{r_{Sn^{+4}}}{r_{O^{-2}}} = \frac{83}{126} = 0.66 \qquad \text{maximum C.N. of } Sn^{+4} = 6$$

$$\frac{r_{O^{-2}}}{r_{Sn^{+4}}} = \frac{126}{83} = 1.52 \qquad \text{maximum C.N. of } O^{-2} = 6$$

Considering the stoichiometry of the salt, the only feasible arrangement is with $C.N._{O^{-2}} = 3$, $C.N._{Sn^{+4}} = 6$; tin dioxide assumes the TiO_2 or rutile structure of. Note that the radius ratio would allow three more tin (IV) ions in the coordination sphere of the oxide ion, but the stoichiometry forbids it.

$$\frac{r_{K^+}}{r_{O^{-2}}} = \frac{152}{126} = 1.21 \quad \text{maximum C.N. of } K^+ = 8$$

$$\frac{r_{O^{-2}}}{r_{K^+}} = \frac{126}{152} = 0.83 \quad \text{maximum C.N. of } O^{-2} = 8$$

Considering the stoichiometry of the salt, the structure must be antifluorite (Fig. 3.3, reversed) with $C.N._{O^{-2}} = 8$, $C.N._{K^+} = 4$.

The radius ratio quite often predicts the correct coordination numbers of ions in crystal lattices. It must be used with caution, however, when covalent bonding becomes important. The reader may have been puzzled as to why beryllium sulfide was chosen to illustrate the radius ratio rule for coordination number 4 instead of zinc sulfide, which was used repeatedly earlier in this chapter to illustrate 4-coordinate structures such as wurtzite and zinc blende. The reason is simple. If ZnS had been used, it would have caused more confusion than enlightenment: It violates the radius ratio rule! Proceeding as above, we have $r_+/r_- = 88$ pm/170 pm = 0.52, indicating a coordination number of 6, yet both forms of ZnS, wurtzite and zinc blende, have a C.N. 4, both cations and anions. If one argues that 0.52 does not differ greatly from 0.41, the point is well taken, but there exist more vexing cases. The radius ratio for mercury(II) sulfide, HgS, is 0.68, yet it crystallizes in the zinc blende structure. In both of these examples the sp^3-hybridized *covalent* bonding seems to be dominant factor. Both ZnS and especially HgS are better regarded as infinite covalent lattices than as ionic lattices.

It should be kept clearly in mind that the radius ratio rules apply strictly only to the packing of hard spheres of known size. As this is seldom the case, it is surprising that the rules work as well as they do. Anions are not "hard" like billiard balls, but polarizable under the influence of cations. To whatever extent such polarization or covalency occurs, errors are apt to result from application of the radius ratio rules. Covalent bonds are directed in space unlike electrostatic attractions, and so certain orientations are preferred.

There are, however, other exceptions that are difficult to attribute to directional covalent bonds. The heavier lithium only marginally obey ther ule, and perhaps a case could be made for C.N. = 4 for LiI (Fig. 6.8). Much more serious, however, is the problem of coordination number 6 versus 8.

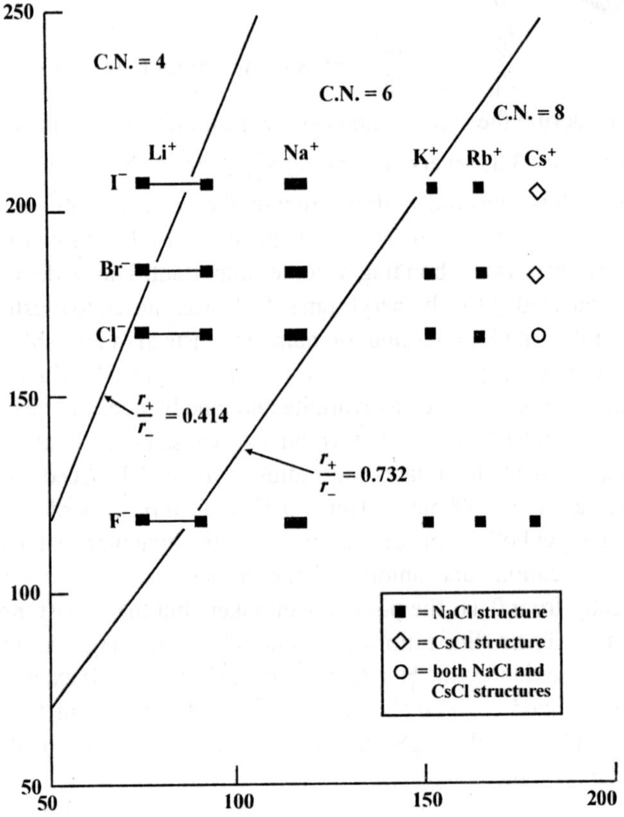

Fig. 6.8 : Actual crystal structure of the alkali halides (as shown by the symbols) constrasted with the predictions of the radius ratio rule: The figure is divided into three regions by the lines r_+/r_- = 0.414 and r_+/r_- = 0.732, predicting coordination number 4 (wurtzite or zinc blende, upper left), coordination number 6 (rock salt, NaCl, middle), and coordination number8 (CsCl, lower right). The crystal radius of lithium, and to a lesser extent that of sodium, changes with coordination number, so both the radii with C.N. = 4 (left) and C.N. = 6 (right) have been plotted.

The relative lack of 8-coordinate structures—CsCl, CsBr, and CsI being the only known alkali metal examples—is commonly found, if hard to explain. There are no 8-coordinate oxides, MO, even though the larger divalent metal ions, such as Sr^{+2}, Ba^{+2}, and Pb^{+2}, are large enough that the radius ratio rule would predict the CsCl lattice. There is no simple explanation for these observations. Some solace can be obtained from the observation that the Madelung constant for C.N. = 8 (1.763) is hardly larger than the value for C.N. = 6 (1.748). This small difference means that the predict lattice energies from the electrostatic model are less than 1% greater for the higher coordination number (see Fig. 6.7, where the lines for NaCl and CsCl almost coincide for $r_+/r_- > 0.732$). Thus, small energies comings from other sources could tip the balance.

What, then, are we to make of this problem? From a pragmatic point of view we can accept the radius ratio rule as a useful, if imperfect, tool in our arsenal for predicting and understanding the behaviour of ionic compounds. From a theoretical point of view it rationalizes the choice of lattice for various ionic or practically ionic compounds. It is failings call our attention to forces in solids other than purely electrostatic ones acting on billiard-ball-like ions.

Lattice energy

Consider a salt MX having the sodium chloride structure, the ions carrying charges z_+e and $z\ e$ where e is the electronic charge and z_+ and z are positive integers (neccssarily the same in this case, but it is desirable to develop the argument in a general way). Each M^{z+} ions is surrounded by six X^{z-} ions at a distance r, tweleve M^{z+} ions at $\sqrt{2}r$, eight X^{z-} ions at $\sqrt{3}r$, six M^{z+} ions at $\sqrt{4}r$, twenty-four X^{z-} ions at $\sqrt{5}r$, and so on. The electrostatic energy change when a M^{z+} ion is brought from infinity to its stable position in a sodium chloride type lattice is therefore given by

$$\frac{6e^2}{r} z_+z_- + \frac{12e^2}{\sqrt{2}r}(z_+)^2 - \frac{8e^2}{\sqrt{3}r} z_+z_-$$

$$+\frac{6e^2}{\sqrt{4}r}(z_+)^2 - \frac{24e^2}{\sqrt{5}r} z_+z_- \ldots.$$

$$= \frac{z_+z_-e^2}{r}6 - \frac{12z_+}{\sqrt{2}z_-} + \frac{8}{\sqrt{3}} - \frac{6z_+}{\sqrt{4}z_-} + \frac{24}{\sqrt{5}} - ....$$

Since the ratio of the charges on the ions (z_+/z_-) is constant for a given type of structure (unity for NaCl, two for CaF_2, etc.), the series in parentheses is a function only of the crystal geometry, and is independent of the values of z_+, z_- and r; it is called the Madelung constant A after its first evaluator, and the electrostatic energy of interaction of a cation with all other ions in the crystal is then $-z_+z_-e^2 A/r$. The environmens of both ions in the sodium chloride structure being indentical, the same expression gives the electrostatic energy of interaction of an anion with al! other ions. The total electrostatic ineraction energy is then given by half the sum of the interaction energies of cations and anions, since addition of the two would take interaction into account twice. Thus for a mole of MX (N pairs of ions, where N is Avogadro's number) the total electrostatic energy change on formation of the latice from ions at infinity is $-z_+z\ e^2\ NA/r$. (There is also a decrease in energyof 2RT representing the mechanical energy liberated when the gaseous ions form a solid, but this is very small relative to the electrostatic energy terms and may be neglected here.)

For the sodium chloride structure, the series for the Madelung constant converges to a value of 1.7476. For the caesium chloride structure, the Madelung constant is 1.7476. It might seem surprising that, for a given interionic distance, the total electrostatic interaction energy is almost the same for sixfold and eightfold coordination, but the slowness with which the series for Madelung constants converge emphasizes how important it is to consider all interactions in the lattice and not just the interaction of an ion and its nearest neighbours. Values for some other Madelung constants are given in Table 6.3. Attention should be drawn to the fact that different values are sometimes given from structures containing ions having different charges (e.g. 5.038 instead of 2.519 for the Madelung constant of the fluorite structure). This practice results from including the charges on the ions in the Madelung constant (i.e. quoting $-z_+z\ e^2\ A$) and is not to be ıecommonded.

Many expressions have been used to take into account the repulsive contribution to the interaction energy. The repulsive energy increases very steeply as interionic distance decreases after the ions have come into contact and is most simply represented by a term + NB/r^n, where B is a constant, the repulsion coefficient, which is approximately proportional to the number of nearest neighbours and n is another constant, called the Born exponent, which can be evaluated from compressibility data. The Born exponent generally

Table 6.3 : Maedlung constants

Structure type	A
CsCl	1.7627
NaCl	1.7476
ZnS (zinc blende)	1.638
ZnS (wurtzite)	1.641
CaF_2	2.519
TiO_2 (rutile)	2.408*
CdI_2	2.191*

* Values for these structures vary slightly according to the ratio of the lattice constants for the unit cell.

lies between 8 and 10, but is lower if both ions have the helium electronic configuration and higher if both have the xenon electronic configuration. The total interaction energy is then given by

$$-\frac{NAz_+z_-e^2}{r}+\frac{NB}{r^n}$$

and the lattice energy (defined as the energy required at 0°K to convert one mole of the salt into gaseous ions infinitely removed from one another) is

$$U=\frac{NAz_+z_-e^2}{r}-\frac{NB}{r^n}$$

Since $dU/dr = 0$ at the equilibrium distance $r = r_0$,

$$0=-\frac{NAz_+z_-e^2}{r_0^2}+\frac{nNB}{r_0^{n+1}}$$

whence

$$B=\frac{Az_+z_-e^2r_0^{n-1}}{n}$$

and

$$U = \frac{NAz_+z_-e^2}{r_0} 1 - \frac{1}{n}$$

This is the Born-Landé expression for lattice energy. More detailed calculations of lattice energies employ more sophisticated expressions for the repulsive energy, and also take into account London or van der Waals forces, which are attributed to synchronized oscillating dipoles in adjacent ions, and zero-point energies; the results are not greatly modified, however, as may be seen from the data for sodium chloride and magnesium oxide in table 6.4. In estimating lattice energies of compounds which are unknown or whose structures are unknown (or are of low symmetry, so that the Madelung constants are unknown), chemists usually employ the simple formula and often assume n = 9; in a comparative treatment a small change in n is relatively unimportant. Physicists, on the other hand, being mainly concerned with the most accurate description of a very few compounds whose properties are known in great detail, naturally prefer more rigorous (and, unfortunately, more difficult to handle) formulae for lattice energies.

Table 6.4 : Contributions to lattice energies

Energy	*NaCl**		*MgO**	
Electrostatic attraction	+ 860	(205.6)	+ 4634	(1108)
Repulsion	− 99	(23.6)	− 699	(167)
London	+ 12	(2.9)	+ 6	(1.5)
Zero-point	− 7	(1.7)	− 18	(4.4)
Total	766	(183.2)	3923	(938)

*Values are in kJ and, in parentheses, in kcal.

Lattice energies derived from formal ionic charges, Madelung constants, Born exponents and interionic distances are often referred to as 'calculated' values to distinguish them from values obtained by means of thermochemical cycles. It should, however, be pointed out that interionic distances measured by X-ray diffraction are themselves experimental data and may conceal departures from ideal ionic behaviour. Furthermore, the real changes on ions may be less than their formal charges, and the evaluation of the non-electrostatic part of the lattice energy is subject to some uncertainty. Nevertheless, the concept of lattice energy is of immense importance in inorganic chemistry, and many examples of its utility will be given later.

We may conclude this section by returning briefly to the subject of interionic distances and ionic radii. The expression for the repulsion constant B may be rewritten in the form :

$$r_0 = \frac{nB}{Az_+z_-e^2}^{1/(n-1)}$$

If we consider a salt existing in the caesium chloride and sodium chloride structures, and assume that, since repulsive forces are significant only over short distances, B is proportional to the number of nearest neighbours, and that $n = 9$.

$$\frac{r_0 \text{ (CsCl form)}}{r_0 \text{ (NaCl form)}} = \frac{8 \times 1.748}{6 \times 1.763} = 1.036$$

For the CsCl and NaCl, RbCl, forms of CsCl RbBr and RBI the ratio of the experimental values of r_0 is found to be 1.027, 1.036, 1.029 and 1.025 respectively; the agreement is at least good enough to suggest that the premise that the repulsive forces are important over much shorter distances than the electrostatic forces is correct. The effect of the change in or on the lattice energy is, however, diminished by the change in the Madelung constant, and the overall conclusion is that there is probably very little difference between the lattice energies of the two forms of the alkali metal halides being discussed in this paragraph. This point could be tested experimentally by, for example, measuring the heats of dissolution of the two forms in water, but no such measurements have yet been made.

The Born-Haber Cycle

One test of whether an ionic model is a useful description of a substance such as sodium chloride is its ability to produce an accurate value of the enthalpy of formation. Note that we cannot make this test simply by measuring the enthalpy of reaction 4.4, or its reverse. The former is possible in principle but not experimentally feasible. The latter is not possible because sodium chloride does not vapourize cleanly to Na^+ and Cl^- but to NaCl, which then dissociates into atoms.

To handle the energy problem, a *thermodynamic cycle*, called the Born-Haber cycle is used. This is illustrated in Fig. 6.9. The basic idea is that the formation of NaCl(s) from the elements, Na(s) + $\frac{1}{2}$$Cl_2$(g), whose enthalpy is by definition the enthalpy of formation of NaCl(s), can be broken down into a series of steps. If the enthalpies of these steps are added, algebraically, the result must be equal to ΔH^o_f according to the Law of Conservation of Energy, the First Law of Thermodynamics. We thus have the equation

$$\Delta H^o_f = \Delta H^o_{vap} + \frac{1}{2}\Delta H^o_{diss} + \Delta H^o_{EA} + \Delta H^o_{ion} + U$$

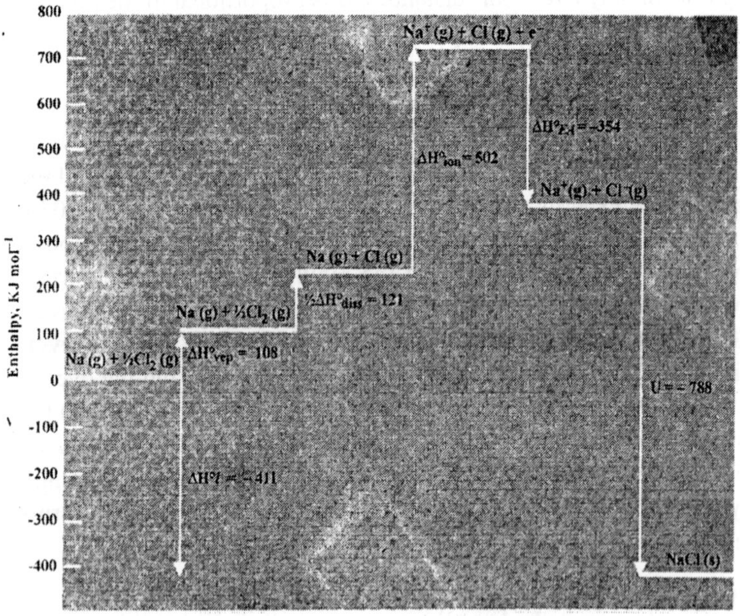

Fig. 6.9 : The Born-Haber Cycle for NaCl

where the enthalpy terms are for vapourization of sodium (ΔH^o_{vap}), dissociation of Cl_2(g) into gaseous atoms (ΔH^o_{diss}), electron attachment to Cl(g) to give Cl^-(g) (ΔH^o_{EA}), ionization of Na(g) to Na^+ (g) + *e* (ΔH°_{ion}) and the formation of NaCl(s) from gaseous ions (U).

More generally, any one of these energies can be calculated if all the others are known. For NaCl all of the enthalpies except *U* in Eq.

4.7 have been measured

$$\begin{array}{rcl} \Delta H^{o}_{f} & & -411 \\ -\Delta H^{o}_{vap} & & -108 \\ -\frac{1}{2}\Delta H^{o}_{diss} & = & -121 \\ -\Delta H^{o}_{EA} & & -354 \\ -\Delta H^{o}_{ion} & & -502 \\ \hline U & = & -788 \text{ kJ mol}^{-1} \end{array}$$

This result is very close (within 2.7%) of the value of U calculated by using the refined ionic model (– 766 kJ mol^{-1}). Actually, even precise calculations give a value that agrees to within less than 1%. This good agreement supports (but does not prove) the idea that the ionic model for NaCl is a useful one.

Often, the Born-Harber cycle, or a similar one, is used differentially. If it is assumed that *U* calculated on the ionic model is correct, the cycle can be used to estimate some other energy term. For example, there is no convenient direct way to measure the enthalpy of formation of the gaseous CN^- ion. From a Born-Haber cycle for NaCN, where values of all the other enthalpies are available and U is calculated, it is found that ΔH_f for CN^-(g) is 29 kJ mol^{-1}.

Frenkel defect occurs in AgBr when a silver ion vacates the octahedral hole and occupies a tetrahedral hold instead. Schottky defects occur when both a cation and an anion are missing from a structure, leaving vacancies.

Defects are localized irregularities, which can have a marked effect on properties of a crystalline substance. For example, defects are important in the photochemical decomposition of solids such as silver salts; in reactions involving heterogeneous catalysis, as occur when Cu_2O catalyzes the oxidation of CO to CO_2; and in the luminescence of compounds used for cathode-ray screen. Zinc sulfide fluoresces blue when exposed to ultraviolet radiation or cathode rays after it is heated above 500°C. This phenomenon appears to be associated with a defect structure which possess vacant sulfur positions. Heating zinc sulfide to 1100°Cwith a trace of copper, which becomes incorporated within the lattice, gives a defect structure which fluoresces yellow-green when irradiated with cathode rays.

Point defects

Another important kind of defect is the *point defect*. These include the *lattice vacancy* and the *lattice interstitial*,illustrated in Fig. 6.10 (a) and b, respectively. A vacancy occurs when an atom, ion, or molecule is missing from the lattice, and an interstitial occurs when a particle is trapped at a location which is not a lattice site.

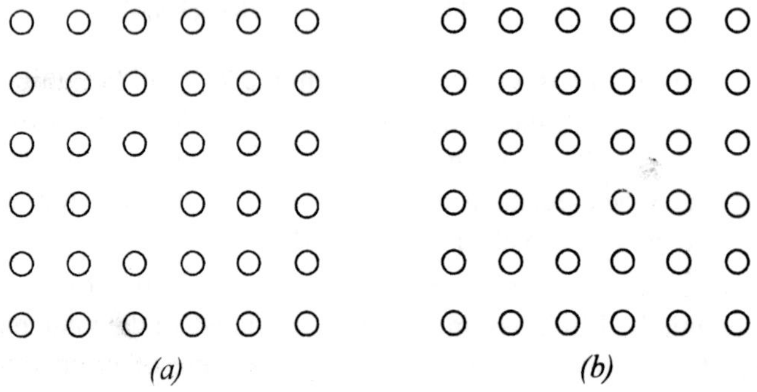

Fig. 6.10 : Point defects. (a) Lattice vacancy. (b) Lattice interstitial.

In ionic solids point defects often occur in two common combinations : the *Schottky defect*, Fig. 6.11(a), is a pair of lattice vacancies of opposite charge, and the *Frenkel defect*, Fig. 6.11, is an ion misplaced in an interstitial position plus the vacancy where it "ought" to be. Note that in each the electrical neutrality of the crystal is preserved, that is, there are still equal numbers of positive and negative charges.

There are several other kinds of point defects known as *colour centers*. These also occur in ionic solids and are namd because their presence gives colours to otherwise colourless crystals. One of these, the F centre (from the German, *Farbenzentrum*, "colour center"), is an electron located at an anion site.

Also classed as point defects are impurity defects, which consist of foreign particles present at lattice sites. If, for example, molten NaCl to which a few tenths of a per cent $CaCl_2$ has been added is crystallized by cooling, the resulting NaCl lattice has an occasional Ca^{2+} where an

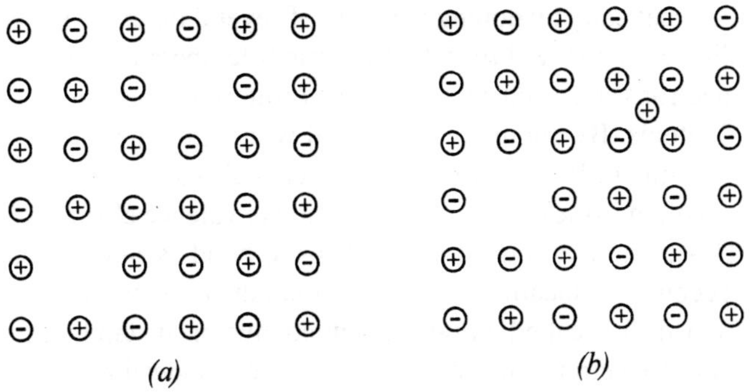

Fig. 6.11 : Schottky and Frenkel defects in ionic solids.
(a) Schottky defect. (b) Frenkel defect.

Na^+ ought to be. In order to preserve electrical neutrality in the crystal, for each Ca^{2+} ion present there will be a cation vacancy (a missing Na^+) somewhere in the crystal. Similarly, ZnS can be prepared with Cl^- ions replacing some of the S^{2-} ions. For every two Cl^- ions a cation vacancy is present. Compounds to which small amounts of impurities have been added are said to have been *doped.* Doping often changes the properties of a substance greatly. The $CaCl_2$-doped sodium chloride mentioned above has, for example, a much higher electrical conductivity than does pure NaCl.

Semiconductors In semiconductors the presence of certain impurity defects has a spectacular effect on electrical conductivity. A *semiconductor* is a substance whose electrical conductivity increases as temperature is increased. (In a metal the opposite is true: increasing the temperature increases the vibrations of the ions in the crystal lattice., and the interferes with motion of the delocalized electrons, reducing conductivity.) In a semiconductor at low temperature, essentially all electrons are localized, bound to specific atoms. Raising the temperature frees some electrons, which become delocalized. as in a metal. As the temperature goes up, more weakly bound electrons are made available to conduct electric current, that is, to move through the crystal.

Silicon is a good example of a semiconductor. At room temperature the conductivity of silicon is very low, but it can be increased dramatically by doping with any of several other elements. In crystalline Si (periodic group IVA) at room temperature all four valence electrons are used in sp^3 hyrbid orbitals to bond to four adjacent Si atoms. By doping the Si crystal with a group V element such as P, As, Sb, or Bi, the structure of the crystal is left unchanged, but every so often in the crystal an atom with five valence electrons is present. This atom uses four of its electrons to bond as Si does, but the fifth becomes delocalized and free to conduct electricity. The group V-doped Si is called an *n*-type semiconductor, n standing for *negative*, since electrons (negative charges) are responsible for the conductivity.

Doping an Si crystal with a group III A element such as B, Al, Ga, or In produces an Si crystal structure in which an occasional atom with only three valence electrons is present. The electron vacancy is called a *hole*. Under the influence of an electric field, an electron can fall into a hole and annihilate it. But this leaves a hole where the electron was, so that it looks as if the hole has moved in a direction opposite to that in which the electron has actually moved. The hole thus appears to have a positive charge and to move through the silicon lattice as if it were a real particle. The group III-doped Si is called a p-type semiconductor, since *positive* holes appear to be responsible for the conductivity.

Electrical conduction in a semiconductor with impurity centers consists of electrons moving in one direction (in an n-type semiconductor) or holes moving in the other (in a p-type semiconductor). This is shown schematically in Fig. 6.12, where the electrons are represented by minus signs and the holes by plus signs. Various combinations of p- and n-type semiconductors are used to make solid-state electronic devices. A solid-state *diode* (Fig. 6.13) is a combination of one of each type of semiconductor. It is used as a rectifier for alternating current, because it will allow current to flow in one direction only. When a voltage is applied in one direction across a diode, as in Fig. 6.13(a), electrons migrate through the n-type semiconductor in one direction and holes, through the p type semiconductor in the other. When they reach the junction beteen the semiconductors they annihilate each other. (Actually, the electron just drops into the hole.) Conduction continues because more electrons are fed into the n-type semiconductor from the left (in the illustration), and more holes are created in the p-type semiconductor on the right as electrons are removed. When the polarity of the applied voltage is reversed, however, electrons and holes tend to move away from each other and from the junction, as shown in Fig. 6.13(b). For most diodes, unless the voltage is high, holes and free electrons cannot be formed at the junction, and so the flow of current through the diode is prevented.

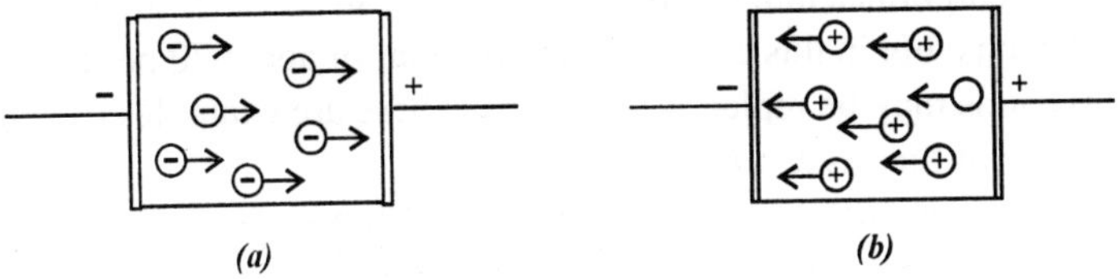

Fig. 6.12 : Conduction in semiconductors (a) n-type (b) p-type

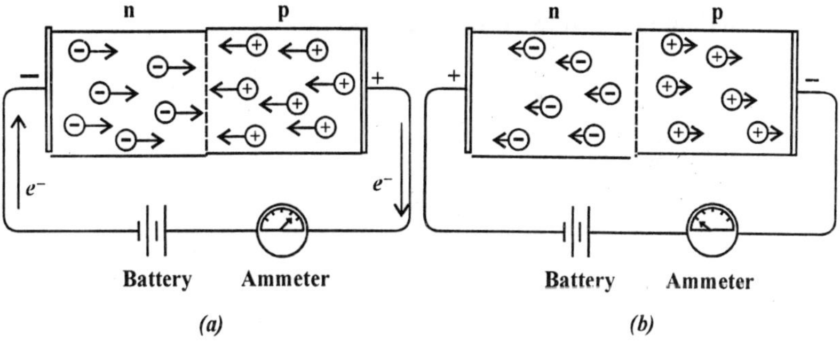

Fig. 6.13 : A solid-state diode.
(a) Current flows through diode. (b) Polarity reserved; no current.

Other applications of semiconductors include the *transistor*, in which a sandwich, either pnp or npn, is used to detect or amplify electrical signals. The *light-emitting diode* (l.e.d.) is a diode in which the energy produced when electrons annihilate holes at the pn junction is released as light. In the *photodiode*, light striking the junction creates equal the numbers of holes and electrons cell ("battery") by producing a voltage across its two elements. Similar to the photodiode, converting radiant energy into electrical energy.

Solid solutions

Impurities may sometimes be incorporated into the host lattice up to very high concentrations. For example, mixtures of potassium chloride and potassium bromide ranging from pure KCl to pure KBr may be recrystallized out of aqueous solution. In each case a single crystalline form with the rock-salt structure is obtained. K^+ ions occupy the cation sites, while Cl^- and Br^- are randomly distributed over the anion sites. The radii of Cl^- (0.181 nm) and Br^- (0.195 nm) are evidently close enough to allow one to substitute for the other. Similarly, I^- ions can be incorporated into the rock-salt structure of AgBr by substituting for Br^- ions, but in this case only a maximum of about 70 per cent of the Br^- ions may be replaced by I^- ions. These are examples of solid solutions of the *substitutional* type.

In *an interstitial solid solution* foreign atoms, ions, or molecules occupy the nooks and crannies, the *interstices*, in the host lattice. One example of this is *austenite*, an alloy in which carbon atoms occupy some of the interstices in a face-centered cubic array of iron atoms. Most of the transition elements will form such solid solutions with small atoms such as H, B, C, and N.

7

Co-ordination Compound

Werner's coordination theory

The basic features of the structures of coordination compounds, as it is still convenient to call them, were elucidated by Alfred Werner over seventy years ago. Werner, in discussing the nature of compounds such as $CoCl_3$, $6NH_3$ (as it was then written), postulated that neutral molecules or oppositely charged ions are grouped or coordinated round a central atom in what he termed the first sphere of attraction, the number of such ligands being the coordination number of the central atom. The latter had two sorts of valency. One, the primary valency, was identified with the oxidation state of the central atom; in $CoCl_3$, $6NH_3$, for example, this would be + 3, the positive charge being balanced by that on the three chloride ions held in the outer coordination sphere. The ammonia molecules were envisaged as being held by secondary valencies round the central atom, having an octahedral distribution in space; the compound was thus formulated as $[Co(NH_3)_6]$ Cl_3 (hexamminecobalt (III) chloride). It was also postulated that the number of coordinated ligands might vary as the central atom was changed, and that such ligands would not show the same reactivity as the same species in the free state.

The chief methods available to Werner for the investigation of complexes were the study of chemical reactivity, electrical conductivity and isomerism. The following examples, taken from his work, illustrate their application. By the action of oxidizing agents on solutions of cobalt (II) chloride in the presence of ammonia, three compounds are formed according to the conditions employed. One of these is the yellow $[Co(NH_3)_6]Cl_3$ already mentioned; all of the chloride in this compound is precipitated by cold aqueous silver nitrate. Another is the pink compound $CoCl_3.5NH_3.H_2O$; all the chloride in this is precipitated by cold silver nitrate, and a salt $Co(NO_3)_3.5NH_3.H_2O$ can be isolated from the reaction. This suggests the pink compound is $[Co(NH_3)_5(H_2O)]$ Cl_3 (aquopentammine-cobalt (III) chloride). The third product, purple $CoCl_35NH_3$, yields only two moles of silver chloride on treatment with cold aqueous silver nitrate, and is thus $[Co(NH_3)_5Cl]$ Cl_2 (chloropentammine-cobalt (III) chloride). In the series of complexes obtainable in the system K^+–$Pt^{II}Cl^-$–NH_3, the values for the molar conductivities in aqueous solution of $K_2[PtCl_4]$, $K[PtCl_3(NH_3)]$, $[PtCl_2(NH_3)_2]$, $[Pt(NH_3)_3Cl]$ Cl and $[Pt(NH_3)_4]$ Cl_2 (268, 107, 1, 116 and 261 $ohm^{-1}cm^2$ respectively) help in formulating the complexes, the contribution to the molar conductivity of any ion $I^{x\pm}$ being found to be about 60 x $ohm^{-1}cm^2$. The octahedral geometry of the cobalt (III) complexes and the planar geometry of the platinum (II) complexes was established with a high degree of probability by the studies of geometrical and optical isomerism, mentioned.

Today there is an overwhelming amount of X-ray diffraction and other data in support of Werner's formulations; we shall now summarize this, and then end the chapter by a short account of all types of isomerism in metal complexes with special reference to recent work.

Coordination Numbers and Geometries

The term coordination number has already been introduced (page 99) in discussing the packing of ions in ionic crystals. It is widely used in discussing the structure of the complexes formed by cations. In addition to the number of ligands surrounding a cation, it is important to know their spatial arrangement, the coordination geometry. We now discuss the coordination numbers, in increasing order, and under each the most geometrical arrangements.

Coordination Number Two. This is relatively rare, occurring mainly with the + 1 ions of Cu, Ag, and Au and with Hg^{2+}. The geometry is linar, and ions such as $[H_3N—Ag—NH_3]^+$, $[NC—Ag—CN]^-$, and $[Cl—Au—Cl]^-$ are examples.

Coordination Number Three. The most important geometries are planar and pyramidal. It is a relatively rare coordination number. Examples are the planar HgI_3^- ion and the pyramidal $SnCl_3^-$ ion. In most instances where the stoichiometry might suggest it, for example, $AlCl_3$, $FeCl_3$, $PtCl_2PR_3$, etc., no 3-coordinate mononuclear species occur. Instead, there are dinuclear species in which two ligands are shared so as to give each cation a coordination number four.

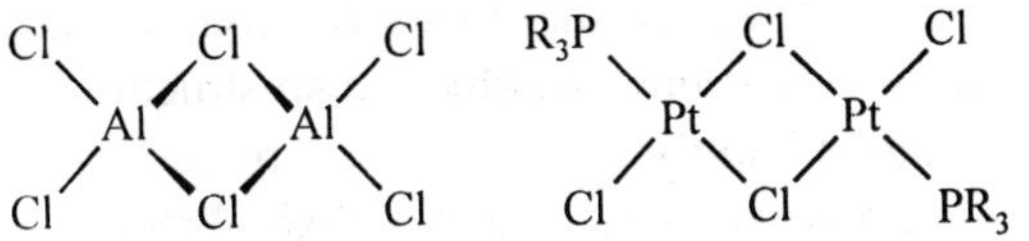

Coordination Number Four. This is one of the most important and gives two main geometries, tetrahedral and square. Tetrahedral complexes are most common, being formed almost exclusively by nontransition metal cations as well as by transition metal ones other than those to the right side of the *d* block. Examples of tetrahedral complexes include $Li(H_2O)_4^+$, BeF_4^{2-}, BF_4^-, $AlCl_4^-$, $FeCl_4^-$, $CoBr_4^{2-}$, ReO_4^-, and many others. Square complexes are particularly common for Cu^{2+}, Ni^{2+}, Pd^{2+}, Pt^{2+}, Au^{3+}, Rh^+, and Ir^+. The cations that characteristically form square complexes are those with eight *d* electrons. These eight electrons form four electron pairs that fill all of the *d* orbitals except $d_{x^2-y^2}$. This orbital together with the *s*, p_x, and p_y orbitals a set of hybrids, dsp^2, directed toward the vertices of a square.

Coordination Number Five. This is less common than 4 or 6, but still very important. There are two symmetrical geometric arrangement. the trigonal bipyramid (**tbp**) and the square pyramid

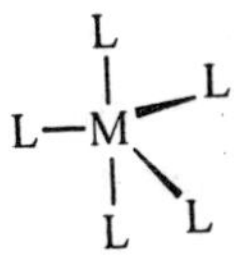

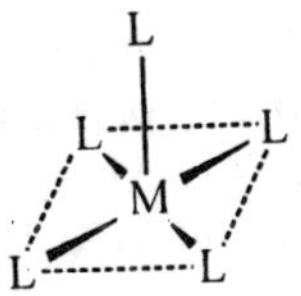

(**sp**), 6 (a) and (b). Although these two may appear quite different, they usually do not differ greatly in energy, and one can be converted into the other by rather small changes in bond angles. For these reasons, many 5-coordinate complexes donot have either structure precisely, but a structure that is intermediate between the two. Moreover, even those that do have a structure approximating closely to one of these ideal forms are *stereo-chemically nonrigid.* This means that the ligands do not remain fixed in their places, but interchange places rapidly.

An interesting illustration of the similar stabilities of the two geometries is afforded by the $Ni(Cn)_5^{3-}$ ion, which forms one crystalline salt in which both the **tbp** and **sp** structures are present.

Coordination Number Six. This is enormously important, since nearly all cations form 6-coordinate complexes. Practically all of these have one geometrical form, the octahedron, 6-V. It is essential to recognize that the octahedron is an extremely symmetrical figure, even though some of the stylized ways of drawing it might not show this clearly. All six ligands, and all six M – L bonds are equivalent in a regular octahedral ML_6 complex.

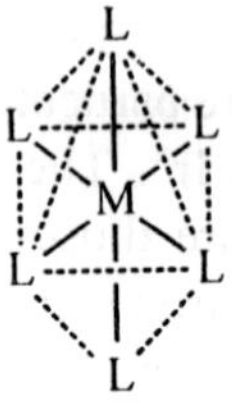

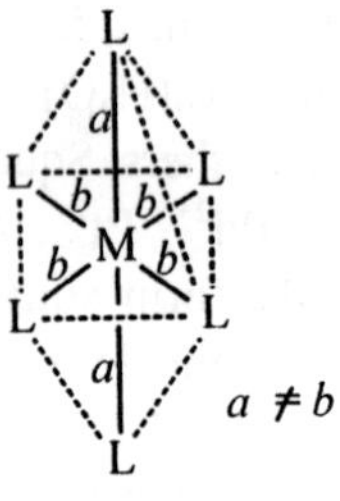

As with other prototype geometries, we continue to describe complexes as "octahedral" even when different kinds of ligands are present and, hence, the full symmetry of the true octahderon cannot be retained. Even in cases where all ligands are chemically the same, octahdera are often distorted, either by electronic effects inherent in the metal ion or by forces in the surroundings. A compression or elongation of one L—M—L axis relative to the other two is called a *tetragonal* distortion, 6-V, whereas a complete breakdown of the equality of the axes gives a *rhombic* distortion, 6-VII. If the octahedron is comporessed or elongated on an axis connecting the centers of two opposite triangular faces, the distortion is called *trigonal*,

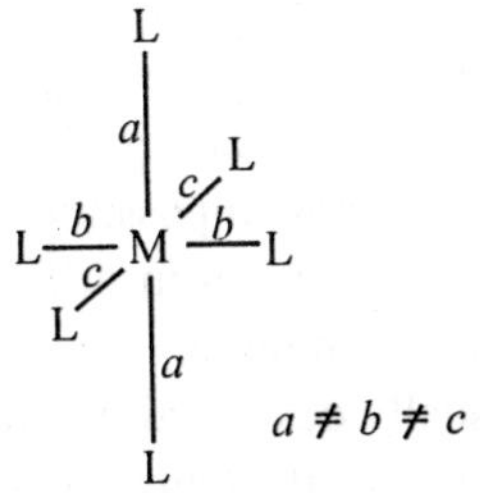

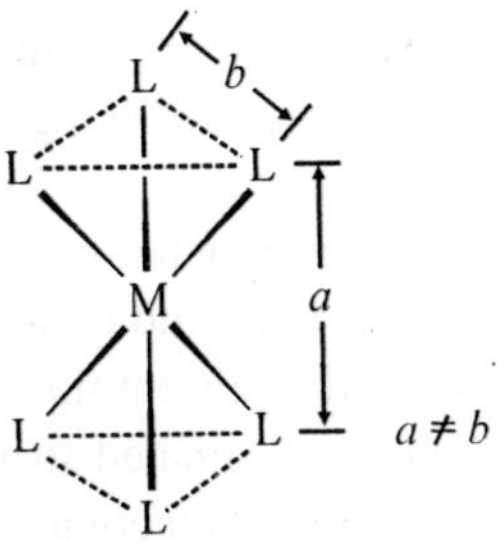

There are a very few cases in which six ligands lie at the vertices of a trigonal prism. The prism is related to the octahedron in a simple way: if one triangular face of an octahedron is rotated 60° relative to the one opposite to it, a prism is formed. The superior stability of the octahedron compared with the prism has, at least, two causes. The most evident is steric: the octahedron allows the ligands to say further away from each other, on the average, than does the prism for any given M—L distance. It is also likely that in most cases the metal ion can form stronger bonds to an octahedral set of ligands. The cases where a trigonal prism is found mostly involve either a set of six sulfur atoms, which may interact directly with each other to stabilize the prism, or some sort of rigid cage ligand which forces the prismatic arrangement.

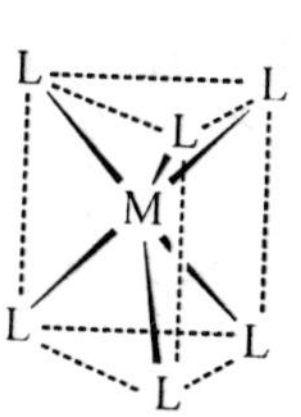

Coordination number seven. For coordination numbers above six the differences between idealized geometries become even smaller than for five-coordination, and where structures are somewhat distorted from basic types by the presence of ring systems it is sometimes almost meaningless to describe them in terms of preconceived geometries. Moreover, for high coordination numbers the metal atoms involved are usually (though not invariably) those of the second and third transition series, and determination of the exact positions of ligand atoms of much lower atomic number is then impossible unless X-ray diffraction studies are supplemented by neutron diffraction work; up to the present time this has rarely been done.

Seven-coordinated structures are usually described in terms of three types: the pentagonal bipyramid the capped trigonal prism (or one-face centered trigonal prism) and the capped octahedron (or one-face centered octahedron). The third of these was first reported many years ago, and was believed to be present in the ions $[ZrF_7]^{3-}$ and the isostructural $[NbOF_6]^{3-}$; more recently, however, it has been found that in the case of $[ZrF_7]^{3-}$ the report was incorrect, and since it now seems that there is no well-authenticated case of the capped octahedral structure, we shall confine attention to the first two.

The pentagonal bipyramidal structure (which is that of iodine heptafluoride) occurs in $[ZrF_7]^{3-}$ (d^0) and the bridged oxalate complex of titanium (III), μ-oxalatobis (oxalato) hexaquodititanium (III) (d^1) (Figs. 7.1. and 7.2 respectively). In the latter, each titanium atom is coordinated by four oxygen atoms from oxalate and three from water; the water molecules occupy one equatorial and two apical positions round each titanium. It occurs also in the $[Fe(EDTA)(H_2O)]^-$ ion (d^5) (in which EDTA represents the ethylenediamine tetraacetate

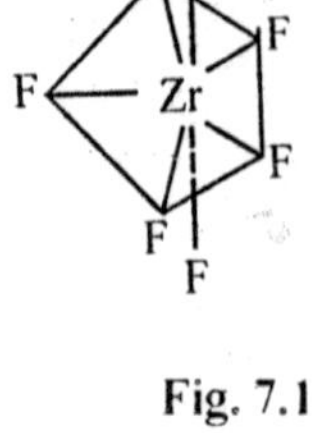

Fig. 7.1

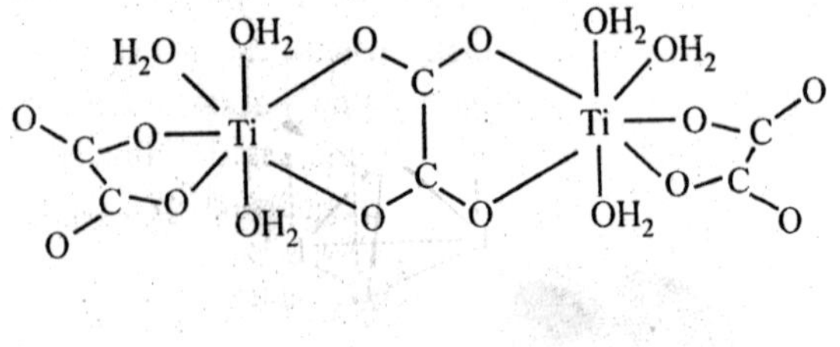

Fig. 7.2

anion), with the water molecule occupying an equatorial position. If, however, the anion is replaced by that of diaminocyclohexane-tetraacetic acid, the structure changes to a capped trigonal prism, presumably owing to the steric effect of the cyclohexane ring. The balance between the two structures must be a fine one, for [Mn(EDTA) $(H_2O)]^{2-}$ (also a d^5 system) is also a capped trigonal prism.

The simplest examples of the capped trigonal prismatic are $[NbF_7]^{3-}$ and $[TaF_7]^{3-}$ (Fig. 7.3). In the former, the positions of the fluorine atoms have been confirmed by neutron diffraction; all lie within the range 1.91-1.96 Å of the metal atom. (The closest appro-

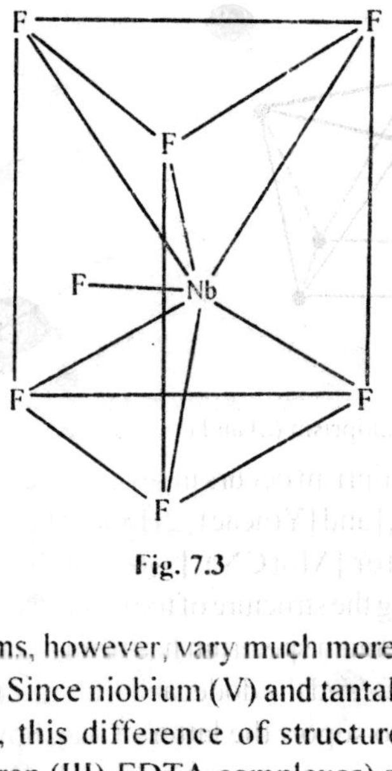

Fig. 7.3

aches of fluorine atoms, however, vary much more widely, extending from 2.36 to 2.91 Å.) Since niobium (V) and tantalum (V) are both d^0 like zirconium (IV), this difference of structure (like that of the manganese (II) and iron (III)-EDTA complexes) shows that for high coordination numbers the isoelectronic principle (that molecules or ions having the same number of electrons have the same structure) is no longer valid. There is, in fact, no reliable general guide to the structures of species of these less common coordination numbers.

Coordination number eight

Of the three basic structures for eight-coordination (the cube, the square antiprism and the dodecahedron) only two commonly occur in complexes. The cube is well kown in ionic salts, presumably because its symmetry lends itself to packing in infinite lattices, but individual molecules and ions usually adopt other structures in which, incidentally, repulsions between pairs of bonded electrons should be less. Recently, however, it has been found that the anion in Na_3 $[PaF_8]$ is an almost perfect cube, and the corresponding compounds of uranium (V) and plutonium (V) are isostructural. The square antiprism and the dodecahedron are shown in Fig. 7.4(a) and (b) respectively; as with the structures for seven-coordination, there appears to be little to choose between them on energetic grounds, and its seems likely that the $[Mo(CN)_8]^{4-}$ ion, for example, can have either structure.

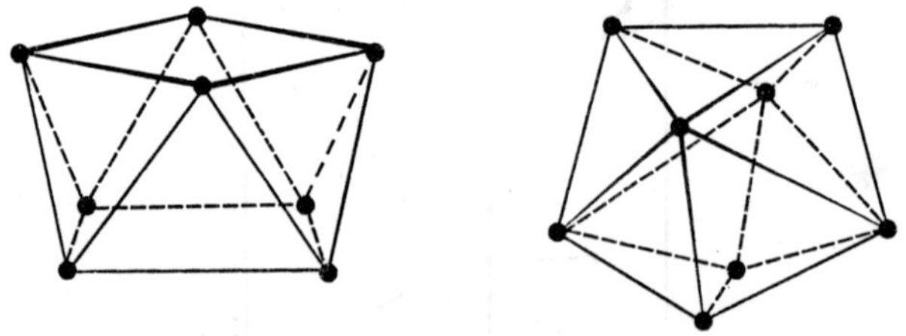

Fig. 7.4 : The square antiprism (a) and dodecahedron (b)

The square antiprism occurs in several d^0 complexes, among them $[TaF_8]^{3-}$, $[Zr(acac)_4]$ and $[Y(acac)_3.2H_2O]$; it has also been reported for $[ReF_8]^{2-}$ (d^1) and for $[Mo(CN)_8]^{3-}$ (d^1) in Na_3 $[Mo(CN)_8].4H_2O$. Evidence concerning the structure of the $[Mo(CN)_8]^{4-}$ ion (d^2) in solution is conflicting: the Raman spectrum has been interpreted on the basis of both the antiprismatic and the dodecahedral structures, though the e.s.r. spectrum appears to suport the latter; it has, however, recently been reported that the anion is a square antiprism in $H_4[Mo(CN)_8].6H_2O$ and in the analogous tungsten compound.

So far the square antiprism has not been found among the structures of compounds of the metals of the first transition series. The dodecahedron, however, occurs in a group of nitrates in which the ligand is bidentate: $Ti(NO_3)_4$ (d^0), $[Mn(NO_3)_4]^{2-}$ (d^5), $[Co(NO_3)_4]^{2-}$ (d^7) and $[Zn(NO_3)_4]^{2-}$ (d^{10}). It occurs also in $[Ti\ (diarsine)_2\ Cl_4]$ (diarsine = *o*-phenylenebis-dimethylarsine), $[Zr(C_2O_4)_4]^{4-}$, $[CrO_8]^{3-}$ (in which four peroxide ions act as bidentate ligands), $[Mo(CN)_8]^{3-}$ in the tetrabutylammonium salt and $[Mo(CN)_8]^{4-}$ in $K_4[Mo(CN)_8].2H_2O$.

Coordination number nine

The only structure found in complexes not containing polydentate chelating groups (which may largely determine the configuration) is the tricapped trigonal prism shown in Fig. 7.5. This occurs in $[Nd(H_2O)_9]^{3+}$ (in which the Nd—O distances are all nearly equal), in several other hydrated rare earth cations (also, incidentally, in $SrCl_2.6H_2O$, in which each strontium ion has three water molecules bonded to it alone and shares six others with adjacent metal ions in a cationic chain) and in $[ReH_9]^{2-}$ (d^0).

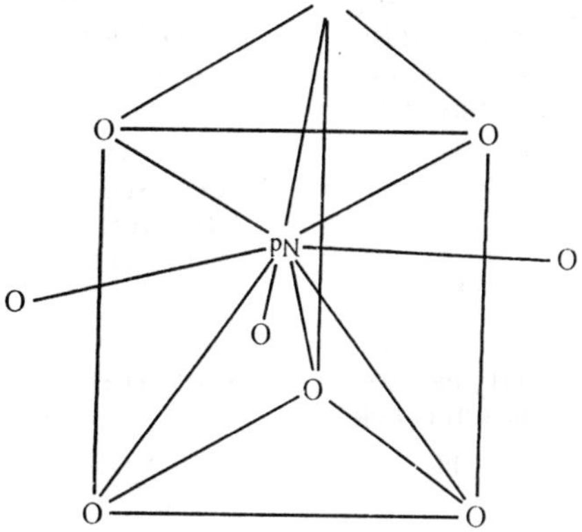

Fig. 7.5 :

Higher coordination numbers

The only higher coordination numbers yet found in complexes are ten, eleven and twelve, and up to the present time coordination numbers above nine in molecules or ions have been identified only for the lanthanides and actinides. Examples of ten-coordination occur in $K_4Th(C_2O_4)_4.4H_2O$ (in which the arrangement of the ligands round the thorium is that of a bicapped square antiprism) and $La(NO_3)_2$. 2dipy, in which the ten-coordination is provided by three bidentate nitrate ions and the two bidentate dipyridyl molecules; in this instance the structure is described as a bicapped dodecahedron

Ligands

One of the best methods for classifying complex compounds is on the basis of the nature of the ligands involved. Generally speaking, the ligands supply the electron density for the covalent bond formed to the transition-metal atom. Thus, it is not surprising that the ligand atoms directly attached to the transition-metal atom are those found at the extreme right of the representative elements, as shown in Table 7.1, where the number of electrons available exceeds the number covalent bonds the ligand atom can form.

Table 7.1: Some Typical ligands

	C	N	O	P
Neutral species	CO, C_2H_4 C_6H_6, CNR	N_2, NH_3, NR_3 $NH_2CH_2CH_2NH_2$, NO, N_2H_4, RCN bipyridyl	R_2O, ROH, R_3PO, R_3AsO, R_2SO, $CH_3C(O)CH_2C(O)CH_3$	–
Ions	CN^-	NO_2^-, N_3^-, NCS^-, NH2	OH^-, O_2^-, O_2^{2-}, CO_3^{2-}, SO_4^{2-} $S_2O_3^{2-}$, RCO_2^-, VO_2^-, $C_2O_4^{2-}$	F^-
		P	S	Cl
Neutral species		PH_3, PR_3, PX_3, $R_2PCH_2CH_2PR_2$	R_2S, $RSCH_2CH_2SR$	–
Ions		PR_2^-	SCN^-, $S_2O_3^{2-}$	Cl^-
	As	Se	Br	
Neutral species		R_2As, $AsCl_3$, $R_2AsCH_2CH_2AsR_2$	R_2Sc	–
Ions		AsR^{2-}	$SeSCN^-$	Br^-
		Sb	Te	I
Neutral species		R_3Sb	T_2Te	–
Ions		SbR_2	–	I^-

The ligand atoms can be associated with neutral molecules such as the trivalent Group V derivatives, MR_3 (*IV*), where the substituent can be organic groups, hydrogen, or halogens (*V*), as well as similar derivatives of the Group VI elements, R_2M. In each of these cases the ligand atom has at least one pair of unshared electrons (1 and 2). Other neutral ligands include oxidized derivatives of the Group V elements (R_3MO), binary oxides (CO and NO), and organic derivatives containing the CN moiety (RCN and RNC). Even organic compounds such as C_2H_4 and C_6H_6 containing unsaturated systems can act as ligands towards transition-metal atoms. The chemistry and properties of the organometallic derivatives

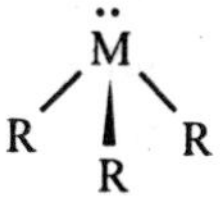

1

2

of the transition metals are properly considered as a distinct part of transition-metal coordination chemistry.

It should not be surprising that virtually all of the anions can act as ligands since these species carry a negative charge, a fact which suggests that an unshared pair of electrons is available. However, a few anions such as the tetraalkyl borates (BR_4^-) must be excluded from this generalization. Inspection of Table 17.5 indicates that many anions are known to act as ligands. Theoretically some anions have two or more possible sites for coordination. For example, the Lewis formulation for CN^- (**3**) (VI), SCN^- (**4**) and NO_2^- (**5**) show two possible coordination sites. In each instance complex compounds

3

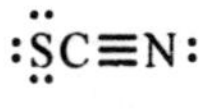

4

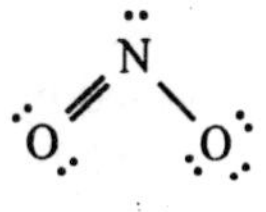

5

in which the metal is coordinated to one or the other site have been isolated. The point is considered in more detail in the next section.

The coordination behaviour of the lower oxidation states of the transition metals, some of the transition metals can exhibit valence states as high as 8+. In some ways oxyanions, in which the transition metals exhibit their highest oxidation states, can be considered as being composed of a highly charged cation surrounded by oxide, O^2, ligands. In this view of theoxidation states M^{7+}, M^{6+}, and M^{5+} give rise to the tetrahedral complex oxyanions MO_4^- (MnO_4^-, ReO_4^-), MO_4^{2-} (CrO_4^{2-}, ReO_4^{2-}, MoO_4^{2-}, WO_4^{2-}), and MO_4^{3-} (VO_4^{3-}). Of course, it is not correct to describe these species as collections of the corresponding ions because the high polarizing power of atoms in his oxidation states leads to significant covalent bonding, but this point of view is often useful in certain discussions.

Polydentate Ligands

Many species have only one point of attachment when they form complex ions with metal atoms; such ligands are called monodentate ligands.* Some ligands could have several lone electron pairs and thus form several co-ordinate-covalent bonds. Such ligands are called polydentate ligands. If, because of geometrical considerations, the lone electron pairs are sufficiently well separated to form bonds to the same metal atom, the ligand is called a chelate. For example, triethylenediamine (**6**) is bidentate but the two amine groups are constrained to be opposite each other and hence can form coordinate bonds to two different metals. Ethylenediamine (**7**), however, can

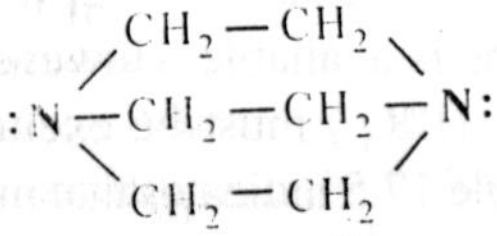

6

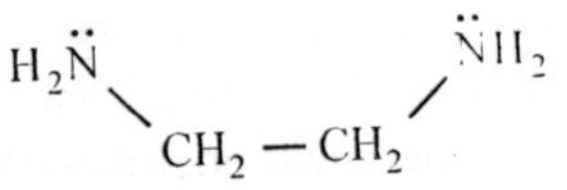

7

also form two metal-nitrogen coordinate bonds, but because of the geometry of the molecule, the bonds are formed to the same atom, giving a five-membered ring. As in the chemistry of carbon compounds, five- and six-membered rings are unstrained and are easily formed by chelating ligands. Although bidentate chelates are the most widely investigated ligands, species with three, four, five, and six coordination sites are also known; examples of such ligands appears in Table 7.2.

Polydentate chelates usually incorporate a two- or three-carbon chain between the ligand atoms, which provides sufficient separation of the latter for formation of five- or six-membered heterocyclic systems. In some instances, such as with carboxylate ions (*VII*), four-membered rings can be formed. In addition, *N, N*-diethylthiocarbamate ion (**8**), CO_3^{2-} (**9**), and NO_3^- (**10**) (*VIII*) are

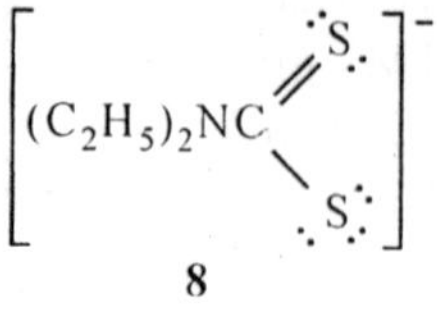

8

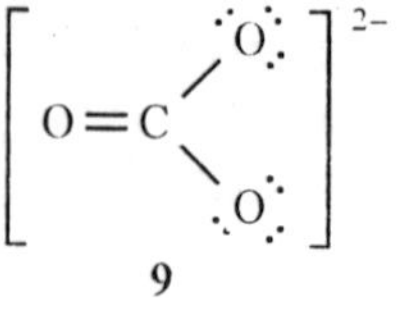

9

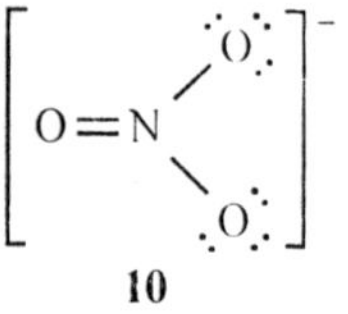

10

known to act as bidentate ligands. As indicated in Table 7.2, chelate ligands may be neutral species or charged species.

Table 7.2 : Some typical ligands

Name	Structure
Blidentate ligands	
ethylenediamine (en)	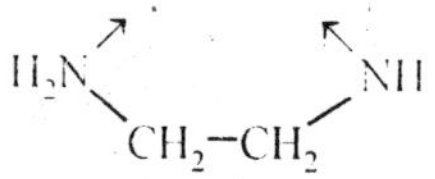
bipyridyl (bipy)	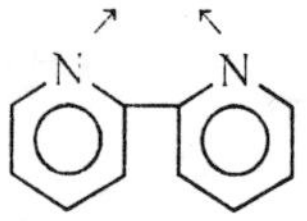
o-phenylenebisdimethylarsine (diars)	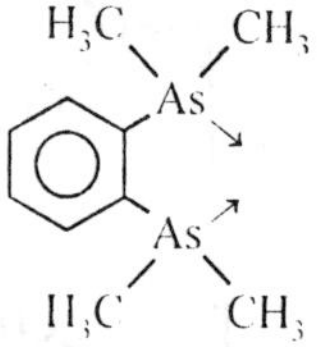
8-quinolinolate ion	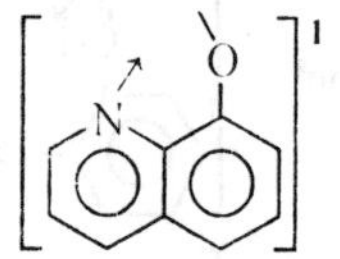
N, N-diethylthiocarbamate ion	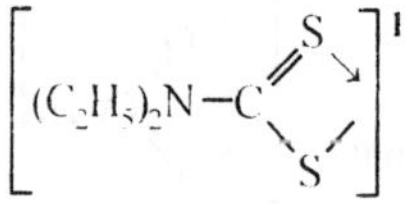
acetylacetonate ion (acac)	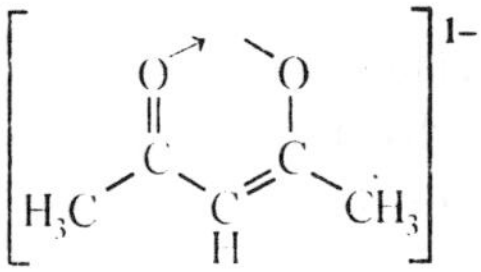

Tridentate ligangs

diethylenetriamine (dien)

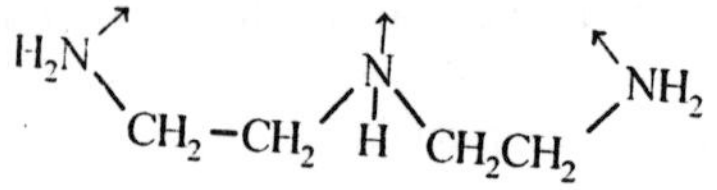

iminodicetate ion

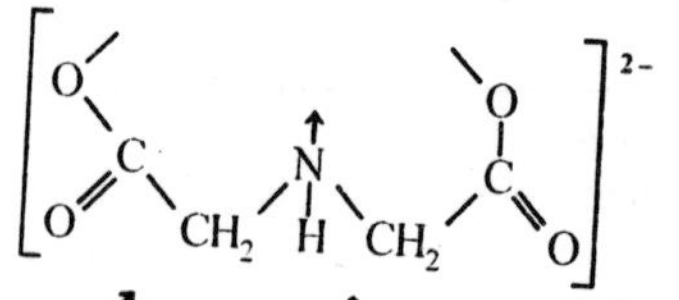

Tetradentate ligands

triethylenetetramine (trien)

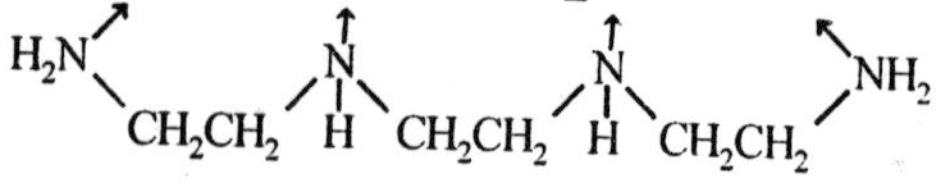

Pentadentate ligands

ethylenediaminetriacetate ion

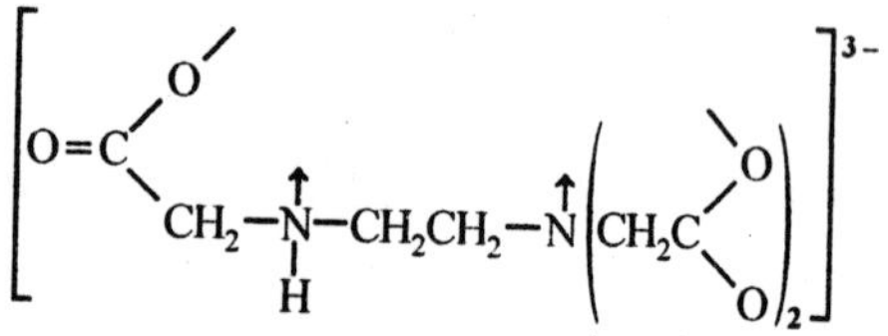

Hexadentate ligands

1,8-bis (salicylideneamino)-3,6-dithiooctate

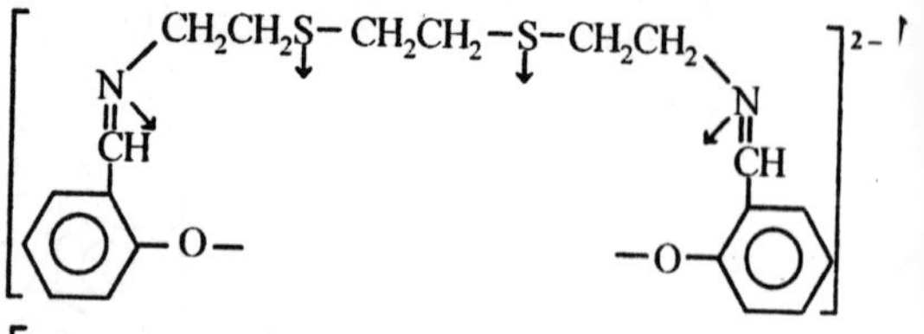

ethylenediaminetetraacetate ion (EDTA)

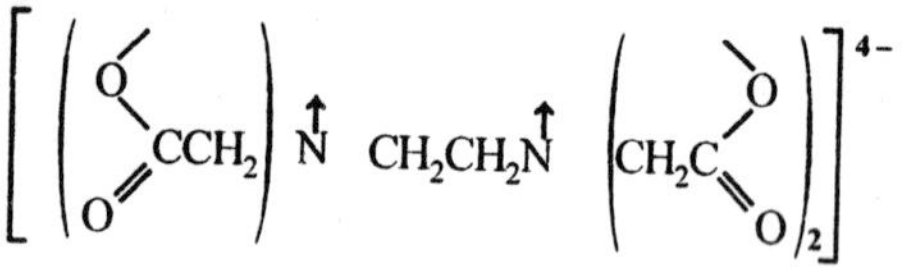

Charged chelates may satisfy one primary and one secondary valence [e.g., acetylacetonate (**11**) (*LX*) and 8-quinolinolate (**12**)], or two secondary and two primary valences (e.g., **8**, **9** and **13**).

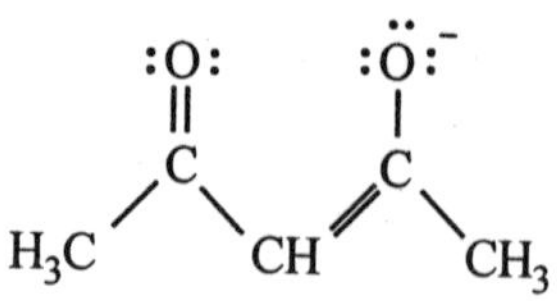

11

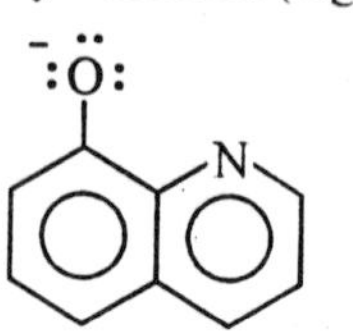

12

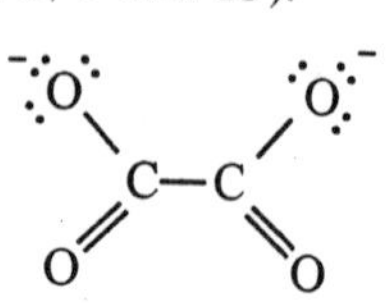

13

Bidentate Ligands. These are the most common of the polydentate ligands. Neutral ones include diamines, diphosphines, diethers, and β-ketoenolate anions, of which the commonest are

$H_2\ddot{N}CH_2CH_2\ddot{N}H_2$ ethylenediamine.en.

$(C_6H_5)_2\ddot{P}CH_2CH_2\ddot{P}(C_6H_5)_2$ diphos

$CH_3\ddot{O}CH_2CH_2\ddot{O}CH_3$ glyme

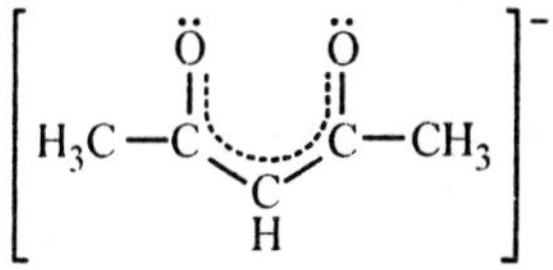

acetylacetonate, acac

all of which form five-membered rings with the metal atom, as well as a number of anions that from four-membered rings, such as

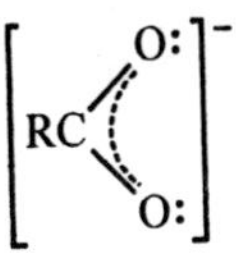

carboxylates

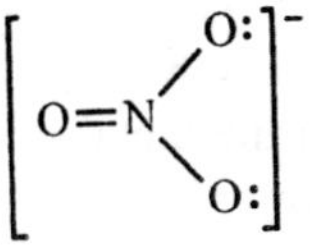

nitrate

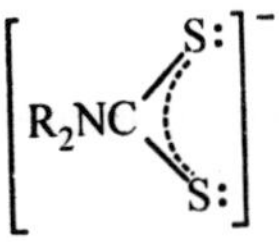

dithiocarbamates

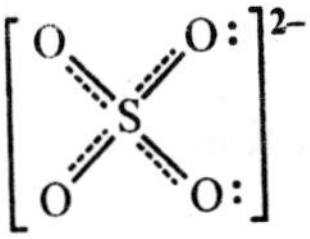

sulfate

and a number of others.

Polydentate Ligands. These include tri-, quadri-, penta-, and hexadentate ligands. Examples of tridentate ligands are

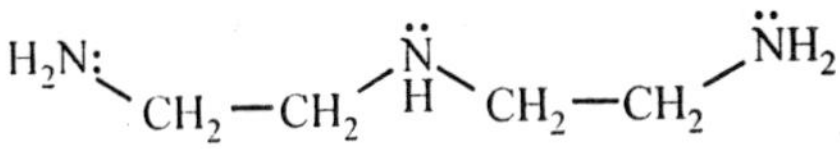

diethylene triamine, dies

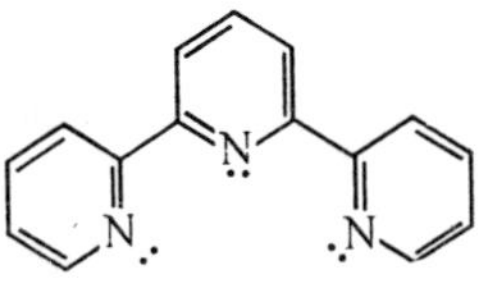

terpyridyl, terpy

Nomenclature. A systematic set of rules for formulating and naming complex ions has been adopted by IUPAC. At this time we will list and give examples of only the most basic of these rules. (Additional details can be found in App. C.)

Formulas of Complexes

1. The formula of a complex shows the central atom first, followed by the ligands. Examples: $Cu(NH_3)_4^{2+}$; $CrCl_6^{3-}$.
2. When the ligands in a complex are not all alike, anion ligands are written before neutral ligands. Examples: $CrCl_2(NH_3)_4^{+}$; $Fe(OH)(H_2O)_5^{2+}$.

Names of Complexes

1. Complex ions are named by specifying first the ligands and then the central atom. (Note that this is the reverse of the order in a formula.)
2. The oxidation state of the central metal atom is indicated with a Roman numeral in parentheses at the end of the name of the metal.
3. Anionic ligands have names ending in *-o*.

   (a) When serving as ligands, anions whose names end in *-ide* have this suffix replaced with *-o*.

Anion	Anion name	Ligand name
Cl^-	Chlor*ide*	Chlor*o*
Br^-	Brom*ide*	Brom*o*
OH^-	Hydrox*ide*	Hydrox*o*
CN^-	Cyan*ide*	Cyan*o*

   (b) When serving as ligands, anions whose names end in *-ate* have this suffix replaced with *-ato*.

Anion	Anion name	Ligand name
SO_4^{2-}	Sul*fate*	Sulf*ato*
$S_2O_3^{2-}$	Thiosulf*ate*	Thiosulf*ato*
$C_2O_4^{2-}$	Oxal*ate*	Oxal*ato*
SCN^-	Thiocyan*ate*	Thiocyan*ato*

4. *Neutral ligands are named as the neutral molecule.* There are some exceptions to this rule, the most important of which are H_2O, named as the *aquo* ligand; NH_3, named as the *ammine* ligand; and CO, named as the *carbonyl* ligand.

5. The numbers of ligands in a complex are specified by using the Greek prefixes *di-*, *tri-*, *tetra*, etc. The prefixes, *bis-* (twice), *tris-* (thrice), and *tetrakis-* (four times) are also used to eliminate ambiguity when the name of a ligand itself contains a Greek prefix. Example: bis (ethylenediamine).
6. *The name of an anionic complex ends in -ate.* This is added to the stem of the English, or sometimes the Latin, name of the metal. Thus in an anionic

Table 7.3 : Name of some complex

Formula	Name
$Co(H_2O)_6^{2-}$	Hexaaquocobalt(II) ioin
$CoCl_4^{2-}$	Tetrachlorocobaltate(II) ion
$Ni(CN)_4^{2-}$	Tetracyanonickelate(II) ion
$Cr(OH)_2(H_2O)_4^+$	Dihydroxotetraaquochromium(III) ion
$Cr(OH)_3(HO)_3$	Trihydroxotriaquochromium(III)
$Cr(OH)_4(H_2O)_2^-$	Tetrahydroxodiaquochromate(III) ion
$CoCl_4(NH_3)_2^-$	Tetrachloodiamminecobaltate(III) ion

complex *chromium* becomes *chromate*, *aluminum* becomes aluminate, and *iron* becomes *ferrate*. (Cationic and neutral complexes are named with the English name of the metal and no suffix.)

Table 7.3 illustrates the use of these rules.

Finally, it should be noted that as usual an ionic *compound* is formulated and written with the cation followed by the anion.

Isomerism

The variety of coordination geometries and the extensive number of ligands with different modes of attachment lead to an interesting display of isomerism among the coordination compounds. Isomerism can be discussed under two general categories: stereoisomers which arise because of the arrangement of ligands about the transition-metal atom, and isomes which occur because ligands can be attached to the metal in different ways.

Stereoisomerism

Complex species which exhibit cooridination number 2 and 3 cannot exhibit stereoisomerism. Four coordinate species which have a tetrahedral arrangement of ligands yield stereoisomers only when all four ligands are different, which, of course, leads to an enantiomorphic pair. Optical isomers are discussed in a subsequent section.

Four-coordinate planar complex ions exhibit geometrical isomerism if they possess equal numbers of two kinds of ligands, ie., Ma_2b_2. The ligands of the same kind can be opposite each other (**33**) or adjacent to each other (**34**), such arrangements being designated *trans* and *cis*, respectively, after the spatial analogy found in the disubstituted

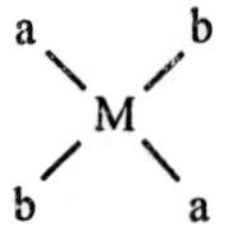

trans

33

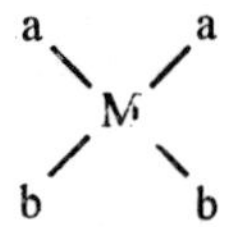

cis

34

ethylenes (CHX=CXH). The stereochemistry of neutral planar complexes such as $PtCl_2.2NH_3$ can be elucidated by dipole-moment measuremens, the *cis* form being nonzero within experimental error. As a subsidiary point, the fact that the compound $PtCl_2.2NH_3$ gives one isomer with a zero dipole moment is evidence that this isomer, at least, cannot be tetrahedral. The correctness of the *cis-trans* assignment in these platinum compounds is verified by several simple experiments. For example, cis-$PtCl_2(NH_3)_2$ reacts readily with the bidentate bivalent oxalate ion (**13**) to give $Pt(NH_3)_2C_2O_4$, which is logically the *cis* isomer since it is generally assumed that bidentate ligands can only bridge adjacent coordination positions. In addition, complex ions containing in three isomeric forms (**35-37**). If the four-coordinate Pt^{2+} complexes were tetrahedral, only two isomers, which would be enantiomorphs, should be possible. *Cis* and *trans* isomers are also possible in planar binuclear complex species.

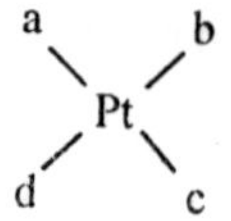

35

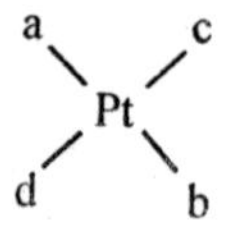

36

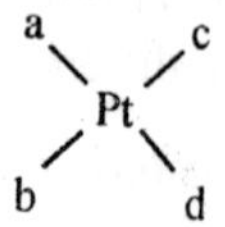

37

For example, platinum forms a complex species with the empirical formula $PtCl_2.PR_3$ which is dimeric in benzene solution. The latter fact leads to a chlorine-bridged formulation which has three possible arrangements of the nonbridged ligands(**38, 39** and **40**). The first two

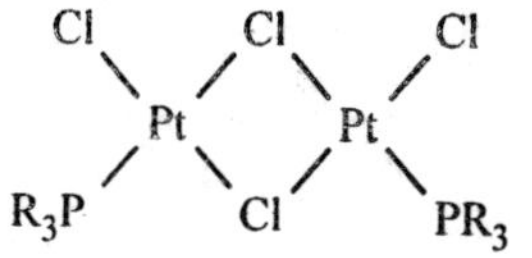

cis

38

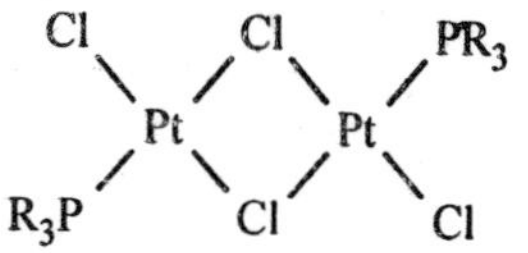

trans

38

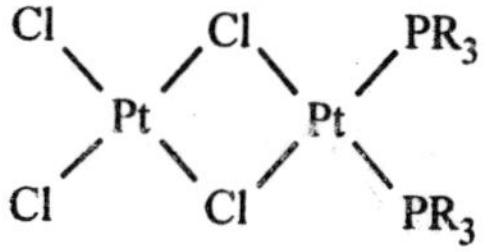

gem

isomers have a *cis-trans* relationship of the ligands, whereas the last is analogous to the structure of 1,1-dichloroethylene, sometimes called *gem-dichloroethylene*.

Bidentate ligands carrying susbtituents on one or more of the heterocyclic ring atoms can form geometrical isomers. For example, two molecules of a methyl-substituted ethylenediamine can be complexed in a planar arrangement in two different ways which correspond to a *cis* orientation (**41**) of the methly groups with respect

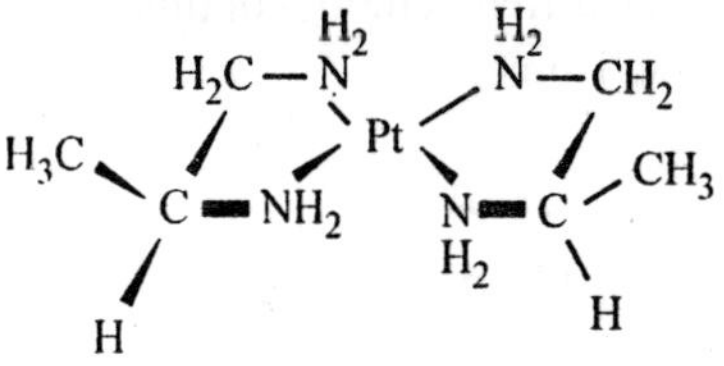

41

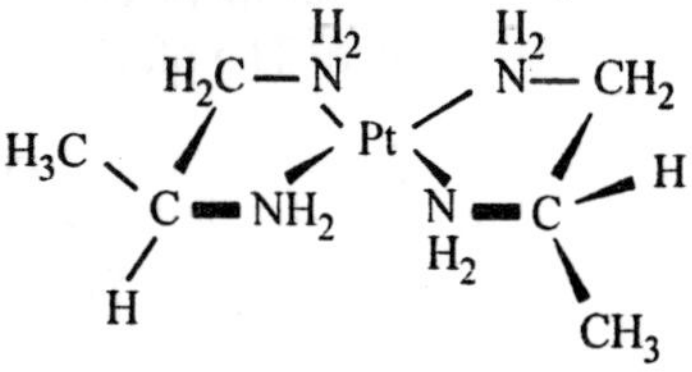

42

to the median plane of the ring atoms; the other possible orientation is *trans* (**42**). There are, in fact, two *cis* and *two trans* isomers with respect to this orientation. It is apparent that very complex arrangements of geometrical isomers can arise if substitution on the ligands is considered.

Ionization Isomerism. Compounds that have the same molecular formula but differ in which anions are coordinated yield different ions when in solution. Examples are

$[Co(NH_3)_4Cl_2]NO_2 \qquad [Co(NH_3)_4Cl_2]^+ + NO_2^-$

$[Co(NH_3)_4Cl(NO_2)]Cl \qquad [Co(NH_3)_4Cl(NO_2)]^+ + Cl^-$

and

$[Coen_2(NO_2)Cl]SCN, \qquad [Coen_2(NO)_2(SCN)]Cl,$

$[Coen_2(SCN)Cl]NO_2 \; en = H_2NCH_2CH_2NH_2$

in these illustrations the square brackets are used to enclose the metal atom and all the ligands that are directly bound to it, that is, are in the coordination shell. This use of square brackets is a way of making this distinction in formulas when necessary and will be found in the research literature. They can, however, be omitted when no confusion would arise, as in $Co(NH_3)_3Cl_3$.

The concept of ionization isomerism provides the key to understanding many simple but otherwise puzzling observations. For example, there are three different substances of the composition $CrCl_3.6H_2O$. One is violet and is $[Cr(H_2O)_6]Cl_3$; it does not lose water over H_2SO_4 and all Cl^- is immediately precipitated by Ag^+ from a fresh solution. $[Cr(H_2O)_5Cl]Cl_2.H_2O$ is green; it loses one H_2O over H_2SO_4 and only two thirds of its Cl content is precipitated promptly. $[Cr(H_2O)_4Cl_2]Cl.2H_2O$ which is also green loses two H_2O over H_2SO_4 and only one third of its Cl content is promptly precipitated.

Linkage Isomerism. Some ligands can bind in more than one way, and often isomeric complexes with the different modes of binding can be isolated. The oldest example is the isomeric pair:

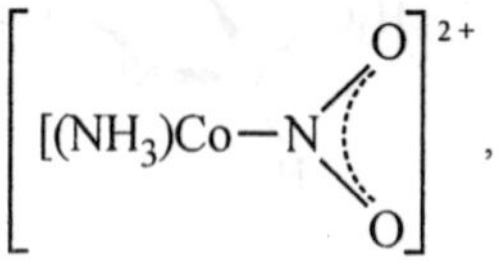

, $\qquad [(NH_3)_5Co{-}O{-}N{=}O]^{2+}$

nitro — nitrito

Other ligands prone to give linkage isomers, or at least to the bind in different ways in different compounds, are SCN^-, which may use either S or N as the donor atom and sulfoxides, $R_2S{=}O$, which may use either S or O as the donor.

Coordination Isomerism

Coordination isomerism occurs when both the anion and cation are complex species with a different distribution of ligands. The same metal ion in the same oxidation state may be present in both ions

$[Cr(NH_3)_6]$ $[Cr(SCN)_6]$ and
$[Cr(NH_3)_4(SCN)_2]$ $[Cr(NH_3)_2(SCN)_4]$
$[Pt(NH_3)_6]$ $[PtCl_4]$ and $[Pt(NH_3)_3Cl]$ $[Pt(NH_3)Cl_3]$,

or the metal may be present in two different oxidation states

$[Pt(NH_3)_4]$ $[PtCl_6]$ and $[Pt(NH_3)_4Cl_2]$ $[PtCl_4]$.

Coordination isomers also may involve two different metals

$[Co(en)_3]$ $[Cr(CN)_6]$ and $[Cr(en)_3]$ $[Co(CN)_6]$
$[Cu(NH_3)_4]$ $[PtCl_4]$ and $[Pt(HH_3)_4]$ $[CuCl_4]$.

In fact, it is possible in some instances to isolated isomers in which the distribution of ligands in the cation and anion vary in a regular manner, as in the series $[Co(en)_3]$ $[Cr(C_2O_4)_3]$, $[Co(en)_2C_2O_4]$ $[Cr(en)(C_2O_4)_2]$, $[Co(en)_2C_2O_4]$ $[Co(en)(C_2O_4)_2]$ and $[Cr(en)_3]$ $[Co(C_2O_4)_3]$.

A variation of coodination isomerism occurs in polynuclear complexes where the ligands can be arranged in differnt ways with resepct of the metal ions. For example, there are two possible

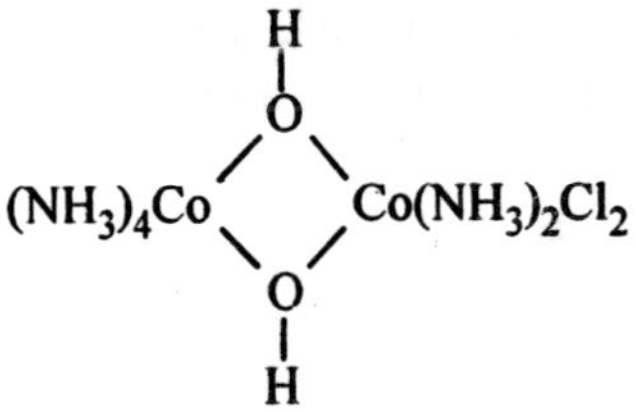

53

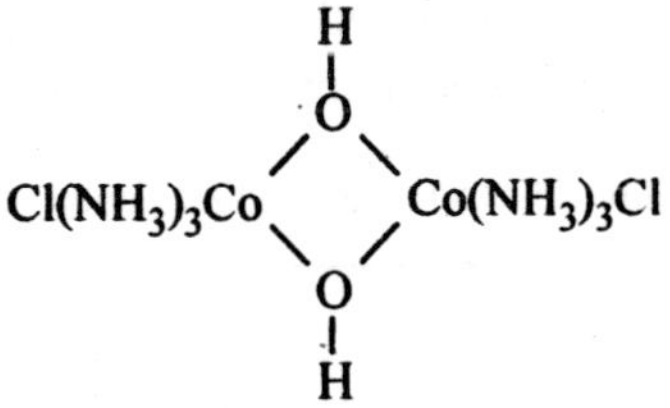

54

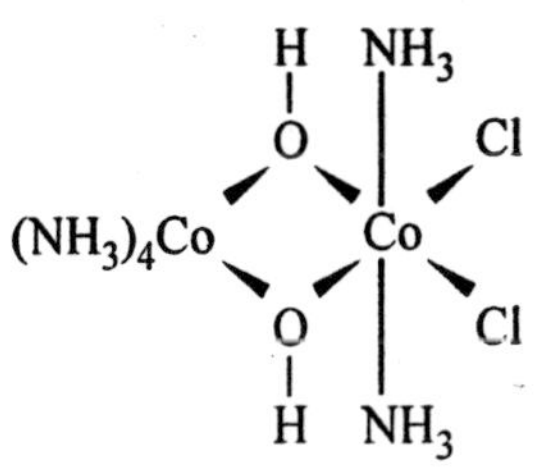

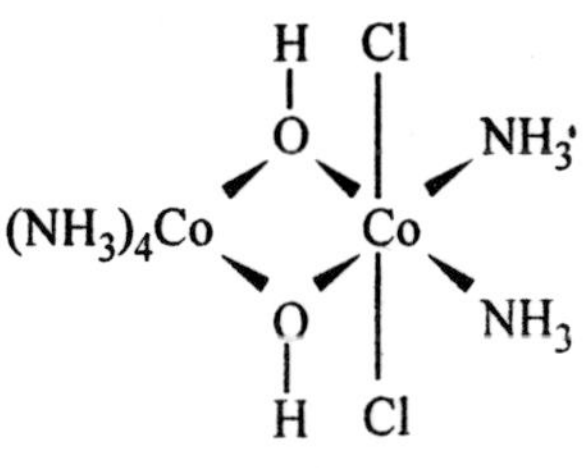

distributions (53 and 54) of chlorine atoms in the binuclear species with the empirical formulation $[Co_2(NH_3)_6^-(OH)_2Cl_2]^{2+}$. Of course three stereoisomers (**55-57**) are possible for this formulation; **57** is optically active.

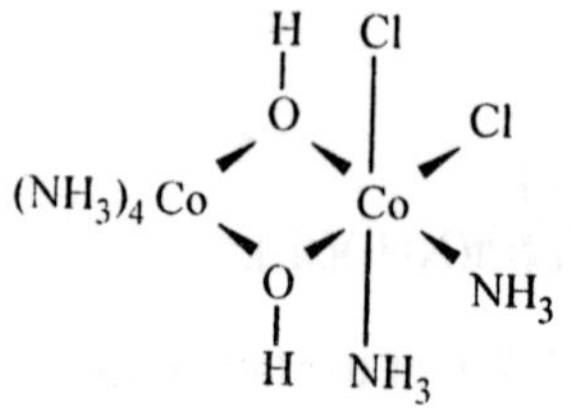

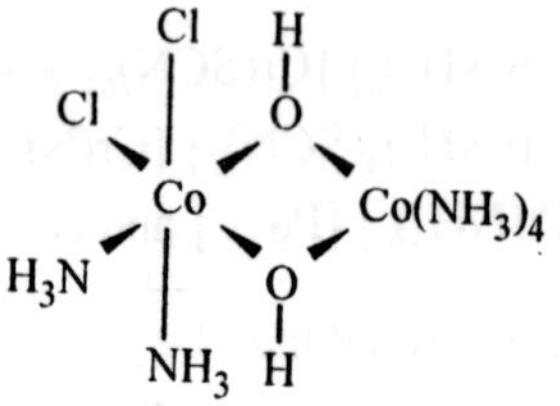

Polymerization Isomers

Polymerization isomers have the same empirical formulas but differ in their molecular weights. The name represents a formalism because there is no polymerization of smaller units into larger units in the usual sense of the term. The differences in the isomers still represent differences in arrangements of ligands and not in numbers of monomers. A relatively simple example of polymerization isomerism occurs in the chemistry of platinum where three different compounds exist with the empirical formula $Pt(NH_3)_2$-Cl_2; their molecular weights correspond to $Pt(NH_3)_2Cl_2$, $[Pt(NH_3)_4]$ $[Pt(NH_3)Cl_3]_3$. Experimental results show that these compounds should be formulated as $Pt(NH_3)_2Cl_2$, $[Pt(NH_3)_4]$ $[PtCl_4]$, and $[Pt(NH_3)_4]$ $[Pt(NH_3)Cl_3]_2$, respectively. In fact, the coordination isomer of the last-named compound, $[Pt(NH_3)_3Cl]_2[PtCl_4]$, also has been identified. All these platinum-containing species are planar complexes, which also leads

Table 7.5 : Octahedral coordination polymers

Molecular formula	Formulation
$Co(NH_3)_3(NO_2)_3$	$[Co(NH_3)_3(NO_2)_3]$
$Co_2(NH_3)_6(NO_2)_6$	$[Co(NH_3)_6]$ $[Co(NO_2)_6]$
	$[Co(NH_3)_4(NsO_2)_2]$ $[Co(NH_3)_2(NO_2)_4]$
$Co_3(NH_3)_9(NO_2)_9$	$[Co(NH_3)_5NO_2]$ $[Co(NH_3)_2(NO_2)_4]_2$
$Co_4(NH_3)_{12}(NO_2)_{12}$	$[Co(NH_3)_6]$ $[Co(NH_3)_2(NO_2)_4]_3$
	$[Co(NH_3)_4(NO_2)_2]_3$ $[Co(NO_2)_6]$
$Co_5(NH_3)_{15}(NO_2)_{15}$	$[Co(NH_3)_5NO_2]_3$ $[Co(NO_2)_6]_2$

to the possibility of *cis-trans* isomers for the first member of the family. Table 7.5 lists some coordination polymers isolated for octahedral cobalt complexes.

Inspection of the results listed in Table 7.5 indicates that all of the possible isomeric species have not been characterized for a given molecular formula.

Theory of Bonding

Valence Bond Theory

The first successful application of bonding theory to coordination compounds was made by Linus Pauling. It is usually referred to as the *valence bond theory* of coordination compounds.[9] It is closely related to the hybridization and geometry of noncomplex compounds as discussed in Chapter 5. From the valence bond point of view the formation of a complex is a reaction between a Lewis base (ligand) and a Lewis acid (metal or metal ion) with the formation of a coordinate covalent (or dative) bond between the ligand and the metal. For example, the ions Ni^{+2}, Pd^{+2}, and Pt^{+2} have d^8 configurations. Complexes of Pd^{II} and Pt^{II} are customarily 4-coordinate, square planar, and diamagnetic, and this arrangement is often found in Ni^{II} complexes as well. Inasmuch as the ground state of these ions is paramagnetic (^3F), the ligands in these diamagnetic complexes must cause pairing of the two odd electrons. Pauling suggested that this occurs via the use of one *d* orbital by the ligands. It is customary to represent the metal orbitals by boxes or circles for the purpose of accounting for the distribution of the metal electrons and those donated by the ligand :

$Pt = [Xe]4f^{14}$ $5d^8$ $6s^2$ $6p^0$ Ground state atom, ^3F

$Pt^{+2} = [Xe]4f^{14}$ $5d^8$ $6s^0$ $6p^0$ Ground state ion, ^3F

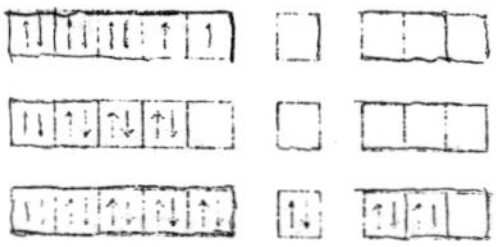

$Pt^{+2} = [Xe]4f^{14}$ Excited state ion, ^1D

$PtCl_4^{-2} = [Xe]4f$ dsp^2 hybrid = square planar geometry

Electrons donated by $4Cl^-$ ions.

In those cases (limited to Ni^{+2}) in which paramagnetic 4-coordinate complexes are known, all five *d* orbitals of the metal must be occupied by the d^8 electrons, eliminating the possibility of involvement of a low-lying *d* orbital (3*d* in Ni) in the ligand-metal bonds. Pauling suggested that these complexes would involve sp^3 hybrids:

Ni = [Ar]	$3d^8$	$4s^2$	$4p^0$	Ground state ion, ^3F
Ni^{+2} = [Ar]	$3d^8$	$4s^0$	$4p^0$	Ground state ion, ^3F

$NiCl_4^{-2}$ = [Ar] sp^3 hybrid = tetrahedral geometry

Electrons donated by $4Cl^-$ ions

According to this interpretation, it should be possible to predict the geometry of a d^8 complex if the magnetic susceptibility is known, namely : diamagnetic = square planar; paramagnetic = tetrahedral. This rule, the "magnetic criterion of bond type," proved exceedingly useful for a number of years, but we now know that it can be misleading, which indicates one of the weaknesses of the simplified VB interpretation of bonding in complexes.

In a similar manner two possible were postulated for 6-coordinate complexes of d^6 ions, such as Fe^{+2}, Co^{+3}. Consider Co^{III} complexes, for examples as well as all others synthesized prior to the formulation of Pauling's VB treatment of coordination were

Co = [Ar]	$3d^7$	$4s^2$	$4p^0$	Ground state atom, ^4F
Co^{+3} = [Ar]	$3d^6$	$4s^0$	$4p^0$	Ground state ion. ^5D

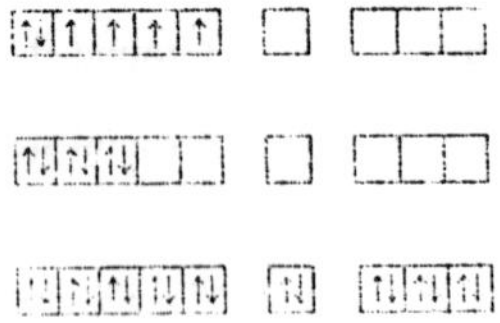

Co^{+3} = [Ar] Excited state ion, ^1I

$Co(NH_3)_4^{+3}$ = [Ar] d^2sp^3 hybrid = octahedral geomery

Electrons donated by $6NH_3$ molecules

The resulting VB description adequately accounts for the properties of the complex, i.e., no unpaired electrons and a perfectly octahedral structure. The discovery that a paramagnetic complex of Co^{III} could be formed—CoF_6^{-3}, having four unpaired electrons—required and adjustment of the theory. Pauling suggested that the flouride ion could bond through the "outer" $4d$ orbitals. Such a hybrid, $3sp^3 4d^2$, would have identical symmetry with the $3d^2sp^3$ hybrids invoked above. The hexafluorocobaltate(III) ion can thus be formulated as :

$$CoF_6^{-3} = [Ar]$$

$3d^6$	$4s$	$4p$	$4d$	
↑↓ ↑ ↑ ↑ ↑	↑↓	↑↓ ↑↓ ↑↓	↑↓ ↑↓ ☐ ☐ ☐	sp^3d^2 hybrid = octahedral geometry

Electrons donated by $6F^-$ ions

Strengths and weakness of the valence bond approach

Until about 30 years ago, the valence bond theory was almost the only applied to coordination compounds by chemists. In the following decade a revolution occurred, and today few inorganic chemists use simple valence bond theory in discussions of these complexes. It is not because valence bond theory is wrong—it isn't. It has simply proved to be more convenient to explain the properties of these molecules is different terms. The valence bond theory of complex compounds in still alive. Pauling is currently interpreting some unusual properties of coordination compounds in valence bond terms. The student should therefore be aware of the technique and the terminology, although the present-day inorganic chemist needs a much broader theoretical viewpoint in order to interpret adequately the experimental data presented by coordination chemistry.

Crystal field theory

In crystal field theory, a complex is regarded as composed of a metal ion surrounded by anions or negative ends of polar ligands, and the bonding is assumed to be purely electrostatic. We shall confine attention to octahederal, tetrahedral, tetragonal and square complexes; the treatment of more complex systems is difficult, and few useful results have yet been obtained.

Octahderal complexes. Suppose six ligands are brought towards the positively charged central metal ion along the x-, y- and z-axis. There will be a net electrostatic attraction, but owing to repulsion between the ligand electrons in the d-orbitals of the metal there will be a general increase in the energy of these orbitals. However, not all d electrons will be affected equally; those in the $d_{x^2-y^2}$ and d_{z^2} orbitals (which are directed along the *x*-, *y*- and *z*-axes) will be repelled more than those in the d_{xy}, d_{yz} and d_{xz} orbitals, which are directed between these axes. The former group of orbitals are, for reasons which will be indicated later, designated e_g orbitals, the latter group t_{2g} orbitals (the respective names d_γ and d_δ are also used sometimes).

It is obvious that the three t_{2g} orbitals are degenerate. This is, however, also true of the two e_g orbitals: the d_{z^2} orbitals can be envisaged as a linear combination of $d_{z^2-x^2}$ and $d_{z^2-y^2}$ orbitals which, by symmetry, must behave as the $d_{x^2-y^2}$ orbital. The relative amounts by which the energies of the t_{2g} and e_g orbitals are affected can be seen by considering the following argument. Suppose a d^{10} cation is placed at the centre of a hollow sphere of radius equal to the metal-ligand distance, with a total charge equal to that of the six ligands spread uniformly over its surface. In this symmetrical environment the five *d* orbitals are still degenerate. Then imagine the total charge on the sphere being collected into six point-charges lying on the surface of the sphere but also at the apexes of the octahedron associated with the *x*-, *y*- and *z*-axes. Such redistribution of charge cannot affect the total energy of the system when all the *d* orbitals are filled symmetrically, and therefore the combined energy of the e_g orbitals must be raised by the same amount as that by which the combined energy of the t_{2g} orbitals is lowered. The difference in energy between the t_{2g} and e_g orbitals is known as Δ_o (the subscript denotes the octahedral environment) or sometimes (for historical reasons) as 10 *Dq*. Thus the three t_{2g} orbitals are each 0.4 Δ_o or 4 *Dq* lower, and the two e_g orbitals each 0.6 Δ_o or 6 *Dq* higher, than the centre of gravity of the set of levels. This pattern of splitting is general for any splitting of energy levels so long as the forces are purely electrostatic and the set of levels being split is well removed from other sets of energy levels with which they might interact.

The relationship between the energy levels of *d* orbitals in the free metal ion, in a spherically symmetrical field and in an octahedral crystal field, is summarized in Fig. 7.6. The differences between the energy levels in the free ion and in a spherically symmetrical field depends upon the nature of the metal ion, the nature of the ligand and the internuclear distance. It is important to understand that although the energies of the *d* orbitals are raised when the ion is put into any negative field, the total energy of the system is lowered by the much larger effect of the electrostatic interaction of the negative charges or dipoles with the positive charge on the cation. The magnitude of Δ_o is usually quite small in comparison with the total energy of interaction, but it is on the splitting of the *d* orbitals that crystal field theory concentrates attention; this point is sometimes forgotten when small variations in Δ_o are being discussed for systems for which the total interaction energies may be quite different.

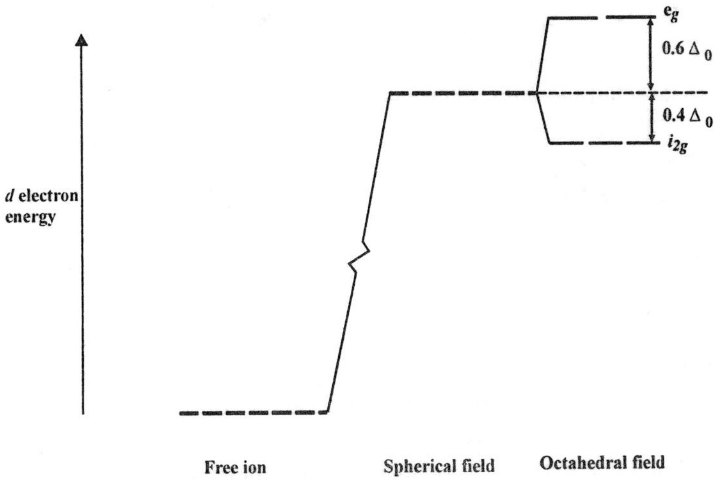

Fig. 7.6 : The effects on *d* orbitals of a spherical and of an octahedral crystal field.

The great significance of Δ_o is that it can often be determined experimentally from the absorption spectrum of the complex. In simple cases the frequency at the maximum of the lowest frequency band in the visible or ultraviolet absorption spectrum corresponds to promotion of an electron from a t_{2g} orbital to an e_g orbital. For many transition metal complexes this band in the visible region is the origin of the colour of the complex; the $[Ti(H_2O)_6]^{3+}$ ion, for example, has a broad band with a maximum at 20,400 cm^{-1} (4900 Å), corresponding to an increase in energy of 244 kJ (58 kcal) $mole^{-1}$ for this promotion. such a transition would be spectroscopically forbidden in the free Ti^{3+} ion (the selection rule for electronic spectra is $\Delta l = \pm 1$, where l the second qauntum number), but occurs in the complex because the vibration of the ligand molecules about the central ion destroys the perfect octahedral symmetry and broadens what would otherwise be a line into a band; the intensity of the absorption, however, remains low (i.e. the colour is pale). The quantity 0.4 Δ_o (8160 cm^{-1}, 98 kJ, or 23 kcal) is called the crystal field stabilization energy (CFSE) or, more generally in order to accommodate the same quantity for non-electrostatic complexes treated by ligand field theory, the legand field stabilization energy (LFSE). For ions containing several electrons, the interpretation of the spectra is often more complicated, for reasons which we shall discuss later, and comparison with values of Δ_o estimated theoretically is often helpful in assigning bands. Nevertheless, a large number of values of Δ_o for different ions and different ligands are now known with a fair degree of certainty.

There is a striking general pattern in the relative values of Δ_o for different ligands. Almost irrespective of the nature of the metal ion, Δ_o increases along the series :

$I^- < Br^- < Cl^- < F^- < OH^- < C_2O_4^{2-} < H_2O <$ pyridine $< NH_3 <$ ethylenediamine $<$ dipyridyl $<$ *o*-phenanthroline $< CN^-$

This is the so-called spectrochemical series. It should be noted that since it is an experimental series it incorporates all effects of the ligands in splitting the *d* orbitals (π bonding, for example), and that its validity in no way depends upon the validity or otherwise of crystal field theory. For a given ligand and a given oxidation state of the central metal ion, it is found that Δ_o varies somewhat for metals in the same transition series (e.g. over the range 8000 cm^{-1} to 14,000 cm^{-1} for $[m(H_2O)_6]^{2+}$ in he first series). For a given ligand, it is higher for the oxidation state of the same metal (e.g. 9400 cm^{-1} for $[Ir(NH_3)_6]^{2+}$ and 13,700 cm^{-1} for $[Fe(H_2O)_6]^{3+}$). Finally, Δ_o is about 30-50 per cent greater for members of the second transition series than for corresponding members of the first transition series, and greater again by the same margin on going from the second to third transition series (e.g. values for $[Co(NH_3)_6]^{3+}$, $[Rh(NH_3)_6]^{3+}$ and $[Ir(NH_3)_6]^{3+}$ are 23,000, 34,000 and 41,000 cm^{-1} respectively).

Let us now consider the interpretation of the magnetic properties of octahedral complexes on the basis of crystal field theory. For the electronic configurations d_1, d_2 and d_3 the t_{2g} orbitals are half-filled progressively and the electrons remain unpaired; for d^4, however, two possibilities exist. If Δ_o is small, the fourth electron goes into an e_g orbital and all electrons are unpaired; if, however, Δ_o is large, it may cause less absorption of energy to pair the fouth electron with one of the t_{2g} electrons than to put it in an e_g orbital. Similarly for d^5, the two possible electronic configurations are $(t_{2g})^3$, $(e_g)^2$, with five unpaired electrons, and $(t_{2g})^5$, with only one unpaired electron. These are the high-spin and low-spin configurations mentioned earlier. Which configuration is found depends upon the total octahedral stabilization energy and the energy of repulsion between the paired electrons. Two kinds of repulsion are actually involved, the electrostatic repulsion between like charges and a quantum mechanical exchange interaction which favours parallel spins for electrons in the same set of *d* orbitals. For the present, however, we shall not separate these effects and shall consider only the total mean pairing energy per electron, *P*, which can be obtained approximately from spectroscopic data. The mean pairing energies per electron vary irregularly from metal to metal (as do the Δ_o values) for a given ligand.

Fig. 7.6 : Shows the occupation of the 3*d* orbitals of Fe^{3+} in a high-spin (weak field) and in a low-spin (strong field) complex (the numerical values for the magnetic moments are discussed in the following chapter). In Fig. 7.7 are shown the high-spin and low-spin electron configurations for d^4, d^5, d^6 and d^7 ions in the octahedral fields; it will be remembered that the distinction does not arise for

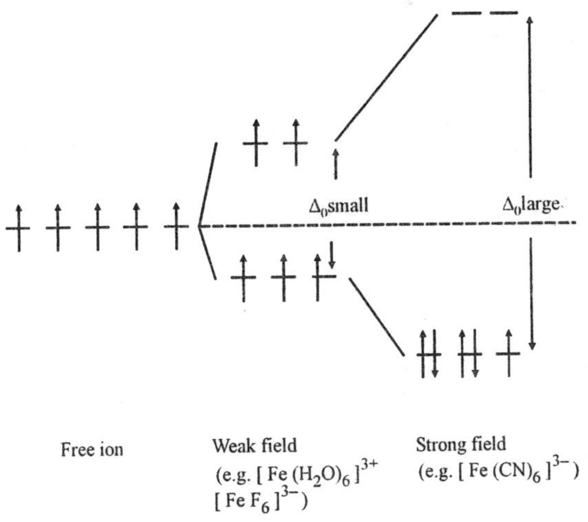

Fig. 7.6. : Occupation of the 3*d* orbitals in weak and strong octahedral fields, illustrated for $Fe^{3+}(3d^5)$ complexes

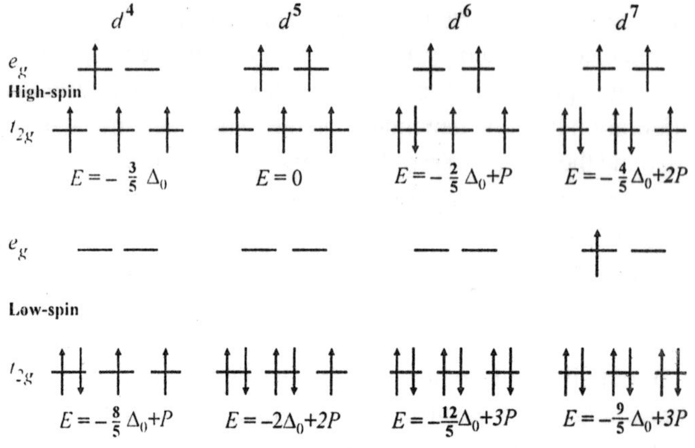

Fig. 7.7 : Configurations, crystal fields stabilization energies, and pairling energies of d^4, d^5, d^6 and d^7 ions in high-spin and low-spin ground states in octahedral fields.

d^1, d^2 and d^3 ions in octrahedral fields, and the same is true for d^8, d^9 and d^{10}, in which the distribution of electrons is $(t_{2g})^6$, $(e_g)^2$, $(t_{2g})^6(e_g)^3$ and $(t_{2g})^6(e_g)^4$ respectively. Also shown in Fig. 15.5 are the energies o the high-spin and low-spin configurations of d^4, d^5, d^6 and d^7 ions relative to the energy of the ion in a spherical field, expressed in terms of Δ_o and P. In each case it can be seen that the high-spin and low-spin configurations have equal energy if Δ_o is equal to P; or, to put the matter differently, whether the complex is of the high-spin or of the low-spin type depends on whether Δ_o is less or greater than P.

Tetragonal Distortion of Octahedral Complexes
(John-Teller Distortion)

The shape of tansition metal complexes is determined by the tendency of electron pairs to occupy positions as far away from each other as possible. This is the same as for the main group compounds and complexes. In addition, the shapes of transition metal complexes are affected by whether the *d* orbitals are symmetrically or asymmetrically filled.

Repulsion by six ligands in an octahedral complex splits the *d* orbitals on the central metal into t_{2g} and e_g levels. It follows that there is a corresponding repulsion between the *d* electrons and the ligands. If the *d* electrons are symmetrical arranged, they will repel all six ligands equally. Thus the structure will be a completely regular octahedron. The symmetrical arrangement of *d* electron are shown in Table 7.5.

All other arrangements have an asymmetrical arrangement of *d* electrons. If the *d* electrons are asymmetrically arranged, they will repel some ligands in the complex more than others. Thus the structure is distorted because some ligands are prevented from approaching

Table 7.5 : Symmetrical electronic arrangements

Electronic configuration	t_{2g}	e_g	Nature of ligand field	Examples
d^0			Strong or weak	$Ti^{IV}O_2$, $[Ti^{IV}F_6]^{2-}$
	[ ][ ][ ]	[ ][ ]	[↑][↑][↑] [↑][ ]	$[Ti^{IV}Cl_6]^{2-}$
d^3				$[Cr^{III}(oxlate)_3]^{3-}$
	[↑][↑][↑]	[ ][ ]	[↑↓][↑↓][↑↓] [↑][ ]	$[Cr^{III}(H_2O)_6]^{3+}$
d^5				$[Mn^{II}F_6]^4$
	[↑][↑][↑]	[↑][↑]	[↑↓][↑↓][↑↓] [↑↓][↑]	$[Fe^{III}F_6]^{3-}$
d^6				$[Fe^{II}(CN)_6]^{4-}$
	[↑↓][↑↓][↑↓]	[ ][ ]		$[Co^{III}(NH_3)_6]^{3+}$
d^8				$[Ni^{II}F_6]^4$
	[↑↓][↑↓][↑↓]	[↑][↑]		$[Ni^{II}(H_2O)_6]^{2+}$
d^{10}	[↑↓][↑↓][↑↓]	[↑↓][↑↓]	Strong or weak	$[Nn^{II}(NH_6)_6]^{2+}$
				$[Zn^{II}(H_2O)_6]^{2-}$

the metal, as closely as others. The e_g orbitals point directly at the ligands. Thus asymmetric filling of the e_g orbitals results in some ligands being repelled more than others. This causes a significant distortion of the octahedral shape. In contrast the t_{2g} orbitals do not point directly at the ligands, but point in between the ligand directions. Thus asymmetric filling of the t_{2g} orbitals has only a very small effect on the stereochemistry. Distortion caused by asymmetric filling of the t_{2g} orbitals is usually too small to measure. The electronic arrangements which will produce a large distortion are shown in Table 7.6.

The two e_g orbitals $d_{x^2-y^2}$ and d_{z^2} are normally degenerate. However, if they are asymmetrically filled then this degeneracy is destroyed, and the two orbitals are no longer equal in energy. If the

Table 7.6 : Asymmetrical electronic arrangements

Electronic configuration	t_{2g}	e_g	Nature of ligand field	Examples
d^4	↑ ↑ ↑	↑	Weak field	Cr(+II).Mn(+III) (high-spin complex)
d^7	↑↓ ↑↓ ↑↓	↑	Strong field	Co(+II).Ni(+III) (low-spin complex)
d^9	↑↓ ↑↓ ↑↓	↑↓ ↑	Either strong or weak	Cu(+II)

dz^2 orbitals contains one more electron than the $d_{x^2-y^2}$ orbitals then the ligands approaching along $+z$ and $-z$ will encounter greater repulsion than the other four ligands. The repulsion and distortion result in elongation of the octahedron along the z-axis. This is called tetragonal distortion. Strictly it should be called tetragonal elongation. This form of distortion is commonly observed.

If the $d_{x^2-y^2}$ orbitals contains the extra electron, then elongation will occur along the x and y axes. This means that the ligands approach more closely along the z-axis. Thus there will be four long bonds and two shorts bonds. This is equivalent to compressing the octahedron along the z-axis, and is called tetragonal compression. Tetragonal elongation is much more common than tetragonal compression, and it is not possible to predict which will occur.

For example, the crystal structure of CrF_2 is distorted rutile (TiO_2) structure. Cr^{2+} is octahedrally surrounded by six F^-; and there are four Cr—F bonds of length 1.98 – 2.01 Å, and two longer bonds of length 2.43 Å. the octahedron is said to be tetragonally distorted. The electronic arrangement in Cr^{2+} is d^4. F^- is a weak field ligand, and so the t_{2g} level contains three electrons and the e_g level contains one electron. The $d_{x^2-y^2}$ orbitals has four lobes whilst the d_{z^2} orbital has only two lobes pointing at the ligands. To minimize repulsion with the ligands, the single e_g electron will occupy the d_{z^2} orbitals. This is equivalent to splitting the degeneracy of the e_g level so that d_{z^2} is of lower energy, i.e. more stable, and $d_{x^2-y^2}$ is of higher energy, i.e. less stable. Thus the two ligands approaching along the + *z* and – *z* directions are subjected to greater repulsion than the four ligands along + x, – x, + y and – y. This causes tetragonal distortion with four short bonds and two long bonds. In the same way MnF_3 contains Mn^{3+} with a d^4 configuration, and forms a tetragonally distorted octahedral structure.

Many Cu(+II) salts and complexes also show tetragonally distorted octahedral structure. Cu^{2+} has a d^9 configuration.

To minimize repulsion with the ligands, two electrons occupy the d_{z^2} orbital and one electron occupies the $d_{x^2-y^2}$ orbital. Thus the two ligands along – *z* and – *z* are repelled more strongly than are the other four ligands.

The examples above show that whenever d_{z^2} and $d_{x^2-y^2}$ orbitals are unequally occupied, distortion occurs. This is known as Jahn-Teller distortion. The Jahn-Teller theorem states that 'Any non-linear molecular systesm in a degenerate electronic state will be unstable, and will undergo some sort of distortion to lower its symmetry and remove the degeneracy. More simply, molecules or complexes (of any shape except linear). which have an unequally filled set of orbitals

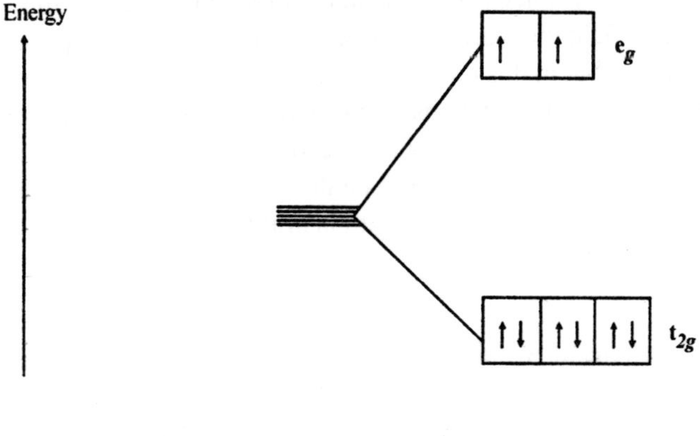

Fig. 7.7 : d^8 arrangement in weak octahedral field.

(either t_{2g} or e_g), will be distorted. In octahedral complexes distortion from the t_{2g} level are too small to be detected. However, distortions resulting from uneven filling of the e_g orbitals are very important.

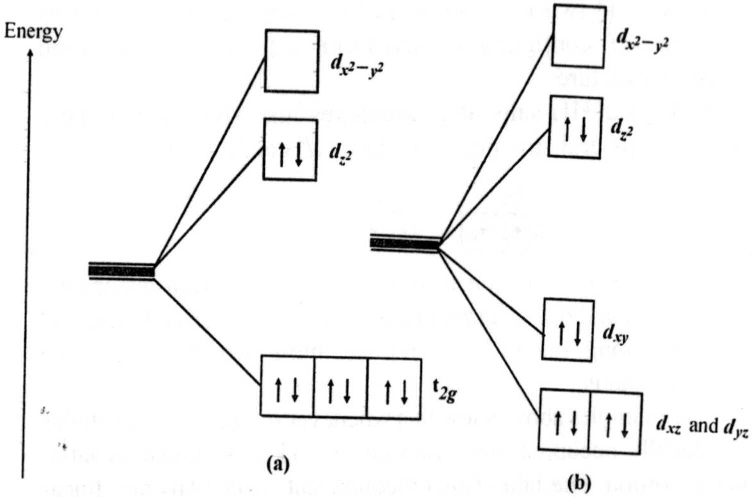

Fig. 7.7 : d^8 arrangement in very strong octahedral field. Tetragonal distortion splits (a) the e_g level; and (b) also splits the t_{2g} level. The d_{xy} orbital is higher in energy than the d_{xz} or d_{yz}. (For simplicity this is sometimes ignored.)

Tetrahedral complexes. The spliting of the *d* orbitals depends markedly upon the symmetry of the ligand field round the central ion. For a tetrahedral field the *d* orbitals are again split into two groups of different energies, but in this case the d_{xy}, d_{xz} and d_{yz} orbitals have the higher energy since their directions lie closer to the ligands than those of the d_{z^2} and $d_{x^2-y^2}$ orbitals. For the same ion and ligand at the same distance as in the octahedral case, it can be shown that Δ_t, the difference between the two sets of orbitals in a tetrahedral field, is given by

$$\Delta_t = \frac{4}{9}\Delta_0$$

(This relationship would not be expected to be quantitatively very useful in practicle, since a change of coordination number from six to four would nearly always be accompanied by a change in inter-atomic distance.) The effects on *d* orbitals of a spherically symmetrical and a tetrahedral crystal field are then as represented in Fig. 7.8.

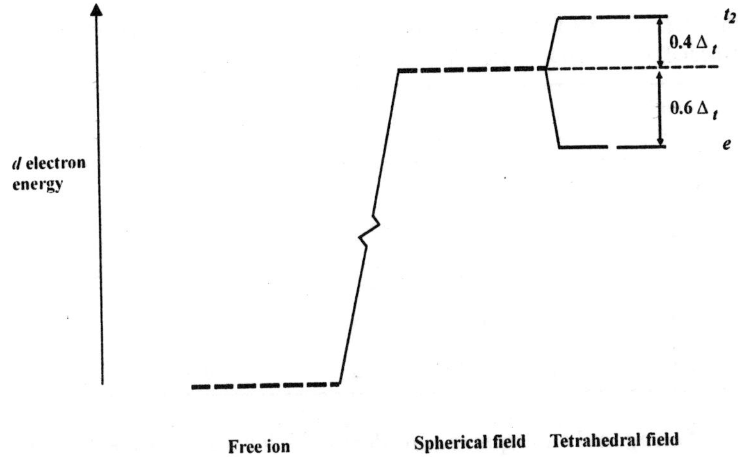

Fig. 7.8 : The effects on d orbitals of a spherical and a tetrahedral field MAIC-Q.

The spectrochemical series follows the same order as for octahedral fields, but low-spin tetrahedral complexes have not yet been obtained; apparently the crystal field stabilization energies are never sufficient to bring about spin-pairing. Nor have substantial distortions yet been found in tetrahedral complexes, but this may be because relatively few have been examined.

Crystal field stabilization energies for high-spin tetrahedral and octahedral configurations of the same species at the same distance are shown in Table 7.8. Two points of interest about the values for tetrahedral fields may be noted. First, the difference between the stabilizations is least for d^0, d^5 and d^{10} (for which it is zero in all cases), and next smallest for d^1, d^2, d^6 and d^7; most of the examples of tetrahedral coordination given in belong to one of these systems. Secondly, since Δ_t is much lower than Δ_o, complexes of the same ion with ligands occupying nearby positions in the spectro-chemical series have their absorption maxima for d-d spectra at lower frequencies and higher wavelengths for tetrahedral complexes than for octachedral complexes. Thus high-spin octahedral cobalt (II) complexes such as $[CO(H_2O)_6]^{2}$ absorb in the blue region of the spectrum and are pink, whilst tetrahedral cobalt (II) complexes such as $[CoCl_4]2^-$ absorb in the red and are blue in colour.

Table 7.8 : Crystal field stabilization energies for high-spin tetrahedral and octahedral configurations (other things being equal)

Numberof d electrons	*CFSE* *Tetrahedral*	*Octahedral*
0, 5	0	0
1, 6	$0.6\ \Delta_t = 0.27\ \Delta_o$	$0.4\ \Delta_o$
2, 7	$1.2\ \Delta_t = 0.53\ \Delta_o$	$0.8\ \Delta_o$
3, 8	$0.8\ \Delta_t = 0.35\ \Delta_o$	$1.2\ \Delta_o$
4, 9	$0.4\ \Delta_t = 0.18\ \Delta_o$	$0.6\ \Delta_o$

Tetragonal and square complexes. We have already mentioned the tetragonal distortion of an octahedral complex arising from the Jahn-Teller effect. Such a distortion has the effect. Such a distortion has the effect of destroying the degeneracy of the e_g orbitals; if, for example, the metal ion has four near and two distant ligands, the d_{z^2} orbital is of lower energy than the $d_{x^2-y^2}$ orbital. If the difference in energy between them is large enough, a d^8 complex (e.g. of Ni^{II}) will be diamagnetic with the electronic configuration $(d_{xz})^2(d_{yz})^2$ $(d_{xy})^2(d_{z^2})^2$. In the limit of tetragonal distortion, two of the ligands are removed to infinity and a square-planar complex results. The relative changes in energy when an octahedral complex is successively distorted are shown in Fig. 7.9, from which we can see that planar complexes of electronic configurations d^7, d^8 and d^9 (the only ones at present known) should have 1, 0 and 1 unpaired electrons respectively. Fig. 7.9, however, does not really tell us much about the likelyhood of forming planar complexes. As we have pointed out before, the crystal field stabilization energy is only a small part of the total interaction energy in the complex. It may also be doubted whether a truly electrostatic system would result in planar coordination, and it is indeed noteworthy that ligands which often give planar complexes are also good π-bonding ligands. We shall threfore not discuss planar coordination further in terms of crystal field theory; fuller treatments are given in more specialized works.

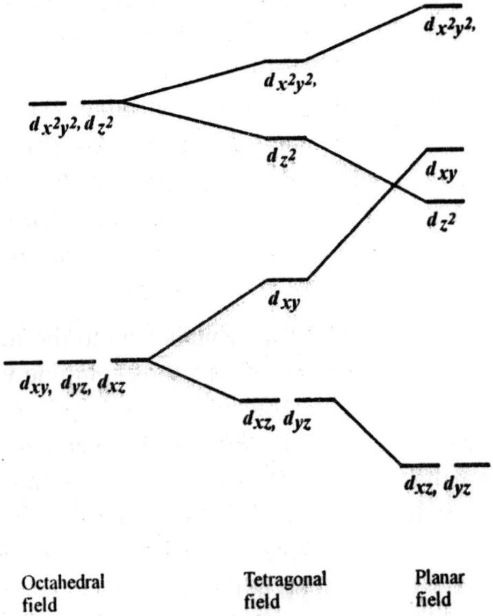

Fig. 7.9 : Crystal field splittings of the d orbitals in octahedral, tetragonal and planar complexes.

Ligand-field theory: An elegant model of the bonding in complex ions is found in the application of molecular-orbital theory to these complexes and is usually called *ligand-field theory*. A complete discussion of the way the atomic orbitals of a central ion and six ligands combine to form molecular orbitals in a complex is not appropriate here, but a capsule description of the process is in order.

Imagine that a complex ion is formed as follows: Six ligands approach a metal ion by moving in toward it along the three cartesian coordinate axes. This will form an octahedral complex with one ligand at each of the six corners of the octahedron and with the metal ion at the centre. As the ligands move in, an inward-pointing orbital of each, say a *p*-orbital, overlaps certain orbitals of the central ion. These are the *ns* orbital, the np_x, np_y, and np_z orbitals, and the $(n-1)d_{x^2-y^2}$ and $(n-1)d_{z^2}$ orbitals. Only these orbitals can provide favourable overlap with the ligand orbitals. [*Favourable overlap means that there is a net reinforcement of the wave functions in the overlap region.* This means that there is a buildup of electronic charge density there; in other words, a bond.] the inward-pointing lobes of the ligand orbitals favourably overalp with the designated *d, s,* and *p,* orbitals, but not with the d_{zy}, d_{xz}, and d_{yz} orbitals. The reason for this that the wave function carries opposite algebraic signs in adjacent lobes of these three orbitals so that no matter what the sign on the ligand orbital lobe is, net reinforcement is impossible. (This just means that reinforcement by one *d*-orbital lobe is offset by cancellation by the other.)

Favourable overlap of atomic orbitals lead to the formation of molecular orbitals. The six metal orbitals thus combine with the six ligand orbitals to form twelve molecular orbitals, six bonding and six antibonding. In the diagram the subsets of molecular orbitals are labeled with shorthand designations for the symmetries of the orbitals, but except for the t_{2g} and e^*_g designations we need not concern ourselves with these. Note, however, that the electron pairs from the ligand orbitals all end up in low-energy, *bonding* molecular orbitals. (Each of these started out as a long pair in, for example, a cyanide ion, a water molecule, a fluoride ion, etc.) These molecular orbitals are all of the σ type because each is symmetrical around the bond axis.

Higher in energy than the bonding orbitals in the complex are the t_{2g} orbitals, which are *nonbonding orbitals*. While electrons in bonding orbitals increase the stability of a molecule (or complex ion) and electrons in antibonding orbitals decrease it, electrons in nonbonding orbitals have no effect on the stability of a collection of atoms. The three t_{2g} orbitals correspond to the d_{zy}, d_{xz}, and d_{yz} orbitals of the original central ion. There are six antibonding orbitals in the complex, each higher in energy than the bonding or nonbonding orbitals. Note that the antibonding orbitals of lowest are the two labeled e^*_g.

In Fig. 22.3 we have not shown any of the electrons in the $(n-1)d$ subshell of the metal ion. There may be various numbers of these, depending on the identity and oxidation state of the ion, and these end up distributed between the t_{2g} and e^*_g subsets of molecular orbitals. We will discuss their precise distribution shortly.

When carried out with the fewest possible approximations, calculations based on the ligand-field model yield excellent descriptions of complex ions. But such calculations are not appropriate for this text. Fortunately, however there is a simplified, yet surprisingly powerful, version of ligand-field theory which is amazingly successful is predicting many of the properties of complex ions. This is called *crystal-field theory*, and for most of the remainder of this chapter we will emphasize the model.[1]

Molecular Orbital Theory

The molecular orbital theory incorporates covalent bonding. Consider a first row transition element forming an octahedral complex, for example

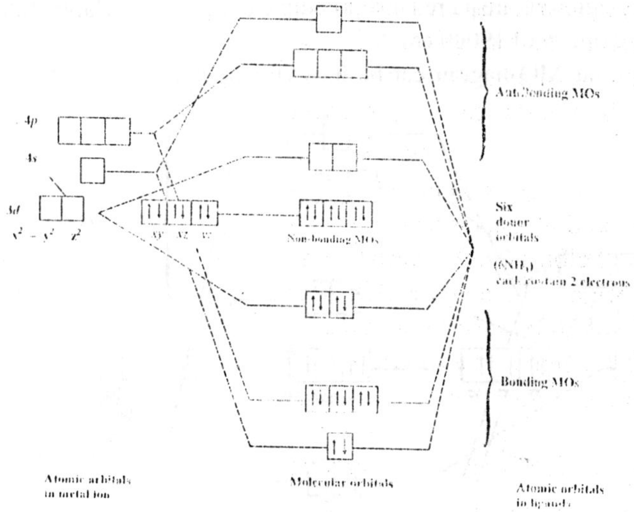

Fig. 7.10 : Molecular orbital diagram for $[Co^{III}(NH_3)_6]^{3+}$

$[Co^{III}(NH_3)_6]^{3+}$. The atomic orbitals on Co^{3+} which are used to make molecular orbitals are $3d_{x^2-y^2}$, $3d_{z^2}$, $4s$, $4p_x$, $4p_y$ and $4p_z$. A $2p$ atomic orbital from each NH_3 containing a lone pair is also used to make molecular orbitals. Thus there are 12 atomic orbitals, which combine to give 12 molecular orbitals (six bonding MOs and six antibonding MOs). The 12 electrons from the six ligand lone pairs are placed in the six bonding MOs. This accounts for the six bonds. The transition metal Co^{3+} has other *d* orbitals, which have so far been ignored. These are the $3d_{xy}$, $3d_{xz}$ and $3d_{yz}$ orbitals. These form non-bonding MOs, and in Co^{3+} they contain six electrons, but contribute nothing to the bonding. The antibonding MOs are all empty. The arrangement is shown in Figure 7.10. We should predict that the complex should be diamagnetic as all the electrons are paired. The complex should be coloured since promotion of electrons from the non-bonding MOs to the antibonding e^*_g MOs is feasible. The energy jump Δ_o is 23000 cm^{-1}. The six non-bonding *d* electrons are paired in this complex because Δ_o is larger than the pairing enegy of 19000 cm^{-1}.

A similar MO diagram can be drawn for the complex $[Co^{III}F_6]^{3-}$.

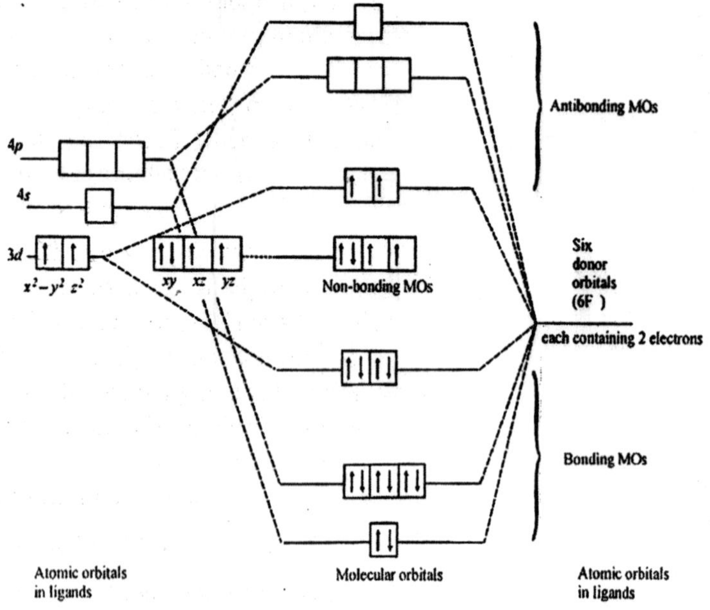

Fig. 7.11 : Molecular orbital diagram for $[CoF_6]^{3-}$

However, the energies of the 2*p* orbitals on F^- are much lower than the energy of the corresponding orbital on N in NH_3. This alters the spacing of the MO energy levels (Figure 7.11). Spectroscopic measurements show that Δ_o is 13000 cm^{-1}. Thus the gap between the non-bonding MOs and the antibonding e^*_g MOs is less than the pairing energy of 19000 cm^{-1}. Thus the non-bonding d electrons do not pair up as in the $[Co(NH_3)_6]^{3+}$ complex because there is a net gain in energy if electrons are left unpaired. Thus $[CoF_6]^{3-}$ has four unpaired electrons and is a high-spin-complex, whilst $[Co(NH_3)_6]^{3+}$ has no unpaired electrons and is a low-spin-complex.

Thus the MO theory explains the magnetic properties and spectra of complexes equally as well as the crystal field theory. Both theories rely on spectra to measure the energy of Δ_o. Either theory may be used depending on which is the most convenient.

The MO theory is based on wave mechanics and so has the disadvanage that enthalpises of formation and bond energies cannot be calculated directly. So far we have considered σ bonding between ligands and the central metal. The MO theory has the great advantage that it is easily extended to cover π bonding. Pi bonding helps to explain how metals in low oxidation states (e.g. $[Ni^0(CO)_4]$) can form complexes. It is impossible to explain any attractive force in such a complex the crystal field theory because the lack of charge on the metal. Pi bonding also helps to explain the position of some ligands in the spectrochemical series. There are two cases:

1. Where the ligands act as π acceptors, by accepting electrons from the central metal. Examples include CO, CN^-, NO^+ and phosphines.
2. Where the ligands act as π donors and transfer charge from ligand to metal in π interactions as well as σ interactions. Pi bonding of this kind commonly occurs in oxoions of metals in high oxidation states, e.g. $[Mn^{VII}O_4]$ and $[Cr^{VI}O_4]^{2-}$.

π acceptors

Ligands such as CO, CN^- and NO^+ have empty π orbitals with the correct symmetry to overlap with the metal t_{2g} orbitals, forming π bonds. This is often described as back bonding. Normally the π orbitals on the ligands are of higher energy than the metal t_{2g} orbitals. No more electrons are added to the scheme as the ligand π orbitals are empty, but the π interaction increases the value of Δ_o. This accounts for the position of these ligands as 'strong field ligands' at the right of the spectrochemical series.

π donors

The ligand has filled π orbitals which overlap with the metal t_{2g} orbitals, giving a π bond. Thus electron density is transferred from the ligand to the metal. The σ bonding also transfers charge to the metal. This type of complex is favoured when the central metal has a high oxidation state, and 'is short of electrons'. The ligand π orbitals are lower in energy than the metal t_{2g} orbitals. Delocalizing π electrons from the ligand to the metal in this way reduces the value of Δ. It is not always clear if π donor bonding has occurred, but it most likely with ligands at the left of the spectrochemical series.

8

Metallic Carbonyls

Introduction

Most transition metals form complexes with a wide variety of uncharged molecules such as carbon monoxide, substituted phosphines, and unsaturated or aromatic hydrocarbons. In many of these, the metal is in zero or another low oxidation state and, π bonding between the metal atom and the ligands is believed to play an important part in stabilizing these complexes. The formation of metal carbon σ and π bonds in a metal carbonyl is shown in Fig. 8.1.

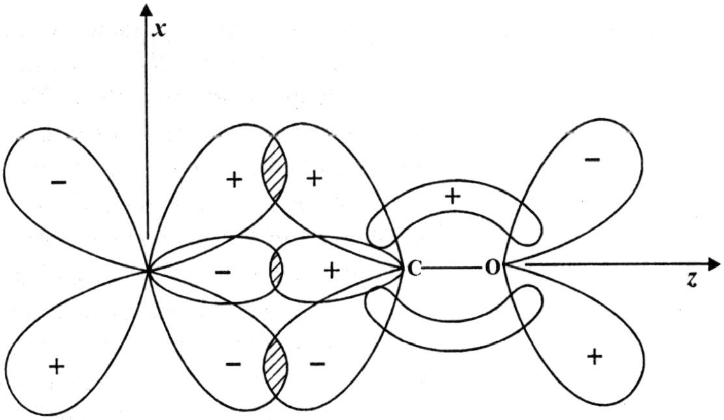

Fig. 8.1 : The formation of α and π bonds between a transition metal and CO. Only one t_{2g} metal orbital, and one π and one π* ligand orbital, are shown

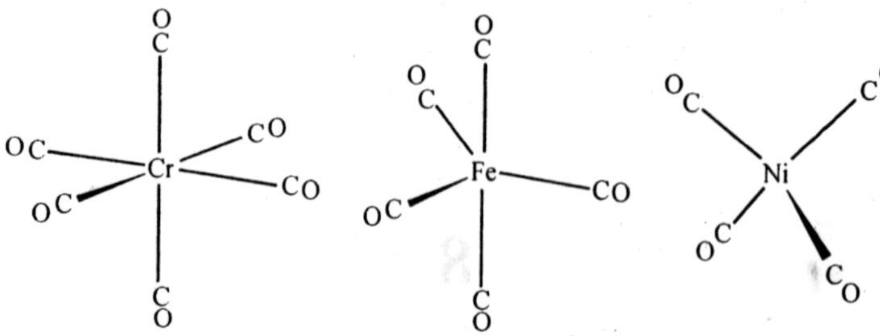

Fig. 8.2 : Structures of simple carbonyls of chromium, iron, and nickel.

Almost all of the transition metals form compounds in which carbon monoxide acts as a ligand. There are three points of interest with respect to these compound: (1) Carbon monoxide is not ordinarily considered a very strong Lewis base, and yet it forms strong bonds to the metals in these complexes; (2) the metals are always in a low oxidation state most often formally in an oxidation state of zero, but also in low positive and negative oxidation states; and (3) the 18-electron Rule is obeyed. Examples of the application of the rule to some simple metal carbonyls are

Cr	=	$24e^-$	Fe	=	$26e^-$	Ni	=	$28e^-$
6CO	=	$12e^-$	5CO	=	$10e^-$	4CO	=	$8e^-$
$Cr(CO)_6$	=	$36e^-$	$Fe(CO)_5$	=	$36e^-$	$Ni(CO)_4$	=	$36e^-$

Metals with odd atomic numbers cannot satisfy the 18-electron Rule by simple addition of carbon monoxide ligands since the resultant moiety will have an odd number of electrons (and hence not 18) no matter how many carbonyls are added.

Preparation Of the simple carbonyls, the nickel and iron compounds can be prepared by direct combination of CO with the finely divided elements

$$Ni + 4CO \rightarrow Ni(CO)_4$$

$$Fe + 5CO \rightarrow Fe(CO)_5$$

Nickel carbonyl forms readily at room temperature under one atmosphere of carbon monoxide; the reaction is reversible and serves as a method for preparing very pure nickel. In contrast, $Fe(CO)_5$ is formed less readily, temperature of 200° at a pressure of 100 atm being required to effect formation.

The other simple carbonyls are prepared by reducing compounds of the transition metal in a carbon monoxide atmosphere. A variety of reducing agents have been used in this process, e.g., aluminum alkyls.

$$CrCl \xrightarrow[CO]{(C_2H_5)_6Al_2} Cr(CO)_6,$$

the active metals, such as sodium

$$MoCl_5 \xrightarrow{Na + CO} Mo(CO)_6,$$

zinc, and magnesium, and hydrogen

$$2CoCO_3 + 2H_2 + 8CO \xrightarrow[150^\circ]{300\ atm.} Co(CO)_8 + 2CO_2 + 2H_2O.$$

These reactions are generally conducted with suspensions in organic solvents such as ether, tetrahydrofuran, or diglyme, using carbon monoxide pressures in the range of 200-300 atm. In some instances carbon monoxide can act as the reducing agent

$$RuI_3 \xrightarrow[1\ atm]{CO} Ru(CO)_2I_2$$

$$Ru(CO)_2I_2 \xrightarrow[CO]{Ag} Ru(CO)_5$$

$$OsO_4 + 9CO \xrightarrow[50\ atm]{100^\circ} Os(CO)_5 + 4CO_2$$

$$Re_2S_7 + 17CO \xrightarrow[200\ atm]{250^\circ} Re(CO)_{10} + 7COS$$

Iron pentacarbonyl can be used to form certain other metal carbonyls because its carbon monoxide ligands are labile

$$MCl_5 \xrightarrow[M = Mo, W]{Fe(CO)_5} M(CO)_6.$$

The polynuclear carbonyls are prepared by thermal or photochemical decomposition of the simple carbonyls

$$2Fe(CO)_5 \xrightarrow{hv} Fe_2(CO)_9 + CO.$$

or from polynuclear carbonylate anions, which are discussed in a subsequent section.

Bonding Initially it might appear unusual for carbon monoxide to act as a sufficiently strong ligand to form stable metal complexes; certainly, up to this point there has been little indication that carbon monoxide exhibits significant basic characteristics. The unusual stability of transition-metal-to-carbon bonds in these complexes has been attributed to multiple-bond formation, as shown schematically in Fig. 8.3. The σ component of this bond involves overlap of the

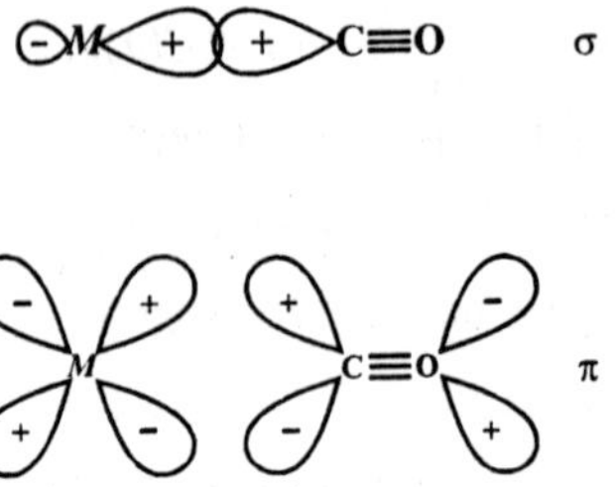

Fig. 8.3. Orbital representation for the s and p components in a metal carbonyl bond. The p component involves antibonding orbitals of the C—O bond order.

filled *sp* hybrid orbital with an appropriately directed vacant orbital on the metal atom. The latter could be an atomic orbital or an appropriate combination of atomic orbitals which form one of the usual hybrids, e.g. sp^3 in the case of $Ni(CO)_4$, d^2sp^3 for $M(CO)_6$ (M=Cr, Mo, W), and dsp^3 for $Fe(CO)_5$. The π component is believed to arise from the interaction of filled metal d orbitals with the antibonding π orbitals of the carbonyl moiety. The formation of the metal-carbon σ bond results in a net increase of electron density on the metal atom, but the π bond results in a net increase of electron density on the metal atom, but the π component lead to a decrease in electron density. The net affect of this process leaves the metal atom little changed, but the electron density donated to the π* carbonyl orbitals decreases the bond order of the C—O bond. In valence-bond terms, **17** and **18** represent the main contributing canonical

$$M \leftarrow C \equiv O:$$

17 18

forms in the description of the wave function for the metal-carbon bond. The predicted decrease in the C—O bond order can be seen in the decrease in ν(CO) from 2155 cm^{-1} for the free ligand to ~2000 cm^{-1} for the complexed carbonyl group. This description of the metal-carbon bond is also reflected in ν(CO) for an isostructural and isoelectronic series in which the charge on the metal is varied regularly. The data in Table 8.1 for two such

Table 8.1 : Carbonyl stretching frequencies in some mononuclear species

Complex	ν(CO) cm^{-1}
$[Mn(CO)_6]^+$	2096
$Cr(CO)_6$	2000
$[U(CO)_6]$	1859
$Ni(CO)_6$	2046
$[Co(CO)_4]^-$	1883
$[Fe(CO)_4]^{2-}$	1788

series of components show the predicted decrease in ν(CO) as the negative charge on the metal increase.

Structure of metal carbonyls[3,4]

The structures of the mononuclear carbonyls $V(CO)_6$, $Cr(CO)_6$, $Fe(CO)_5$ and $Ni(CO)_4$ have been established by X-ray diffraction, infrared spectroscopy and (expect for the vanadium compound) electron diffraction; and $W(CO)_6$. All of the hexacarbonyls have regular octahedral structures, and nickel carbonyl has a regular tetrahedral structure. In the trigonal bipyramidal molecule of iron pentacarbonyl, however, the most recent electron diffraction study suggests that the Fe—C apical bond is slightly shorter than the equatorial one, the lengths being 1.806 and 1.833 Å respectively; it is interesting to note that in trigonal bipyramidal molecules of typical elements it is usually the equatorial bond that is shorter, but the significance of this observation is not clear.

X-ray diffraction studies have shown that in the solid state $Mn_2(CO)_{10}$, $Fe_2(CO)_9$ and $Co_2(CO)_8$ have the structures shown in Figs. 8.4, 8.5 and 8.6 respectively; $Te_2(CO)_{10}$ and $Re_2(CO)_{10}$ are isomorphous with $Mn_2(CO)_{10}$. The structure of $Mn_2(CO)_{10}$ may be described as two staggered octahedra sharing one corner; in each of the octahedra the four equatorial Mn—CO bonds are bent inwards

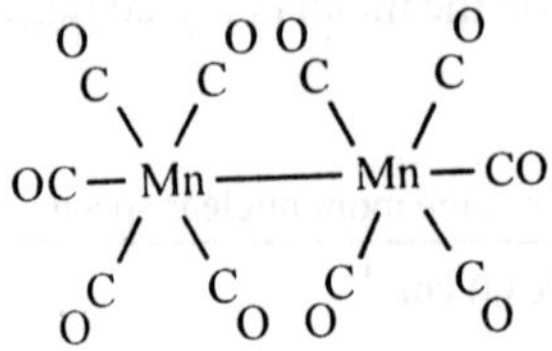

Fig. 8.4

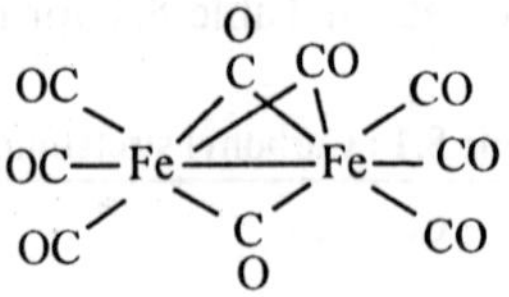

Fig. 8.5

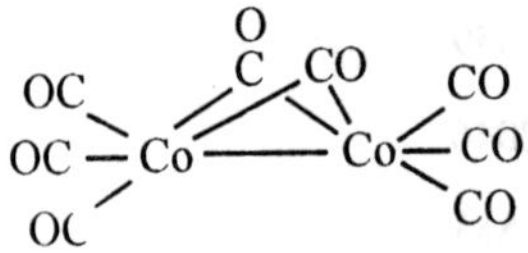

Fig. 8.6

slightly. In $Fe_2(CO)_9$ the structure consists of two octahedra sharing a face; since the compound is diamagnetic it is usually inferred that a metal-metal bond is present. This does not necessarily follow from the diamagnetism, since coupling of unpaired spins through multicentre iron-carbonyl-iron bonding could occur: but the shortness of the Fe—Fe distance probably indicates a metal-metal bond; Mn—Mn in $Mn_2(CO)_{10}$, in which there must be a metal-metal bond, is as great as 2.92 Å. A puzzling property of $Fe_2(CO)_9$, in view of its molecular structure, is its insolubility in all organic solvents. Di-cobalt octacarbonyl has approximately the structure of $Fe_2(CO)_9$ with one of the bridging carbonyls removed; again the diamagnetism of the compound and Co—Co distance (2.52 Å) are generally taken to indicate a metal-metal bond.

Metal-carbon distances in $Fe_2(CO)_9$ and $Co_2(CO)_8$ fall into two groups, metal-bringing carbonyl distances being about 0.1 Å longer than metal-terminal carbonyl distances being about 0.1 Å longer than metal-terminal carbonyl distances. Such a difference is compatible with the concept of two-electron donation by terminal carbonyls, though the possible existence of π bonds of different strengths makes quantitative interpretation impossible. That the extent of π bonding to terminal and bridging carbonyls is different is clearly shown by the carbonyl stretching frequencies. Carbon monoxide itself has a stretch frequency of 2143 cm^{-1}; neutral metal carbonyls known to have no bridging carbonyl groups have stretching frequencies in the range 2125-2000 cm^{-1}; and $Fe_2(CO)_9$ and $Co_2(CO)_8$, in addition to showing bands in this region, also show carbonyl absorption at 1830 and 1860 cm^{-1} respectively.

Of the three trinuclear carbonyls at present known, crystalline $Os_3(CO)_{12}$ and $Ru_3(CO)_{12}$ have the structure shown in Fig. 8.7, whilst $Fe_3(CO)_{12}$ has the structure shown in Fig. 8.8

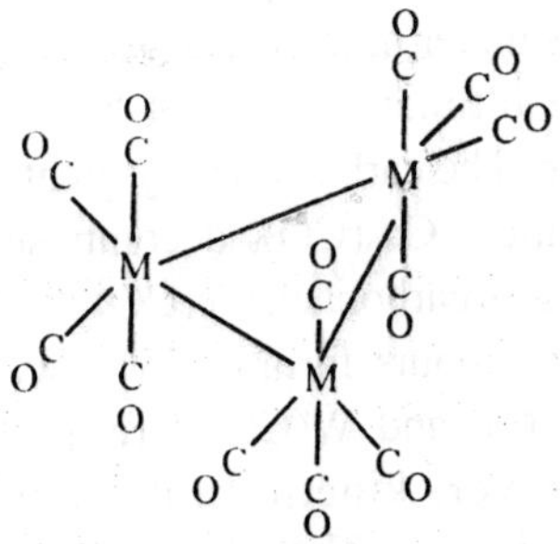

Fig. 8.7

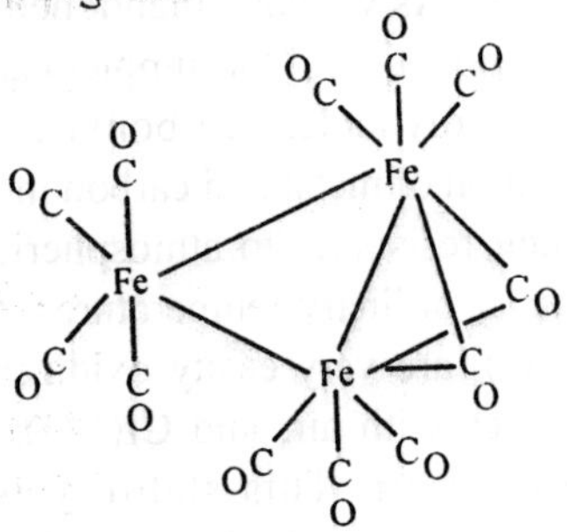

Fig. 8.8

A similar change of structure occurs among the tetracarbonyls of members of the cobalt group : $Ir_4(CO)_{14}$ has the non-bridged structure shown in Fig. 8.9, whilst $Co_4(CO)_{12}$ and $Rh_4(CO)_{12}$ have the structure shown in Fig. 8.10. The only hexametalcarbonyl whose structure has yet been determined is $Rh_6(CO)_{16}$, which contains a Rh_6 octahedron, two terminal carbonyl groups on each metal atom and four carbonyl groups each bridging three metal atoms (Fig. 8.11).

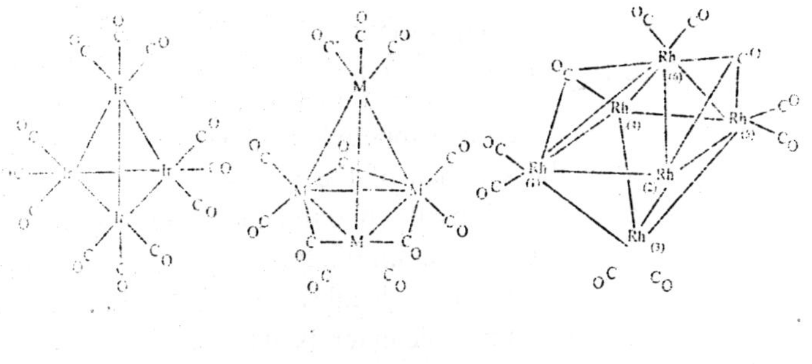

Fig. 8.11. (Four terminal and two bringing CO groups omitted)

Physical and chemical properties of the carbonyls[1]

The metal carbonyls are all crystalline solids with the exceptions of $Fe(CO)_5$, $Ru(CO)_5$, $Os(CO)_5$ and $Ni(CO)_4$, which are liquids. They melt or decompose at low temperatures, and are either insoluble in water (e.g. $Ni(CO)_4$ or decomposed by it (e.g. $Fe(CO)_5$). With the exception of $Fe_2(CO)_9$, all are more or less soluble in organic solvents. As we have mentioned already, thermal or photochemical decomposi-tion of the simpler carbonyls under mild conditions often yields polynuclear carbonyls, but the ultimate action of heat is formation of metal and carbon monoxide. The carbonyls vary greatly in their resistance to atmospheric oxidation: $Co_2(CO)_8$ decomposes in air at ordinary temperatures, forming a carbonate; $Fe(CO)_5$ and $Ni(CO)_4$ are very easily oxidized, their vapours forming explosive mixtures with air; and $Cr(CO)_6$, $Mo(CO)_6$ and $W(CO)_6$ are quite stable in air. This stability is, however, kinetic rather than thermodynamic character; the bond energies in $Cr(CO)_6$, $Fe(CO)_5$ and $Ni(CO)_4$ (obtained from the heats of combustion, the bond energies in carbonyl groups being assumed to be the same in all three compounds) are 113, 117 and 146 kJ (27, 28 and 35 k cal) respectively.

The exchange of carbon monoxide with some metal carbonyls has been studied using ^{14}CO; $Ni(CO)_4$ and $Co_2(CO)_8$ exchange rapidly, and $Fe(CO)_5$, $Cr(CO)_6$ and $Mn_2(CO)_{10}$ very slowly. The rate of exchange of the nickel compound is independent of the concentration of ^{14}CO and depennds on the first power of that of $Ni(CO)_4$; the mechanism is thus

$$Ni(CO)_4 \rightarrow Ni(CO)_3 + CO \qquad \text{(slow)}$$
$$Ni(CO)_3 + 14CO \rightarrow Ni(CO)_3\,(^{14}CO) \qquad \text{(last)}$$

Another ligand such as triphenylphosphine may replace ^{14}CO without change in the kinetics of the reaction. The intermediate $Ni(CO)_3$ has been isolated, by the photolysis of $Ni(CO)_4$ in an inert matrix at low temperatures. A point of interest about the exchange of CO in $Co_2(CO)_8$ is that all eight molecules are evquivalent in reactivity in solution; this is understandable in terms of the equilibrium between the ordinary (bridged) structure and the non-bridged structure under these conditions.

One of the earliest known reactions of iron pentacarbonyl was its conversion into the $[Fe(CO)_4]^{2-}$ anion by the action of aqueous sodium hydroxide. All neutral metal carbonyls can be converted into anions, a molecule of carbon monoxide being eliminated; it can be considered as being replaced by two electrons (in a doubly charged anion such as $[Fe(CO)_4]^{2-}$, one electron and a hydrogen atom (as in $[HFe(CO)_4]^-$) or two hydrogen atoms (in the uncharged carbonyl hydride, $H_2Fe(CO)_4$. There are now, indeed, more carbonylate anions than neutral carbonyls into cations (e.g. $Fe(CO)_5$ into $[HFe(CO)_5]^+$) and other cations can be obtained in different ways. The action of chlorine, bromine or iodine on some carbonyls give carbonyls halides; for example,

$$2MNo(CO)_6 + 2Cl_2 \rightarrow [Mo(CO)_4Cl_2]_2 + 4CO$$

$$Mn_2(CO)_{10} + Br_2 \rightarrow 2BrMn(CO)_5$$

$$Fe(CO)_5 + I_2 \rightarrow I_2Fe(CO)_4 + CO$$

The recent preparation of a carbonyl fluoride by the reaction

$$4Ru(CO)_5 + 4F_2 \rightarrow [Ru(CO)_3F_2]_4 + 8CO$$

(a similar reaction occurs between $Mo(CO)_6$ and fluorine) is of particular interest, since it has been widely believed that carbonyl fluorides do not exist. Most other reactions of metal carbonyls involve replacement of carbon monoxide by Lewis bases (e.g. tertiary phosphine, PF_3, pyridine, NO, *o*-phenanthroline and ethylenedioamine) or organic molecules or radicals; the latter will be discussed in the following chapter. Meanwhile, we shall complete the description of the inorganic chemistry of the carbonyls by discussing in turn anionic and cationic carbonyl complexes, Lewis base complexes, carbonyl hydrides,

Carbonyl Halides : Halogen derivatives of the metal carbonyls have been prepared by reaction of the transition-metal carbonyls with the appropriate halogens. In some cases direct addition of the halogen molecule occurs to give a compound with a higher coordination number.

$$Fe(CO)_5 + X_2 \rightarrow Fe(OX)_5X_2$$

which then decomposes by loss of carbon monoxide

$$Fe(CO)_5X_2 \rightarrow Fe(CO)_4X_2 + CO.$$

The halogen adducts formed by mixed carbonyl derivatives of the Group VIB metals containing bi- and tridentate ligands are sufficiently stable to be isolated; the properties of some of the adducts are given in Table 8.3. Such adducts can exist as neutral, anionic, or cationic species, but they all appear to be heptacoordinate.

Table 8.3 : The formulas and colours of some stable, mixed carbonyls derivatives of the Group VIB elements

Compound[a]	Colour
$[Mo(CO)_3 (dipy)Br_2]$	deep yellow
$[W(CO)_3 (dipy)I_2]$	orange
$[Cr(CO)_2 (diars)_2Br]Br$	pale yellow
$[Cr(CO)_2 (diars)_2I]I_3$	brown
$(C_5H_5NCH_3)$ $[Mo(CO)_4I_3]$	yellow
$(C_5H_5NCH_3)$ $[Mo(CO)_4I_3]$	yellow
[a]	Dipy = 2,2'-dipyridyl; diars = *o* -phenylenebisdimethylarsine.

Metal-metal bonds in polynuclear carbonyls are often cleaved when these substances are treated with halogens.

$$2BrMn(CO)_5 \xrightarrow{120^\circ} Mn(CO)_4Br_2 + 2CO.$$

The monohalo derivatives are monomeric and the polyhalo derivatives are often dimeric with halogen bridges, as indicated in Fig. 8.12. Irradiation of metal hexacarbonyls in the presence of halogens yields seven-coordinate dimeric complexes which are presumably halogen bridged :

$$2M(CO)_6 + 2X_2 \xrightarrow{uv} [M(CO)_4X_2]_2 + 4CO$$

$$M = W, Mo; X = Br, I.$$

The halogen bridges can often be broken by treating the dimers with a strong ligand such as pyridine

$$[Mn(CO)_4I_2]_2 + 4py \rightarrow 2Mn(CO)_3(py)I + 2CO$$

which yields monomeric species. Mixed metal carbonyl halides have also been prepared by the action of carbon monoxide on metal halides

$$n\text{RuI}_3 + 2n\text{CO} \rightarrow [\text{Ru(CO)}_2\text{I}_2]_n + \frac{n}{2}\text{I}_2$$

$$2\text{PtCl}_2 + 2\text{CO} \rightarrow [\text{Pt(CO)Cl}_2]_2.$$

Carbonylate Anions

Carbonylate anions are formed when carbonyls are treated with aqueous or alcoholic alkali hydroxide or with amines, sulfoxides, or other Lewis bases, when metal—metal bonds are cleaved with sodium, or when certain carbonyls are refluxed with salts in an ether. Illustrative examples are

$$\text{Fe(CO)}_5 + 3\text{NaOH(aq)} \rightarrow \text{Na[HFe(CO)}_4\text{](aq)} + \text{Na}_2\text{CO}_3\text{(aq)} + \text{H}_2\text{O}$$

$$\text{Co}_2(\text{CO})_8 + 2\text{Na/Hg} \xrightarrow{\text{THF}} 2\text{Na}[\text{Co(CO)}_4]$$

$$\text{Mn}_2(\text{CO})_{10} + 2\text{Li} \xrightarrow{\text{THF}} 2\text{Li}[\text{Mn(CO)}_5]$$

$$2\text{Co}^{2+}\text{(aq)} + 11\text{CO} + 12\text{OH}^- \xrightarrow{\text{KCN(aq)}} 2[\text{Co(CO)}_4]^- + 3\text{CO}_3^{2-} + 6\text{H}_2\text{O}$$

The stoichiometries of the simpler carbonylate ions obey the noble-gas formalism. Most of them are readily oxidized by air. The alkali-metal salts are soluble in water, from which they can be precipitated by large cation such as $[\text{Ph}_4\text{As}]^+$.

There are also many polynuclear species; those of iron have been much studied. They are obtained by reactions such as

$$\text{Fe}_2(\text{CO})_9 + 4\text{OH}^- \rightarrow [\text{Fe}_2(\text{CO})_8]^{2-} + \text{CO}_3^{2-} + 2\text{H}_2\text{O}$$

$$\text{Fe(CO)}_5 + \text{Et}_3\text{N} \xrightarrow{\text{H}_2\text{O},\ 80} [\text{Et}_3\text{NH}][\text{HFe}_3(\text{CO})_{11}]$$

An important general reaction of carbonylate anions or substituted carbonylate ions is with halogen compounds. By this reaction, metal—carbon or metal—metal bonds can be formed. Typical examples are

$$\text{Mn(CO)}_5^- + \text{ClCH}_2\text{CH{=}CH}_2 = (\text{CO})_5\text{MnCH}_2\text{CH{=}CH}_2 + \text{Cl}^-$$

$$\text{Fe(CO)}_4^{2-} + 2\text{Ph}_3\text{PAuCl} \rightarrow (\text{Ph}_3\text{PAu})_2\text{Fe(CO)}_4 + 2\text{Cl}^-$$

$$\text{Co(CO)}_4^- + \text{Mn(CO)}_5\text{Br} \rightarrow (\text{OC})_4\text{CoMn(CO)}_5 + \text{Br}^-$$

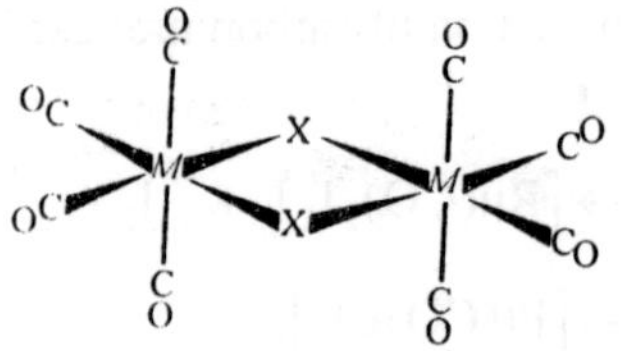

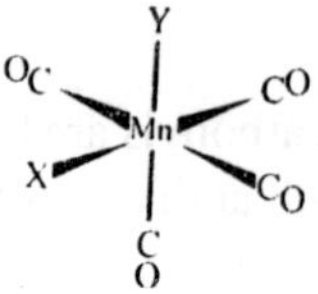

Fig. 8.12 : The structures of $[M(CO)_4X]_2$ (M = Mn. Re: X = halogen) and *trans*-$Mn(CO)_4XY$.

Finally, anionic carbonyl halides are obtained when metal carbonyls are treated with ionic halides

$$M(CO)_6 + R_4N^+X^- \rightarrow R_4N^+ [M(CO)_5X]^- + CO$$
$$M = Cr.\ Mo.\ W$$
$$Mn2(CO)_{10} + 2R_4N^+X^- \rightarrow (R_4N^+)_2 [Mn_2(CO)_8X_2]^{2-} + 2CO$$

The metal carbonyl halides are useful reagents for the preparation of compounds containing mixed metal-metal bonds

$$(CO)_5MnI + Na[Mr] \rightarrow (CO)_5MnMR + NaI$$
$$Mr = Co(CO)_4^-.$$

Carbonyl Hydrides (IX) Acidification of aqueous solutions containing carbonylate anions liberates carbonyl hydrides

$$Na[Co(CO)_4] + H^+ \xrightarrow{aq} HCo(CO)_4 + Na^+$$

$$NaHFe(CO)_4 + H^+ \xrightarrow{aq} H_2Fe(CO)_4 + Na^+.$$

These substances can also be prepared directly from hydrogen

$$Mn_2(CO)_{10} + H_2 \xrightarrow[200°]{200\text{ atm}} 2HMn(CO)_5$$

$$Co + 4CO + \frac{1}{2}H_2 \xrightarrow[150°]{50\text{ atm}} HCo(CO)_4.$$

The carbonyl hydrides are generally unstable, volatile liquids, as indicated by the data in Table 8.4. These susbstances are sparingly soluble in water in which they act as acids; in this respect the nomenclature seems to be anomalous.

Table 8.4 : The properties of some carbonyl hydrides

Compound	Colour	Mp, °C	pK_a	$\tau(^1H)$. ppm[a]
$HMn(CO)_5$	colourless	−25	~7.1	17.5
$H_2Fe(CO)_4$	yellow	−70	~4.4	21.1
$H_2Fe_3(CO)_{11}$	red	—	—	25
$HCo(CO)_4$	yellow	−26	strong	20

Table 8.5 : Some Representative Metal Carbonyls and Carbonyl Hydrides

Compound	Colour and form	Structure	Comments
A. *Mononuclear Carbonyls*			
$V(CO)_6$	Black crystals; dec. 70°; sublimes in vacuum	Octahedral	Yellow-orange in solution; paramagnetic ($1e^-$)
$Cr(CO)_6$ $Mo(CO)_6$ $W(CO)_6$	Colourless crystals; all sublime in vacuum	Octahedral	Stable to air; dec. 180-200°
$Fe(CO)_5$	Yellow liquid; mp − 20° bp 103°	*tbp*	Action of uv gives $Fe_2(CO)_9$
$Ru(CO)_5$	Colourless liquid; mp − 22°	tbp (by ir)	Very volatile and difficult to prepare
$Ni(CO)_4$	Colourless liquid; mp − 25° bp 43°	Tetrahedral	Very toxic; musty smell; flammable; decomposes readily to metal
B. *Polynuclear Carbonyls*			
$Mn_2(CO)_{10}$	Yellow solid mp 151° Subl. 50° (10^{-2} mm)		The Mn—Mn bond is long (2.93 Å) and $Mn_2(CO)_{10}$ is reactive
$Fe_2(CO)_9$	Gold solid mp 100° decomp.		Very insoluble and nonvolatile
$Fe_3(CO)_{12}$	Green-black solid mp 140-150 decomp.		Moderately soluble

$Rh_4(CO)_{12}$	Brick red solid mp 150° decomp. Subl. 65° (10^{-2} mn)		Useful reagent for many carbonyl rhodium compounds
C. *Carbonyl Hydrides*[a]			
$HmN(CO)_5$	Colourless liquid mp – 25°	Octahedral	Stable at 25°C, weak acid τ = 17.5
$H_2Fe(CO)_4$	Yellow liquid, colourless gas mp – 70°	Uncertain	Decomp. – 10°. Weak acid τ = 21.1
$H_2Fe_3(CO)_{11}$	Dark red liquid	Uncertain	
$HCo(CO)_4$	Yellow liquid, colourless liquid, mp – 20°	Distorted *tbp*	Decomp. above mp strong acid τ = 20

9

Hydrogen

Element	Atomic	Isotopes	Atomic Number	Electron configuration
H	1.008	1. 2. 3	1	1

Occurrence of hydrogen. In the order of abundance, the element hydrogen stands eighth among the elements. It makes up only about 0.1 per cent of the earth's crust, but since it constitutes about 11 per cent a water and is a constituent of all plant and animal tissues, hydrogen makes up about 1 per cent of all the matter on the earth. In all of these materials, hydrogen occurs in the compound form. Because of the large excess of oxygen in the atmosphere and the great reactivity of hydrogen toward oxygen, only about of $\frac{1}{300}$ of 1 per cent of the atmosphere is elementary hydrogen.

Periodic Classification

Hydrogen is the first element in the Periodic Table. With atomic number = I, the nucleus of the most abundant isotope is one proton with one valency electron outside it; the mass number is therefore I.

There is a naturally occurring isotope of mass number = 2. This has a neutron as well as one proton in the nucleus. Of course, the atomic number remains 1 and this isotope occupies the same position *(isos topos)* as ordinary hydrogen in the Periodic Table. However, the doubled mass accentuates the physical difference between the isotopes, which become of great interest and are given special names. The heavier isotope is called *deuterium* ,${}_1^2H = D$, and the lighter, sometimes, protium, ${}_1^1H = H$. On earth, the weight ratio $\frac{H}{D}$ is roughly constant $\left(\text{about}\frac{6000}{1}\right)$ about I in all sources of hydrogen, so that roughly 1 in 12000 atoms of hydrogen is deuterium. Hydrogen, therefore, consists mostly of H_2 molecules with some D_2 and HD molecules. A third isotope, artificial and radioactive (by β decay), has a mass number = 3.${}_1^3H$ is known as *tritium* and has 2 neutrons and 1 proton in the nucleus. While diffusion and thermal diffusion methods may be used to separate the isotopes, the technical production of water rich in deuterium, heavy water, D_2O and HOD, is accomplished by electrolysis in Norway where there is abundant cheap hydroelectric power. The $\frac{H}{D}$ ratio in the gas discharged is less than in water, causing enrichment of the electrolyte liquor. In 1933, using about 500 gallons of water from commercial electrolysis cells (already enriched), Lewis obtained 83 ml of 99% pure D_2O.

Deuterium and hydrogen (protium) will differ only in properties dependent on mass, enabling diffusion methods to be used in their separation. Chemically, the relatively large difference in mass (D = 2, H = 1) leads to a noticeable difference in reaction rates, a heavier atom being slower than a lighter atom with the same kinetic energy. Farkas found that chlorine in the light at 30°C combines with hydrogen at three times the rate of its reaction with deuterium. The melting-point of deuterium, D_2, is 4.70°C higher and the boiling point 3.2°C higher than for hydrogen, H_2. The maximum density for heavy water, D_2O, deuterium oxide, occurs at 11.22°C compared with 4.08°C for H_2O; furthermore, D_2O has a freezing point 3.8°C higher and a boiling-point 1.4°C higher than water.

Laboratory Preparation of Hydrogen

For the preparation of hydrogen in laboratory quantities, the action of sulfuric acid on zinc is frequently employed. The apparatus shown in Figure 9.1 may be used. Acid is added through the thistle tube and reacts with the zinc placed in the flask. The hydrogen gas is collected by displacement of water in which it is relatively insoluble:

$$Zn + H_2SO_4, \rightarrow ZnSO_4 + H_2 \uparrow$$

This reaction may be catalyzed by the addition of a few drops of copper sulfate solution. Zinc displaces the less active copper:

$$Zn + CuSO_4 \rightarrow ZnSO_4 + Cu$$

The finely divided copper deposits on the zinc and acts as a catalyst for the reaction.

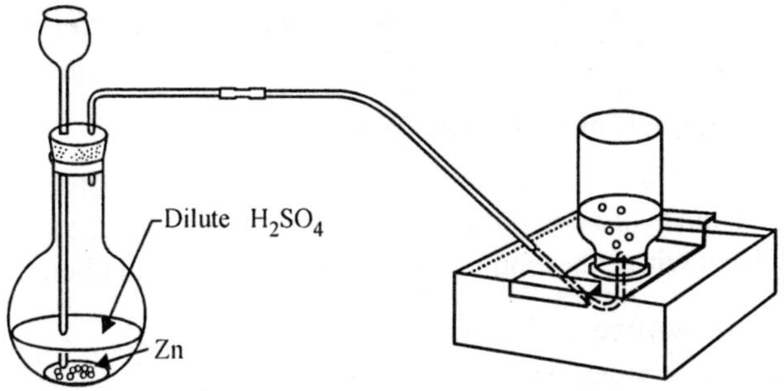

Fig. 9.1 : Apparatus for generating hydrogen.

Action of water on metals

Those metals near the top of the activity series are sufficiently active to displace hydrogen from water. If a small piece of sodium metal is dropped into water, a very vigorous reaction ensues with the liberation of hydrogen gas. The experiment must be carried out with a great deal of caution, as the action may throw sodium or sodium hydroxide around.

Sodium displaces only one hydrogen atom in the water molecule, forming sodium hydroxide and hydrogen. Sodium hydroxide is a base and turns red litmus blue:

$$2\ Na + 2\ H_2O \rightarrow \underset{\text{sodium hydroxide}}{2\ NaOH} + H_2 \uparrow$$

Potassium, barium, and calcium displace hydrogen from water in a similar manner; in each case the metal hydroxide (base) and hydrogen are produced. These first metals in the activity series are sufficiently active to displace hydrogen from ice water:

$$\text{Metal (K, Ba, Ca, Na)} + H_2O \rightarrow \text{Metal hydroxide} + H_2\uparrow$$

The next nine metals in the activity series, magnesium, aluminum manganese, zinc, chromium, cadmium, iron, cobalt, and nickel, while not active enough to displace hydrogen from cold water, will act with water in the form of steam. These metals displace all of the hydrogen in the water molecule and yield the oxide of the metal used and free hydrogen :

$$\text{Metal} + H_2O\ \text{(steam)} \rightarrow \text{Oxide of Metal} + H_2\uparrow$$

Lead and tin, which follow in the series, are too inactive to act with water at any temperature to any noticeable degree.

Preparation of hydrogen from bases

Certain metals will displace hydrogen from certain bases (metallic hydroxides) to yield hydrogen. For example, zinc or aluminum will act with either sodium or potassium dioxide to liberate hydrogen :

Manufacture of Hydrogen

Action of steam on coke or hydrocarbons

The first process involving coke is carried out by passing steam over white-hot coke at a temperature of about 1200°C. The reaction is endothermic and the temperature falls to 800°C. A blast of air is now turned on to raise the temperature to 1200°C again; the two reactions are :

$$C + H_2O \rightarrow \underbrace{CO + H_2}_{\text{water gas}} \qquad \text{(endothermic)}$$

$$2C + (O_2 + 4N_2) \rightarrow \underbrace{2CO + 4\,N_2}_{\text{produce gas}} \ \text{(exothermic)}$$

The water gas is now treated with more steam at a temperature of about 450°C in the presence of a special iron oxide catalyst; the carbon monoxide is oxidised to carbon dioxide and an extra molecule of hydrogen is formed:

$$\underbrace{CO + H_2}_{\text{water gas}} + H_2O \rightarrow CO_2 + 2H_2 \qquad \text{(shift reaction)}$$

Carbon dioxide is removed by scrubbing with water under pressure or by treatment with a hot solution of potassium carbonate.

The above method is now being replaced by a very similar process which uses cheap refinery hydrocarbons in place of coke. The first stage of the process is operated at a temperature of about 900°C with a nickel catalyst, when a mixture of carbon monoxide and hydrogen (synthesis gas) is formed. The second stage (the shift reaction) is exactly similar to the water gas/steam reaction above, and carbon dioxide is removed by the same technique. Traces of carbon monoxide still remaining are removed by absorption in an ammoniacal solution of copper (I) formate.

$$C_nH_{2n+2} + nH_2O \rightarrow nCO + (2n + 1)\, H_2$$

synthesis gas

$$nCO + nH_2O \rightarrow nCO_2 + nH_2 \text{ (shift reaction)}$$

Other methods of producing hydrogen

Hydrogen is produced as a by-product during the cracking of hydrocarbons, and in the manufacture of chlorine and sodium hydroxide by electrolysing a concentrated solution of sodium chloride.

Properties of hydrogen

(a) Physical

Hydrogen is **a** colourless, odourless gas and is relatively insoluble in water; 100 ml. of water dissolves about 2 ml. of hydrogen gas under ordinary conditions of temperature and pressure. It is the lightest known substance; therefore it has the smallest density. Hydrogen gas can be liquefied and solidified only with considerable difficulty; the gas liquefies at temperatures below –253°, and solidifies at –259°.

Certain metals such as platinum and palladium have the property of adsorbing large volumes of hydrogen, often adsorbing many hundred times their own volume of the gas. If the metal is finely divided, thus increasing its surface area, more gas will be adsorbed. This hydrogen appears to be more active-than ordinary hydrogen in many reactions and is referred to as "activated hydrogen gas." This is the basis for the use of platinum as a catalyst for many reactions in which hydrogen is one of the reacting substances.

Hydrogen gas *diffuses* more rapidly than any other gas-by diffusion we mean that it mixes rapidly with other gases. All gases will diffuse or mix with one another, and the speed of diffusion is a function of the density of the gases- the lower the density, the more rapid the rate of diffusion. This property may be demonstrated by the following experiment. A beaker of hydrogen gas is placed over a porous clay pipe arranged as shown in Figure 9.2, The hydrogen gas passes into the pipe faster than the air can diffuse out, since hydrogen is the lighter gas and therefore diffuses more rapidly. As a result the increased pressure inside the pipe and bottle forces liquid up through the jet.

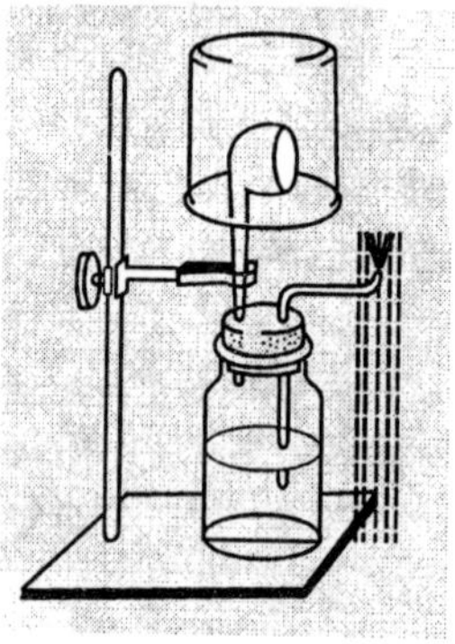

Fig. 9.2 : Apparatus to show the relative rates of diffusion of air and hydrogen.

(b) Chemical

A mixture of hydrogen and oxygen gases may be heated moderately without any apparent action taking place; however, if the mixture is ignited by a spark, combination takes place with explosive violence to form water :

$$2\,H_2 + O_2 \rightarrow 2\,H_2O$$

This action may be used as a basis for testing the purity of a sample of hydrogen. A small amount of the gas is ignited in a test tube; if a slight explosion occurs, the gas contains air or oxygen and is therefore impure. Pure hydrogen burns quietly with an almost colourless flame and does not support combustion.

When hydrogen gas is passed over heated copper oxide in a test tube, the copper oxide gradually changes from black to a copper-coloured powder (Figure 9.3). Small drops of liquid condense on the sides of the test tube. The reddish deposit remaining in the test tube is metallic copper, and the liquid formed on the sides of the tube is water:

$$CuO + H_2 \rightarrow Cu + H_2O$$

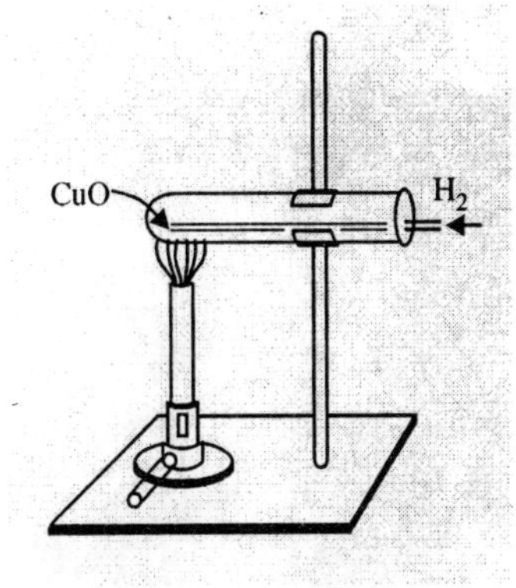

Fig. 9.3 : Reduction of cupric oxide by hydrogen gas.

This reaction is another example of displacement, since the hydrogen has taken the place of the copper in the copper oxide. In such a chemical change, hydrogen is sometimes referred to as a *reducing* agent, since the compound has been reduced or changed to an element. Oxygen or those substances which give up oxygen in a chemical change are referred to as *oxidizing* agents; hence copper oxide in the above case is termed the oxidizing agent. The copper oxide may be said to be reduced and the hydrogen to be oxidized, and the entire process may be referred to as an oxidation-reduction reaction. Oxidation and reduction are simultaneous processes; one never takes place without the other.

Certain other metallic oxides may be reduced with hydrogen to form the free metal. Some of the less common metals—tungsten and molybdenum, for example—are prepared in a very pure condition in this manner. In general, the cost of production with hydrogen as a reducing agent is relatively high, and cheaper reducing agents are employed in the commercial preparation of most common metals.

Hydrogen combines directly with many nonmetals; for example,

$$H_2 + S \rightarrow \underset{\text{hydrogen sulfide}}{H_2S}$$

$$H_2 + Cl_2 \rightarrow \underset{\text{hydrogen chloride}}{2\,HCl}$$

Active Hydrogen

Atomic hydrgen

The hydrogen molecule can be dissociated into atoms using high energy sources, such as a discharge tube containing hydrogen at low pressure, or a high current density arc at high temperature. This dissociation is highly endothermic:

$$H_2 \rightarrow 2H \ \Delta H = +435.9 \text{ kJ/mol of } H_2$$

Many metals are able to catalyse the recombination of hydrogen atoms, e.g. platinum and tungsten, which results in the liberation of the same quantity of energy as is needed to effect the dissociation. This effect is used in the atomic hydrogen blowlamp for welding metals. As is to be expected, atomic hydrogen is a very powerful reducing agent, e.g. it will reduce metallic oxides and chlorides to metals, and oxygen to hydrogen peroxide.

Nascent hydrogen

When hydrogen is bubbled into an iron (III) solution no reduction takes place; if the reaction is carried out by adding a mixture of an iron (III) salt in dilute sulphuric acid to zinc an immediate reduction to iron (II) ions takes place. It was at one time thought that an explanation of these observations required the participation of nascent hydrogen (hydrogen at the instant of formation). This notion is now known to be false, since zinc would be expected to be a better reducing agent than hydrogen on account of its higher position in the electrode potential series; indeed, zinc itself effects the reduction without the need for dilute acid:

$$2Fe^{3+}\ Zn \rightarrow 2Fe^{2+} + Zn^{2+}$$

It should be thermodynamically possible for hydrogen gas to reduce iron (III) ions to the iron (II) state, since its standard electrode potential (arbitrarily fixed at zero) is more negative than that for the system Fe^{3+}/Fe^{2+} (E^{ϕ} = + 0.76v). The fact that it does not do so must be due to a high energy of activation for the reaction, a consequence of the fact that the bond uniting two hydrogen atoms together is a very strong one.

Hydride type and the Periodic Table

A metal is an element which forms simple positive ions (cations) by the loss of electrons. However, it is difficult to decide whether some elements are best classed as metals or non-metals. The strong metals form ionic hydrides. They are metal elements with low electronegativity (highly electropositive) found in Groups I and II. Beryllium and magnesium, the less electropositive members of Group II, form hydrides which are not of this simple typc.

	I	II
Metals forming	Li	
ionic hydrides	Na	
	K	Ca
	Rb	Sr
	Cs	Ba

Calcium hydride is the least ionic. Hydrides of metals below sodium in Group I are highly reactive.

	III	IV	V	VI	VII
Elements forming covalent hydrides	B	C	N	0	F
		Si	P	S	Cl
		Ge	As	Se	Br
		Sn	Sb	Te	I
		Pb	Bi	Po	

Non-metals form covalent hydrides. But so do certain metals and metalloids. The stability of the hydrides falls off with increase in atomic number. In Group V, BiH_3 was first detected as a radioactive gas in small yield from the action of acid on an alloy of magnesium and a radioactive bismuth isotope. This hydride must be placed in the progression NH_3, PH_3, AsH_3, SbH_3 and BiH_3, where the hydrides prove progessively less stable with the development of metallic characteristics. The hydride type, and particularly the stability of the covalent hydride, is a good criterion for grading the elements.

Scandium Subgroup

Element	Symbol	Atomic weight	Atomic number
Scandium ...	Sc	44.96	21
Yttrium	Y	88.92	39
Lanthanum ..	La	138.92	57
Actinium	Ac	227	89

Ionic (salt-like) hydrides

In reactions with certain strong metals, a hydrogen atom receives an electron to form the negative hydrogen ion, the hydride ion, H^-. Lithium and calcium react:

$$[Li]^+[H_o^x]^- \quad LiH$$
$$[Ca]^{2+}[H_o^x]_2^- \quad CaH_2$$

Atomic number	*Electronic configuration* *Before*	*After*
I	H I	2 (He)
3	Li 2, I	2 (He)
20	Ca 2, 8, 8, 2	2, 8, 8, (Ar)

These hydrides are colourless solids, of low volatility and with ionic lattices. They are known as the saline, or salt-like, hydrides. Although fusion brings slow decomposition, rapid electrolysis yields hydrogen at the anode in accordance with Faraday's Laws. The metal appears at the cathode. A few reactions may be mentioned, using lithium hydride and writing ionic equations. With water, lithium hydroxide and hydrogen are formed:

$$H^- + H_2O \rightarrow H_2\uparrow + OH^-$$

Anhydrous ammonia yields the amide:

$$H^- + NH_3 \rightarrow H_2 + NH_2^-$$

Carbon dioxide gives the formate, $HCOO^-Li^+$; carbon monoxide, the formate and carbon:

$$H^- + {}_2CO \rightarrow H.C00 + C$$

and sulphur dioxide, the ditmonite

$$2u^- + 2SO_2 \rightarrow S_2O_4^{2-} + u_2$$

Covalent hydrides

By far the greatest number of hydrides come under this classification and, except for a few, they are gaseous at ordinary temperature. They are formed by the elements from Group IVB to Group VIIB in the Periodic Table.

Other hydrides

Certain non-transition metals form non-volatile hydrides which are not ionic and are relatively unstable, examples are BeH_2, MgH_2, AlH_3, ZnH_2, CdH_2, HgH_2, CuH.

(Metal-like) Interstitial Hydrides

A group of elements, constituting nearly all the transition metals, reacts with hydrogen under a variety of conditions to form substances the structures of which have been termed interstitial. Originally the term interstitial compound implied that the arrangement of metal atoms was essentially the same as that in the pure metal, with hydrogen atoms entering the interstices. Even though this is not generally the case, the term is still used. In contrast with the saline hydrides, which are colourless and occupy less volume than the metal from which they are formed (Table 9.1), the metal-like hydrides occupy a greater volume than the parent metal, are metallic in appearance, and often possess variable composition.

The metallic hydrides can be prepared by direct combination, usually at elevated temperatures. Thus, platinum, palladium, and iron are permeable to hydrogen at elevated temperatures; tantalum absorbs hydrogen to yield a brittle product. Hydrogen can also be absorbed under more moderate conditions. For example, metallic cathodes used in the electrolysis of aqueous as well as nonaqueous solutions often absorb very large quantities of hydrogen; in some instances (for example, iron) electrolytically deposited metals will also absorb hydrogen.

Table 9.1 : Change in density between some metals and their hydrides (14-16)

Hydride	Density change, %
LiH	+ 52.8
NaH	+ 44
CaH_2	+ 10
BaH_2	+ 20
$LaH_{2.76}$	− 12.8
$CeH_{2.69}$	− 17.5
$TaH_{0.76}$	− 9.1
$VH_{0.56}$	− 6.7

Although many examples of metallic hydrides are known which possess a stoichiometry suggesting the formation of true compounds (Table 9.2), there is evidence that these susbstances are not compounds in the usual sense.

Table 9.2 : A comparison of the structures of some metals and of their hydrides

Hydride	Structure[a] of metal atoms in	
	Hydride	Pure metal[b]
TiH_2	Fluorite[c]	bcc, hcp
ZrH, TiH	Zinc b lende[d]	bcc, hcp
Pd_2H	Zinc blende[e]	cccp
Zr_4H	Zinc blende[f]	bcc, hcp
CuH	Wurtzite	ccp
LaH_2, PrH_2, NdH_2	Fluorite[c]	hcp
CeH_2	Flurite[c]	ccp, hcp
ThH_2	Distorted fluorite[c]	ccp
CrH	Wurtzite	bcc
CrH_2	Fluorite	bcc

b Key : bcc, body-centered cubic; hcp, hexagonal closest-packed; ccp, cubic closest-packed

c All tetrahedral holes in the cubic-closest-packed structure of metal atoms filled.

d One-half of ccp tetrahedral holes filled.

e One-fourth of ccp tetrahedral holes filled.

f One-eight of ccp tetrahedral holes filled.

Thus, the hydrogen-zirconium phase diagram shows the existence of α-zirconium (hexagonal closest-packed) which can dissolve several per cent of hydrogen; β-zirconium (body-centered cubic) which is stable at high temperatures and can dissolve up to 50 atom % hydrogen; and the compound ZrH_2, which exists in two modifications (face-centered cubic and tetragonal) (*V*). As a second example, although well defined stoichiometries for the lanthanum dihydrides and trihydrides have been reported, there are also substances reported with variable composition in the range $MH_{1.8}$ to $MH_{3.0}$.

The fundamental difference between the ionic hydrides and the metallic hydrides is clearly illustrated by the way the corresponding metals absorb hydrogen (Fig 9.4) The metal which form the saline hydrides show extended temperature ranges in which hydrogen absorption does not occur until conditions favourable for the formation of the hydride are reached; at this point the expected stoichiometric amount of hydrogen is "absorbed." In contrast, some of the metals that form metallic hydrides absorb hydrogen in amounts that are stoichiometrically correct, whereas for others hydrogen absorption *increases* as the temperature is raised. Little evidence has been found for compound formation in the metallic hydrides.

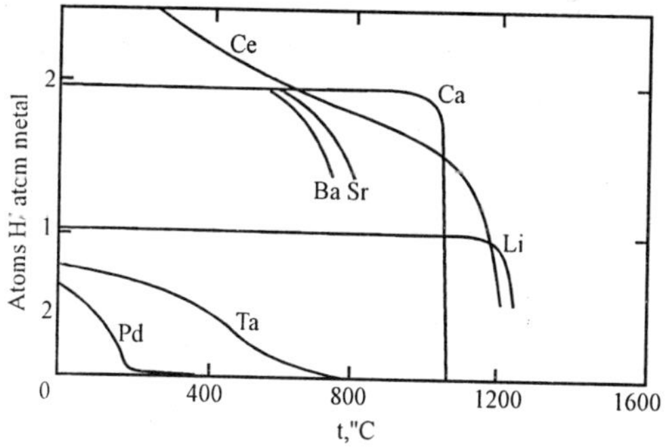

Fig. 9.4 : The absorption of hydrogen by various metals (8).

The nature of the hydrogen atoms bound in the metallic hydrides is not clearly understood. Some of these substances are good reducing agents. suggesting the presence of hydridic or, indeed, atomic hydrogen. The rapid diffusion of hydrogen in metals such as palladium indicates that perhaps unbound hydrogen atoms are present. The constitution of these systems may well vary with composition. For example, up to a composition of $PdH_{0.5}$ the palladium-hydrogen system has the properties associated with a conductor, but at higher concentrations of hydrogen it is a semiconductor. In the same range of stoichiometries the paramagnetism decreases from the value for the pure metal to a minimum at a composition corresponding to $PdH_{0.6}$. It is apparent that a unified description of the metallic hydrides is not presently available.

The Hydrogen Bond

When hydrogen is bonded to another atom, X, mainly F, O, N, or Cl such that the X-H bond is quite polar with H bearing a partial positive charge, it can interact with another negative or electron-rich atom, Y, to form what is called a hydrogen bond (H-bond), written as

$$X\text{-}H\text{----}Y$$

Although the details are subject to variation, and controversy, it is generally believed that typical hydrogen bonds are due largely to electrostatic attraction of H and Y. The X-H distance slightly longer, but this bond remains essentially a normal 2-electron bond. The H---Y distance is generally much longer than that of a normal covalent H-H bond.

In the case of the very strongest hydrogen bonds, the X to Y distance becomes quite short and the X-H and Y-H distances come close to being equal. In these cases there are presumably covalent and electrostatic components in both the X-H and Y-H bonds.

Experimental evidence for hydrogen bonding came first from comparisons of the physical properties of hydrogen compounds. Classic examples are the apparently abnormally high boiling points of NH_3, H_2O, and HF such as heats of vaporization provided further evidence for association. Although physical properties reflecting association are still a useful tool in detecting hydrogen bonding, the most satisfactory evidence for solids comes from X-ray and neutron-diffraction crystallographic studies, and for solids, liquids, and solutions from infrared and nuclear magnetic resonance spectra.

Structural evidence for hydrogen bonds is provided by the X to Y distances, which are shorter than the expected van der Waals contact when a hydrogen bond exists. For instance in crystalline $NaHCO_3$ there are four kinds of O----O distances between HCO_3^- ions with value of 3.12, 3.15, 3.19, and 2.55 Å. The first three are about equal to twice the van der Waals radius of oxygen, but the last one indicates a hydrogen bond, O-H---O. When an X-H group enters into hydrogen bonding, the X-H stretching band in the infrared spectrum is lowered in frequency, broadened, and increased in integrated intensity. These changes afford a very useful means of studying H-bonding in solution.

Uses of Hydrogen

Until this century only small quantities of hydrogen were required as a fuel in the form of town gas and water gas, for filling, balloons and in the oxyhydrogen blowlamp for welding. Nowadays, however, large quantities of the gas employed in the processes listed below :

(a) Manufacture of ammonia by the Haber process. This is used in turn to manufacture nitric acid, which can then be converted into explosives, dyestuffs and nitrogenous fertilizers.

(b) Manufacture of hydrogen chloride and hydrochloric acid. The hydrogen and chlorine are reacted together; the products is hydrogen chloride, which forms hydrochloric acid with water.

(c) Manufacture of organic chemicals, e.g. methanol, CH_3OH. This compound is obtained by reacting carbon monoxide with hydrogen at 400°C and 300 atmosphere pressure in the presence of zinc oxide and chromium (III) oxide.

(d) Manufacture of margarine. Groundnut oil is reacted with hydrogen in the presence of a nickel catalyst to give a solid fat.

(e) Extraction of some metals from their oxides. Molybdenum and tungsten are extracted by reduction of their oxides with hydrogen.

(f) Liquid hydrogen has been used as a rocket fuel.

Acids and Bases

Early Ideas

As early as 1663 Robert Boyle noted that a class of chemical compounds existed that had a similar set of properties, different from those of other compounds. These substances are called acids, and among their properties can be listed the following:

(a) Sour taste
(b) Ability to change the colour of vegetable dyes (indicators)
(c) High solvent power (dissolving ability)
(d) Tendency (for some) to react with certain metals to evolve hydrogen.

Other chemical compounds, called alkalis, were noted for their ability to produce a 'soapy feel' in water, affect vegetable dyes, but above all to neutralise acids to produce a salt and water. It soon became apparent that other substances shared this last property with alkalis—this oxides and hydroxides of metals insoluble in water—and the term base, introduced by Rouellie in 1754, was used to cover all compounds that on reaction with an acid formed a salt and water only.

The Arrhenius Theory of Acids and Bases

In 1884 Arrhenius suggested that substances which contained hydrogen and gave hydrogen ions in aqueous solution were acids; similarly, bases were substances which produced hydroxyl ions when dissolved in water. It is now known that the hydrogen ion is hydrated in solution in the hydroxonium ion, H_3O^+, and indeed some recent work has suggested that hydration may proceed further to give $H^+(H_2O)_4$ or $H_9O_4^+$ ions. In this book hydrogen ions in aqueous solution will be represented as H_3O^+ ions.

The ionisation of hydrogen chloride and nitric acid, both covalent compounds, in water can be represented thus :

$$HCl + H_2O \rightarrow H_3O+ + Cl^-$$

$$HNO_3 + H_2O \rightarrow H_3O^+ + NO_3^-$$

(the chloride and nitrate ions are almost certainly hydrated)

The dissociation of a base in aqueous solution follows a similar course and, although hydration of cation and anion takes place, it is easier to write down the simple ions. Thus the dissociation of potassium and sodium hydroxides, both existing in the form of ions in the solid state (electrovalent compounds), can be represented as :

$$K^+OH^- + \text{water} \rightarrow K^+ + OH^-$$

$$Na^+OH + \text{water} \rightarrow Na^+ + OH^-a$$

The neutralisation of an acid by a base, for example the reaction between hydrochloric and acid and sodium hydroxide in aqueous solution, can be represented thus :

$$H_3O^+ + Cl^- + Na^+ + OH^- \rightarrow 2H_2O + Cl^- + Na^+$$

or, since chloride and sodium ions occur on both sides of the equation more simply as :

$$H_3O^+ + OH^- \rightarrow 2H_2O$$

The Arrhenius concept of acids and bases thus allows the process of neutralisation to be expressed by one simple equation. Furthermore, it predicts that interaction between dilute strong acids, and bases, i.e. those that are completely dissociated in solution, should evolve a constant amount of heat. This prediction is borne out in practice when it is found that one gramme equivalent of any strong acid neutralises one gramme equivalent of strong base in dilute solution with the evolution of 57.3 kJ.

Bronsted-Lowry Acids and Bases

Two kinds of definitions that scientists use, and that you have used perhaps without knowing their names are called operational and conceptual definitions. An **operational definition** is based on *directly observable properties* or effects. Acids can be defined operationally as substances with a sour taste that neutralize bases and turn litmus paper red. Similarly, bases can be defined operationally in terms of their properties.

A **conceptual definition** is based on the *interpretation of observed facts.* Acids can be defined conceptually as substances that produce hydrated hydrogen ions as the only positive ions in their aqueous solutions. Bases also can be defined conceptually as substances that produce hydroxide ions as the only negative ions in their aqueous solutions.

The conceptual definitions presented earlier in this chapter are based on Arrhenius's nineteenth-century interpretation of observed facts. You come now to a newer interpretation, that of Bronsted and Lowry.

In 1923, J.H. Bronsted (1879-1947), a Danish chemist, and T. M. Lowry (1874-1936), an English chemist, proposed independently a new conceptual definition of an acid. The concept is known as the Bronsted-Lowry theory. According to this theory, an acid is a substance, either molecule or ion, that can *donate* a proton (a hydrogen atom without its single electron) to another substance. A base, according to the theory, is any substance that can *accept* a proton from another substance. Thus, an acid is a **proton donor,** and a base is a **proton acceptor.**

For an illustration of these definitions, see the equation for the ionization of hydrogen chloride gas in water solution (Equation 14). Look again at Figure 9.5, which shows the reaction with models.

$$Hcl(g) + H_2O(l) \rightarrow H_3O^+(aq) + Cl^-(aq) \qquad \text{(Eq. 14)}$$

acid	base	hydronium	chloride
(proton	(proton	ion	ion
donor)	acceptor)		

With electron-dot symbols, Equation 14 can be written :

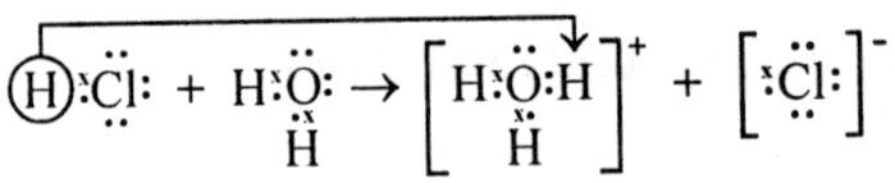

In this reaction, the HCl molecule donates a proton (H^+) to the water molecule, thus forming a hydronium ion. (The proton, with its single positive charge, gives the hydronium ion its positive charge.) By accepting the proton, water acts as a Bronsted-Lowry base.

Another example of the Bronsted-Lowry definition is the reaction for the ionization of ammonia gas. Ammonia gas is very soluble in water and ionizes slightly in solution. See Figure 9.5.

Ionization of Ammonia Gas, NH_3

$$NH_3(g) + H_2O(l) \rightarrow NH_4^+(aq) + OH^-(aq) \qquad \text{(Eq. 15)}$$

Base	acid	acid	base

With electron-dot symbols. Equation 15 can be written :

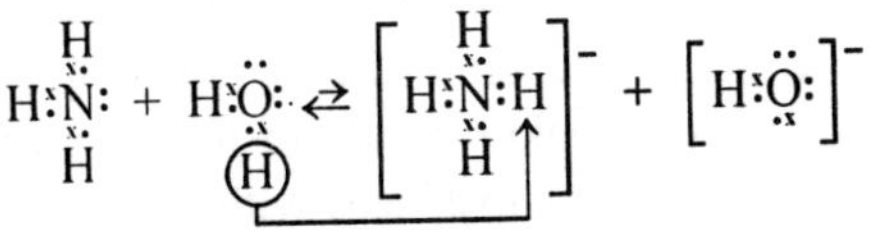

In this reaction, a water molecule donates a proton to an ammonia molecule. The loss of a proton, H^+, from a water molecule leaves behind a hydroxide ion, OH^-. The gain of a proton by an ammonia molecule, NH^3, produces an ammonium ion, NH_4^+. In the reverse reaction, the NH_4^+ donates a proton, which the OH^- ion accepts. Hence. NH_1^+ is an acid in the reverse reaction, and OH^- is a base.

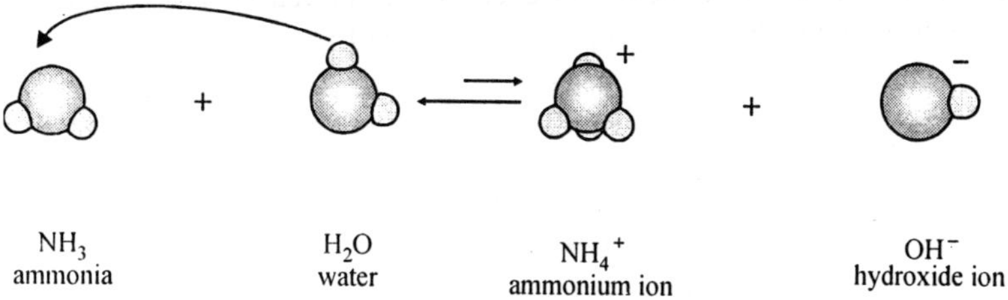

Fig. 9.5 : In a reversible reaction favoring the existence of the molecules rather than the ions, ammonia gas reacts with water to form ammonium ions and hydroxide ions.

Arrhenius, Bronsted-Lowry Theory Compared

	Theory	
	Arrhenius	**Bronsted-Lowry**
Definition of an acid	Any substance that releases H^+ ions as the only positive ion in aqueous solutions	Any substance that donates a proton
Definition of a base	Any substance that releases OH^- ions as the only negative ion in aqueous solutions	Any substance that accepts a proton

Fig. 9.6 : The acid-base theories of Arrhenius and Bronsted-Lowry

Notice that in the ionization of ammonia (equation 15), water functions as an acid, but in the ionization of hydrogen chloride

(Equation 14), it functions as a base. Whether water donates or accepts protons depends on what substance it reacts with. Substances with a strong tendency to donate protons (that is, substance with loosely held protons because of weaker bonds to hydrogen atoms) make water a proton acceptor. But substances with a strong attraction for protons make water a proton donor.

The Lewis Theory. A concept of acids and bases which is more general in its application is that introduced by G. N. Lewis. According to the Lewis theory, any substance which an unused pair of electrons (such as NH_3) may be considered as a base, while any substance which may attach itself to such an available pair of electrons would be classed as an acid. For example, in the reaction of an acid anhydride with a basic anhydride to form a salt :

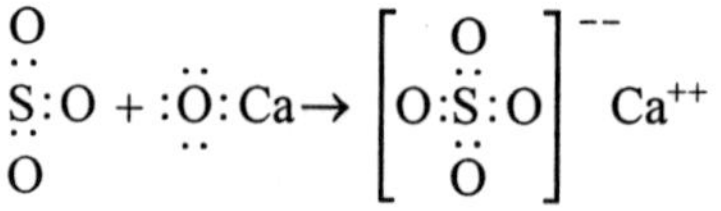

Fig. 9.6

CaO would be considered a base and SO_3 an acid. In the Lewis theory, a base is *an electron pair donor* and an acid an *electron pair acceptor*.

If we apply the Lewis theory to the ionization of acids, for example, to

$$HCl + H_2O \rightarrow H_3O^+ + Cl^-$$

we would classify H_2O as a base, since it donates a pair of electrons to H^+.

The Lux-Flood definition

Lux originally proposed a different definition of acids and bases which was extended by Flood. Instead of using protons, or ions characteristic of the solvent, they defined acids as oxides which accept oxygen, and bases as oxides donate oxygen. Thus :

$$CaO + CO_2 \rightarrow Ca^{2+} [CO_3]^{2-}$$

$$SiO_2 + CaO \rightarrow Ca^{2+} [SiO_3]^{2-}$$
$$6Na_2O + P_4O_{10} \rightarrow 4Na_3^{+} [PO_4]^{3-}$$
acid base

This system is very useful in dealing with anhydrous reactions in fused melts of oxides, and other high temperature reactions such as are found in metallurgy and ceramics.

This theory has an inverse relationship to aqueous, chemistry, since Lux-Flood acids are oxides which react with water, giving bases in water, and Lux-Flood bases react with water, giving acids.

$$Na_2O + H_2O \rightarrow 2NaOH$$
$$P_4O_{10} + 6H_2O \rightarrow 4H_3PO_4$$

The Usanovich definition

This defines an acid as any chemical species which reacts with bases, gives up cations, or accepts anions or electrons. Conversely a base is any chemical species which reacts with acids, gives up anions or electrons, or combines with cations. This is very wide definition and includes all the Lewis acid-base type of reactions, and in addition it includes redox reactions involving the transfer of electrons.

Hard and soft acids and bases

Metal ions may be divided into two types depending on the strength of their complexes with certain ligands.

Type (a) metals include the smaller ions from Groups 1 and 2, and the left side of the transition metals, particularly when in high oxidation states, and these form the most stable complexes with nitrogen and oxygen donor (ammonia, amines, water, ketones, alcohols), and also with F^- and Cl^-.

This empirical classification was useful in predicting the relative stabilities of complexes. Pearson extended the concept into a broad range of acid-base interactions. Type (a) metals are small and not very polarizable. Perarson called these metals hard acids, and the ligands hard bases. In a similar way, type (b) metals and the ligands they prefer are larger and more polarizable, and he called these soft acids and soft bases. He stated the relationship *hard acids prefer to react with hard bases, and soft acids react with soft bases.* This definition takes in the usually accepted acid-base reactions (H^+ strong acid,

OH^- and NH_3 strong bases), and in addition a great number of reactions involving the formation of simple complexes, and complexes with π bonding ligands.

Table 9.3 : Some hard and soft acids and bases

Hard acids	*Soft acids,*
H^+,	Pd^{2+}, Pt^{2+}, Cu^+, Ag^+, Au^+, Hg^{2+},
Li^+, Na^+, K^+,	$(Hg_2)^+$, Tl^+
Be^{2+}, Mg^{2+}, Ca^{2+}, Sr^+,	$B(CH_3)_3$, B_2H_6, $Ga(CH_3)_3$, $GaCl_3$,
Al^{3+}, BF_3, $Al(CH_3)_3$, $AlCl_3$,	$GaBr_3$, GaI_3
Sc^{3+}, Ti^{4+}, Zr^{4+}, VO^{2+}, CrI2,	[Fe(CO)5], $[Co(CN)_5]^{3-}$
MoO^{3+}, WO^{4+},	
Ce^{3+}, Lu^{3+},	
Co_2, SO_3	
Hard bases	*Soft bases*
NH_3, RNH_2, N_2H_4	H^-, CN^-, SCN^-, $S_2O_3^{2-}$, I^-, RS^-,
H_2O, ROPH, R_2O	R_2S, CO, B_2H_6, C_2H_4, R_3P, $P(OR)_3$
OH_-, NO_3^-, ClO_4^-, CO_3^{2-}, SO_4^{2-}–,	
PO_4^{3-}, CH_3COO^-, F^-, Cl^-	

10

Elements of Zero Group : Noble Gases—Group VIIIA (18)

1. Introduction

The electronic configuration of rare-gas atoms consists of filled *ns* and *np* orbitals, giving each atom a closed - shell structure. Classically such electronic configurations have been considered the epitome of an inert electronic structure; for example, many of the Lewis arguements concerning the nature of ionic and covalent compounds, as well as derivations of this theory, are based on the supposed inertness of such configurations. Since 1962 a considerable number of compounds containing chemically bonded rare gas atoms have been reported. As might be expected from their electronic configurations, shown in Table 10.1, the rare gases are monatomic, a conclusion supported by the ratio of their heat capacities (Cp/Cv),which is the value predicted (1.667) for a monatomic gas. The physical properties of the rare gases given in Table 10.1 show trends which are essentially ideal because no permanent dipolar or electrostatic interactions occur among the species which constitute the sample. Thus, the boiling and melting points vary regularly with increasing atomic number, as do many other physical properties.

The fact that rare-gas atoms can be condensed to give liquid and solid phases at low tempreatures suggests that weak interactive forces are operative in the spherically symmetrical, closed -shell electronic systems. It was suggested by London that electronic motions in a symmetrical molecule or atom give rise to fluctuating dipoles which time average to zero (1). These London forces cause the rare gases to form condensed phases at temperatures which are sufficiently low that the kinetic energy is not an important factor. The London, or dispersion, forces theoretically are directly proportional to the polarizability of the atoms, which is a measure of how loosely the elections in the system are held, and inversely proportional to their ionization potential; This prediction correlates well with the data in Table 10.1 which show that the polarizabilities of the rare gases increase with increasing atomic weight but the ionization potenial decreases in the same order. Thus, we would expect the London forces to increase with increasing atomic number, which should also be the order of increasing melting or boiling points.

Table 10.1 :

	Atomic Number	Electronic Configuration	First Ionisation Energy (kJ/mol)	Atomic Radius (A)	M.P. °C
He	2	2 $1s^2$	2372	1.2	–271.4 at 30 atm.
Ne	10	2.8 $...2s^2 2p^6$	2080	1.6	–248.7
Ar	18	2.8.8 $...3s^2 3p^6$	1519	1.9	–189.2
Kr	36	2.8.18.8 $...4s^2 4p^6$	1351	2.0	–157
Xe	54	2.8.18.18.8 $...5s^2 5p^6$	1170	2.2	–111.5
Rn	86	2.8.18.32.18.8 $...6s^2 6p^6$	1037	–	–71

The noble gases, except for helium, have eight electrons in their outer shell. Since they form chemical compounds less readily than other elements, this closed shell of eight electrons must be an extremely stable one. Theories of valency have been developed according to which it is suggested that elements react to achieve a stable noble gas configuration either by loss and gain of electrons, or by electron sharing. Although many exceptions are known where electronic configurations other than that of a noble gas are achieved, these simple ideas lead to an understanding of a large range of diverse chemical structures.

The discovery of the noble gases

Although the observation of a line in the sun's spectrum as early as 1868 led workers to postulate the existance of an unknown element in the sun's atmosphere, the isolation in 1889 of helium from the mineral cleveite by heating was not recognized as related phenomenon. The first definitive work was by Lord Rayleigh, who noticed a discrepancy between the density of "chemical nitrogen" and that of "atmospheric nitrogen." The former was obtained by chemically removing nitrogen from various nitrogen oxides, ammonia, or other compounds. The latter was obtained by removal of oxygen, carbon dioxide and water vapour from air. The careful work necessary to establish this difference has often been pointed to quite rightly, as an example of the importance of precise measurements. Unfortunately, too often the emphasis has been upon the number of significant figures rather than the realization by Rayleigh and Ramsay that the difference was chemically significant. Their arguments concerning the significance of the ratio of specific heats of the noble gases[3] ($C_p/C_v = 1.66$) and their rebuttal of various arguments advanced by their critics shows as much chemical insight as the gas density argument, if not more.[4]

Ramsay and Rayleigh succeeded in isolating all of the noble gases except radon and in showing that they were inert to all common reagents. They also discovered the identity of alpha particles and ionized helium.

Helium

This gas was first discovered by spectroscopic examination of light from the sun; it is formed by nuclear fusion of hydrogen atoms in stellar reactions (p.161). It occurs, together with natural gas, in various parts of the USA and Canada , some natural gas wells containing as much as 7 percent of the element. It has probably accumulated as a result of natural radioactive decay processes. Because it is very light and non-inflammable it has been used to inflate the tyres of large aircraft, thus increasing their maximum payload by several hundred pounds. An oxygen/helium mixture is used by drivers, since helium is less soluble in the blood than nitrogen, which is notorious for causing decompression sickness, or the 'bends'. It is used to provide inert atmospheres during certain welding operations, and seems to have a bright future in gas-cooled nuclear reactors since it does not become radioactive and is non-corrosive. Liquid helium is used in low temperature research.

Argon

Argon is extracted from the atmosphere, one useful source being Haber process plants, where it tends to accumulate in the recycled gases. It is used in electric lamps to suppress the evaporation of the filament, thus allowing higher temperatures to be used, with a consequent increase in efficiency. Like helium, it is used to provide an inert atmosphere for welding operations.

Neon, Krypton and xenon

These noble gases are extracted from liquid air along with argon; they can be separated by fractional distillation and selective absorption on charcoal. Neon is used in shop signs and street lamps; since the characteristic red glow produced when an electrical discharge is passed through neon at low pressure is able to penetrate fog, it finds a use in devices for illuminating airfields. Krypton is used in discharge tubes; there are no significant uses for xenon.

Radon

This gas is radioactive and is not found in the atmosphere. The most stable isotope of radon is 222/86**R**n, which is obtained during the decay of radium chloride solution.

$$222/86\text{Ra} \rightarrow 222/86\ \text{Rn} + 4/2\ \text{He}$$

It has been used in the treatment of cancer because of its radioactivity.

Compounds of the Elements(I)

The extremely stable eletronic configurations of the rare gases, suggest that only highly electronegative atoms or group can compete successfully with these atoms for their electrons to form either ionic or covalent species. Before the successful preparation of the first rare- gas compound, several suggestions had been made concerning the possibility of chemically combining rare-gas atoms. An unsuccessful attempt was made to combine xenon with fluorine under the influence of an electric discharge but 30 years later similar experiments proved successful in addition, stable rare-gas -halogen compounds were predicted on theoretical grounds before they were prepared.

Although we might expect the heavier rare gases to accommodate more than four electron pairs in their valence shells, it is highly unlikely that the neutral atoms are sufficiently electronegative to form a bond to a Lewis base. Thus, using the electronegativities of the elements and the ionization potentials of the rare gases as crude indicators of possible compound formation, we might expect that the most probable candidates for such compounds are the largest rare gases, radon and xenon, and the most electronegative of the elements, fluorine (4.0), oxygen(3.5) nitrogen (3.0), and chlorine (3.0). From a practical standpoint radon is excluded from these considerations because all of its isotopes are radioactive and have relatively short half-lives. In fact, the majority of the known rare- gas compounds contain xenon bonded to fluorine or oxygen. Krypton forms fewer and less stable compounds; evidence exists for unstable rare-gas compounds containing chlorine, but no nitrogen-containing compounds have been characterized. Thus far stable compounds containing the rare-gas atom in 1+,2+,4+, and 6+ formal oxidation states have been prepared.

Ionic Derivative

In the early 1960s Neil Bartlett was studying the properties of platinum hexafluoride, an extremely powerful oxidizing agent. In fact, by merely mixing oxygen with platinum hexafluoride it is possible to remove one election from the oxygen molecule and isolate the product:

$$O_2 + PtF_6 \rightarrow O^+_2\ [PtF_6]$$

Barlett realized that the first ionization energy of molecular oxygen, 1180 kj mol^{-1}(12.2eV), is almost identical to that of xenon, 1170kj mol^{-1}(12.1eV). Furthermore, the dioxygenyl cation should be roughly the same size as a Xe^+ion, and hence the lattice energies of the corresponding compounds should be similar. He mixed xenon and platinum hexafluoride with the result that an immediate reaction took place with formation of a yellow solid.[10]Based on the volumes of gases that reacted, Bartlett suggested the formulation $Xe^+[PtF_6]^-$ However, the reaction is not so simple as was once thought , and the product has been formulated variously as $Xe[Pt_2F_6]_2$,$Xe[PtF_6]x$ (1<+<2), or $FXe+[PtF_6]-+FXe+[Pt_2F_{11}]-$If a large excess of inert SF_6

is added as a diluent, pure Xe+PtF-6 may be isolated. Despite the experimental difficulties in characterizing the products of the reaction there was no doubt whatever that a reaction had taken place and the myth of inertness was shattered forever.

It is of some interest to note that several groups were thinking about the possibility of compound formation by the noble gases at the time of Bartlett's discovery. R.Hoppe's research group in Germany had analyzed bond energies of related nonmetal compounds (see below) and concluded that XeF_2 and XeF_4 should exist. However, they did not attempt any preparations untill after the publication of Bartlett's work. Clifford and Zeilenga in the United States were reinvestigating the earlier attempts to synthesize xenon fluorides, Working with Pyrex apparatus presented difficulties due to contamination from silicon, but they later isolated a product believed to be Xe_2SiF_6. Meanwhile, workers at Oak Ridge Laboratories were investigating the curious lack of xenon in the gas above molten fluoride reactor fuels. Fission of UF_4 in this fuel should have produced both Kr (which was observed) and Xe (which was not). With 20-20 hindsight it is easy for us to ascribe this discrepancy to the formation of xenon fluorides.

Once the initial breakthrough had been made, a flood of experiements followed, and within a year eight compounds were known and many physical measurements had been made.

Randon has been oxidized in anhydrous HF using the chlorine fluorides, bromine fluorides, IF_7, and NiF_6^{2-} ion (14). Electromigration studies of the stable solutions suggest that the radon is present as either Rn^2+ or RnF^+.

Covalent Compounds. The xenon fluorides can be prepared by direct fluorination of xenon under a variety of conditions

$$Xe + F_2\ XeF_2 \quad (3)$$

$$XeF_2 + F_2\ XeF \quad (4)$$

$$XeF_4 + F_2\ XeF_6 \quad (5)$$

including heat, ultraviolet irradiation, and electric discharge; some properties of the known xenon fluorides are given in Table 10.2. Xenon difluoride is readily fluorinated [Eq.(4)], hence it must be removed from the reaction zone if a relatively pure sample is required. In practice, a mixture of xenon and fluorine is circulated through a closed loop containing a section heated to 400°and cooled to –50°. Relatively pure XeF_2 condenses in the latter section. Xenon difluoride can be also prepared by using O_2F_2at -118°as a fluorinating agent (15), Xenon difluoride has been reported as the product formed in the high- voltage discharge (6000V) of a mixture of $CF_4$and xenon. Other fluorine - containing compounds, such as SiF_4;CF_3 OF, and FSO_3F have been suggested or reported to act as fluorine sources in the preparation of xenon fluorides.

Table 10.2 : Some properties of the rare-gas fluorides.

Oxidation state	Compound	Mp. °C	Vapor pressure, torr at 25°C	Color	
				Solid	Gas
2+	XeF_2	129	3.8	white	colorless
	KrF_2	<0 (subl.)	—	white	—
4+	XeF_4	~114	3	white	colorless
6+	XeF_6	49.5	28	white (< 42°C)	yellow
				yellow (> 42°C)	
	$XeOF_4$	−46.2	28	white	colorless

Xenon tetra fluoride forms readily as the major product when fluorine and xenon are allowed to react and XeF_2 is not removed.Indeed, it is possible to form this compound by allowing sunlight to fall on a mixture of the elements sealed in a quartz vessel. The preparation of xenon hexafluoride requires more severe conditions, i.e., a high proporation of fluorine (F2:Xe+20:1), high pressures (50-500atm), elevated temperatures (200-700°C) and extended periods of heating.

Krypton is less reactive with PtF_6 or RhF_6 and does not react with fluorine under ultraviolet irradiation. However KrF_2 is formed when a mixture of the elements is subjected to an electric discharge at 84°K or irradiated with an election beam at 123°K.Krypton difluoride is markedly less stable than XeF_2. decomposing rapidly at 0°C. Krypton tetrafluoride has been prepared as an unstable colourless solid by subjecting the elements to an electric discharge at low temperatures.

The preparation of radon fluorides has been established using radio-chemical techniques. Neither the stoichiometry nor the constitution of these fluorine-containing compounds has been established because of the radiochemical instability of radon.

The reaction between xenon and chlorine under the influence of a micro-wave discharge has been studied. The infrared spectra of the products,condensed at 200K, contain evidence for a linear, symmetrical compound, $XeCl_2$(16). A similar experiment revealed the existence of $KrCl_2$ in a low temperature matrix (17).

Xenon difluoride is a white solid which has a sufficiently high vapour pressure to permit its sublimation under mild conditions. The compound exhibits characteristic bands in the mass spectrum (XeF_2+,XeF_2++, and Xe+) and in the infrared region (549cm-1) which can be used to monitor the formation or loss of this species.

Xenon defluoride is decomposed by water. aqueous KOH, and sulfuric acid. A 0.15 M solution of XeF_2 can be prepared in water, in which it has a half-life of about seven hours at 0°C. In alkaline solution elemental xenon is rapidly formed

$$XeF_2 + 2OH^- \rightarrow Xe + \tfrac{1}{2}O_2 + 2F^- + H_2O, \qquad (6)$$

It reacts with iodide ion

$$XeF_2 + 2I^- \rightarrow Xe + I_2 + 2F^-, \qquad (7)$$

Hydrogen

$$XeF_2 + H_2 \rightarrow Xe + 2HF, \qquad (8)$$

and ammonia

$$\tfrac{3}{2}XeF_2 + 4NH_3 \rightarrow \tfrac{3}{2}Xe + 3H_4F + \tfrac{1}{2}N_2 \qquad (9)$$

to give elemental xenon. The reaction with hydrogen at 200° [Eq.(8)] has been used to establish the composition of Xef2. Xenon difluoride can fluorinate nitrogen oxides

$$2NO + XeF_2 \rightarrow 2FNO + Xe \qquad (10)$$

$$2NO_2 + XeF_2 \rightarrow 2O_2NF + Xe. \qquad (11)$$

Electron-pair-repulsion arguments for xenon difluoride, which is isoelectronic with ICl_2^-, suggest that the molecule should be linear (1).

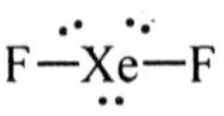

1

Xenon tetrafluoride is a volatile white solid which apparently can exist in two modifications. This substance can be distinguished from XeF_2 by spectroscopic means since its mass spectrum contains the species XeF_3+ and its infrared spectrum exhibits a band at 590cm-1. Xenon tetrafluoride is reduced by hydrogen

$$XeF_4 + 2H_2 \rightarrow Xe + 4HF \quad (12)$$

mercury

$$XeF_4 + 4Hg \rightarrow Xe + 2Hg_2F_2, \quad (13)$$

xenon

$$XeF_4 + Xe \rightarrow 2XeF_2, \quad (14)$$

Table 10.3 : The structural parameters of the rare-gas fluorides.

Compound		Molecular shape	M–F bond distances, A
XeF_2	(solid)	linear	2.00
	(gas)	linear	1.977
KrF_2	(gas)	linear	1.889
XeF_4	(solid)	planar	1.95
	(gas)	planar	1.94
KrF_4	(gas)	planar	1.94
XeF_6	(gas)	planar	1.89
XeF_6	(gas)	distorted octahedral	1.91
$XeOF_4$[a]		tetragonal pyramid	1.900

[a] d(xe-O)=1.703 Å.

and water

$$XeF_4 + 2H_2O \rightarrow Xe + 4HF + O_2. \quad (15)$$

The reaction with water is pH-dependent; in acid or neutral solution 50% of the oxygen and 70% of the xenon expected on the basis of Eq.(15) are obtained. However, in 6N NaOH the gaseous products are formed in accordance with Eq.(15). Xenon trioxide has been isolated from the hydrolysis of XeF_4, suggesting a rather complex disproportionation process. Aqueous iodide ion is oxidized by XeF_4.

$$XeF_4 + 4I^- \rightarrow Xe + 2I_2 + 4F^- \quad (16)$$

Attempts to dissolve XeF_4 in ethanol , tetrahydrofuran, dioxane, or cyclopentadiene lead to a very vigorous reaction, although the compound apparently dissolves smoothly in CS_2.CCI_4, dimethylformamide, cyclohexane, or diethyl ether; the ether solution exhibits strong oxidizing characteristics. Solutions of XeF_4 in trifluoroacetic anhydride, after treatment with trifluroroacetic acid, yield a yellow solid reported to be $(CF_3CO_2)_4Xe$.

Neither alkali-metal fluorides nor SbF_5form complexes with XeF_4,When XeF_4 is dissolved in SbF_5 a gas is evolved and a yellow complex (mp63°C) with the stiochiometry $XeF_2.2SbF_5$ is formed. A similar -straw-coloured tantalum complex ($XeF_2.2TaF_5$, mp81°) forms when XeF_4 is added to TaF_5.

Electron -pair repulsion arguments suggest xenon tetrafluoride should be a planar molecule (2) ; the data in Table 10.3 verify this prediction. A mixed

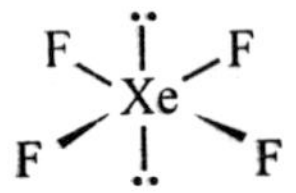

2

$XeF_2.XeF_4$lattice complex is known which contains both linear and squareplanar molecules.

Xenon hexafluoride, a pale- yellow solid, is the least volatile and least stable of the xenon fluorides. A violent reaction occurs when the compound is added to water, but under controlled conditions, e.g., moist air or low temperatures, hydrolysis proceeds smoothly. Xenon trioxide can be obtained by evaporating the excess water in such experiments. Incomplete hydrolysis of XeF_6 yields $XeOF_4$ (mp46.2°).Aqueous XeF_6 solutions react with I-to give iodine [cf.Eq.(&)].

Xenon hexafluoride or its derivatives form more complexes than do the other xenon fluorides. The reaction of xenon, fluorine, and SbF_5 yields three compounds $XeF_6.2SbF_5$(mp 108°), XeF_62SbF_5 (mp205°), and $2XeF_6.SbF_5$ (mp 1000), which have different x-ray powder patterns. Xenon hexafluoride react with Lewis acids producing compounds such as XeF_6. BF_3, $XeF_6.AsF_5$, and $2XeF_6.VF_5$; it reacts with alkali-metal fluorides, giving species formulated as $MXeF_7$. The latter are thermally unstable and yield octafluoroxenates

$$2MXeF_7 - M_2XeF_8 + XeF_6. \qquad (18)$$

Rubidium and cesium octafluoroxenates are stable to 400° They do, however, hydrolyze in aqueous solutions to give xenon-containing oxyanions. Xenon oxytetrafluorides also yield adducts with some alkali-metal fluorides, e.g.MF.$XeOF_4$(M = K and Cs) and 4RbF.2$XeOF_4$, as well as with Lewis acids, 2SbF_5.$XeOF_4$.

Incomplete electron-diffraction data suggest that the structure of XeF_6 appears to be a distorted octahedron, although some spectroscopic data have been interpreted in terms of a symmetrical octahedral arrangement. The electron-pair-repulsion model favours the latter structure, which is isoelectronic with IF_7; a process similar to that suggested for the interconversion of the two possible structures for IF_7, has been suggested to explain the available structural information on XeF_6 (22). Xenon oxytetrafluoride, on the other hand, exhibits a square-pyrimidal arrangement of atoms in agreement with its Lewis formulation.

Oxygen Derrivatives

Derivatives of Xe(11)"Although xenon fluorides react more or less vigorously with water. the controlled hydrolysis of these compounds leads to reasonably stable solutions containing xenon- oxygen species. Xenon difluoride dissolves in water to give a transient yellow colour; some have suggested this arises from the formation of a very unstable xenon oxide.

$$XeF_2 + H_2O = XeO + 2HF \quad (19)$$

which eventually yields xenon

$$XeO = Xe + \frac{1}{2} O_2 \quad (20)$$

The unidentified species present in these solutions appears to be reasonably stable in that it can be extracted into CCI_4; XeF_2 are good oxidizing agents and can oxidize Ag_2 +, CI– to CI_2, IO_4–, BiO_2– to BrO_4–,CO_2 + to Co_3 +, and Ce_3 + to Ce_4 +. in acetonitrile XeF_2 does not act as an oxidizing agent unless traces of HF are present; under these conditions it will oxidize iodine to IF_7. The active species in this reaction is through to be XeF + or Xe_2F_3 + (23).

Xenon difluoride reacts with very strong oxyacids HOR [R = CIO_3 (24), SO_2F (24) TeF_5 (25-27), and -CO-CF_3 (27), to give derivatives that are formally related to $Xe(OH)_2$

$$XeF_2 + HOR = FXeOR + HF \quad (21)$$

$$XeF_2 + 2HOR \rightarrow Xe(OR)_2 + 2HF \quad (22)$$

$Xe(OTeF_5)_2$ (mp35-370) is thermally stable to 130°, above which temperature it decomposes

$$Xe(OTeF_5)_2 \rightarrow Xe + \tfrac{1}{2}O_2 + O(TeF_5)_2 \qquad (23)$$

The trifluoroacetate $[Xe(O_2CCF_3)_2]$ a dark yellow substance, is markedly less stable ; it decomposes by thermal or mechanical shock at -20°C, forming xenon, CO_2 , and hexafluoroethane quantitatively. The reaction of xenon difluoride with $Xe(OTeF_5)_2$ gives a mixed derivative in quantitative yield

$$XeF_2 + Xe(OTeF_5)_2 \rightarrow 2XeOTeF_5 . \qquad (24)$$

$XE(OTeF_5)_2$ forms a yellow 1:1 adduct (mp, 160°C) with arsenic pentafluoride, which has been formulated as $[XeOTeF_5]^+$ $[AsF_6]^-$ on the basis of its Roman spectrum. Xenon difluoride will, however, displace $FXeOTeF_5$ form this compound, suggesating that the relative order of F- donor strengths is XeF_2 > $FXeOTeF_5$ (28).

The crystal structure of $Xe(OTeF_5)_2$ shows a planar arrangement of groups about the linear O-Xe-O unit with trans TeF5 moieties (25, 26). The linear arrangement of the O–X–O unit is consustent with the linear structure observed for XeF_2.

Xenon Trioxide and *Its Derivatives* (II). Xenon tetrafluoride undergoes dispoportionation when it is hydrolyzed

$$3XeF_4 + 6H_2O \rightarrow XeO_3 + 2Xe + \tfrac{3}{2}O_2 + 12HF. \qquad (25)$$

The product, XeO_3, can be isolated as a dangerously explosive white crystalline compound by careful evaporation of the solvent. Xenon trioxide can also be obtained from the hydrolysis of $XeOF_4$. The Raman Spectra of concentrated aqueous solutions of XeO_3 are consistent with the presence of a pyramidal molecule, but other species are also present Xenon trioxide is a weak base, the equilibrium

$$XeO_3 + OH^- \; f \;\; HXeO_4^- \qquad (26)$$

being impotant at pH11. Crystalline acid salts of xenic acid ($MHXeO_4$, M = Na, K, Rb, Cs) have been obtained by removal of solvent from a mixture containing equivalent amounts of the alkali hydroxide and XeO_3.

Xenic acid is also a powerful oxidizing agent. It can oxidize the same species as does an aqueous XeF_2 solution, and is sufficiently powerful to oxidize Br– to BrO_3^-, Mn_2+ to MnO_4^-, and Pu_3 + to Pu_4 +. The advantage of using xenic acid as an oxidizing agent lies in the fact that the reduction product is a gas.

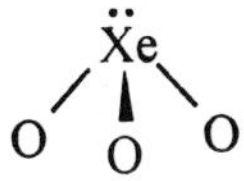

3

Reaction of the hexafluoride with water is violent, the final product being XeO_3.

$$XeF_6 + 3H_2O = XeO_3 + 6HF$$

Xenon complexes

Xenon difluoride will react with fluoride ion acceptors to form complexes of three types: $2XeF_2.\ MF_5$ (M = As, Ru, Os,Ir,Pt), $XeF_2.\ MF_5$ (M = As,Ru, Os, Ir, Pt, Ta, Nb) and $XeF_2.2MF_5$ (M = Sb, Ru, Ir, Pt, Ta, Nb). Some are formed directly from the difluoride and the pentafluoride of the element in question, but in a number of cases BrF3 is used as a solvant. In the 2 : 1 series the arsenic compound is found by single- crystal X-Ray analysis to be $[Xe_2F_3]^+[AsF_6]^-$, and Raman and infrared spoectra show the rest of the group to be similarly constituted. There is spectroscopic evidence that members of the second series should be formulated as $[XeF]^+[MF6]^-$ while, in the third series, the antimony compound has been shown by X-ray analysis to be $[XeF]^+\ Sb_2F11]^-$, with a flurine-bridged anion. Spectroscopic evidence again shows the other adducts of this type to be similarly constituted. The platinum metal compounds in each group are isomorphous.

It will be recalled that the first noble gas compound to be described was formulated as $XePtF_6$ by Bartlett. Reaction of xenon with PtF_6 and related hexafluorides gives products which vary in composition in the range $Xe(MF_6)1\text{-}2$, and it is uncertain how these should be formulated and even if all are single substances.

Xenon tetrafluoride is reported to form an adduct with SbF_5, which is a good fluoride ion acceptor , but its nature is not fully established. It does not accept fluoride ions as does XeF_6, and , as mentioned earlier, this difference is useful in separating XeF_6 from

XeF_4. The hexafluoride forms numerous adducts with fluoride ion acceptors (e.g.$XeF_6.AsF_5$;$XeF_6.BF_3$;$XeF_6.PtF_5$), and the structures of the arsenic and platinum derivatives are known , both containing the XeF_5+ ion, which is also present in the cubic form of solid $XeF_6(XeF_5+F^-)$. There are also other adducts which have different stochiometries. The alkali metal fluorides act as fluoride ion donors in forming highly reactive complexes of two types ($M^IF.XeF_6$ and $2M^IF.XeOF_6$) with the hexafluoride, while NOF gives $2NOF.XeF_6$. There can be little doubt that these contain XeF_7^- or XeF_8^- ions, through their structures have not been reported. A few complexes of $XeOF_4$ with fluoride ion acceptors are also known (e.g.$XeOF_4.2SbF_5$), which are probably $XeOF_3^+$ salts, and , in addition, there are some complexes with fluoride ion donors (e.g.Csf.$XeOF_4$; 3KF.$XeOF_4$), in which $XeOF_5^-$ or more complex fluorine-bridged ions are likely to be formed.

Other xenon compounds. Xenon dichloride has been obtained by passing a xenon- chlorine mixture containing a very large excess of xenon through a microwave discharge and condensing the product on a caesium iodide window at 20°K, so as to study the infrared spectrum. It appears to be unstable at room temperature and has not been prepared in macroscopic quantities. Both the di-and the tetrachloride, as well as $XeBr_2$, have been detected by Mossbauer spectroscopy as products of the β-decay of ^{129}I in $^{129}ICl_2^-$, $^{129}ICl_4^-$ and $^{129}IBr_2^-$.

Monosubstituted derivatives of XeF_2 are formed when xenon difluoride is treated with HSO_3F, $HClO_4$, $HOTeF_5$ or HOC(O) CF_3, the driving force for the reaction being

$$XeF_2 + HL \rightarrow FXeL + HF$$

provided by the highly exothermic formation of HF. All of these ligands, it will be noted , are highly electronegative. By using two molecular proportions of the anhydrous acid it is also possible to make the disubstituted derivatives. $Xe(OSO_2F)_2$, $Xe(ClO_4)_2$, $Xe(OTeF_5)_2$ and $Xe(OCOCF_3)_2$. The monosubstituted products are unstable and those of the second type even more so, though all have been analyzed.

Chemical Properties Of The Noble Gases

The noble gases were isolated and discovered because of their lack of reactivity. For a long time it was thought that they really were chemically inert. Before 1962, the only evidence for compound formation by the noble gases was some molecular ions formed in discharge tubes, and clathrate compounds.

Molecular ions formed under excited conditions

Several molecular ions such as He^{+}_{2}, HeH^{+}, HeH^{2}+ and $Ar^{+} \pm 2$ are formed under high energy conditions in discharge tubes. They only survive momentarily and are detected spectroscopically. Neutral molecules such as He_2 are unstable.

Clathrate compounds

Clathrate compounds of the noble gases are well known. Normal chemical compounds have ionic or covalent bonds. However, in the clathrates atoms or molecules of the appropriate size are trapped in cavities in the crystal lattice of other compounds. *Though the gases are trapped, they do not form bonds.*

If an aqueous solution of quinol (1,4- dihydroxybenzene) is crystallized under a pressure of 10-40 atmospheres of Ar,Kr or Xe , the gas becomes trapped in cavities of about 4Å diameter in the β-quinol structure. When the clathrate is dissolved, the hydrogen bonded arrangement of β-quionol breaks down and the noble gas escapes. Other small molecules such as O_2, SO_2, H_2S,MeCN and CH_3OH form clathrates well as Ar, Kr, and Xe.The smaller noble gases He and Ne do not form clathrate compounds because the gas atoms are small enough to escape from the cavities. The composition of these clathrate compounds corresponds to 3 quinol: 1 trapped molecule, though normally all the cavities are not filled.

The gases Ar,Kr, and Xe may be trapped in cavities in a similar way when water is frozen under a high pressure of the gas. These are clathrate compounds, but are more commonly called 'the noble gas hydrates'. They have formulae approximating to $6H_2O$:1 gas atom. He and Ne are not trapped because they are too small. The heavier noble gases can also be trapped in cavities in synthetic zeolites, and samples have been obtained containing up to 20% of Ar by weight. Clathrates provide a convenient means of stroing radioactive isotopes of Kr and Xe produced in nuclear reactors.

11

S-Block Elements : Alkali Metals Group IA (1)

11.1. Physical data of Group alkali metal

	Atomic Number	Electronic Configuration	First Ionisation (KJ/mol)	Standard Electrode potential (volts)	Atomic Radius (Å)	Ionic Radius (Å)	M.p. °C	B.p. °C
Li	3	2.1 $1s^2 2s^1$	520	– 3.04	1.33	0.60	179.1	1340
Na	11	2.8.1 $...2s^2 2p^6 3s^1$	494	– 2.71	1.57	0.95	97.5	885
K	19	2.8.8.1 $...3s^2 3p^6 4s^1$	418	– 2.92	2.03	1.33	63.5	775
Rb	37	2.8.18.8.1 $...4s^2 4p^6 5s^1$	403	– 2.99	2.16	1.48	39.0	690
Cs	55	2.8.18.18.8.1 $...5s^2 5p^6 6s^1$	374	– 3.02	2.35	1.69	28.5	670

The elements lithium, sodium, potassium, rubidium, caesium and francium are called the alkali metals. Not much is known about the last-named, since it is radioactive and all its isotopes are exceedingly short-lived; it is formed during the radioactive decay of actinium.

The metals are extremely reactive and electropositive (prone to lose electrons) and exist in combination with other elements or radicals as positive ions, e.g. Na^+Cl^-, $(K^+)_2SO_4^{2-}$; they are therefore never found in the free state in nature. They all adopt the body-centred cubic structure in which each atom is surrounded by eight nearest neighbours with six more atoms only slightly further distant (Fig. 11.1).

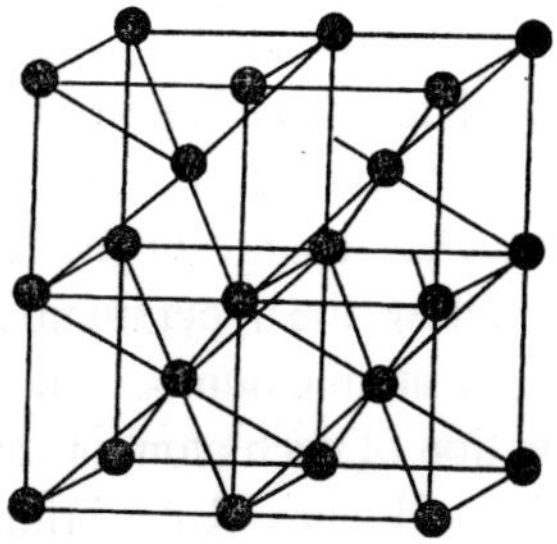

Fig. 11.1. The body-centred cubic structure of the alkali metals

All alkali metal atoms have one electron in the outer shell preceded by a closed shell containing eight electrons—except for lithium which has a closed shell of two. In chemical combination this single electron is very readily transferred, giving a unipositive metal ion with the stable electronic configuration of a noble gas, e.g. Na^+(2.8) is isoelectronic with Ne(2.8). Compounds of the alkali metals are therefore generally predominantly ionic and exist as high melting-point solids in which as many ions of opposite charge surround each other as possible (e.g. sodium chloride and caesium chloride structures. These coumpounds are generally water soluble and white—unless the anion happens to be coloured.

A bonding involving the alkali metals

Atoms of the alkali metals are relatively very large and the single outer valency electron is easily attracted away from the controlling influence of the relatively distant nucleus, leaving an ion of inert gas electronic configuration and unit positive charge. The electronegativity decreases from lithium, showing in a quantitative way the decline in affinity of the elements for electrons as atomic number rises. The electronic configurations of the atoms are the simplest found and this is reflected in the regularity of their properties. The alkali metals are characterized by their unit ionic valency.

Except for certain compounds of lithium, the compounds formed by the alkali metals are ionic and generally have the typically ionic attributes, notably, relatively high melting—and boiling-points with low volatility, good solubility in water and poor solubility in organic solvents. The ionic nature of the compounds is confirmed by X-ray methods. The formation of hydrated ions may be explained by the formation of co-ionic bonds between the central ion and the oxygen atoms of water molecules or simply, by close electrostatic attraction.

An atom of an alkali metal acquires the electronic configuration of the nearest inert gas by the loss of one electron. The resulting cation retains the nuclear charge of the parent alkali metal. In the illustrations which follow, the upper number of in each case refers to the nuclear charge, the atomic number, and the lower numbers to the electronic configuration of each atom or ion.

The application of Fajans' Rules brings a deeper understanding of the varying degrees of ionic character encountered in compounds formed the alkali metals. The lithium cation, being the smallest in the series, has the greatest power of deformation and covalent traits appear in certain lithium compounds. Lithium hydroxide is a weaker base than the other hydroxides. Of the halides, the iodide ion is the largest and most easily polarized. Lithium iodide is the least ionic of the alkali metal halides.

Electronic configurations

Inert gases	*Ions*		
2	3	Li^+F^-	lithium fluoride
He	Li^+	Na^+Cl^-	sodium chloride
2	2	K^+Br^-	potassium bromide
10	11	Rb^+I^-	rubidium iodide
Ne	Na^+	Cs^+H^-	caesium hydride
2, 8	2, 8	$(Li^+)_2O^{2-}$	lithium monoxide
18	19	$(Na^+)_2(O_2^{2-})_2$	sodium peroxide
Ar	K^+	$K^+(O_2^{2-})$	potassium hyperoxide
2, 8, 8	2, 8, 8	$(Rb^+)_2S^{2-}$	rubidium sulphide
36	37	Cs^+OH^-	caesium hydroxide
Kr	Rb^+	$Li^+ClO_4^-$	lithium perchlorate
2, 8, 18, 8	2, 8, 18, 8	$Na^+NO_3^-$	Sodium nitrate
54	55	$K^+IO_2^-$	potassium iodate
Xe	Cs^+	$(Rb^+)_2SO_4^{2-}$	sodium sulphate
2, 8, 18, 18, 8	2, 8, 18, 18, 8	$(Cs^+)_2CO_3^{2-}$	caesium carbonate
simple valency	I		

Note: to avoid ambiguity, certain ions have been enclosed in brackets.

when the metals are separately heated in oxygen are, in family order, lithium monoxide, Li_2O, sodium peroxide, Na_2O_2, and for the others, hyperoxides, KO_2, RbO_2 and CsO_2. The first two are colourless, or nearly so, but the others are deep yellow in colour and are odd electron structures stabilized in the anion by resonance. It is argued that the small cations, by their polarizing power, prevent the dispersal of negative charge which would occur during the progressive increase in proportion of oxygen from monoxide to hyperoxide,

$$\underset{\text{monoxide ion}}{O^{2-}} \xrightarrow{\frac{1}{2}O_2} \underset{\text{peroxide ion}}{O_2^{\;2-}} \xrightarrow{O_2} \underset{\text{hyperoxide ion}}{2(O_2^{\;-})}$$

Again, polyhalide complex ions are not formed by the first two metals, but have been isolated for the others. Caesium forms CsI_3 and CsI_4, rubidium forms RBI_3, while for potassium, only a hydrate, $KI_3.H_2O$, has been isolated.

Occurrence, extraction and uses of the alkali metals

Lithium occurs in alumino-silicate rocks, e.g. spodumene, $Li^+Al^{3+}(SiO_3^{\;2-})_2$ sodium in rock salt, Na^+Cl^-, and Chile saltpetre, $Na^+NO_3^{\;-}$, and potassium in carnallite, $K^+Cl.Mg^{2+}(Cl^-)_2.6H_2O$, and saltpetre, $K^+NO_3^{\;-}$. Compounds of rubidium and caesium are obtained in small amounts during the recrystallisation of naturally occurring salts of the other alkali metals. Francium is produced during the radioactive decay of actinium.

The metals are extracted by electrolysis of their fused chlorides, other halides being added to lower the melting point and thus economise on electrical power. Sodium is used to a much larger extent than the other alkali metals, primarily because sodium compounds are more abundant. It is extracted by the electrolysis of fused sodium chloride to which a little calcium chloride has been added (Downs process). A graphite anode and steel cathode are employed; chlorine is a valuable co-product.

Cathode **Anode**

$$Na^+\text{discharged} \longleftarrow Na^+Cl^- \longrightarrow Cl^-\text{discharged}$$

$$2Na^+ + 2e \rightarrow 2Na \qquad\qquad 2Cl^- \longrightarrow 2Cl\cdot + 2e$$

$$2Cl\cdot \longrightarrow Cl_2$$

Lithium is used to a slight extent as a component in alloys; in the form of lithium aluminium hydride, $Li^{+}AlH_4^{-}$, it finds a use in organic research laboratories, since this compound is a selective reducing agent, e.g. it will reduce a carboxylic acid grouping to an alcohol but generally will leave $C \equiv C$ and $C = C$ bonds intact. Sodium vapour is used in road lighting. An alloy with potassium is used in a sealed system for withdrawing heat from nuclear reactors, e.g. the Dounreay fast breeder. Over 100 000 tonnes of sodium a year were alloyed with lead for the manufacture of lead tetraethyl, which raised the 'octane rating' of petrol. It reacts at red heat with ammonia and carbon to form sodium cyanide, which is used in the extraction of silver and gold. It has also been used as an alternative to magnesium in the extraction of titanium. In the laboratory it is used as a reducing agent, either alone or as an amalgam.

Potassium is sometimes used as a laboratory reducing agent and caesium, when alloyed with aluminium and barium, is used in the construction of photoelectric cells; but, apart from a few such special uses, these metals and rubidium possess few advantages over sodium, which is cheaper.

Chemical Properties

As a family the alkali metals are the most reactive of the metals; lithium stands apart from the others of the series in terms of reactivity and the substances with which it reacts.

The alkali metals react to varying degrees with oxygen, hydrogen, and the halogens. In the case of oxygen the nature of the product depends upon the alkali metal. Thus lithium does not react below 100°; however, at 200° LiO_2 is formed. Sodium gives a mixture of Na_2O and Na_2O_2. Potassium yields predominantly KO_2 at 200°, but K_2O_2 at higher temperatures. Under certain conditions cesium can form any one of a series of suboxides, Cs_7O, Cs_4O, Cs_7O_2, and Cs_3O. The Rb/Rb_2O phase diagram shows the formation of Rb_3O, which is a copper-coloured substance.

Hydrogen and the halogens yield the corresponding hydrides and halides when heated with the alkali metals:

$$2M + H_2 \rightarrow 2MH \qquad (1)$$

$$2M + X_2 \rightarrow 2MX. \qquad (2)$$

In general, under a given set of conditions the reactivity increases with increasing atomic weight of the alkali metal. For example, lithium reacts only superficially with liquid bromine but potassium explodes. Lithium is unusual in that it reacts readily with gaseous nitrogen to form the nitride

$$6Li + N_2 \rightarrow 2Li_3N. \quad (3)$$

whereas no reaction is observed for the other alkali metals (9). Thus, all the alkali metals react with air, lithium predominantly with the nitrogen present and the remaining members of the series with oxygen.

The alkali metals react with simple molecules containing acidic hydrogen atoms to give the corresponding metal derivatives; hydrogen in usually formed in the process:

$$H_2O_{(l)} + M \rightarrow MOH + \frac{1}{2}H_2 \quad (4)$$

$$C_2H_5OH_{(l)} + M \rightarrow C_2H_5OM + \frac{1}{2}H_2 \quad (5)$$

$$2H_2S + 2M \rightarrow M_2S + MSH + \frac{3}{2}H_2 \quad (6)$$

$$C_2H_2 + M \rightarrow C_2HM + \frac{1}{2}H_2 \quad (7)$$

$$NH_{3(g)} + M \rightarrow MNH_2 + \frac{1}{2}H_2 \quad (8)$$

$$NH_{3(g)} + 2Li \rightarrow Li_2NH + H_2 \quad (9)$$

The reaction of lithium with gaseous ammonia proceeds at 300-400°C and two products ($LiNH_2$ and Li_2NH) can be formed in the process (*VI*). It should be recalled that alkali-metal amides can also be formed by the catalyzed decomposition of the blue alkali-metal ammonia solutions. Indeed, all of the reactions given in Eqs. 4–9 occur quite readily with solutions of the alkali metals in liquid ammonia.

$$O^{2-} \xrightarrow{\frac{1}{2}O_2} O_2^{2-} \xrightarrow{O_2} 2O_2^-$$

All three oxide ions are unstable in the pressure of water. e.g.

$$(Li^+)_2O^{2-} + H_2O \longrightarrow 2Li^+OH^-$$

or $$O^{2-} + H_2O \longrightarrow 2OH^-$$

$$(Na^+)_2O_2^{2-} + 2H_2O \longrightarrow 2Na^+OH^- + H_2O_2$$

or $$O_2^{2-} + 2H_2O \longrightarrow 2OH^- + H_2O_2$$

$$2K^+O_2^- + 2H_2O \longrightarrow 2K^+OH^- + H_2O_2 + O_2$$

or $$2O_2^- + 2H_2O \longrightarrow 2OH^- + H_2O_2 + O_2$$

The above three reactions involve the abstraction of protons from water molecules; the oxide ions thus function as strong bases in the Bronsted Lowry sense.

Hydrides, M^+H^-

Hydrides have been synthesized by union of the elements for all of the alkali metals. Lithium reacts at red heat, with incandescence, to form the most stable hydride of the series, m.p. 680°C:

$$2Li + H_2 \rightarrow 2LiH$$

The other metals react in the region of 400°C. Sodium hydride melts with decomposition above 700°C while the others dissociate below their melting-points.

All are colourless, crystalline solids with the sodium chloride lattice in which hydrogen is present as the negative anion, H^-. Structurally, these hydrides are classified as saline, or salt-like.

Lithium hydride merits closer study. It is formed as a transparent, vitreous, deliquescent mass. It melts without decomposition and on electrolysis yields lithium at the cathode and hydrogen at the anode in quantities required by Faraday's Laws,

$$2H^- - 2e \rightarrow H_2 \qquad 2Li^+ + 2e \rightarrow 2Li$$

In cold water a vigorous reaction occurs with the formation of hydrogen and the hydroxide, involving essentially the hydride ion:

$$H^- + H_2O \rightarrow H_2 + OH^-$$

Lithium nitride, Li_3N, exists as colourless transparent crystals with a hexagonal structure. It is made by passing nitrogen over lithium; the reaction takes place slowly at room temperature, and more rapidly as the temperature is raised.

Lithium nitrate reacts readily with water to give lithium hydroxide and ammonia. If heated in a current of hydrogen it is reduced to lithium hydride; the reaction is reversible.

Lithium nitride forms double nitrides (which may be solid solutions) with certain other metallic nitrides. The double nitrides may be obtained (*i*) by heating lithium nitride with the second metal in an atmosphere of nitrogen or ammonia, (*ii*) by heating the lithium alloy of the second metai in an atmosphere of ammonia or nitrogen, or (*iii*) by heating a mixture of lithium nitride and the nitride of the second metal.

$(Li, Co)_3N$ is a black, brittle substance, decomposed by water or cold acids to give ammonia, nitrogen, hydrogen, and free cobalt. The double nitrides of nickel and copper behave similarly. Li_5SiN_3, Li_5TiN_3, Li_5GeN_3 are yellow, easily hydrolysed, salt-like substances with fluorite lattices. LiMgN, LiZnN, $LiAlN_2$, and the oxynitrides Li_5MN_3, $2Li_2O$ where M = Ti, Si, Ge, are also known.

Sodium nitride and potassium nitride. Neither sodium nitride nor potassium nitride can be made by the direct combination of the metal and molecular nitrogen under any conditions. These nitrides can be made:

(*i*) by the action of active nitrogen on a film of the metal (continued action produces the azide; if, however, potassium azide is heated in vacuo, 20% remains behind as the nitride);

(*ii*) by heating an intimate mixture of the metallic azide and the metal, prepared by the evaporation of a mixed solution of the metal and the azide in liquid ammonia.

Sodium nitride is red at room temperature, and darkens on heating to become black at 300°. It begins to dissociate at 300°. It is not affected by air or oxygen at room temperature, but it burns in oxygen if warmed. When treated with water (in which it is insoluble) it liberates ammonia. At slightly elevated temperatures it is reduced by hydrogen to sodium hydride and ammonia.

Carbonates, $(M^+)_2CO_3^{2-}$, and Hydrogen-carbonates (bicarbonates), $M^+HCO_3^-$

Carbonic acid is a very weakly ionized dibasic acid and forms both normal and acid salts:

H_2CO_3 carbonic acid	$KHCO_3$ potassium hydrogencarbonate, (bicarbonate)	K_2CO_3 potassium carbonate

The thermal stability of the carbonate and the existence, or not, of a solid hydrogencarbonate are criteria of the degree of electropositive character of the combined metal.

Except for lithium carbonate, the carbonates of the alkali metals are not decomposed by heat. Lithium carbonate yields lithium monoxide and carbon dioxide:

$$Li_2CO_3 \rightarrow Li_2O + CO_2 \uparrow$$

The solubilities of the carbonates increase with atomic number. The carbonates of potassium, rubidium and caesium are extremely soluble (solubility = 1.33 g in 100 g water at 25°C). Lithium carbonate may be precipitated using solutions of ammonium carbonate and lithium sulphate, or other convenient salts of suitable concentration:

$$2Li^+ + CO_3^{2-} \rightarrow Li_2CO_3$$

Lithium carbonate bears some resemblance to carbonates of Group II. The other carbonates are prepared by the action of carbon dioxide on the hydroxide:

$$2OH^- + CO_2 \rightarrow CO_3^{2-} + H_2O$$

Solid hydrogencarbonates are formed by all of the alkali metals except lithium, the hydrogencarbonate of which is known only in solution.

Hydrogencarbonates are formed by the action of carbon dioxide and water on the carbonate in the cold, and are decomposed by heat:

$$CO_3^{2-} + CO_2 + H_2O \rightleftharpoons 2HCO_3^-$$

Carbonates and hydrogencarbonates liberate carbon dioxide with acids stronger than carbonic acid:

$$2H^+ + CO_3^{2-} \rightarrow H_2O + CO_2 \uparrow$$
$$H^+ + HCO_3^- \rightarrow H_2O + CO_2 \uparrow$$

Both carbonates and hydrogencarbonates are hydrolysed giving solutions on the alkali side of neutrality. The carbonates are more alkaline than the hydrogencarbonates, which are classed as acid salts:

$$CO_3^{2-} + H_2O \rightleftharpoons HCO_3^- + OH^-$$
$$CO_3^- + H_2O \rightleftharpoons H_2CO_3 + OH$$

The greater alkalinity of carbonate solutions results in the precipitation of a basic carbonate, while hydrogencarbonates give a normal carbonate, with solutions of the salts of some metals:

$$Zn^{2+} + 2HCO_3^- \rightarrow ZnCO_3 \downarrow + CO_2 \uparrow + H_2O$$

$$3Zn^{2+} + CO_3^{2-} + 4OH^- + 2H_2O \rightarrow ZnCO_3.2Zn(OH)_2.2H_2O \downarrow$$

Sodium carbonate, soda, soda ash, $(Na^+)_2CO_3^{2-}$

The common hydrates are:

$Na_2CO_3.10H_2O$	decahydrate
$Na_2CO_3.H_2O$	monohydrate.

Sodium carbonate is freely soluble in water; solubility = 21.4 g in 100 g water at 25°C. On crystallization, a hydrate is obtained, the composition of which depends on the temperature. The anhydrous compound is formed by heating a hydrate or the hydrogencarbonate to constant weight. It melts at 854°C. The solubility curve of sodium carbonate is shown in Figure 11.2. Below 32.5°C, the decahydrate separates; above 37.5°C, the monohydrate is formed and, in the interval, a hepathydrate may be obtained. Solid sodium carbonate monohydrate is conveniently obtained by evaporation of the solution at the boiling point. Anhydrous sodium carbonate slowly becomes hydrated in damp air. The decahydrate, which forms large translucent crystals, commonly used domestically as washing soda, effloresces in air, becoming coated with the monohydrate, seen as a white powdery coating.

In volumetric analysis, sodium carbonate is used for the standardization of strongly ionized acids. If necessary, absorbed moisture should be expelled by heating, with stirring to avoid caking, and the anhydrous sodium carbonate cooled in a desiccator.

In quantitative analysis, sodium carbonate solution is boiled with mixtures containing metals which form insoluble products, before testing for anions. This avoids possible misinterpretation of observations: thus, in the test for sulphate ions, addition of barium chloride solution to a solution of a lead(II) salt would produce a white precipitate of lead(II) chloride, to be confused possibly with the anticipated precipitate of barium sulphate. Fusion mixture, consisting of 'equimolecular' proportions of sodium carbonate and potassium carbonate, is used as a flux in charcoal block reductions. The powdered substance is mixed with three times its bulk of fusion mixture and heated in the reducing blow-pipe flame. Carbonates of heavy metals are decomposed to the oxide as the residual ions pass into the block. Further action may occur with the oxide being reduced to metal. By using a mixture of carbonates, the fusion point of the flux is lowered as much as possible.

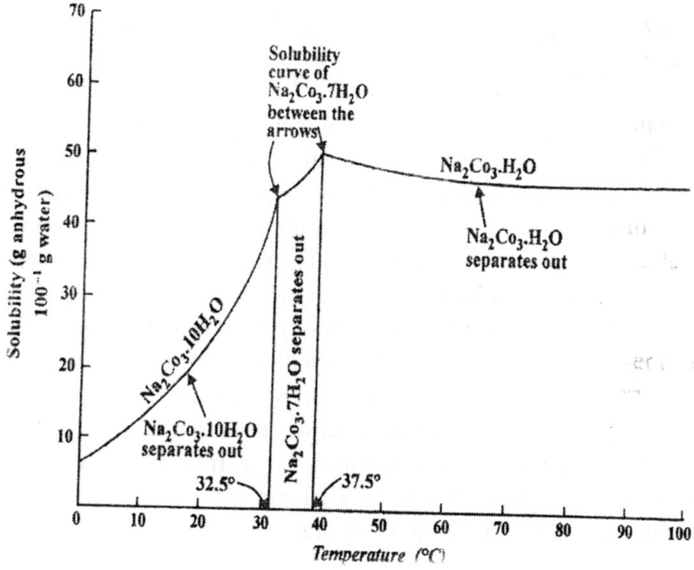

Fig. 11.2 : Solubility Curve of Sodium Carbonate. Note that the solubility of the monohydrate falls with the rise in temperature

The manufacture of sodium carbonate forms a major part of the Alkali Industry.

Potassium carbonate, pearl-ash, potash, $(K^+)_2CO_3^{2-}$

There are three hydrates: $K_2CO_3.2H_2O$, $2K_2CO_3.3H_2O$ and $K_2CO_3.H_2O$.

Potassium carbonate cannot be produced commercially by the Solvay process because potassium hydrogencarbonate is rather too soluble to be precipitated. It is obtained from potassium chloride, which occurs in salt deposits, by the Precht process. Hydrated magnesium carbonate is added to a concentrated solution of potassium chloride and carbondioxide is blown in with the resulting precipitation of a double salt:

$$2KCl + 3(MgCO_3.3H_2O) + CO_2 \rightarrow$$
$$2(KHCO_3.MgCO_3.4H_2O)\downarrow + MgCl_2$$

Treatment with magnesium oxide regenerates magnesium carbonate and potassium carbonate dissolves:

$$2(KHCO_3.MgCO_3.4H_2O)\downarrow + MgO\downarrow \rightarrow$$
$$3(MgCO_3.3H_2O)\downarrow + K_2CO_3$$

After filtration, evaporation yields potassium carbonate. It is freely soluble in water, 113 g dissolving in 100 g water at 25°C. The commercial product is known as pearl-ash.

In the laboratory, potassium carbonate solution could be prepared by boiling a hydrogencarbonate solution, obtained by the action of carbon dioxide on strong caustic potash solution. Evaporation and crystallization yields a hydrate. Usually, the dihydrate separates from solution but others with less water of crystallization may be prepared. Above 130°C, the anhydrous salt is formed. It melts at 900°C.

Potassium carbonate is deliquescent and is used as a desiccating agent for drying organic liquids and to reduce the solubility of ethanol. Moist solid potassium carbonate will absorb carbon dioxide to form the hydrogencarbonate. Alternatively, when carbon dioxide is passed into a concentrated aqueous solution of potassium carbonate, the hydrogencarbonate is precipitated. After filtration, it is left to dry. On heating it decomposes to the carbonate, carbon dioxide and steam.

Potassium carbonate may be extracted from various animal and vegetable products including wood ash. It is a by-product of beet sugar manufacture. Potassium carbonate is used in the production of hard glass and soft soap.

Sodium nitrate, Chile saltpetre, $Na^+NO_3^-$

Sodium nitrate is mined in a rainless desert region of Chile. Deposits of caliche, containing about 30% of sodium nitrate, are broken up by means of explosives and treated with water. Sodium nitrate is crystallized out. The iodate-bearing mother liquors are the most important world source of iodine. Sodium nitrate is a white solid, somewhat hygroscopic and very soluble in water. At 25°C, the solubility is 91.8 g in 100 g water. Sodium nitrate melts at 316°C and decomposes readily above 550°C. It forms highly combustible mixtures with organic materials, acetates, cyanides, thiocyanates, hypophosphites and sodamide. The elements carbon, phosphorus and sulphur form carbonate, phosphate and sulphate explosively on heating.

Sodium nitrate may be used as a nitrogenous fertilizer, and in the manufacture of explosives, flares and fireworks, the production of lead glass, for curing meat, and in chemical processes, including nitration and the Lead Chamber production of sulphuric acid.

Potassium nitrate, nitre, saltpetre, $K^+NO_3^-$

Potassium nitrate is obtained on the commercial scale by mixing boiling concentrated solutions of sodium nitrate and potassium chloride from natural sources. Sodium chloride separates as the least soluble pairing of ions and is filtered off. On cooling to about 20°C, potassium nitrate becomes the least soluble salt and crystallizes out. Potassium nitrate is much more soluble in hot water than cold, while the solubility of sodium chloride is little affected by change of temperature. At 25°C, the solubility of potassium nitrate is 31.6 g in 100 g water.

Potassium nitrate is a white crystalline solid, melting at 340°C and similar in its reactions to sodium nitrate but preferred in the manufacture of gunpowder and pyrotechnics because it does not absorb moisture from the atmosphere. Gunpowder contains potassium nitrate, charcoal and sulphur in the approximate proportions, 6 : 1 : 1 by weight. The mixture is ground to a fine powder when moist and slowly dried out. The release of a great volume of hot expanding gases, nitrogen and oxides of carbon, produces the disruptive power. It is exploded by mechanical shock.

Sodium hydrogen carbonate is much less soluble in water than potassium hydrogen carbonate and both solutions are alkaline by hydrolysis, but not to the same extent as the corresponding carbonates:

$$HCO_3^- + H_2O \rightleftharpoons H_2CO_3 + OH^-$$

Because sodium hydrogen carbonate solution is less alkaline than sodium carbonate solution, the former is often used to precipitate normal carbonates which would otherwise be formed as basic carbonate, e.g.

$$Cu^{2+} + 2HCO_3^- \longrightarrow CuCO_3 + H_2O + CO_2$$

The hydrogen carbonate ions tend to be held together in crystal structures by hydrogen bonding (p. 67) giving layers of polymeric anions. e.g. the partial structure of sodium hydrogen carbonate is:

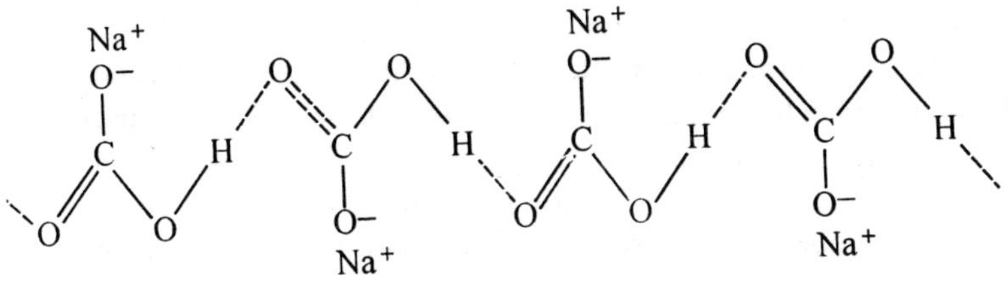

The halides of sodium and potassium, M^+X^-

The fluorides $M^+ F^-$

The fluorides can be obtained by reacting aqueous solutions of the carbonate or hydroxide with an aqueous solution of hydrofluoric acid. Polythene vessels must be used since the acid attacks glass.

$$Na^+OH^- + H^+F^- \longrightarrow Na^+F^- + H_2O$$

If an excess of acid is used, it is possible to isolate the solid hydrogen fluorides $Na^+HF_2^-$ and $K^+HF_2^-$. The hydrogen fluoride ion contains onc molecule of hydrogen fluoride, hydrogen bonded to the very electronegative fluoride ion:

$$(F \text{——} H \text{ - - - - } F)^-$$

The hydrogen fluorides decompose on heating to give the normal fluorides and hydrogen fluoride.

The chlorides, M^+Cl^-

Sodium chloride occurs extensively in sea water; as a solid it is found as rock salt, which is coloured owing to the presence of impurities. In Great Britain, salt deposits occur below ground level and the salt is brought to the surface as a solution in water. Pure sodium chloride can be obtained by passing hydrogen chloride gas through a saturated solution of the impure salt, when it is precipitated—the common ion effect.

The solubility of sodium chloride in water is affected very little by rise in temperature: 35.5 g at 0°C and 39 g at 100°C. When added to ice, it is possible to achieve a temperature of – 23°C and an ice-salt mixture is often used as a cheap freezing mixture in the laboratory.

Sodium chloride is a vital raw material in many industrial process, and the British alkali industry is sited in Cheshire where there are extensive salt deposits. It is used in the manufacture of sodium, sodium hydroxide, chlorine, sodium carbonate, sodium hypochlorite, sodium chlorate and in the preparation of many sodium salts. It is also used to produce the sodium silicate glaze on earthenware vessels. Its more familiar uses include the flavouring and curing of foods.

Potassium chloride is far less abundant than sodium chloride and is found in sea water and in the Stassfurt (Germany) deposits as carnallite, $K^+Cl^-.Mg^{2+}(Cl^-)_2.6H_2O$. It is extracted from the latter source by fractional crystallization; potassium chloride is less soluble in water at ordinary temperatures than magnesium chloride and can be crystallised out.

Potassium chloride is used to counteract potassium deficiency in soil and in the production of potassium hydroxide.

The bromides, M^+Br^-

Potassium bromide is more important than sodium bromide and is used as a sedative and in the production of silver bromide for use in photography. It can be made by adding bromine to a warm concentrated solution of potassium hydroxide. A mixture of potassium bromide and potassium bromate is formed, which can be crystallised out. On heating with charcoal the bromate is reduced to the bromide:

$$3Br_2 + 6OH^- \longrightarrow 5Br^- + \underset{\text{bromate}}{BrO_3^-} + 3H_2O$$

$$2K^+BrO_3^- + 3C \longrightarrow 2K^+Br^- + 3CO_2$$

The iodides, M^+I^-

Potassium iodide is more important than the corresponding sodium salt and its aqueous solution is used in the laboratory for dissolving iodine, which is not very soluble in water, as the polyiodide, $K^+I_3^-$. Sodium iodide also forms a polyiodide.

Potassium iodide can be obtained by a process similar to the one used for making potassium bromide, bromine being replaced by iodine.

Coordination Behaviour

Crystallographically, the coordination numbers of the alkali metals in the predominantly ionic compounds are governed by the radius-ratio rules; the smaller ions (Li^+, Na^+, K^+, Rh^+) give octahedral structures while the larger Cs^+ ion can attain a coordination number of eight in some of its compounds. In General, the alkali-metal ions do not form particularly stable coordination compounds. The structures of some solid hydrates as well as compounds incorporating very easily polarizable legends such as ammonia give an insight into the coordination behaviour of these metals.

Lithium usually exhibits a coordination of 4, although occasionally under special conditions a coordination number of 6 may be observed. For example, the crystal structure of $LiClO_4.3H_2O$ shows that lithium ions are linked in linear chains by water molecules, each lithium being associated with six octahedrally disposed water molecules. The perchlorate ions are interpersed between the chains. A similar arrangement of oxygen atoms about the lithium atom occurs in $LiIO_3$ (*15*). Four-coordinate lithium is undoubtedly present in the very stable ammoniate $LiI{\cdot}4NH_3$; the solid-state structure of $LiOH{\cdot}H_2O$ is polymeric (Fig. 11.3) with each lithium being tetrahederally

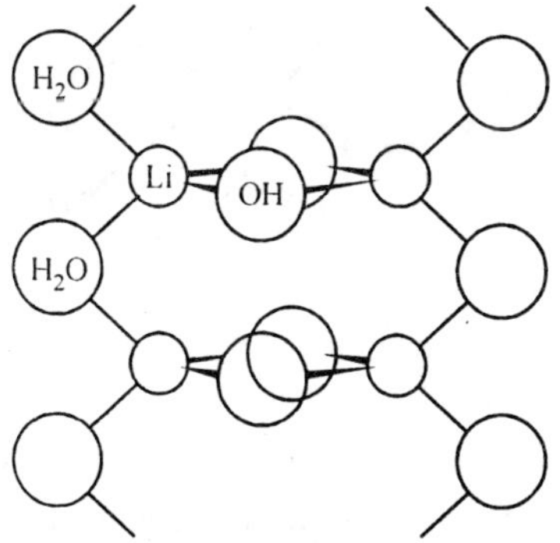

Fig. 11.3 : $LiOH{\cdot}H_2O$ crystallizes in a double than chain of lithium atoms with water molecules forming bridges between the lithium atoms in the chain and OH units holding the chains together.

surrounded by oxygen atoms. Each polymeric chain is attached to an adjacent chain by hydrogen bridges (Chapter 6). Lithium acetylacetonate [the salt of the enol form of acetylacetone, CH_3—CO—CH=C(OH)CH_3] form a dihydrate in which lithium has achieved a coordination number of 4 (Fig. 11.4).

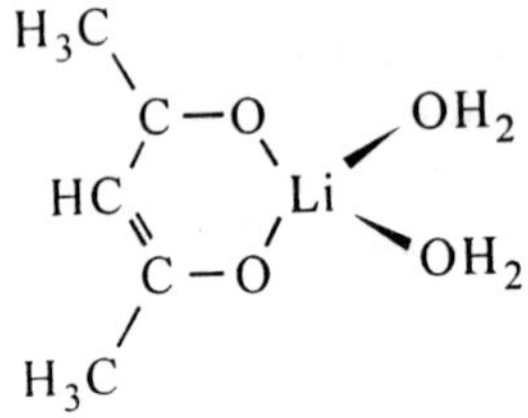

Fig. 11.4 : Lithium acetylacetonate crystallizes with two water molecules of hydration. The metal ion is tetrahderally surrounded by oxygen atoms.

The alkali metals larger than lithium usually exhibit higher coordination numbers. For example, sodium, potassium, and rubidium iodides form reasonably stable ammoniates containing six ammonia molecules ($MI{\cdot}6NH_3$); sodium acetylacetonate forms a tetrahydrate in which the sodium ion is octahedrally surrounded by oxygen atoms (Fig. 11.5). Salicylaldehyde, which is acidic, forms compounds with potassium, rubidium, and cesium in which

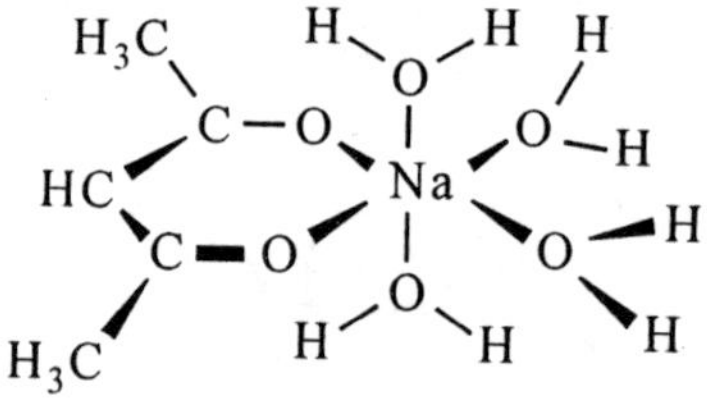

Fig. 11.5 : Sodium acetylacetonate forms a tetrahydrate in which the metal atom is octahedrally surrounded by oxygen atoms.

these metals acquire octahedral coordination, as shown in Fig. 11.6. The formulation of these compounds appears to be MOC_6H_4—CHO·2HOC_6H_4—CHO in which one salicyladehyde molecule has been converted into an anion and two are present in their molecular forms. The fact that the heavier alkali metals form 6-coordinate structures

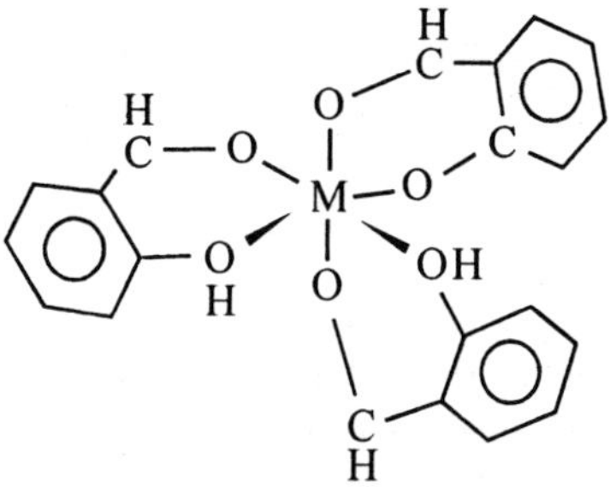

Fig. 16.6 : The heavier alkali metals form compounds with salicyladehyde in which the metal atoms is octahedrally surrounded by oxygen atoms. M = K, Rb, or Cs.

does not exclude the formation of structures with a lower coordination number. Thus, NaI also forms a 4-coordinate ammoniate ($NaI \cdot 4NH_3$) which has been shown to have a tetrahedral arrangement of ammonia molecules around the sodium ion; salicyladehyde also forms 4-coordinate structures with all of the alkali-metal ions. In addition, all the alkali metals form acetylacetonates that are dihydrates and presumably contain 4-coordinate cations similar to that shown in Fig. 11.4. The hydrated compounds exhibit solubility properties of covalent coumpounds; for example, $Na(acac)_2 \cdot 2H_2O$ is soluble in benzene.

Solutions of the Alkali Metals in Liquid Ammonia

The solutions of the alkali metals in liquid ammonia (Table 11.2) are dark blue. The densities of the solutions are lower than that of liquid ammonia. The behaviour of the solutions under the ultramicroscope, and many of their other properties, show that the solutions are not colloidal. At – 33°C the conductivity of a saturated solution of sodium in liquid ammonia is 0.5×10^{14} mhos, which is half that of metallic mercury. The mobility of the anion is 280 times that of the cation. Several theories have been advanced to explain the properties of medium and dilute solutions of the alkali metals in ammonia. It was now generally accepted that the system is ionic. The cations are solvated metal cations $[Na(NH_3)_n]^+$, and the anions are solvated electrons. The paramagnetism of the solution decreases with increasing concentration, from which it may be inferred that the solvated electrons may associate in pairs.†

On evaporation of the ammonia the alkali metals (except lithium) are deposited as copper-coloured spongy solids which probably consist of crystals of the metal together with adsorbed liquid. Lithium crystallises as the ammoniate Li, $4NH_3$, and in this resembles the alkaline earth metals which crystallise as hexa-ammoniates such as Ca, $6NH_3$.

Table 11.2 : Solutions of the Alkali Metals in Liquid Ammonia

Metal	Solubility, gm in 100 gm of ammonia at 0°	Notes on the equilibrium	Period required for formation of the amid, XNH_2‡
Li	9.41§	Forms two liquid phases below 60°	Several days
Na	18.7	Eutectic at – 110° and 13% Na	Several days
K	32.7		1 hour
Rb	Sol.		30 min
Cs	Sol.		15 min

‡ These changes are catalysed by iron, platinum, ferric oxide, or by light of wave length 2150-2500 Å.

§ The heat of solution at – 33° is 9.65 kcal/gm atom.

The alkali metals in liquid ammonia solution react with ammonium salts (which are acids in the solvent system,) to form salts. Ammonia and hydrogen are normally liberated:

$$2NH_4Cl + 2Na = 2NaCl + 2NH_3 + H_2.$$

Ammonium nitrate, however, produces the nitroxylate:

$$NH_4NO_3 + 3K = NH_3 + K_2NO_2 + KOH,$$

and there is no evolution of hydrogen or any other insoluble gas.

The anomalies of lithium—the diagonal relationship with magnesium

In general the differences between the first and second members of a particular periodic group are more pronounced than those between the second and third; in this respect lithium is no exception and its abnormally high negative electrode potential has already been noted.

Table 11.3 : Additive Reactions of the Alkali Metals in Liquid Ammonia Solution

Metal	Reagent	Product
Na	N_2	No reaction
KP	K_3P_2	
Li, Na	P	Li_2, Na_2P
KAs	K_3As, NH_3 and K_2As_4, NH_3	
Na	Sb	Na_3Sb_5, Na_3Sb_6, Na_3Sb_7
Na	Bi	Na_3Bi, Na_3Bi_3, $Na_3Bi_{5\text{-}7}$
Li	O_2	Li_2O, Li_2O_2
Na	O_2	Slow stream of oxygen: Na_2O, which is ammonolysed to $NaNH_2$, which is further oxidised to $NaNO_2$ and NH_3. If $NaNH_2$ is slowly added to liquid ammonia through which oxygen is passing rapidly a yellow product is obtained of composition Na_2O_2 and NaO_2
K	O_2	As for sodium
Li	S	Li_2S, Li_2S_2, Li_2S_x
Na	S	Na_2S, Na_2S_x
Na	Se	Na_2Se, Na_2Se_2, Na_2Se_6
K	Te	K_2Te, K_2Te_3
Na	Pb	NaPb, $NaPb_2$, Na_4Pb_9
Na	Hg	$NaHg_8$
Rb	S	Rb_2S_x $x = 2, 3$, or 5 but not 4 or 6
Cs	S	Cs_2S_x $x = 2, 3, 5$ or 6 but not 4
All	Graphite	Compounds of ideal formulae $C_{12}M(NH_3)_2$ and $C_{25}M(NH_3)_2$. The first consists of alternate layers of graphite and alkali metal ammoniate; the second of two layers of graphite to one of ammoniate

In many aspects of its chemistry lithium shows similarities with magnesium; thus it only forms a monoxide on heating in oxygen, its carbonate is fairly readily decomposed to the oxide and carbon dioxide, and the nitrate on decomposition gives the oxide, nitrogen dioxide and oxygen. Other similarities with magnesium include the insolubilities of its carbonate and phosphate in water, the formation of a carbide and nitride by direct combination of the elements, and a strong tendency for its ions to hydrate in aqueous solution. Its halides, in particular, are soluble in many organic solvents, indicating that they possess appreciable covalent character. In all these respects lithium shows marked differences from the other alkali metals.

Many of the resemblances to magnesium can be explained as being due to these two elements having very similar electropositivities. Since electropositivity increases from top to bottom in any periodic group and decreases from left to right across a particular period, the increase in electropositivity in going down one place in the Periodic Table (magnesium is one place lower down than lithium) is compensated for by the decrease which occurs in moving across a period from left to right (magnesium is one place to the right of lithium).

The appreciable covalent character of several compounds of lithium depends on the fact that its lone valency electron is not completely transferred during chemical reaction; this is readily understood in terms of the small size of the lithium atom.

Hydrolysis of alkali metal salts

The alkali metal halides, nitrates and sulphates are neutral in aqueous solution; but other salts, for instance hydrogen carbonates, carbonates, cyanides and sulphides, show an alkaline reaction in solution. This is because the latter group of anions behave as strong bases and are able to abstract protons from solvent water molecules, thus giving rise to hydroxyl ions:

$$CO_3^{2-} + H_2O \rightleftharpoons HCO_3^- + OH^-$$

$$HCO_3^- + H_2O \rightleftharpoons H_2CO_3 + OH^-$$

$$CN^- + H_2O \rightleftharpoons HCN + OH^-$$

$$S^{2-} + H_2O \rightleftharpoons HS^- + OH^-$$

$$HS^- + H_2O \rightleftharpoons H_2S + OH^-$$

When alkali metal cyanides and sulphides are warmed with water volatile hydrogen cyanide and hydrogen sulphide are driven off as gases; the hydrolysis therefore goes to completion and solutions of alkalis remain.

Summary to trends in alkali metal

All the alkali metal are extremely electropositive, increasingly so with increasing atomic number; they therefore tend to form predominantly electrovalent compounds and these are generally soluble in water. Lithium, because of the small size of its atom and unipositive ion, tends to form compounds in which there is some degree of covalent character, and in many respects behaves like its diagonal neighbour in the Periodic Table—magnesium.

The alkali metals from hydroxides, which are strong bases, and salts, which are either neutral in solution (those of a strong acid, e.g. the nitrates) or alkaline (those of a weak acid, e.g. the carbonates). Their salts are generally stable to heat; thus the carbonates are not decomposed in the Bunsen flame and strong heating is required to decompose the nitrates to the nitrites and oxygen. Only the hydrogen carbonates of the alkali metals are sufficiently stable to exist as solids. Although many of the alkali metal salts are hydrated in the solid state, there is little tendency for the alkali metal ions to hydrate strongly in aqueous solution. Lithium is exceptional in this respect and is sufficiently hydrated in aqueous solution to have the highest negative electrode potential of this group of elements.

12

S-Block Elements : The alkaline earth metals—Group IIA (2)

Some physical data of alkaline earth elements

	Atomic Number	Electronic Configuration	Ionisation energy First	(kJ/mol) Second	Standard Electrode Potential (volts)	Atomic Radius (Å)	Ionic Radius (Å)	M.p. °C	B.p. °C
Be	4	2.2 $1s^22s^2$	899	1757	– 1.70	0.89	0.31	1280	2970
Mg	12	2.8.2 ....$2s^22p^63s^2$	738	1451	– 2.37	1.36	0.65	650	1100
Ca	20	2.8.8.2 ....$3s^23p^64s^2$	589	1145	– 2.87	1.74	0.97	850	1490
Sr	38	2.8.18.8.2 ....$4s^24p^65s^2$	549	1064	– 2.89	1.91	1.13	760	1380
Ba	56	2.8.18.18.8.2 ....$5s^25p^66s^2$	502	965	– 2.90	1.35	710	1640	
Ra	88	2.8.18.32.18.8.2 ....$6s^26p^67s^2$							

Introduction

The alkaline-earth metals resemble the alkali metals in many of their properties. There are, of course, differences which arise from the different number of valence electrons in the two groups of elements, but taking these into account there is a remarkable uniformity in the behaviour of the two families. The ground-state configuration of the alkaline-earth metals consists of a rare-gas core with two electrons in the next *s* orbital (Table 12.1). A comparison of the first ionization potentials of the alkaline-earth metals and the alkali metals, suggests that the 1 + oxidation state might be easily attained for the former group. In reality, the 1 + valence is a chemically unimportant state for the alkaline-earth metals; only under unusual conditions have 1 + ions been observed in this family. For example, the electrolysis of conducting aqueous solutions using beryllium electrodes or of pyridine solutions using magnesium electrodes yields highly reducing solutions at the anode which are thought to certain Be^+ and Mg^+ respectively.

Table 12.1 : Some properties of the alkaline-earth metals and their ions

Element[a]	*Ionisation potentials eV Electronic configuration*	*I*	*II*	*Standard Electro- III*	*electrode potential negativity*	*Ionic radius V*	ΔH_{hyd} *Å*
Be (4)	[He] $2s^2$	9.32	153.85	1.5	1.69	0.31	– 24.8
Mg (12)	[Ne] $3s^2$	7.64	80.21	1.2	2.37	0.65	– 20.2
Ca (20)	[Ar] $4s^2$	6.11	51.21	1.0	2.87	0.94	– 16.6
Sr (38)	[Kr] $5s^2$	5.69		1.0	2.89	1.10	– 15.2
Ba (56)	[Xe] $6s^2$	5.21		0.9	2.90	1.29	– 13.7
Ra (88)	[Rn] $7s^2$	5.10			2.92	1.50	

a Nuclear charge Z given in parentheses
b Estimated

Even though the removal of a second electron from an alkaline-earth metal ion requires approximately twice the energy as is required for the first electron, stable compounds containing 2 + ions are formed. Apparently this increase in energy is more than compensated by lattice formation or by the solvation of these ions. The ions in this family are isoelectronic with the corresponding 1 + ions of the alkali metals. The increase in nuclear charge in going from an alkali metal to the next alkaline-earth metal gives rise to a small ion; since the charge on the alkaline-earth ions is also greater their charge densities are greater than those of the alkali-metal ions. A consideration of the data in Table 12.1 strongly suggests that the simple compounds of the alkaline-earth metals should be even more ionic than the corresponding alkali-metal compounds. In comparable compounds the alkaline-earth ions would be less polarizable than the alkali-metal ions. However, the data indicate that higher electric fields would be generated by the alkaline-earth ions; that is, these ions would be more polarizing in their compounds than the isoelectronic alkali-metal ion. Thus, a clear understanding of the structures and energetics of the ionic compounds of the alkaline-earth metals can be obtained from the radius-ratio rules and the simple electrostatic arguments. In considering the structures of the ionic halides we must, of course, recognize that each alkali-metal cation has two counter ions present. This stoichiometry, which arises from the electronic configuration of the metal, requires a more complex structure (e.g., the flourite or rutile structures) than those observed for the alkali halides. However, many binary compounds

Table. 12.2 : The charge density on the alkali and alkaline-earth ions

Ion	*Charge density* $C\,Å^{-3} \times 10^{-19}$
Li^+	1.22
Na^+	0.42
K^+	0.16
Cs^+	0.12
Rb^+	0.08
Be^{2+}	25.6
Mg^{2+}	2.78
Ca^{2+}	0.92
Sr^{2+}	0.57
Ba^{2+}	0.36

of the alkaline-earth metals containing divalent ions exhibit the common sodium-chloride structure in which the cation exhibits a coordination number 6. Beryllium forms a sufficiently smaller ion so that the lower 4-coordinate wurtzite or zinc-blende structures are preferred.

Although the alkaline-earth metals are more difficult to vaporize and ionize than the alkali metals, the higher hydration

Table 12.3 : The crystal structures of some of the simple binary compounds of the alkaline-earth elements

Types of compound	Structures				
	Be[a]	*Mg*[a]	*Ca*[a]	*Sr*[a]	*Ba*[a]
MO	wurtzite (2530)	NaCl (2802)	NaCl (2587)	NaCl (2430)	NaCl (1923)
MS	zinc blende (—)	NaCl (d.)	NaCl (d.)	NaCl (> 2000)	NaCl (1200)
MSe	zinc blende (—)	NaCl (—)	NaCl (—)	NaCl (—)	NaCl (—)
MTe	zinc blende (—)	zinc blende (—)	NaCl (—)	NaCl (—)	NaCl (—)
MF_2	cristobalite (880)	rutile (~ 1350)	fluorite (~ 1350)	fluorite (~ 1300)	fluorite (1285)
MCl_2	chain (430)	$CdCl_2$[b] (715)	fluorite (773)	fluorite (870)	flourite (960)

a Melting point in °C given in pictures.

b A layer-type structure containing octahedral $CdCl^6$ groups linked by sharing edges to form infinite layers. Relatively weak forces exist between layers.

energies associated with the former ions lead to standard electrode potentials for both groups of elements that are nearly the same.

Of the alkaline-earth metals, radium is the least abundant, comprising ~ 10^{-10}% of the igneous rocks on earth; all the isotopes of radium are radioactive, ^{226}Ra, with a half-life of approximately 1600 years, being the most stable. The known chemical properties of radium are consistent with those of the heavier alkaline-earth metals. We shall address our discussion in this chapter only to remaining alkaline-

Table 12.4 Group II (Alkaline-Earth Metals) : Physical Data

	Be	Mg	Ca	Sr	Ba
Atomic number	4	12	20	38	56
Electronic configuration	2, 2	2, 8, 2	2, 8, 8, 2	2, 8, 18, 8, 2	2, 8, 18, 18, 8, 2
Atomic weight	9.0122	24.305	40.08	87.62	137.34
Atomic radius/Å	0.89	1.36	1.74	1.91	1.98
Ionic radius/Å	0.31	0.66	0.99	1.12	1.34
Atomic volume/ml mol $^{-1}$	4.85	14.00	26.08	34.01	38.26
Standard electrode potential (H Scale: 25°C)/volts	– 7.70	– 2.34	– 2.87	– 2.89	– 2.90
Boiling-point/°C	1500	1107	1187	1366	1537
Melting-point/°C	1283	651	851	800	850
Electronegativity	1.5	1.2	1.0	1.0	0.9
Density/g ml $^{-1}$ (at 20°C)	1.85	1.74	1.55	2.60	3.59
Abundance (% by weight)	0.006	2.1	3.63	0.042	0.039

earth metals. In this discussion it should be noted that beryllium, like lithium within the alkali-metal family, exhibits properties that are markedly different from the heavier members of its family.

A general survey of bonding involving the alkaline-earth metals

The atoms of alkaline earth metals, except for beryllium, which, although it is in Group II, is not strictly an alkaline-earth according to the original meaning of the term, are comparatively large and readily lose both valency electrons to give ions of inert gas electronic configuration and double positive charge. No cations of unit positive charge are ever formed. The electronegativities fall with increase in atomic number, the cations becoming increasingly stable. The alkaline-earth metals form ionic compounds in which they have an ionic valency of two. This is the Group Valency.

Ionic bonding in compounds is usually associated with high melting-points and boiling-points, good solubility in water and poor solubilities in organic solvents, although conclusions based on solubilities alone should be judged with caution. Many compounds of the alkaline-earth metals are characterized by these properties and their ionic structure may be shown by X-ray methods. However, it is interesting to note that the alkaline-earth metals form many insoluble compounds, although most are undoubtedly of ionic structure.

Ions of the alkaline-earth metals have twice the net positive charge and are smaller than the corresponding ions of the adjacent alkali metals. According to Fajans' Rules, the cations of the alkaline-earth metals will have a greater polarizing effect on latent covalency in certain compounds. The hydroxides are relatively weaker and certain salts are slightly hydrolyzed. The salts are more prone to thermal decomposition. The tendency to re-capture electrons diminishes with increasing ionic radius, because the electrical attraction between nucleus and electron falls off with distance (inverse square law). Barium forms completely ionic compounds while beryllium, which lies on the diagonal separating metals on the left from non-metals on the right in the Periodic Classification, forms both covalent and ionic compounds, but the latter have a strong tendency always to revert to the covalent condition. The strong polarizing effect of the beryllium cation, much more marked than that of lithium, is due to the inadequate shielding of the relatively tiny nucleus by the duplet of electrons. Consider the halides, which are always good for the diagnosis of covalency formation: the anhydrous halides of beryllium have comparatively low melting-points, are non-conducting when fused are readily sublime, dissolve in many organic solvents and dissolve in water with hydrolysis. The covalent tendencies of beryllium are also shown by the hydrolysis of oxosalts, the ease of formation of basic salts and the amphoteric nature of the hydroxide. Magnesium is intermediate in position between beryllium and the main body of alkaline-earth metals but its compounds are predominantly ionic and are written with ionic structures, as are those of the remaining metals of this family. Magnesium halides are hydrolyzed to the basic salts on evaporation of aqueous solutions.

In covalent compounds, as in ionic, beryllium exerts the group valency. Therefore, in covalency the inert gas electronic configuration is not reached. Beryllium chloride may be used in place of aluminium chloride or boron trifluoride in the Friedel-Crafts reaction, there being a tendency to take up a halide ion leaving a positively charged organic ion, a carbonium ion.

The tendency to form complexes decreases with increases in atomic number. Beryllium forms many complexes but these do not concern us here. Magnesium forms complexes with inorganic and organic compounds of oxygen and to a lesser extent, with nitrogen. Increasing atomic number brings greater size and less attraction for molecules with lone pairs of electrons available for donation. Notable among the complexes are hydrated cations and ammines. Magnesium salts may have as many as twelve molecules of water of crystallization. Ammines, which are complexes of ammonia, are usually formed by the chlorides, bromides and iodides, those of magnesium being the best defined. Examples of magnesium chloride hydrates and ammines follow:

$MgCl_2.xH_2O$ where $x = 12, 8, 6$ or 4

and $MgCl_2.yNH_3$ where $y = 6, 4$ or 2

In assigning structures for complexes, co-ionic bonds are drawn between the central ion and the electron-donating atom, oxygen or nitrogen, of the molecules being taken on. Alternatively, water molecules may be regarded as held by electrostatic attraction.

Electronic configurations

inert gases	*ions*		
10	12	$Mg^{2+}O^{2-}$	magnesium oxide
Ne	**Mg^{2+}**	$Mg^{2+}(NO_3^-)_2$	magnesium nitrate
2, 8	2, 8	$Mg^{2+}NH_4^+PO_4^{2-}$	magnesium ammonium phosphate
18	20	$Ca^{2+}\begin{matrix}C^-\\ \vert\vert\vert\\ C^-\end{matrix}$ 4	calcium carbide
Ar	**Ca^{2+}**	$Ca^{2+}(H^-)_2$	calcium hydride
2, 8, 8	2, 8, 8	$(ca^{2+})_2(PO_4^{3-})_2$	calcium phosphate
36	38	$Sr^{2+}(Cl^-)_2$	strontium chloride
Kr	**Sr^{2+}**	$Sr^{2+}(OH^-)_2$	strontium hydroxide
2, 8, 18, 8	2, 8, 18, 8	$Sr^{2+}CO_3^{2-}$	strontium carbonate
54	56	$Ba^{2+}(O_2^{2-})$	barium peroxide
Xe	**Ba^{2+}**	$Ba^{2+}S^{2-}$	barium sulphide
2, 8, 18, 18, 8	2, 8, 18, 18, 8	$Ba^{2+}SO_4^{2-}$	barium sulphate

simple valencey 2

Note: to avoid ambiguity certain ions have been enclosed in brackets.

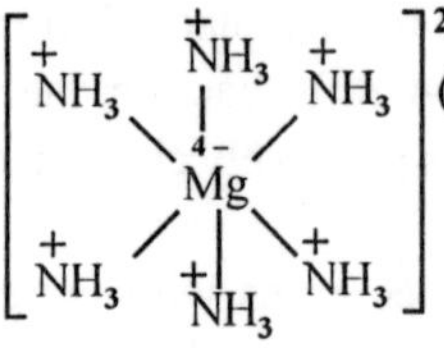

hexammine formed by magnesium chloride $Mg(NH_3)_6Cl_2$

$BeCl_2$
beryllium chloride

$BeCl_2 \cdot {}_4H_2O$
hydrated
beryllium chloride

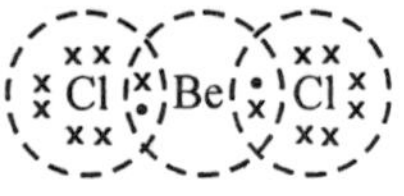

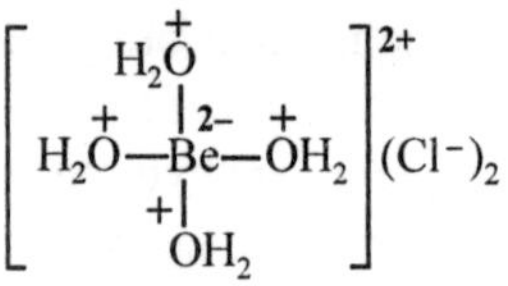

Electronic configuration acquired by beryllium : 2, 4
(no dipole moment
linear structure)

2, 8 (Ne)

$BeSO_4$
anhydrous
beryllium sulphate
$Be^{2+}SO_4^{2-}$

$BeSO_4, {}_4H_2O$
hydrated
beryllium sulphate
$[Be(OH_2)_4]^{2+}SO_4^{2-}$

Electronic configuration acquired by beryllium : 2 (He)

2, 8 (Ne)

Cations of the alkaline-earth metals, formed by the ejection of two electrons from each atom, have the electronic configuration of the nearest inert gas, but the nuclear charge of the original metal. In the illustrations above, the upper number in each case refers to the nuclear charge, the atomic number, ands the lower numbers to the electronic configuration of each atom or ion.

Occurrence, extraction and uses of the Alkaline-earth metals

Beryllium occurs as beryl, $3BeO.Al_2O_3.6SiO_2$, magnesium as magnesite, $Mg^{2+}CO_3^{2+}$, kieserite, $Mg^{2+}SO_4^{2-}.H_2O$ and carnallite, being extracted on an increasing scale; the central atom in the chlorophyll molecule is magnesium. Calcium, barium and strontium occurs as carbonates and sulphates. The presence of magnesium and calcium compounds in natural waters is the principal cause of hardness.

The metals are extracted by electrolysis of their fused halides, other halides being added to lower the melting point and in the case of beryllium to make the melt a better electrical conductor. Magnesium has more uses than the other Group IIA metals; its extraction from sea water involves the precipitation of magnesium hydroxide by the addition of slaked lime (obtained from oyster shells—principally calcium carbonate), conversion into magnesium chloride by heating in a stream of hydrogen chloride and finally electrolysis of the fused chloride in the presence of other halides. A steel-lined tank as cathode and a graphite rod as anode are used; magnesium is discharged at the cathode and floats on the molten electrolyte. It is protected from atmospheric oxidation by a blanket of town gas.

$$\underset{\text{in sea water}}{Mg^{2+}} + Ca^{2+}(OH^-)_2 \longrightarrow Mg^{2+}(OH^-)_2 + Ca^{2-}$$

$$Mg^{2+}(OH^-)_2 + 2HCl \longrightarrow Mg^{2+}(Cl^-)_2 + 2H_2O$$

Cathode **Anode**

$$Mg^{2+}\text{discharged} \longleftarrow Mg^{2+}(Cl^-)_2 \longrightarrow Cl^-\text{discharged}$$

....

$$Mg^{2+} + 2e \rightarrow Mg \qquad\qquad 2Cl^- \longrightarrow 2Cl\cdot + 2e$$

$$2Cl\cdot \longrightarrow Cl_2$$

Beryllium is used for making containers for atomic fuel since it absorbs very few neutrons and does not become radioactive. It is also used as a window material in X-ray apparatus since it is transparent to X-rays. An alloy with copper has a higher tensile strength than any other non-ferrous material.

Magnesium is used extensively in the construction of many light alloys, principally in conjunction with aluminium. It is also used to afford cathodic protection to structures such as ships' hulls and buried pipelines; the greater reactivity of magnesium ensures that this metal corrodes in preference to the main steel structure. Magnesium is also used as a reducing agent in the extraction of some metals.

Calcium, strontium and barium metals have no very extensive important uses.

Physical Properties

Beryllium exhibits exceptional physical properties within this series of elements. Unlike the alkali metals, the alkaline-earth metals vaporize to monomeric species; that is, there is no indication that diatomic or polyatomic species are formed.

Chemical Properties

The chemical behaviour of beryllium, the first member of the alkaline-earth family of elements, is markedly different from that of the other members of this series, although qualitatively each member gives essentially the same type of reaction. This behaviour is similar to that observed for the alkali metals.

The alkaline-earth metals react readily with the oxidizing gaseous elements to form binary compounds containing the metal in the 2 + oxidation state:

$$2M + O_2 \rightarrow 2MO \qquad (1)$$

$$M + X_2 \rightarrow MX_2 \qquad (2)$$

$$X = F, Cl, Br, I \qquad (3)$$

$$M + S \rightarrow MS. \qquad (4)$$

In general these reactions occur more readily with the heavier members of these series and the more powerful oxidizing agents. For example, beryllium does not react with cold Cl_2, Br_2, or I_2 but a vigorous reaction occurs if these systems are heated; magnesium reacts very readily with gaseous F_2 and Cl_2 at room temperature, whereas I_2 vapour reacts markedly less vigorously at 600°. Nitrogen reacts with all of the elements to form the corresponding nitrides

$$3M + N_2 \rightarrow M_3N_2, \quad (5)$$

but the reaction with beryllium requires a markedly higher temperature (1000°C) than is needed for the other members of the series. Hydrogen reacts with all of the alkaline-earth metals except beryllium to form the corresponding hydrides

$$M + H_2 \rightarrow MH_2, \quad (6)$$

$$M = Mg, Ca, Sr, Ba, \quad (7)$$

which are salt-like in character. The alkaline-earth metals, except beryllium, are sufficiently strong as reducing agents to react with many oxygen-containing compounds to give oxides; thus, magnesium reacts with B_2O_3 to give a mixture of MgO and Mg_3B_2.

All of the alkaline-earth metals except beryllium react with water to give hydrogen and the corresponding hydroxide:

$$M + 2H_2O \rightarrow M(OH)_2 + H_2. \quad (8)$$

The reaction with magnesium proceeds slowly because of the presence of a relatively impervious layer of MgO; amalgamation of the metal prevents the formation of this protective coating and the reaction proceeds vigorously. Alcohols react with the alkaline-earth metals in a similar manner:

$$M + 2ROH \rightarrow M(OR)_2 + H_2. \quad (9)$$

In the case of magnesium, elevated temperatures or activation by iodine is required to make this reaction occur rapidly. Although water has no action on beryllium, dilute aqueous acids

$$Be + 2HCl \xrightarrow{aq} Be^{2+}_{(aq)} + 2Cl^- + H_2. \quad (10)$$

or concentrated sodium hydroxide solution will dissolve the metal to give hydrogen.

Aqueous solutions of beryllium salts are extensively hydrolyzed. Since most beryllium salts can be prepared as very stable hydrates, $BeX_2.4H_2O$ (Table 12.5), it is reasonable to assume that these substances are best formulated as $[Be(H_2O)_4]X_2$ on the basis of their properties.

Table 12.5 : The numbers of water molecules of crystallization associated with alkaline-earth metal compounds

Type of compound	*Molecules of water per forumla unit*				
	Be	*Mg*	*Ca*	*Sr*	*Ba*
MCl_2	4	12, 8, 6, 4	6, 4, 2, 1	6, 2, 1	2
MBr_2	4	10, 6, 4	6	6, 2	2, 1
MI_2	—	10, 8, 6	8, 6	6, 2, 1	6, 2, 1
MCO_3	4	5, 3, 1	6, 1	0	0
$M(NO_3)_2$	4	9, 6, 2	4, 3, 2	4	0
MSO_4	4.2	12, 7, 6, 1	2, 1/2	0	0
$M(ClO_4)_2$	4	6, 4, 2	2	4, 2	3, 1

For example, $Be(ClO_4)_2.4H_2O$ loses no water until the temperature at which it begins to decompose completely. Beryllium chloride dissolves readily in water with the evolution of heat. The compound $BeCl_2.4H_2O$ can be readily isolated from this mixture. This compound does not lose water after several months exposure to P_2O_5 under high vacuum. Although a detailed understanding of the nature of aqueous solutions containing divalent beryllium is not available, the properties of these solutions, e.g. high viscosity, low cation mobility, highly hydrated than other divalent species. Depending upon the counter-ion present, several classes of species are thought to be present in aqueous solutions of beryllium salts. For example, if the counter-ion is chloride or sulfate a series of complex equilibria incorporating $[Be.aq]^{2+}$, $[Be_2(OH).aq]^{3+}$, $[Be_3(OH)_2.aq]^{4+}$, $[Be_3(OH)_3.aq]^{3+}$, $[Be_3(OH)_4.aq]^{2+}$, $[Be_6(OH)_8.aq]^{4+}$, and counter ion does not appear in these species. On the other hand, hydrolysis of Be^{2+} in aqueous fluoride solutions gives the mononuclear species $BeF_n^{(2-n)+}$ ($n = 1, ..., 4$).

In general, the alkaline-earth metals heavier than beryllium are not as extensively hydrated in solution. Thus, like beryllium perchlorate, $Mg(ClO_4)_2$ forms a tetrahydrate $[Mg(ClO_4)_2.4H_2O]$; indeed, it is possible to prepare several other hydrates of $Mg(ClO_4)_2$. All of these compounds readily lose water at 250° to form the anhydrous salt.

The marked difference in behaviour of Be^{2+} when compared with the corresponding ions of the other members of this family can be undoubtedly correlated with the large difference in charge density of the ions. Although there is little direct evidence concerning the nature of the alkaline-earth metal ions in aqueous solution, the trend in ionic mobilities (Be, 30; Mg, 55.5; calcium, 59.8; strontium, 59.8; barium, 64.2 at 25°) indicates that divalent beryllium is the most solvated of these ions.

Hydrolysis of the Alkaline-earth metal salts

The soluble salts of strong acids (e.g. chlorides and nitrates) of strontium and barium show a neutral reaction in aqueous solution, and those of the magnesium and calcium are only very slightly acidic. The corresponding salts of beryllium are hydrolysed in solution and invariably show an acid reaction.

This behaviour is very much in line with the progressive increase in ionic size on moving down this particular group. Although the majority of beryllium compounds are essentially covalent, the hydrated ion $[Be(H_2O)_4]^{2+}$ exists in solution and is much smaller than the magnesium ion. Because of its small size, the beryllium ion is able to attract water molecules very strongly and thus weaken the hydrogen-oxygen bonds in bonded water molecules; other solvent water molecules are thus able to act as Bronsted-Lowry bases and abstract protons:

$$[Be(H_2O)_4]^{2+} + H_2O \rightleftharpoons [Be(H_2O)_3(OH)]^+ + H_3O^+$$

The hydrogen carbonates and sulphides of these metals, like those of the alkali metals, show an alkaline reaction in aqueous solution and for exactly the same reason.

Although not classified as salts, hydrides, carbides and nitrides of these metals are hydrolysed by water, since the hydride, carbide, and nitride ions are strong bases:

$$H^- + H_2O \longrightarrow OH^- + H_2$$
$$(C \equiv C)^{2-} + 2H_2O \longrightarrow 2OH^- + HC \equiv CH$$
$$N^{3-} + 3H_2O \longrightarrow 3OH^- + NH_3$$

Anomalous Behaviour of Beryllium

Be differs from the rest of the group for three reasons.

1. It is extremely small, and Fajans' rules state that small highly charged ions tend to form covalent compounds.

2. Be has a comparatively high electronegativity. Thus when beryllium reacts with another atom, the difference in electronegativity is seldom large, which again favours the formation of covalent compounds. Even BeF_2 (electronegativity difference 2.5) and BeO (electronegativity difference 2.0) show evidence of covalent character.
3. Be is in the second row of the periodic table, and the outer shell can hold a maximum of eight electrons. (The orbitals available for bonding are one 2*s* and three 2*p* orbitals). Thus Be can form a maximum of four conventional electron pair bonds, and in many compounds the maximum coordination number of Be is 4. The later elements can have more than eight outer electrons, and may attain a coordination number of 6 using one *s*, three *p* and two *d* orbitals for bonding. Exceptions occur if multi-centre bonding occurs, as for example in basic beryllium acetate, when higher coordination numbers are obtained.

Thus we should except Be to form mainly covalent compounds, and commonly have a coordination number of 4. Anhydrous compounds of Be are predominantly two-covalent, and BeX_2 molecules should be linear.

Electronic structure of beryllium atom in the ground state

Electronic structure of beryllium atom in excited state

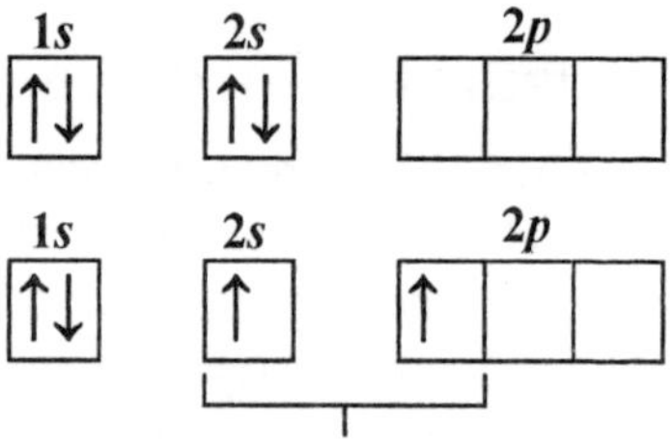

two unpaired electrons can form bonds with two X atoms - linear molecule (*sp* hybridization)

In fact linear molecules exist only in the gas phase, as this electronic arrangement has not filled the outer shell of electrons. In the solid state four-fold coordination is always achieved. There are several ways by which this can be achieved.

1. Two ligands that have a lone pair of electrons may form coordinate bonds using the two unfilled orbitals in the valence

shell of Be. Thus two F^- ions might coordinate to BeF_2, forming $[BeF_4]^{2-}$. Similarly deithyl ether can coordinate to Be (+ II) in $BeCl_2$, forming $[BeCl_2(OEt_2)_2]$.

2. The BeX_2 molecules may polymerize to form chains, containing bridging halogen groups, for example $(BeF_2)_n$, $(BeCl_2)_n$. Each halogen forms one normal covalent bond, and uses a lone pair to form a coordinate bond (Figure 11.3c).
3. $(BeMe_2)_n$ has essentially the same structure as $(BeCl_2)_n$, but the bonding in the methyl compound is best regarded as three-centre two-electron bonds covering one Me and two Be atoms.
4. A covalent lattice may be formed with a zinc blende or wurtzite structure (coordination number 4), for example by BeO and BeS.

In water beryllium salts are extensively hydrolysed to give a series of hydroxo complexes of unknown structure. They may be polymeric and of the type:

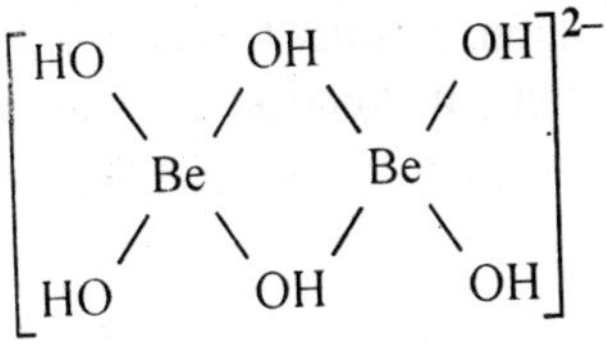

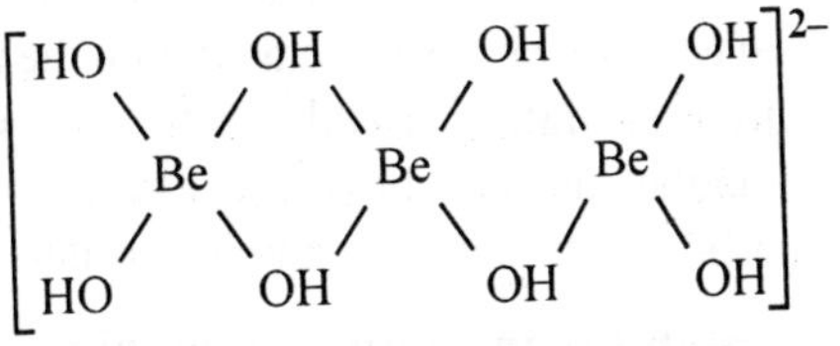

If alkali is added to these solutions the polymers break down to give the simple mononuclear beryllate ion $[Be(OH)_4]^{2-}$, which is tetrahderal. Many beryllium salts contain the hydrated ion $[Be(H_2O)_4]^{2+}$ rather than Be^{2+}, and the hydrated ion too is a tetrahedral complex ion. Note that the coordination number is 4. Forming a hydrated complex increases the effective size of the beryllium ion, thus spreading the charge over a larger area. Stable ionic salts such as $[Be(H_2O)_4]SO_4$. $[Be(H_2O)_4](NO_3)_2$ and $[Be(H_2O)4]Cl_2$ are known.

Beryllium salts are acidic when dissolved in pure water because the hydrated ion hydrolyses, producing H_3O^+. This happens because the Be—O bond is very strong, and so in the hydrated ion this weakens the O—H bonds, and hence there is a tendency to lose protons. The initial reaction is

$$H_2O + [Be(H_2O)_4]^{2-} \rightleftharpoons [Be(H_2O)_3(OH)]^{-} + H_3O^{-}$$

but this may be followed by further polymerization, involving hydroxo-bridges structures $[Be_2OH]^{3+}$, $[Be_3(OH)_4]^{3+}$. In alkaline solutions $[Be(OH)_4]^{2-}$ is formed. The other group 2 salts do not interact so strongly with water, and do not hydrolyse appreciably.

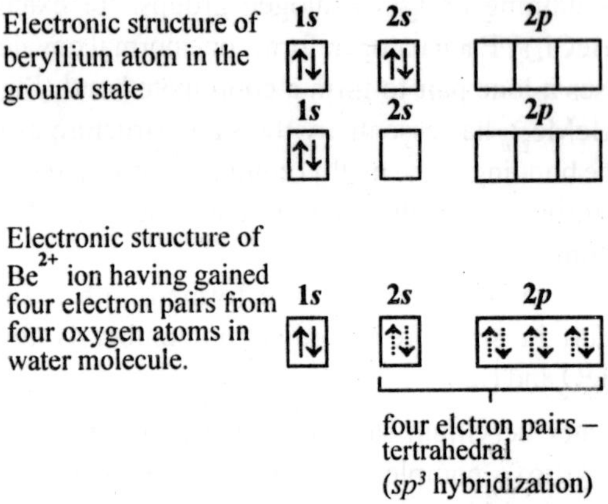

Beryllium salts rarely have more than four molecules of water of crystallization associated with the metal ion, because there are only four orbitals available in the second shell of electrons, whereas magnesium can have a coordination number 6 by using some 3*d* orbitals as well as 3*s* and 3*p* orbitals.

Solubility in Liquid Ammonia

The alkaline-earth metals dissolve in liquid ammonia to form the blue solutions containing solvated electrons. The properties of these solutions, when their concentrations are expressed in terms of equivalents of alkaline-earth metal, parallel those of the alkali metals. For example, the visible spectra of dilute solutions containing equivalent amounts of alkali metal and alkaline-earth metal are indistinguishable. This observation strongly supports the suggestion that the band at ~ 14.500 Å arises from transitions associated with the solvated electron and that the alkaline-earth metal atoms dissolve and lose two electrons in liquid ammonia:

$$M + (2x + y)NH_3 \rightarrow M(NH_3)_y^{2+} + 2e(NH_3)_x^{-}. \qquad (1)$$

In the case of calcium, evaporation of the metal-ammonia solutions yields a substance with the formula $Ca(NH_3)_6$. Attempts to dissolve magnesium in pure ammonia give no indication of forming a blue solution; rather, upon prolonged contact with boiling ammonia, hydrogen is evolved. However, if a strong base such as $NaOCH_3$ or KNH_2 is present, a blue solution readily forms (*3*). This observation can be understood if one recognizes that Mg^{2+} has a relatively high charge density. Thus, if any magnesium does dissolve according to Eq. (1), the Mg^{2+} formed would be expected to interact sufficiently strongly with the solvent to give ammonolysis products:

$$Mg(NH_3)_x^{2+} + NH_3 \rightleftharpoons Mg(NH_3)_{x-1}^{2+}(NH_2)^- + NH_4^+. \quad (2)$$

Reactions analogous to Eq. (2) are common in water with small highly charged ions such as Al^{3+}. Thus, the dissolution of magnesium would produce solvated electrons [Eq. (1)] and NH_4^+ [Eq. (2)], which would, of course, react to form hydrogen:

$$NH_4^+ + e(NH_3)_x^- \rightarrow (x+1)NH_3 + \frac{1}{2}H_2. \quad (3)$$

In the presence of a strong base the equilibrium in Eq. (2) can be suppressed to the point where the NH_4^+ concentration is negligible. Under these conditions the reaction described by Eq. (3) does not occur and the solution turns blue. There are also reports that under favourable conditions beryllium can be made to dissolve in liquid ammonia. Based upon observations on the dissolution of magnesium, we might expect that similar arguments would be valid for the dissolution of beryllium since the charge density of Be^{2+} is significantly greater than that of Mg^{2+}.

Oxide Compoudi of alkaline-earth metals

	covalent	*ionic*				
Monoxides	[Beo]	$M^{2+}O^{2-}$	[MgO]	[CaO]	[SrO]	BaO
	covalent structure	white solids				
Peroxides		$M^{2+}(O_2^{2-})$ white solids	$(MgO + MgO_2)$	CaO_2	SrO_2	[BaO_2]

The alkaline-earth metals react with oxygen on heating to form the monoxides. Further in oxygen results in the formation of a peroxide with barium only at atmospheric pressure, but with strontium also under considerable pressure. The final products obtained on heating the elements in oxygen under atmospheric pressure are indicated by the framed formulae. However, the peroxides are best obtained by neutralization using hydrogen peroxide on the hydroxides or by using sodium peroxide and a solution of the appropriate salt, the octahydrate being the product for calcium, strontium and barium:

$$Ca^{2+} + O_2^{2-} + 8H_2O \rightarrow CaO_2.8H_2O \downarrow$$

Magnesium peroxide has a variable composition.

Thermal decomposition of the carbonates may be used for the preparation of all monoxides, except that of barium:

$$SrCO_3 \rightarrow SrO + CO_2 \uparrow$$

The thermal decomposition of barium carbonate is facilitated by heating with carbon, when a very porous, highly reactive form of the oxide results:

$$BaCO_3 + C \rightarrow BaO + 2CO \uparrow$$

Alternatively, nitrates or hydroxides may be decomposed by heat. The oxides are strongly basic white solids, ionic in structure and melting at very high slightly but enough for the solution to turn litmus blue. The others are slaked by water, reacting with great vigour when freshly prepared, barium oxide becoming incandescent. Excess moisture escapes as steam.

Magnesium oxide, magnesia, $Mg^{2+}O^{2-}$

The melting-point of magnesium oxide is 2800°C.

Magnesia is used in the production of magnesite and chrome-magnesite bricks for the steel and other metallurgical industries where a basic furnace lining is required. In granular form, it is used for the basic process in the open hearth furnace in steel-making. The raw materials required in the production of magnesia are dolomite, $CaCO_3.MgCO_3$, sea water, which contains magnesium chloride and sulphate (more accurately, Mg^{2+}, Cl^- and SO_4^{2-}) and coal as fuel. Dolomite is burnt with coal to yield 'Dolime':

$$CaCO_3.MgCO_3 \rightarrow CaO.MgO + 2CO_2 \uparrow$$

This is slaked, made into a slurry and residues from the coal removed. The slurry is mixed with sea water from which calcium hydrogencarbonate has been removed as a precipitate of carbonate by an earlier addition of a little 'Dolime'. As the least soluble combination of ions present, magnesium hydroxide is precipitated:

$$CaO.MgO + 2H_2O + Mg^{2+} \rightarrow 2Mg(OH)_2 \downarrow + Ca^{2+}$$

A thick, creamy slurry settles, is filtered and heated above 1600°C to give magnesium oxide of not less than 90% purity.

Calcium oxide, quicklime, $Ca^{2+}O^{2-}$

Calcium oxide melts at 2572°C.

Quicklime is made in great quantity by heating limestone, calcium carbonate, to 900-1200°C in a lime kiln, fired by burning producer-gas. Thermal dissociation or calcium carbonate becomes thermal decomposition because the carbon dioxide is swept away:

$$CaCO_3 \rightleftharpoons CaO + CO_2 \uparrow$$

Quicklime is usually obtained as white lumps. Addition of water yields slaked-lime, calcium hydroxide:

$$CaO + H_2O \rightarrow Ca(OH)_2$$

As the reaction proceeds, the mass swells and cracks, becoming very hot and steaming, finally crumbling to a fine white bulky powder. In dry air, quicklime keeps indefinitely. It is asked just before use in making mortar and plaster in the building trade. In the cold, quicklime reacts with acids but not with acidic oxides. Industrially, it is used in metallurgical processes to remove acidic oxides as fusible slags:

$$\underset{\text{silica}}{SiO_2} + CaO \rightarrow \underset{\text{calcium silicate}}{CaSiO_3}$$

$$\underset{\text{phosphorus pentoxide}}{P_4O_{10}} + 6CaO \rightarrow \underset{\text{calcium phosphate}}{2Ca_3(PO_4)_2}$$

Quicklime is used in Gossage's Process for making caustic soda and in the production of bleaching powder. Soda-lime is made by slaking quicklime with concentrated aqueous caustic soda. It is used in the laboratory as an absorbent for carbon dioxide, and waste acidic gases, and for decarboxylation of salts of carboxylic acids. Quicklime is used to dry ammonia, which reacts with the usual drying agents, and sometimes, to dry alcohol.

At red heat, calcium oxide reacts with chlorine to yield calcium chloride and oxygen:

$$2CaO + 2Cl_2 \rightarrow 2CaCl_2 + O_2$$

Barium peroxide, $Ba^{2+}O_2^{2-}$

When barium oxide is heated to a dull redness in air or oxygen, barium peroxide is formed in a reversible reaction, accompanied by evolution of heat:

$$2BaO + O_2 \rightleftharpoons 2BaO_2$$

Further heating, or reduction in pressure, results in decomposition. In Brin's Oxygen process, now obsolete, barium peroxide was formed by heating barium oxide in air under increased pressure and the oxygen extracted drawn off under reduced pressure.

Barium peroxide is a dense off-white powder used for bleaching straw and silk, and in the laboratory preparation of hydrogen peroxide. The commercial form is relatively inactive and a much more reactive, bulky, porous and crystalline octahydrate results from dissolving it in dilute ice-cold hydrochloric acid followed by neutralization with alkali:

$$BaO_2 \downarrow + 2H^+ \rightarrow Ba^{2+} + H_2O$$
$$Ba^{2+} + H_2O_2 + 2OH^- + 6H_2O \rightarrow BaO_2.8H_2O \downarrow$$

Barium peroxide liberates hydrogen peroxide with acids, being a salt of hydrogen peroxide and therefore a true peroxide. Carbon dioxide displaces oxygen from barium peroxide:

$$2BaO_2 + 2CO_2 \rightarrow 2BaCO_3 + O_2$$

With potassium hexacyanoferrate(III), potassium barium hexacyanoferrate(II) is formed and oxygen evolved,

$$BaO_2 + 2K_3Fe(CN)_6 \rightarrow K_6Ba[Fe(CN)_6]_2 + O_2$$

This reaction may be employed for the quantitative estimation of barium peroxide.

Hydroxides

$Be(OH)_2$ is amphoteric, but the hydroxides of Mg, Ca, Sr and Ba are basic. The basic strength increases from Mg to Ba, and Group 2 shows the usual trend that basic properties increase on descending a group.

Solution of $Ca(OH)_2$ and $Ba(OH)_2$ are called lime water and baryta water respectively, and are used to detect carbon dioxide. When CO_2 is bubbled through these solutions, they become 'turbid' or 'milky' due to the formation of a suspension of solid particles of $CaCO_3$ or $BaCO_3$. If excess CO_2 is passed through these 'milky' solutions then the turbidity disappears as soluble bicarbonates form with the excess CO_2. Baryta water is rather too sensitive as it gives a positive test for CO_2 by exhaling breath on it, whereas with lime water, breath (or other gas) must be blown through the solution as bubbles.

$$Ca^{2+}(OH^-)_2 + CO_2 \rightarrow \underset{\substack{\text{insoluble}\\ \text{white precipitate}}}{CaCO_3} + H_2O \xrightarrow[]{\substack{\text{Excess}\\ CO_2}} \underset{\text{soluble}}{Ca^{2+}(HCO_3^-)_2}$$

The bicarbonates of Group 2 metals are only stable in solution. Caves in limestone regions often have stalactites growing down from the roof, and stalagmites growing up from the floor. Water percolating through the limestone contains some $Ca^{2+}(HCO_3^-)_2$ in solution. The soluble bicarbonate decomposes slowly into the insoluble carbonate, and this results in the slow growth of the stalactites and stalagmites.

$$Ca^{2+}(HCO_3^-)_2 \rightarrow CaCO_3 + CO_2 + H_2O$$

Sulphates, $M^{2+}SO_4^{2+}$

Magnesium sulphate differs from the others in being freely soluble in water:

	Solubility g anhydrous salt in 100 g water	
$MgSO_4$	35.8	(20°C)
$CaSO_4$	0.11	(25°C)
$SrSO_4$	0.011	(18°C)
$BaSO_4$	0.00022	(25°C)

This determines the most convenient methods of preparation. Magnesium sulphate is prepared by the neutralization of dilute sulphuric acid using metals, oxide, hydroxide or carbonate, excess of which may be filtered off. The solution, which is not hydrolysed, deposits the heptahydrate, $MgSO_4.7H_2O$. The other sulphates are prepared by precipitation. In testing for sulphate ions in solution, the addition of barium chloride solution and hydrochloric acid, the eliminate carbonate ions, etc., which would also cause precipitation, yields a white precipitate of barium sulphate. An interesting application of the difference in solubilities of sparingly soluble sulphates of calcium and strontium appears in qualitative analysis: strontium sulphate is precipitated by addition of a saturated solution of the more soluble calcium and magnesium ions, associated with sulphate and chloride ions, accounts for the permanent hardness of water.

Magnesium sulphate, magnesium tetraoxo-sulphate, $Mg^{2+}SO_4^{2-}$

Magnesium sulphate was discovered in spring water at Epsom, Surrey, in 1695. The colourless, crystalline heptahydrate, $MgSO_4.7H_2O$ is known as Epsom salts. Magnesium sulphate also occurs in natural salts deposits at Stassfurt and elsewhere which may contain:

kieserite	$MgSO_4.H_2O$
schönite	$K_2SO_4.MgSO_4.6H_2O$
and kainite	$K_2SO_4.MgSO_4.MgCl_2.6H_2O$
	(or, more simply, $KCl.MgSO_4.3H_2O$)

Epsom salts is made commercially by reacting magnesite or dolomite with sulphuric acid or by re-crystallization of the naturally occurring salt. It is used as a purgative and in the tanning and dyeing industries.

Calcium sulphate, calcium tetraoxosulphate, $Ca^{2+}SO_4^{2-}$

Calcium sulphate occurs naturally as anhydrite, $CaSO_4$, and as the dihydrate, gypsum, $CaSO_4.2H_2O$, various forms of which are known as selenite (transparent), satin spar (fibrous) and alabaster (opaque).

When gypsum is heated to 97°C, loss of water occurs and the hemihydrate results, a change which is essentially:

$$2(CaSO_4.2H_2O) \rightleftharpoons (CaSO_4)_2.H_2O + 3H_2O$$

The product is plaster of Paris. Addition of water reverses the change and a solid mass of interlocking crystals of gypsum results, accompanied by a slight expansion and generation of heat. Plaster of Paris gives sharply defined castings which are cheap and easy to make. If surface pores in the object are filled by application of paraffin wax dissolved in a volatile solvent, such as petrol, the casting formed from it has a smooth, hard, ivory-like surface. Plaster of Paris is widely used in surgery for keeping parts of the body rigidly in position. Setting takes up to 15 minutes and may be accelerated by addition of salt or delayed by the use of borax of alum.

Further heating to 200°C yields anhydrous calcium sulphate which also takes up water but stronger heating above 400°C produces a 'dead-burnt' product which is re-hydrated only very slowly. Hard-wearing plaster surfaces (Parian Cement, Estrich's Plaster) use the slow-setting form which sets without expansion, or plaster of Paris and alum (Keene's Cement). Stucco is plaster of Paris made with glue.

Calcium sulphate is used as a filling in the manufacture of glazed paper and for the production of sulphuric acid.

Strongly heated to above 960°C, calcium sulphate begins to dissociate:

$$2CaSO_4 \rightleftharpoons 2CaO + 2SO_2 \rightleftharpoons 2SO_2 + O_2$$

The relatively more volatile sulphur trioxide may be displaced by heating calcium sulphate with silica, the reaction commencing at 870°C:

$$CaSO_4 + SiO_2 \rightarrow \underset{\text{calcium silicate}}{CaSiO_3} + SO_2 \uparrow$$

This is similar to the displacement of phosphorus pentoxide from calcium phosphate in the production of phosphorus.

Strontium sulphate, stronium tetraxosulphate, $Sr^{2+}SO_4^{2-}$

Strontium sulphate melts at 1605°C.

Strontium sulphate is found naturally as celestine or celestine. It is a white anhydrous solid.

Barium sulphate, barium tetraoxosulphate, $Ba^{2+}SO_4^{2-}$

Barium sulphate melts at 1580°C.

Barium sulphate is found naturally as heavy spar or barytes. It is a white solid and very insoluble in water but as it is an important source of barium compounds, methods have been devised to use in preparations.

On boiling barium sulphate with a concentrated solution of sodium carbonate, some of the less soluble barium carbonate is formed:

$$BaSO_4 \downarrow + CO_3^{2-} \rightleftharpoons BaCO_3 \downarrow + SO_4^{2-}$$

The precipitate is filtered off, washed free of sodium sulphate and boiled with fresh sodium carbonate solution. The cycle is repeated. Alternatively, barium sulphate is fused with 4-6 times its weight of sodium carbonate, cooled and sodium sulphate extracted with boiling water. The barium carbonate resulting from either treatment is dissolved in acid.

Barium sulphate may be reduced to the sulphide by heating to a dull redness in a fireclay crucible with powdered charcoal for two hours:

$$BaSO_4 + 4C \rightarrow BaS + 4CO\uparrow$$

The solid product, barium sulphate, is extracted with boiling water and decomposed with acid.

Barium sulphate is used in paper-making to improve opacity and weight. Nearly all white and white tinted paints, oil and water, contain the white pigment lithopone, which is a mixture of barium sulphate and zinc sulphide. Barium sulphide is formed by reduction of barytes with coke in a rotary inclined kiln heated by producer gas, purified zinc sulphate solution:

$$\underset{\text{(hydrolysed)}}{Ba^{2+} + S^{2-}} + Zn^{2+} + SO_4^{2-} \rightarrow BaSO_4 \downarrow + ZnS \downarrow$$

The mixture, which is not blackened in sulphurous town air, is filtered off, dried and processed further for incorporation in paints.

Barium sulphate begins to dissociate slightly near the melting-point.

Halides, $M^{2+}(X^-)_2$

The fluorides are all sparingly soluble and prepared by mixing solutions containing the required ions:

$$Ca^{2+} + 2F^- \rightarrow CaF_2 \downarrow$$

Calcium fluoride, formed here as a white gelatinous precipitate, is the most important alkaline-earth fluoride. Generally, the other halides are very soluble, those of barium having the lowest solubility. Chlorides are prepared by the neutralization of hydrochloric acid, on addition of hydroxide or carbonate, excess being filtered off. For magnesium chloride, the oxide or metal may be used. The most stable hydrates formed by the chlorides are:

$MgCl_2.6H_2O$	$CaCl_2.6H_2O$
$SrCl_2.6H_2O$	$BaCl_2.2H_2O$

The hexahydrates are very soluble in water and deliquescent. On heating, the magnesium salt is somewhat hydrolysed to the oxide chloride, formerly oxychloride, while calcium chloride is less hydrolysed and strontium chloride and barium chloride not at all:

$$MgCl_2.6H_2O \rightarrow Mg(OH)Cl + HCl\uparrow + 5H_2O\uparrow$$
$$Mg(OH)Cl \rightarrow MgO + HCl\uparrow$$

The tendency of the halides to form ammines and the amount of hydration declines with increases in atomic number of the metal.

Calcium fluoride, $Ca^{2+}(F^-)_2$

Calcium fluoride is sparingly soluble (solubility = 0.0025 g in water at 18°C). It melts at 1330°C.

Calcium fluoride occurs naturally as fluorite or fluorspar. The blue variety, used ornamentally, is known as Blue John. Calcium fluoride is used in the production of hydrofluoric acid by treatment with concentrated sulphuric acid:

$$CaF_2 + H_2SO_4 \rightarrow CaSO_4 + 2HF\uparrow$$

Magnesium chloride, $Mg^{2+}(Cl^-)_2$

The common hydrate is the hexahydrate $MgCl_2.6H_2O$, although combination with 12, 8, 6, 4 and 2 molecules of water of crystallization may be encountered.

Magnesium and chloride ions are present in sea water and natural brines. Magnesium chloride is found in salt deposits formed by the evaporation of salt takes; thus, carnallite has the composition $KCl.MgCl_2.6H_2O$.

Anhydrous magnesium chloride cannot be prepared by direct evaporation of the aqueous solution because of hydrolysis. It is formed by passing dry chlorine over a mixture of magnesium oxide and carbon.

$$MgO + C + Cl_2 \rightarrow MgCl_2 + CO\uparrow$$

It melts at 715°C. At 20°C, 54.1 g anhydrous magnesium chloride dissolves in 100 g water. Other general methods of preparation include the action of dry chlorine or dry hydrogen chloride on the heated metal. Alternatively, the hexahydrate may be heated in a stream of hydrogen chloride, the high concentration of which reduces hydrolysis, or a double salt, formed by evaporation of a solution of magnesium chloride and ammonium chloride in equi-molar quantities, may be heated:

$$MgCl_2.NH_3Cl.6H_2O \rightarrow MgCl_2 + NH_4Cl + 6H_2O\uparrow$$
$$MgCl_2 + NH_4Cl \rightarrow MgCl_2 + NH_3\uparrow + HCl\uparrow$$

Calcium chloride, $Ca^{2+}(Cl^-)_2$

The common, deliquescent hydrate has six molecules of water of crystallization, $CaCl_2.6H_2O$, and melts at 30°C. A tetrahydrate and dihydrate may also be prepared.

Calcium chloride is formed in great quantity as a by-product of the Ammonia-Soda process. There are few demands for it. Previously fused calcium chloride is a useful drying agent but may not be used for ammonia or alcohols, with which it forms compounds normally given as $CaCl_2.8NH_3$, $CaCl_2.4CH_3OH$ and $CaCl_2.6C_2H_5OH$.

At 20°C, 81.5 g anhydrous calcium chloride saturates 100 g water. The anhydrous salt melts at 782°C.

Barium chloride, $Ba^{2+}(Cl^-)_2$

Barium chloride may be made in the laboratory from barium sulphate by conversion into the carbonate or sulphide followed by reacting the product with hydrochloric acid. Commercially, calcium chloride is heated with barytes and coke, the produce being extracted with water and purified:

$$BaSO_4 + 4C + CaCl_2 \rightarrow BaCl_2 + CaS + 4CO\uparrow$$

It crystallizes with two molecules of water of crystallization. Barium chloride dihydrate is neither deliquescent nor hygroscopic and yields the anhydrous salt. without hydrolysis, on heating. Anhydrous barium chloride melts at 955°C. At 20°C, 37.2 g saturates 100 g water.

Sulphides, $M^{2+}S^{2-}$

The sulphides may be prepared by union of the elements or by reduction of the sulphate. Calcium, strontium and barium sulphates may be reduced by prolonged heating with wood charcoal in a furnace, the reduction of barium sulphate being the most difficult:

$$CaSO_4 + 4C \rightarrow CaS + 4CO\uparrow$$

Calcium sulphide is sparingly soluble, strontium sulphide dissolves readily in warm water while barium sulphide is readily soluble in cold water, all passing into solution as the hydrogensulphide and hydroxide formed by hydrolysis:

$$Cas + H_2O \rightleftharpoons Ca^{2+} + HS^- + OH^-$$
$$HS^- + H_2O \rightleftharpoons H_2S + OH^-$$

Magnesium sulphide is rapidly decomposed by water according to the same scheme.

Calcium hydrogensulphide (hydrosulphide), $Ca(HS)_2$, is made industrially by acting on milk of lime with hydrogen sulphide:

$$OH^- + H_2S \rightleftharpoons SH^- + H_2O$$

It is used for removing hair from ox-hides and in depilatories generally.

Nitrates

Nitrates of the metals can all be prepared in solution and can be crystalized as hydrated salts by the reaction of HNO_3 with carbonates, oxides or hydroxides. Heating the hydrated solids does not give the anhydrous nitrate because the solid decomposes to the oxide. Anhydrous nitrates can be prepared using liquid dinitrogen tetraoxide and ethyl acetate. Beryllium is unusual in that it forms a basic nitrate in addition to the normal salt.

$$BeCl_2 \xrightarrow{N_2O_4} Be(NO_3)_2.2N_2O_4 \xrightarrow[\text{under vacuum}]{\text{warm to } 50°C} Be(NO_3)_2$$

$$\xrightarrow{125°C} \underset{\substack{\text{basic}\\\text{beryllium}\\\text{nitrate}}}{[Be_4O(NO_3)_6]}$$

Basic beryllium nitrate is covalent and has an unusual structure. Four Be atoms are located at the corners of a tetrahedron, with six NO_3^- groups along the six edges of the tetrahedron, and the (basic) oxygen at the centre. The structure is of interest partly because beryllium is unique in forming a series of stable covalent molecules of formula $[Be_4O(R_6)]$ where R may be NO_3^-, $HCOO^-$, CH_3COO^-, $C_2H_5COO^-$, $C_6H_5COO^-$, etc. Thus, basic beryllium nitrate is one of a series of similar molecules. The structure is also of interest because the NO_3^- groups act as bidentate ligands in forming a bridge between two Be atoms.

Carbides

When BeO is heated with C at 1900-2000°C a brick red coloured carbide of formula Be_2C is formed. This is ionic, and has an anti-fluoride structure, *i.e.* like the CaF_2 structure except that the positions of the metal ions and anions are interchanged. It is unusual because it reacts with water, forming methane, and is thus called a methanide.

$$Be_2C + 4H_2O \rightarrow 2Be(OH)_2 + CH_4$$

Group 2 metals typically form ionic carbides of formula MC_2. The metals Mg, Ca, Sr and Ba form carbides of formula MC_2, either when the metal is heated with carbon in an electric furnace, or when their oxides are heated with carbon. CaC_2 made in this way is a grey coloured solid, but is colourless when pure. BeC_2 is made by heating Be with ethyne.

On heating, MgC_2 changes into Mg_2C_3. This contains C_3^{4-}, and it reacts with water to form propyne $CH_3C \equiv CH$ (methyl acetylene).

$$Ca + 2C \xrightarrow{1100°C} CaC_2$$

$$CaO + 3C \xrightarrow{2000°C} CaC_2 + CO$$

The MC_2 carbides all have a distorted sodium chloride type of structure. M^{2+} replaces Na^+ and $C \equiv C^{2-}$ replaces Cl^-. The C_2^{2-} ions are elongated, not spherical like Cl^-. Thus the axis along which the C_2^{2-} ions lie is lengthened compared with the other two axes, and this is called tetragonal distortion. At temperatures above 450°C the C_2^{2-} ions adopt random positions rather than being aligned in this way, and the unit cell then becomes cubic rather than tetragonal.

Calcium carbide is the best known. It react with water, liberating ethyne (formerly called acetylene), and is thus called an acetylide.

$$CaC_2 + 2H_2O \rightarrow Ca(OH)_2 + C_2H_2$$

At one time this reaction was the main source of ethyne for oxy-acetylene welding. Production of CaC_2 peaked at 7 million tonnes/ year in 1960 but had declined to 4.9 million tonnes in 1991 because ethyne is now obtained from processing oil.

CaC_2 is an important chemical intermediate. When CaC_2 is heated in an electric furnace with atmospheric dinitrogen at 1100°C, calcium cyanamide CaNCN is formed. This is an important reaction because it is one method of fixing atmospheric dinitrogen. (The Haber process for NH_3 is another method of fixing dinitrogen.)

$$CaC_2 + N_2 \rightarrow CaNCN + C$$

The cyanamide ion $[N{=}C{=}N]^{2-}$ is isoelectronic with CO_2, and has the same linear structure. CaNCN is produced on a large scale, particularly in locations where there is cheap electricity. At one time about 1.3 million toones/year was produced, though this is now increasing. Cyanamide is widely used (particularly in SE Asia and the Far East) as a slow acting nitrogenous fertilizer, as it hydrolyses slowly over a period of months. CaNCN has the advantage over more soluble nitrogenous fertilizers such NH_4NO_3 or urea, in that it is not washed away with the first rainstorm.

$$CaNCN + 5H_2O \rightarrow CaCO_3 + 2NH_4OH$$

Other important industrial uses of CaNCN are the manufacture of cyanamide H_2NCN which is used to make urea and thiourea, and of melamine which forms hard plastics with formaldehyde.

$$CaNCN + H_2SO_4 \rightarrow H_2NCN + CaSO_4$$

$$CaNCN + CO_2 + H_2O \rightarrow \underset{\text{cyanamide}}{H_2NCN} + CaCO_3$$

$$H_2NCN + H_2O \xrightarrow{\text{pH<2 or > 12}} H_2N.CO.NH_2\,(\text{urea})$$

$$H_2NCN + H_2S \longrightarrow H_2N.CS.NH_2\,(\text{thiourea})$$

$$\underset{\text{cyanamide}}{H_2NCN} \xrightarrow{\text{pH 7-9}} \underset{\text{dicyanamide}}{NCNC(NH_2)_2} \xrightarrow{\text{pyrolysis}} \quad ????$$

It is interesting that BaC_2 also reacts with N_2, but it forms a cyanide $Ba(CN)_2$, not a cyanamide $(NCN)^{2-}$.

Hard waters Most natural waters with the exception of rain water contain dissolved mineral matter. This mineral matter is derived from the rocks and soil and contains calcium, magnesium, or ions salts as chlorides, sulfates, or bicarbonates. Waters containing any of these salts are objectionable for certain household and industrial uses and are termed *hard waters*. If soap is added to hard water, a precipitate forms before a lather can be obtained. This residue may stain the cloth with which it comes in contact, and of course the soap is wasted. Soap is a sodium or potassium salt of complex organic acids—stearic, palmitic, or sodium or oleic acids. We may represent the formula for soap as NaSt, in which St stands for the stearate acid radical. If calcium ions are present in water, the reaction with soap is as follows:

$$Ca^{++} + 2NaSt \rightarrow Ca(St)_2 \downarrow + 2Na^{+}$$

Temporary water hardness Waters containing bicarbonates of calcium, magnesium, or iron are said to possess *temporary hardness*, since the hardness may be removed simply by boiling the water. Thus, when calcium bicarbonate is heated, the following reaction takes place:

$$Ca(HCO_3)_2 \rightarrow CaCO_3 \downarrow + H_2O + CO_2$$

Calcium carbonate precipitates and since the calcium ions are removed, the water will no longer act with soap. Such a water from which the ions responsible for the precipitation of soap are removed is termed *soft* water. Decomposition of bicarbonates is the cause of scale in tea kettles in which considerable water has been boiled. Water containing bicarbonates is also objectionable for industrial uses, as it will form boiler-scale deposit (carbonates of calcium or magnesium) on the inside of boiler pipes. Since the scale is a poor conductor of heat, boiler efficiency is reduced.

Temporary hardness may be removed by adding the proper amount of slaked lime, $Ca(OH)_2$, or household ammonia, NH_4OH:

$$Ca(HCO_3)_2 + Ca(OH)_2 \rightarrow 2CaCO_3 \downarrow + 2H_2O$$

$$Ca(HCO_3)_2 + 2NH_2OH \rightarrow CaCO_3 \downarrow + (NH_3)_2CO_3 + 2H_2O$$

Permanent hardness Permanent hardness in water results from chlorides or sulphates of metals such as calcium or magnesium. Boiling does not remove these salts; a substance called a *water softener* must be added to the water to form a precipitate with these salts. Common water softeners are soda, Na_2CO_3, and borax, $Na_2B_4O_7$. Either of these will precipitate calcium or magnesium carbonate or borate from solution as an insoluble substance. Typical reactions of these water softeners are as follows:

$$Ca^{++} + CO_3^{--} \rightarrow CaCO_3 \downarrow$$

$$Mg^{++} + B_4O_7^{--} \rightarrow MgB_4O_7 \downarrow$$

In some cases, both soda and slaked lime are used to soften water, the slaked lime to remove temporary hardness and the soda to remove permanent hardness.

Zeolites Another method of softening water utilizes *zeolites*, minerals that contain sodium aluminium silicate. If water is allowed to filter through this coarse-grained material, sodium ions are displaced by calcium and magnesium ions (this is called *ion exchange*):

$$Na_2 \text{ Zeolite} + Ca^{++} \rightarrow \text{Ca Zeolite} + 2Na^{+}$$

Sodium ions left in solution are not objectionable, since they do not make the water hard. In this process, the sodium zeolite may be regenerated with a concentrated sodium chloride solution. By mass action, the above reaction is reversed and the original zeolite re-formed:

$$\text{Ca Zeolite} + 2NaCl \rightarrow CaCl_2 + 2Na_2 \text{ Zeolite}$$

An artificial zeolite called Permutit is also effective as a water softener.

Organometallic Compounds (II)

The similarity of the electronegativities of calcium, barium, and strontium with the alkali metals suggests that their simple alkyl derivatives (MR_2) would be predominantly ionic in character. Although extensive work has not been done in this area, the data available support this suggestion. For example, the dimethyl derivatives of calcium, strontium, and barium are reported to be white crystalline compounds that decompose under vacuum at 400°, are rapidly hydrolyzed, and burn readily when exposed to air.

In contrast, the markedly greater electronegativities of beryllium and magnesium suggest that the organic derivatives of these elements should be more covalently bonded. This prediction correlates with the observed properties of the methyl derivatives of these elements. For example, when heated under vacuum the compounds $Mg(CH_3)_2$ and $Be(CH_3)_2$ sublime before the melt.

The organometallic derivatives of the alkaline-earth elements can, in general, be prepared by the methods described for the alkali organometallic compounds. Undoubtedly the most useful of the organometallic derivatives from the standpoint of synthesis is the so-called Grignard reagent, the mixture formed when magnesium metal is allowed to react with organic halides in either solution:

$$RX + Mg \xrightarrow{(C_2H_5)_2O} RMgX.$$

Magnesium occupies a rather position with respect to the reactivity of the organometallic compounds it forms. Metals that are much more electronegative than magnesium form more reactive derivatives and, of course, are more difficult to handle. Accordingly, the Grignard reagent has achieved a position of great synthetic usefulness because of the relative simplicity of synthesis and the ease with which it can be used. In addition to the reactions that Grignard reagents undergo with organic functional groups, they, together with the organolithium compounds, are useful intermediates for the preparation of a variety of other organometallic compounds by halogen-exchange reactions:

$$AsCl_3 + 3CH_3MgI \rightarrow As(CH_3)_3 + 3MgClI$$
$$SiCl_4 + 4CH_3MgI \rightarrow Si(CH_3)_4 + 4MgClI.$$

The nature of the Grignard reagent in solution has been the source of considerable controversy (III). Either solutions containing Grignard reagents become very viscous if the solvent in evaporated; however, it is virtually impossible to remove all the ether from these solutions. This behaviour suggests that the solvent in chemically combined and that association of the species present occurs as the solvent is removed. Molecular-weight determinations of several Grignard reagents by the boiling-point method verify that association occurs in these solution (6). The results show that the degree of association ranges from about 1 to 4. Since ether is a Lewis base, these results suggest that species such as those shown in Figure 12.1 may be permit.

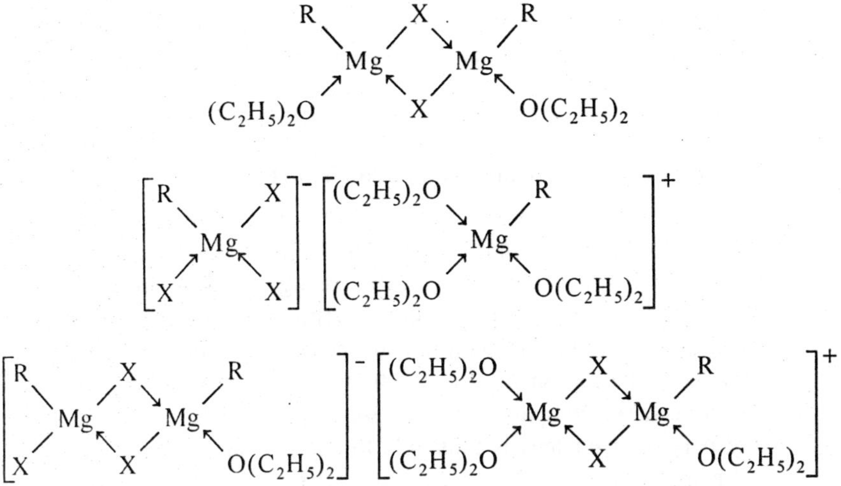

Fig. 12.1 : Solutions of the Grignard reagent are associated and conduct an electric current; magnesium is found in both the cationic and anionic species. Shown here are some species which may be present in these solutions.

Addition of dioxane to ether solutions of the Grignard reagent precipitates dioxane complexes of magnesium halides; dialkylmagnesium compounds can be isolated from the remaining solution. Ether solutions containing equivalent amounts of R_2Mg and MgX_2 behave in the same way as solutions prepared from RX and magnesium. With no additional information one might be tempted to suggest that equilibria of the type

$$2RMgX \rightleftharpoons R_2Mg + MgX_2$$

are present in solution. However, tracer experiments indicate that practically no exchange occurs between $MgBr_2$ and $(C_2H_5)_2Mg$ in ether solution. Electrical conductivity data show that ions are present in Grignard solution and that both anions and cations contain magnesium; these observations are consistent with the species shown in Fig. 12.1. Whatever the nature of the species that are present in Grignard solutions, they react as if the carbon atom is negative with respect to magnesium ($^{\delta-}C—Mg^{\delta+}$). The nature of the products formed when

Grignard reagents react can be understood on the basis of the attack of the negative carbon atom on the positive part of the other reagent, although this is not necessarily the mechanism of the reaction IV):

$$\overset{\delta-}{R}-\overset{\delta+}{Mg}X+\overset{\delta-}{Cl}-\overset{\delta+}{Si}Cl_3 \rightarrow RSiCl_3 + MgXCl.$$

Summary of trends in alkaline-earth metals

The elements of this group are very electropositive, although less so than corresponding members of the alkali metal group. Electropositivity increases with increasing atomic number. They therefore form predominantly ionic compounds, although beryllium is exceptional in forming compounds in which the bonding is usually covalent. The reason for this can be traced to the small size of the beryllium atom compared with the sizes of the other Group IIA atoms; the beryllium nucleus is thus able to exert a considerable force of attraction on its outer two valency electrons of beryllium is compared with that of aluminium, its diagonal neighbour.

The Group IIA metals form hydroxides which are strong bases, except that of beryllium which is amphoteric; the solubility of these hydroxides increases with increasing atomic number. Their salts are fairly stable to heat but less so than the corresponding salts of the alkali metals. Thus the carbonates can be decomposed into the oxides and carbon dioxide at high temperature, and the nitrates on decomposition give the oxides, nitrogen dioxide and oxygen; the stability to heat increases with increasing atomic number. The hydrogen carbonates exist in solution but any attempt to isolate them as solids leads to decomposition into the carbonates.

The solubilities of the Group IIA salts show a definite trend but in some cases this shows up with increasing atomic number as an increase and in others and a decrease in solubility. The fluorides increase in solubility with increasing atomic number, but the chlorides, carbonates and sulphates decrease in solubility in this order. In order to explain these observations one must consider changes of lattice and hydration energies for comparable salts. The fluorides increase in solubility in the order Mg < Ca < Sr < Ba because the lattice energy decreases more rapidly along the series than does the hydration energy. The reverse is true for the chlorides, carbonates and sulphates because in these compounds the lattice energy of comparable salts decreases less rapidly than hydration energy.

As the atomic and ionic sizes of the Group IIA metals increase with increasing atomic number there is an increase in chemical reactivity, and calcium, strontium and barium are sufficiently reactive to force the hydrogen atom to accept an electron and form ionic hydrides. Ionic carbides and nitrides can be formed by direct combination and in this respect there is no parallel to be found in Group IA, except for lithium which has a small enough ions to form a stable carbide and nitride lattice.

The Group IIA metals form no oxides containing the superoxide ion, (p. 176) presumably because their ions are too small to form a stable crystal lattice—their ions are very much smaller than those of the corresponding alkali metals. The larger ions of strontium and barium are able to form peroxide like the alkali metals and for the same reason, but pressure is required.

The tendency for the Group IIA metal ions to form complexes decreases with increasing atomic number; thus of the heavier ions only calcium forms a complex with ethanol. The calcium tetraethanolate ions presumably involves dative bonding from the oxygen atoms of ethanol molecules, which the four molecules of ethanol arranged tetrahedrally (theory of electron pair repulsion (Fig. 12.2).

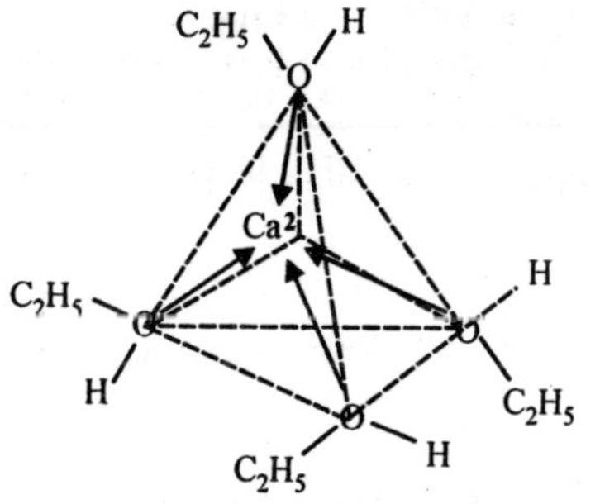

Fig. 12.2 : The structure of the calcium tetraethanolate ion

Magnesium, having a smaller ion than calcium, forms complexes more readily and the solubility of its bromide and iodide in alcohols, ketones and ethers is probably due to the formation of dative bonded complexes involving bonds between the magnesium ion and the oxygen of the organic compound.

13

p-Block Elements : boron, aluminium, gallium, indium and thalium—Group IIIA (13)

Some physical data of Group IIIA elements

	Atomic Number	Electronic Configuration	Ionisation energy (kJ/mol)			Standard Electrode Potential (volts)	Atomic Radius(Å)	Ionic Radius (Å)	M.p.	B.p.
			First	Second	Third	$M^{3+} + 3e \longrightarrow M$		M^{3-}	°C	°C
B	5	2.3 $1s^2 2s^2 2p^1$	800	2427	3659		0.80	(0.20) Estimated Value	2030	3930
Al	13	2.8.3 ....$2s^2 2p^6 3s^2 3p^1$	578	1816	2744	– 1.66	1.25	0.50	660	2450
Ga	31	2.8.18.3 ....$3s^2 3p^6 3d^{10} 4s^2 4p^1$	579	1979	2952	– 0.52	1.25	0.62	29.8	2240
In	49	2.8.18.18.3 ....$4s^2 4p^6 4d^{10} 5s^2 5p^1$	558	1817	2693	– 0.34	1.50	0.81	156.2	2050
Tl	81	2.8.18.32.18.3 ....$5s^2 5p^6 5d^{10} 6s^2 6p^1$	589	1970	2866	+ 072	1.55	0.95	303.6	1470

All these elements exhibit a group valency of three, but because of the very large input of energy that is necessary to form the 3-valent ions—the sum of the first three ionisation energies—their compounds when anhydrous are either essentially covalent or contain an appreciable amount of covalent character. Boron never forms a B^{3-} ion since the enormous amount of energy required to remove three electrons from a small atom cannot be repaid with the formation of a stable crystal lattice, even with the most electronegative fluoride atom.

The electronic configuration of the boron and aluminium atoms are similar inasmuch at the penultimate shell has a noble gas configuration, whereas the penultimate shell of the gallium, indium and thallium atoms contains eighteen electrons.

Boron, which is non-metallic, and aluminium, which is clearly metallic, are best considered separately. Gallium, indium and thallium are weakly metallic and can profitably be studied as a group; in many aspects their chemistry is similar to that of aluminium.

Boron

Occurrence, extraction and uses of boron

Boron occurs principally as borates, e.g. sodium tetraborate (borax), $(Na^+)_2B_4O_7^{2-}.10H_2O$, in which the boron atom is part of an anionic complex.

Boron can be obtained as an amorphous brown powder by treating borax with hydrochloride acid, igniting the boric acid, H_3BO_3, to give the oxide, B_2O_3, and finally reducing the latter with magnesium at a high temperature:

$$B_2O_3 + 3Mg \rightarrow 2B + 3Mg^{2+}O^{2-}$$

It is used in the construction of high impact-resistant steel and, since it absorbs neutrons, in reactor rods for controlling atomic reactions.

A crystalline form of boron can be obtained by the thermal decomposition of boron tri-iodide on a tantalum filament:

$$2BI_3 \rightarrow 2B + 3I_2$$

Properties of boron

Amorphous boron is a very reactive element combining directly with oxygen, sulphur, nitrogen and the halogens to give respectively an oxide, sulphide, nitride and a halide. The oxide, sulphide and nitride are giant molecules with covalent bonds running completely through the structures.

$$4B + 3O_2 \rightarrow 2B_2O_3$$
$$4B + 6S \rightarrow 2B_2S_3$$
$$4B + N_2 \rightarrow 2BN$$
$$2B + 3Cl_2 \rightarrow 2BCl_3$$

At red heat it will reduce steam to hydrogen :

$$2B + 3H_2O \rightarrow B_2O_3 + 3H_2$$

Diborane, B_2H_6

This compound, which is an inflammable and very reactive gas, is the simplest hydride of boron. It can be made by the reduction of boron trichloride with lithium aluminium hydride and must be handled in vacuum systems which employ mercury valves, since it attacks tap-grease.

$$4BCl_3 + 3Li^+AlH_4^- \rightarrow 2B_2H_6 + 3Li^+Cl^- + 3AlCl_3$$

The structure of this molecule immediately poses problems since the number of valency electrons is twelve, yet at least fourteen are required for conventional single covalent bonded structures similar to ethane, i.e. the molecule is electrons deficient. Physical evidence shows that the structure can be considered to involve something resembling boron—boron double bonds, the extra two electrons required for this being provided by two of the six hydrogen atoms which bridge the boron atoms as protons :

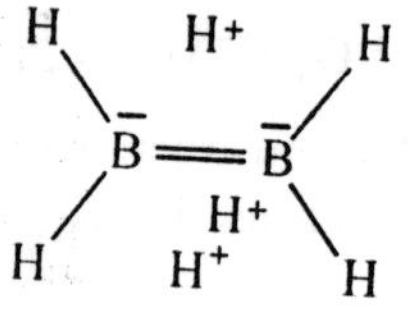

A more sophisticated but more difficult theory envisages the two bridging hydrogen atoms forming bonds with both boron atoms.

Other hydrides of boron are B_4H_{10}, B_5H_9, B_5H_{11} and $B_{10}H_{14}$; they are all electron deficient.

Borohydrides.[11] Closely related to the boranes is a large group of ionic species known collectively as the hydroborates, the simplest of which has the formula $[BH_4]^-$: salts of this anion would be designated as tetrahydromonoborates (– 1). The name borohydride is, however, in common use and will be adopted here to avoid confusion. Other hydroboron ions containing more than one boron atom will be described later.

Borohydrides are formed by a number of the metallic elements, the most important being the alkali metal compounds, which can be made by reaction of the alkali metal hydride with diborane (6) in one of the higher ethers (e.g. diglyme, $CH_3O(CH_2)_4OCH_3$). Formally this can be regarded as a Lewis acid-base interaction between BH_3 and H^- to give $[H_3B \leftarrow H]^-$. There are, however, a number of the better preparative methods that avoid the isolation of B_2H_6, one being the reaction of sodium hydride with methyl borate in tetrahydrofuran (THF) where the sodium salt is precipitpated:

$$4NaH + 4B(OCH_3)_3 \xrightarrow{THF} NaBH_4 + 3Na[B(OCH_3)_4]$$

Aluminium borohydride, the first compound of this class to be characterized, is formed when trimethylaluminium and diborane (6) react at 80°C:

$$Al(CH_3)_3 + 2B_2H_6 \rightarrow Al(BH_4)_3 + B(CH_3)_3$$

Alkyls of a number of other elements (e.g. Li, Be, Mg, Zn and Ga) undergo a similar reaction to give the metal borohydrates. Metathetical reactions are also widely used, as is shown by the following examples:

$$NaBH_4 + LiCl \xrightarrow{iso-Pr\,NH_2} LiBH_4 + NaCl$$

$$2NaBH_4 + BeCl_2 \xrightarrow{125°} Be(BH_4)_2 + 2NaCl$$

$$3NaBH_4 + AlCl_3 \xrightarrow{ether} Al(BH_4)_2 + 3NaCl$$

$$2Al(BH_4)_3 + MF_4 \xrightarrow{20°} M(BH_4)_4 + 2AlF_2(BH_4)\ (M = U, Th)$$

$$8LiBH_4 + 2TiCl_4 \xrightarrow{20°} 2Ti(BH_2)_3 + 8LiCl + B_2H_6 + H_2$$

Boron trioxide, B_2O_3

Boron trioxide can be obtained by burning in oxygen or by fusing orthoboric acid :

$$2H_3BO_3 \rightarrow B_2O_3 + 3H_2O$$

It is usually obtained as a glassy material whose structure consists of randomly oriented three-dimensional networks of BO_3 groups, each oxygen atom uniting two boron atoms:

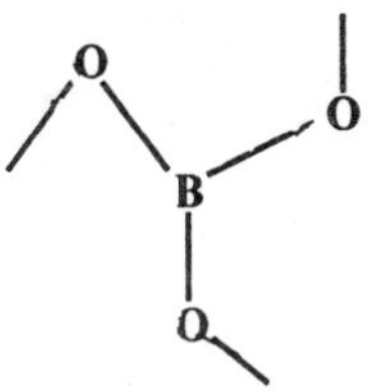

Boron trioxide in an acidic oxide and slowly reacts with water, forming orthoboric acid. When fused with metallic oxides it forms borate glasses which are often coloured; this is the basis of the borax bead test in qualitative analysis.

Orthoboric acid, H_3BO_3

Orthoboric acid is formed when the boron halides are hydrolysed or when dilute hydrochloric acid is added to a solution of borax:

$$B_4O_7^{2-} + 2H^+ + 5H_2O \rightarrow 4H_3BO_3$$

It is obtained as a white solid on subsequent crystallization.

Orthoboric acid is a weak monobasic acid and in aqueous solution the boron completes its octet by removing OH^- from water molecules:

$$B(OH)_3 + 2H_2O \rightarrow B(OH)_4^- + H_3O^+$$

It therefore functions as a Lewis acid and not as a proton donor.

When heated, orthoboric acid first forms metaboric acid (empirical formula HBO_2) and then boron trioxide.

The-structure of orthoboric acid is based on the planar $B(OH)_3$ unit, individual units being joined together by hydrogen bonds to give a two-dimensional sheet. It is now known that the hydrogen bonds linking the oxygens of different $B(OH)_3$ groups are unsymmetrical, i.e. they are not midway between the oxygen atoms. The layers of two-dimensional sheets are held together by weak van der Waals' forces, and this accounts for the cleavage of the solid into flakes.

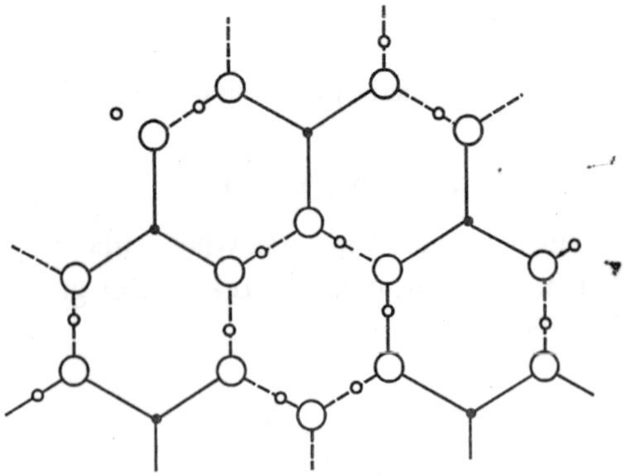

Fig. 13.1 : The structure of orthoboric acid

Borate Salts : Boric and metaboric acids form a variety of salts with many metal ions. Anhydrous metal borates can be prepared by fusion of the metal oxide with B_2O_3,

$$Na_2O + B_2O_3 \rightarrow Na_2B_2O_4.$$

Hydrated salts can be obtained by crystallizing the compounds from aqueous solutions. Large quantities of boron are found naturally as hydrated sodium borates [$(Na_2B_4O_7).10H_2O$, borax] or as calcium salts [$(CaB_3O_4(OH)_3.H_2O$, colemanite]. A survey of the stoichiometries of the borates shows that no discernible pattern exists. However, structural studies indicate that many of the simpler borates consist of BO_3 units linked in several different ways, as shown in Fig. 13.2.

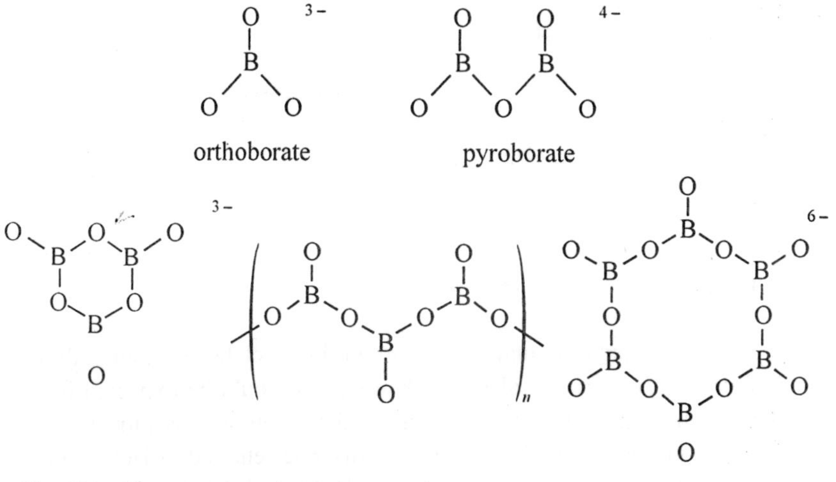

Fig. 13.2 : The structures of some borate anions.

Orthoborates. Discrete anions which contain a B : O ratio of 1 : 3 are known as orthoborates. In such compounds (e.g., $InBO_3$ and $Ca_3(BO_3)_2$) the orthoborate ion is planar.

Pyroborates. Pyroborates are compounds such as $Mg_2B_2O_5$ containing the ion $B_2O_5^{4-}$. In these ions the BO_3 groups share an oxygen atom, but the ions, in general, are not perfectly planar.

Metaborates. Metaborates possess structures where each BO_3 unit shares two oxygen atoms. Basically, the ring structures and chain structures shown in Fig. 13.2 are possible. It should be recalled that the basic ring structure exhibited by some metaborates is the same as that found in metaboric acid. In addition, the dehydration of monomethylboric acid, $CH_3B(OH)_2$, gives a cyclic trimer, (CH_3BO_3), which has also this same basic boron-oxygen ring structure (Fig. 13.3). All of the atoms except calculated distance expected from single-bond radii (1.54 Å).

The anions containing a B : O ratio of 2 : 3 (which corresponds to BO_3 groups sharing three oxygen atoms) are not yet known.

Modern Inorganic Chemistry

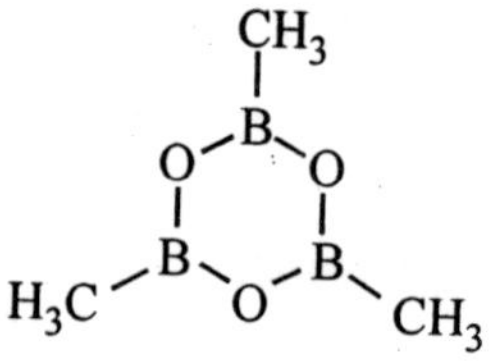

∠BOB = 112°
d(B—O) = 1.39 Å

Fig. 13.3 : Structural parameters for $(CH_3BO)_3$

Compounds containing oxyanions in which boron atoms exhibit four-coordination are also well known. As would be expected from previous arguments, oxygen atoms in these compounds surround the boron in a tetrahderal arrangement, discrete tetrahedral BO_4 groups being present in compounds such as BPO_4 and $BAsO_4$. Tetrahedral BO_3OH groups are found in compounds like $CaBSiO_4(OH)$ and $CaB_3O_4(OH)_3.H_2O$. Tetrahedral $B(OH)_4$ groups are found in minerals such as $Na_2ClB(OH)_4$ and $CaClB(OH)_4$.

Complex Borates. The ortho-, meta-, and pyroborate anions all contain three-coordinate boron atoms. Oxyanions containing only four-coordinate boron atoms are not known. However, a series of oxyanions containing both four-and three-coordinate boron atoms have been found. For example, the ion $B_5O_{10}^{5-}$, found in the compound $KH_4B_5O_{10}.2H_2O$, exhibits the structure shown in Fig. 13.4 (a), in which a central four-coordinate boron atoms; the planes of the six-membered B—O rings are perpendicular to each other. The cyclic structure shown

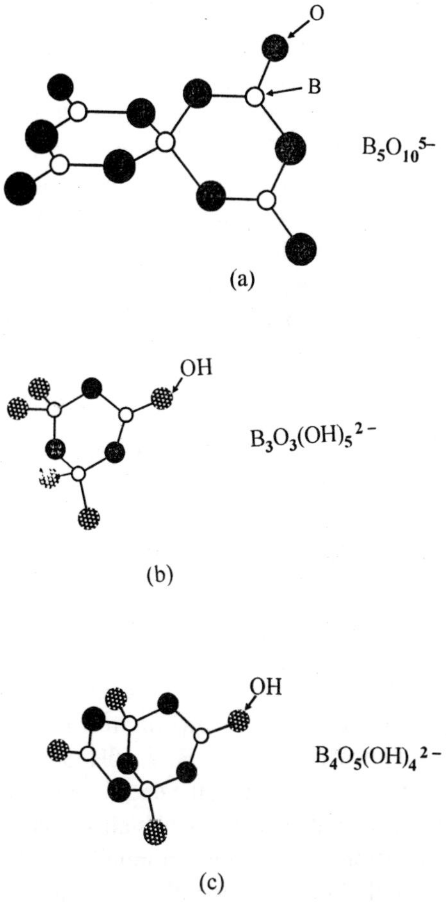

Fig. 13.4 : The structures of some complex borate ions: the hydrogen atoms are not shown in (b) and (c).

in Fig. 13.4b, containing two four-coordinate and one three coordinate boron atoms, exists in the anion $B_3O_3(OH)_5^{2-}$ which is present in the compound $CaB_3O_3(OH)_5.4H_2O$. A related ion is found in borax, which has the empirical formula $Na_2B_4O_7.10H_2O$. Actually the anion present in this compound is related to the $B_3O_3(OH)_5^{2-}$ anion; it is formed by attaching a bridging planar BO_3 group sharing oxygen atoms with the four-coordinate BO_4 species in the ring to give the structure shown in Fig. 13.4c. Thus, borax should be more properly formulated as $Na_2[B_4O_5(OH)_4].8H_2O$.

The average B—O bond length in crystalline borates containing a planar BO_3 group is very close to 1.36 Å. In contrast, the B—O bond distance increases to an average value of 1.48 Å for tetrahedral four-coordinate boron atoms. The change in bond length suggests that there is a change in bond order in going from a planar three-coordinate structure to a four-coordinate structure. Oxygen atoms in these structures possess two lone-pair electrons which can interact with the empty p_z orbital on a planar sp^2 hybridized boron atom (1) in much the same way as the bond shortening which occurs for planar boron compounds containing B—F and B—Cl bonds. The increase

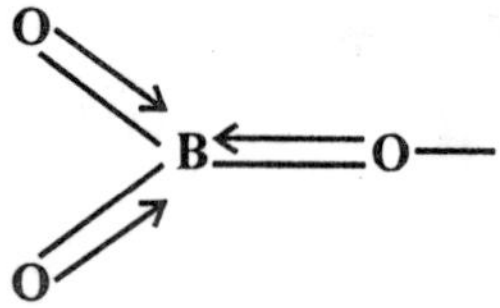

(1)

in bond length for tetrahedral species correlates with the fact that boron atoms possessing four bonds require the use of sp^3 hybrid orbitals, which does not leave a low-lying orbital available for back-coordination.

The two most important naturally occurring compounds of boron are both hydrates of disodium tetraborate, kernite, $Na_2B_4O_7.4H_2O$, and tincal (borax), $Na_2B_4O_7.10H_2O$, which are mined in southern California, U.S.A. Borax, the decahydrate, $Na_2B_4.10H_2O$, is obtained by extraction with water and crystallization. Borates find a great many applications, being incorporated into heat-resistant borosilicate glass and fibreglass, porcelain enamels and ceramics, detergents, adhesives and sizes, fluxes and fireproofing preparations. Borax is a colourless, crystalline solid which is much more soluble in hot water than cold. A solution of borax is hydrolyzed in water:

$$B_4O_7^{2-} + 7H_2O \rightleftharpoons 2OH^- + 4H_2BO_2$$

Borax may be titrated with an acid provided an indicator is used which shows its 'alkaline' colour to boric acid. Hydrochloric acid is frequently standardized against borax using methyl orange (pH range : 2.9-4.0) as indicator:

$$Na_2B_4O_7 + 2HCl + 5H_2O \rightarrow 2NaCl + 4H_2BO_2$$

This is similar to the titration of carbonates, where methyl orange is unaffected by weak carbonic acid. In qualitative analysis, the borax bead test gives characteristic colours with certain cations. When borax is heated in a small loop in a platinum wire, it swells and forms a colourless, transparent bead, possibly containing metaborate and boric oxide:

$$Na_2B_4O_7 \rightarrow 2NaBO_2 + B_2O_3$$

The cold bead is moistened and a minute quantity of the unknown substance is taken up, too much giving a dark, opaque bead in the subsequent heating. The bead is heated in the reducing zone of the bunsen flame and finally, in the oxidizing zone. Coloured borates are formed; for copper:

$$CuO + B_2O_3 \rightarrow Cu(BO_2)_2$$

Under oxidizing conditions, copper gives green (hot) and blue (cold), manganese gives amethyst (violet), cobalt, blue, and nickel, reddish-brown (all cold) coloured beads. Orthoboric acid may be obtained by adding concentrated hydrochloric acid to a hot strong solution of borax. On cooling, it crystallizes out. Orthoboric acid is used as a mild antiseptic and in glazes. Addition of the calculated quantity of sodium hydroxide to borax yields the metaborate:

$$Na_2B_4O_7 + 2NaOH \rightarrow 4NaBO_2 + H_2O$$

Boron halides[9]

Boron forms halides of three types, the simple trihalides of the type BX_3 (X = F, Cl, Br, I), polyboron halides and a monofluoride, BF, since it is closely associated with the short-lived difluoride of silicon.

Boron has an electronegativity of 1.9 on the Pauling scale, so that the polarity of the boron to halogen is $B^+—X^-$ in each instance. On this basis it would be expected that the acceptor properties of boron in the trihalides would decrease in the order $BF_3 > BCl_3 > BBr_3 > BI_3$. Fluorine should exercise the greatest electron-withdrawing effect and so lead to the greatest positive charge on boron, which would favour the acceptance of electrons from a donor. There is, however, much evidence that the stability of many comparable adducts of the trihalides is in the order $L \rightarrow BI_3 > L \rightarrow BBr_3 > L \rightarrow BCl_3 > L \rightarrow BF_3$. Thus, for example, values of $\Delta H°$ for 1 : 1 adduct formation between pyridine and a boron halide in nitrobenzene solution are – 143, – 189 and – 217 kJ (– 34.2, – 45.2 and – 51.8 kcal) mole^{-1} for BF_3, BCl_3 and BBr_3 respectively.

A suggested explanation for this anomaly is that $p_\pi - p_\pi$ back-bonding occurs, involving the transfer of electron density from the halogen to a vacant *p* orbital on the boron atom, and that this is at a maximum for BF_3 and decreases along the series. This occurs in the absence of a ligand and results in a decrease in B—X bond length below the value expected for a single bond. Its effect on the interaction with a ligand is to reduce the positive charge on boron and so reduce its acceptor function. If the B—X bonds have some double bond character this will also increase the energy required to change the hybridization on boron from sp^2, which it has in the planar trihalide molecules, to sp^3, which characterizes the 1 : 1 adduct. This, in turn, would be directly reflected in the energetics of adduct formation.

Formation of the tetrahaloborates represents a special case of donor-acceptor interaction, with the halide ion acting as an electron donor. The most familiar example is the tetrafluoroborate ion BF_4^-, which is formed readily from BF_3 and a metallic fluoride. The B—F bond in BF_4^- is longer than in BF_3, in keeping with the fact that the $2p_z$ orbital of boron is no longer vacant, but forms part of the sp^3 hybrid, so that back-bonding cannot occur. It has been suggested that electrostatic repulsion between the fluorine atoms may also contribute to bond lengthening.

The remaining tetrahaloborates are more difficult to prepare and require a large cation for stability in the solid state. The BCl_3 will react with an alkyl ammonium chloride in chloroform solution forming the BCl_4^- salt. Similarly BCl, when heated in a sealed tube with KCl, RbCl or CsCl, using nitrobenzene as solvent, also yields tetrachloroborates. The most general reaction, however, is that between the boron trihalide and an alkyl ammonium chloride in anhydrous liquid hydrogen chloride. This can also give mixed haloborates; for example,

$$Me_4NCl + BF_3 \xrightarrow{HCl} Me_4NBF_4Cl$$

Anhydrous HF does not yield HBF_4 with BF_3, but an aqueous solution of the acid is obtained when BF_3 dissolves in aqueous HF, just as SiF_4 gives H_2SiF_6; neither acid can be isolated pure and is attempted. All of the tetrahaloborates other than salts of BF_4^- are rapidly decomposed by water.

Some boron trifluoride adducts exist in an ionized form, the dihydrate $BF_3.2H_2O$, for example, having a substantial to the ions H_3O^+ and $[BF_3OH]^-$. In other adducts with fluorides, fluoride ion transfer can be shown to occur. Thus X-ray analysis of $SF_4.BF_3$ shows it to be $SF_3^+BF_4^-$, while in the case of $NOF.BF_3$ the presence of the tetrafluoroborate ion may be detected by infrared spectroscopy.

Exchange reactions take place in mixtures of boron halides and these, too, can be detected spectroscopically; for exchange,

$$BCl_3 + BBr_3 \rightleftharpoons BBrCl_2 + BBr_2Cl$$

Such a reaction is best explained by postulating an intermediate in which the two boron atoms are joined by two B-(Hal)-B bridges, with alternative modes of breakdown. It has also been shown by ^{11}B n.m.r. spectroscopy that BCl BrI is present in a mixture of the three halides. Exchange between alkyl boranes and boron halides can occur in the presence of a Lewis base catalyst; for example,

$$R_3B + 2BF_3.O(C_2H_5)_2 \rightleftharpoons 3RBF_2 + 2(C_2H_5)_2O$$

No attempt will be made here to describe the many applications of the boron halides and their derivatives in organic chemistry, where they can act as catalysts for esterifications, Friedel-Crafts acylations and alkylations, certain polymerization and other types of reaction.

Nitrogen Derivatives of Boron

The atom air B—N is isoelectronic with C_2. Therefore, compounds containing boron bonded to nitrogen theoretically can exhibit the range of structure types encountered in carbon chemistry. Thus, compounds containing tetra-, tri-, and dicoordinate boron atoms attached to an aminoid nitrogen atom are isoelectronic with the corresponding C_2 system (Fig. 13.5). Although the BN analogues of the triple-bonded carbon system are not yet known, many examples of the other two possibilities exist.

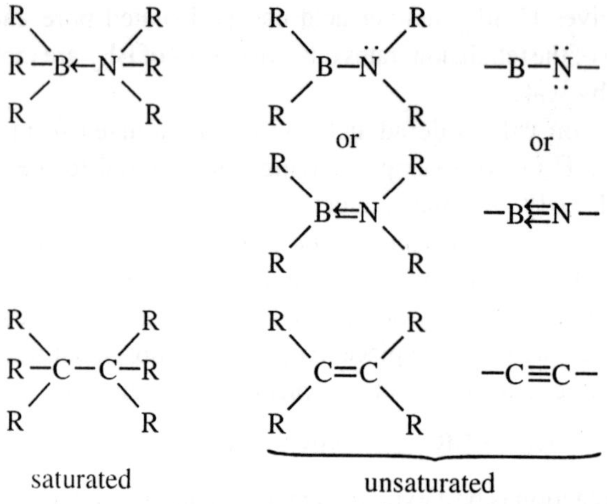

Fig. 13.5 : Boron-nitrogen compounds form isoelectronic structures with carbon compounds.

Boron Nitride. Boron nitride, $(BN)_x$, can be prepared by the action of nitrogen or ammonia on elemental boron at 1000° or by heating borax with ammonium chloride:

$$Na_2B_4O_7.10H_2O + 2NH_4Cl \rightarrow 2NaCl + 2BN + B_2O_3 + 14H_2O$$

The compound melts at 3000° under pressure. It is inert to many common fused alkali.

Boron nitride exists in two modifications. The form normally obtained from usual preparative procedures exhibits a layer-type structure, shown in Fig. 13.6 (a), reminiscent of graphite. Each layer consists of a hexagonal arrangement of alternating boron and nitrogen atoms similar to the hexagonal arrangement of carbon atoms in the graphite structure. However, in contrast to the graphite structure, the boron atoms in one layer of the BN structure are above the nitrogen atoms of the next layer. The B—N distance is 1.446 Å, which is shorter than that expected for a single bond between these atoms (1.54 Å).

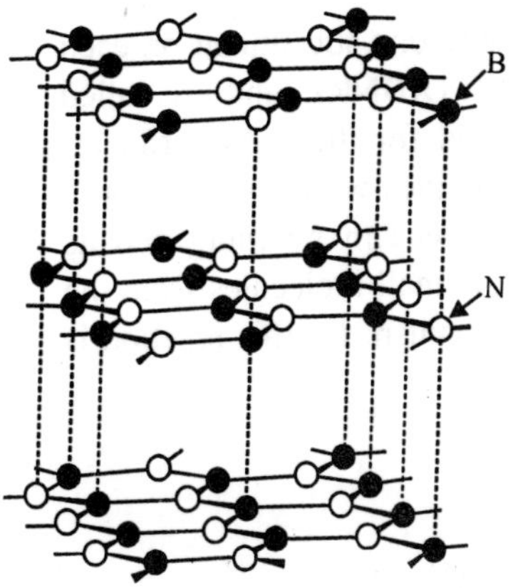

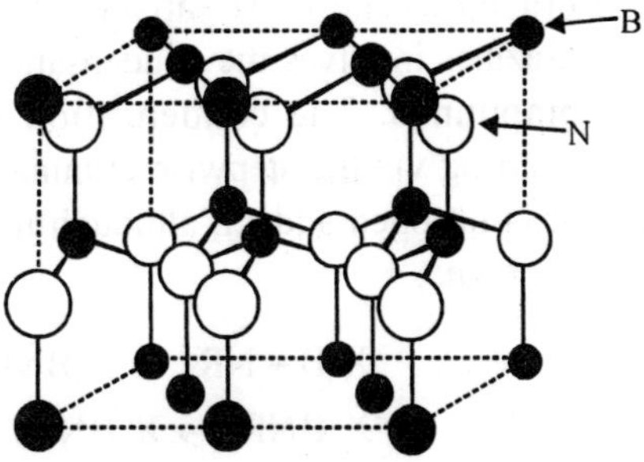

Fig. 13.6 : Boron nitride crystallizes in a graphite-like structure (a) as well as in the zinc-blende structure (b). Boron phosphide and arsenide also crystallize in the zinc-blende structure.

A comparable decrease in the carbon-carbon distances in the graphite sheets (1.42 Å) is observed over the carbon-carbon single-bond distance (1.54 Å). Again in contrast to the graphite structure which has a larger interlayer distance (3.35-3.45 Å), the interlayer distance in BN is the same as the distance between boron and nitrogen atoms within a layer. The hexagonal form of boron nitride can intercalate foreign atoms in the same way as graphite. However, in spite of the structural and chemical resemblance to graphite of the hexagonal form of boron nitride, the physical properties are very different. For example, boron nitride is a white substance and a very good insulator.

When hexagonal boron nitride (density 2.25 g cm^{-3}) is heated to 1400-1800° at 85,000 atm, it is transformed to a more dense form (3.45 g cm^{-3}) which has the zinc-blende structure shown in Fig. 13.6 (b). This form, called borazon, is isostructural and isoelectronic with diamond and exhibits many of the properties of diamond. Boron reacts with phosphorus and arsenic—also Group IV elements—to yield a phosphate (BP) and an arsenide (BAs) as well as with Group IV elements to yield binary compounds which also have the zinc-blende structure shown in Fig. 13.6.

Borazines (VI)

Boron-nitrogen species carrying only one substituent on each atom are known, but the exist as trimers, $(XBNR)_3$, and are called borazines. The most satisfactory method for the preparation of borazines involves the condensation of amines with trivalent boron compounds. The condensation reactions can be imagined as occurring via the stepwise elimination of small molecules from the boron-nitrogen adduct, although the mechanism for this reaction is unknown:

$$X_3B + NR_3 \rightarrow X_3BNR_3$$
$$X_3BNR_3 \rightarrow X_3BNR_2 + RX$$
$$3X_2BNR_2 \rightarrow (XBNR)_3 + 3RX$$

The condensation reaction invariable gives a symmetrically substituted borazine, and in some cases it is possible to obtain unsymmetrical derivatives. It is not necessary to isolate the intermediates in the preparation of borazines, but in several instances the boron-nirtogen adduct X_3BNR_3, as well as the aminoborane (X_2BNR_2), have been isolated before their conversion to the corresponding borazine. The addition compound can be formed by directed combination, the displacement of hydrogen halide from an ammonium salt,

$$[R'_3NH]^+X^- + R_3B \rightarrow R_3BNR'_3 + HX,$$

or by the displacement of hydride ion from a borohydride with an ammonium salt,

$$[R'_3NH]^+X^- + MBH_4 \rightarrow H_3BNR'_3 + H_2 + MX,$$

Conversion of the adduct to the borazine occurs most readily when RX is a hydrogen halide, which can be eliminated by pyrolysis of the pure adduct, by refluxing the adduct in a solvent or by treatment with a strong base like triethylamine. Vigorous pyrolytic conditions (~ 450°) must be used if RX is an alcohol or a hydrocarbon,

$$3(n-BuO)_2BNHR \rightarrow (n-BuOBNR)_3 + 3n-BuOH$$

$$3(CH_3)_3B + 3NH_3 \rightarrow (CH_3BNH)_3 + 6CH_4.$$

The molecular structures for a number of borazines have been determined by diffraction methods. The data show that these compounds possess a cyclic structure of alternating boron and nitrogen atoms. All of the atoms in the ring lie in a plane, within experimental error. The boron-nitrogen bond distances are about 1.42 Å, which is considerably shorter than the B—N distance in the boron nitrogen adducts R_3BNR_3. Moreover, all the bond distances in the ring are equal. These results strongly suggest that the borazine cyclic trimer is best described as containing a B—N multiple bond rather than

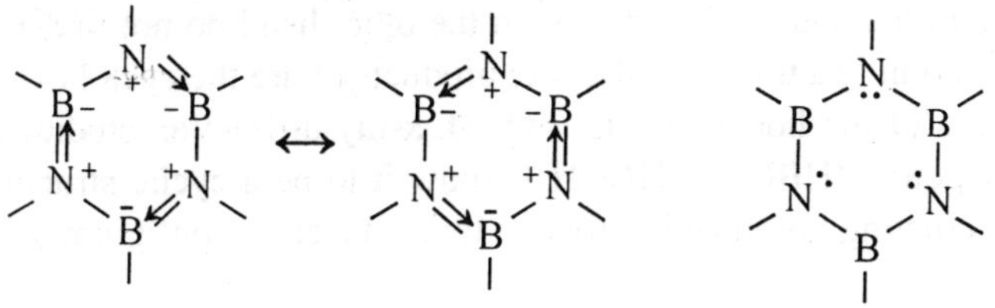

a single bond. The single-bonded structure implies that borazine is a cyclic training in which the nitrogen atoms exhibit a tetrahedral arrangement of electron pairs; under these circumstances the ring should be puckered. The two possible valence-bond structures that exist with multiple boron nitrogen bonds are reminiscent of the resonance structures for benzene. The electronic formulations shown in are a consistent with the known structural data for the borazine nucleus. Presumably all the ring atoms use sp^2 hybridized orbitals to form the three σ - bonds necessary. The unhybridized p orbitals, which are normal to the plane of the ring, form a π system containing three electron pairs which were originally associated with the nitrogen atoms.

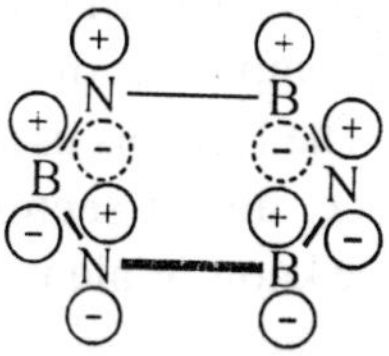

Although the borazine ring has many similarities to the isoelectronic benzene rign, the reactions of the two systems are markedly different. For example, the borazine readily undergoes addition reactions with molecules containing potentially acidic hydrogen atoms such as hydrogen halides, water, methanol, and carboxylic acids, to give borazanes:

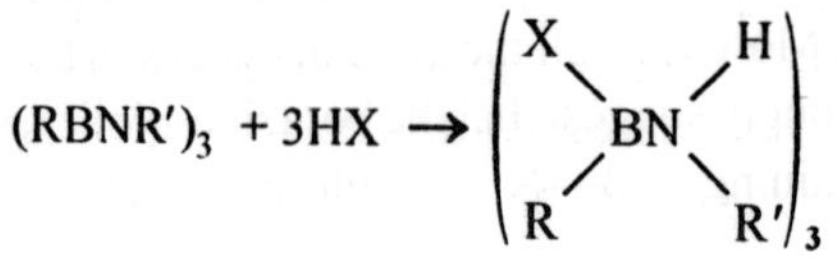

Benzene and its derivatives on the other hand do not undergo such addition reactions readily. The products of are the cyclohexane analogues of the borazine ring. Indeed, x-ray diffraction studies of the compound $(HBCH_3)_3(HNCH_3)_3$ show it to be a cyclic structure with alternating boron and nitrogen atoms in a chair configuration.

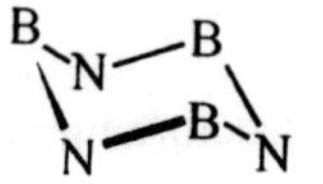

28

Borazines undergo substitution reactions primarily at the boron sites. Borazines carrying a hydrogen atom at the boron site are reducing agents; that is, such hydrogen atoms are hydridic in nature. B-Hydrido- and B-halogeno-borazines react with Grignard reagents, yielding B-substituted borazines. Substitution reactions rarely occur at the nitrogen sites, and unsymmetrically N-substituted borazines are relatively rare.

Aluminium

Although the element itself was isolated only a little more than a hundred years ago, compounds of aluminium have beem known and used for centuries. The Romans used the word *alum* to describe any astringent substance, and it seems likely that the compound of aluminium which we know as alum today was included in this group.

Because of the activity of aluminium and the stability of its compounds. efforts to liberate the element were unsuccessful until

1827 when Wöhler succeeded in producing it by the reduction of aluminium chloride with metallic potassium:

$$AlCl_3 + 3K \rightarrow Al + 3KCl$$

However, the cost of the method prohibited its production in any quantity, and for many years the metal was only a curiosity. It was recognized that the element would be extremely useful if it could be produced economically, and many attempts were made to develop cheaper processes of manufacture. The substitution of metallic sodium for the potassium reduced the cost from \$ 160 to \$ 10 per pound. The development of electrolytic methods for the preparation of sodium reduced the cost further to \$ 4 a pound. The development of electrolytic methods for the preparation of sodium reduced the cost further to \$ 4 a pound. Efforts to produce aluminium electrolytically from water solutions in a manner similar to that for many other metals failed; a coating of aluminium oxide was produced almost immediately which stopped the passage of the current. Likewise, the use of carbon at a reducing agent was not successful, as aluminium carbide was produced instead to the free metal. The problem was finally solved in 1886 by an American, Charles Martin Hall, while working as a student at Oberlin College. He found that electrolysis of aluminium oxide could be effected if the oxide was dissolved in melting *cryolite*, Na_3AlF_6, a double fluoride salt found in Greenland. When Hall's process was industrialized, the price of aluminium quickly dropped to about thirty cents a pound.

Occurrence. Aluminium is the most abundant of the metals and stands third in abundance among the chemical elements, the percentage in the earth's crust being 7.4. It does not occur in the free state because of its activity, but its compounds are spread widely over the earth's surface. It is a consistent of most siliceous soils and rocks such as *clay*, $2H_2O{\cdot}Al_2O_3{\cdot}2SiO_2$; *feldspar*, $KAlSi_3O_8$ or $NaAlSi_3O_8$, and *mica*, $KH_2Al_3(SiO_4)_3$ or $NaH_2Al_3(SiO_4)_3$. The ore of alminium is *bauxite*, $Al_2O_3{\cdot}2H_2O$, large deposits of which are found in France, British and Dutch Guiana, and in the states of Arkansas and Alabama. Bauxite is the only ore of aluminium at present because of the difficulties of extracting the metal from other naturally occurring compounds. As mentioned above the double fluoride salt cryolite, Na_3AlF_6, is important as a solvent in the electrolytic production of metallic aluminium.

Metallurgy. The process for the preparation of aluminium today is essentially the same as that developed by Hall in 1886. Purified dehydrated bauxite is dissolved in molten Na_3AlF_6, which acts as a solvent for the Al_2O_3. In modern practice, AlF_3 and CaF_2 are added to lower the melting point and increase fluidity. The mixture, when melted, is a good conductor of the electric current, and electrolysis results in the decomposition of the Al_2O_3 to form metallic aluminium and oxygen. The electrolysis is carried out in carbon-lined vasts which act as cathodes. The anodes are carbon rods (Figure 31.7) which dip into the electrolyte.

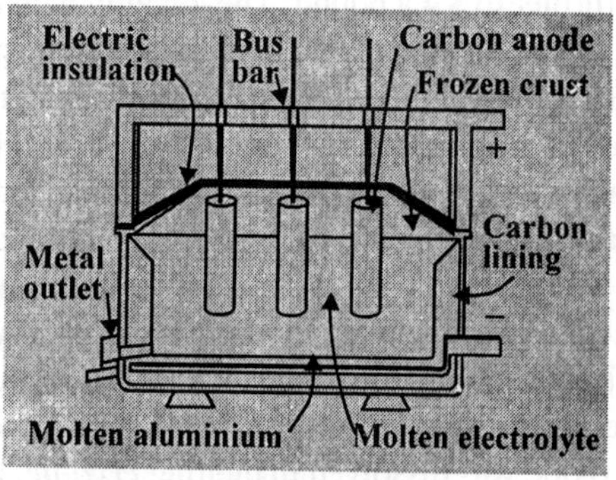

Fig. 13.7 : Preparation of aluminium by electrolysis (Hall process).

At the beginning of the process, the carbon anodes are brought near the bottom of the carbon-lined *vat* to produce an arc to melt the cryolite. Electrolysis of the Al_2O_3, then begins with aluminium ions moving toward the sides and bottom of the vat, and the oxygen ions moving toward the carbon anodes. At the cathode, aluminium ions are discharged and metallic liquid aluminium settles to the bottom of the vat where it may be drawn off. Oxygen, which is produced at the anodes, rapidly oxidizes the latter. About 0.6 pound of carbon is used for every pound of aluminium produced; hence the carbon electrodes must be frequently replaced. The resistance to the passage of the current produces enough heat to keep the cryolite bath in a molten condition. The aluminium metal thus produced is above 99 per cent pure, which is adequate for most purposes.

Principal costs of this method are the electric energy required and the treatment of the bauxite ore to prepare pure Al_2O_3. The bauxite ore is ground and treated with sodium hydroxide under pressure to convert the aluminium oxide to an aluminate:

$$Al_2O_3 + 2NaOH \rightarrow 2NaAlO_2 + H_2O$$

Iron oxide, which is nearly always present as an impurity, is not amphoteric and hence is insoluble in the alkaline solution and may be filtered off. The solution of sodium aluminate is diluted with water and heated to hydrolyze the salt and precipitate aluminium hydroxide:

$$NaAlO_2 + 2H_2O \rightarrow Al(OH)_3 \downarrow + NaOH$$

The hydroxide is filtered off and ignited to the oxide, which may then be dissolved in the cryolite bath. The caustic soda is recovered.

Properties and Uses of Aluminium

When freshly cut, aluminium has a silvery appearance, but it soon becomes superficially oxidized and assumes a dull white luster. The metal is very light and processes high tensile strength. Weight for weight it is twice as good conductor of electricity as copper. Although aluminium is a relatively reactive metal, the tenacious coating of the oxide which forms on it prevents further atmospheric corrosion. When heated in the air finely divided aluminium burns with a brilliant light. Because of the formation of a protective coat of oxide, aluminium does not decompose water. It is readily dissolved by hydrochloric acid and sulphuric acid, but is rendered passive by nitric acid. Aluminium dissolves in concentrated alkalies with the formation of aluminates and hydrogen.

$$2Al + 2OH^- + 6H_2O \rightarrow 2Al(OH)_4^- + 3H_2 \uparrow$$

This behaviour is in accord with the amphoteric character of aluminium hydroxide. When heated, aluminium combines directly with the halogens, nitrogen, carbon, and sulphur.

The fact that aluminium is an excellent conductor of heat, together with its light weight and resistance to corrosion, accounts for its use in the manufacture of cooking utensils. The most important uses of aluminum are in the airplane and other transportation industries, which depend upon the lightness, toughness, and tensile strength of the metal. Aluminum is also used in the manufacture of electrical transmission wire, as a paint pigment, and, in the form of foil, as a wrapping material.

Aluminium is one of the best reflectors of heat and light, including the wavelengths in the ultraviolet. For this reason it is used as an insulating material, and as a mirror in reflecting telescopes. The 200-inch mirror in the world's largest telescope at Mount Palomar, California, is coated with aluminum.

About half of the aluminum produced in this country is converted to alloys for special uses. **Duralumin** in an alloy containing aluminium, copper, manganese, and magnesium. It is light but nearly as strong as steel, so it is useful in the construction of aircraft. **Aluminum bronzes** contain copper, and occasionally some silicon, manganese, iron, nickel, and zinc. These light-weight alloys have high tensile strength and great resistance to corrosion, so they are used extensively in the manufacture of crankcases and connecting rods for gasoline engines. Alnico is a magnetic alloy containing 50 per cent iron, 20 per cent aluminum, 20 per cent nickel, and 10 per cent cobalt, which will lift more than 4,000 times its own weight of iron. It is made by pressing together the constituent metals in powder form and heating the mixture just below the melting point. This is an example of **powder metallurgy**—a branch of science which is growing rapidly in importance.

When powdered aluminum and iron(III) oxide are mixed and ignited by means of a magnesium fuse a vigorous and highly exothermic reaction occurs.

$$2Al(s) + Fe_2O_3(s) \rightarrow 2Fe(s) + Al_2O_3(s) + 203{,}500 \text{ cal}$$
$$(\Delta H^\circ_{298} = -\,203{,}500 \text{ cal})$$

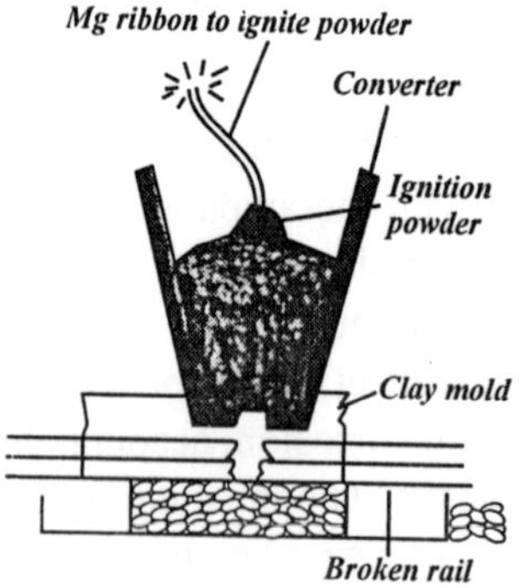

Fig. 13.8 : Thermite welding of a rail. Molten metal, mostly iron, flows out of the converter at the bottom and welds the broken rail without having to remove it from the roadbed.

This is known as the **thermite**, or **Goldschmidt**, reaction. The temperature of the reaction mixture rises to about 3,000°C, so the iron and aluminium oxide become liquid. The process is frequently used in welding large pieces of iron or steel. Thermite bombs were used for incendiary purposes during World War II, because of the high temperature generated by the thermite reaction and because the reaction is not readily quenched by water. The thermite reaction is used in the reduction of metallic oxide (such as MoO_3 and WO_3) which are not readily reduced by carbon, or others (such as MnO_2 and Cr_2O_3) which do not give pure metal when reduced by carbon. The reaction is also of particular value when carbon-free alloys, such as ferrotitanium, are desired.

The halides of aluminium

Aluminium fluoride, $Al^{3+}(F^-)_3$

This compound can be made by the direct combination of aluminium and fluorine. It is the only ionic halide of aluminium and is sparingly soluble in water.

Aluminium chloride, Al_2Cl_6

This is made by passing hydrogen chloride or chlorine over heated aluminium; since it is readily attacked by the moisture in the air it must be prepared under anhydrous conditions, the tube of soda lime being used to keep out moisture as well as to absorb excess chlorine (Fig. 13.9).

$$2Al + 3Cl_2 \rightarrow Al_2Cl_6$$
$$2Al + 6HCl \rightarrow Al_2Cl_6 + 3H_2$$

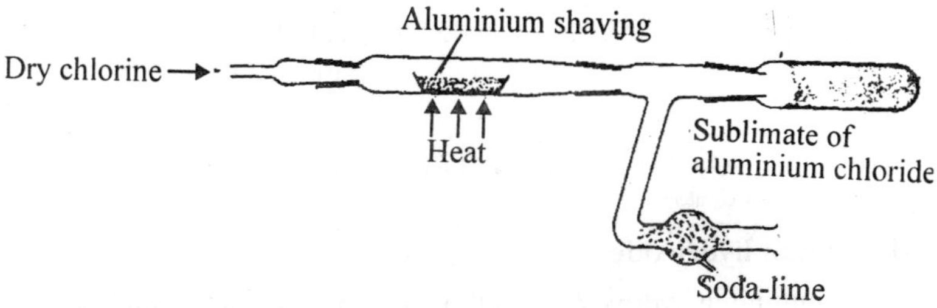

Fig. 13.9 : Preparation of anhydrates aluminium chloride

It can also be obtained by passing a stream of chlorine over a mixture of aluminium oxide and charcoal heated to about 1000°C:

$$(Al^{3+})_2(O^{2-})_3 + 3C + 3Cl_2 \rightarrow Al_2Cl_6 + 3CO$$

When pure, aluminium chloride is a white solid which sublimes at a temperature of about 180°C. Molecular weight determinations in solution, e.g. in benzene, and in the vapour state indicate that the molecule has the formula Al_2Cl_6, unlike the halides of boron which are monomeric. The aluminium atoms complete their octets by dative bonding from two chlorine atoms, the arrangement of the chlorine atoms about each aluminium atom being roughly tetrahedral:

Equation

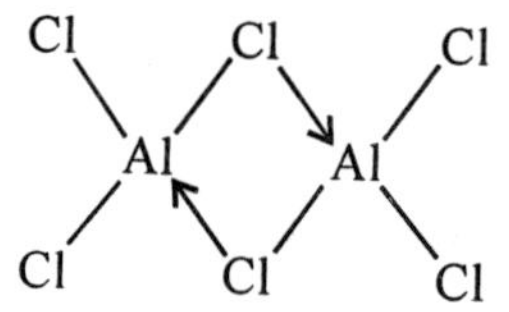

At high temperatures aluminium chloride exists as the monomer $AlCl_3$ in which the aluminium atom has a sextet of electrons; as predicted this molecule is planar and symmetrical:

Equation

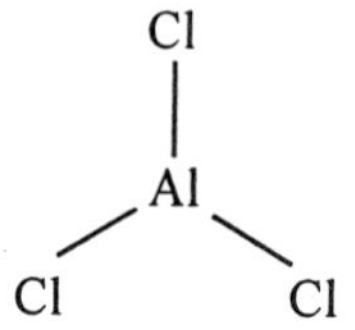

When aluminium chloride is added to water it reacts exothermically to give hydrated aluminium ions, $[Al(H_2O)_6]^{3+}$ and chloride ions, Cl^-. The energy needed to break the Al—Cl covalent bonds is derived from the high heat of hydration of the small highly charged Al^{3+} ion.

$$\underset{\text{covalent}}{Al_2Cl_6} + 12H_2O \rightarrow \underset{\text{ionic}}{2[Al(H_2O)_6] + (Cl^-)_3}$$

Aluminium hydroxide

Aluminium hydroxide, $Al(OH)_3$, is formed as a white flocculent precipitate when ammonium hydroxide is added to the solution of an aluminium salt. The hydroxide is amphoteric, since it dissolves in both strong bases and acids:

$$Al(OH)_3 + 3H^+ \rightarrow Al^{+++} + 3H_2O$$

$$Al(OH)_3 + 3OH^- \rightarrow Al(OH)_6^{---} \text{ or } AlO_3^{---} \cdot 3H_2O$$

In the first reaction, aluminium act as a metal and the hydroxide is neutralized by the acid to produce Al^{+++}. In the second reaction, the aluminium acts as a nonmetal and dissolves in the base to form aluminate ions, AlO_3^{---}. Meta-aluminate ions, AlO_2^-, may be formed by using a lesser amount of base, according to the reaction.

$$Al(OH)_3 + OH^- \rightarrow Al(OH)_4^- \text{ or } AlO_2^- \cdot 2H_2O$$

If aluminium hydroxide is precipitated in a solution containing an organic dye, the hydroxide adsorbs the colour and removes it from solution. This adsorptive property of aluminium hydroxide is used advantageously in the dye industry. Certain fabrics and materials will not adsorb a dye directly and the addition of a *mordant* is necessary—something which will fix the dye to the cloth. Aluminium hydroxide acts in this capacity if precipitated with the fibers of the cloth.

Aluminium hydroxide also finds use, because of its adsorptive capacity for suspended matter, in the purification of water.

Lithium aluminium hydride, $LiAlH_4$

When finely divided lithium hydride reacts with an ethereal solution of aluminium chloride, lithium aluminium hydride is formed in solution. This compound has outstanding reducing powers in organic chemistry. Excess aluminium chloride yields aluminium hydride, $(AlH_3)_n$, of uncertain constitution:

$$4LiH + AlCl_3 \rightarrow LiAlH_4 + 3LiCl$$

$$3LiAlH_4 + AlCl_3 \rightarrow 4AlH_3 + 3LiCl$$

Carboxylic acids may be reduced to be corresponding alcohols with high yields by reduction in ethereal solution and subsequent hydrolysis by 10% sulphuric acid or caustic soda solution. Certain linkages are not affected and this simplifies syntheses. Simple examples illustrate these reactions:

$$\underset{\text{phenylacetic acid}}{C_6H_5.CH_2.COOH} \rightarrow \underset{\text{2-phenyl ethanol}}{C_6H_5.CH_2CH_2OH}$$

$$\underset{\text{cinnamaldehyde}}{C_6H_5.CH{=}CH.CHO} \rightarrow \underset{\text{cinnamyl alcohol}}{C_6H_5.CH{=}CH.CH_2OH}$$

Lithium aluminium hydride may be used to prepare hydrides: silicon tetrachloride is reduced to silane,

$$LiAlH_4 + SiCl_4 \rightarrow LiCl + AlCl_3 + SiH_4 \uparrow$$

Finally, carbon dioxide is reduced to methanol in 81% yield by lithium aluminium hydride.

The alums

The alums are double salts of general formula $M^+M^{3+}(SO_4^{2-})_2.12H_2O$, which form isomorphous crystals and in which M^+ and M^{3+} are respectively univalent and tervalent. In structure, the alums consist of simple ions, being not complexes, but double salts. Potash alum or potassium alum is the common alum, with the formula $KAl(SO_4)_2.12H_2O$ which, for convenience, may be written $K_2SO_4.Al_2(SO_4)_3.24H_2O$. The univalent cation may belong to the alkali metals or be silver, thallium or ammonium. Aluminium may be replaced as the tervalent metal by the ions: Cr^{3+}, Mn^{3+}, Fe^{3+} or Co^{3+}. In solution the simple the simple ions disperse.

The alums may be prepared by mixing the appropriate sulphates in proportions of the formula weight of the anhydrous salt and a evaporating to crystallization. But certain alums may be prepared by specially adapted methods. The alums are isomorphous, appearing as octahedral crystals which form overgrowths and solid solutions.

Aluminium oxide (alumina), Al_2O_3

Aluminium oxide occurs naturally as bauxite which is contaminated with silica and iron(III) oxide impurities. When freshly prepared it reacts readily with dilute acids and strong alkalis to give salts. It is thus amphoteric, unlike the oxide of boron which is exclusively acidic.

$$(Al^{3+})_2(O^{2-})_3 + 6H^+ \rightarrow \underset{\text{hydrated}}{2Al^{3+}} + 3H_2O$$

$$(Al^{3+})_2(O^{2-})_3 + 2OH^- + 3H_2O \rightarrow \underset{\text{aluminate ion}}{2Al(OH)_4^-}$$

It is used for dehydrated alcohols to the corresponding alkenes, e.g.

$$\underset{\text{ethanol}}{CH_3-CH_2-OH} \xrightarrow[300°C]{\text{Alumina}} \underset{\text{ethylene}}{H_2C=CH_2} + H_2O$$

As a finely divided powder it is frequently used as the absorbing stationary phase in column chromatography, e.g. for separating chlorophylls.

A very hard crystalline form of aluminium oxide called corundum occurs naturally; it is used as an abrasive. It can be made heating amorphous aluminium oxide to about 2000°C.

Aluminium forms a series of mixed oxides with other metals, some of these occurring naturally as semi-precious stones. These include ruby (Cr^{3+}), and blue sapphire (Co^{2+}, Fe^{2+}, Ti^{4+}); these are now produced synthetically and find an application as bearings in watches.

Aluminium carbide, Al_4C_3 (aluminium methide)

This covalent solid can be made by direct combination of aluminium and carbon at high temperature. It is hydrolyzed by water, with the evolution of methane:

$$Al_4C_3 + 12H_2O \rightarrow 4Al(OH)_3 + 3CH_4$$

Aluminium sulphide, Al_2S_3

This is made by heating together aluminium powder and finely divided sulphur; the reaction can be exceedingly violent. It is covalent and is readily hydrolyzed by water with the evolution of hydrogen sulphide:

$$Al_2S_3 + 6H_2O \rightarrow 2Al(OH)_3 + 3H_2S$$

Aluminium triethyl, $(C_2H_5)_3Al$

This compound is made by heating aluminium powder, ethene and hydrogen under high pressure:

$$2Al + 6C_2H_4 + 3H_2 \rightarrow 2(C_2H_5)_3Al$$

It is spontaneously inflammable in air and explosively reactive towards water. The liquid is dimeric and two ethyl groups act as bridging units:

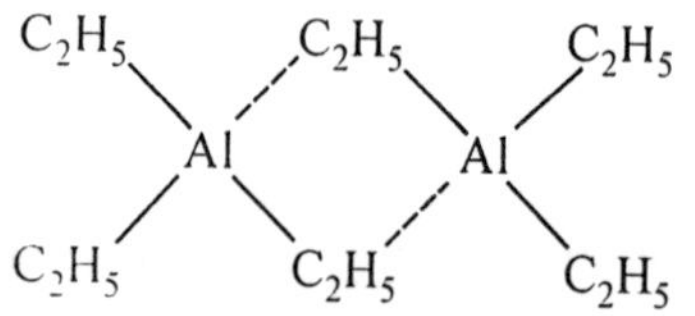

Aluminium triethyl in conjunction with titanium(IV) chloride is the catalyst in one industrial process for converting ethylene into polythene. A similar type of catalyst system is used in the polymeriztaion of propene into polypropene (polypropylene).

Ceramics. Clay is a product of the weathering of complex silicates, particularly feldspar, $KAlSi_3O_8$ or $NaAlSi_3O_8$, which is the most abundant class of all minerals of the earth's crust. Although rock silicates are in general quite insoluble, ground waters, being dilute solutions of carbonic acids, have in geological time chemically altered them. Feldspar is progressively altered to the mineral *mica* and thence to *kaolin* according to such chemical changes as

$$3KAlSi_3O_8 + H_2CO_3 \rightarrow \underset{\text{mica}}{KH_2Al_3(SiO_4)_3} + 6SiO_2 + K_2CO_3$$

$$2KH_2Al_3(SiO_4)_3 + H_2CO_3 + 3H_2O \rightarrow \underset{\text{kaolin}}{3H_4Al_2Si_2O_9} + K_2CO_3$$

The potassium carbonate is dissolved by water and the clay may be carried in colloidal suspension in streams and finally deposited in beds. Ordinary clays are quite impure—usually containing considerable quantities of iron. Kaolin, also called white clay or china clay, is a clay of high purity. When clay is mixed with water, a very plastic mass is formed with may be molded into any desired shape. This property of clay is the basis of the *ceramic* industries which produce ricks, tile, pottery, stoneware, and porcelain. After the molding and drying of the clay, the product is baked. Most of the residual molecular water is lost, leaving a very hard, resistant, highly porous product. If the porosity is undesirable, the product may be glazed by producing a very thin film of fused matter on the surface. A cheap glaze can be produced by throwing salt on the object while it is still hot; a thin liquid coating of sodium aluminium silicate forms, which becomes compact and impervious when cooled. Bricks and tiles are made from the cheaper and impure forms of clay, while china and porcelain are produced from kaolin.

Gallium, indium and thallium

Gallium, indium and thallium are obtained by electrolysis of aqueous solutions of their salts. They are soft, white and fairly reactive metals, combining on heating with many non-metals such as the halogens and sulphur. Gallium resembles aluminium in reacting with sodium hydroxide solution: it is a liquid over a remarkable range of temperature (30°-2240°C) for which no adequate explanation is yet possible.

The halides

The trifluorides are high melting-point ionic solids which resemble aluminium fluoride. Gallium(III) chloride is a dimeric covalent solid which is rapidly hydrolyzed by water and which dissociates into the monomer on heating; it is thus very similar in behaviour to aluminium chloride. Indium(III) chloride, however, is monomeric and, since it conducts electricity in the fused state, is appreciably ionic. Thallium(III) chloride decomposes on heating to thallium(I) chloride, TlCl, in which the metal exhibits a valency of one.

The oxides

The 3-valent oxides are similar to aluminium oxide but become progressively less difficult to reduce with increasing atomic number. The oxides are more basic in the same order; thus gallium(III) oxides is less amphoteric than aluminium oxide and thallium(III) oxide is exclusively basic.

1-valent compounds

Gallium and indium both form chlorides with the empirical formula XCl_2 in which they appear to be 2-valent. Physical evidence, however, shows that the correct formulation of these chlorides is $X^{I}(X^{III}Cl_4)$, in which both gallium and indium exhibit a valency of one. The tendency towards univalency is even more pronounced in the case of thallium and is due to the reductance of the two *s*-electrons of the heavier elements to be transferred or to participate in covalent bond formation. It is called the *inert pair effect.*

Some 1-valent compounds of thallium resemble corresponding compounds of the alkali metals; thus the oxide. Tl_2O, and the hydroxide. TiOH, are strong bases. Other compounds are similar to those of silver, thus thallium(I) chloride is only sparingly soluble in water and is sensitive to light.

14

p-Block Elements : (carbon, silicon, germanium, tin and lead)—Group IVA (14)

Some physical data of Group IVA elements

	Atomic Number	Electronic Configuration	Atomic Radius (Å)	Ionic Radius (Å) M^{2+}	Ionic Radius (Å) M^{4+}	M.p. °C	B.p. °C
C	6	2.4 $1s^2 2s^2 2p^2$	0.77			38.50 (subl.)	
Si	14	2.8.4 ....$2s^2 2p^6 3s^2 3p^2$	1.17		0.41	1410	2680
Ge	32	2.8.18.4 ....$3s^2 3p^6 3d^{10} 4s^2 4p^2$	1.22	0.93	0.53	937	2830
Sn	50	2.18.18.4 ....$4s^2 4p^6 4d^{10} 5s^2 5p^2$	1.41	1.12	0.71	232	2687
Pb	82	2.8.18.32.18.4 ....$5s^2 5p^6 5d^{10} 6s^2 6p^2$	1.54	1.20	0.84	327	1751

All these elements exhibit a group valency of four, but because an enormous amount of energy is needed to remove four electrons from their atoms, they form compounds which are predominantly covalent. Similarly the gain of four electrons to give 4-valent anion is energetically impossible.

Germanium, tin and lead form 2-valent compounds in which the two *s*-electrons are inert (inert pair effect). The stability of this state relative to the 4-valent state increase steadily from germanium to lead, i.e. 2-valent germanium compounds tend to be strongly reducing and revert to the 4-valent state, while for lead this is the predominant valency state. 2-valent compounds of tin and lead are often predominantly ionic.

Carbon is non-metallic and so too is silicon; germanium has properties of both metals and non-metals (it is a metalloid), while the elements tin and lead are definitely metallic. There is a smooth transition from non-metallic to metallic properties on passing down the series silicon, germanium, tin and lead, but the first member carbon differs considerably from silicon. The chemistry of silicon is very similar to that of boron, its diagonal neighbour in the Periodic Table.

The chemistry of carbon is dominated by its tendency to form chains and rings of carbon atoms in which other atoms, particularly hydrogen, play an important part. The chemistry of silicon is completely different and here the important feature is the formation of silicon-oxygen bonds which are present in the giant molecule silica, and in the polymeric silicate anions.

Carbon

Carbon in Nature. Carbon is found in nature both in the elemental state and in numerous compounds. Native carbon occurs in the form of two simple substances, diamond and graphite. A third simple substance consisting also of carbon, namely, charcoal, can be obtained only artificially. However, there are substances in nature very close in composition to charcoal. Such, for instance, are various kinds of mineral coal, forming huge deposits at many points of the globe. Some mineral coals contain as much as 99 per cent carbon.

Carbon compounds are very widespread. Besides mineral coal, the depths of the earth contain large accumulations of mineral oil, which is a complex mixture of various carboniferous compounds, mainly hydrocarbons. Salts of carbonic acid occurs in immense quantities in the earth's crust, particularly the calcium salt, which frequently forms whole mountains of limestone and chalk. The air always contains carbon dioxide. Finally, plants and animal organisms consists of substances, the main constituent of which is carbon. Thus, carbon is one of the most abundant elements on the earth, although its total content in the earth's crust has been estimated at not over 0.35 per cent by weight.

Carbon occupies quite a specific position among other elements with regard to the abundance and diversity of its compounds. The number of carbon compounds already known is well over a million, while the compounds of all the rest of the elements taken together do not total more than 25 or 30 thousand.

The diversity of carbon compounds is due to the ability of its atoms to unite with one another into long chains or rings consisting sometimes of scores or even hundreds of atoms.

Allotropy of carbon

Elemental carbon is known in three allotropic modification: *diamond*, crystallizing in the isometric system; *graphite*, belonging to the hexagonal system, and *amorphous carbon*, usually called coal or charcoal. The latter modification is apparently but a peculiar cryptocrystalline form of graphite. It had not been established for certain to the present day whether an amorphous modification of carbon really exists or not.

The chemistry of carbon and silicon are best considered separately. Germanium, tin and lead can profitably be studied as a group.

Carbon Cycle

Some of the planets are thought to be surrounded by a layer of carbondioxide, and it is probable that all deposits of carbon compounds in the earth originally came from carbon dioxide in the atmosphere. Plants use atmospheric carbon dioxide for growth and animals feed on plants. The decomposition of death animals and vegetation, followed by geological upheavals which subjected the decaying matter to great temperatures and pressures for long periods of time in the absence of oxygen, almost certainly accounts for coal, oil and natural gas. The carbonate minerals were formed by the sedimentation of the shells of microscopic sea animals, followed by compression of the layers and the redistribution of the oceans. Pure carbon has two allotropes and both of these—diamond and graphite—are found naturally.

The percentage of carbon dioxide in the atmosphere remains constant at a figure of about 0.03 per cent by volume by the operation of a carbon cycle in nature.

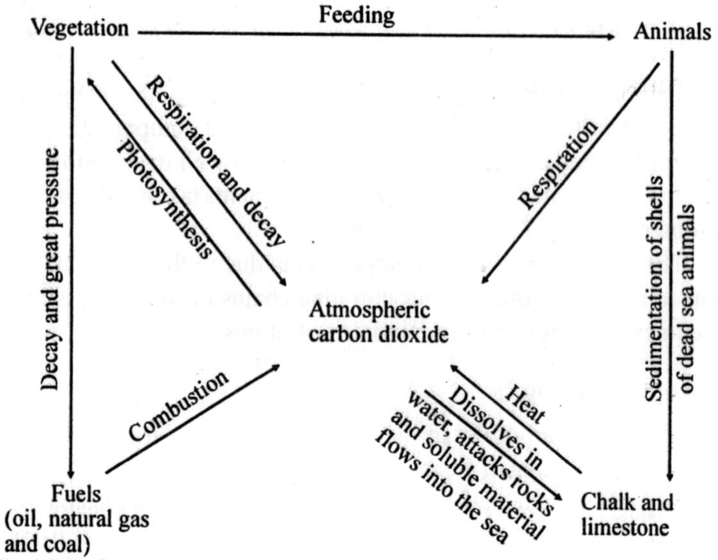

Fig. 14.1 : Carbon cycle.

Diamond shares with a synthetic compound, boron nitride, the distinction of being the hardest substance known.* When pure, it is transparent and colourless. It is much less reactive chemically than other forms of carbon, although when heated strongly in air, diamond will burn. In sharp contrast to graphite, diamond does not conduct the electric current, but because it is an excellent conductor of heat, diamond feels cold to the touch. X-rays readily penetrate diamond, but paste or glass imitations are opaque to x-rays. This and the hardness make good tests for real diamond. It is difficult to believe that a substance so different from graphite can, nevertheless, be the same chemical element. But the truth of this is easily proved. Diamond may be burned in oxygen to yield nothing but carbon dioxide and a trace of ash. Exactly the same result is obtained by burning graphite. *Diamond is an allotrope of graphite, an allotrope with striking differences in properties and a fantastic difference in price.* Diamonds cost, weight for weight, up to two hundred and fifty million times as much as graphite.

Examination of diamond crystals by x-ray diffraction methods shows that the structure of diamond is quite different from that of graphite. Instead of being arranged in parallel layers, each carbon atom is closely bound by electron sharing to four other carbon atoms. This tightly knit, interlocking arrangement is doubtless responsible for the extreme hardness and general lack of reactivity of diamond in comparison with other forms of carbon

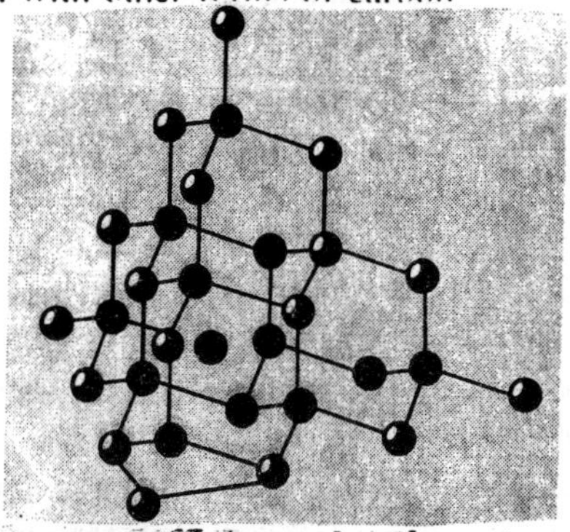

Fig. 14.2 : Arrangement of Carbon atom in diamond.

Graphite

Graphite is mined in Mexico, Siberia, Austria and Ceylon, but industrial applications have now outstripped this supply and it is manufactured synthetically (see below).

Its density, 2.3 g cm^{-3}, is less than that of diamond; it is also very soft and slippery and again these properties are readily understood in terms of its structure. Each carbon atom in graphite is surrounded by three more in the same plane, the bonding is covalent and the angle between each covalent bond is 120°. The structure extends in two dimensions and is thus a sheet of carbon atoms arranged in regular hexagons. Individual sheets are held together one above the other only by weak van der Waals' forces, the separation between the sheets being very much greater than the distance between carbon atoms within each sheet.

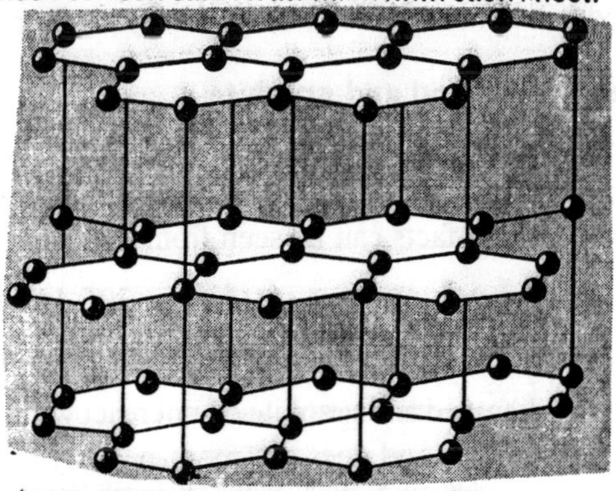

Fig. 14.3 : Arrangement of carbon atoms in graphite

Graphite is less dense than diamond because its structure is less compact. Since the weak forces holding the individual sheets together are easily broken, the sheets are easily moved relative to one another and this accounts for the slippery nature of graphite. Within each sheet every carbon atom is only bonded to three neighbours and thus alternate single and double bonds must be present to complete the octet of each carbon atom. (Since the C—C bond distances in each sheet are identical, resonance structures must be invoked.) The presence of double bonds mean that mobile electrons are present, and these can readily be transferred from one hexagon to another by the application of an electric field; graphite, unlike diamond, is therefore a conductor of electricity.

Graphite is used as a lubricant either as a power or an dispersion in oil. Mixed with clay it is used in 'lead' pencils. It is chemically very inert and since it conducts electricity it is used as the anode material in many electrolytic processes. Crucibles made of graphite are not attacked by dilute acids or by fused alkalis.

Massive graphite is used in foundry operations, especially for making crucibles which will withstand very high temperatures. It is also for electrodes for arc lights, electric furnaces, and is spectrographic analysis. One of the most spectacular is in the uranium reactor for the production of atomic power.

Colloidal graphite has many applications. Because of its greasy feel, the suitability of graphite as a lubricant has been recognized for a long time. Where high temperatures are involved, graphite may be used as a "dry" lubricant under conditions which would destroy an oil or grease.

The manufacture of diamond and graphite

Both diamond and graphite are thermodynamically unstable in the presence of oxygen and, of the two allotropes, graphite is more stable than diamond. These facts can be seen from the equations:

$$\text{C(Diamond)} + O_2(g) \rightarrow CO_2(g) \quad \Delta H^{\phi}_{298} = -395.4 \text{ kJ/mol}$$

$$\text{C(Graphite)} + O_2(g) \rightarrow CO_2(g) \quad \Delta H^{\phi}_{298} = -393.4 \text{ kJ/mol}$$

The reason why diamond and graphite do not react with oxygen at room temperature and diamond does not spontaneously transform itself into graphite is because high activation energies oppose these reactions.

The allotropy of carbon is referred one form (graphite) is always thermodynamically more stable than the other (diamond).

Graphite is manufactured by the Acheson process which requires cheap electrical power. A mixture of coke and sand is maintained at a very high temperature for about 24 hours in an electric furnace. Silicon carbide is the probable intermediate which decomposes into graphite and silicon, the latter substance volatilising away:

$$SiO_2 + \underset{\text{coke}}{3C} \rightarrow SiC + 2CO$$

$$SiC \rightarrow Si + \underset{\text{graphite}}{C}$$

Although graphite is more stable than diamond at ordinary temperatures and pressures, in the region of 3000°C and at extremes of pressure (60-120 thousand atmospheres) an equilibrium between graphite and diamond exists. However, even under such extremes the transformation of graphite into diamond is still extremely slow and a catalyst is required to speed up the rate of attainment of equilibrium. Since 1955 diamonds have been successfully made from graphite by utilising high temperatures and extremes of pressure in the presence of a rhodium catalyst. Only small stones have yet been produced but they are of vital importance in industry.

Impure form of carbon

Amorphous carbon (charcoal). If carboniferous compounds are heated without access of air, carbon is evolved from them in the form of a black mass called amorphous carbon or charcoal. It has been proved by X-raying that such carbon consists in many cases of tiny crystals, invisible even under the microscope, but having the same structure as graphite. It is possible, however, that along-side these crystals charcoal contains really amorphous modifications of carbon.

Charcoal is one of the most refractory substances. Charcoal can be observed to melt only if a very intense electric current is passed through a graphite rod. The drops of molten charcoal thus formed solidify into pure graphite. The melting point of charcoal is about 3,500°C.

Amorphous carbon is insoluble in ordinary solvents but dissolves in many molten metals such as iron, nickel and platinum, Upon cooling it precipitates out of these solutions as graphite. The specific gravity of charcoal varies between 1.8 and 2.1.

Charcoals differ considerably in properties on substance from which they are obtained and on the method of their preparation. Besides they always contain impurities which greatly influence their properties. The most important technical types of amorphous carbon are: *coke*, *wood charcoal*, *bone charcoal* and *carbon black*.

Coke is obtained by dry distillation of mineral coal. It is used chiefly in metallurgy for reducing metals from their ores.

Wood charcoal is obtained by heating wood out of contact with air. Formerly this used to be done in a very crude manner: the wood would be piled up in large heaps, "bonfires," covered with turf, burnt, but the greater part would turn into charcoal. Nowadays charcoal is usually prepared in large iron retorts, making it possible to collect the valuable products of dry distillation, namely, methyl alcohol, acetic acid, etc.

Charcoal is used in metallurgy, in forge-smithing, for the preparation of gunpowder, as well as for the absorption of vapours and gases.

Bone charcoal is prepared by charring degreased bones. It contains from 7 to 11 per cent carbon, about 80 per cent calcium phosphate and other mineral salts. A distinguishing feature of bone charcoal is its very high absorptive capacity, especially with respect to organic dyes, and it is used to remove various kinds of colouring matter from solutions.

Carbon black (lampblack) is the purest form of amorphous carbon. It is prepared in industry by burning pitch, turpentine or other substances rich in carbon in a limited supply of air. It is widely used as black paint (India ink, printer's ink, etc.) and as an important component of rubber in the rubber industry.

Chemical properties of carbon

Carbon in any form will react with oxygen at a sufficiently high temperature to give dioxide; in a deficiency of oxygen, carbon monoxide is formed as well. Charcoal will combine directly with sulphur, some metals and fluorine.

$$C + 2S \rightarrow CS_2$$

$$Ca + 2C \rightarrow Ca^{2+}(C \equiv C)^{2-}$$

$$C + 2F_2 \rightarrow CF_4$$

It will reduce steam, forming water gas and many oxides of metals; these reductions are of industrial importance.

It is not attacked by dilute acids, but concentrated nitric acid and sulphuric acid are reduced if warmed with carbon according to the equations:

$$C + 4HNO_3 \rightarrow 2H_2O + 4NO_2 + CO_2$$
$$C + 2H_2SO_4 \rightarrow 2H_2O + 2SO_2 + CO_2$$

Carbon monoxide, CO

Carbon monoxide, together with carbon dioxide, is formed when carbon or carbonaceous matter is burnt in a deficiency of air or oxygen. It is also produced when carbon dioxide is reduced by red hot carbon; the importance of this reversible reaction in connection with the extraction of metals has been discussed previously.

$$C + CO_2 \rightleftharpoons 2CO$$

A mixture of carbon monoxide and hydrogen (synthesis gas) is formed during the first stage in the industrial production of hydrogen.

In the laboratory it can be obtained by dehydrating formic acid with concentrated sulphuric acid:

$$HCOOH \xrightarrow{-H_2O} CO$$

If oxalic acid is dehydrated in the same way, carbon dioxide is formed as well:

$$(COOH)_2 \xrightarrow{-H_2O} CO + CO_2$$

Carbon monoxide is a colourless, odourless gas which burns in air with a blue flame, forming carbon dioxide. It is exceedingly poisonous, combining with the haemoglobin in the blood more readily than oxygen so that normal respiration is impeded very quickly. Ordinary gas masks are no protection against the gas, since it is not readily absorbed by active charcoal. In the presence of air, a mixture of manganese(IV) oxide and copper(II) oxide catalytically oxidises it to carbon dioxide, and this mixed catalyst is used in the breathing apparatus worn by rescue teams in mine disasters.

Carbon monoxide is a powerful reducing agent, being employed industrially in the extraction of iron and nickel:

$$Fe_2O_3 + 3CO \rightarrow 2Fe + 3CO_2$$
$$NiO + CO \rightarrow Ni + CO_2$$

It reacts with many transition metals, forming volatile carbonyls; the formation of nickel carbonyl followed by its decomposition is the basis of the mond process for obtaining very pure nickel:

$$Ni + 4CO \xrightarrow{90°C} Ni(CO)_4 \; Ni + 4CO$$

The transition metal carbonyls are covalent, the carbon atom of the carbon monoxide molecule being bonded to the metal atom. The structure of nickel carbonyl, which is a tetrahedral molecule (to minimise electron pair repulsion), is given on the opposite page; it is a resonance hybrid of the two forms:

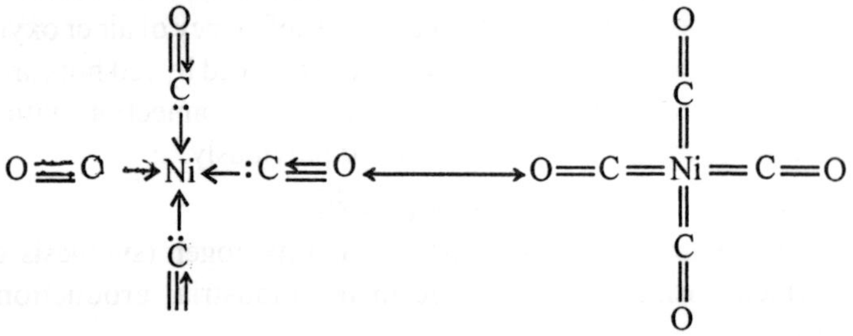

In addition to reacting with oxygen, carbon monoxide combines with sulphur to give carbonyl sulphide and with chlorine in the presence of light to give carbonyl chloride (phosgene). The latter is an exceedingly poisonous gas.

$$CO + S \rightarrow \underset{\text{carbonyl sulphide}}{COS}$$

$$CO + Cl_2 \rightarrow \underset{\text{carbonyl chloride}}{COCl_2}$$

Although carbon monoxide is not a true acid anhydride since it does not react with water to produce an acid, it reacts under pressure with fused sodium hydroxide to give sodium formate:

$$Na^+OH^- + CO \rightarrow HCOO^-Na^+$$

Formic acid is produced from the sodium salt by the addition of dilute hydrochloric acid.

With hydrogen under pressure and in the presence of a zinc oxide/chromium(III) oxide catalyst it reacts to give methanol; this reaction is of industrial importance.

$$CO + 2H_2 \xrightarrow{ZnO/Cr_2O_3} CH_3OH$$

Carbon monoxide is readily absorbed by an ammonical solution of copper(I) chloride to give $CuCl.CO.2H_2O$. It reduces an ammonical solution of silver nitrate to silver (black) and, in the absence of other gaseous reducing agents, this serves as a test for the gas. It can be estimated by reaction with iodine pentoxide, the iodine which is produced quantitatively being titrated with standard sodium thiosulphate solution.

$$5CO + I_2O_5 \rightarrow I_2 + 5CO_2$$

Carbon monoxide is considered to have a structure involving the three resonance forms:

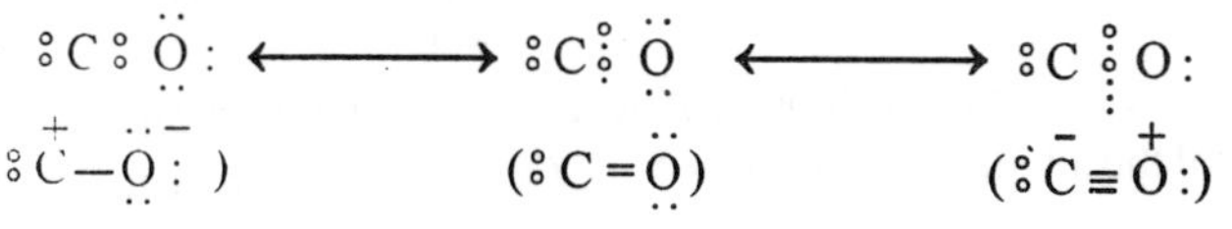

Carbon dioxide

Carbon dioxide is present in the air to the extent of 3.5 parts in 10,000. This amount is constantly being added to by burning carbon compounds, decay, fermentation, respiration, and volcanic gases. In turn, some of the carbon dioxide of the air is being absorbed by plants, dissolved by water, and fixed into such minerals as calcium or magnesium carbonate. There are some gas wells in the central United States and Mexico from which carbon dioxide issues at considerable pressure. Escaping carbon dioxide from ground waters in a cave near Naples, Italy, builds up such a concentration that a small dog walking into the quiet air of the cave will suffocate. A man walking upright, however, will have little discomfort, since carbon dioxide is about 1.4 times as heavy as air, and the concentration at the cave floor is much higher than at the five-foot level. Carbon dioxide is not a poison, but in a moderately high concentration a person suffers from lack of oxygen and inability to get rid of carbon dioxide from the body.

Preparation. (*a*) Carbon dioxide is the product of the complete burning of carbon or its compounds. Most vaporizable or gaseous carbon compounds, when mixed with air, form an explosive mixture. Occasionally, serious explosions occurs when a suspension of finely divided solid carbon compounds in air (coal dust, flour dust, and so on) is subjected to a spark or flame. Precautions must be taken in mills that are grinding organic material so that sparks or flame do not contact the organic dust-laden air. The slow oxidation processes of rotting and respiration form carbon dioxide as a product.

(*b*) Fermentation of carbohydrate compounds results in the formation of alcohol and carbon dioxide. Yeast cells secrete an organic compound called "zymase," which hastens the fermentation process. Such an organic substance, which is produced by living cells and which catalyzes specific chemical changes is called an *enzyme*:

$$\underset{\text{glucose}}{C_6H_{12}O_6} \xrightarrow{\text{zymase}} \underset{\text{alcohol}}{2C_2H_5OH} + 2CO_2 \uparrow$$

(*c*) Carbon dioxide is produced when metal carbonates (except those of the alkali metals such as Na_2CO_3 or K_2CO_3) are heated. Heating limestone to produce lime constitutes a large industry:

$$CaCO_3 \rightarrow CaO + CO_2$$

This chemical change requires a temperature of about 900°. The reaction is reversible, so a draft is used to keep removing the CO_2.

(*d*) The most convenient laboratory method of preparing carbon dioxide is by action of an acid on a carbonate or bicarbonate:

$$CaCO_3 + 2H^+ \rightarrow Ca^{++} + H_2O + CO_2$$

$$Na_2CO_3 + 2H^+ \rightarrow 2Na^+ + H_2O + CO_2$$

$$NaHCO_3 + H^+ \rightarrow Na^+ + H_2O + CO_2$$

Physical Properties

Carbon dioxide gas is 1½ times as heavy as air and, since it is noninflammable, may be used to blanket a flame and extinguish it. A 4 per cent carbon dioxide atmosphere over burning wood, even if 17 per cent oxygen is present. will smother the flame. It is moderately soluble in water under ordinary conditions and markedly more soluble under pressure. Its solubility in enhanced by the combination:

$$CO_2 + H_2O \rightleftharpoons H_2CO_3$$

Carbonic acid. which has a slightly sour or biting taste, is used in carbonated or soda water.

Carbon dioxide may be liquefied by pressure and is transported as a liquid in steel cylinders. If the nozzle of such a cylinder is opened, with the nozzle end lower than the rest of the cylinder, liquid carbon dioxide is ejected into the air. The liquid under the reduced pressure immediately boils. Since a boiling liquid absorbs its heat of vapourization from the surroundings, a marked cooling takes places. If the liquid is allowed to squirt into a cloth bag so that the cooling is confined, some of the carbon dioxide liquid will cool to the solid state. Solid carbon dioxide, which commercially is called "dry ice," is made in this fashion. The solid carbon dioxide gasifies in air at a temperature of about – 79°.

A convenient low-temperature mixture may be prepared by making a mush of solid carbon dioxide and ether. Its use may be emonstrated as follows:

Into a crystallizing dish containing a mush of ether and carbon dioxide, place a small matchbox or pillbox. Fill the box with mercury. As the mercury begins to solidify, insert a small wooden stick so that the mercury freezes around it. The box mold may then be torn from the solid mercury. One now has mercury hammer with which nails may be driven.

Chemical properties

Carbon dioxide is a very stable substance except at exceedingly high temperatures, where it decomposes:

$$2CO_2 \rightarrow 2CO + O_2$$

Carbon dioxide from the atmosphere is constantly dissolving in water over the earth's surface; thus, natural water is really a dilute solution of carbonic acid. This dilute acid has much to do with the disintegration of rocks. Minerals are acted on by the dilute acid, and today there are enormous deposits of limestone, $CaCO_3$, and dolomitic limestone, $CaCO_3 \cdot MgCO_3$.

Carbon dioxide is tested for on the basis of the chemical change:

$$Ca(OH)_2 + CO_2 \rightarrow CaCO_3 \downarrow + H_2O$$

A calcium hydroxide solution (lime water), when subjected to carbon dioxide, becomes milky as calcium carbonate precipitates. If carbon dioxide is added to a fine suspension of calcium carbonate, the milkyness will clear up, since calcium acid carbonate, which is water soluble is formed:

$$CaCO_3 + H_2CO_3 \rightarrow Ca(HCO_3)_2$$

This chemical change is a most important one in the problem of water softening. Naturally occurring limes one is constantly being slowly dissolved by carbonic acid in ground water to form soluble calcium acid carbonate (also called calcium bicarbonate).

Carbonyl chloride, phosgene, $COCl_2$

When chlorine and carbon monoxide are passed together through animal charcoal at 30-40°C, phosgene is formed:

Equation $$CO + Cl_2 \longrightarrow O{=}C\begin{matrix}\diagup Cl\\ \diagdown Cl\end{matrix}$$

It is a colourless gas (b.p. 8°C), very poisonous and with a smell of musty hay. Carbonyl chloride is the acid chloride of carbonic acid, which is formed on hydrolysis:

Equation $$O{=}C\begin{matrix}\diagup Cl\\ \diagdown Cl\end{matrix} + 2H_2O \longrightarrow O{=}C\begin{matrix}\diagup OH\\ \diagdown OH\end{matrix} + 2HCl$$
$$\downarrow$$
$$H_2O + CO_2$$

Ammonia gives carbamide, or urea:

Equation $$O{=}C\begin{matrix}\diagup Cl\\ \diagdown Cl\end{matrix} + 2NH_3 \longrightarrow O{=}C\begin{matrix}\diagup NH_2\\ \diagdown NH_2\end{matrix} + 2NH_4Cl$$

Carbonyl sulphide, carbon oxysulphide, SCO (COS)

Carbon monoxide and sulphur vapour react to form carbonyl sulphide:

$$CO + S \rightarrow SCO$$

It is usually prepared by the hydrolysis of thiocyanic acid, formed by the action of strong sulphuric acid on potassium thiocyanate:

$$KSCN + H_2SO_4 \rightarrow KHSO_4 + HSCN$$

$$HSCN + H_2O_4 \rightarrow S{:}C{:}O + NH_3 \xrightarrow[H_2SO_4]{} (NH_4)_2SO_4$$

Carbonyl sulphide is a colourless, very poisonous gas (b.p. – 50°C), which is hydrolyzed slowly to carbon dioxide and hydrogen sulphide, and burns in air to carbon dioxide and sulphur dioxide. Normally it has an objectionable odour, but is odourless when pure.

Carbonic acid

When carbon dioxide dissolves in water the greater part of it is only loosely hydrated. This hydrated species is in equilibrium with carbonic acid, hydrogen ions, and hydrogen carbonate and carbonate anions:

$$CO_2 + aq. \rightleftharpoons CO_2aq. \rightleftharpoons H_2CO_3 \rightleftharpoons H^+ + HCO_3^- \rightleftharpoons 2H^+ + CO_3^{2-}$$

Although pure carbonic acid cannot be isolated, solid carbonates are plentiful and the Group IA metals form solid hydrogen carbonates.

Carbonates contain discrete planar ions in which the three carbon-oxygen bond lengths are identical; the ion is thus a resonance hybrid containing equal contributions from the three forms:

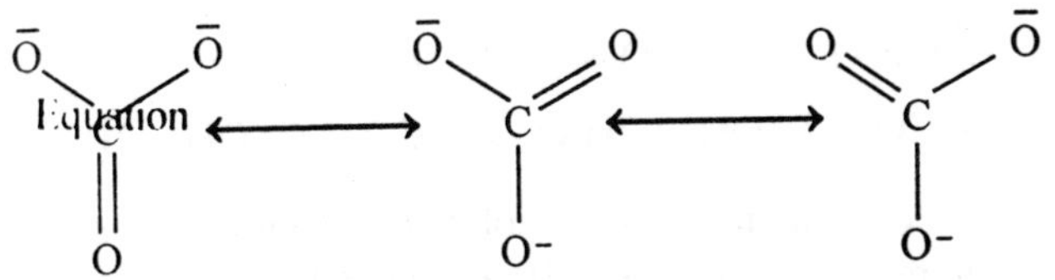

Silicon does not form an analogous discrete silicate anion, SiO_3^{2-}, because of the reluctance of the larger silicon atom to form a multiple bond with oxygen. Silicates contain polymeric silicate anions, some of which contain the anion which approximates to the empirical formula SiO_3^{2-}.

Carbonates

The carbonates of the Group I metals, except lithium, and ammonium carbonate are the only soluble carbonates met here. The only solid hydrogen carbonates encountered are those of sodium and potassium. Soluble carbonates may be prepared by the action of carbon dioxide on the alkali:

$$2OH^- + CO_2 \rightarrow H_2O + CO_3^{2-}$$

Excess of carbon dioxide yields the hydrogen-carbonate:

$$CO_3^{2-} + H_2O + CO_2 \rightarrow 2HCO_3^-$$

For the removal of carbon dioxide from gases, caustic potash is used because the products are freely soluble in water. One very important use of carbon dioxide is in the Ammonia-Soda process for the manufacture of sodium carbonate. Hydrogencarbonates of calcium and magnesium cause temporary hardness of water. Insoluble carbonates may be precipitated by mixing solutions containing the required ions:

$$Ba^{2+} + CO_3^{2-} \rightarrow BaCO_3 \downarrow$$

Solutions of both carbonates and hydrogen-carbonates are hydrolyzed, the former being the more alkaline:

$$CO_3^{2-} + H_2O \rightleftharpoons HCO_3^- + OH^-$$
$$HCO_3^- + H_2O \rightleftharpoons H_2CO_3 + OH^-$$

Where both normal carbonates and mixtures of carbonate and hydroxide, called basic carbonates, exist, the former are generally precipitated by addition of sodium hydrogencarbonate solution and the latter by sodium carbonate solution:

$$Zn^{2+} + 2HCO_3^- \rightarrow ZnCO_3\downarrow + CO_2\uparrow + H_2O$$

$$3Zn^{2+} + CO_3^{2+} + 4OH^- + 2H_2O \rightarrow ZnCO_3.2Zn(OH)_2.2H_2O\downarrow$$

Addition of sodium carbonate solution to solutions containing the ions of aluminium, chromium(III) and iron(III) precipitates very insoluble hydroxides with evolution of carbon dioxide and not the carbonates, which have so far not been prepared:

$$3CO_3^{2-} + 3H_2O + 2Al^{3+} \rightarrow 2Al(OH)_3\downarrow + 3CO_2\uparrow$$

The thermal decomposition of carbonates and hydrogen-carbonates and the action of acids have been described. Weak metals form carbonates which are readily decomposed by heat while the strongest metals of Group I and II form carbonates which cannot be decomposed by heat. Only the metals of Group I, except lithium, form hydrogencarbonates which are readily isolated as solids. In the laboratory, carbonates and hydrogencarbonates are recognized by the action of heat and of dilute acids. The only soluble carbonates normally encountered are those of ammonium, sodium and potassium. The only hydrogencarbonates used in the laboratory are those of sodium and potassium. A mixture of a carbonate and hydrogencarbonate in solution may be identified by addition of magnesium sulphate solution. An immediate precipitation of white magnesium carbonate reveals the presence of carbonate ions while further precipitation on boiling the filtrate is due to the decomposition of magnesium hydrogencarbonate to magnesium carbonate, similar to the removal of temporary hardness and confirms the presence of hydrogencarbonate. Alternatively, calcium chloride solution is used, the initial precipitate of calcium carbonate removed and hydrogencarbonate ions neutralized by addition of ammonium carbonate with a resultant precipitation of calcium carbonate.

Cyanates and cyanic acid, HOCN

Potassium cyanate is formed when red lead is fused with potassium cyanide,

$$4KCN + Pb_3O_4 \rightarrow 4KOCN + 3Pb$$

It is a colourless, crystalline solid. When mixed with ammonium sulphate and heated, urea and potassium sulphate are formed. The transformation of ammonium cyanate into urea was achieved by Wöhler in 1828 and was the first preparation from inorganic compounds of an animal product:

$$2KOCN + (NH_4)_2SO_4 \rightarrow K_2SO_4 + 2NH_4OCN \rightleftharpoons 2NH_2.CO.NH_2$$

This was a breakdown in the supposed barrier between the inorganic and the organic, the latter being associated with the concept of a vital force. Acidification of a cyanate yields cyanic acid but this is hydrolyzed immediately:

$$HOCN + H_2O \rightarrow NH_3 + CO_2$$

Pure cyanic acid is a colourless liquid obtained by heating cyanuric acid, $(HOCN)_3$, which may in turn be prepared from urea.

Thiocyanates and thiocyanic acid, HSCN

Potassium thiocyanate is obtained when potassium cyanide is fused with sulphur,

$$KCN + S \rightarrow KSCN$$

It is a colourless, deliquescent, crystalline solid. Thiocyanic acid is obtained by distillation of potassium thiocyanate with dilute sulphuric acid at low temperatures, strong acid yielding carbonyl sulphide, etc.

$$2KSCN + H_2SO_4 \rightarrow K_2SO_4 + 2HSCN$$

Thiocyanic acid behaves as a tautomeric mixture of two forms, H—S—C≡N, and S=C=N--H, both giving a resonance hybrid thiocyanate ion,

$$\overline{S}-C\equiv N \leftrightarrow S=C=\overline{N}.$$

Suboxides of carbon

Malonic anhydride, C_3O_2, is formed by the dehydration of malonic acid with phosphorus pentoxide at 140°C in *vacuo*:

Equation $$CH_2\begin{cases}COOH\\COOH\end{cases} - 2H_2O \rightarrow C_2O_2$$

It is a colourless gas (b.p. 6.8°C), with an objectionable odour. At 200°C, it decomposes:

$$C_3O_2 \rightarrow CO_2 + 2C$$

The structure is O=C=C=C=O. C_5O_2 has been reported. Mellitic anhydride is $C_{12}O_9$. A sulphur analogue, C_3S_2, of malonic anhydride has been made in small quantities.

Cyanogen, (CN)$_2$

Cyanogen is evolved on mixing strong solutions of potassium cyanide and copper(II) sulphate when copper(II) cyanide, precipitated initially, breaks down to cyanogen and copper(I) cyanide:

$$Cu^{2+} + 2CN^- \rightarrow Cu(CN)_2\downarrow$$

$$2Cu(CN)_2\downarrow \rightarrow 2CuCN\downarrow + \begin{matrix}C\equiv N\uparrow\\|\\C\equiv N\end{matrix}$$

It may also be formed by thermal decomposition of mercury(II) cyanide:

$$Hg(CN)_2 \rightarrow Hg + (CN)_2$$

Cyanogen is a colourless, extremely poisonous gas (b.p. – 21°C), which burns with a violet flame to form carbon dioxide and nitrogen:

$$(CN)_2 + 2O_2 \rightarrow 2CO_2 + N_2$$

It is soluble in water and slowly hydrolyzed, to give a range of products:

$$\begin{matrix} C\equiv N \\ | \\ C\equiv N \end{matrix} + 4H_2O \longrightarrow \begin{matrix} COO^-NH_4^+ \\ | \\ COO^-NH_4^+ \end{matrix}$$

ammonium oxalate

$$(CN)_2 + H_2O \rightarrow \underset{\text{hydro-cyanic acid}}{HCN} + \underset{\text{cyanic acid}}{HCNO}$$

With caustic alkali, a mixture of the cyanide and cyanate results:

$$2OH^- + (CN)_2 \rightarrow CN^- + CNO^- + H_2O$$

Cyanogen will react with certain heated metals to form cyanides:

$$2K + (CN)_2 \rightarrow 2KCN$$

Because of the superficial resemblance of these reactions to those of the halogens, cyanogen has been called a pseudo-halogen.

Carbides. Several classes of binary compounds containing carbon are known. Generally speaking the carbides are compounds formed by the more electropositive elements, compounds of carbon with nonmetals such as the halogens and the first two members of the Group V and VI elements and of hydrogen are generally not considered in discussions of the carbides.

Preparation. Binary compounds containing carbon can be prepared by three general methods. Direct reaction often occurs when a mixture of carbon and an element is heated to temperatures in excess of about 2200°C. The reduction of metal oxides, or other compounds, with carbon at high temperatures also yields carbides. The third general method involves heating a sample of the metal (often electrically, as in the form of a wire) with a hydrocarbon such as CH_4 at low pressures.

In addition to these reactions, binary carbon compounds of copper, silver, gold, magnesium, zinc, and cadmium can be prepared by passing acetylene into solutions containing compounds of these metals. The alkyl derivatives of the last two named elements react with acetylene in organic solvents

$$R_2M + HC\equiv CH \rightarrow 2RH + MC_2$$

M = Cd, Zn, Mg.

Ammonical solutions of the common metal salts of the remaining elements also react with acetylene

$$2M^+ + HC \equiv CH \rightarrow M_2C_2 + 2H^+$$

In both cases the reactions reflect the acid character of acetylene. Since undoubtedly the metal is present as a complexed cation and the concentration of protons (and perhaps free acetylene) is very small in the basic media in which such reactions occur. The binary comounds formed in the reactions sometimes are called acetylides; they are extremely sensitive to heat and mechanical shock.

Structure and Reactions. It has been difficult to establish a consistent basis for the classification of carbides. As we shall see, a structural basis for classification does not always reflect consistent chemical properties. Generally, the carbides as a group are divided into salt-like (or ionic) carbides. Interstitial carbides, and covalent carbides.

The Interstitial Carbides, formed by the transition elements, are highmelting and very hard. Basically, there are two groups of compounds which fall into this classification. The truly interstitial compounds are formed by metals which have atomic radii greater than ~1.3 Å. Atoms of this size can form close-packed structures with interstices sufficiently large to accept carbon atoms. The elements titanium (1.47 Å), zirconium (1.60 Å), hafnium (1.59 Å), vanadium (1.35 Å), niobium (1.47 Å), tantalum (1.47 Å), molybdenum (1.40 Å), and tungsten (1.41 Å) form carbides with the general formulas MC and M_2C possessing interstitial structures. On the other hand, there is a group of carbides formed by transition elements with radii smaller than ~ 1.3 Å which have been classified historically as interstitial, but which do not exhibit the basic close-packed structure of metal atoms. The carbides of chromium (1.29 Å), manganese (1.37 Å), iron (1.26 Å), cobalt (1.25 Å), and nickel (1.25 Å) fall into this category. The formulas for these compounds (e.g. Cr_7C_3, Cr_3C_2, Mn_4C, Mn_3C, Mn_5C_2, Mn_7C_3, Fe_3C, Co_2C) cannot be explained in terms of the atoms existing in their normal valence states. Indeed, the structural data available suggest that carbon-carbon interactions occur in these systems. For example, the structure Cr_3C_2 shows a zig-zag carbon chain running through the crystal with a C—C distance of 1.66 Å. Only slightly greater than that observed in aliphatic carbon compounds (1.54 Å) .

The truly interstitial carbides are chemically inert. For example, TiC is impervious to attack by water or aqueous HCl even at 600°C. The borderline interstitial carbides are markedly more reactive. Thus, Mn_3C yields mixtures of hydrocarbons (CH_4 and C_2H_6) as well as hydrogen when it reacts with water.

The *Covalent Carbides* are formed when carbon is bonded to elements with nearly the same electronegativity. The most common examples of this type of compound are SiC and B_4C. Silicon carbide—commonly called carborundum—crystallizes in a large number of polymorphos, all of which are modifications of the zinc-blende or wurtzite structure.

Carbides are conveniently classified in terms of the bonding present into salt-like (ionic), interstitial and covalent carbides.

Salt-like (ionic) carbides

The reactive metals, i.e. those of Groups IA and IIA, form carbides which are colourless solids and non-conductors of electricity. With the exception of beryllium carbide, they are decomposed by water, with the liberation of ethyne:

$$Ca^{2+}(C \equiv C)^{2-} + 2H_2O \rightarrow Ca^{2+}(OH^-)_2 + H—C \equiv C—H$$

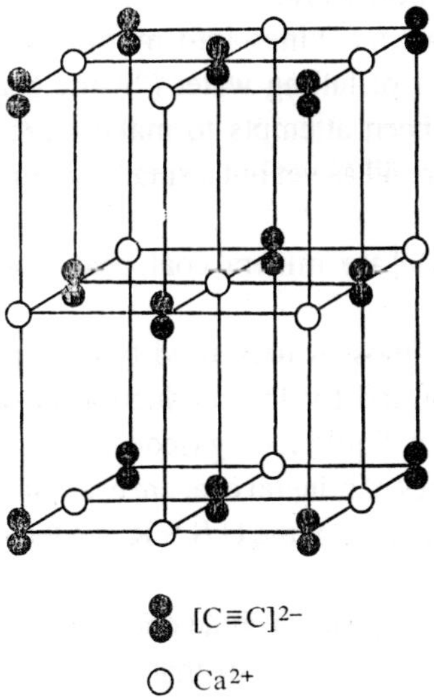

Fig. 14.4 : The structure of calcium carbide

Many of these carbides have a structure in some ways resembling that of sodium chloride, one cell axis being elongated to accommodate the $(C \equiv C)^{2-}$ ions (tetragonal symmetry). Notice that the co-ordination number of each ion is six, the same as in the structure of sodium chloride, but that the axis along which the acetylide ions lie is elongated (Fig. 14.4)

Beryllium and aluminium carbides give methane on hydrolysis with water and, although normally included in this category, are best thought of as a transition from true ionic to essentially covalent carbides. The carbides of the lanthanides are considered to be ionic.

Fuel and Types of Fuel

Amorphous carbon, as charcoal and coke, mineral coal, and many carbon compounds, play a very important part of modern life as sources of various kinds of energy. When coal and carboniferous compounds are burnt, heat is evolved which can be employed for heating, preparing food and for many industrial processes. Most of the heat obtained, however, is converted into other kinds of energy and is used to do mechanical work.

True, to put various machines into motion we use, besides thermal power, the power of falling water ("white coal") and wind power; there have also been attempts to make direct use of solar energy; however, these are all as yet but a small fraction of the power obtained by burning fuel.

The chief types of fuel are: mineral coal, peat, wood, mineral oil and natural gas.

Mineral coal is the most widely used type of fuel. It is used both directly for burning and for the production of more valuable types of fuel namely, coke, liquid fuel, gaseous fuel.

The world's resources of mineral coal are estimated at 7,900,000 million tons. The coal resources of the U.S.S.R. are the second largest in the world.

The coal mined in tsarist Russia could not cover the requirements of even the underdeveloped industry of that time, and mineral coal was imported additionally from abroad. The only source of coal in the whole country at that time was the Donets Basin. Russia occupied the sixth place in world coal mining.

After the October Revolution great changes took place in the coal industry. During the pre-war five-year plan periods new coal basins were put into operation in Central Asia, in the Caucasus, in Eastern Siberia. Coal mining was started in a number of other districts as well. The output of coal in 1959 amounted to 506.5 million tons. By 1965 it will be 600 to 612 million tons per year. But the share of coal in this country's fuel production will gradually decrease, as the output of oil and natural gas makes headway.

Mineral coal is the remains of the vegetable kingdom which existed on our planet in the long past periods of its life. The older the coal, the more carbon it contains.

Three main types of mineral coal are distinguished:

1. *Anthracite*, the oldest of mineral coals. It has a high density and bright lustre, contains an average of 95 per cent carbon and gives a large amount of heat when burnt. The resources of anthracite in U.S.S.R. constitute about 5.5 per cent of its total coal resources.

2. *Bituminous* coal is a compact black substance containing between 75 and 90 per cent carbon. It is the most widely used to all mineral coals.

3. *Brown* coal or lignite contains 65 to 70 per cent carbon. Is brown in colour. The youngest of all mineral coals, is often bears traces of the structures of the wood from which it originated. Brown coal is very hygroscopic, has a high ash content (from 7 to 38 per cent), and for that reason is used only as local fuel.

Recently larger and larger quantities of brown coal have been employed as a raw material for the production of valuable types of liquid fuel, gasoline and kerosene. For this purpose the powdered coal is *hydrogenated*. i.e., heated under high pressure with hydrogen in the presence of catalysts. Under the action of hydrogen the very complex organic molecules split apart and form simpler molecules of saturated hydrocarbons (see § 169.) The products resulting from the hydrogenation of coal are again split and converted into gasoline and kerosene.

Of great importance in many districts of the U.S.S.R. is the use of peat.

Peat is a product of the first stage of formation of mineral coals. It forms in immense quantities at the bottoms of marshes from the dead parts of marsh lichens. In explored peat resources the U.S.S.R.

is the richest country in the world. The carbon content of peat does not exceed 65 per cent. Its chief disadvantage as a fuel is its high ash content. Peat is burnt as a local fuel at a number of large power stations.

Dry distillation of peat results in *peat coke*, which contains very little sulphur, making it suitable for smelting high quality iron. Besides, upon dry distillation peat yields a number of valuable chemical products.

Wood occupies a secondary place in the total fuel balance.

In recent years the use of wood as fuel in industry has been continuously decreasing, as timber can be utilized to greater advantage for other purposes.

Mineral oil has been used widely as fuel ever since the end of the XIX century when the internal-combustion engine was invented. This engine burns oil and makes it possible to utilize a much greater part of heat evolved during the combustion of the fuel that the steam engine. Oil products, gasoline and kerosene, are used as fuel for tractors, automobiles, aeroplanes, etc. The great importance of internal combustion engines in modern life makes oil, and especially its products, a fuel of prime importance. The explored oil resources of the Soviet Union are the largest in the world.

The Russian oil industry began to develop extensively under Soviet power. By 1940 the production of oil had increased 3.4 times compared to 1913. In 1959 the oil output of the Soviet Union was 129.5 million tons, while by 1965 it is to reach 230 to 240 million tons per year.

Oil is not only a convenient and highly calorific fuel, but also a most important raw material for the production of a wide range of chemical product. Of great importance as a raw material for the chemical industry are also casing head and refinery gases.

Natural gas, consisting of methane and other saturated hydrocarbons, is a very cheap and convenient fuel. The importance of natural gas is evident from the fact that in the course of 12 years (1940-52) its extraction in the capitalist countries increased threefold and reached 259,000 million cubic metres, whilst the coal output of these countries hardly increased at all during the same period.

Natural gases began to be utilized in our country practically only after the war. In 1947 the Saratov-Moscow gas line, 843 kilometres long, was completed, and this increased the supply of gaseous fuel to the capital by about five times. In 1948 the Dashava-Kiev gas line was put into operation, bringing gas to the Ukrainian capital from Dashava, a very rich source of natural gas in the Carpathian foothills.

During recent years a great number of new gas mains have been built in the Soviet Union. The largest is the Stavropol-Moscow gas main. Huge deposits of natural gas have been discovered in the Ukraine, the Krasnodar Territory, Middle Asia and a number of districts of Siberia.

The gas industry is one of the most rapidly developing branches of the country's heavy industry. In 1955, 10,400 million cubic metres of gas were produced by coal and slate processing, while in 1959 this figure had risen to 37,200 million cubic metres. In 1965 the total gas output is to reach 150,000 million cubic metres.

Natural gas is an exceedingly valuable and economical raw material for the organic synthesis industry and for the manufacture of many synthetic products. A number of new chemical plants now in construction are to be based on natural gas; besides, several operating chemical enterprizes are being switched to this raw material.

Equal quantities of fuel give different quantities of heat when burnt. Therefore, to evaluate the quality of a fuel, its heat value is determined, this being the maximum amount of heat evolved upon complete combustion of one kilogram of the fuel.

Table 14.1 : Elemental Composition and Heat Values of Various Kinds of Fuel

	Content, per cent			
Fuel	**Carbon**	**Hydrogen**	**Oxygen (nitrogen, sulphur)**	**Heat value Cal./kg.**
Wood, air-dry.....	50	6	44	4,500
Peat, air-dry.....	59	6	35	5,400
Brown coal.....	69	5.5	25.5	6,700
Bituminous coal.....	82	4.3	13.7	8,400
Anthracite.....	95	2.2	2.8	8,100
Charcoal.....	100	—	—	8,080
Mineral oil.....	85.5	14.2	0.3	10,500

The various types of fuel consist mainly of the three elements: carbon, hydrogen and oxygen. It may be considered that the higher the carbon and hydrogen content in the fuel, the greater its heat value.

The average composition of the organic mass of various kinds of fuel and their heat values are given in Table 14.1.

Gaseous Fuel

Of essential importance for many industries is not only the heat value of a fuel, but also the highest temperature that can be obtained with the fuel, it burnt rationally. This temperature is called the pyrometric burning effect and, besides the calorific power of the fuel, it depends to a considerable extent on the physical state of the latter.

Experience shows that for complete combustion of a solid fuel, approximately one and a half times the amount of air needed theoretically must be introduced into the furnace where it is being burnt. This excess air absorbs part of the heat evolved and greatly lowers the pyrometric burning effect.

A higher pyrometric effect is obtained when a liquid fuel, such as crude oil, is burnt. The oil is introduced into the stoker by means of special atomizers, called *burners*, which mix it thoroughly with air, so that it does not require such as an excess of the latter as solid fuel.

But only gaseous fuel can be mixed perfectly with air and therefore it burns up completely with almost the theoretical quantity of air and gives the highest pyrometric effect.

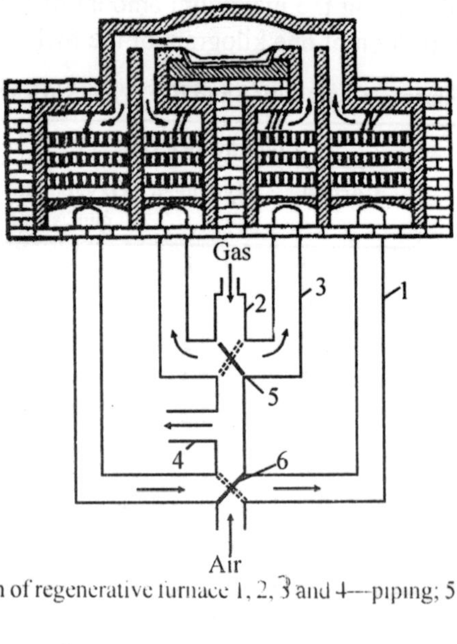

Fig. 14.5 : Diagram of regenerative furnace 1, 2, 3 and 4—piping; 5 and 6—valves

Another important advantage of gaseous fuel is the possibility of utilizing the heat of the combustion products formed for pre-heating the fuel and the air mixed with it, thus, increasing the temperature developed during combustion still more.

The heat of the hot combustion products is utilized in regenerative furnaces, a diagram of which is shown in Fig. 14.5. The furnace consists of four regenerative chambers filled with refractory bricks laid in squares.

Air is introduced into regenerator IV through pipe 1, and the fuel gas into regenerator III through pipes 2 and 3. The gases mix under the roof of the furnace and burn. The combustion products pass through regenerators I and II, transferring their heat to the bricks in these chambers. The cooled gases are discharged through pipe 4 into the chimney stack. If valves 5 and 6 are turned into the position indicated by dotted lines on the diagram, the direction of the gases changes: now the air and the fuel gas pass into regenerators I and II, where they are preheated, while the combustion products are discharged through regenerators III and IV.

In regenerative furnaces the temperature may rise as 1,800°C. In such furnace not only iron and steel, but even platinum, can be melted.

Of course, part of the fuel must be expended on converting the solid fuel into gaseous. Calculation show that about one-third the total heat that the fuel can yield is lost in this way. However, this loss is made up for by the high temperature developed in burning the resulting gases, especially if it is taken into account that they can be produced from such materials as peat, slack coal, etc., which will not give high temperatures if burned in ordinary furnaces.

The most important types of gaseous fuel are: natural gas, producer gas, water gas and coal or coke-oven gas.

Producer gas is obtained by blowing air through a layer of red-hot coal. This process is carried out in tall cylindrical furnaces, called producers (Fig. 14.6). The fuel, say coal, is dumped into the producer at the top, while air is forced in at the bottom. As the coal burns, carbon dioxide forms in the lower part of the producer and upon rising is reduced by the hot coal to carbon monoxide; the latter is discharged from the furnace together with the nitrogen of the air, which has not taken part of the reaction through an opening in the side wall of the producer. A mixture of these two gases, containing theoretically two volumes of nitrogen per volume of carbon monoxide (or more precisely 34.7 per cent CO and 65.3 per cent N_2) is known as producer gas. Producer gas usually contains a little carbon dioxide (2.5 to 5 per cent).

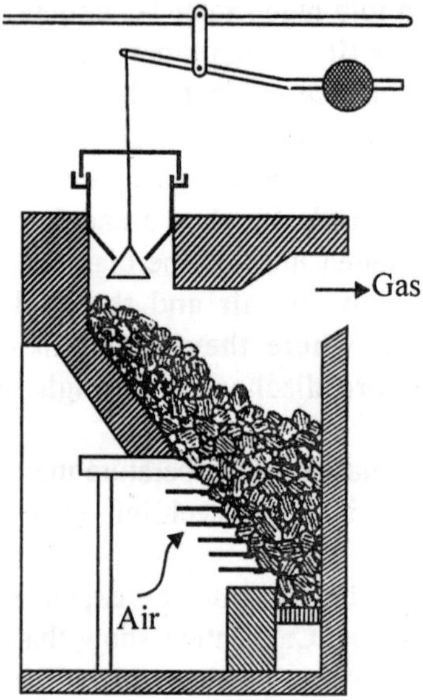

Fig. 14.6 : Diagram of producer

The reactions taking place in the producer may be expressed by the following equations:

$$C + O_2 = CO_2 + 97.7 \text{ Cal.}$$
$$CO_2 + C = 2CO - 38.3 \text{ Cal.}$$

More heat is liberated during the first of these reactions than is absorbed during the second, so that in the long run some excess heat is obtained, which keeps the coal in the producer hot.

Water gas is made by passing steam through a layer of red-hot coal in a producer:

$$C + H_2O = CO + H_2 - 28.1 \text{ Cal.}$$

As can be seen from the equation, water gas consists of equal volumes of carbon monoxide and hydrogen. As both component parts are combustible, water gas gives a higher temperature, when burnt, than producer gas.

As the formation of water gas is accompanied by absorption of heat, the passing of steam through the coal in the producer soon makes it cold. Therefore the production of water gas is usually alternated with the production of producer gas, making it is possible to keep the coal at red heat.

Water gas is used not only as a gaseous fuel but also as a raw material for the preparation of hydrogen for the synthesis of ammonia.

A gas close in composition to water gas can be obtained by passing a mixture of steam and oxygen through the producer.

Coke-oven or *coal gas* is the gas obtained by heating bituminous coal to a high temperature out of contact with air. Large quantities of this gas are used pure or mixed with natural gas in industry, in laboratories and for domestic purposes.

Coal gas is a mixture of various combustible gases. Its composition depends on the raw materials, but on the average it may be expressed by the following figures:

Hydrogen........................	55 p.c.
Methane........................	30 p.c.
Carbon monoxide........................	4 p.c.
Other hydrocarbons........................	3 p.c.
Incombustible impurities (CO_2, N_2, O_2)........................	8 p.c.

When using coal gas or its mixtures with natural gas for domestic purposes, it should be handled with care, as it contains a considerable percentage of carbon monoxide. Special precipations should be taken to prevent leakage. When the gas is not burning, the gas cocks should always be turned off.

Underground gasification of coal. Coal can be converted into gaseous fuel by gasification right in the seams in which it occurs (under the earth).

The possibility of underground gasification of coal in Russia was first suggested by D. Mendeleyev in 1888 after a trip to the

Donets Basin, where he made a thorough study of the state of the coal industry.

In an article entitled "The Future Might Lying on the Banks of the Donets," published in the journal "Severny Vestnik," Mendeleyev wrote: "....There will probably even come a time when coal will no longer be extracted from the earth but will be converted right there, in the earth, into combustible gases which will be sent over long distances through pipes."

Later Mendeleyev repeatedly returned to the idea of underground gasification of coal. In particular, with respect to the underground fires of coal seams occurring at that time in the Urals, he wrote: "....As to these coal seam fires, it seems to me that they could be utilized by getting them under control and directing them to make the coal burn like in a producer, i.e., with limited supply of air. Then carbon monoxide should form and producer gas would be produced in the seams."

For a number of years Mandeleyev persistently put forth his ideas of underground gasification, deeply convinced of the possibility of their realization. But all Mendeleyev's appeals to coal manufacturers to experiment with underground coal gasification were in vain. His suggestions were considered too fantastic and unreal at that time.

The English chemist Ramsay worked on the problem of underground gasification of coal between 1910 and 1915.

However, only in the U.S.S.R. has D. Mendeleyev's in idea begun to be realized. Numerous experiments carried out in the Donets and Moscow Coal basins have shown the technical possibility of solving this problem.

Occurrence of silicon

Silicon is second only to oxygen in abundance, constituting approximately 26 per cent of the earth's crust. Since practically all the silicon occurs combined with oxygen as SiO_2 (or as compounds of SiO_2 with metal oxides), this compound comprises over 50 per cent of the weight of all matter available to us.

The outer zone of the earth is composed largely of what are called *siliceous rocks*. Rocks that have probably resulted from solidification of liquid on or near the earth's surface are known as igneous rocks. The weathering of these rocks, erosion, and deposition by water have resulted in vast deposits of *sedimentary* rocks such as sandstone, shale, and limestone. As new intrusions of hot liquid rocks have penetrated the earth's crust, heat and marked pressure effects have caused alteration of igneous and sedimentary rocks to *metamorphic* rocks. When a sandstone is subjected to marked heat and pressure effects, it crystallizes to a more compact rock called quartzite; limestone by heat and pressure is altered to the metamorphic rock, marble.

The chemistry of igneous rocks is the chemistry of many complex silicates. Liquid, glassy masses which originally covered all or part of the earth (some lavas still exist in the earth) are metal oxides dissolved in, or combined with, silicon dioxide. When a water solution of many salts freezes, one finds many different salts crystals throughout the ice.

Similarly, a magma (molten rock), on solidifying, contains crystalline compounds throughout the quartz, SiO_2, or the SiO_2 may be part of the compound.

Granite is a good example of an igneous rocks. A molten mass of SiO_2, Al_2O_3, K_2O, Na_2O, Fe_2O_3, and H_2O may solidify to the minerals *feldspar*, $K_2O{\cdot}Al_2O_3{\cdot}6SiO_2$ or $Na_2O{\cdot}Al_2O_3{\cdot}6SiO_2$; *mica*, $K_2O{\cdot}2Al_2O_3{\cdot}Fe_2O_3{\cdot}6SiO_2{\cdot}2H_2O$; and *quartz*, SiO_2. Close examination of a good granite building stone will reveal cream-coloured feldspar, black flaky mica, and white quartz. As varying compositions of molten magmas solidify or are altered chemically by weathering, hundreds of different minerals may form. Other common silicates are *wollastonite*, $CaSiO_3$; *asbestos*, $Mg_3Ca(SiO_3)_4$; *talc*, $H_2Mg_3(SiO_3)_4$; zircon, $ZrSiO_4$; *clay*, $2H_2O{\cdot}Al_2O_3{\cdot}2SiO_3$. These types of siliceous rock comprise most of our surface rock and soil particles.

Preparation

Silicon may be prepared by the reduction of silicon dioxide with an active metal like magnesium or aluminium at an elevated temperature:

$$SiO_2 + 2Mg \rightarrow Si + 2MgO$$

More commonly, silicon is prepared by reducing the oxide with carbon in an electric furnace:

$$SiO_2 + 2C \rightarrow Si + 2CO\uparrow$$

The proportions of the reactants must be carefully regulated to prevent the formation of silicon carbide.

Properties and uses

Silicon as ordinarily prepared is a brownish powder but may exist also as hard gray crystals with a metallic luster. The element is relatively inactive at ordinary temperatures. When heated with oxygen, it combines readily to form the dioxide; however, the action soon stops as a tightly adhering protective coating forms on the surface and prevents further action. The elements readily reacts with the halogens at elevated temperatures to form volatile halides; for example,

$$Si + 2Cl_2 \rightarrow SiCl_4\uparrow$$

Silicion alloyed with iron as *ferrosilicon* finds extensive application in the manufacture of steel as a deoxidizing agent for dissolved gases. The alloy is prepared by heating together ferric oxide, silicon dioxide, and carbon. Silicon in small amounts in steel gives a softer, easily magnetized steel used in magnets and transformers. When silicon is alloyed with iron in larger amounts, a product is obtained which is highly resistant to the action of acids and chemicals. *Duriron* (15 per cent Si, 85 per cent Fe) is used for laboratory drain pipes and sinks. A fragment of pure silicon (or germanium) is the vital element in transistors—new electronic devices which give promise of cheaper television sets, tiny radios improved electric motors, and desk-size electronic brains. A wafer of elemental silicon is used in the conversion of solar energy to electrical energy in the solar battery.

Silicon is not attacked by acids, but does react as a nonmetal with strong bases to liberate hydrogen:

$$Si + 4OH^- \rightarrow SiO_4^{----} + 2H_2\uparrow$$

A small amount of industrial hydrogen is produced in this way.

The hydrides of silicon (silanes)

A mixture of hydrides of silicon, containing up to six silicon atoms in a chain, can be obtained by treating magnesium silicide with dilute hydrochloric acid:

$$Mg_2Si + H^+ \rightarrow Mg^{2+} + SiH_4 + H_3Si - SiH_3, \text{etc.}$$

The hydrides can be separated by fractional distillation, and individual isomers have been separated by vapour phase chromatography.

The first member of the series, monosilane, is conveniently made by the reduction of silicon tetrachloride with lithium aluminium hydride:

$$SiCl_4 + Li^+AlH_4^- \rightarrow SiH_4 + Li^+Cl^- + AlCl_3$$

The hydrides are gases or volatile liquids, the boiling points rising with increasing molecular weight. Their stability to heat decreases with increasing chain length (reluctance of silicon atoms to catenate). They are all spontaneously inflammable in air and are hydrolyzed by water—more rapidly in the presence of alkali. This reactivity has been attributed to the availability of *d*-orbitals for bonding with attacking species, which are to available to the carbon atom.

$$SiH_4 + 2O_2 \rightarrow SiO_2 + 2H_2O$$

$$SiH_4 + 2OH^- + H_2O \rightarrow SiO_3^{2-} + 4H_2$$

Silicon dioxide (silica), SiO_2

Silicon dioxide occurs in a number of different crystalline forms, but each form contain silicon atoms bonded tetrahedrally to four oxygen atoms by means of single Si—O bonds. Silicon dioxide is thus quite unlike carbon dioxide in its structure, and once again the reluctance of silicon to form multiple bonds (this time with oxygen atoms) is noted.

The three crystalline forms of silicon dioxide are quartz, tridymite and cristobalite, each of which has a high and low temperature modification. Quartz is the most thermodynamically stable of the three forms, but there is no tendency for the other two forms to transform themselves into quartz, since the migration of atoms in solids is a process requiring very high activation energies.

The structure of crystobalite can be visualised in terms of the diamond structure, carbon atoms being replaced by silicon atoms with one oxygen atom midway between two silicons. In quartz and tridymite spiral structures are present and these forms have optical isomers (right- and left-handed spirals). Since all three forms of silicon dioxide have covalent bonds extending in three dimensions, they are giant molecules and high melting-point solids. The empirical formula corresponds to SiO_2 because each oxygen atom is shared equally by two silicon atoms (Fig. 14.7).

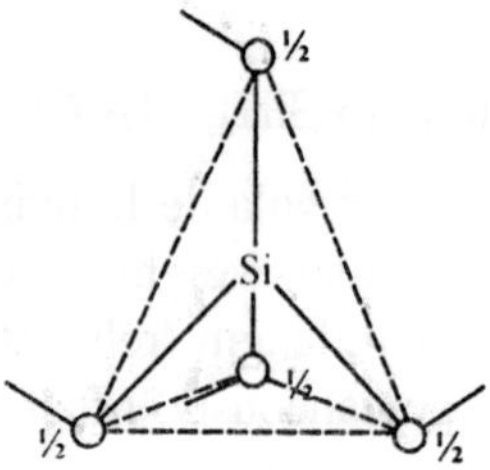

Fig. 14.7 : In silicon dioxide each silicon atom is surrounded tetrahedrally by oxygen atom

Colourless quartz is called rock crystal; purple quartz is the gem amethyst. It is used in the construction of special lenses and prisms, since it is transparent to both infrared and ultraviolet radiation, and its ability to control radio frequencies is utilized in the construction of electronic equipment, i.e. radar devices. Since its industrial applications have now outstripped its natural production, single crystals of quartz are 'grown' artificially from other forms of silica. Sand is essentially small grains of quartz with impurities; it is formed by the weathering of silicate rocks.

When molten silica is allowed to cool, the substance called silica glass is obtained. It consists of randomly oriented chains, sheets and three-dimensional networks of SiO_4 groups, cf. boron trioxide glass. It has a softening point higher than that of ordinary glass and a very low coefficient of expansion. It can, therefore, be heated and cooled rapidly without cracking. It is used in the construction of special chemical apparatus since it is not attacked by acids, except hydrofluoric acid, nor by alkalis unless they are fused. Its main disadvantage is that it is more brittle than ordinary glass and it is expensive.

Silica is non-volatile acidic oxide (although it does not react with water and forms complex mixtures of silicates only when fused with alkalis or alkali carbonates) and finds numerous applications, e.g. the extraction of alkali phosphorus, the manufacture of concrete and glass and in the anhydrite process for producing sulphur dioxide. When heated with coke it forms silicon carbide (empirical formula SiC), one of the hardest man-made substances, which is used as an abrasive.

Silicic Acids

Orthosilicic acid, H_4SiO_4, is an extremely weak acid. It cannot be formed from its acid anhydride, SiO_2, by hydration, due to the fact that silica is very nearly insoluble in water. However, the addition of a strong mineral acid to a solution of an alkali metal silicate results in the formation of orthosilicic acid.

$$SiO_4^{4-} + 4H_3O^+ \rightarrow \underline{H_4SiO_4} + 4H_2O$$

The orthosilicic acid first appears as a colloidal dispersion and then very shortly forms a colloidal gelatinous mass. In this form orthosilicic acid adsorbs certain ions to the extent that they cannot be removed by washing with water. It is though that this is the mechanism by which colloidal silicic acids present in soils retain ions of soluble salts which are essential to plant growth.

The loss of water from orthosilicic acid by heating results in silica being formed as the final product of dehydration. Although there is no evidence to support the mechanism, it is thought that the dehydration is stepwise, with the formation of **metasilicic acid** (H_2SiO_3), **disilicic acid** ($H_6Si_2O_7$), **trisilicic acid** ($H_4Si_3O_8$), and perhaps other polysilicic acids as intermediates. Salts of these and other polysilicic acids are found among the naturally occurring silicates.

Silica gel is obtained when gelatinous orthosilicic acid is dehydrated until it contains only a small percentage of moisture. This gel has an open, porous structure with a large surface area per unit of mass. It has a great tendency to adsorb gases and to catalyze certain chemical reactions involves substances in the gaseous state.

Silicates. Silicates, the metal derivatives of the hypothetical silicic acid H_4SiO_4, can be classified roughly into water-soluble or water-insoluble substances. Water-soluble silicates have the general composition $aM_2O{\cdot}bSiO_2{\cdot}cH_2O$, where M is an alkali metal. The insoluble silicates are naturally occurring minerals. Water-soluble silicates are prepared by fusing varying proportions of silica with sodium carbonate at about 1300°. Among the crystalline compounds of this type that have been characterized are Na_2SiO_3 (sodium metasilicate), $Na_2Si_2O_5$ (sodium discilicate), Na_4SiO_4 (sodium orthosilicate), $Na_6Si_2O_7$ (sodium pyrosilicate), as well as the hydrates $Na_2SiO_3{\cdot}nH_2O$ (n = 5, 6, 8, 9). Naturally occurring silicates consist of a wide array of compounds incorporating a variety of linkages for the SiO_4 units present.

A consideration of the structural relationships is perhaps the best basis for discussing the oxyanions of silicon. As is the case with the modifications of silica, the fundamental structural unit in a silicate is the tetrahedral SiO_4 moiety. All silicates can be formally derived from this unit by sharing one or more concerns. No silicates are known where two silicon atoms are joined by more than one oxygen atom; that is, only corners of SiO_4 tetrahedra are shared, never edges or faces.

Orthosilicates. Discrete SiO_4 units are found in compounds such as M_2SiO_4 where M is a divalent metal ion such as magnesium, manganese, or iron. These compounds are, in formal sense, the salts of orthosilicic acid. The SiO_4 units are clustered about the cation, making it six-coordinate; in the case of Be_2SiO_4, where the cation is smaller, the clustering of SiO_4 units leads to four-coordination of the cation. Another description of the structures of the orthosilicates considers that the oxygen atoms are arranged in an approximate hexagonal close-packed structure with silicon atoms in half of the tetrahedral holes (forming the SiO_4 units) and divalent metal ions such as magnesium, manganese, or iron in the octahedral holes; if the metal ion is sufficiently small (e.g. Be^{2+}), it can occupy the other half of the tetrahedral holes.

Many minerals have the basic orthosilicate structure; for example, olivine, which has the general formula $9Mg_2SiO_4{\cdot}Fe_2SiO_4$, exhibits the Mg_2SiO_4 structure, but one in ten of the divalent metal atoms occupying the octahedral holes is iron. Garnets are a group of orthosilicates with the general formula $M_3M'_2(SiO_4)_3$, where M = Ca^{2+}, Mg^{2+}, or Fe^{2+} and M' = Al^{3+}, Cr^{3+}, or Fe^{3+}. The SiO_4 units are packed about M to give eight-coordination and about M' to give six-coordination.

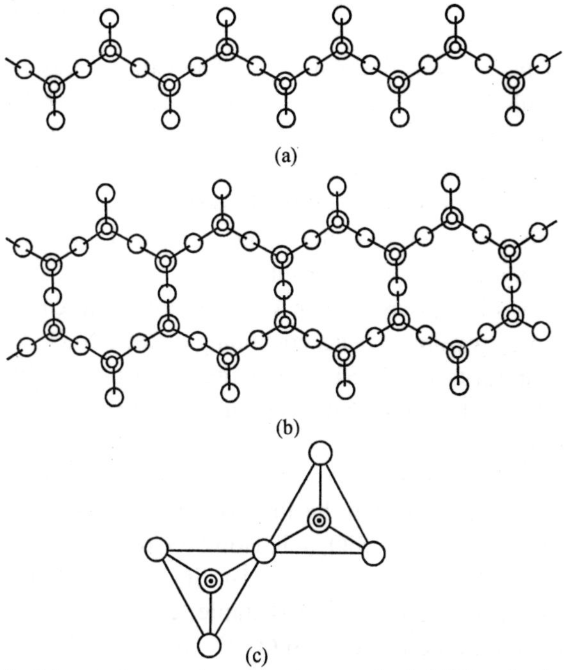

Fig. 14.8 : Silicate anions can be formed from tetrahedral SiO_4 units sharing (*a*) one corner to form a chain or (*b*) two corner to form a double chain. The $Si_2O_7^{6-}$ ions (*c*) is the simplest condensed silicate.

Chain Silicates. The two types of silicate chain structure shown in Fig. 14.8 are known. Simple chain with the composition $(SiO_3)^{2-}$ consist of tetrahedral SiO_4 units joined to form infinite chains. The synthetic metasilicates such as Na_2SiO_3 and Li_2SiO_3 possess an infinite chain structure for the anions, as do the minerals classified as pyroxene [$MgSiO_3$, enstatite; $CaMg(SiO_3)_2$, diopside; $NaAl(SiO_3)_2$, jadeite; $LIiAl(SiO_3)_2$, spodumene]. The chains in such structures are held together by the cations lying between them so that they achieve an appropriate coordination number. For example, the chains in enstatite are arranged to give Mg^{2+} a coordination number of 6, but when Mg^{2+} is replaced by the larger Ca^{2+} ion, as in diopside, the structure accommodates to give the larger Ca^{2+} ion a coordination number of 8.

The other common, infinite-chain structure shown in Fig. 14.8 involves a double chain of SiO_4 units crosslinked so that every other SiO_4 unit share two of the four oxygen atoms. Such structures are known for that class of minerals called amphiboles, which contain anions of the composition $(Si_4O_{11})^{6-}$. The cations present in such minerals link the adjacent double chains to each other by achieving the appropriate coordination number using the nonbridging oxygen atoms in the chains. The various asbestos minerals are amphiboles; the fibrous nature of such minerals reflects the relatively weak electrostatic forces between the chains compared with the stronger forces holding the SiO_4 units together.

A few silicates are known with short discrete chain lengths. For example, two-units chains containing discrete $Si_2O_7^{6-}$ ions exist in the structures of $Sc_2Si_2O_7$ and hemimorphite, $Zn_3(Si_2O_7)\cdot Zn(OH)_2\cdot H_2O$. This anion consists of the two SiO_4 tetrahedra sharing a corner (Fig. 14.9).

Cyclic Silicates. Theoretically there could be a very large number of cyclic silicates in which the first SiO_4 unit in a chain shares an oxygen atom with the last SiO_4 unit. However, only two cyclic anions, shown in Fig. 14.8, are known to exist. $Si_3O_9^{6-}$ occurs in $BaTiSi_3O_9$ (benitoite), $Na_2ZrSi_3O_9$. $2H_2O$ (catapleite), and $Cu_6Si_6O_{18}\cdot 6H_2O$ (dioptase). The cyclic structure $(Si_6O_{18})^{12-}$ exists in $Be_2Al_2Si_6O_{18}$ (beryl, emerald). The ring anions are arranged in sheets; the metal atoms lie between the sheets holding the rings in different sheets together. When large cations are present, they are six-coordinate as in benitoite. The smaller cations such as $B3^{2+}$ are four-coordinate.

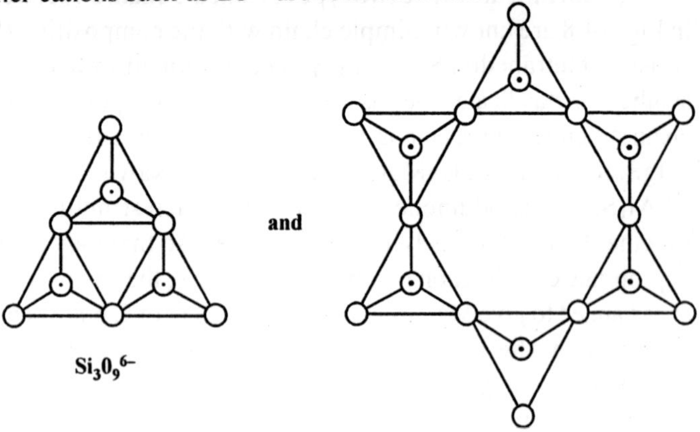

$Si_30_9^{6-}$ and $Si_30_{18}^{12-}$

(a) (b)

Fig. 14.9 : Cyclic silicates can be formed if SiO_4 tetrahedra share two oxygen atoms. (*a*) The smallest cyclic silicate is $Si_3O_9^{6-}$. (*b*) $Si_6O_{18}^{12-}$ is a common cyclic silicate ion.

Layer Silicates. When each SiO_4 tetrahedron share three oxygen atoms with adjacent tetrahedra the infinite sheet structure shown in Fig. 14.10 is formed, with the composition $(Si_2O_5)^{2-}$. The sheet arrangement for compounds containing the empirical unit $Si_2O_5^{2}$ is

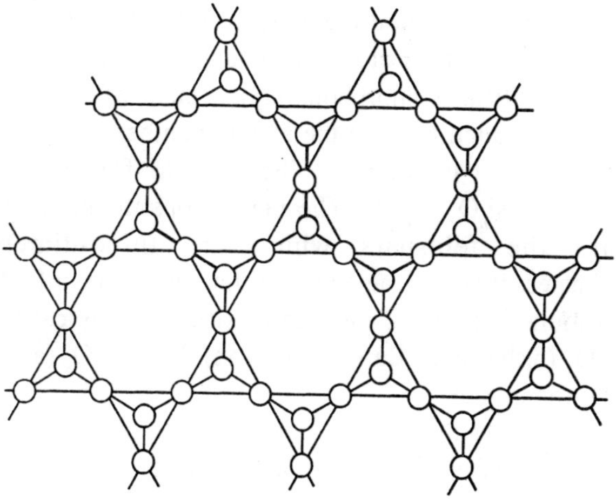

Fig. 14.10 : When SiO_4 units share three oxygen atoms, an infinite sheet can be formed.

the most common, although recently a discrete ion with this stoichiometry has been isolated in the compound $[Ni(C_2H_4NH_2)_3]Si_2O_5 \cdot 8.7 H_2O$. The anion in this substance is $Si_6O_{16}^{6-}$ and contains a prismatic arrangement of silicon atoms, each having three bridging and the terminal oxygen atom.

There are two possible arrangements of oxygen atoms of the SiO_4 units in a sheet. All of the oxygen atoms can be directed towards one side of the sheet or half can be on one side and half on the other. Many silicates and compounds containing aluminosilicates, in which an appropriate cation has been substituted for silicon, exhibit this layer structure. The counter cations lie between the infinite sheets, holding them together by relatively weak electrostatic forces. Minerals which are found in such layered structures are the kaolins $[Al_2(OH)_4Si_2O_5]$, talc $[Mg_2(Si_2O_5)_2 \cdot Mg(OH)_2]$, and the micas. Kaolins are part of the general class of minerals known as clays; these minerals are soft, easily hydrated, and exhibit cation-exchange properties. Talc of soft, cleaves very easily, and can be used as a dry lubricant. Micas have very nearly perfect cleavage parallel to the layer framework.

Three-Dimensional Silicate. If each SiO_4 tetrahedron shares all of its corners, a three-dimensional network is formed. The possible structures, which are, of course, neutral are described in the section on silica. Isomorphous replacement of aluminium for silicon yields an anion because the atomic number of aluminium is one less than that of silicon. The general formula of such structures is $[AlSi_nO_{2(1+n)}]^{-1}$. The major constituents of igneous rocks are the feldspars which are three-dimensional aluminosilicates. It should be apparent that an almost infinite array of compositions is possible for aluminosilicates. Typical feldspars are $KAlSiO_3$ (orthoclase), $BaAl_2O_8$ (celsian), $NaAlSi_3O_8$ (albite), and $CaAl_2Si_2O_8$ (anorthite).

A more interesting class of three-dimensional aluminosilicates is the zeolites. Unlike feldspars, these substances have a more open structure, which permits them to absorb liquids and gases rather easily. Generally speaking, three families of zeolites are known. In the first type of zeolite the AlO_4 and SiO_4 tetrahedra are linked into a three-dimensional open framework in which four- and six-membered rings predominate, as shown in Fig. 14.11; chabazite ($Na_2CaAl_2Si_4O_{12}\cdot 6H_2O$), gemelinite ($Na_2CaAl_2Si_4O_{12}\cdot 6H_2O$), and erionite ($CaMgNa_2K_2\,Al_2Si_6O_{16}\cdot 6H_2O$) are examples of this class.

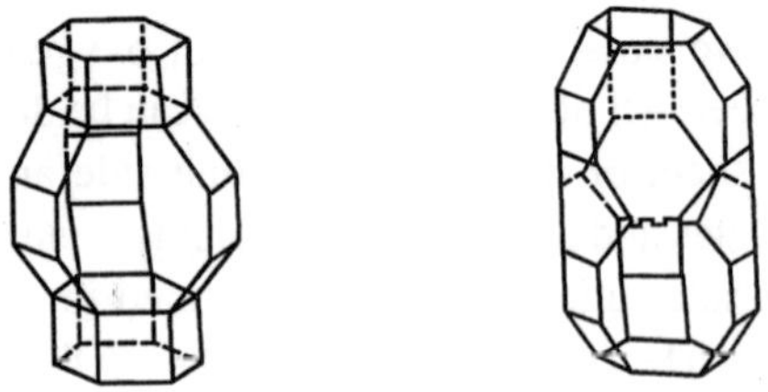

(a) (b)

Fig. 14.11 : Depending upon their composition, zeolites can be produced with difficult sized cavities: (a) erionite ($CaMgNa_2K_2Al_2Si_6O_{16}\cdot 6H_2O$) and (*b*) gemelinite ($Na_2CaAl_2Si_4O_{12}\cdot 6H_2O$)

Silicon Carbide

When a mixture of sand and a large excess of coke is heated in an electric furnace, *silicon carbide* (**carborundum**), SiC, is produced according to the equation

$$SiO_2(s) + 3C(s) + 149{,}300\ cal \rightarrow SiC(s) + 2CO(g)$$

$$(\Delta H^\circ_{298} = +149{,}300\ cal)$$

The product comes from the furnace in the form of blue-black irridescent crystals, nearly as hard as diamonds and very stable at high temperatures. The crystals are crushed, the particles are screened to uniform size, mixed with a binder of clay or sodium silicate, molded into various shapes such as grinding wheels, and fired. Silicon carbide is used as an abrasive for cutting, grinding, and polishing.

Silicon carbide exists in three different crystalline forms; yet in each of these form carbon and silicon atoms have alternate positions, and each atoms is surrounded tetrahedrally by four others. One crystalline form has the diamond structure. In other to rupture a crystal of silicon carbide, one must break a number of strong covalent bonds. The high decomposition temperature (above 2,200°), the extreme hardness, the brittleness, and the chemical inactivity of silicon carbide are in accord with a diamond-like structure.

Glass

Glassy substances not infrequently occur in nature. Of these the commonest is obsidian, from which many primitive peoples made weapons and ornaments. The earliest artificial production of glass is lost in antiquity Pliny ascribes the discovery of glassmaking to accident, thus:

The story is, that a ship, laden with nitre, being moored upon this spot [near the mouth of the river Belus in Syria], the merchants, while preparing their repast upon the seashore, finding no stones at hand for supporting their cauldrons, employed for the purpose some lumps of nitre which they had taken from the vessel. Upon its being subjected to the action of the fire, in combination with the sand of the seashore, they beheld transparent streams flowing forth of a liquid hitherto unknown; this, it is said, was the origin of glass.

The earliest known glaze is that found on stone beads from Egypt, dating about 12,000 B.C. By the year 1000 B.C. very beautiful and fairly complicated glass objects were being made in several parts of the world. Glassmaking is thus one of the oldest of civilized arts. Beginning about the eleventh century the city of Venice became a centre for glass manufacture and Venetian glass is still prized. The beauty of its design has never been surpassed. From Venice the art of glassmaking spread throughout the world.

Many substances may be obtained in the "glassy" condition. Pure silica for instance, may occur as a glass or as a crystalline material. We shall limit our discussion to commercial glasses, particularly the ordinary "window pane" variety. Commercial glass is always a mixture rather than a pure chemical compound. The composition varies within considerable limits, depending on the kind of glass desired and the formula used by the glass manufacture. Silicon dioxide is a constituent of particularly all glasses, and, in window glass amounts to about 70 per cent of the whole. The other constituents are alkali oxides and alkaline earths such as sodium oxide, calcium oxide, and magnesium oxide. A typical analysis of glass might be as follows: SiO_2, 70 per cent; CaO, 10 per cent; MgO, 2.5 per cent; Na_2O, 15 per cent; other oxides such as Fe_2O_3, Al_2O_3, etc., 2.5 per cent. These substances can scarcely be thought of as existing in the form of oxides in the glass. They are, rather, combined to form a mixture of, say, calcium silicate, magneisum silicate, and so forth. These several silicates are mutually dissolved in each other and in the excess of silicon dioxide. Glass is, therefore, a complicated product, the composition and character of which has not been understood until quite recent years, although the actual manufacture of glass is one of the oldest of chemical industries.

Thanks to studies by x-ray diffraction methods, the essential difference between a glassy substance and the same substance in the crystalline state is now clear. In crystals, it will be recalled, the atoms are regularly arranged throughout the whole mass. In glasses the atoms may be combined to each other in the same fashions as in crystals, but the arrangement of the atoms in space is much more nearly random. The regular rows and planes which characterize atomic arrangement in crystals are absent in glasses.

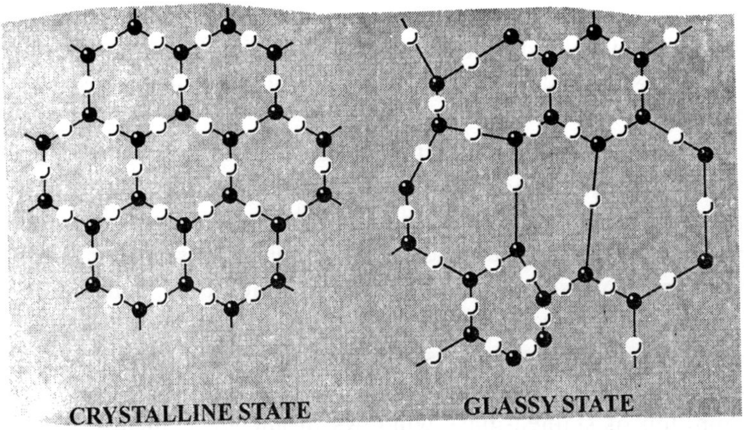

Fig. 14.12 : A glass differs from a crystal in the lack of long-range ordered arrangement in the former, although the immediate neighbours of each atom may be the same in each case. The illustrations are diagrammatic only—the third dimension being omitted for simplicity in drawing.

Glasses are very like liquids in respect to the arrangement of atoms. Glasses may be thought of as solutions which have been subjected to a very great degree of supercooling so that the solution can no longer flow like a liquid. We might define a **glass** as an amorphous substance or mixture of substances which has been supercooled from the fused condition until it is for all practical purposes rigid. Glass which is very old, or which has been subjected to certain heat conditions, tends to crystallize, or devitrify, which is exactly what would be expected of a substance in a metastable, super cooled condition.

Commercial production of glass consists of mixing together clear white sand with the appropriate amounts of alkali and alkaline earth compounds. These latter may be added in the form of the oxides, but carbonates and sulphates and other compounds are often used because these are converted to oxide at the high temperature of the furnace. As the temperature is raised, the various ingredients begin to interact and to dissolve in each other. Finally, at temperatures ranging up to about 1500°C, the mixture becomes a clear, free-flowing liquid, which may be poured out and allowed to solidify to glass. There are, of course, innumerable technical details in the production of a glass with the desired properties, free from undesirable colours, or air bubbles, and free from excessive internal strains. These details and many other interesting properties of glass will be found in works on the subject.

The most outstanding properties of glass are probably its resistance to corrosion and its transparency. No substance may, strictly speaking, be considered insoluble in water, but glass possesses this property to a high degree; hence its almost universal use for containers for such diverse substances as milk, medicine, and acids. Comparatively few substances attack glass. Of sodium hydroxide, is another example. But glass resists the ordinary action of the air and of water for indefinitely long periods, as it witnessed by the good condition of stained-glass windows in ancient cathedrals. The transparency of glass is, of course, the basis for its use for windows, lamp bulbs, and optical apparatus. What condition leads to transparency in a chemical substance is not completely known, but glass certainly possesses to a remarkable degree the two desirable properties of transparency and corrosion resistance.

Glass is subject to a condition known as "strain." If the sample is heated or cooled irregularly, there is set up a tendency for the glass to shatter. This is probably due to unequal expansion or contraction of the glass under changing temperature. In all large glass objects it is necessary to anneal the sample. Annealing consists of heating the glass to a temperature somewhat below its softening point, and then gradually lowering the temperature.

Annealing is generally accomplished in a few hours, but in special cases may extended over months. Freedom for strain is indicated by the use of polarized light, and special glass apparatus is often so tested before being sold.

One other property of glass will be mentioned, and that is its low electrical conductivity. Glass has a measurable electrical conductivity, but this is so exceedingly small that for all practical purposes glass may be considered an excellent electrical insulator.

A large variety of glasses are made for special purposes. A few of these will be described. So-called lead glasses are made by melting lead oxide or lead silicate with alkalis and silica. The lead content may average about 15 per cent, although much higher concentrations are known and glasses containing up to 92 per cent lead oxide have been made. Such glass is as dense as cast iron. The lead glasses have a high refractive index and are important for optical lenses and prisms. The finest cut-glass tableware is made of lead glass.

The name "**Pyren**" is well known as a brand of glass. This word is a trade mark of the Corning Glass Works, covering various glass having valuable electrical, thermal, and corrosion-resistant properties. A common Pyrex glass has the percentage composition: SiO_2, 80.5; B_2O_3, 12.9; Na_2O, 3.8; K_2O, 0.4; Al_2O_3, 2.2. Borosilicate glasses, of which the above is an example, have unusual resistance to rapidly changing temperature, and they are specially valuable and prized for laboratory apparatus.

In 1934 a type of Pyrex glass was cast into a huge mirror for the Mount Palomar Observatory of the California Institute of Technology. This disc is 200 in, in diameter, and its production by the Corning Glass Works is unquestionably the most ambitious undertaking in the history of glass manufacture. The disc was successfully cast on the second try, and, after literally years of careful fabrication and grinding, the mirror is now in operation at the observatory. The telescope, of which the mirror is part, extends to a fantastic distance the observable part of the universe.

An interesting glass approaching pure silica in its properties has been developed by the Corning Glass Works. This product, known as Vycor, is made by treating a special borosilicate glass with acid. The acid leaches out most of the boron oxide and other substances, leaving a sort of skeleton of almost pure silica. When this material is heated, it fuses into a product almost as resistant to heat and corrosion as pure silica, but, of course, very much easier to fabricate. Vycor products are especially valuable for laboratory equipment which must stand high temperatures.

A few other special glasses will be mentioned. All the glasses described above contain large properties of silica. A glass has been devised which contains the oxide of the rare-earth element lanthanum, together with small amounts of other oxides, but no silica. Such glass has an exceptionally high refractive index and is valuable for lens systems in optical instruments.

Glass is often coloured. The various colours are produced by added coloured compounds, or by the use of colloidal material. For instance, chromic oxide Cr_2O_3, can be used to give a green colour;

cobalt oxide gives a blue. The exact colour often depends upon the way in which the colouring material is added and how the glass is later heated. Beautiful ruby glass is made by adding gold to the glass melt. The gold is dispersed in the colloidal condition. Red glass may also be made with copper, or more frequently with selenium. Most red glass is made by adding selenium and cadmium fulfide. Translucent, or opal, glasses are made by adding powdered substances which do not dissolve in the glass melt. Fluorspar and cryolite are often used for this purpose.

One of the most remarkable of glasses is photosensitive, like photographic film. The material is also a product of the Corning Glass Works. The developed picture is actually inside the glass, rather than on the surface. In some cases it is possible to dissolve always the unexposed glass, leaving a perfect glass image of the object photographed.

The structures of germanium, tin and lead

Tin exhibits allotropy, three crystalline forms being known with the transition temperatures as shown below

$$\underset{\text{'grey'}}{\alpha\text{–tin}} \xrightleftharpoons{13.2°C} \underset{\text{'white'}}{\beta\text{–tin}} \xrightleftharpoons{161°C} \gamma\text{–tin} \xrightleftharpoons{232°C} \text{liquid tin}$$

Although β-tin should change into α-tin at a temperature below 13.2°C. unless some α-tin is added to catalyze the reaction.α-Tin has the diamond structure whereas both β- tin and γ-tin are metallic(an approach to close packing of the atoms). Since α-tin has a more open structure than β-and γ-tin its density is considerably less than the densities of the other two allotropes.

This type of allotropy in which two allotropes are equally stable at the transition temperature is referred to as enantiotropy. The allotropy of carbon at ordinary temperatures and pressures is a different phenomenon–monotropy–and has been discussed previously.

Compounds of germanium, tin and lead

Unlike carbon and silicon, germanium, tin and lead can exhibit valencies of 2 and 4; in the 2-valent state the two s electrons are inert, taking no part in chemical bonding. 2-valent germanium compounds are strongly reducing and revert to the 4-valent state; the same is true of 2-valent tin compounds but to a lesser extent. Lead is predominantly 2-valent and here the 4-valent state tends to be strongly oxidizing.

4-valent compounds of these elements are generally predominantly covalent (a few such as SnF_4 contain ionic boding), whereas many of the 2-valent compounds are essentially ionic.

Occurrence, extraction of Germanium

Germanium occurs to the extent of about 6 per cent by weight in the ore germanite, (Cu, Ge, Fe, Zn, Ga)(S, As). When the ore is strongly heated with hydrochloric acid germanium (IV) chloride, $GeCl_4$, distils off. After purification by fractional distillation it is hydrolyzed by water to germanium (IV) oxide, GeO_2, from which the metal is obtained by reduction with carbon or hydrogen.

$$GeCl_4 + 2H_2O \rightarrow GeO_2 + 4HCl$$
$$GeO_2 + 2H_2 \rightarrow Ge + 2H_2O$$

Very pure germanium is obtained by zone refining; it is a semiconductor and like silicon is used in the construction of transistors.

Properties of germanium

Germanium is hard and brittle, exhibiting the group valency of 4.

$$Ge \xrightarrow[\text{heat}]{O_2} GeO_2$$
$$Ge \xrightarrow[\text{heat}]{S} GeS_2$$

The halides of germanium

Germanium forms all four tetrahalides. Germanium (IV) chloride can be made by passing chlorine over the heated metal. It is a fuming colourless liquid, readily hydrolyzed by water, with the evolution of hydrogen chloride. Digermanium hexachloride, Ge_2Cl_6, which contains the weak Ge–Ge bond is also known.

Germanium (II) chloride, $GeCl_2$, is a colourless solid and is probably partially ionic. It is obtained by passing germanium (IV) chloride vapour over heated germanium, but it is unstable with respect to the tetrachloride and begins to disproportionate at about 75°:

$$GeCl_4 + Ge \rightarrow 2GeCl_2$$

The oxides of germanium

Germanium reacts with oxygen when heated to give germanium (IV) oxide which is a white solid; this oxide can also be obtained by the action of concentrated nitric acid on germanium.

It is less acidic than silicon dioxide, although it reacts with basic oxides to give germanates (IV) which are often isomorphous with the silicates.

$$(Na+)_2O^{2-} + GeO_2 \rightarrow (Na^-)_2GeO_3^{2-}$$

Germanium (II) oxide, GeO, has been obtained by reduction of Ge^{IV} in solution. It is unstable in the presence of oxygen, reverting to germanium (IV) oxide.

$$2GeO + O_2 \rightarrow 2GeO_2$$

Germanium oxide GeO is a black crystalline powder. It is very unstable. It is prepared by carefully reducing germanium dioxide with magnesium or metallic germanium. It reacts with the hydrohalic acids forming dihalides.

Germanium hydroxide Ge $(OH)_2$ can be precipitated by alkalis or ammonia from a solution of germanium chloride :

$$GeCl_2 + 2NaOH = Ge(OH)_2 + 2NaCl$$

Germanium tetrachloride $GeCI_4$ is prepared by heating germanium in a stream of chlorine or by passing hydrogen chloride through a heated suspension of germanium dioxide in concentrated hydrochloric acid:

$$GeO_2 + 4\ HCI \rightarrow GeCI_4 + 2\ H_2O$$

The reaction is reversible and goes from right to left in an excess of water.

Germanium tetrachloride is a colourless mobile liquid with a specific gravity of 1.874 and a boiling point of 83°C. In properties it resembles silicon tetrachloride.

Germanium disulphide GeS_2 is known to occur in two modifications–amorphous and crystalline. The moisture of the air decomposes GeS_2, liberating hydrogen sulphide.

$$GeS_2 + 2\ H_2O = GeO_2 + 2\ H_2S$$

Germanium disulphide readily forms stable thiosalts.

Germanium hydrides. Treatment of germanium chloride with sodium amalgam in a stream of hydrogen, or decomposition of a germanium-magnesium alloy by acids results in germanium tetrahydride GeH_4. It is a colourless gas which, like arsenic hydride, decomposes when heated to form a metallic mirror. The melting point of germanium tetrahydride is–166°C and its boilling point–88°C.

During the preparation of the simplest germanium hydride, small quantities of its homologues, Ge_2H_6 and Ge_3H_8, are also formed.

Tin

There is some evidence to indicate that tin and lead were regarded as identical metals by early writers. Although lead and copper vessels are common in Roman and Greek relics, comparatively little tin has been found.

Since the metal does not occur in the free state, the ancients must have obtained it from its ores, probably by heating the oxide mineral in a wood fire.

Occurrence

The only ore of tin of importance is the oxide, SnO_2, which is called *cassiterite*. At one time England led in the production of tin, but at the present time most of the ore comes from British Malaya, Bolivia, and the Dutch East Indies.

Metallurgy

The fact that the metallurgy of tin is relatively simple undoubtedly accounts for its use in early times; the method of smelting by simply heating in a wood fire was probably discovered by accident. The oxide mineral is readily reduced with hot carbon, and the low melting metal flows from the waste rock into a collecting crucible.

$$SnO_2 + 2\,C \rightarrow Sn + 2\,CO$$

The liquid tin is cast into blocks. It is refined by heating the blocks to a temperature which melts the tin and leaves most impurities as solid matter. The tin may also be refined by electrolysis, by making impure tin the anode and pure tin sheet the cathode in a cell in which sulfuric and fluosilicic acids are used as the electrolyte.

Properties

Tin is a silver-white metal of moderate hardness, and its melting point of 232° is quite low. It exists in at least two allotropic forms

$$\underset{\text{gray}}{Sn} \overset{18^\circ}{\rightleftharpoons} \underset{\text{White}}{Sn}$$

White tin is distinctly crystalline, and if the metal is bent back and forth, it produces a creaking noise, called "tin cry," due to the crystals slipping over one another. The gray variety is formed when the white form is subjected to low temperatures over a long period of time. When the change is once started, it progresses rapidly with a change in appearance from white metallic to grayish powder. Tin buttons on coats of Russian soldiers have been known to crumble to a gray powder in winter. The action is termed "tin disease" or "tin pest."

The metal appears just above hydrogen in the activity series. It does not corrode or tarnish in the air, and dilute acids act on the metal only slowly. Hydrochloric and sulfuric acids act slowly on the metal to produce stannous salts:

$$Sn + 2\,H^+ \rightarrow Sn^{++} + H_2$$

Concentrated nitric acid acts vigorously on the metal to form metastannic acid:

$$Sn + 4\,HNO_3 \rightarrow H_2SnO_3 \downarrow + 4\,NO_2 + H_2O$$

Uses-One of the principal uses of tin is in the production of tin plate. In the process, sheet iron which has been cleaned of oxide coating by "pickling" in acid is dipped into molten tin, and a thin layer of tin is coated onto the iron surface. The tin acts as a protective coating for the iron, which is the more active metal. If the surface of tin plate is scratched so that the iron is exposed, corrosion takes place rapidly, probably because of an electric couple set up between the two metals. Most of the tin plate is used in the manufacture of cans for storage of foods; utensils of various kinds are also made from tin plate. Tin pipes are used for conveying distilled water, and tin foil is a common wrapping material. Considerable tin is used in the production of alloys: *bronze* (Cu, Sn); *pewter* (Cu, Pb, Sb, Sn); *solder* (Pb, Sn); and *type metal* (Pb, Sn, Sb). The distribution of tin used in the United

States for various commercial products is approximately as shown in Figure 14.13.

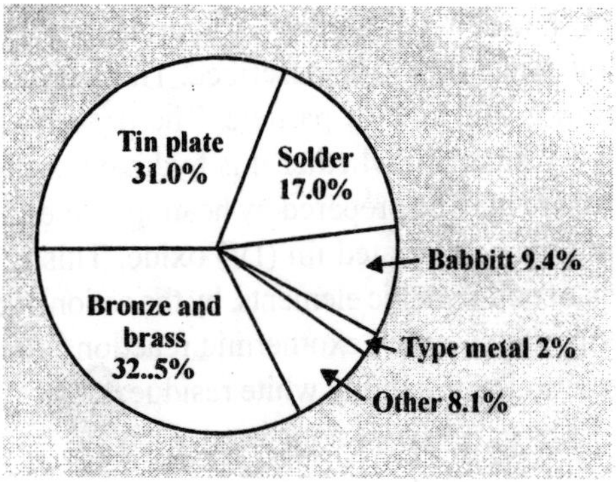

Fig. 14.13 :

Tin is a comparatively expensive metal. A "tin can" is really an iron can with a very thin coating of protective tin on the surface. Similarly, a "tinsmith" works with iron sheeting, which may have a coating of tin, zinc, or other metals. Much zinc, aluminum, or lead foil now is substituted for the more expensive tin foil.

Compounds of tin—Tin forms two series of compounds, stannous and stannic. The higher valence form, stannic, behaves more as a nonmetal than the stannous form; hence stannic salts are much less common than stannous. Both stannous and stannic hydroxides are amphoteric, since they are soluble in acid and base; the addition of sodium hydroxide to the hydroxides of tin produces stannites and stannates, respectively:

$$Sn(OH)_2 + 2\,OH^- \rightarrow Sn(OH)_4^- \text{ or } SnO_2^{-}.2H_2O$$
$$Sn(OH)_4 + 2\,OH^- \rightarrow Sn(OH)_6^{--} \text{ or } SnO_3^{--}.3H_2O$$

Sn^{II} and Sn^{IV} : the oxides of tin

Tin forms oxides in which it exerts the principal valency of 4 and the subsidiary valency of 2. Tin (II) oxide (stannous oxide) smoulders when exposed to air and may become incandescent, forming tin (IV) oxide, (stannic oxide)

$$2SnO + O_2 \rightarrow 2SnO_2$$

Tin (IV) oxide is found as the ore, cassiterite or tin-stone. Hydrated forms of both oxides exist. Tin (II) hydroxide has not been isolated as such and is probably best regarded as hydrated tin (II) oxide. The same is true of tin (IV) hydroxide, known more frequently as stannic acid, but better regardes as hydrated tin (IV) oxide of which two forms α and β, will be described. These probably differ in degree of hydration and size of particle. The properties of tin (IV) oxide depend very much on whether it is hydrated or dried.

Tin (IV) oxide may be prepared by heating molten tin in oxygen or air or igniting the β- hydrated tin (IV) oxide. This is formed in a way reminiscent of non-metallic elements, by the action of concentrated nitric acid on tin, the complex exothermic reaction proceeding with increasing vigour to leave a bulky white residue:

$$Sn + {}_4HNO_3 \rightarrow$$
$$SnO_2 \text{hdrated} + {}_4NO_2 \uparrow + {}_2H_2O$$

Tin (IV) oxide is a white powder which is reduced to the metal on heating with carbon.

$$SnO_2 + {}_2C \rightarrow Sn + {}_2CO \uparrow$$

and reacts with fused alkalis to form the stannate (IV)

$$SnO_2 + {}_2OH^- \rightarrow \underset{\text{stannate (IV)}}{SnO_3^{2-}} + H_2O$$

The reaction expressed in this way is similar to those which occur with carbon dioxide or silica and alkali. But sodium stannate (IV), $Na_2SnO.3H_2O$, used commercially as a mordant in dyeing, cannot be dehydrated and is thought to be $Na_2Sn(OH)_6$ the hexahydroxostannate (IV). The simpler ion is retained usually in equation. Tin (IV) oxide, as prepared above, does not react with dilute acids. It dissolve in hot concentrated sulphuric acid from which it is precipitated on addition of water. Tin (IV) oxide is used to manufacture white glass, white tiles and glazes and as a polishing medium.

Addition of aqueous ammonia to a solution of tin (IV) chloride, or of dilute hydrochloric acid to sodium stannate (IV) solution precipitates α-hydrated tin (IV) oxide:

$$SnCl_4 + {}_4NH_3 + {}_2H_2O \rightarrow$$
$$4(NH_4^+ + Cl^-) + SnO_2 \text{ hydrated} \downarrow$$

$$SnO_3^{2-} + 2H^+ \rightarrow$$

complex
as above

$$SnO_2 \text{ hydrated} \downarrow + H_2O$$

This form of tin (IV) oxide is much more amphoteric. Hydrochloric acid forms tin (IV) chloride. As this compund contains some (hydrated) tin (IV) ions, it is considered to be a salt:

$$SnO_2 \text{ hydrated} + {}_4HCI \rightarrow SnCI_4 + {}_2H_2O$$

At red heat, α-hydrated tin (IV) oxide loses water.

Unlike tin (IV) oxide, which is assigned the ionic structure, $Sn^{4+}(O^{2-})_2$, tin (II) oxide has a covalent macromolecular structure. It is a dark powder, variously describes as brown, black and olivegreen, formed when tin (II) oxalate is heated in the absence of air.

$$\begin{matrix} COO \\ | \\ COO \end{matrix} Sn \rightarrow SnO + CO \uparrow + CO_2 \uparrow$$

Alternatively, the hydrated oxide ('tin (II hydroxide') may be precipitated by addition of aqueous ammonia to tin (II) chloride solution, collected, dried and heated, all operations being done in an inert atmosphere. Tin (II) oxide is amphoteric: tin (II) salts are formed with acids, and stannate (II) with alkalis:

$$SnO + {}_2H+ \rightarrow H_2O + Sn^{2+}$$

$$SnO + OH^- \rightarrow \underset{\text{stannate(II)}}{HSnO_2-}$$

With concentrated alkali, disproportionation occurs followed by formation of sodium stannate (IV):

$${}_2SnO + Sn \rightarrow SnO_2$$

$$SnO_2 + {}_2OH^- \rightarrow SnO_3^{2-} + H_2O$$

The action of alkali, and sometimes acids, on so-called amphoteric oxides is not as straightforward as it might seem. The oxide may pass into colloidal solution, the micelles being stabilized by adsorption of ions of electrolyte. This is called *peptization*.

Sn^{II} and Sn^{IV}: sulphides of tin

Both tin(IV) sulphide (stannic sulphide), SnS_2, and tin (II) sulphide (stannous sulphide), SnS, may be formed by union of the elements. On heating above 265°C, tin (II) sulphide slowly disproportionates into a mixture of tin and tin (IV) sulphide, the sulphide of the principal valency state:

$$2SnS \rightarrow Sn + SnS_2$$

The sulphides are more usually prepared by precipitation using hydrogen sulphide and solutions of the chlorides, slightly acidified to reduce hydrolysis,

$$SnCl_4 + {}_2H_2S \rightarrow SnS_2 \downarrow {}_4HCl$$

$$SnCl_2 + H_2S \rightarrow SnS \downarrow {}_2HCl$$

Tin (II) sulphide is brown and the tin (IV) compound, yellow.

Tin (IV) sulphide dissolves in warm (colourless) ammonium sulphide solution to form ammonium thiostannate (IV), $(NH_4)_2SnS_3$:

$$SnS_2 + S^{2-} \rightarrow SnS_3^{2-}$$

It is usually more convenient to use yellow ammonium (poly) sulphide, $(NH_4)_2S.S_x$. Tin (II) sulphide will not dissolve in colourless ammonium sulphide but will react with warm yellow ammonium sulphide, being oxidized to the tin (IV) state and forming ammonium thiostannate (IV):

$$SnS + S^{2-}.S \rightarrow SnS_3^{2-}$$

Addition of acid precipitates the yellow tin (IV) sulphide:

$$SnS_3^{2-} + {}_2H^+ \rightarrow SnS_2 \downarrow + H_2S$$

(compare carbonates)

When yellow ammonium sulphide is used for the separation of Group II into the Groups IIa and IIb of the Qualitative Analysis Tables, tin (II) sulphide is converted into the tin (IV) compound. Alternatively, separation of the sulphides which form thiosalts may be effected by boiling with lithium hydroxide reagent, which dissolves the sulphides of arsenic, antimony and tin, although with the last-name some hydrolysis occurs. The reagent contains potassium nitrate to prevent dispersal of sulphides into the colloidal condition:

$$_2SnS + _2OH^- \rightarrow HSnO_2^- + HSnS_2^-$$

stannate(II) thio-
stannate(II)

$$_3SnS_2 + 6OH^- \rightarrow SnO_3^{2-} + _2SnS_3^{2-} + _3H_2O$$

stannate(IV) thio-
stannate(IV)

Acidification precipitates sulphides in the original valency state.

Both tin (II) sulphide and tin (IV) sulphide dissolve rapidly in concentrated hydrochloric acid with liberation of hydrogen sulphide.

$$SnS + _2HCl \rightarrow SnCl_2 + H_2S \uparrow$$

$$SnS_2 + _4HCl \rightarrow SnCl_4 + _2H_2S \uparrow$$

Mosaic gold for gilding is manufacutred by heating a mixture of tin filings, sulphur and ammonium chloride to give a mass of tin (IV) sulphide.

The halides of tin

Like germanium, tin forms all four tetrahalides. Tin (IV) chloride, $SnCl_4$, is made by passing chlorine over the metal. It is a colourless liquid which fumes in moist air, owing to hydrolysis. It is not completely hydrolyzed by water unless the solution is dilute and in the presence of a little water it is possible to obtain the solid hydrate $SnCl_4.5H_2O$. This compound has ionic character and presumbly contains the $[Sn(H_2O)_4]^{4+}$ ion.

Tin (IV) fluoride, unlike the chloride, is a white solid and is considered to be essentially ionic $Sn^{4+}(F^-)_4$.

Tin (II) chloride can be obtained as an anhydrous solid on passing hydrogen chloride over heated tin, or as the dihydrate by reacting tin with concentrated hydrochloric acid:

$$Sn + 2HCl \rightarrow SnCl_2 + H_2$$

$$Sn + 2HCl + 2\,H_2O \rightarrow SnCl_2.2H_2O + H_2$$

Although more stable with respect to the tetrachloride than the corresponding dichloride of germanium, it is quite a powerful reducting agent, e.g. it will reduce mercury (II) salts to mercury and iron (III) salts to the iron (II) state:

$$2HgCl_2 + Sn^{2+} \rightarrow Hg_2Cl_2 + Sn^{4+} + 2Cl^-$$

White solid

$$Hg_2Cl_2 + Sn^{2+} \rightarrow 2Hg + Sn^{4+} + 2Cl-$$

$$2Fe^{3+} + Sn^{2+} \rightarrow Sn^{4+} + 2Fe^{2+}$$

Both the anhydrous chloride and the dihydrate are hydrolyzed by water the solution becoming milky with the formation of a basic salt.

$$SnCl_2 + H_2O \rightleftharpoons Sn(OH)Cl + HCl$$

The solution can be kept clear in the presence of hydrochloric acid with which the chloride forms $SnCl_3^-$ and $SnCl_4^{2-}$ complexes. Tin is also added to prevent aerial oxidation.

Anhydrous tin (II) chloride is an angular covalent molecule as expected, since two of the outer electrons of the tin atom are not involved in bonding. The bond angle is 95°, which is consistent with the theory that repulsion between lone pair-bonding pair electrons is greater than that between bonding pair-bonding pair electrons.

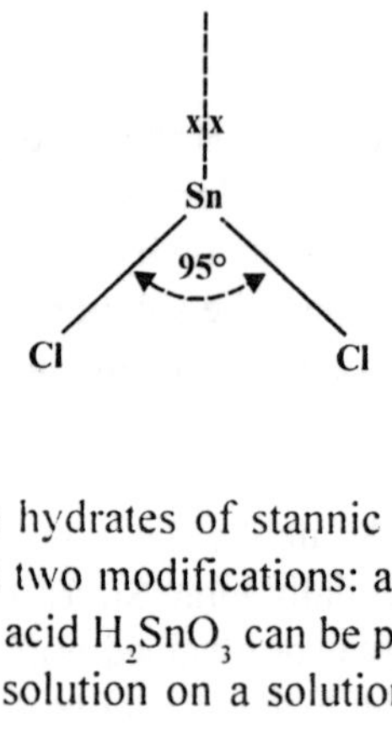

Stannic acids- The hydrates of stannic oxide are known as stannic acids and exist in two modifications: as α-stannic acid, and β-stannic acid. α-Stannic acid H_2SnO_3 can be prepared by the action of an aqueous ammonia solution on a solution of stannic chloride $SnCl_4$.

The formation of the white precipitate which separates out is usually represented by the equation

$$SnCl_4 + 4\ NH_4OH = \downarrow H_2SnO_3 + {}_4NH_4Cl + H_2O$$

As the precipitate is dried it gradually loses water until pure stannic oxide remains. Thus, no acid or any definite composition is obtained. Therefore, the above formula for α-stannic acid is but the simplest of its possible formulas. It would be more correct to represent the composition of this acid by the formula $mSnO_2.nH_2O$.

α-Stannic acid dissolves readily in alkalis, forming salts which contain the complex anion $[Sn(OH_6)^{--}$ and are called stannates:

$$H_2SnO_3 + 2NaOH + H_2O = Na_2[Sn(OH)_6]$$

Sodium stannate separates out of solution as crystals of a composition which may also be expressed by the formula $Na_2SnO_3.3H_2O$. This salt is used as a mordant in dyening and as a filler for silk. Silk fabrics treated before dyeing with solutions of tin compounds sometimes contain as much as 50 per cent of their total weight in tin.

Acids also dissolve α-stannic acid to form salts of tetravalent tin.

For example:

$$H_2SnO_3 + 4HCI_4 \rightleftharpoons SnCI_4 + 3H_2O$$

In an excess of hydrochloric acid $SnCI_4$ adds two HCI molecules to form complex chlorostannic acid $H_2[SnCI_6]$. The ammonium salt of this acid $NH_4[SnCI_6]$ is used for the same purposes as sodium stannate.

β-Stannic acid is obtained as a white powder by the action of concentrated nitric acid on tin. Its composition is just as indefinite as that of α-stannic acid. Contrary to α-stannic acid, it dissolves neither in acids nor in alkalis. But it can be transferred into solution as a stannite by fusion with alkalis. If kept in contact with the solution from which it was precipitated α-stannic acid also gradually passes into β-stannic acid.

Lead

Lead and objects have been found in Egyptian tombs, and the element is mentioned several times in the Old Testament. Uses of the metal in early times included backing for glass mirrors, coffins, weights, net sinkers, and arrow tips. In the excavated ruins of Pompeii, which was covered by volcanic ash in 79 A,D., lead water pipe is seen in place as used so many years ago.

Occurrence

Various estimates place the percentage of lead in the earth's crust at about 0.0005 percentage. which is much smaller than the percentage of such elements as zirconium, cerium, and vanadium, which we usually think of as being rare elements. There has been marked concentration of the limited natural occurrence of lead by percolating earth water and it occurs chiefly as PbS, *galena. Cerussite*, $PbCO_3$, and *anglesite*, $PbSO_4$, are less common ores but are found in sufficient quantities to warrant mining operations.

Lead ores are widely distributed over the earth's surface, the metal is produced on all of the five continents. In North America, principal deposits are in the United States in Missouri, Kansas, Oklahoma, Arizona, and Idaho. In Europe, Spain is one of the principal sources. Australia is the largest producer in the British Commonwealth

Metallurgy

Crushed lead ores are usually concentrated by gravity and flotation processes. The concentrate is then subjected to a partial roasting process in which part of the lead sulfide is changed to lead oxide and part to lead sulfate:

$$2\,PbS + 3\,O_2 \rightarrow 2\,PbO + 2\,SO_2$$
$$PbS + 2O_2 \rightarrow PbSO_4$$

The product of the roasting operation is mixed with limestone to act as a flux in the removal of siliceous impurities and with coke to act as a reducing agent on the oxides present. The mixture is smelted in a small blast furance. Although the reactions are somewhat complicated, the principal ones may be represented by the equations:

$$2\,PbO + C \rightarrow 2\,Pb + CO_2$$
$$PbS + 2\,PbO \rightarrow 3\,Pb + SO_2$$
$$PbSO_4 + PbS \rightarrow 2\,Pb + 2\,SO_2$$

The product of the smelting is crude lead or lead bullion which contains small amounts of gold, silver, copper, and other metals. The metal may be partially refined by heating and stirring in the presence of air which oxidizes copper, arsenic, and antimony. The resulting oxides float on the surface of the metal and may be removed by skimming. To further purify the lead and to remove valuable metals, two methods are employed: (1) Betts process and (2) Parkes process.

In Betts process, crude lead is made an anode in an electrolytic cell in which a sheet of pure lead is used as the cathode; the electrolyte is *fluosilicic acid*, H_2SiF_6. The process is similar to the refining of copper; lead dissolves from the anode and is plated out on the pure cathode. Gold, silver, and copper form anode mud, while other metallic ions remain in solution

Properties

Lead is one of the most dense of the common metals with a specific gravity of 113. It is soft and malleable but has little tensile strength. A freshly cut surface has a bright metallic appearance but on standing becomes coated with a thin coherent layer of oxide and a basic carbonate. Because of a protective covering on its furface, the metal is relatively inactive and resistant to corrosion.

In the activity series it appears above hydrogen, but the action of acids on it is very slow. With hydrochloric and sulfuric acids, a coating of lead chloride or lead sulfate quickly covers the surface, thus preventing further action. It is readily soluble in hot nitric acid, with the formation of the nitrate salt and oxides of nitrogen:

$$3\ Pb + 8\ HNO_3 \rightarrow 3\ Pb(NO_3)_2 + 2\ NO + 4\ H_2O$$

$$\text{or}\ \ 3\ Pb + 8H^+ + 2NO_3^- \rightarrow 3\ Pb^{++} + 2\ NO + 4\ H_2O$$

The metal very slowly reacts with water which contains air, forming a loose deposit of lead hydroxide:

$$2\ Pb + 2\ H_2O + O_2 \rightarrow 2\ Pb(OH)_2$$

Because of this action, lead pipes should not be used for carrying drinking water; lead compounds are poisonous.

Uses

Because of its relative inactivity and the ease with which it may be worked, lead is used extensively in the manufacture of a wide variety of articles. One of its principal uses is the manufacture of plates for storage batteries. Sheet lead is used extensively in lining sinks and vats, and for conduits for corrosive chemicals. Other uses include lead shot, cable coverings, lead foil, alloys, lead pipes, and production of white lead for paints. Principal alloys are pewter, solder, type metal, and bearing metal.

Compounds of lead—Several oxides of lead are known, the more important ones being the monoxide PbO, red lead Pb_3O_4; and lead dioxide, PbO_2.

PbII and PbIV: the oxides of lead: lead monoxide, PbO: red lead, Pb_3O_4: lead dioxide, PbO_2; (lead sesquioxide, Pb_2O_3)

The principal valency state of lead is two, while the group valency, four, has become subsidiary. Lead (II) oxide (monoxide), PbO, is the most stable oxide, the others reverting to it on heating. Red lead, minium, trilead tetroxide, Pb_3O_4, may be regarded as mixed oxide, $_2PbO.PbO_2$, dilead (II) lead (IV) oxide, but is better assigned the constitution, lead (II) orthoplumbate (IV), $(Pb^{2+})_2PbO_4^{4-}$. The plumbate (IV) state is much more stable than the lead (IV). Lead (II) oxide is prepared in the laboratory by heating the nitrate, carbonate, 'hydroxide' (hydrated oxide) or any other oxide to above 600°C. Commercially, a yellow form, massicot, is obtained by gently heating lead in air while fusion produces litharge, a reddish-yellow form, which is then ground to a powder:

$$2Pb + O_2 \rightarrow {}_2PbO$$

Red lead is a scarlet powder obtained by heating massicot in air to 400°C; on further heating above 470°C, litharge results:

$$6PbO + O_2 \underset{470°C}{\overset{400°C}{\rightleftharpoons}} 2Pb_3O_4$$

Lead (IV) oxide (dioxide) cannot be prepared from the monoxide and oxygen because it decomposes when heated to 300°C:

$$2PbO_2 \rightarrow 2PbO + O_2 \uparrow$$

Dilead trioxide, lead sesquioxide, Pb_2O_3, is much less important. It is an orange powder. There is some uncertainty about its exact composition but there is experimental evidence for regarding it as lead (II) metaplumbate (IV), $Pb^{2+}PbO_3^{2-}$.

Lead (II) oxide is an amphoteric oxide. It is insoluble in water, but reacts with both acids and alkalis. Because most salts of lead are insoluble, the nature of the acid used is important. Dilute nitric acid gives lead (II) nitrate and boiling dilute hydrochloric acid, lead (II) chloride, which though sparingly soluble in the cold is moderately soluble in hot solution.

$$PbO + {}_2H^+ \rightarrow H_2O + Pb^{2+}$$

Plumbate (II) is formed in reaction with alkali:

$$PbO + OH^- \rightarrow HPbO_2^-$$
$$HPbO_2^- + OH^- \rightarrow PbO_2^{2-} + H_2O$$

Lead (II) hydroxide, $Pb(OH)_2$, has not been isolated: addition of aqueous ammonia to a solution of lead (II) nitrate yielding a hydrated oxide, $(PbO)_2.H_2O$, as a white precipitate. It is amphoteric, but much less acidic than the corresponding tin (II) compound. Lead (II) oxide is easily reduced to the metal on heating with carbon, carbon monoxide and hydrogen, the charcoal block reduction being very convenient:

$$PbO + C \rightarrow Pb + CO$$
$$PbO + CO \rightarrow Pb + CO_2$$
$$PbO + H_2 \rightarrow Pb + H_2O$$

Lead (II) oxide is used in the manufacture of lead plate accumulators, as a drier in varnishes, for the production of lead compounds and in the manufacture of optical glass

Lead (IV) oxide is a puce-coloured solid, insoluble in water and nitric acid. It is precipitated by boiling a solution of lead(II) acetate with sodium hypochlorite solution, the colour of the mixture changing from orange to dark brown. The precipitate is filtered at the pump, washed with hot water and dried at 120°C in an oven:

$$(CH_3COO)_2Pb + ClO^- + H_2O \rightarrow PbO_2\downarrow + 2CH_3COOH + Cl^-$$

Lead (IV) oxide is a strong oxidizing agent. On heating, it evolves oxygen, leaving litharge. At 0°C, with concentrated hydrochloric acid, some lead (IV) chloride, lead tetrachloride, is formed but soon loses chlorine on warming. At ordinary temperatures, chlorine is evolved immediately leaving lead (II) chloride (compare MnO_2):

$$PbO_2 + {}_4HCl \rightarrow PbCl_4 + {}_2H_2O$$
$$PbCl_4 \rightarrow PbCl_2 + Cl_2$$

Oxygen is liberated when lead (IV) oxide is heated with concentrated sulphuric acid and lead (II) sulphate remains (compare MnO_2):

$${}_2PbO_2 + {}_2H_2SO_4 \rightarrow {}_2PbSO_4 + {}_2H_2O + O_2$$

Lead (II) sulphate is formed with incandescence on warming lead (IV) oxide in sulphur dioxide, or with sulphur:

$$PbO_2 + SO_2 \rightarrow PbSO_4$$

Lead dioxide and nitric acid will oxidize manganese (II) salts to permanganic acid, HMnO4:

$$2Mn^{2+} + {}_4H^+ + {}_5PbO_2 \rightarrow$$
$$2MnO_4^- + {}_2H_2O + {}_5Pb^{2+}$$

Lead (IV) oxide reacts with strong alkalis to form the plumbate (IV):

$$PbO_2 + {}_2OH^- \rightarrow PbO_3^{2-} + H_2O$$

Plumbates of the type Ca_2PbO_4 (calcium orthoplumbate(IV)) and Na_2PbO_3, sodium metaplumbate (IV) corresponding to $Pb(OH)_4$ or H_4PbO_4 and H_2PbO_3, both being hypothetical, are knwon. Calcium plumbate is used in priming paints for iron and steelwork to resist rusting. Many plumbates (IV) are hydrated and being isomorphous with stannates (IV) may be assigned similar formulae: $Na_2PbO_3.{}_3H_2O$ becomes $Na_2[Pb(OH)_6]$ the hexahydroxoplumbate (IV). Some, but not sodium plumbate (IV), decompose on dehydration. The general preparation of the plumbate (IV) requires fusion of lead (IV) oxide and the appropriate base. They are decomposed by carbon dioxide. Because of the formation of certain lead (IV) salts from lead (IV) oxide with acids under carefully regulated conditions, lead (IV) oxide is considered to be weakly basic as well as acidic and is, therefore, classified as an amphoteric oxide. Lead (IV) oxide is used in the grids of lead plate accumulators.

Red lead, trilead tetroxide, Pb_3O_4, is a scarlet powder. Its commercial production and decomposition has been described. In its reactions it may be regarded as a mixed oxide, ${}_2PbO.PbO_2$, or as lead (II) orthoplumbate (IV), Pb_2PbO_4. With nitric acid, lead (II) nitrate and the puce dioxide are formed:

$$Pb_2PbO_4 + {}_4HNO_3 \rightarrow$$
$${}_2Pb(NO_3)_2 + Pb(OH)_4$$

or $$\downarrow$$

$${}_2PbO.PbO_2 + {}_4HNO_3 \rightarrow$$
$${}_2Pb(NO_3)_2 + {}_2H_2O + PbO_2$$

Warm concentrated hydrochloric acid is oxidized to chlorine, lead (II) chloride remaining:

$$_2PbO.PbO_2 + {}_4HCl \rightarrow {}_2PbCl_2 + {}_2H_2O + PbO_2$$

and $$PbO_2 + {}_4HCl \rightarrow PbCl_2 + {}_2H_2O + Cl_2 \uparrow$$

$$Pb_3O_4 + 8HCl \rightarrow {}_3PbCl_2 + {}_4H_2O + Cl_2 \uparrow$$

Warm concentrated sulphuric acid yields oxygen and white lead (II) sulphate:

$$_2({}_2PbO.PbO_2 + {}_2H_2SO_4 \rightarrow$$
$$_2PbSO_4 + {}_2H_2O + PbO_2)$$
$$_2PbO_2 + {}_2H_2SO_4 \rightarrow$$

Adding, $$_2PbSO_4 + {}_2H_2O + O_2 \uparrow$$

$$_2Pb_3O_4 + 6H_2SO_4 \rightarrow$$
$$6PbSO_4 + 6H_2O + O_2 \uparrow$$

Red lead may be reduced to lead on heating with carbon, carbon monoxide or hydrogen. Red lead is used as a priming paint for wood and steel work, although other primers (zinc chromate and calcium plumbate (IV)) have been developed for steel, and paints containing a suspension of aluminium metal are used to seal the surface of resinous woods and hardboard. Mixed with linseed oil, it is used by plumbers as a sealing compound for gas pipe connections. As an oxidizing agent, red lead is incorporated in some matchheads and it is also used in the production of optical glass.

The halides of lead

Lead forms a tetrafluoride and a tetrachloride; the former is a solid and appreciably ionic, $Pb^{4+}(F^-)_4$, while the latter is a covalent liquid, $PbCl_4$, readily hydrolyzed by water and easily decomposed into lead (II) chloride, $PbCl_2$, and chlorine. The tetrabromide and tetraiodide do not exist, presumably because bromine and iodine are not sufficiently strong oxidizing agents to convert Pb^{II} to Pb^{IV}.

The stable valency state of lead is 2 and all four dihalides can be obtained as sparingly soluble solids by adding halide ions to a soluble lead (II) salt, e.g. the nitrate:

$$Pb^{2-} + 2X^- \rightarrow PbX_2 \; (X^- = F^-, Cl^-, Br^- \text{ or } I^-)$$

The fluoride is ionic, $Pb^{2+}(F^-)_2$, and the chloride, bromide and iodide contain some ionic character which decreases with increasing atomic number of the halogen. Except for the iodide, which is yellow, they are white solids.

Lead (II) chloride and lead (II) iodide are of importance in qualitative analysis and their formation serves as tests for Pb^{2+} in solution. Both are much more soluble in hot water than in cold and the chloride dissolves in concentrated hydrochloric acid with the formation of the $PbCl_4^{2-}$ complex ion:

$$PbCl_2 \rightleftharpoons Pb^{2+} + 2Cl^-$$

$$+$$

$$4Cl^- \quad \text{From conc. hydrochloric acid}$$

$$\updownarrow$$

$$PbCl_4^{2-}$$

Lead (IV) sulphide (PbS_2) has not been prepared but a lead (II) disulphide ($Pb^{2+}S.S^{2-}$) has been made. However, lead (II) sulphide, PbS, is the important sulphide. It is found naturally as *galena.*

Lead (II) sulphide may be produced by union of the elements but more usually as black precipitate obtained on passing hydrogen sulphide into a solution of lead salt:

$$Pb^{2+} + H_2S \rightarrow PbS \downarrow + {}_2H^+$$

In qualitative analysis, some lead passes through from the initial precipitation of lead (II) chloride in Group I to Group II. It is precipitated as the sulphide in the presence of hydrochloric acid, the precipitate appearing red or yellow at first, due to $PbS.PbCl_2$ and then black as PbS is formed. Paint containing lead compounds will darken in the sulphurous atmosphere of our towns due to the formation of lead (II) sulphide.

Lead (II) sulphide is insoluble in water, colourless and yellow ammonium sulphide, and in alkali. In these respects it differs from sulphides of tin

Concentrated nitric acid oxidizes it to lead (II) sulphate, but at lower concentrations, lead (II) nitrate and sulphur result. Hydrogen peroxide oxidizes lead (II) sulphide to the sulphate, which is white:

$$PbS + {}_4H_2O_2 \rightarrow PbSO_4 + {}_4H_2O$$

This reaction has been applied to the restoration of darkened oil paintings.

Pb^{IV}: oxocompounds

Compounds formed between quadrivalent lead and oxoacids are rare and very unstable. Lead (IV) sulphate, $Pb(SO_4)_2$, is formed as yellow crystals by the electrolysis of strong sulphuric acid below 30°C using a lead anode. It is hydrolyzed by water to lead (IV) oxide (the dioxide) and acid:

$$Pb(SO_4)_2 + {}_2H_2O \rightarrow PbO_2 \downarrow + {}_2H_2SO_4$$

A lead (IV) pyrophosphate lead (IV) diphosphate, PbP_2O_7, has been prepared by the reaction between phosphoric acid and lead (IV) oxide at 300°C. However, the easiest compound to prepare is lead tetra-acetate or lead (IV) acetate, $Pb(O.OC.CH_3)_4$. It is made by the addition of red lead in small portions to warm glacial acetic acid. On cooling, colourless crystals separate. Lead (II) acetate is also formed and remains in solution:

$$Pb_3O_4 + 8CH_3COOH \rightarrow$$
$$Pb(O.OC.CH_3)_4 + 2Pb(O.OC.CH_3)_2 + {}_4H_2O$$

Experiments have shown that lead tetra-acetate is virtually covalent. It is hdrolyzed on warming with water to the dioxide and acetic acid.

$$Pb(O.OC.CH_3)_4 + {}_2H_2O \rightarrow$$
$$PbO_2 \downarrow + {}_4CH_2.COOH$$

Lead (IV) acetate is a strong oxidizing agent, used in organic chemistry to oxidize glycols with adjacent hydroxyl groups smoothly to aldehydes or ketones.

Pb^{II}:oxosalts

The only common soluble salts of lead are the nitrate and the acetate, formed by the addition of litharge, 'hydroxide' or carbonate to a slight excess of hot dilute acid followed by evaporation to yield colourless crystals.

Lead (II) nitrate, $Pb(NO_3)_2$, is much more soluble in hot water than in cold, 59.6 g dissolving in 100 g water at 25° C and 134 at 100°

C. On heating, lead (II) nitrate decomposes with decrepitation, forming nitrogen dioxide, oxygen and lead (II) oxide:

$$Pb(NO_3)_2 \rightarrow {}_2PbO + {}_4NO_2 \uparrow + O_2 \uparrow$$

The crystals are anhydrous, unusual for a heavy metal nitrate.

Lead (II) acetate, $Pb(O.OC.CH_3)_2.{}_3H_2O$, is known as sugar of lead because of its sweet taste, although it is poisonous. It is soluble compound of lead and is largely covalent.

Insoluble salts of lead may be precipitated by mixing solutions containing the necessary ions.

Lead (II) sulphate, $PbSO_4$, occurs naturally as anglesite: it is only very slightly soluble in water (solubility = o.oo45 g in 100 g water at 25°C) but readily soluble in ammonium acetate solution due to the removal of lead(II) ions as covalent lead (II) acetate and complex acetates.

Lead (II) chromate, $PbCrO_4$, is a yellow solid, sold as a pigment, chrome yellow. It is less soluble even that the sulphate and will not dissolve in ammonium nitrate solution. A basic lead (II) chromate, the pigment chrome red is formed with hot alkali:

$${}_2PbCrO_4 + {}_2OH^- \rightarrow$$
$$PbCrO_4.PbO + CrO_4^{2-} + H_2O$$

Normal lead (II) carbonate, $PbCO_3$, is formed as a white precipitate on adding sodium hydrogencarbonate solution to lead (II) nitrate solution:

$$Pb^{2+} + {}_2HCO_3^- \rightarrow PbCO_3 \downarrow + H_2O + CO_2 \uparrow$$

Sodium carbonate solution, which contains a greater concentration of hydroxyl ions due to hydrolysis, forms a basic carbonate:

$${}_3Pb^{2+} + {}_2CO_3^{2-} + {}_2OH^- \rightarrow Pb(OH)_2.{}_2PbCO_3 \downarrow$$

Cerussite, $PbCO_3$, is a naturally occurring form of lead (II) carbonate.. Basic lead carbonate or white lead is used in many white and coloured paints. It is prepared by the action of acetic acid vapour on lead in an atmosphere containing carbon dioxide at about 60°C, a basic acetate being initially formed. Various processes have been devised to achieve a fine product of good covering power. The Dutch process uses vinegar as a source of acetic acid vapour and fermenting dung to generate carbon dioxide and warmth. White lead darkens in the sulphurous atmosphere of towns, due to the formation of lead (II) sulphide.

The Lead Storage Battery. The oxidizing properties of tetravalent lead and its transition into the more stable bivalent state are the basis of the design and operation of widely used lead storage batteries.

Electric storage batteries are devices for accumulating electric power for use at some future moment. Power is accumulated by passing electric current through the battery, thus giving rise to a chemical process during which electrical energy is transformed into chemical; the storage battery is then said to be charged. The charged battery may subsequently be used as a galvanic cell. Whereupon the reaction by which it was charged proceeds in the opposite direction, and the chemical energy accumulated in the storage battery is converted back into electrical ; as this energy is used up, the battery is said to be discharged.

In the simplest case a lead storage cell* consists of two lead grids, the spaces of which are filled with a putty of lead oxide and water. The plates are immersed in a rectangular glass vessel containing dilute sulphuric acid of specific gravity between 1.15 and 1.20 (22 to 28 per cent H_2SO_4).

Owing to the reaction

$$PbO + H_2SO_4 = PbSO_4 + H_2O$$

the lead oxide is presently converted into lead sulphate. If direct current is now passed through the device by connecting one plate to the negative and the other to the positive pole of a current source, the battery will be charged, the following processes taking place at the electrodes:

$$\overset{+II}{PbSO_4} + 2e^- \xrightarrow{\text{cathode}} \overset{0}{Pb} + SO_4''$$

$$\overset{+II}{PbSO_4} - 2e^- + 2H_2O \xrightarrow{\text{anode}} \overset{+IV}{PbO_2} + 4H + SO_2''$$

Adding up these equations we get the summary equation of the reaction of charging a storage battery:

$$2\ PbSO_4 + 2\ H_2O = Pb + PbO_2 + 4\ H^{\cdot} + 2\ SO_4''$$

Thus, as the current is passed through, the lead sulphate at the cathode turns into a spongy mass of metallic lead and that at the anode, into dark brown lead dioxide.

When this process is complete, the storage battery is fully charged. Completion of charging is indicated by the water beginning to decompose vigorously, liberating hydrogen at the cathode and oxygen at the anode (the accumulator is said to "boil")

If the plates of a charged storage battery are connected by means of a conductor, current arises in the latter, the electrons moving from the lead plate to the lead dioxide plate. The appearance of current is due to the following. Some Pb ions pass into solution from the lead plate charging the latter negatively. The electrons liberated at the lead plate pass over to the PbO_2 and reduce the tetravalent lead into bivalent. As a result $Pb^{\cdot\cdot}$ ions form at both plates and combine with the $SO^{4''}$ ions in solution into insoluble lead sulphate, discharging the battery.

The processes taking place during the discharge of a storage battery can be repesented by the following scheme:

Negative Electrode

$$\overset{0}{Pb} - 2e^- + SO_4'' = \overset{+II}{PbSO_4}$$

Positive Electrode

$$\overset{+IV}{PbO_2} + 2e^- + 4H^{\cdot}\ SO_4'' = \overset{+II}{PbSO_4} + 2H_2O$$

Addition of the above equations makes it obvious that the reaction taking place during discharge of the battery is the reverse of the charge reaction. Therefore both processes may be expressed by a single equation:

$$2\ PbSO_4 + 2\ H_2O \underset{\text{discharge}}{\overset{\text{charge}}{\rightleftharpoons}} Pb + PbO_2 + 4\ H^{\cdot} + 2\ SO_4''$$

When a storage battery is discharged the concentration of the sulphuric acid gradually decreases. owing to the consumption of H-ion and SO_4''-ion and the formation of water. Therefore. the degree of discharge of a storage battery can be judged by the specific gravity of the acid, measured by means of a densimeter.

The e.m.f. of a lead storage battery equals two volts and under normal load remains alomst unaltered throughout its entire period of operation. If the voltage begins to fall, the storage battery must be recharged.

Paints

Ordinary paints contain three essential ingredients; (1) a "vehicle," which is essentially a quick-drying oil; (2) a "body," or opaque substance of high covering power; and (3) a "pigment," if a colour other than white is desired. Linseed oil is used as the vehicle or carrier of the body of the paint, which is usually white lead. The oil oxidizes in the air to form a hard, tough film which adheres to the surface of wood or other material and protects it. Linseed oil is usually boiled with lead and manganese oxides which act as catalysts and hasten the time of drying. White lead, $(PbCO_3)_2$, is a substance much used as a paint body because of its high covering power. The main objection to its use is the darkening action by hydrogen sulfide which produces black lead sulfide. Titanium oxide is now largely replacing white lead as a body material in paints.

Interior paints contain lithopone, a $BaSO_4$-ZnS mixture, as a body since lithopone is not darkened by hydrogen sulfide. If a coloured paint is desired, a pigment is added to the lithopone. Common pigments are red lead, lead chromate, Prussian blue, iron oxide, chromium oxide, and carbon. In addition, a filler is frequently added to prevent the body from being spread too thin.

In recent years, so-called "water soluble" or "rubber latex" paints for interior use have become very popular. In these preparations, pigment and body are incorporated into an emulsion containing rubber-like polymerization products of styrene and butadiene. These emulsions may be thinned with water.

High-gloss paints or enamels may contain resins (resins from polymerization of glycerine and phthalic anhydride) in a wide variety of oil combinations. For example, enamel used for automobile finishes contains soybean oil, alkyd resins, and resins produced from polymerization of urea or melamine with formaldehyde.

15

p-Block Elements : (Nitrogen, Phosphorus, Arsenic, Antimony and Bismuth)—Group VA (15)

	Atomic	Electronic	Atomic	Ionic Radius	M.p.	B.p.
	Number	Configuration	Radius (A)	M3+	°C	C°
N	7	2.5 $1s^2 2s^2 2p^3$	0.74		-210	-1958
P	15	2.8.5 $...2s^2 2p^6 3s^2 3p^3$	1.10		44.1 (white)	280.5 (white)
As	33	2.8.18.5 $...3s^2 3p^6 3d^{10} 4s^2 4p^3$	1.21	0.69		610 (sublimation)
Sb	51	2.8.18.8.5 $...4s^2 4p^6 4d^{10} 5s^2 5p^3$	1.41	0.90	630.5	1380
Bi	83	2.8.18.32.18.5 $...5s^2 5p^6 5d^{10} 6s^2 6p^3$	1.52	1.20	271	1450

General Features of the Nitrogen Group. The nitrogen group includes the typical elements of the fifth group of the Periodic Table, nitrogen and phosphorus, as well as arsenic, antimony and bismuth, elements with atoms of similar structure in the odd series of the large periods. Together these five elements make up the main subgroup of the fifth group called also the nitrogen group.

The elements of the nitrogen group, having five electrons in the outer layer of their atoms, may be characterized on the whole as non-metals. However, their capacity for gaining electrons is much weaker than that of the corresponding elements of the sixth and seventh groups. Owing to the presence of five outer electrons, the highest positive valency of the nitrogen group elements equals five, and the highest negative valency, three. Owing to their relatively lower electronegativity, the bond between hydrogen and the elements of the nitrogen group is less polar than that between hydrogen and the elements of the sixth and seventh groups. Therefore, the hydrogen compounds of the nitrogen group elements do not split off hydrogen ions in aqueous solution and hence do not possess acidic properties.

The physical and chemical properties of the elements of the nitrogen group vary with increasing atomic number. But since the non-metallic properties of nitrogen are less pronounced than those of oxygen, and the more so than of fluorine, further weakening of these properties in passing to the subsequent elements results in rapid growth of metallic properties. The latter become perceptible already in arsenic; antimony possesses metallic and non-metallic properties in approximately equal degree, while in bismuth metallic properties predominate considerably over non-metallic.

Nitrogen

Nitrogen seems first to have been characterized by a Scottish scientist, D. Rutherford (not to be confused with Lord Rutherford, the physicist, who lived over a hundred years later). D. Rutherford published a Latin manuscript in 1772; the manuscript contained this passage:

By the respiration of animals, healthy air is not merely rendered mephitic [that is, charged with carbon dioxide], but it also suffers another change, for, after the mephitic portion is absorbed by a solution of caustic alkali, the remaining portion is not rendered salubrious, and, although it occasions no precipitate in limewater, it nevertheless extinguishes flame, and destroys life.

Rutherford's "phlogisticated air," as he called it, was studied by Scheele and by Lavoisier. The latter called it *azote*, by which name it is still known in France. The name "nitrogen" was suggested by J. A. C. Chaptal in 1823. Derivation of the name is from the Greek word *nitron* (saltpeter), from which nitrogen may be obtained.

About 80 percent of the atmosphere consists of nitrogen. Every square mile of the earth's surface has above it millions of tons of nitrogen. This atmospheric nitrogen constitutes an inexhaustible source. Nitrogen is indispensable for fertilizers and for explosives. Toward the end of the nineteenth century it was feared that available supplies of nitrogen for these uses would soon be exhausted unless means were found for extracting and using nitrogen from the air. But about 1913 there was developed a process for the use of atmospheric nitrogen, and since that time the supply has been limited only by the capacity of the ammonia and nitric acid manufacturing plants. No shortage of raw material has since occurred or ever will occur. Production capacity in the world is millions of tons per year.

Other sources of nitrogen are saltpeter or potassium nitrate, KNO_3, the supply of which is very limited, and Chile saltpeter, or sodium nitrate, $NaNo_3$. Extensive deposits of this latter substance are found in northern Chile. They constitute an important source of nitrogen. Still other substances which yield appreciable supplies of nitrogen compounds are coal and animal residues.

It should be pointed out that nitrogen is an essential constituent of all living matter. The cells of animal tissues contain nitrogen in complicated compounds called proteins. Nitrogen is not only necessary for animal and plant growth, but it is produced by the decay of animal and plant tissues.

Laboratory preparation. (a) *From the air*. This involves separating the two main constituents of air-nitrogen and oxygen, either by physical means or by removing the oxygen by chemical means. Since oxygen is quite an active substances and nitrogen is much less active chemically, the difference in activity may be used as a basis for separation. Phosphorus readily combines with oxygen in the air to form phosphorus pentoxide. Since nitrogen does not combine with phosphorus, an effective means of separation is as follows: A piece of phosphorus is burned in air which is enclosed in a bottle inverted over water. The oxygen is removed by combination with the phosphorus. The phosphorus pentoxide dissolves in the water, forming phosphoric acid. The reactions are

$$4P + 5\,O_2 \rightarrow 2\,P_2O_5$$
$$P_2O_5 + 3H_2O \rightarrow 2H_3PO_4$$

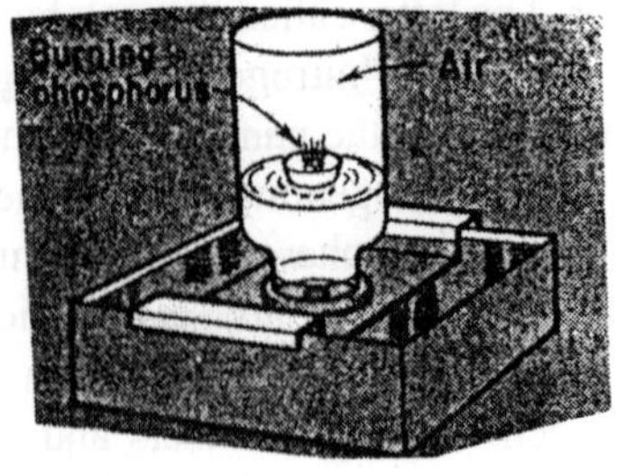

Nitrogen prepared by this method is not pure, as it contains a small amount of water vapoor and the rare gases present in the atmosphere. Certain metals, such as heated copper, may also be used to remove the oxygen from the air in much the same manner as phosphorus.

(b) From certain compounds of nitrogen. Chemically pure nitrogen is conveniently prepared by heating together solutions of sodium nitrite and ammonium chloride:

$$NaNO_2 + NH_4CI \ ?\ NaCI + NH_4NO_2$$
$$\downarrow$$
$$N_2 + 2H_2O$$

This is a double decomposition reaction between two salts; the ammonium nitrite formed, since it is unstable toward heat, breaks down to yield nitrogen and water. A combination of any nitrite salt and any ammonium salt would give a reaction similar to that above.

If ammonia is passed over hot copper oxide, the following reaction takes place:

$$3CuO + 2NH_3 \ ?\ 3Cu + 3H_2O + N_2$$

Cupric oxide acts as an oxidizing agent, ammonia as a reducing agent.

Commercial preparation. Commercial nitrogen is prepared by liquefaction of air and subsequent fractional distillation. The nitrogen obtained by this process contains about one per cent of other substances, mainly argon. Argon is also inactive chemically, so its presence does not interfere in the use of the nitrogen.

Pure nitrogen is a colourless gas with no odour, very slightly soluble in water. It is a little lighter than air, one litre of it weighing 1.25 gr. When cooled strongly under high pressure nitrogen becomes a liquid which boils at –195.8° and solidifies into a snow-like mass at –210°C.

Nitrogen is conspicuous for its chemical inertness. At ordinary temperatures it is almost incapable of forming compounds. When heated it combines quite readily with some metals, such as lithium, magnesium, calcium, titanium. Finally, at very high temperatures nitrogen combines directly with oxygen and hydrogen.

Animals placed in an atmosphere of nitrogen soon perish, not as a result of nitrogen poisoning, but due to the absence of oxygen. This property accounts for the old name of nitrogen, *azote*, from the Greek "azoos," meaning lifeless. The present English name "nitrogen" is derived from the Latin "nitrogenium," meaning "originating saltpetre."

The nitrogen molecule consists of two atoms (N_2). These atoms are tied together very stably, which accounts for the chemical inertness of nitrogen under ordinary conditions.

Nitrogen forms a large number of compounds, which are prepared indirectly and easily undergo various changes.

Free nitrogen is used in electrical engineering to fill "half-watt" lamps. But the greater part of nitrogen obtained from the air is used for the preparation of synthetic ammonia and calcium cyanamide.

Nitrogen Cycle in Nature. During the decay of organic substances a considerable part of the nitrogen contained in them turns into ammonia which is subsequently oxidized into nitric acid by nitrifying bacteria, which live in the soil. The nitric acid thus produced reacts with the salts of carbonic acid contained in the soil, for instance, with $CaCO_3$, to form calcium nitrate:

$$2\,HNO_3 + CaCO_3 = Ca(NO_3)_2 + CO_2 + H_2O$$

But during decay part of the organic nitrogen always escapes as elemental nitrogen into the atmosphere. Nitrogen is liberated also during the combustion of organic substances, when wood, coal, peat or other fuels are burnt. Besides, there are bacteria which are capable, in a limited supply of oxygen, of abstracting oxygen from nitrates, liberating free nitrogen. As a result of the activities of these denitrifying bacteria, part of the combined nitrogen passes from the form assimilable by green plants (nitrates) into an inaccessible form (free nitrogen). Thus, by no means all the nitrogen contained in dead plants is returned to the soil; some of it is constantly evolved in the free form and is therefore lost for the plants.

The continuous depletion of mineral nitrogen compounds should have long since resulted in complete cessation of life on the earth, were it not for the existence in nature of other processes replenishing the nitrogen losses. These processes include primarily the electric discharge occurring in the atmosphere, always leading to the formation of a certain amount of nitrogen oxide; the latter combines with water giving nitric acid, which is transformed into nitrates in the soil. Another source which keeps continuously supplementing the nitrogen compounds in the soil is the vital activities of the so-called *nitrobacteria*, capable of assimilating atmospheric nitrogen. Some of these bacteria inhabit the roots of plants of the bean family causing the formation of characteristic swellings, called "nodules." Hence, the name *nodule bacteria*. Assimilating atmospheric nitrogen, *nodule bacteria* convert it into nitrogen compounds, and plants, in their turn, convert the latter into proteins and other complex substances. That is why bean plants, contrary to others, thrive excellently in soils which are almost devoid of nitrogen compounds.

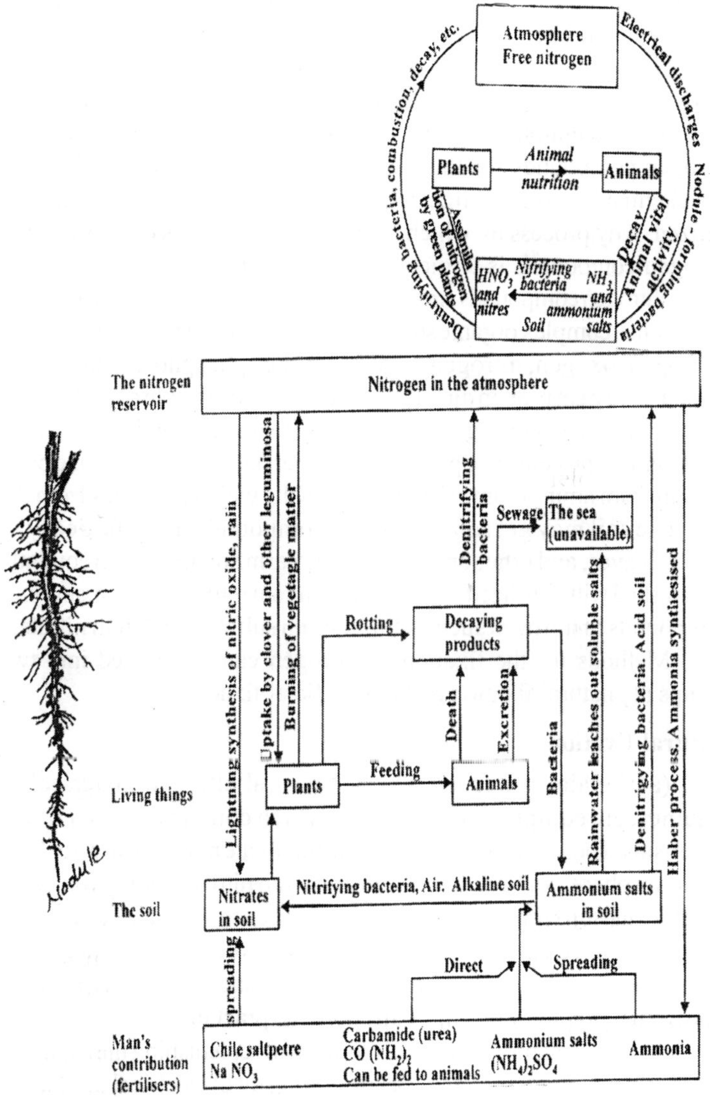

Fig. 15.1 : The nitrogen cycle.

The activities of bacteria which assimilate atmospheric nitrogen are the main reason for the quantity of combined nitrogen in the solid remaining more or less constant despite its losses due to the decomposition of nitrogen compounds. This decomposition is balanced by the formation of new nitrogen compounds, and thus there is a continuous nitrogen cycle in nature (Fig.15.1).

Nitrogen in the combined state–that is, in compounds–is far more useful and important than in the free state. Animals and plants (with few exceptions) cannot use free nitrogen but must depend on nitrogen compounds for this essential element. It is the problem of the chemist to convert free nitrogen into compounds that can be utilized. Any process by which elementary nitrogen is converted into nitrogen compounds is termed *nitrogen fixation.*

Nitrogen compounds are essential to plant and animal life in building up proteins, complex organic substances containing the elements carbon, hydrogen, oxygen, nitrogen, and sometimes phosphorus and sulfur. Animals must eat protein directly, as the body is incapable of building up proteins from other nitrogen compounds. Beans, peas and meat contain considerable protein. The plant, on the other hand, can synthesize plant protein from ammonium salts or nitrate salts which it derives from the soil. The soil may get some nitrogen compounds from the decay of plants, leaves, and other organic matter. Farm soil must be frequently replenished with fertilized in the form of animal refuse matter, manure, or compounds of nitrogen such as ammonium sulfate or calcium nitrate.

Methods for the fixation of nitrogen can be divided into two groups (1) natural fixation and (2) artificial fixation.

Natural fixation

(a) Certain plants such as peas, beans, alfalfa, and clover, called legumes, are equipped with nodules which contain nitrogen fixing organisms. These organisms are able to assimilate nitrogen directly from the atmosphere, converting free nitrogen into nitrogen compounds. To replenish the soil with nitrogen, legumes may be grown on the depleted soil and plowed under, in which case decomposition occurs, furnishing fixed nitrogen to the soil. A system of crop rotation, using leguminous plants for maintaining soil fertility, is now an approved practice.

(b) One of nature's principal methods of replenishing the supply of nitrogen compounds in the soil occurs as a result of electrical discharges in air and subsequent rainfall. At the high temperature associated with lightning bolts or other arc discharges, nitrogen and oxygen gases combine to form nitrogen oxides. These oxides with air moisture form nitrous and nitric acids which come to the earth in rainfall. A person out in a thunder shower is being subjected to a shower of very dilute nitric and nitrous acids. As these acids act on soil minerals, soluble nitrogen-containing salts are formed in soil waters. For example, as the very dilute nitric acid acts on limestone, $CaCO_3$, calcium nitrate is formed this salt is a common soil mineral. Considering the bulk of the atmosphere as a factory for HNO_3 production, and the geologic time of operation, we can account for millions of tons of nitrate compounds formation.

Artificial fixation

Man's methods of nitrogen fixation will be discussed in detail in the next chapter. The important processes are: (1) the Haber; (2) the Birkeland-Eyde; (3) the cyanamide.

The nitrogen cycle. It might occur to us that since nitrogen is continually being fixed from the air by natural and artificial methods, the supply of nitrogen in the atmosphere would decrease. Careful analysis of the air, however, shows that this is not the case; the percentage of nitrogen in the air remains nearly constant because a balance between free and combined nitrogen is maintained as a result of decomposition of nitrogen compounds, yielding elementary nitrogen. Although the process involves more steps than indicated, the diagram represents the essential changes taking place.

We may start with elementary nitrogen and trace some of the possible changes. A nitrogen atom in the free state may be fixed by (1) legumes into plant protein, (2) lightning discharges into nitrate salts, or (3) artificial processes into ammonium or nitrate salts. A plant may then utilize the salts for production of plant protein. The plant may serve as food for animal in which case the original nitrogen atom may become a part of animal protein; or degradation of the plant may again produce ammonium salts. Animal protein, in undergoing decay or degradation, may be converted either into ammonium salts or into free nitrogen which finds its way back into the atmosphere where it is again ready to start another cycle of changes.

AMMONIA AND AMMONIUM SALTS

Decay of nitrogenous matter in the absence of air produces ammonia. The decomposition may be brought about with the aid of heat or in the presence of putrefying bacteria. Ammonia at one time was produced by distilling animal horns and hoofs: hence the name "spirits of hartshorn". As a result of the action of bacteria on nitrogen compounds, small quantities of ammonia gas may be present in air. Because of the great solubility of ammonia in water, however, it rapidly finds its way by the action of rain and snow to the soil where it may be converted into other compounds.

Laboratory preparation of ammonia

(a) *From ammonium salts.* Ammonia is most conveniently prepared in the laboratory by the action of a strong base on an ammonium salt; for example:

$$NH_4CI + NaOH \rightarrow NH_4OH^1 + NaCI$$
$$NH_4OH \rightarrow NH_3 \uparrow + H_2O$$

The reaction is caused to go to completion by the use of heat, since ammonium hydroxide is unstable and breaks down to give water and ammonia which escapes as a gas and is collected by the displacement of air. (See Figure 15.2.) Since the gas is extremely soluble in water, it cannot be collected by water displacement. The reaction given above is a typical one for the preparation of ammonia. Any ammonium salt and strong base may be employed. $Ca(OH)_2$ is usually instead of sodium hydroxide because it is cheaper.

$$(NH_4)_2SO_4 + Ca(OH)_2 \rightarrow CaSO_4 + 2\,NH_4OH$$
$$2\,NH_3 \rightarrow\ +2\,H_2O$$

(b) From nitrides. The action of water on nitride produces ammonia:

$$Mg_3N_2 + 6\,H_2O \rightleftharpoons 3\,Mg(OH)_2 + 2\,NH_3 \uparrow$$

These reactions of nitrides are of theoretical interest only.

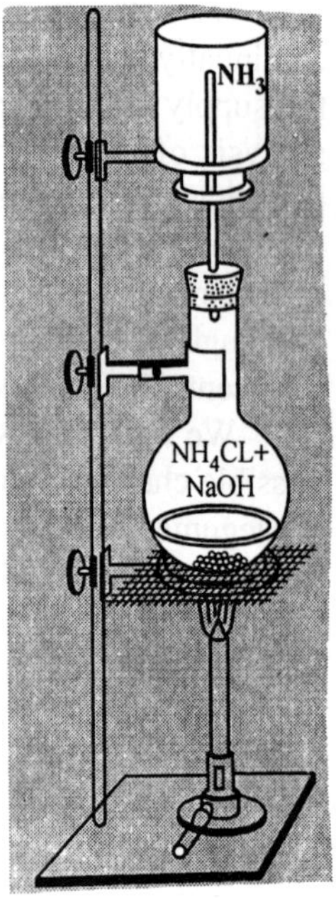

Fig. 15.2 : Ammonia may be collected by up-ward displacement of air.

Commercial preparation of ammonia

(a) *From coal.* Coal, a product of decomposition of vegetation, contains a small percentage of combined nitrogen. When coal is heated n the absence of air in the production of coke, an ammoniacal liquor is produced as a by-product. Ammonia is recovered from this liquor by distillation and subsequent absorption in water. The ammonium hydroxide solution thus produced may be converted into ammonium salts by the addition of acids. Most of the ammonium sulfate used as a fertilizer is prepared in this manner.

(b) *Haber process.* When nitrogen and hydrogen gases are heated together in the presence of certain catalysts, some ammonia is formed according to the reaction:

$$N_2 + 3\,H_2 \rightleftharpoons 2\,NH_3 + 24{,}000 \text{ cal.}$$

The reaction is very incomplete at ordinary conditions, with equilibrium established when the ammonia content is less than one percent. Since the reaction is exothermic, it is evident from a consideration of Le Chatelier's principle that a low temperature is favourable to the formation of ammonia. However, if the temperature is below about 500°, the combination of nitrogen and hydrogen occurs too slowly to be practical as a commercial means of preparation. Applying Le Chatelier's principle, we note that an increase of pressure will shift the equilibrium to the right–that is, toward the formation of ammonia. An increase of pressure form 1 atmosphere at 500° to 200° atmospheres at 500° increase the equilibrium yield of ammonia in the above reaction from 0.13 percent to 17.6 percent. The process is ordinarily carried out at a temperature of approximately 500° and at a pressure of 200 atmospheres in the presence of a catalyst mixture such as Fe_3O_4 and $K_2O\,Al_2O_3$. Ammonia is liquefied from the equilibrium mixture by rapid refrigeration of the gas mix, and the residual N_2 and H_2 is counterpassed back to the high-pressure catalyst chamber. Since large supplies of nitrogen compounds are essential in making explosives, this process was vital to Germany during World War I. Germany's access to the Chilean nitrates was cut off by the British blockade, and had Germany been unable to utilize nitrogen from the air to produce explosives, the war would have lasted only a few months. Haber, a German chemist, perfected the process just World War I began. In World War II, Germany again had to rely on synthetic nitrogen compounds.

(c) The cyanamide process

During the war years of 1914-1918, the United States built several plants–the largest at Muscle Shoals, Alabama–for the production of synthetic ammonia by the cyanamide process. This process depends for its success upon the development of cheap electric power. The first step in the preparation of ammonia consists in heating lime, CaO, and carbon together in an electric furnace to a temperature of approximately 3000°. Calcium carbide is formed:

$$CaO + 3\ C \rightarrow CaC_2 + CO$$

Nitrogen obtained from the liquefaction of air is passed over the crushed calcium carbide at a temperature of about 1000°. Calcium cyanamide, $CaCN_2$, is produced:

$$CaC_2 + N_2 \rightarrow CaCN_2 + C$$

The cyanamide is then treated with steam under pressure, forming ammonia and calcium carbonate:

$$CaCN_2 + 3\ H_2O \rightarrow CaCO_3 \downarrow + 2NH_3$$

Cyanamide itself may be used directly as a fertilizer, since this last reaction proceeds slowly in soil.

Physical properties of ammonia

Ammonia is a colourless gas with a strong, irritating odour. The mucous membranes are attacked by the gas, and inhalation of undiluted gas can seriously affect the lung tissue. The gas is mild stimulant in small quantities, being the active component of smelling salts.

Extremely soluble in water: I volume of water dissolves 1.300 volumes of ammonia at 0°
Density: 0.771 g/liter; about ? the density of air
Critical temperature: 132°
Boiling point: –33.4° at atmospheric pressure
Heat of vaporization: 297 cal/g

This last figure is exceeded only by water. Because of its high heat of vapourization, ammonia is useful in refrigeration. Liquid ammonia is a useful solvent, dissolving many polar compounds and thus acting like water.

Chemical properties. *Ammonia as a base.* Ammonia acts as a base because the unshared pair of electrons in the ammonia molecule can combine with a proton, forming the ammonium.

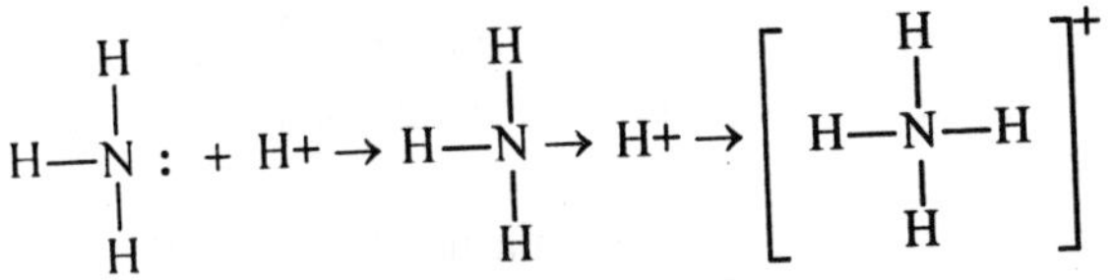

NH_3 + H+ NH:

Ammonia can take protons from water, formating a weakly alkaline solution containing hydroxide ions.

$$NH_3 + H_2O \rightleftharpoons NH_4^+ + OH \rightleftharpoons NH_4OH \quad \text{(ammonium hydroxide)}$$

Since the ionization takes place only to a small degree, the solution is weakly alkaline. Ammonium hydroxide reacts with acids to form ammonium salts.

$$NH4+ + OH^- + H_3O^+ + Cl^- \rightarrow NH^{4+} + Cl^- + 2H_2O$$

Ammonium chloride

$$2NH4+ + 2OH + 2H_3O + SO_4 \rightarrow 2NH + SO_4 + NH_2O$$

Ammonium sulfate

Ammonia gas is readily liberated from these salts by the action of strong alkali such as sodium hydroxide.

$$NH4+ + Cl^- + Na^+ + OH^- \rightarrow NH_3 \rightarrow + H_2O + Na^+ + Cl^-$$

Since the gaseous ammonia is basic, its presence can be detected by holding a place at basic red - paper above the solution and nothing that be turns blue. The presence of ammonia in the vapour is a test for an ammonium salt. Ammonia gas can be prepared conveniently in the laboratory by this reaction.

If ammonium chloride, a white salt, is heated in a test tube, it disappears from the bottom of the tube but reappears as a coating on the upper, cooler parts of the tube. In this process, there is no evidence of the substance passing through a liquid stage. This occurs because ammonium chloride breaks up into ammonia and hydrogen chloride when it is heated.

$$NH4Cl \rightarrow NH3 + HCl \qquad \text{(dissociation)}$$

In the cooler portions of the tube the reverse reaction occurs, and ammonium chloride deposits on the tube.

$$NH_3 + HCl \rightarrow NH_4Cl \qquad \text{(combination)}$$

Since the solid ammonium chloride is converted to a gas without an intermediate liquid stage, this process is analogous to the sublimation of such compounds as mercuric chloride and arsenic trioxide.

Metal complex ions. Ammonia has the property of attaching itself to certain metallic ions forming complex ions such as $Cu(NH_3)_4{}^{++}$. When ammonium hydroxide is added slowly to a solution of a copper salt, a precipitate is formed according to the reaction

$$Cu^{++} + 2Cl^- + 2NH_4OH \rightarrow Cu(OH)_2 \rightarrow + 2NH4+ + 2Cl^-$$

As more ammonium hydroxide is added, the precipitate dissolves, forming a deep-blue solution containing $Cu(NH_3)4{++}$

$$Cu(OH)_2 + 4NH_4OH \rightarrow Ca(NH_3)_4{}^{++} + 2OH + 4H_2O$$

Among the common metals, ZN^{++}, Cd^{++}, Fe^{++}, Co^{++}, Ni^{++}, and Ag^+ *form ammonia complex ions*, the bivalent ions forming $M(NH_3)_4$ and Ag_4 forming $Ag\,(NH_3)2+$. This reaction is of great importance in analysis of metal ions.

Ammonia is a reducing agent, the ostward -process. Ammonia cannot be reduced since nitrogen is in its lowest oxidation state, 3–; however, it can be oxidized fairly readily by burning it in oxygen or in air. In the latter case catalysis are necessary to make the reaction proceed at a rapid rate. The oxidation of ammonia is the first step in the preparation of nitric acid, one of the important acids in the chemical industry. The process by which it is carried out is the Ostwald process, and the fundamental reaction is

$$4NH_3 + 5O_2 \rightarrow \underset{\text{Nitric}}{4NO} + 6H_2O + \text{heat}$$

A platinum catalyst is used to increase the rate of reaction when air is used as the source of oxygen. A mixture of ammonia and air is passed through a fine platinum gauze at an initial temperature of about 700°. As the reaction proceeds the heat evolved maintains the high temperature. This process provides practically all of the nitric oxide of commerce.

Liquid ammonia as a solvent

Liquid ammonia is ionized to a slight extent:

$$2NH3 \rightarrow NH4+ + NH2^{-}$$

cf. $$2H2O \rightarrow H3O+ + OH^{-}$$

Whereas in water acids are substances that provide hydroxonium ions and alkalis are substances that provide hydroxyl ions, in liquid ammonia acids give rise to ammonium ions, e.g. ammonium salts, and basis give rise to amide ions, e.g. ionic amides. Thus in liquid ammonia an ammono-acid neutralises an ammono-base to give a salt and ammonia (an exact parallel with aqueous systems), e.g.

$$\underset{\text{ammono-acid}}{NH_4^+Cl^-} + \underset{\text{ammono-base}}{Na^+NH_2^-} \rightarrow \underset{\text{salt}}{Na^+Cl^-} + \underset{\text{ammonia}}{2NH_3}$$

or $$NH_4^+ + NH_2^- \rightarrow 2NH_3$$

Ionic nitrides such as lithium nitride function as bases in liquid ammonia:

$$3NH_4^+Cl^- + (Li^+)_3N^{3-} \rightarrow 3Li^+Cl^- + 4NH_3$$

or $$3NH_4^+ + N^{3-} \rightarrow 4NH_3$$

More parallels with aqueous systems exist; for instance hydrolysis and ammonolysis are essentially very similar processes:

$$SiCl_4 \xrightarrow{\text{hydrolysis}} Si(OH)_4 \xrightarrow{\text{heat}} SiO_2 + 2H_2O$$

$$3SiCl_4 \xrightarrow{\text{ammonolysis}} 3Si(NH_2)_4 \xrightarrow{\text{heat}} Si_3N_4 + 8NH_3$$

Hydrazine, N_2H_4

Hydrazine is a colourless fuming covalent liquid and is made by the oxidation of ammonia with sodium hypochlorite in the presence of gelatine (this is thought to complex with heavy metal ions and so prevent them from catalyzing the oxidation of ammonia to nitrogen). The reaction proceeds in two stages:

$$NH_3 + OCl^- \rightarrow OH^- + NH_2Cl$$

$$NH_2Cl + NH_3 + OH^- \rightarrow N_2H_4 + Cl^- + H_2O$$

Hydrazine is an endothermic compounds which burns in air to give nitrogen and steam with a vigorous evolution of heat. This fact accounts for the interest shown in it as a potential rocket fuel. It is weaker base than ammonia but, because it contains two lone pairs of electrons (one lone pair on each nitrogen atom), it is bifunctional; it thus forms two chlorides, for example $(NH_2–NH_3^+)Cl^-$ and $(NH_3^+–NH_3^+)(Cl^-)_2$.

Aldehydes and ketones condense with hydrazine to give hydrazones, e.g.

$$R-\overset{\displaystyle H}{\overset{|}{C}}=O+H_2N-NH_2 \longrightarrow \underset{\text{a hydrazone}}{R-\overset{\displaystyle H}{\overset{|}{C}}=N-NH_2}+H_2O$$

A derivative of hydrazine, 2, 4- dinitrophenylhydrazine, forms yellow crystalline condensation products with aldehydes and ketones in the same manner, and a determination of their melting points affords a useful method of determining the these compounds.

Hydrazoic acid, HN_3, and azides

Hydrazoic acid is a colourless, highly explosive liquid. It can be made by treating sodamide with nitrous oxide at about 190°C; sodium azide is formed, which reacts with dilute sulphuric acid to give hydrazoic acid:

$$2Na^+NH_2^- + N_2O \rightarrow Na+N_3^- + Na^+OH^- + NH_3$$
$$2Na^+N_3^- + (H+)_2SO_4^{2-} \rightarrow 2HN_3 + (Na^+)_2SO_4^{2-}$$

Hydrazoic acid is covalent and its molecule is asymmetrical:

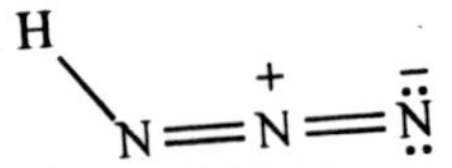

Covalent azides such as lead azide, $Pb(N_3)_2$, are explosive and are used as detonators, whereas ionic azides, e.g. sodium azide, $Na^+N_3^-$, decompose non-explosively when heated into the metal and nitrogen. The azide ion is linear as expected.

The halides of nitrogen

These are of no great importance, but it is of interest to note that nitrogen trifluoride, NF_3, is covalent gas resistant to hydrolysis, whereas the trichloride, NCl_3, is a covalent and highly explosive yellow oil readily decomposed by water into ammonia and hypochlorous acid. This hydrolysis thought to proceed by a stepwise replacement of chlorine, the lone pair on the nitrogen atom bonding to a hydrogen atom of a water molecule:

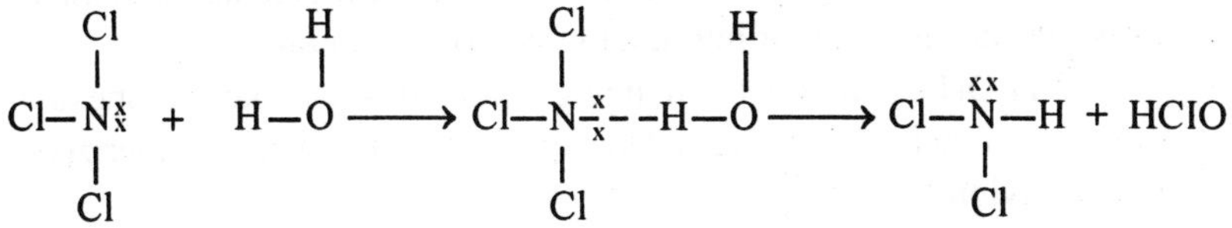

$$NHCl_2 + H_2O \rightarrow NH_2Cl + HClO$$
$$NH_2Cl + H_2O \rightarrow NH_3 \quad + HClO$$

Hydroxylamine NH_2OH

Hydroxylamine forms colourless crystals tat melt at 33°C. It is thermally unstable and decomposes into NH_3, N_2, HNO_2 and N_2O easily. It explodes if heated strongly. It is usually handled in aqueous solution, or as one of its salts, since these are more stable than free NH_2OH.

Hydroxylamine is a weaker base than is ammonia or hydrazine. Salts contain the hydroxylammonium ion [NH3OH]+.

$$NH_2OH + HCl \rightarrow [NH_3OH]^+Cl^-$$
$$NH_2OH + H_2SO_4 \rightarrow [NH_3OH]^+HSO\text{-}4$$

The appropriate reduction potentials suggest that hydroxylamine should disproportionate. It disproportionates slowly in acidic solutions:

$$\overset{-1}{4[NH_2OH\,H]+}\ ?\ \overset{+1}{N_2O} + \overset{-III}{2NH+4} + 2H^+ + 3H_3O$$

and rapidly in alkaline solutions:

$$3\overset{-1}{NH_2OH} \rightarrow \overset{0}{N_2} + \overset{-III}{NH_3} + 3H_2O$$

Both NH_2OH and its salts are very poisonous, and they are also strong reducing agents.

Hydroxylamine is manufactured by reducing nitrites, or from nitromethane:

$$NH_4NO_2 + NH_4HSO_3 + SO_2 + 2H_2O \rightarrow [NH_3OH]^+HSO\text{-}4 + (NH_4)_2SO_4$$
$$CH_3NO_2 + H_2SO_4 \rightarrow [NH_3OH]^+HSO\text{-}4 + CO$$

Hydroxylamine has donor properties (like NH_3 and N_2H_4): the N atom can form coordinate bonds, and can complex with metals. In addition it adds easily to double bonds in organic molecules, and thus provides an easy way of introducing N atoms into molecules.

NH_2OH is manufactured in large quantities to make cyclohexanone oxime, which is converted to caprolactam and then polymerized to give nylon-6.

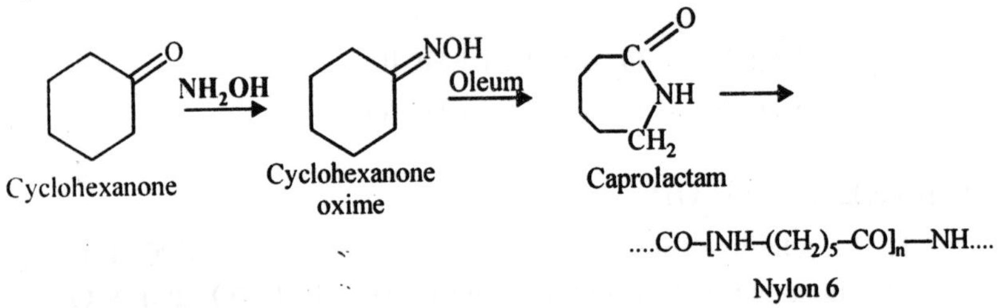

Fig. 15.2 : Nylon-6

Dinitrogen oxide, nitrous oxide, N_2O

Discovered by Priestley (1772), dinitrogen oxide is a colourless gas (b.p.–88°) with a sweet taste and pleasant odour. It is readily liquefied under pressure, the critical temperature being about 36°C. Dinitrogen oxide, mixed with oxygen, is still used as an anaesthetic by dentists, and for minor operations, and is sometimes called 'laughing gas 'on account of the exhilaration and sometimes, hysteria it produces in small quantities. Dinitrogen oxide is fairly soluble in water, I volume of water dissolving I volume of gas at 6°C. It is a neutral oxide although it may be regarded as the anhydride of hyponitrous acid, from which it is formed on heating, but the reaction cannot be reversed and dinitrogen oxide does not react with alkali to form salts:

$$\underset{\text{hyponitrous acid}}{H_2N_2O_2} \rightarrow H_2O + N_2O$$

The dinitrogen oxide molecule is linear and unsymmetrical, with a very small dipole moment, explained by its being a resonance hybrid of two structures each of which would be highly polar in opposite senses:

$$\left\{ \overline{N} = \overset{+}{N} = O \leftrightarrow N \equiv \overset{+}{N} - \overline{O} \right\}$$

Dinitrogen oxide may be prepared by thermal decomposition of ammonium nitrate:

$$NH_4NO_3 \rightarrow N_2O \rightarrow + {}_2H_2O\ ?$$

The decomposition is exothermic, starting at I85°C and becoming violent at 250°C. Impurities in the gaseous product include nitrogen oxide, nitrogen dioxide, nitrogen, oxygen and carbon dioxide. To reduce the possibility of an explosion, a mixture of ammonium sulphate and sodium nitrate may be heated, sodium sulphate remaining as the residue. The gas is often collected over hot water on account of its solubility, but rapid collection over cold water is much more convenient. Very pure dinitrogen oxide is prepared by other methods. Solution of sodium nitrite and hydroxylammonium chloride are mixed and warmed.

$$\overset{+}{NH_3}OH + NO_2^- \uparrow N_2O\ ? + {}_2H_2O$$

Nitrates may be reduced: nitric acid is reduced by tin(II) chloride in the presence of hydrochloric acid:

$${}_4Sn^{2+} + {}_2NO_3^- + IO^{H+} \rightarrow$$
$${}_4Sn^{4+} + N_2O \uparrow + {}_5H_2O$$

Dinitrogen oxide is chemically unreactive unless heated when it becomes a strong oxidizing agent, giving up its oxygen:

$$2N_2O \rightarrow 2N_2 + O_2;\ \ \Delta H = -2 \times 7^1\,kJ$$

Dinitrogen oxide supports the combustion of substances which are hot enough to decompose it, the proportion of oxygen given being greater than that in air. Strongly burning sulphur forms sulphur dioxide, and phosphorus gives the pentoxide, leaving nitrogen:

$${}_2N_2O + S \rightarrow SO_2 + {}_2N_2$$
$${}_{10}N_2O + {}_4P \rightarrow P_4O_{10} + {}_{10}N_2$$

Dinitrogen oxide re-kindles a glowing splint and is decomposed by heated metals such as copper and iron:

$$N_2O + Cu \rightarrow CuO + N_2$$

Dinitrogen oxide could easily be confused with oxygen. Points of distinction are the sweet taste and solubility, combustions in dinitrogen oxide leave residual nitrogen in equal volume and, unlike oxygen, it does not react with an alkaline solution of pyrogallol, nor does it give brown fumes of nitrogen dioxide with nitrogen oxide.

Nitrogen pentoxide, dinitrogen pentoxide, N_2O_5

When anhydrous nitric acid is dehydrated by distillation with phosphorus pentoxide, nitrogen pentoxide and metaphosphoric acid are formed:

$$P_4O_{10} + {}_4HNO_3 \rightarrow {}_4HPO_3 + {}_2N_2O_5 \uparrow$$

decomposition at 32.5°C. It rapidly decomposes when heated:

$$_2N_2O_5 \rightarrow 2N_2O_4 + O_2$$

Nitrogen pentoxide is nitric anhydride: it is hygroscopic, uniting vigorously with water to form nitric acid:.

$$N_2O_5 + H_2O \rightarrow {}_2HNO_3$$

Nitrogen pentoxide is a powerful oxidizing agent

It is thought that the structure of the gaseous molecule is a resonance hybrid of the form as

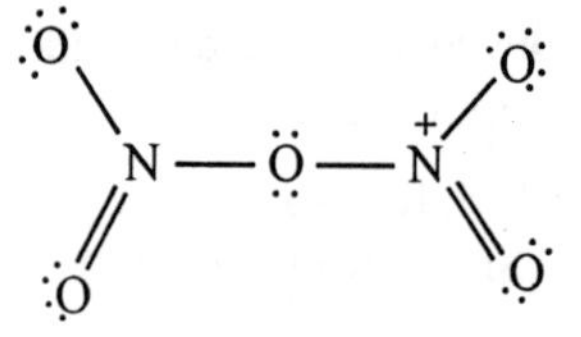

Nitric Oxide Nitric oxide (NO) is formed when acid solutions of nitrates or nitrites are reduced by metals

$$8HNO_3 + 3Cu \rightarrow 3Cu(NO_3)_2 + 4H_2O + 2NO$$

$$2KNO_3 + 6Hg + 4H_2SO_4 \rightarrow 2NO + 3Hg_2SO_4 + K_2SO_4 + 4H_2O$$

or other reducing agents

$$2KNO_3 + 4H_2SO_4 + 6FeSO_4 \rightarrow 3Fe_2(SO_4)_3 + 2NO + K_2SO_4 + 4H_2O$$

$$2HI + 2HNO_2 \rightarrow I_2 + 2NO + 2H_2O.$$

The large-scale preparation of NO involves the oxidation of ammonia above 500° in the presence of platinum

$$4NH_3 + 5O_2 \rightarrow 4NO + 6H_2O.$$

Direct combination of the elements occurs under energy-rich conditions

$$N_2 + O_2 \rightleftharpoons 2NO$$

but the equilibrium is unfavourable even at high temperatures about 5 volume percent of NO being formed at 3200°C.

Nitric oxide is thermodynamically unstable at room temperatures, the liquid or compressed gas slowly undergoing disproportionation,

$$4NO \rightarrow N_2O_3 + N_2O.$$

Since NO possesses an odd electron it might be expected to undergo radicaltype reactions. Indeed, small quantities of NO inhibit chain reactions. Although it generally behaves as expected for a radical species, the degree of reactivity is markedly depressed. Nitric oxide can act as an oxidizing agent;

It supports combustion of carbon compounds as well as elemental carbon and phosphorus at elevated temperatures. In addition it can be reduced by $LiAIH_4$ (to hyponitrous acid), SO_2 (to N_2O), and Cr_2^+ (to NH_2OH). Nitric oxide is also reactive towards oxidizing agents. It reacts immediately with oxygen to form nitrogen dioxide

$$2NO + O_2 \rightarrow 2NO_2,$$

and with the halogens to give nitrosyl halides

$$2NO + X_2 \rightarrow 2XNO$$
$$X = F, CI, Br.$$

In the reaction with fluorine, an additional product can be formed, F_3NO. Aqueous permanganate ion converts NO quantitatively into $NO3^-$.

The physical properties of NO, indicate a rather unusual behaviour for a radical species. Contrary to the behaviour of most radicals, it is not coloured in the gaseous phase, although the pure liquid and solid are reported to be blue. The relatively high value of the entropy of vapourization (Trouton's constant = 27.1) suggests that association occurs in the liquid phase. Nitric oxide in the liquid state is paramagnetic but solid NO is diamagnetic at low temperatures suggesting that intermolecular coupling occurs under these conditions. However an unusual combination of factors found in the NO molecules confuses the obvious interpretation of the temperature dependence of the apparent magnetic moment. The odd electron has an orbital moment about the molecular axis which may be coupled with the spin momentum to give a diamagnetic ground state of the molecule. The paramagnetic excited state is about 350 cal higher in energy than the ground state. The thermal energy available at room temperature (600 cal mole^{-1}) is sufficient to keep most of the molecules in the excited spin state, which is paramagnetic. As the sample is cooled, the distribution of molecules shifts toward the diamagnetic state. Thus the temperature dependence of the magnetic susceptibility is not necessarily attributable to the dimerization process.

$$2NO \rightleftharpoons N_2O_2.$$

Nitric oxide is unusual in that its molecule contains an odd number of electrons, so the octet cannot be completed round the nitrogen atom. The structure of the molecule is a resonance hybrid of the two forms:

Only in the liquid and solid has the tendency to dimerization been observed and loose units, of the type below, been shown to exist:

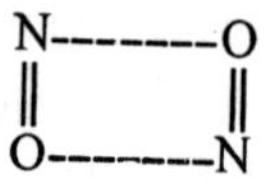

The loss of one electron from the nitric oxide molecule should produce a positive ion isoelectronic with the nitrogen molecule. This entity, the nitrosonium ion NO^+, does indeed exist in compounds such as $NO^+ClO_4^-$ and $NO^+HSO_4^-$.

Dinitrogen trioxide, N_2O_3

This oxide only exists in the pure condition as a pale blue solid at low temperatures. The liquid, which is also pale blue, and the vapour contain dinitrogen trioxide in equilibrium with nitric oxide and nitrogen dioxide, the dissociation being about 90 percent complete at room temperature:

$$N_2O_3 \rightarrow NO + NO_2$$

A convenient way of making the unstable liquid is to condense equal volumes of nitric oxide and nitrogen dioxide at –20°C. When this same mixture is dissolved in sodium hydroxide almost pure sodium nitrite is formed; Dinitrogen trioxide is thus the acid anhydride of the unstable nitrous acid.

$$NO + NO_2 + 2Na{+}OH{-}\ ?\ 2Na^{+}NO_2^{-} + H_2O$$

The structure of the oxide is thought to involve resonance between forms such as:

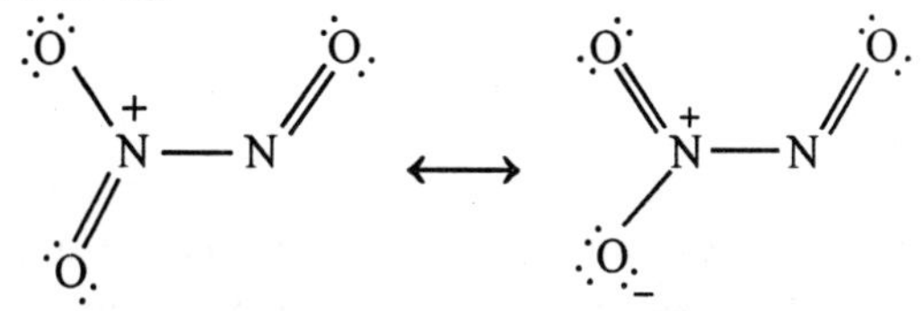

Nitrogen dioxide, NO_2 (dinitrogen tetroxide, N_2O_4)

Nitrogen dioxide exists in equilibrium with the dimer N_2O_4, the extent of the dimerization decreasing with increasing temperature. At low temperature, in the solid state, the oxide exists almost entirely as the pale yellow N_2O_4; as the temperature is increased the colour darkens as NO_2, which is black, is produced. Dissociation into NO_2 is about 90 percent complete at 100°C and 100 percent complete at 150°C, when the colour is black. Further increase in temperature results in a loss of colour as the nitrogen dioxide decomposes into nitric oxide and oxygen.

$$N_2O_4 \underset{\text{cool}}{\overset{\text{heat}}{\rightleftharpoons}} 2NO_2 \underset{\text{cool}}{\overset{\text{heat}}{\rightleftharpoons}} 2NO + O_2$$

Nitrogen dioxide can be prepared by the action of heat on most metallic nitrates, but lead nitrate is usually used since it does not contain any water of crystallization. Oxygen is evolved as well but the two gases can be separated by passing them through a freezing mixture of ice and salt; nitrogen dioxide condensed as a pale yellow liquid, b.p.22°C.

$$2Pb(NO_3)_2 \longrightarrow 2PbO + 4NO_2 + O_2$$

Nitrogen dioxide is a dense poisonous gas and very soluble in water, forming a mixture of nitrous and nitric acids, of which it is the mixed anhydride:

$$2NO_2 + H_2O \longrightarrow HNO_2 + HNO_3$$

Nitrous acid then decomposes gradually into nitric acid and nitric oxide:

$$3HNO_2 \longrightarrow HNO_3 + 2NO + H_2O$$

With alkalis it forms a mixture of the nitrite and nitrate:

$$2NO_2 + 2OH^- \longrightarrow NO_2^- + NO_3^- + H_2O$$

Because nitrogen dioxide decomposes into nitric oxide and oxygen at about 150°C it will oxidise strongly-burning substances such as magnesium, phosphorus and even charcoal. At room temperature it will oxidise hydrogen sulphide to sulphur, sulphurous acid to sulphuric acid and iodide ions to iodine, and is itself reduced to nitric oxide:

$$H_2S + NO_2 \longrightarrow NO + H_2O + S$$

$$SO_2 + H_2O + NO_2 \longrightarrow H_2SO_4 + NO$$

$$2I^- + H_2O + NO_2 \longrightarrow I_2 + 2OH^- + NO$$

Nitrogen dioxide, like nitric oxide, has a molecule containing an odd number of electrons, it structure being angular and a resonance hybrid of the two forms:

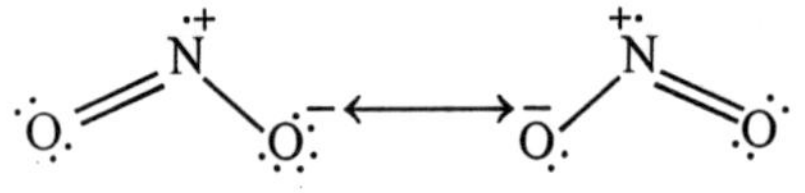

The dimer N_2O_4 has a planar structure, the odd electron on each of the nitrogen atoms of the NO_2 molecules pairing to form a weak N–N bond:

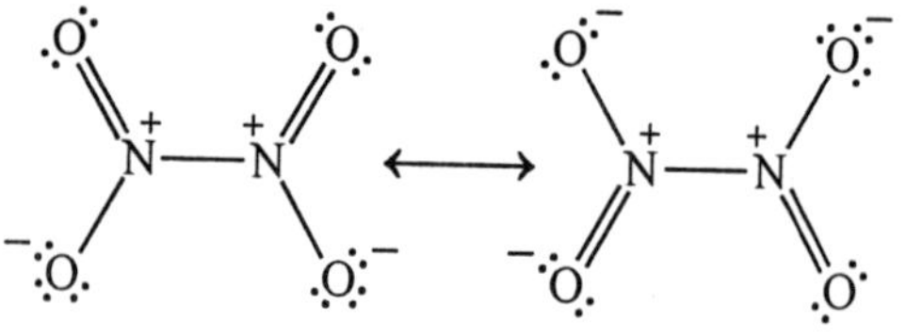

The molecule is a resonance hybrid since all four N–O bonds are equivalent.

Liquid dinitrogen tetroxide will ionise slightly into NO^+ and NO_3^- ions:

$$N_2O_4 \rightleftharpoons NO^+ + NO_3^-$$

This ionization is increased by adding substances capable of removing either the NO^+ or NO_3^- ions. Thus zinc nitrate, for example, will dissolve in the liquid to form the complex anion, $[Zn\ (NO_3)_4]^{2-}$, which is present in the nitrosonium salt:

$$(NO+)_2[Zn(NO_3)_4]^{2-}$$

Nitrogen Oxyacids

Three nitrogen-containing acids have been well characterized. Dinitrogen oxide is formally the anhydride of hyponitrous acid, which has the empirical formula HON. On the same basis, nitrous acid (HNO_2) and nitric acid (HNO_3) can be hydrolytically related to N_2O_3 and N_2O_5, respectively. There are no acids known which correspond to the other common oxides of nitrogen, NO and NO_2. These substances undergo disproportionation when they react with water to form compounds containing nitrogen in different oxidation states

$$4NO + H_2O \xrightarrow{OH^-} 2HNO_2 + N_2O$$

$$2NO_2 + H_2O \longrightarrow HNO_3 + HNO_2$$

Hyponitrous Acid (V). Hyponitrous acid ($H_2N_2O_2$) cannot be prepared by the reaction of N_2O with water, although a formal relationship between the acid and its anhydride exists. The compound is best prepared by the oxidation of hydroxylamine with nitrous acid, organic nitrates, or metal oxides such as CuO, Ag_2O, or HgO. In general the yields are small in these processes. However, a better yield of acid can be obtained by treating dry silver hyponitrite with HCI gas in ether. Filtration and cvaporation of the solution yields a white, crystalline, explosive compound. Cryoscopic measurements indicate that the acid and its esters are dimeric, $R_2N_2O_2$, and that the diethyl ester has a dipole moment near zero. These results suggest that the acid is best formulated with a –N = N– bond, the nitrogen substituents being in

RON
||
NOR

the trans orientation. Solutions of hyponitrous acid evolve N2O

$$H_2N_2O_2 \rightarrow H_2O + N_2O.$$

Hyponitrous acid is a weak dibasic acid in water ($K_1 = 9 \times 10^{-18}$, $K_2 = 1 \times 10^{-11}$ at 25°) but alkali and alkaline-earth metal salts can be prepared. The salts, like the free acids, are explosive

$$3Na_2N_2O_2 \rightarrow 2NaNO_2 + 2Na_2O + 2N_2.$$

Nitrous acid, HNO_2 and nitrites

Nitrous acid is a weakly ionized unstable acid, known only in solution. It is prepared when required by addition of dilute acid to well-cooled sodium nitrite solution:

$$H^+ + NO_2^- \rightarrow HNO_2$$

A pale blue solution is formed and brown nitrogen dioxide is observed as the acid breaks down:

$$_3HNO_2 \rightarrow HNO_3 + H_2O + {}_2NO$$
$$_2NO + O_2 \rightarrow 2NO_2$$

The dissociation constant of nitrous acid is 4.5 X IO^{-5} mole $litre^{-1}$ at room temperature, and a decimolar solution has pH = 2.17 and is 6.7% ionized.

Nitrous acid gives two series of covalent organic compounds. The reaction between sodium nitrite and an alkyl halide yields the alky1 nitrite, an ester, while silver nitrite yields a mixture of the nitrite and an isomeric nitro-compound:

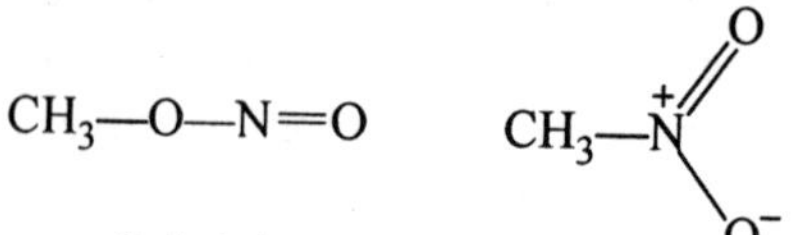

methyl nitrite **nitromethane**

Undissociated nitrous acid is a tautomeric mixture (not a resonance hybrid) of two forms:

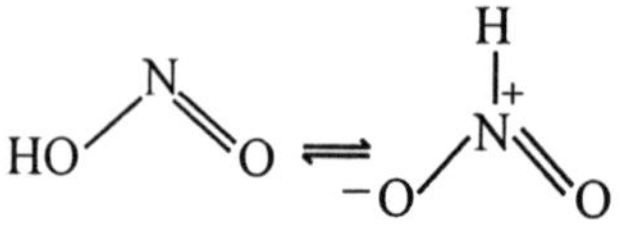

$(C_5H_5\overset{+}{N} \equiv N)\overline{C}l$ below 10°C with aromatic primary amines. Otherwise, the $-NH_2$ group is replaced by –OH. Both reactions are important in organic chemistry.

Nitrous acid may be oxidized to nitric acid, and nitrites to nitrates, by strong oxidizing agents:

$$NO_2^- + H_2O - 2e \rightleftharpoons NO_3^- + 2H$$

Nitric Acid

Nitric acid is the most important compound of nitrogen owing to its diverse applications in the national economy.

Nitric acid is used in large quantities for the manufacture of nitrogen fertilizers and organic dyes. It is used as an oxidant in many chemical processes, is employed in the manufacture of sulphuric acid by the lead-chamber process, serves for dissolving metals, for the preparation of nitrates, is used in the manufacture of cellulose varnishes, cinema film and in a number of other chemical processes. Nitric acid is also employed for the production of smokeless gunpowder and explosives used widely in mining and various earth work (construction of canals, dams, etc.), as well as for military purposes.

Industrial Preparation of Nitric Acid. Nitric acid can be manufactured by three methods which will be described in the chronological order of their use in industry.

1. *Preparation of nitric acid from Chile saltpetre.* The oldest method of preparation of nitric acid, employed as far back as the middle of the XVII century, consists in heating Chile saltpetre with concentrated sulphuric acid:

$$NaNO_3 + H_2SO_4 = NaHSO_4 + HNO_3$$

The vapours of nitric acid liberated upon heating are passed into a water-cooled receptacle, where they condense into a liquid. If gently heated in an excess of sulphuric acid, the reaction leads to the formation of the acid salt. But if saltpetre is taken in sufficient quantity and the mixture is heated more strongly, the normal salt results:

$$2\,NaNO_3 + H_2SO_4 = Na_2SO_4 + 2\,HNO_3$$

However, in this case a considerable amount of nitric acid is lost during the reaction due to decomposition. For this reason the process is adjusted to take place according to the first equation.

Up to the beginning of the present century this was the only process used commercially for the manufacture of nitric acid. It has since been almost completely replaced by other methods based on the fixation of atmospheric nitrogen.

2. *Production of nitric acid from air by the arc method.* This method, suggested by Birkeland and Eyde, and first employed industrially in Norway in 1905, is based on the direct union of nitrogen with oxygen, as described in 140. The high temperature required for this reaction is obtained by means of an electric arc fed by a powerful source of electric current. If the flame of the electric arc is brought between the two poles of a strong electromagnet (Fig. 15.3) it becomes disc-shaped, greatly increasing the surface of the flame. Such a fiery disc, up to three metres in diameter, is struck in a special furnace, built of refractory brick, in which the temperature rises as high as 3,000 or 3,500°C. Air is blown into the furnace through channels in its walls. Coming into contact with the flame of the electric arc, the air is strongly heated and part of it combines into nitric oxide.

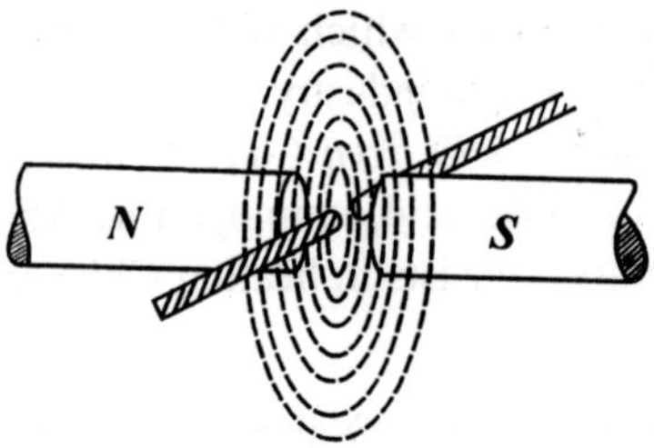

Fig. 15.3 : Electric are in magnetic field.

The gases discharged from the furnace, containing 2 to 3 percent nitric oxide, are cooled rapidly to 1,000 or 1,100°C to keep the nitric oxide formed from decomposing back into nitrogen and oxygen. When the gas mixture is further cooled, the nitric oxide combines with oxygen to form nitrogen dioxide, which is absorbed by water and changes into nitric acid.

The manufacture of nitric acid by the above method requires a great amount of electric power, and it developed mostly in Norway, which possesses vast resources of cheap electric power. By 1925 the world production of nitrogen compounds by the arc process had reached 42,000 tons per year (recalculated for nitrogen). Lately, however, the arc process has been employed on a considerably smaller scale, as it was found that nitric acid couid be manufactured more economically by the oxidation of synthetic ammonia.

3. *Production of nitric acid by oxidation of ammonia.* The most important method of nitric acid production nowadays is the catalytic oxidation of ammonia by the oxygen of the air. It was indicated, when describing the properties of ammonia, that the latter burns in oxygen, forming water and free nitrogen. But in the presence of catalysts the oxidation of ammonia may take a different course. If, for instance, a mixture of ammonia and air in definite proportions is passed through a red-hot platinum wire gauze, which serves as a catalyst, at 750°C the ammonia is transformed almost quantitatively into nitric oxide:

$$4\ NH_3 + 5\ O_2 = 4\ NO + 6\ H_2O + 215.8\ \text{Cal.}$$

The nitric oxide is easily transformed into nitrogen dioxide which combines with water to form nitric acid.

The catalytic oxidation of ammonia into nitric acid has been known for a long time, but only in the early XX century was this process used successfully for the oxidation of ammonia the catalyst was platinum. At present an alloy of platinum and rhodium containing five to ten percent rhodium is used more often. The yield of nitric oxide is 96 to 98 percent.

Fig.15.4 is a diagram of an ammonia oxidation plant operated at atmospheric pressure. Air, freed by filtration from mechanical impurities, is blown by means of fan 1 into heat exchanger 2 where it is heated by the gases discharged from the contact apparatus to a temperature of about 300°C, after which it passes into mixer 3. Fan 4 blows gaseous ammonia into the same mixer. From here the ammonia air mixture, containing 10 to 11 percent ammonia, enters contact apparatus 5, consisting of two hollow cones placed base to base. Several screens of thin platinum-rhodium wire, used as the catalyst, are fixed horizontally in the wide part of the apparatus. In the contact apparatus the ammonium is oxidized into nitric oxide. The hot gas mixture discharged from the contact apparatus goes through the heat exchanger, where it is cooled to 500-500°C, and then passes consecutively through a steam boiler (to utilize the heat), special coolers and, finally, absorption towers, in which mainly the nitric oxide is oxidized into nitrogen dioxide, and the nitric acid is formed.

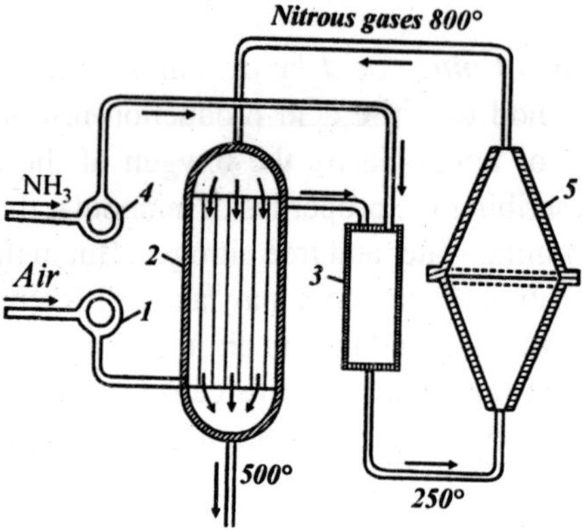

The acid obtained in the towers usually contains about 50 to 55 percent HNO_3. It is concentrated by distillation with sulphuric acid, which serves as a means of holding back the water. Besides, a method has been developed lately for the manufacture of concentrated nitric acid by the reaction between liquid nitrogen dioxide and water (or dilute nitric acid) in the presence of oxygen at 50 atm. pressure and a temperature of 75°C. The reaction proceeds according to the following summary equation:

$$2\,N_2O_4 + O_2 + 2\,H_2O = 4\,HNO_3 + 14.2\text{ Cal.}$$

By this method, called "direct synthesis," 98 percent nitric acid is obtained directly.

At present most of the vast quantities of nitric acid produced is obtained by the oxidation of ammonia. And since most of the ammonia is prepared by synthesis from the elements, this methods in the long run, like the previous one, is based on the fixation of atmospheric nitrogen.

The founder of the Russian home manufacture of synthetic nitric acid was I. Andreyev, an engineer, who in 1914 first raised the question of the organization in Russia of nitric acid production by the oxidation of ammonia obtained during the carbonizaton of coal.

Andreyev made a thorough study of the oxidation of ammonia over catalysts prepared under his supervision, and built a pilot plant. The experimental data obtained at this plant were used in designing the first Russian synthetic nitric acid plant which was started up in July 1917. The experience gained at that plant was widely used by Soviet engineers in designing new plants built during the years of the first five-year plans.

Physical properties of nitric acid

Pure nitric acid is a colourless liquid, b.p.86°C. Its vapour consists of molecules which are planar; resonance is involved:

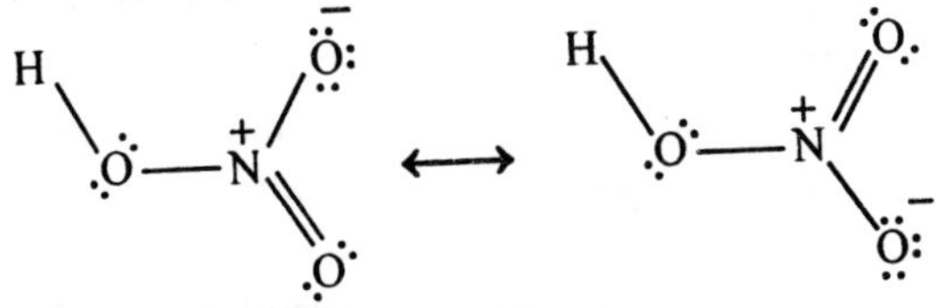

The pure liquid, however, is conducting and some ionization occurs:

$$2HNO_3 \rightleftharpoons NO_2^+ + NO_3^- + H_2O$$

cf. $$2H_2O \rightleftharpoons H_3O^+ + OH^-$$

Ionization of nitric acid is much enhanced in the presence of a strong proton acceptor, and in aqueous solution it behaves as a strong acid:

$$HNO_3 + H_2O ? H_3O^+ + NO_3^-$$

Chemical properties of nitric acid

In aqueous solution, nitric acid shows many of the reactions of a strong acid, e.g. it displaces carbon dioxide from a carbonate and neutralizes a base to form a salt and water. With the possible exception of magnesium under certain conditions, it will not react with metals to evolve hydrogen, because it is an oxidizing agent.

Concentrated nitric acid decomposes photochemically and gradually assumes a yellow colour due to the presence of dissolved nitrogen dioxide:

$$4HNO_3 \longrightarrow 4NO_2 + 2H_2O + O_2$$

The same reaction occurs, but at a much faster rate, when the acid is heated.

Concentrated nitric acid in the presence of concentrated sulphuric acid reacts with many organic compounds. For instance, glycerol reacts with this mixture to give glyceryl trinitrate (nitroglycerine); the temperature must be kept below 30°C or the reaction will become uncontrollable:

$$\begin{array}{l} CH_2-OH \\ \quad | \\ CH-OH \\ \quad | \\ CH_2-OH \\ \textbf{Glycerol} \end{array} + 3HNO_3 \longrightarrow \begin{array}{l} CH_2-O-NO_2 \\ \quad | \\ CH-O-NO_2 \\ \quad | \\ CH_2-O-NO_2 \\ \textbf{Glyceryl trinitrate} \end{array} + 3H_2O$$

Glyceryl trinitrate is an explosive which detonates violently with slight shock. It decomposes exothermically, being transformed from a liquid to a disordered mixture of gases, i.e. there is a large decrease in free energy. Dynamite is made by absorbing glyceryl trinitrate in Kieselguhr; it retains the explosive qualities of glyceryl trinitrate but is much safer to handle.

Benzene and its derivatives are nitrated by the same reagents, the reacting species being the nitronium ion, NO_2^+. It is thought that the NO_2^+ ion first attacks the benzene ring, with the formation of a positively charged complex; this then loses a proton to give the nitro compound, e.g.

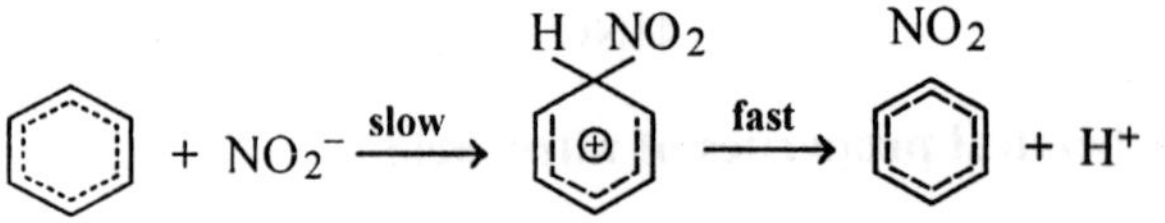

The purpose of the sulphuric acid is to increase the NO_2^+ concentration beyond that already present in the pure nitric acid; the following two ionizations have been postulated:

$$2HNO_3 \rightleftharpoons NO_2^+ + NO_3^- + H_2O$$

$$HNO_3 + H_2SO_4 \rightleftharpoons NO_2^+ + HSO_4^- + H_2O$$

If these equations are essentially correct the addition of nitrate ions, NO_3^-, should reduce the concentration of NO_2^+ by reversal of equation (1) above (the mass action effect). This is proved to be true, since nitration in the presence of nitrate ions is retarded. The existence of the NO_2^+ ion has been conclusively established in salts such as $NO_2^+ClO_4^-$.

The usual reduction products of nitric acid are nitric oxide (for the dilute acid) and nitrogen dioxide (for the more concentrated acid), although others are possible, such as nitrous oxide and ammonia. These redox reactions are discussed below.

(a) Oxidation of non-metals by nitric acid

Non-metals such as sulphur and phosphorus give their highest oxides when treated with concentrated nitric acid; these then react with the water produced to give acids:

$$S + 6HNO_3 \longrightarrow H_2SO_4 + 6NO_2 + 2H_2O$$

$$P + 5HNO_3 \longrightarrow H_3PO_4 + 5NO_2 + H_2O$$

The same type of reaction often occurs with weak metals; thus tin is oxidized to tin (IV) oxide:

$$Sn + 4HNO_3 \rightleftharpoons SnO_2 + 4NO_2 + 2H_2O$$

(b) Oxidation of metals by nitric acid

Most metals react with nitric acid to form nitrates; nitric oxide tends to predominate when the acid is dilute, and nitrogen dioxide if the concentration of the acid is increased, e.g.

$$3Cu + 8HNO3 \longrightarrow 3Cu^{2+} + 6NO_3^- + 4H_2O + 2NO \text{ (50 percent acid)}$$

$$\text{or } 3Cu \longrightarrow 3Cu^{2+} + 6e \quad \text{(oxidation)}$$

$$8HNO_3 + 6e \longrightarrow 6NO_3^- + 4H_2O + 2NO \quad \text{(reduction)}$$

$$Cu + 4HNO_3 \longrightarrow Cu^{2+} + 2NO_3^- + 2H_2O + 2NO_2 \quad \text{(conc.acid)}$$

$$Cu \longrightarrow Cu^{2+} + 2e \quad \text{(oxidation)}$$

$$4HNO_3 + 2e \longrightarrow 2NO_3^- + 2H_2O + 2NO_2 \quad \text{(reduction)}$$

In very rare instances, using very dilute nitric acid and a fairly electropositive metal, the reduction of nitric acid can proceed further to give nitrous oxide or hydrogen, e.g.

$$4Zn + 10HNO_3 \longrightarrow 4Zn^{2+} + 8NO_3^- + 5H_2O + N_2O$$
$$Mg + 2HNO_3 \longrightarrow Mg^{2+} + 2NO_3^- + H_2$$

By employing Devarda's alloy (45 percent Al, 5 percent Zn, 50 percent Cu) in alkaline solution, nitric acid and nitrates can be converted quantitatively into ammonia:

$$4Zn + NO_3^- + 7OH^- + 6H_2O \longrightarrow 4Zn(OH)_4^{2-} + NH_3$$

Nitric acid of any concentration has no effect on very unreactive metals such as gold and platinum. The concentrated acid has no effect either on aluminium, iron or chromium; this is unexpected. It is thought that the surface of these three metals is first attacked, a thin impervious oxide film being formed which stops further action; the phenomenon is referred to as passivity.

(c) Oxidation of some compounds by nitric acid

Compounds of metals that exhibit variable valency can often be oxidized from a lower to a higher valency state with nitric acid; for instance an acidified solution of an iron (II) salt is converted into the iron (III) state by dilute nitric acid:

$$3Fe^{2+} + 4H^+ + NO_3^- \longrightarrow 3Fe^{3+} + 2H_2O + NO$$

Other compounds oxidized by nitric acid include hydrogen sulphide and hydrogen chloride:

$$3H_2S + 2HNO_3 \longrightarrow 4H_2O + 2NO + 3S$$
$$3HCl + HNO_3 \longrightarrow Cl_2 + NOCl + 2H_2O$$

The former reaction is a nuisance in qualitative analysis, since the formation of sulphur is easily mistaken for a metallic sulphide and is difficult to remove; nitric acid should be avoided here whenever possible. The latter reaction is the basis of the solvent action of aqua regia (3 volumes of concentrated hydrochloric acid and 1 volume of concentrated nitric acid). Aqua regia is capable of dissolving gold and platinum, which are resistant to attack by nitric acid. It is thought that the metals are first oxidised by the nitric acid (the reaction being catalyzed by the nitrosyl chloride NOCl) and the cations are removed as fast as they are formed by complexing with chloride ions, so preventing the reverse reaction. e.g.

Uses of nitric acid

Nitric acid is used in the manufacture of nitrogenous fertilizers, e.g. ammonium nitrate and sodium nitrate. Reaction with organic compounds leads to the production of explosives, e.g. glyceryl trinitrate and trinitrotoluene, and intermediates used in the manufacture of dyes.

Aquaregia

A mixture consisting of one volume of nitric and three volumes of hydrochloric acids is called aqua regia. Aqua regia dissolves certain metals which do not dissolve in nitric acid, including gold. Its action is due to the fact that nitric acid oxidizes hydrochloric acid, liberating free chlorine and forming nitrosyl chloride NOCl:

$$\overset{+V}{HNO_3} + 3\,HCl = Cl_2 + 2\,H_2O + \overset{+III}{NOCl}$$

Nitrosyl chloride is an intermediate reaction product and decomposes into nitric oxide and chlorine:

$$2\,NOCl = 2\,NO + Cl_2$$

The chlorine liberated combines with the metals forming metal chlorides, and therefore, when metals are dissolved in aqua regia they are converted to chlorides, and not nitrates:

$$Au + 3\,HCl + HNO_3 = AuCl_3 + NO + 2\,H_2O$$

Salts of Nitric Acid. The salts of nitric acid are known under the generic name of nitrates. They are usually obtained by the action of nitric acid on the corresponding metals or metal oxides.

Nitrates are very soluble in water. When heated dry all nitrates decompose, liberating oxygen, the nitrates of the most active metals splitting off only part of the oxygen and passing into nitrites, salts of nitrous acid:

$$2\,KNO_3 = 2\,KNO_2 + O_2$$

But the nitrates of the majority of other metals decompose into the metal oxide, oxygen and nitrogen dioxide. For example:

$$2\,Cu(NO_3)_2 = 2\,CuO + 4\,NO_2 + O_2$$

Splitting off oxygen readily, nitrates are very powerful oxidants at high temperatures. On the other hand, their aqueous solutions are almost devoid of oxidizing properties. The most important are the nitrates of sodium, potassium, ammonium and calcium, which are sometimes known by the common name of *nitres*.

Sodium nitrate or sodium nitre $NaNO_3$, known also as Chile saltpetre, occurs in large quantities in nature only at one point of the globe, in the Republic of Chile, when it received its name.

Potassium nitrate or potassium nitre KNO_3, known also as saltpetre, found in nature in minor quantities, is mostly prepared artificially by the exchange reaction between sodium nitrate and potassium chloride.

Potassium nitrate is used for the manufacture of black gunpowder, which is a specially treated mixture of saltpetre, sulphur and charcoal. It usually contains 75 percent saltpetre, 10 percent sulphur and 15 percent charcoal.

The combustion of gunpowder is a complicated process involving several reactions. The gases formed during this process, of which the most important are nitrogen, carbon dioxide and carbon monoxide, occupy a volume approximately 2 thousand times as large as that of the powder burnt, which accounts for its explosive power. Besides gaseous substances, the combustion of gunpowder results in solid substances, namely, K_2CO_3, K_2SO_4, K_2S, etc., which form smoke and leave a deposit in the gun barrel. Under the action of the moisture of the air K_2S decomposes partly, liberating hydrogen sulphide. That is why the odour of hydrogen sulphide can always be detected coming out of the muzzle of a gun after using black gunpowder. Sodium nitrate is not used to prepare gunpowder as it is very hygroscopic.

Ammonium nitrate or ammonium nitre NH_4NO_3 is prepared by neutralizing nitric acid with ammonia. Is used as a fertilizer. Forms explosive mixtures, known as ammonals, with various combustible substances. Ammonals are used extensively for industrial blasting.

Calcium nitrate or calcium nitre $Ca(NO_3)_2$ has been manufacture -d in large quantities since the beginning of the present century by neutralizing nitric acid with lime; is used as a fertilizer.

Phosphorus

Although there is a great deal of similarity between phosphorus and nitrogen in the types of compounds which they form, these two elements are very different in chemical and physical properties. The greater atomic radius of the phosphorus atom makes it impossible for two phosphorus atoms to combine through a triple bond as do nitrogen atoms. One of the reasons given for the low chemical activity of elementary nitrogen is the presence of the triple bond uniting the two atoms of nitrogen. A great deal of energy is required to break this bond and free atoms for chemical reaction. Since the triple bond is absent in phosphorus molecules [A-11], much less energy is required to make individual atoms of phosphorus available for chemical reaction. As a result, phosphorus is a much more active element than nitrogen.

Occurrence of phosphorus

Like nitrogen, phosphorus is essential to the animal kingdom. The bony structure of the body contains a large proportion of calcium phosphate. Phosphorus compounds are also present in the nuclei of the cells of the body. Phosphorus is an essential constituent of the soil, being transferred to the cells of the plant. It is found dominantly in the seeds of plants.

Phosphorus is never found free in nature because of its ready combination with oxygen. The principal natural form is **phosphate rock** which is practically pure calcium phosphate, $Ca_3(PO_4)_2$. This mineral of great importance in the fertilizer industry, and each year millions tons of mined principally in the United States and North Africa

Preparation

Phosphorus is prepared by heating a mixture of phosphate rock, sand and coke to a high temperature in an electric furnace. Heat is produced by the resistance to the passage of current between the electrodes near the bottom of the furnace. The reaction probably takes place in two steps:

$$2Ca_3(PO_4)_2 + 6SiO_2 \rightarrow 6CaSiO_3 + P_4O_{10}$$

$$P_4O_{10} + 10C \rightarrow P_4 \uparrow + 10CO$$

Molten calcium silicate (slag) collects at the bottom and may be drawn off. Phosphorus leaves the furnace as a vapour; the vapour is condensed to a liquid which is further cooled and cast into sticks.

Properties

Phosphorus as prepared by this process is a soft, white, waxlike solid which gradually turns yellow in colour. When exposed to air, oxidation takes place slowly with the evolution of heat, and the element spontaneously ignites at a temperature of 40° to 45°. The glowing action of phosphorus in the dark is due to the light emitted during the slow oxidation.

The element is insoluble in water but readily dissolves in carbon disulfide and certain other organic solvents. For storage and handling it is placed under water. Great care should be exercised in handling the element; it should always be picked up with tongs or forceps, never with the fingers, since warmth from the fingers may be sufficient to raise the temperature to the kindling point and cause spontaneous ignition. Phosphorus burns and painful and slow to heal.

In contrast to nitrogen, phosphorus is fairly active, combining readily with oxygen, the halogens, and certain metals. Phosphorus burns brightly in the air, producing dense white fumes of the trioxide and pentoxide. In moist air these fumes form droplets of phosphorus and phosphoric acids which appear as a fog or mist. Because of this behaviour, the element is used in production of smoke screens in warfare.

When yellow phosphorus is heated in the absence of air to a temperature of 250-300°, it changes to a modification called *red* phosphorus. Red phosphorus is nonpoisonous in contrast to the extremely poisonous character of the yellow form; it ignites in air only at a high temperature and is insoluble in carbon disulfide.

Uses Because of its low kindling temperature, phosphorus finds use as a medium for starting fires. It is used in incendiaries and tracer bullets, and its principal use is in the manufacture of matches. At one time yellow phosphorus was used in matches; its use, however, is now prohibited, since factory workers inhaling air containing poisonous phosphorus vapour developed a dreadful disease causing decay of the bones of the jaw. A compound of phosphorus—phosphorus sesquisulfide P_4S_3—has been found to be as effective as yellow phosphorus in match manufacture, and since it is nonpoisonous, workers may handle it in safety.

The essential ingredients of a match are (1) phosphorus sesquisulfide, which ignites at a low temperature; (2) a combustible substance such as paraffin, rosin, or antimony sulfide; (3) an oxidizing agent like potassium chlorate or lead dioxide; (4) glue to bind the ingredients; (5) wood stick. For the "strike anywhere" match, the above ingredients constitute. The head; friction with a rough surface produces enough heat to ignite the phosphorus compound and initiate the reaction of the combustible substance and oxidizing agent. The flame produced is finally trasmitted to wood stick. The "safety" match is the same as the "strike anywhere" match except that it contains no phosphorus in the head. Instead, the phosphorus compound is placed on the striking surface of the box along with ground glass and glue, and friction causes a bit of the phosphorus to ignite the ingredients of the head.

Allotropic forms of Phosphorus

Phosphorus vapour consists of molecules P_4. It condenses to crystalline white phosphorus, which probably also consists of molecules P_4. White phosphorus is metastable. There are at least four forms which are more stable, to which white phosphorus changes under appropriate conditions. Among these are violet phosphorus of which the molecular constitution is unknown, black phosphorus which has a giant molecule structure, and two other forms known as red phosphorus and scarlet phosphorus.

White phosphorus is metastable with regard to the other allotropes. The heats of solution of the three allotropes in bromine show the differences in their energy content:

	Specific gravity	Heat of solution per gm atom in bromine
Black phosphorus	2.70	38.30 kcal
Violet phosphorus	2.34	42.50 kcal
White phosphorus	1.83	59.48 kcal

White phosphorus is poisonous, and chemically it is extremely active.

If cubic white phosphorus is cooled to—76.9°, or if it is subjected to a pressure of 12,000 atm, a hexagonal crystalline form is obtained. The cubic and hexagonal crystals of white phosphorus are probably polymorphs, in both of which the P_4 molecule is present.

Violet phosphorus, or α-metallic phosphorus, is chemically inert. In air it neither glows nor ignites, and it is not oxidized. There is no solvent for it and hence it is non-toxic. It is a non-conductor of electricity.

Black phosphorus, or β**-metallic phosphorus,** is chemically inert. In does not ignite at 400° in air. It conducts electricity.

Black phosphorus has a layer structure in which each atom forms three strong bonds with its nearest neighbours. The figure shows one layer idealized. The actual bond angles are 100° and not 90° as shown

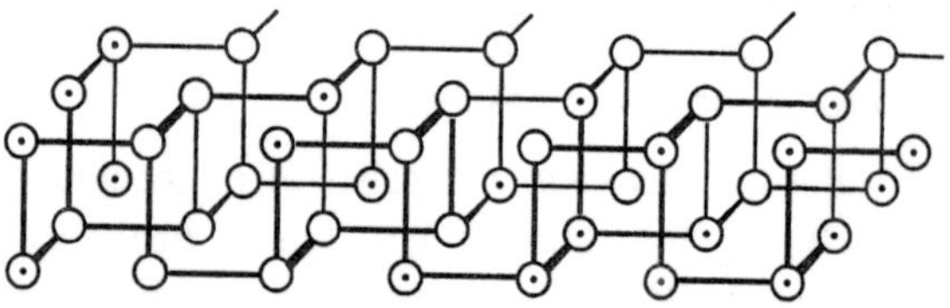

Diagram

It heated to 550° black phosphorus changes to the red variety.

Red phosphorus is tasteless, smell-less, and non-toxic. It is chemically inert compared with white phosphorus. Red phosphorus volatiles to molecules P_2. Red phosphorus sublimes at 413°, but volatilized red phosphorus condenses to white phosphorus. The volatilization of red phosphorus proceeds irreversible because the proportion of P_2 in equilibrium with P_4 is less than one part in 10^7.

It is possible that red phosphorus is a modification of violet phosphorus, since on prolonged heating the colour darkens to violet, and the sp. gr. rises to 2.34, and the m.ps. of the two varieties are the same.

Scarlet phosphorus is said to be obtained pure by heating PBr_3 with mercury at 240°;

$$2PBr_3 + 3Hg = 2P + 3HgBr_2$$

It is probably another modification of violet phosphorus. It is oxidized only very slowly and is non-poisonous. It dissolves in alkali with the evolution of phosphine.

The phosphorus molecules

The molecular weight of phosphorus vapour below 700° corresponds to the molecule P_4. Electron diffraction experiments have established that each atom of the molecule is at the corner of a regular tetrahedron. The distance P—P is 2.21. each phosphorus atom appears to be exerting three bonds at 60° to one another.

Above 700° the molecule begins to dissociate into molecules P_2; 500° only about 1 in 10^7 molecules in the vapour is P_2, but at 1700° about 50% of the P_4 molecules are dissociated. At higher temperatures some phosphorus is present as single atoms. $P_4 = 2P_2$; ΔH=30.0 kcal. P2 = 2P; ΔH = 42.2 kcal. The bond energy P—P is 53 kcal.

Molecular weight determinations show that phosphorus is present at P_4 in its solutions in benzene and in carbon disulphide. The results of Raman spectra and X-Ray experiments show that liquid phosphorus consists of P_4 molecules.

From the experimental evidence available it seems that the molecule P_4 is present in crystalline, liquid, and gaseous white phosphorus. The structure of the molecule P_2 is unknown. The phosphorus atom does not form simple π bonds, and hence it seems unlikely that P_2 is the analogue of the nitrogen molecule $N \equiv N$.

Phosphate Fertilizers

The salts of phosphoric acid are of especially great importance in agriculture. As was mentioned above, phosphorus is contained in the proteins of plants. Therefore, it is just as important an element for plants as nitrogen. Plants receive their phosphorus from the soil, in which it is contained as phosphoric acid salts. But the phosphorus content in soils is very low and a deficiency lowers the crop yield, especially of grain cultures and tubers (sugar beet). To increase the crops various phosphate fertilizers are introduced into the soil.

Natural phosphorus compounds, phosphorites and apatites, contain phosphorus in the form of the insoluble tertiary phosphate $Ca_3(PO_4)_2$, which is difficult for plants to assimilate. To obtain readily assimilated fertilizers, phosphorites are treated chemically to convert the normal salt into an acid one. This is how the most important phosphate fertilizers, called superphosphate and precipitate, are prepared.

To produce *superphosphate*, finely ground natural phosphorite is mixed with enough sulphuric acid to total two molecules of H_2SO4 for every molecule of $Ca_3(PO_4)_2$. The mixture is vigorously mixed and charged into special continuous action chambers, where the following reaction is completed:

$$Ca_3(PO_4)_2 + 2H_2SO_4 = 2CaSO_4 + Ca(H_2PO_4)_2$$

The reaction results in a mixture of gypsum and the primary phosphate $Ca(H_2PO_4)_2$, which is comparatively soluble in water. This mixture, ground or granulated, is known as superphosphate.

The manufacture of superphosphate is one of the largest branches of the basic chemical industry, closely connected with the production of sulphuric acid. Most of the sulphuric acid produced is used for the manufacture of superphosphate. The output of superphosphate only in the capitalist countries amounted to 22 million tons in 1953.

Precipitate is a phosphate fertilizer containing secondary calcium phosphate $Ca_2(HPO_4)_2$ or $CaHPO_4$, which is insoluble in water but dissolves in the acids contained in soils.

To prepare precipitate free phosphoric acid is first evolved from phosphorite by treating the latter with a greater quantity of sulphuric acid than is necessary for the production of superphosphate :

$$Ca_3(PO_4)_2 + 3H_2SO_4 = \downarrow 3CaSO_4 + 2H_3PO_4$$

The phosphoric acid solution is decanted from the precipitate, which contains gypsum and other insoluble admixtures, and milk of lime, i.e., a suspension of lime in water, is added to it in sufficient quantity to form the secondary phosphate :

$$H_3PO_4 + Ca(OH)_2 = \downarrow CaHPO_4.H_2O + H_2O$$

The crystalline precipitate is separated from the liquid and carefully dried, so as not to drive off the water contained in the crystals. This salt is assimilated readily by plants if it has not lost its water of crystallization.

A third type of phosphate fertilizer, produced lately in large quantities, are the ammonium phosphates, known as *ammophoses*. The most important of these is *diammophos* $(NH_4)_2HPO_4$. Ammophoses are complex fertilizers, containing nitrogen besides phosphorus.

Before the Revolution the phosphate fertilizer industry was in its embryonic stage in Russia. In 1913 the output of the few existing superphosphate plants totalled only 70 or 80 thousand tons.

The construction of new plants began only in 1925-26 and immediately assumed a large scale. During the pre-war five-year periods a number of large plants for the production of phosphate fertilizers were erected. By 1938 the U.S.S.R. occupied the first place in Europe for the production of superphosphate.

The production of mineral fertilizers now occupies first place in the output of the chemical industry of the U.S.S.R.

In 1955 the total output of all mineral fertilizers (including phosphate fertilizers amounted to 906 million tons. It will grow even higher during the Sixth Five-Year Plan. The output of various mineral fertilizers in 1960 will reach 19.6 million tons, including a considerable increase in the production of the most effective fertilizers. For instance, the yearly output of concentrated phosphate fertilizers will be about one million tons.

The hydrides of phosphorus

Phosphine, PH_3, the phosphorus analogue of ammonia, can be obtained by heating a concentrated solution of sodium hydroxide with white phosphrous:

$$P_4 + 3OH^- + 3H_2O \rightarrow PH_3 + \underset{\text{hypophosphiteion}}{3H_2PO_{2^-}}$$

It is evolved as a very poisonous gas with an unpleasant smell. As made by the above method it is spontaneously inflammable, forming phosphorus pentoxide contact with air; this is due to the presence of P_2H_4 of phosphorus vapour impurities. When the gas is pure it is not spontaneously inflammable, but it will readily oxidize to phosphorus pentoxide when ignited:

$$4PH_3 + 8O_2 \rightarrow P_4O_{10} + 6H_2O$$

Phosphine can also be obtained by hydrolysis of calcium phosphide:

$$P^{3-} + 3H_2O \rightarrow PH_3 + 3OH^-$$

The phosphine molecule is pyramidal like the molecule of ammonia but, unlike ammonia, it is only slightly soluble in water. This is probably due to the fact that the phosphorus atom is not sufficiently electronegative to hydrogen bond to the hydrogen of a water molecule. The presence of hydrogen bonding in ammonia and its absence in phosphine also explains why phoshpine, which is a heavier molecule than ammonia, has a lower boiling point.

The lone pair of electrons of the phosphine molecule can bond to a proton to form the phosphonium ion, PH_{4+}, which is tetrahedral like the ammonium ion e.g.

$$PH_3 + HI \rightarrow \underset{\text{phosphonium iodide}}{PH_{4^+}I^-}$$

Phosphonium iodide, which is one of the most stable salts containing the phosphonium ion, is less thermally stable than ammonium salts because phosphine is less basic than ammonia, and decomposition occurs at about 60°C:

$$PH_{4^+}I^- \rightarrow PH_3 + HI$$

Unlike ammonium salts it is also decomposed by water and this is a convenient method of making pure phosphine:

$$PH_{4^+}I^- + H_2O \rightarrow PH_3 + H_3O^+ + I^-$$

Phosphine is a much more powerful reducing agent than ammonia, converting silver and copper (II) salts in solution to their phosphides. Which subsequently react to give the free metals, e.g.

$$6Ag^+ + PH_3 + 3H_2O \rightarrow 6Ag + H_3PO_3 + 6H^+$$

Disphosphine, P_2H_4, the phosphorus analogue of hydrazine, is formed in trace amounts during the preparation of phosphine. As mentioned above it is spontaneously inflammable; it is also devoid of basic properites.

Phosphorous trihalides, PX_3

All the four trihalides exist and, except for the triflouride, can be made by direct synthesis. The triflouride is a gas, the trichloride and tribromide are liquids, and the tri-iodide is a solid. Their molecules have the expected pyramidal structure.

Phosphorus trifluroide, PF_3, a colourless gas, is made:

(i) by the action of arsenic trifluoride on phosphorus trichloride:

$$PCI_3 + AsF_3 = PF_3 + AsCI_3$$

(ii) by warming phosphorus tribromide with zinc fluroide:

$$2PBr_3 + 3ZnF_2 = 2PF_3 + 3ZnBr_2,$$

(iii) by heating copper phosphide with lead fluoride:

$$2Cu_3P + 3PbF_2 = 2PF_3 + 6Cu + 3Pb.$$

The gas is hydrolyzed by water to hydrofluoric acid and phosphorous acid, but it does not fume in air. It does not attack glass except at high temperatures. It is decomposed by sparking:

$$5PF_3 = 3PF_5 + 2P.$$

A mixture of the gas and oxygen explodes when sparked:

$$2PF_3 + O_2 = 2POF_3,$$

but phosphorus trifluoride does not combine directly with sulphur. Phosphorus trifluoride can replace carbon monoxide in transition mental carbonyls, and it reacts with haemoglobin.

Phosphorus trichloride, PCI_3, a colourless liquid, is made by passing a stream of dry chlorine over white or red phosphorus (not heated), and condensing the product in an ice-cold receiver. The apparatus is filled with carbon dioxide before the introduction of the phosphorus, and water is excluded by the use of calcium chloride tubes. The crude product is contaminated with chlorine, which is removed by standing the liquid over white phosphorus. Phosphorus trichloride is finally purified by distillator. Phosphorus trichloride can also be obtained by passing gaseous hydrogen chloride over phosphorous oxide P_4O_6:

$$6HCI + P_4O_6 = 2H_3PO_3 + 2PCI_3.$$

Phosphorus trichloride fumes in air and reacts violently with water:

$$PCI_3 + 3H_2O = H_3PO_3 + 3HCI.$$

If phosphorus trichloride is treated with water in the presence of iodine it yields hypophosphoric acid. If treated with ammonia at low temperatures phosphorus trichloride forms PCI_3, $5NH_3$, which on heating gives ultimately $(P_2N_5)_x$. Phosphorus trichloride behaves as an unsaturated compound, talking up chlorine to give PCI_5, oxygen to give $POCI_3$, and sulphur to give $PSCI_3$. It also acts as a reducing agent when treated with:

(i) sulphur trioxide:

$$PCI_3 + SO_3 = POCI_3 + SO_2,$$

(ii) hot concentrated sulphuric acid:

$$PCI_3 + 2H_2SO_4 = SO_3HCI + SO_2 + 2HCI + HPO_3,$$

(iii) sulphur chloride:

$$3PCI_3 + S_2CI_2 = PCI_5 + 2PSCI_3.$$

When phosphorus trichloride is oxidized the 3 co-ordinate phosphorus atom becomes 4 co-ordinate, as in the compounds $[PCI_{4^+}][PCI_{6^-}]$, $POCI_3$, $PSCI_3$, or HPO_3.

The chlorine atoms in phosphorus trichloride can be successively replaced by fluorine by the action of AsF_3, SbF_3, or SbF_5, or by passing the vapour over heated calcium fluoride. The chlorine atoms can be replaced by iodine by the action of potassium iodide.

Phosphorus tribromide, PBr_3, a colourless liquid, is formed by the action of bromine on red phosphorus: the reaction is violent. It fumes in the air. If oxygen is passed into boiling phosphorus tribromide, combination to form $POBr_3$, takes place very easily, and an explosion may result. With excess bromine, phosphorus tribromide forms the yellow pentabromide.

If hydrolyzed in the presence of iodine it yields hypophosphoric acid.

Phosphorus tri-iodide, PI_3, a dark red solid, is made by the action of iodine on white phosphorus in carbon disulphide solution, or by the action of potassium iodide on phosphorus trichloride. It is hydrolyzed by water to phosphorous acid.

The pentahalides, PX_5

Except for the pentaiodide, all the pentahalides have been prepared. Phosphorus pentafluoride is a gas; the pentachloride and pentabromide are solids.

Phosphorus pentachloride is prepared by passing chlorine through a flask into which phosphorus trichlorides is dripping. Since it dissociates into the trichlorides and chlorine very readily, the experiment is conducted in an ice-cooled apparatus.

$$PCl_3 + Cl_2 = PCl_5$$

Like the trichloride it is attacked by compounds containing the hydroxyl group, e.g.

$$PCl_5 + H_2O \rightarrow POCl_3 + 2HCl$$

$$POCl_3 + 3H_2O \rightarrow H_3PO_4 + 3HCl$$

$$CH_3COOH + PCl_5 \rightarrow CH_3COCl + POCl_3 + HCl$$

In the vapour state the phosphorus pentachloride molecule has a trigonal bipyramidal structure; in the solid state it is ionic, having the structure $(PCl_{4^+})(PCl_{6^-})$

Phosphorus pentafluoride, PF_5, is a colourless gas. It is made:

(i) by heating a mixture of phosphoric oxide and calcium fluoride:

$$6P_2O_5 + 5CaF_2 = 2PF_5 + 5Ca(PO_3)_2,$$

(ii) by mixing arsenic trifluroide with phosphorus pentachloride in a freezing mixture:

$$3PCl_5 + 5AsF_3 = 3PF_5 + 5AsCl_3,$$

(iii) by heating phosphorus pentachloride with the fluoride of a divalent metal (Ca, Ba, Zn, Pb); if an ammonium or alkaline fluoride is used the complex $M(PF_6)$ is formed instead,

(iv) by warming phosphorus fluorobromide to 15°:

$$5PF_3Br_2 = 3PF_5 + 2PBr_5.$$

Phosphorus pentafluoride fumes in air and is decomposed by water:

$$PF_5 + H_2O = POF_3 + 2HF$$

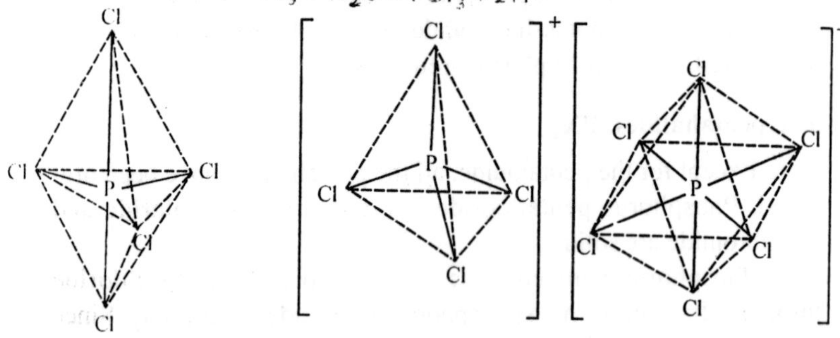

Fig. 15.5.

(a) *The structure of phosphorus pentachloride vapour*
(b) *The structure of solid phosphorus pentachloride*

Phosphorus pentabromide, PBr_5, exists as yellow crystals, which undergo partial dissociation on melting. It is made by allowing phosphorus tribromide to react with bromide. With excess of bromine it forms a heptabromide, PBr_7, which is probably a polybromide $[PBr_{4+}][Br_{3-}]$.

Oxides of Phosphorus, Arsenic and Bismuth

They form fewer oxides than does nitrogen, presumably because of the inability of these elements to form $p\pi$-$p\pi$ double bonds.

Trioxides

Phosphorus trioxides is dimeric and should be written P_4O_6, not P_2O_3. P_4O_6 has four P atoms at the corners of a tetrahedon, with six O atoms along the edges, each O being bonded to two P atoms. The structures of As_4O_6 and Sb_4O_6 are similar to this. Bi_2O_3 is ionic. The structure of P_4O_{10} is shown in Fig. 14.10. Since the P—O—P angle is 127° the O atoms are strictly above the edges, but it is more convenient to draw them on the edges.

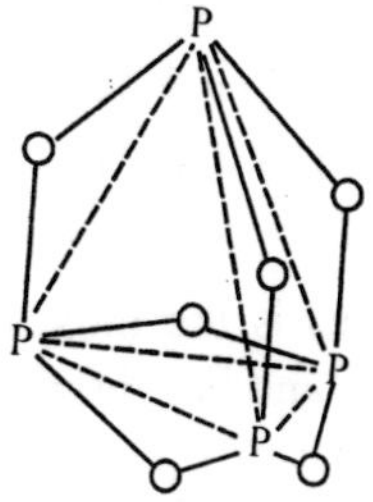

Fig. 15.6 : Structure of phosphorus trioxide P_4O_6.

Because yellow phosphorus is more reactive than is N_2, phosphorus oxides (unlike nitrogen oxides) can all be obtained by burning phosphorus in air.

$$P_4 + 3O_2 \xrightarrow{\text{limited supply of air}} P_4O_6$$

P_4O_6 is formed by burning phosphorus in a limited supply of air. It is a soft white solid (m.p. 24°C, b.p. 175°C). It is removed from the reaction mixture and is purified by distillation. (Higher oxides are formed in a plentiful supply of air.) P_4O_6 will burn in air, forming P_4O_{10}.

$$P_4O_6 + 2O_2 \rightarrow P_4O_{10}$$

As_4O_6 and Sb_4O_6 are obtained by burning the metals in air or dioxygen, since they have less tendency to form higher oxides. Heating the sulphide minerals As_4S_4 (realgar) or As_2S_3 (orpiment) in air also gives As_4O_6. Both As_4O_6 and Sb_4O_6 are very poisonous. Bi_2O_3 is not dimeric like the others.

The basicity of oxides and hydroxides usually increases on descending a group. P_4O_6 is acidic and hydrolyses in water, forming phosphorus acid. (This is considered in more detail later.) Arsenious oxide As_4O_6 is sparingly soluble in water, and Sb_4O_6 is insoluble, As_4O_6 and Sb_4O_6 are both amphoteric since they react with alkali. forming arsenites and antimonites, and with concentrated HCl.

forming arsenic and antimony trichlorides. In the past, various copper arsenites were used as brilliant green pigments. The best known are Scheele's green $Cu_2As_2O_5$ and Paris green $[(CH_3COO)Cu_2(AsO_3)]$. They are seldom used nowadays because they are toxic, and even worse, in damp places bacteria and moulds can produces poisonous volatile substances such as AsH_3 and $As(CH_3)_3$. Bi_2O_3 is wholly basic

$$P_4O_6 + 6H_2O \rightarrow 4H_3PO_3$$

$$As_4O_6 + 12NaOH \rightarrow 4Na_3AsO_3 + 6H_2O$$

$$As_4O_6 + 12HCI \rightarrow 4AsCI_3 + 6H_2O$$

Pentoxides

Phosphorus pentoxides is the most important oxides, and is quite common. It is dimeric and has the formula P_4O_{10}, not P_2O_5. Its structure is derived from that of P_4O_6. Each P atom in P_4O_6 forms three bonds to O atoms. There are five electrons in the outer shell of a P atom. Three electrons have been used in bonding, and the other two comprise a lone pair, which is situated on the outside of the tetrahedral unit. In P_4O_{10} the lone pairs on each of the four P atoms form a coordinate bond to an oxygen atom.

Measurement of the P—O bond lengths shows that the bridging bonds on the edges are 1.60Å but the coordinate bonds on the corners are 1.43Å. Thr bridging bonds compare with those in P_4O_6 (1.65Å) and are normal single bonds, The bonds on the corners are much shorter than a single bond, and are in fact double bonds. These double bonds are different in origin from the 'usual' double bonds such as that in ethene which arises from $p\pi$-$p\pi$ overlap with one electron coming from each C atom. The second bond in P = O is formed by $p\pi$-$d\pi$ back bonding. A full p orbital on the O atom overlaps sideways with an empty *d* orbital on the P atom. Thus it differs from the double bond in ethene in two respects:

1. A *p* orbital overlaps with a *d* orbital, rather than *p* with *p*.
2. Both electrons come from one atom, and hence the bond is a 'dative bond'.

A similar type of back bonding is founds in the carbonyls.

Phosphoric oxide, P_4O_{10}, is a white solid. It is made by burning phosphorus in excess of air or oxygen. The crude product may contain some P_4O_6. This can be removed by sublimation in a current of ozonized oxygen.

The solid is polymorphic. The hexagonal form is obtained when phosphorus is burnt. On heating, this form changes successively into two other forms of increasing density and diminishing volatility.

Phosphoric oxide absorbs moisture giving first cyclotetra-phosphoric acid and then the pyro-and orth- phosphoric acids. All forms absorb water vapour at the same rate, but the hexagonal form is usually employed as a drying agent. It is more efficient than any other drying agent, and is an efficient at 90° as at 0°. Phosphoric oxide is used for making acid oxides by withdrawing the elements of water from the parent acid. For example:

$$H_2SO_4 - H_2O = SO_3,$$

$$2HNO_3 - H_2O = N_2O_5,$$

$$2HClO_4 - H_2O = Cl_2O_7$$

On the addition of liquid water to phosphoric oxide the hexagonal form reacts violently, the orthorhombic very slowly even at 100°, and the tetragonal form produces a gel which slowly liquefies. Phosphoric oxide reacts with ethyl alcohol thus:

$$P_4O_{10} + 6C_2H_5OH = 2(C_2H_5O)_2PO.OH + 2C_2H_5OPO(OH)_2.$$

Phosphoric oxide is very stable, even at high temperatures. The formula P_4O_{10} (mol. wt. 142) is approximately in accordance with the experimental value of the molecular weight of the vapour (150) between 670° and 1100°. On evidence from electron diffraction the structure is similar to that P_4O_6, with an extra oxygen atom on each phosphorus atom.

P_2O_6. If a mixture of phosphoric oxide vapour and oxygen is sparked, a deep violet solid of high oxidizing power is formed. This contains up to 5% of an oxide P_2O_6. If heated to 130° the solid is decolourized, and loses oxygen and oxidizing power, Aqueous solution of the solid oxidise potassium iodide to iodine, and the manganous ion to the permanganate ion. The reaction of P_2O_6 suggest that it is the anhydride of peroxydiphosphoric acid, and that its structure is:

Other oxides

The oxides P_4O_7, P_4O_8 and P_4O_9 are very uncommon. In these oxides one molecule contains P atoms in both the oxidation states (+III) and (+V). P_4O_7 is best made by dissolving P_4O_6 in tetrahydrofuran and reacting it with the correct amount of dioxygen. Heating P_4O_6 under vacuum in a sealed tube gives a mixture of red phosphorus and the oxides P_4O_7, P_4O_8 and P_4O_9. These oxides have structures in between those of P_4O_6 and P_4O_{10}, in that they have one, two or three apical O atoms attached to P atoms. Hydrolysis with water thus yields a mixture of oxoacids in both oxidation states, phosphoric acid P(+V) and phosphorous acid P(+III).

$$\left.\begin{matrix} P_4O_8 \\ P_4O_9 \end{matrix}\right] \xrightarrow{+H_2O} H_3PO_4 + H_3PO_3$$

orthophosphoric acid   orthophosphorous acid

The Oxyacids of Phosphorus

The oxyacids of phosphorus may be grouped in three classes.

(i) Acids having molecules which contain one atom of phosphorus which is in the tetrahedral valence state. The acids in this class are orthophosphoric acid, H_3PO_4, phosphorous acid, H_3PO_3, and hypophosphorus acid, H_3PO_2. If the acids are written in the ionized form the formulae are:

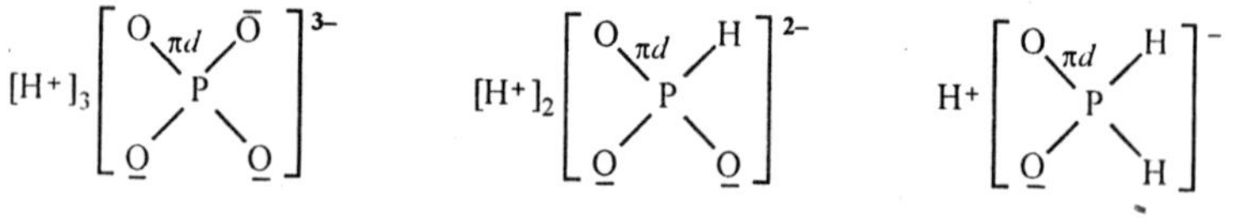

The series may be continued, thus:

phosphonium iodide

In all five compounds the valence state of the phosphorus atom is the same; oxidation of, say, phosphorus acid to phosphoric acid consists in the replacement of a hydrogen atom by an oxygen atom and not in any change in the state of the phosphorus atom.

(ii) Hypophosphoric acid, cyclo-phosphoric acids.

(iii) Di-, tri-, and cyclo-phosphoric acids.

It will be convenient to give accounts of the acids in the order hypophosphorus acid, phosphorus acid, hypophosphoric acid, and orthophosphoric acid, and orthophosphoric acid; the poly-nuclear phosphoric acids can then be discussed immediately after orthophosphoric acid.

Hypophosphorus acid, H_3PO_2, is a solid, m.p. 27°. It is forrmed by the following methods:

(i) The oxidation of phosphine by iodine and water. The reaction is said to be quantitative:

$$PH_3 + 2I_2 + 2H_2O = H_3PO_2 + 4HI.$$

The hydriodic acid may be removed by evaporation.

(ii) The decomposition of barium hypophosphite with the calculated quantity of sulphuric acid. After filtering off the $BaSO_4$ the solution is evaporated at 130°, and cooled in a freezing mixture, when it crystallizes. The barium hypophosphite is made by heating white phosphorus with barium hydroxide. Excess of barium hydroxide is removed by carbon dioxide, and the filtrate is evaporated until barium hypophosphite crystallizes.

Hypophosphorus acid undergoes disproportionation on when heated to 130°:

$$3H_3PO_2 = 2H_3PO_3 + PH_3,$$

and at 200° the phosphorous acid further disproportionates into phosphine and phosphoric acid.

Hypophosphorus acid can be reduced by dilute sulphuric acid and zinc to phosphine, but normally it is a strong reducing agent. It is used in the reduction of aromatic diazonium compounds to the parent aromatic compound, for example:

$$H_2O + C_6H_5N_2Cl + H_3PO_2 = H_3PO_3 + C_6H_6 + N_2 + HCl.$$

Hypophosphorus acid reduces solutions of the salts of Au, Ag, Pt, Hg, Bi, to metal, and precipitates an interstitial hydride of copper from the solution of a cupric salt:

$$3H_3PO_2 + 3H_2O + 2CuSO_4 = 2Cu,H + 3H_3PO_3 + 2H_2SO_4.$$

Hypophosphorus acid is monobasic. The acid and its salts (with the exception of the bismuth and thorium salts) are easily soluble in water, and the alkali metal salts are soluble in alcohol too. The rubidium and caesium salts are hygroscopic. They decompose at 140°. The salts when boiled with alkali evolve hydrogen:

$$NaH_2PO_2 + 2NaOH = 2H_2 + Na_3PO_4.$$

The monobasicity of hypophosphorus acid indicates that one hydrogen atom in the molecule is differentiated from the other two. The results of the determination of the crystal structure of the nickel salt [Ni, $2H_2O$][H_2PO_2]$_2$ show that the anion is tetrahedral. The structure assigned to the acid is:

Structure 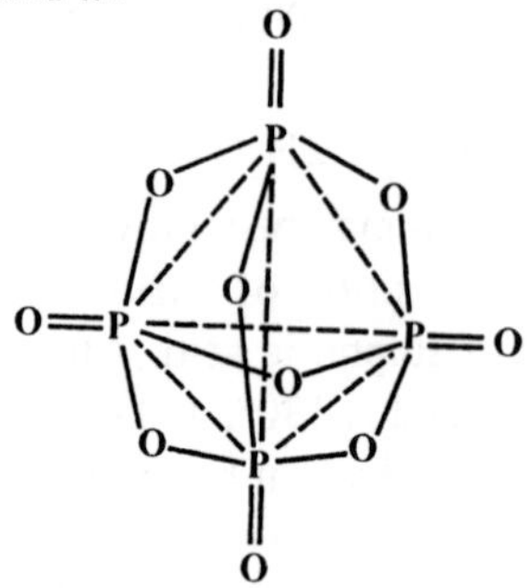

P_4O_{10}

Orthophosphorus acid, H_3PO_3

This is the most important acid based on phosphorus trioxide, and can be obtained from the latter by the addition of water:

$$P_4O_6 + 6H_2O \rightarrow 4H_3PO_3$$

It can also be obtained by hydrolysis of phosphorus trichloride:

$$PCl_3 + 3H_2O \rightarrow H_3PO_3 + 3HCl$$

The pure acid is a deliquescent solid and in aqueous solution it is moderately strong. Since is dibasic its structure is thought to contain only two hydroxyl groups, although its formation by hydrolysis of phosphorus trichloride suggests three. It is thought that an equilibrium exists of the type shown below, which is very largely displaced to the right:

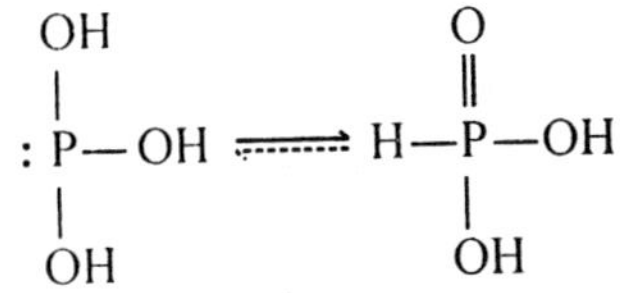

The phosphorus atom is therefore essentially 5-covalent.

The Phosphoric Acids

There are three principal acids, orthophosphoric acid, H_3PO_4, pyrophosphoric acid, $H_4P_2O_7$, and metaphosphoric acid, HPO_3. Since about 1940 our knowledge of the properties of the metaphosphates has been extended by a number of studies, and something has been perceived of the variability and complexity of the constitutions of the substances previously described by this name. Much of the evidence on the constitution of the metaphosphates cannot be regarded as conclusive, but the deductions which can be made from it are consistent, and in accord with our knowledge of the silicates and other oxyacids. The classical nomenclature does not satisfactorily express the relations between the acids suggested by recent work. In the scheme below the strucutres of the acids are drawn, and the suggested new names are attached to them. There are two principal basic structures: (i) strucutres of unbranched chains composed of PO_4 tetrahedra, linked by oxygen atoms held in common,

(ii) cyclic structures in which either three or four PO_4 tetrahedra are linked in the form of a ring, again by oxygen atoms held in common.

Structures of the phosphoric acids in the non-dissociated state:

Unbranches chains:

H_3PO_4

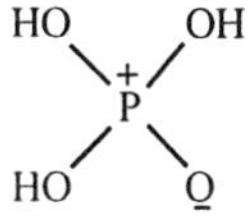

Orthophosphoric acid

Orthophosphoric acid

$H_4P_2O_7$

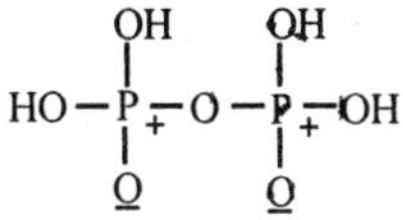

Diphosphoric acid + Pyrophosphoric acid

Diphosphoric acid + Pyrophosphoric acid

$H_5P_3O_{10}$

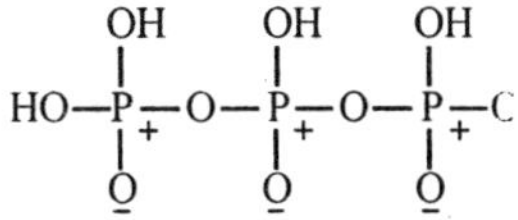

Triphosphoric acid

Triphosphoric acid

The general formula for this series is $H_{4+n}P_{2+n}O_{7+3n}$. *n* is written 0 for diphosphoric acid, and -1 for orthophosphoric acid. ‡

Cyclic strucutures:

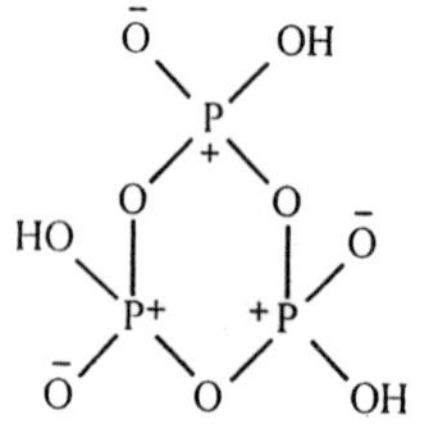

Cyclotriphosphoric acid +
Trimetaphosphoric acid

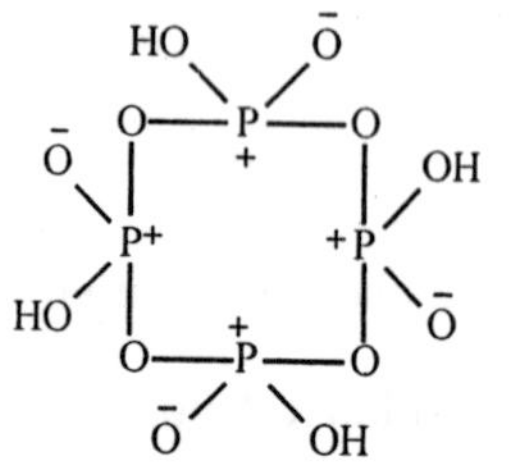

Cyclotetraphosphoric acid +
Tetrametaphosphoric acid

Cyclotriphosphoric acid†
Trimetaphosphoric acid

Cyclotetraphosphoric acid†
Tetrametaphosphoric acid.

The alkali metal salts of orthophosphorus acid, e.g. $Na^{+}H_2PO_{3^-}$ and $(Na^{+})_2HPO_{3^{2-}}$, are soluble in water; the majority of other salts are insoluble. The acid and its salts are strong reducing agents, the free acid decomposing into phosphine—which is a powerful reducing agent itself—

$$4H_3PO_3 \rightarrow PH_3 + 3H_3PO_4$$

Orthophosphoric acid, H_3PO_4,

Orthophosphoric acid is manufactured by heating rock phosphate with sulphuric acid; a purer product is obtained by reacting phosphorus pentoxide with water and then boiling.

$$(Ca^{2-})_3(PO_{4^{3-}})_2 + 3H_2SO_4 \rightarrow 3Ca^{2+}SO_{4^{2-}} + 2H_3PO_4$$

$$P_4O_{10} + 6H_2O \rightarrow 4H_3PO_4$$

Orthophosphoric acid is a deliquescent crystalline soild, although it is usually encountered as a concentrated and very viscous aqueous solution; this high viscosity is due to the presence of hydrogen bonding which links individual molecules of the acid into very large aggregates.

The acid is tribasic, and relatively weak, neutralization with alkalis taking place in stage. When an M/10 solution of sodium hydroxide is added to an M/10 solution of the acid, methyl orange changes colour at the stage corresponding to conversion into sodium dihydrogen orthophosphate, $Na^+H_2PO_4^-$:

$$Na^+OH^- + H_3PO_4 \rightarrow H_2O + Na^+H_2kPO_4 \ (pH\ 4.4)$$

If phenolphthalin is used as indicator, twice as much alkali is needed before the indicator changes colour and this corresponds to the formation of disodium hydrogen orthophosphate, $(Na^+)_2HPO_4^{2-}$:

$$2Na^+OH^- + H_3PO_4 \rightarrow 2H_2O + (Na^+)_2HPO_{4^{2-}} \ (pH\ 9.6)$$

The third stage leading to the formation of trisodium orthophosphate, $(Na^+)_3PO_{4^{3-}}$ cannot be realized in practice since this salt is extensively hydrolyzed in aqueous solution, i.e. the orthophosphate ion, PO_4^{3-}, is a strong base:

$$(Na^+)_3PO_4^{3-} + H_2O \rightleftharpoons Na^+OH^- + (Na^+)_2HPO_4^{2-}$$

or $$PO_4^{3-} + H_2O \rightleftharpoons OH^- + HPO_4^{2-}$$

Most phosphates are insoluble in water but they usually dissolve in the presence of a strong acid such as hydrochloric acid. This is due to the formation of weakly dissociated phosphoric acid which allows the equilibrium below to move over to the right, e.g.

$$\underset{\text{solid}}{(Ca^{2+})_3(PO_4^{3-})_2} \rightleftharpoons \underset{\text{solution}}{3Ca^{2+} + 2PO_4^{3-}} +$$

6H+ from l

$$\downarrow\uparrow$$

$$2H_3PO_4$$

Metaphosphoric acid, HPO_3,

Metaphosphoric acid is obtained by the dehydration of orthophosphoric acid at 316°C. It is a glassy polymeric solid and a mixture of several acids, with the empirical formula HPO_3.

$$H_3PO_4 \rightarrow HPO_3 + H_2O$$

Although the strucutres of the various metaphosphoric acids are still uncertain, mixtures of metaphosphates have been separated by chromatographic methods. These metaphosphate ions are built up from PO_4 units and are cyclic as opposed to linear. The trimetaphosphate ion, $P_3O_9^{3-}$, is show below:

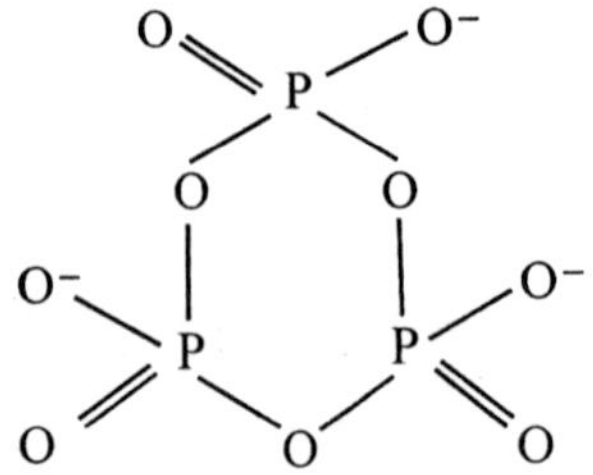

Sodium metaphosphate, empirical formula $Na^+PO_3^-$, is formed when sodium dihydrogen orthophosphate is heated:

$$Na^+H_2PO_4^- \rightarrow Na^+PO_3^- + H_2O$$

It is a mixture, containing cyclic metaphosphate ions of differing ring size.

Pyrophosphoric or Diphosphoric acid, $H_4P_2O_7$, exists as white, granular crystals, m.p. 61°. It is made by the following methods:

(i) By heating orthophosphoric acid at 220°:

$$2H_3PO_4 = H_4P_2O_7 + H_2O.$$

(ii) From sodium diphosphate, which is obtained by heating secondary sodium orthophosphate above 240°:

$$2Na_2HPO_4 = Na_4P_2O_7 + H_2O.$$

The sodium diphosphate is dissolved in water, and by the addition of lead nitrate, lead diphosphate is precipitated. This can be decomposed by hydrogen sulphide to give the free acid.

(iii) By heating a mixture of orthophosphoric acid and phosphoryl chloride:

$$5H_3PO_4 + POCl_3 = 3H_4P_2O_7 + 3HCl.$$

The residue is cooled for some time in a vaccum desiccator at -10°. Crystals of $H_4P_2O_7$ separate.

If diphosphoric acid is boiled with water, or if the aqueous solution stands for some time, it takes up water to yield orthophosphoric acid.

In aqueous solution diphosphoric acid is tetrabasic. The most familiar salts of diphosphoric acid are the acid salts such as $Na_2H_2P_2O_7$, and the salts formed by the complete replacement of hydrogen atoms in the acid, such as $Na_4P_2O_7$; the existence of salts such as $NaH_3P_2O_7$ and $Na_3HP_2O_7$ is doubtful.

Some important phosphates

All naturally occurring phosphates are orthophosphates, the most abundant of these being rock phosphate, $(Ca^{2+})_3(PO_4^{3-})_2$, the source of phosphorus and other phosphates. Rock phosphate is mostly consumed by the fertilizer industry in the manufacture of 'superphosphate of lime', 'triple superphosphate' and 'nitrophos'—a combined phosphatic and nitrogenous fertilizer.

$$(Ca^{2+})_3(PO_{4^{3-}})_2 + 2H_2SO_4 \rightarrow Ca^{2+}(H_2PO_4^-)_2 + 2Ca^{2+}SO_{4^{2-}}$$

'Superphosphate of lime'

$$(Ca^{2+})_3(PO_{4^{3-}})_2 + 4H_3PO_4 \rightarrow 3Ca^{2}(H_2PO_4^-)_2$$

'Triple superphosphate'

$$(Ca^{2+})_3(PO_{4^{3-}})_2 + 4HNO_3 \rightarrow Ca^{2+}(H_2PO_4^-)_2 + 2Ca^{2+}(NO_{3^-})_2$$

'Nitrophos'

Other phosphatic fertilizers are ammonium dihydrogen orthophosphate and diammonium hydrogen orthophosphate, which also counteract nitrogen deficiency.

Commerically importat sodium phosphates include disodium hydrogen pyrophosphate—an ingredient in baking powder—and disodium hydrogen orthophosphate, tetrasodium pyrophosphate and sodium metaphosphate—used in the processing of cheese. Condensed phosphates of sodium are used in the softening of water (p. 295).

Organic phosphates are intimately bound up with life processes, being involved in photosynthesis and protein synthesis. The latter process is still poorly understood but it is thought that deoxyribonucleic acid (DNA) provides the information as to how the amino acids are linked together into protein chains, another orgainc phosphate called ribonucleic acid (RNA) carrying this information to the protein 'factories'.

Phosphazenes and Cyclophosphazenes (Phosphonitrilic Compounds)

Nitrogen and phosphorus show only a slight tendency to catenate *by themselves*. The maximum chain length for nitrogen is three in the azide N_3^- ion, and two for phosphorus in a few compounds such as P_2H_4 and $(Me_2)(S)P—P(S)(Me_2)$. A few ring compounds exist with four, five or six P or As atoms joined together.

In contrast to this, N and P may bond *together,* forming a large number of phosphazenes. In these the P atom is in the oxidation state (+V) and N is in the (+III) state. The compounds are formally unsaturated. Thus monophosphazines may be made by reacting an azide with PCl_3, POR_3 or $P(C_6H_5)_3$:

$$PCl_3 + C_6H_5N_3 \rightarrow Cl_3P = NC_6H_5 + N_2$$
$$P(C_6H_5)_3 + C_6H_5N_3 \rightarrow (C_6H_5)_3P = NC_6H_5 + N_2$$

Diphosphazenes can be made as follows:

$$3PCl_5 + 2NH_4Cl \rightarrow [Cl_3P = N - PCl_2 = N - PCl_3]^+Cl^- + 8HCl$$

However, N and P catenate together, forming an interesting series of polymers.

$$nPCl_5 + nNH_4Cl \xrightarrow{120°-150°C} (NPCl_2)n + 4nHCl$$

(ring compounds cyclophosphazenes)

$$\text{and } Cl_4P(NPCl_2)nNPCl_3$$

(chain compounds polyphosphazenes)

This reaction produces a mixture of ring compounds $(NPCl_2)_n$ where $n = 3,4,5,6$....and fairly short linear chains. The most common rings (n = 3 and 4) contain six or eight atoms. The former are flat and the latter exist in 'chair' and 'boat' conformations.

A large number of chian compounds are known. These range from short chains P_2NCl_7, $P_3N_2Cl_9$, $P_4N_3Cl_{11}$ to those with up to 10^4 units

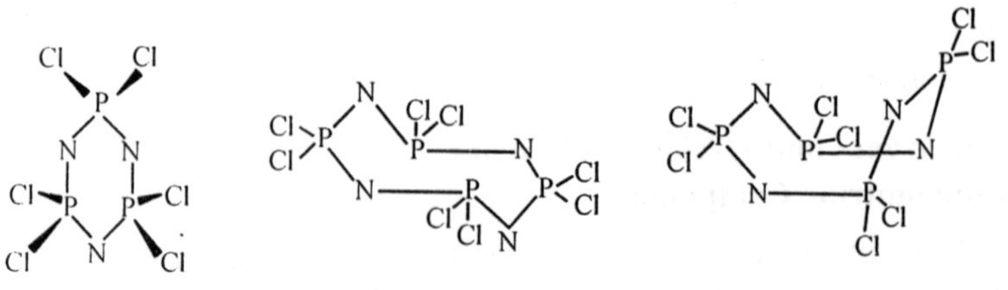

Fig. 15.7 : some cyclophosphazene compounds.

[—N=PCl_2-] linked together. These compounds were originally called phosphonitrilic halids, but are now named systematically poly (chlorophosphazenes).

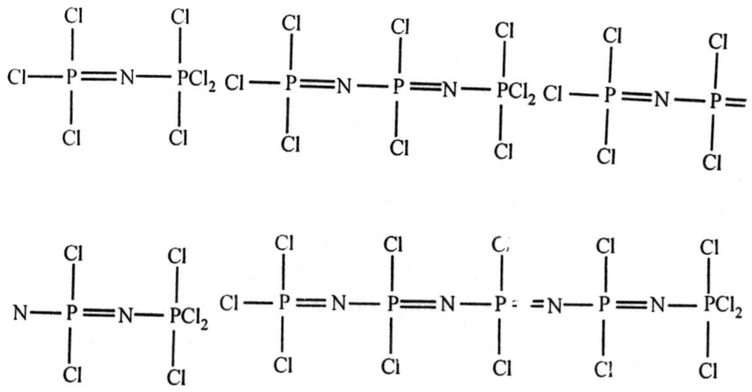

Fig. 15.7 : Some polyphosphazene chain compounds.

The chlorine atoms are reactive, and most reactions of chlorophosphazenes involve replacement of Cl by groups such as alkyl, aryl, OH, OR, NCS or NR_2. Alkyl or aryl groups may be introduced using lithium or Grignard reagents. Substitution may be complete, or partial, and in the latter case many different isomers are possible.

$$[NPCl_2]_3 + 6CH_3MgI \rightarrow [NP(CH_3)_2]_3 + 3MgCl_2 + 3MgI_2$$

$$[NPCl_2]_3 + 6C_6H_5Li \rightarrow [NP(C_6H_5)_2]_3 + 6LiCl$$

$$[NPCl_2]_3 + 6NaOR \rightarrow [NP(OR)_2]_3 + 6NaCl$$

$$[NPCl_2]_3 + 6NaSCN \rightarrow [NP(SCN)_2]_3 + 6NaCl$$

Similar compounds are formed with Br and F. The largest rings formed contain 34 atoms in the chlorides and 12 atoms in the bromides. Some of the long chain polymers are rubber-like, and those with perfluoroalkoxy side groups $[NP(OCH_2CF_3)_2]n$ resemble polythene.

There are many potential uses for the high molecular weight phosphazenes, as rigid plastics, expanded foam, and fibres, since they are waterproof and fireproof, and are unaffacted by petrol, oil, and solvents. They also form flexible plastics which are useful for fuel hoses and gaskets since they retain their elasticity at low temperatures. The phosphazenes are at present far too expensive for general use. Thin films of poly (aminophosphazene) are used in hospitals to cover severe burns and other extensive wounds since they prevent the loss of fluids and keep germs out.

There are two main points of interest in those P—N compounds:

1. The nature of the bonding is not understood. In all of these phosphazene compounds the (apparent) P—N and P=N bonds are equivalent. Their bond lengths are 1.56-1.59 Å. Which is much shorter than the usual single bond distance of 1.77 Å. Thus the bonding in these compounds is not adequately represented by a system of alternate single and double bonds, nor can it be explained by $p\pi$-$p\pi$ bonding and delocalization similar to that in benzene or graphite. It has been suggested that a coordinate bond is formed between a filled sp^2 orbital on N and the empty $3d_{x2-y2}$ orbital on P. This is similar to the $p\pi$-$d\pi$ bonding in the oxides of phosphorus except that in the phosphazenes the π bonds are delocalized over the whole molecule, giving pseudoaromatic character. There are objectivions to this because of the size and energy of the *d* orbitals. Alternatively the singly occupied p_z orbital on N may form a three-centre bond with the d_{xz} and d_{yz} orbitals on the two adjacent P atoms.

2. The polyphosphazenes form a very extensive series of polymers. Carbon based polymers are the most extensive, then the silicones, then the phosphazenes.

Arsenic

Occurrence and preparation. Although arsenic is found in small quantities in the free state, its principal occurrence is as sulfides; *arsenopyrite,* FeAsS; *orpiment,* As_2S_3; and *realgar,* As_2S_2. Small quantities of arsenic are usually associated with most sulfide ores; as a matter of fact, nearly all of the arsenic produced is recovered as a by-product from the flues of smelters utilizing sulfide ores of zinc, copper, and lead. During the roasting process, the arsenic is converted to volatile arsenious oxides, As_2O_3, which is condensed in the flues as a finely divided solid. The element is easily obtained from the arsenious oxide by reduction with carbon:

$$As_2O_3 + 3C \rightarrow 2As + 3CO$$

Properties and Uses of Arsenic

Elemental arsenic exists in several allotropic modifications. Ordinary gray arsenic is monatomic (As). It is metallic in appearance and sublimes at 615°C, forming tetraatomic molecules, As_4, which are tetrahedral in structure. When arsenic vapour is cooled rapidly an unstable yellow crystalline allotrope is produced, which also consists of As_4 molecules. It is soluble in carbon disulfide. Arsenic vapour is yellow in colour, has the odour of garlic, and is very poisonous. Arsenic is a member of an intermediate class of elements which possess to some degree the properties of both metals and nonmetals; hence it is a metalloid. Arsenic ignites when heated, producing white cloud of As_4O_6. It is slowly attacked by hot hydrochloric acid in the presence of air, forming arsenic(III) chloride, $AsCl_3$. Hot nitric acid readily oxidizes arsenic to arsenic acid, H_3AsO_4. Arsenic combines directly with sulfur, the halogens, and many metals. The trihalides (AsX_3) are either liquids or low-melting solides with properties suggestive of covalent rather than ionic bonding, as observed in the absence of saltlike character.

Certain alloys contain arsenic as hardening agent. Examples are bronze and lead shot.

Arsenic, AsH_3, is the only well-characterized hydride of arsenic. It is a poisonous, colourless gas with the odour of garlic. Arsine is prepared by the following methods:

(i) Excess of arsenic is slowly added to sodium dissolved in liquid ammonia, and the bright yellow solution of sodium arsenide so obtained to decomposed with ammonium bromide:

$$Na_3As + 3NH_4Br = (NH_4)_3As + 3NaBr,$$
$$= H_3As + 3Nh_3 + 3NaBr.$$

(ii) Aluminium arsenide (made by the combination of the elements, intimately mixed and fired with an aluminium-barium peroxide primer) is treated with warm water:

$$AlAs + 3H_2O = Al(OH)_3 + AsH_3.$$

This method gives almost pure arsine.

(iii) Sodium arsenide is treated with water:

$$Na_3As + 3H_2O = 3NaOH + H_3As.$$

(iv) Zinc arsenide (prepared by heating the elements together in equivalent proportions) is treated with dilute hydrochloride acid:

$$Zn_2As_2 + 6HCl = 3ZnCl_2 + 2AsH_3.$$

(v) Sodium arsenite is heated with sodium formate (dried at 210°):

$$Na_3AsO_3 + 3HCOONa = 3Na_2CO_3 + AsH_3.$$

Arsine cannot be made by the direct union of molecular hydrogen and arsenic, but it is formed when zinc in contact with any material containing arsenic is acted upon with dilute sulphuric acid, or when a solution of an arsenic compound is electrolyzed using a cathode of metallic mercury.† The formation of arsine by the action of nascent hydrogen on an arsenic compound is the basis of Marsh's test for arsenic.

Arsine decomposes into its elements on gentle heating and, if moist, on exposure to light. It is not oxidized by air, but it reduces many oxidizing agents. It reduces dilute silver nitrate solution to metallic silver:

$$6AgNo_3 + 3H_2O + AsH_3 = H_3AsO_3 + 6Ag + 6HNO_3.$$

If concetrated silver nitrate solution is used, the initial product is the soluble complex, $Ag_3As,3AgNo_3$, which by a reaction in which the silver arsenide reduces the silver nitrate, yields metallic silver when treated with water:

$$Ag_3As, 3AgNO_3 + 3H_2O = 6Ag + 3HNO_3 + H_3AsO_3.$$

Arsine reduces mercuric chloride to mercuric arsenide in three successive stages (Gutzeit's test): $AsH(HgCl)_2$, yellow; $As(HgCl)_3$, brown; As_2Hg, black. Arsine is not basic, and it forms no compounds analogous to the phosphonium compounds. Arsine reacts with sodium, potassium, or potassamide dissolved in liquid ammonia, to give derivatives such as $NaAsH_2$.

Two solid arsenic hydrides are reported. One is made by the reduction of arsenious compounds, as for example, the reduction of an acidified solution of arsenious chloride by stannous chloride in ether:

$$2AsCl_3 + 2HCl + 4SnCl_2 = As_2H_2 + 4SnCl_4.$$

The arsenic hydride is precipitated as a brown powder. The other solid arsenic hydride is made by the oxidation of arsine with stannic chloride:

$$4AsH_3 + 5SnCl_4 = As_4H_2 + 10HCl + 5SnCl_2.$$

The chemical individually of these solid hydrides of arsenic awaits clear demonstration.

Arsenic Trichloride, $AsCl_3$

The recommended preparation involved refluxing arsenic trioxide with disulphur dichloride in a stream of chlorine:

$$2As_4O_6 + 3S_2Cl_2 + 9Cl_2 \rightarrow 8AsCl_3 + 6SO_2$$

Arsenic trichloride may be distilled over. Alternatively, arsenic is heated in a stream of chlorine. A solution of arsenic trichloride results from boiling arsenic trioxide with concentrated hydrochloric acid:

$$As_4O_6 + 12HCL \rightleftharpoons 4AsCl_2 + 6H_2O$$

Arsenic trichloride is a volatile liquid (b.p. 130°C) which fumes in air and is very poisonous. Because of its volatile nature, care must be exercised in using it. It has the same pyramidal structure as phosphorus trichloride but differs in being only partiaily (reversibly) hydrolyzed:

$$4AsCl_3 + 6H_2O \rightleftharpoons As_4O_6 + 12HCL$$

Arsenic trifluoride, AsF_3, is made by treating arsenious oxide with a mixture of powdered fluorspar and concentrated sulphuric acid:

$$As_4O_6 + 12HF = 4AsF_3 + 6H_2O.$$

It is hydrolyzed by water, to arsenious oxide and hydrofluoric acid.

Arsenic trichloride, $AsCl_3$, is formed:

(i) by burning arsenic in chlorine (the reaction occurs spontaneously),

(ii) by heating arsenious oxide in chlorine:

$$11As_4O_6 + 12Cl_2 = 8AsCl_3 + 3(As_4O_{10}, 2As_4O_6),$$

(iii) by distilling a mixture of arsenious oxide, sodium chloride, and concentrated sulphuric acid; the distillate is freed from excess of chlorine by distillation over powdered arsenic,

(iv) by refluxing arsenious oxide with sulphur chloride, passing chlorine through the mixture, and distilling:

$$2As_4O_6 + 3S_2Cl_2 + 9Cl_2 = 8AsCl_3 + 6SO_2.$$

Arsenic trichloride is hydrolysed by water:

$$4AsCl_3 + 6H_2O = As_4O_6 + 12HCL.$$

The reaction is reversible, as is shown by reaction (iii) above. Arsenic trichloride is a good ionizing solvent; the dielectric constant is 12.8 at 20°. Antimony pentachloride, a non-conducting liquid, becomes conducting when dissolved in arsenic trichloride. Arsenic trichloride reacts with arsine to give arsenic and hydrogen chloride.

Arsenic tribromide, $AsBr_3$, is made by treating arsenic with a solution of bromine in carbon disulphide.

Arsenic tri-iodide, AsI_3, is made:

(i) by treating arsenic with a solution of iodine in carbon disulphide,

(ii) by precipitating a solution of arsenious oxide in hot hydrochloric acid with potassium iodide:

$$AsCl_3 + 3KI = AsI_3 + 3KCl.$$

It is less readily hydrolysed than are the other arsenic halides, but it is unstable. The solid and its solutions are slowly oxidised by air. The crystal has a layer structure; the iodine atoms are arranged on a close-packed hexagonal lattice, and the arsenic atoms are in two-thirds of the octahedral holes between layers of iodine atoms.

Sulphides of arsenic, As_4S_4, As_4S_6, As_2S_5, thioarsenites and thioarsenates

Arsenic sulphide, As_4S_4, is the red-orange mineral realgar, and may be formed by fusing arsenic trioxide with sulphur in the correct proportions out of contact with air or by heating the elements together:

$$As_4O_6 + 7S \rightarrow As_4S_4 + 3SO_2$$

$$4As + 4S \rightarrow As_4S_4$$

It inflames readily with oxidizing agents and is used in 'Bengal Fire' flares in pyrotechny, giving arsenic trioxide and sulphur dioxide. With alkalis and sulphide solutions it forms thioarsenites with separation of arsenic.

Arsenic trisulphide, As_4S_6, has been shown to have a similar structure to the trioxide, As_4O_6. It is found as the mineral orpiment (corruption of Latin, *auri pigmentum* = gold paint) which is golden-yellow in colour. Arsenic trisulphide may be formed by heating the trioxide with sulphur in the correct proportions out of contact with air or by passing hydrogen sulphide into a solution of arsenic trichloride or arsenic trioxide in hydrochloric acid, or an arsenite:

$$As_4O_6 + 9S \rightarrow As_4S_6 + 3SO_2$$

$$_4AsCl_3 + 6H_2S \rightleftharpoons As_4S_6 +_{12} HCl$$

Although arsenic trisulphide will not react with concentrated hydrochloric acid while hydrogen sulphide is present, the reaction is reversed by prolonged boiling when both H_2S and $AsCl_3$ are volatilized away. Arsenic trisulphide burns to the trioxide and sulphur dioxide on heating in air. It is an acidic sulphide. Arsenic trisulphide is insoluble in water but dissolves readily in alkalis to form the arsenite ($LiAsO_2$) and thioarsenite ($LiAsS_2$) and mixed compounds (LiAsOS):

$$As_4S_6 + 4OH^- \rightarrow AsO_{2^-} + 3AsS_{2^-} + 2H_2O$$

Arsenic pentasulphide, As_2S_6, may be prepared by heating the trisulphide with sulphur in the calculated proportions, by heating the elements out of contact with air, or by acidification of ammonium thioarsenate. It is a yellow solid. Passage of hydrogen sulphide into a hot 5^N hydrochloric acid containing quinquevalent arsenic appears to give reduction to the tervalent state, with precipitation of sulphur and arsenic trisulphide:

$$H_3AsO_4 + 2HCl \rightarrow HAsO_2 + 2H_2O + Cl_2$$

$$H_3AsO_4 + H_2S \rightarrow HAsO_2 + 2H_2O + S\downarrow$$

$$4HAsO_2 + 6H_2S \rightarrow As_4S_6 \downarrow + 8H_2O + Cl_2$$

Arsenic pentasulphide is acidic, dissolving in aqueous alkali and carbonates to form mixed arsenated and thioarsenates and in solution of sulphides to form thioarsenates; with ammonium sulphide:

$$As_2S_5 + 3S^{2-} \rightarrow 2AsS_{4^{3-}}$$

(tetra)

thiorarsenate

The molecular weight of arsenious oxide vapour corresponds to As_4O_6 up to 800°, and to As_2O_3 above 1800°. The molecular weight determined cryoscopically in nitrobenzene solution corresponds to As_4O_6. X-ray analysis shows that the structure of the molecule in the solid is similar to that P_4O_6; the distance As—O is 1.80. Electron diffraction analysis shows that the molecules in the vapour have the same structure as in the As_4O_6 tetrahedron. The angle OAsO is 100 ± 5°, and the angle AsOAs is 126 ± 3°. These values are the same as those of the corresponding angles in the molecule P_4O_6. In the octahedral form of arsenious oxide the As_4O_6 groups are in positions corresponding to those of carbon atoms in a diamond lattice.

Arsenic oxide, As_4O_{10}, is a white deliquescent solid. The conversion of arsenious oxide to arsenic oxide by direct oxidation is not quantitative, even by treatment with oxygen under pressur. Arsenic oxide is best prepared by heating crystalline arsenic acid above 200°:

$$4(H_3AsO_4)_3, H_2O = 3As_4O_{10} + 22H_2O.$$

Arsenic oxide melts at a red heat, when it begins to decompose into arsenious oxide and oxygen. It is an oxidizing agent (unlike P_4O_{10}). It liberates chlorine from hydrogen chloride. With phosphorus pentachloride it reacts according to the equation:

$$As_4O_{10} + 10PCl_5 = 10POCl_3 + 4AsCl_3 + 4Cl_2.$$

Arsenic oxides dissolves slowly in water to form arsenic acid. The vapour of arsenic oxide is wholly dissociated into arsenious oxide and oxygen. No investigation of the solid appears to have been made.

An oxide of arsenic, As_2O_4, is known. It is probably arsenious arsenate, $As[AsO_4]$.

Arsenious acid, H_3AsO_2.

The solution to arsenious oxide in water is faintly acidic and is *tribasic* (in this aspect it differs from phosphorus acid which is *dibasic*). Its salts as Na_3AsO_3, are known as arsenites which may be either *ortho, meta* or *pyro*:

$$As_2O_3 + 3H_2O \rightleftharpoons 2H_2AsO_3$$

$$\text{Ortho} - \text{arsenite} \rightarrow Na_3AsO_3$$

$$\text{Meta} - \text{arsenite} \rightarrow NaAsO_2$$

$$\text{Pyro} - \text{arsenite} \rightarrow Na_4As_2O_5.$$

Alkali arsenites are soluble in water, but the other arsenites are more or less insoluble. They are soluble in concentrated hydro-chloric acid.

An arsenite gives (i) a brownish precipitate with $FeCl_3$ in neutral solutions, (ii) a pellowish precipitate of silver arsenite, Ag_3AsO_3 with silver nitrate, and (iii) a bright precipitate with copper sulphate. This is known as **Scheele's green** which is used as an insecticide. Its composition is uncertain. It may be cupric arsenite $CuHAsO_3$ or $Cu_3(AsO_2)_2H_2O$. When verdigris (basic copper acetate) is boiled with arsenious acid and acetic acid, **Paris green** is obtained, of the composition $(CuC_2H_3O_2)$. $Cu_3(AsO_2)_2$. It is also used as an insecticide.

Arsenic acid, H_3AsO_4

The acid is obtained when arsenious oxide is dissolved in *aqua regia* or nitric acid. It can also be prepared by passing a current of chlorine in arsenious oxide suspended in water.

$$As_2O_3 + 5H_2O + 2Cl_2 \rightarrow 2H_3AsO_4 + 4HCl$$

$$As_2O_3 + 2HNO_3 + 2H_2O \rightarrow 2H_3AsO_4 + N_2O_3$$

$$As_2O_3 + 5H_2O + 2Cl_2 \rightarrow 2H_3AsO_4 + 4HCl$$

$$As_2O_3 + 2HNO_3 + 2H_2O \rightarrow 2H_3AsO_4 + N_2O_3$$

The crystals of *ortho-arsenic acid* $2H_3AsO_2.H_3O$ melt at 100° and when heated to 180°C, they change to pyro arsenic acid, $H_4As_2O_7$ and on further heating are transformed to meta arsenic acid, $HAsO_3$. Both meta and pyro-acids change in solution to the ortho acid.

The salts of arsenic acid are known as arsenates, and they closely resemble phosphates with which they are isomorphous. The three ortho-arsenates, Na_3AsO_4, Na_2HAsO_4 and NaH_2AsO_4 are known. Magnesium and manganese arsenates on heating strongly are converted to pyro-arsenates, $Mg_2As_2O_7$ and $Mn_2As_2O_7$ just similar to tpyro-Phosphates. The normal ortho-arsenates exist as solides and in solution, but the pyro- and meta-arsenates exist only as solids and are prepared by heating the disodium and dihydrogen arsenates:

$$2Na_2HAsO_4 \rightarrow Na_4As_2O_7 + H_2O$$

$$NaH_2AsO_4 \rightarrow NaAsO_3 + H_2O$$

Arsenious oxide, As_4O_6, exists in two crystalline forms and also an as amorphous glass. It is made by burning arsenic in air. The familiar form is a white solid consisting of octahedral crystals which are produced when the vapour condenses at room temperature, or when aqueous or hydrochloric acid solution of arsenious oxide is allowed to crystallize.† The octaderal form sublimes freely above 135°. The m.p. is 275° under pressure, the b.p. is 465°. The octahedral form is metastable under ordinary conditions and tends to change into the monoclinic form (m.p. about 315°) with which it is in equilibrium at -13°. The change is very slow. It may be brought about by heating the octahedral form with water at 180° for about 40 hours. The monoclinic form is also obtained when the vapour is condensed above 250°, or by cooling a boiling saturated solution of the oxide in sodium arsenite solution.

Arsenious oxides is sparingly soluble in water and, as it is not easily wetted, it dissolves slowly. The solubility is 2% at 25% and 8.2% at 98°. The solution is weakly acid, and perhaps contains H_3AsO_3. Arsenious oxide forms arsenites with alkalis and with sodium carbonate, but with concentrated hydrochloric acid it gives arsenite trichloride. With concentrated sulphuric acid it gives arsenyl sulphate $(AsO)_2SO_4$, and with sulphur trioxide at 100° arsenious sulphate $As_2(SO_4)_3$.

Arsenious oxide is reduced to arsenic when heated with charcoal or potassium cyanide, or when treated with a mixture of hydrochloric acid and stannous chloride, or with hydrochloric acid and copper foil:

$$As_4O_6 + 12HCl + 12Cu = 4As + 12CuCl + 6H_2O.$$

If the reaction is carried out in a test tube the arsenic is deposited as a grey film on the copper foil. Se, Hg, Sb, Bi compounds also give grey deposits when thus treated. When the copper foil coated with arsenic is washed, dried, and heated in a tube open to the air a white sublimate of arsenious oxide is obtained on the cooler parts of the tube. If the same procedure is carried out on the deposits of the other metals no sublimate is obtained. These reactions are the basis of Reinsch's test for arsenic.

Arsenious oxide is, however, usually thought of as a reducing agent, and it is oxidized to arsenate by ozone, H_2O_2, halogens, hypochlorites, ferric chloride, nitric acid, and potassium dichromate solution. The reduction of diodine is reversible unless the acidity of the solution is kept low, say, by the addition of potassium bicarbonate:

$$8KHCO_3 + As_4O_6 + 4I_2 = As_4O_{10} + 8KI + 8CO_2 + 4H_2O.$$

It seems, therefore, that the reducing properties are to be associated with the arsenite ion, rather than with the free acid. Arsenious oxide, or perhaps the arsenite ion, reduces Fehling's solution, precipitating cuprous oxide.

Arsenious oxide is highly poisonous, and is the more dangerous because the vapour has neither taste nor smell. The lethal dose for a man is about 0.1 gm. The recommended antidote is freshly prepared ferric hydroxide made by adding magnesia to a solution of ferric chloride. The ferric hydroxide is said to adsorb the arsenious oxide. Arsenious oxide is a very weak bactericide.

Antimony

Occurrence and preparation The chief source of antimony is the mineral *stibnite*, Sb_2S_3, although small quantities of the element are found in the free state. Antimony may be recovered from stibnite by heating with finely divided iron:

$$Sb_2S_3 + 3Fe \rightarrow 3FeS + 2Sb$$

The molten products separate into two layers with the lighter ferrous sulfide layer on top.

Stibnite may also be roasted to the oxide, followed by reduction with carbon:

$$2Sb_2S_3 + 9O_2 \rightarrow 2Sb_2O_3 + 6SO_2$$
$$Sb_2O_3 + 3C \rightarrow 2Sb + 3CO$$

Properties and Uses of Antimony

Antimony is a lustrous silver-white metal, brittle, and readily pulverized. It tarnishes only slightly in dry air but does oxidize slowly in moist air. The metal is readily attacked by the halogens, phosphorus, and sulfur. It is attacked, but not dissolved, by not nitric acid, forming Sb_4O_6. It is slowly dissolved by hot concentrated sulfuric acid forming $Sb_2(SO_4)_3$ and evolving SO_2. The fact that the metal and its oxides do not react with nitric acid to give the nitrate indicates a lack of marked base-forming properties. However, the metallic nature of the element is more pronounced than that of arsenic, which lies ;immediately above it in Group VA.

Lead hardened by the addition of 10 to 20 per cent of antimony is suitable for shrapnel, bullets, and bearings. Because of its resistance to corrosion by acids, it is used in making storage battery plates. Babbitt metal is an antifriction alloy of antimony with tin and copper, extensively used in machine bearings. Since antimony, lime bismuth, expands on freezing it is uused as a constituent of type metal, to which it confers this property. When forming a piece of type the freezing type metal expands and fills the mold, thus providing type that gives a sharp and distinct imprint.

Stibine and Its Derivatives

Antimony hydride, stibine, SbH_3, is a poisonous gas, b.p. -18°, m.p. -88°, which condenses to a white solid. It has an unpleasant smell. Like phosphine and like arsine, stibine cannot be made by the direct union of the elements.

Stibine is obtained by the action of dilute sulphuric or hydrochloric acid on a Group II metal in the presence of antimony or a compound of antimony. The antimony is best introduced as an alloy of the Group II metal. The most satisfactory Group II metals for the purpose are magnesium and zinc. Calcium or barium may be used, but the yields of stibine are much smaller. Very little stibine is obtained by the action of acids on alloys of antimony with the alkali metals, or with aluminium (cf. Fieitmann's test for arsenic). The reduction of an antimony compound by hydrochloric acid acting on tin in contact with platinum produces metallic antimony.

The purest stibine is made by the action of hydrochloric acid on an alloy of magnesium and antimony in the proportions 2:1. The gas is washed with water, dried, and cooled in liquid air, which solidifies it. Pure gaseous stibine is evolved on warming, and may be collected over mercury.

Stibine is fairly soluble in water, and more soluble in carbon disulphide. It is endothermic and although fairly stable when dry, it decomposes into its elements when heated or sparked. It is attacked by air or oxygen to form water and antimony. It burns with a grey flame giving white fumes of antimonious oxide. It reacts with silver nitrate first to form a precipitate of Ag_3Sb:†

$$SbH_3 + 3AgNO_3 = Ag_3Sb + 3HNO_3.$$

This precipitate is then oxidized by excess of silver nitrate to metallic silver,

$$Ag_3Sb + 3AgNO_3 + 1\frac{1}{2}H_2O = 6Ag + \frac{1}{4}Sb_4O_6 + 3HNO_3,$$

with the precipitation of antimonious oxide, Sb_4O_6. These reactions are parallel to those of arsine, with the important practical difference that arsenic is finally present as the soluble As_4O_6, and antimony is finally present as the insoluble Sb_4O_6. If a mixture of arsine and stibine has been passed through dilute silver nitrate solution, the arsenic is present in solution, and the antimony in the precipitate.

Antimony trichloride, antimony (III) chloride, $SbCl_3$

Finely powdered antimony burns spontaneously in chlorine to form the trichloride. However, the compound is usually prepared by heating antimony trisulphide with concentrated hydrochloric acid:

$$Sb_2S_3 + 6HCl \rightarrow 2SbCl_3 + 3H_2S\uparrow$$

The solution is evaporated to crystallization. Antimony trichloride is a soft white solid ('butter of antimony'), m.p. 73°C and deliquescent. It is hydrolyzed reversibly by cold water to a white precipitate of antimony oxide chloride:

$$SbCl_3 + H_2O \rightleftharpoons SbOCl\downarrow + 2HCL$$

It is unlikely that simple Sb3+ ions exist in solution, considerable hydrolysis taking place. Well defined oxide chlorides, SbOCl and $Sb_4O_5Cl_2$, which re-dissolve in excess acid, have been identified. On boiling in water, antimony trioixde is formed but will re-dissolve in concentrated hydrochloric acid to form antimony trichloride:

$$4SbOCl + 2H_2O \rightarrow Sb_4O_6 + 4HCl$$

the whole hydrolysis being

$$_4SbCl_3 + 6H_2O \rightleftharpoons Sb_4O_5 +_{12} HCl$$

Solutions of antimony trichloride contain added hydrochloric acid to reduce hydrolysis. Antimony oxide chloride, formerly antimonyl or antimony oxychloride, is only slightly soluble in water but dissolves in benzene, carbon disulphide and chloroform, indicating that it has some covalent nature.

Antimony trifluoride, SbF_3, is made :

(i) by dissolving antimonious oxide in aqueous acid, and evaporating the solution,

(ii) by distilling antimony with mercuric fluoride.

In aqueous solution antimony trifluoride is much less hydrolyzed than the other antimony halides, but the solution has an acid reaction. When the aqueous solution is evaporated some hydrogen flouride volatilizes, and a basic antimony fluoride remains. A complex salt, K_2SbF_5, may be made by dissolving antimony trifluoride in hydrofluoric acid containing potassium fluoride in solution.

Antimony tribromide, SbBr3, is made by the direct combination of the elements. In aqueous solution it is hydrolyzed to SbOBr, and $Sb_4O_5Br_2$. It forms complex anions containing a high proportion of bromine. The potassium salts containing these anions are made by adding bromine and antimony tribromide to a solution of potassium bromide. $KSbBr_6$, $KSbBr_7$, and $KSbBr_9$ are known. These salts are black or violet-brown. They lose bromine easily when dry; in aqueous solution they are easily hydrolyzed to antimonic oxide, Sb_4O_{10}. The constitutions of these salts are probably of the perbromide type:

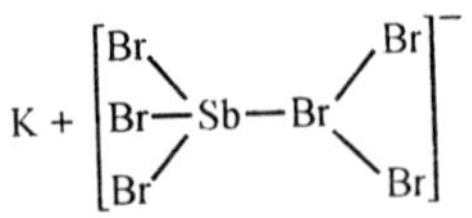

Antimony tribromide forms many complex salts, such as $K_3Sb_3Br_9(NO_3)_3$, $K_3Sb_3Br_9(SCN)_3$, and $NaSb_3Br_9(N_3)$. Dark red, almost black, crystals of the bromo-antimonic acid, $HSbBr_6$, $3H_2O$.

can be obtained from a solution of antimony tribromide and bromine in concentrated hydrobromic acid. Antimony tribromide is monomeric in solution in benzene.

Antimony tri-iodide, SbI_3, is made by dissolving powdered antimony in a solution of iodine in carbon disulphide. When the solution is evaporated the antimony tri-iodide crystallizes in red plated. The crystal structure is similar to that of AsI_3. The vapour of antimony tri-iodide is scarlet

Sulphides of antimony : Sb_2S_3, (Sb_2S_4, Sb_2S_5)

Antimony trisulphide exists in two forms: orange precipitated by passage of hydrogen sulphide through a solution of antimony (III) chloride, and black, the naturally occurring stibnite, obtained by heating the elements in the correct proportion out of contact with air:

$$2SbCl_3 + 3H_2s \rightleftharpoons Sb_2S_3 + 6HCl$$

$$2Sb + 3S \rightarrow Sb_2S_3$$

In the first case, prolonged boiling in the solution used for its preparation converts the orange form to black. On heating, antimony trisulphide burns in air or oxygen to the oxide and sulphur dioxide, the combustion finding use in match heads and fireworks. It dissolves in concentrated hydrochloric acid to give the trichloride and hydrogen sulphide. Antimony trisulphide, will not dissolve in ammonium carbonate solution but is similar to arsenic trisulphide in its reaction with ammonium sulphide and alkalis: parallel equations may be written, showing the formation of thioantimonate(III) and (V).

There is some doubt about the chemical individuality of Sb_2S_4 and Sb_2S_5. 'Antimony Pentasulphide' is made commercially by acting on boiling alkali with sulphur and antimony trisulphide: sodium antimonate (V) is precipitated and thioantimonate(V) in the filtrate is decomposed by acid:

$$2SbS_4^{3-} + 6H^+ \rightarrow Sb_2S_5 \downarrow 3H_2S \uparrow$$

It is used to produce 'red rubber'. Antimony 'pentasulphide' is golden yellow in colour.

Antimonoius oxide, Sb_2O_3, is a white solide existing in two forms. The octahedrat form, senarmonite, is stable up to 570°. The orthorhombic form, valentinite, is stable from 570° to the m.p. about 650°, b.p. 1560°. The strucutures of the two forms are different. The molecular weight of antimonious oxide in the vapour at 1560°

corresponds to the formula Sb_4O_6. The octahedral form is made up of Sb_4O_6 groups, each occupying the position of a carbon atom in the diamond lattice. It is therefore probably that Sb_4O_6 in the vapour, and in the octahedral crystalline form, has an orbital strucutre corresponding to that of P_4O_6. If so, the antimony atom in the oxide in these forms is in the tetrahedral valence state.

The orthorhombic form, however, consists of long chains formed of puckered rings:

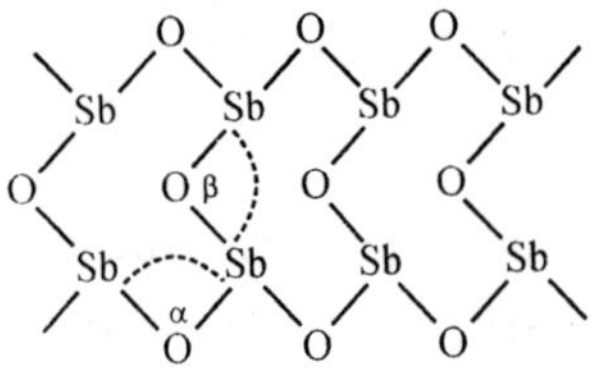

The angles of the structure are: $\alpha = 115°$, $\beta = 129°$. The three angles O—Sb—O at any given antimony atom are 79°, 92° and 100°. Hence the chain is not constructed of regular tetrahedra, each with three shared corners, and this is evidence of the inert pair configuration for the antimony atom in this form of antimonious oxide.

The presence of antimony atom in the tetrahedral valence state in the octahedral form of antimonious oxide would justify the inclusion of the oxide. It is convenient, however, to discuss antimonious oxide with the other antimonious compounds in the present Section.

Antimonious oxide is made:

(i) by heating antimony or antimonious sulphide in air,

(ii) by hydrolyzing antimony trichloride with sodium carbonate solution,

(iii) by digesting the oxychloride, SbOCl, with sodium carbonate solution,

(iv) by passing steam over red-hot antimony.

It becomes yellow on heating (the liquid is yellow), and pale buff on cooling. If heated in air it begins to combine with oxygen at 300° to give Sb_2O_4, but it is easily reduced to antimony by heating with hydrogen or carbon. In reacts with sulphur on heating to give Sb_2S_3. It is almost insoluble in water. It is soluble in hydrochloric acid to form $SbCl_3$, in tartaric acid to form $Sb_2(C_4H_4O_6)_3$, $6H_2O$, in hot concentrated sulphuric acid to yield crystalline $Sb_2(SO_4)_3$ on cooling, and in nitric acid to give $Sb(NO_3)_3$. It dissolves in sodium hydroxide solution to form sodium antimonite (see below). A solution of Sb_2O_3 in tartaric acid, neutralized by sodium carbonate, can be titrated against iodine:

$$Sb_2O_3 + 2H_2O + 2I_2 = 4HI + Sb_2O_5.$$

Antimony dioxide, Sb_2O_4, is a white solid. It is obtained as the ultimate product when metallic antimony, its oxides, or its sulphides, are heated in air between 300° and 900°. It is conveniently made by oxidizing antimonious sulphide with nitric acid, and igniting the product in air at a dull red heat.

When heated it becomes yellow. It is infusible and non-volatile. If heated above 900° it loses oxygen and forms antimonious oxide. It is easily reduced when heated with carbon or hydrogen. It is insoluble in water, but it reddens blue litmus. It is resistant to acids, but it dissolves in hydrochloric acid to give antimony trichloride if a reducing agent, such as hydriodic acid, is present.

BISMUTH

Occurrence and Metallurgy of Bismuth

Bismuth is most often found in the free state in nature. It sometimes occurs in combination, as **bismuth ocher,** Bi_2O_3, and **bismuth glance,** Bi_2S_3. It is produced in the United States as a by-product of the refining of other metals, particularly lead. Ores containing the free metal are treated by heating them in inclined iron pipes, whereupon the metals melts and flows away from the gangue. The oxide and sulphate ores are roasted and then heated with charcoal. As the bismuth is set free, it melts and collects beneath the less dense material.

Properties and Uses

Bismuth is distinctly a metallic element, which hard and brittle and has a silvery luster with a reddish tinge. Bismuth burns when highly heated in air to form the trioxide, Bi_2O_3, and combines directly with the halogens to form halide salts, although these reactions with the halogens are not vigorous. Unlike the other members of the group V*A* elements, bismuth dissolves in nitric acid to form nitrate salt.

Bismuth is used in the preparation of several low-melting alloys such as *Rose'metal* (Sn, Pb, Bi), melting point 94°, and *Wood's metal* (Pb, Cd, Sn, Bi), melting point 71°. These alloys are used in electrical fuses, automatic fire sprinklers, safety plugs for boilers, automatic fire alarms, and so on.

Bismuthine, BiH_3, is made by dissolving an alloy Bi_2Mg_3 in hydrochloric acid, and freezing out by liquid air the bismuthine from the hydrogen evolved. The efficiency of the preparation in terms of bismuth dissolved is only 1 in 20,000. The quantitites of bismuthine prepared have been small, and the knowledge of its chemistry is limited. By consideration of the boiling points of the hydrides of the other elements in Group V, that of bismuthine is judged to be about +20°, and so at room temperature bismuthine is an easily condensible gas. It is decomposed by gentle heating, by alkali, and by concentrated sulphuric acid. It reacts with silver nitrate solution, but not with hydrogen sulphide solution.

Bismuth trifluoroide, BiF_3, is made by evaporating a solution of Bi_2O_3 in hydrofluoric acid. (Fluorine does not easily react with bismuth even when heated.) It is insoluble in water. It dissolves in concentrated hydrofluoric acid with the formation of complexes, such as $H_3(BiF_6)$. Bismuth trifluoride is infusible, and scarcely volatile at a redheat. It has the properties of a salt.

The crystal has the (YF_3) structure in which the metal atom has nine nearest fluorine neighbours.

Bismuth trichloride, bismuth(III) chloride, $BiCl_3$

The normal methods for the preparation of salts of metals may be used. Union of the elements and the action of hydrochloric acid on the trioxide are methods similar to those described above:

$$2Bi + 3Cl_2 \rightarrow 2BiCl_3$$

$$Bi_2O_3 + 6HCl \rightarrow 2BiCl_3 + 3H_2O$$

Bismuth trichloride is a colourless crystalline solid, m.p. 232°C and deliquescent. It is hydrolyzed reversible by water with precipitation of white bismuth oxide chloride, re-dissolving in concentrated hydrochloric acid:

$$BiCl_3 + H_2O \rightleftharpoons BiOCl + 2HCl$$

Hydrolysis does not go further even on boiling with a large excess of water. Solution of bismuth trichloride contain added hydrochloric acid to reduce hydrolysis.

Bismuth tribromide, $BiBr_3$, is made by warming bismuth with bromine. In general it is similar to the chloride. It is soluble in alcohol, acetone, and ether, and in xylene. With water it gives BiOBr.

Bismuth tri-iodide, BiI_3, is made:

(i) by warming bismuth with iodine,

(ii) by adding potassium iodide solution to a solution of bismuth trichloride,

(iii) by adding bismuth oxide to a solution of iodine in stannous chloride saturated with hydrogen chloride.

It is less soluble in water than bismuth chloride or bromide. It is hydrolyzed to red BiOI only when its aqueous solution is boiled. It is soluble in benzene, toluene, xylene, and methylene iodide. It dissolves in hydriodic acid to form the acid $HBiI_4$, and in alkali metal iodides to form red salts such as $KBrI_4$.

Bismuth trioxide, Bi_2O_3, is a pale yellow powder, m.p. 820°. It is formed by heating the metal, the hydroxide, the basic carbonate, or the nitrate, to redness in air. It solidifies from the fused state to the monoclinic α form. At 704° the α modification changes to a tetragonal β form with a large evolution of heat. Bismuth trioxide sublimes at 1010° and is stable up to 1750°. Bismuth trioxide has a giant moleucle structure which may be regarded as the fluorite structure with one-quarter of the anions removed. This structures is almost coincident with a three-dimensional development of the Sb_4O_6 molecule with the units linked through the oxygen atoms.

A hydrated form of bismuth trioxide is formed as a white gelatinus precipitate on adding alkali hydroxide or ammonia to bismuth nitrate solution. It becomes crystalline on standing. Its formula has not been established, but it is conventionally written $Bi(OH)_3$. When heated to 100° the hydrated oxide loses water to form BiO(OH), and subsequently Bi_2O_3. The hydrated oxide is insoluble in water. It is very slightly soluble (10^{-4} molar) in concentrated sodium hydroxide solution. It dissolves in acids to form bismuth salts.

Higher oxides of bismuth, and the bismuthates. Bismuth oxides can be made to take on additional oxygen when treated with oxidizing agents such as chlorine, or air in the presence of alkalis.

The substances so obtained lose oxygen very easily, and not one has been characterized as a chemical individual.

On fusing bismuth trioxide with KOH in air a brown mass of potassium bismuthate is obtained. This material is a strong oxidizing agent, and it oxidises quantitatively a cold solution of a manganous salt in dilute nitric acid to a solution of permanganatic acid. Use is made of this reaction in the detection and estimation of manganese. It was at one time thought that potassium bismuthate must be analogous to the phosphates arsenates, and antimonates, K_3BiO_4, and consequently that the anion contains pentavalent bismuth. More recently the bismuthates are considered to be analogous to the ferrites and the chromites in which the metal in the anion is trivalent, but 8 co-ordinate. On this analogy potassium bismuthate should be formulated $K_5(Bi(OH)_8)$, $4H_2O$, or $K_4(Bi(OH)_7,H_2O),3H_2O$.

Bismuth trisulphide, Bi_2S_3, is obtained as a grey crystalline solid by fusing together bismuth and sulphur. If hydrogen sulphide is passed into an acid solution of a bismuth salt, bismuth trisulphide is obtained as a very dark brown precipitate. The borwn precipitate dissolves in nitric acid and in boiling concentrated hydrochloric acid, but not in alkalis or yellow ammonium sulphide. It dissolves in concentrated K_2S solution, or in K_2S melt, to yield the crystalline salt, potassium thiobismuthite, $KBiS_2$. Bismuth trisulphide is isomorphous with antimony trisulphide, and has a similar structure.

Bismuth nitrate is made by dissolving bismuth oxide, or bismuth basic carbonate, in dilute nitric acid. When the solution is evaporated, deliquescent crystals separate having the constitution $Bi(NO_3)_3$, $5H_2O$. Bismuth nitrate is easily hydrolyzed to the basic nitrate, $BiONO_3$.

Bismuth nitrate forms double nitrates of the formula $3M(NO_3)_2$, $Bi(NO_3)$, $24H_2O$. M'' may be Zn, Cd, Co, or Ni. Bismuth may be replaced by lanthanum or by a rare earth metal. The whole series is isomorphous.

Bismuth sulphate, $Bi_2(SO_4)_3$, is made by dissolving bismuth trioxide in concentrated sulphuric acid. It can be heated fairly strongly without decomposition, but above 400° it breaks down to basic salts and bismuth oxide. The dihydrate and the heptahydrate are known. Bismuth sulphate is isomorphous with the sulphates of yttrium, lanthanum, and praseodymium.

16

P-Block Elements (Oxygen, Sulphur, Selenium and Tellurium)—Group VIA (16)

Some physical data of Group VIA elements

	Atomic Number	Electronic Configuration	Atomic Radius (A)	Ionic Radius(A) X^{2-}	M.p. °C	B.p. °C
O	8	2.6 $1s^2 2s^2 p^4$	0.74	1.40	-218.9	-182.96
S	16	2.8.6 $....2s^2 2p^6 3s^2 3p^4$	1.04	1.84	118.95*	444.6
Se	34	2.8.18.6 $.....3s^2 3p^6 3d^{10} 4s^2 4p^4$	1.17	1.98	217.4+	684.8
Te	52	2.8.18.18.6 $....4s^2 4p^6 4d^{10} 5s^2 5p^1$	1.37	2.21	449.8	1390
Po	84	2.8.18.32.18.6 $....5s^2 5p^6 5d^{10} 6s^2 6p^4$	1.52			

* For monoclinic sulphur
+ For grey selenium

Properties

The Group VIA elements show the usual gradation from non-metallic to metallic properties with increasing atomic number that occurs in any periodic group. Oxygen and sulphur are non-metals, selenium and tellurium are semiconductors and polonium is metallic.

These elements can enter into chemical combination and complete their octets by gaining two electrons to form the 2-valention, e.g.O^{2-}, S^{2-}, bonds, e.g., the hydrides H_2O, H_2S, H_2Se, H_2Te and H_2Po.

The two heavier members of this group can form the 4-valent cation X^{4+}(the inert pair effect), e.g. there is evidence of the presence of Te^{4+} ions in the dioxide TeO_2 and of Po^{4+} ions in the dioxide, PoO_2, and sulphate, $Po(SO_4)_2$

Because sulphur, selenium, tellurium and polonium have vacant d orbitals that can be utilized without too great an energy change, they are able to form covalent compounds in which the octet of electrons is expanded; for instance, the valencies of sulphur in H_2S, SCI_4 and SF_6 are two, four and six respectively. Oxygen, in common with other first row members of the Periodic Table, cannot expand its octet.

Oxygen exists in the form of discrete molecules, a double bond uniting two oxygen atoms together, O = O. The atoms of the other Group 6B elements do not multiple bond to themselves and sulphur, in particular, shows a strong tendency to catenate, puckered S_8 rings being present in rhombic and monoclinic sulphur (the two main allotropic forms of this element). There are two forms of selenium corresponding in structure to rhombic and monoclinic sulphur in which Se_8 rings are present. These forms, however, are readily converted into a 'metallic' form of the element called grey selenium which contains infinite spiral chains of selenium atoms (Fig.16.1). As far as is known, there is only one form of tellurium which has the same structure as grey selenium. Polonium is truly metallic, exhibiting a co-ordination number of six in both its allotropic forms.

Since the chemistry of oxygen and sulphur have little in common, these two elements are treated separately. A brief account of the chemistry of selenium, tellurium and polonium, in so far as in illustrates group trends, is included.

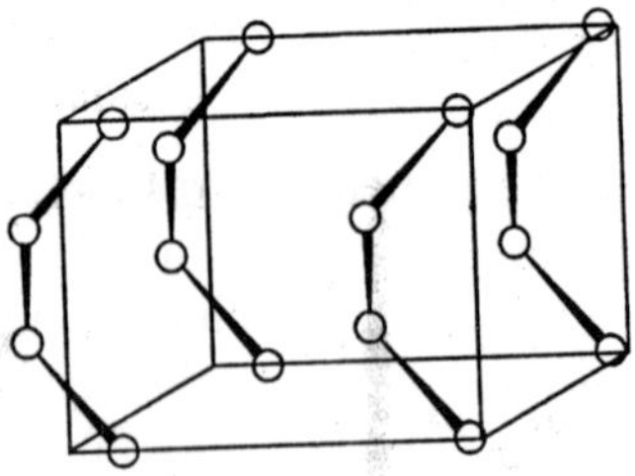

Fig. 16.1 : The spiral chain structure of selenium

Oxygen

Occurrence and preparation of oxygen

Oxygen occurs in the atmosphere to the extent of about 21 per cent by volume (23 per cent by weight). This percentage remains constant by the operation of the highly complex process termed photosynthesis, whereby green plants exposed to sunlight utilise carbon dioxide (one product of respiration and combustion) and water to build up carbohydrate molecules, at the same time releasing oxygen. The element is present in the earth's crust and in water to the extent of about 50 per cent and 89 per cent by weight respectively. It is an essential ingredient in all living matter (together with carbon, hydrogen and nitrogen) and is of prime importance in respiration and combustion processes. Although only slightly soluble in water, enough oxygen dissolves to support marine life.

Laboratory preparation (a) By heating certain oxides. Oxygen is conveniently prepared in the laboratory by repeating Priestley's experiment of heating certain substances. Mercuric oxide, a reddish powder, readily gives up its oxygen when heated, leaving metallic mercury as a residue (Figure 16.2):

$$\underset{\text{mercuric oxide}}{2\,HgO} \rightarrow \underset{\text{mercury}}{2Hg} + \underset{\text{oxygen}}{O_2}$$

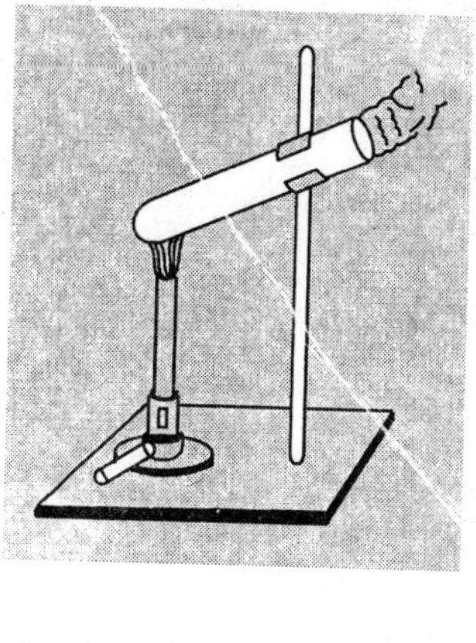

Fig. 16.2 : Heating mercuric oxide yields mercury and oxygen.

Silver oxide, gold oxide, and platinum oxide react similarly, yielding the metal and liberating oxygen when heated. These four metals, mercury, silver, gold and platinum are called the noble metals. Certain dioxides are unstable toward heat, breaking down to give an oxide of the metal and free oxygen, for example,

$$2PbO_2 \rightarrow 2PO + O_2$$

lead dioxide — lead oxide

$$2\,BaO_2 \rightarrow 2BaO + O_2$$

barium dioxide — barium oxide

We must not assume from the above reactions that all metallic oxides give up oxygen when heated. As a matter of fact, most of the oxides of metals are stable and undergo no change when heated.

(b) *By heating certain salts.* Potassium chlorate, a white crystalline substance composed of the elements potassium, chlorine, and oxygen, gives up all of its oxygen when heated above its melting point, 368°C., according to the equation

$$2\,KClO_3 \rightarrow 2KCl + 3\,O_2$$

potassium chlorate — potassium chloride

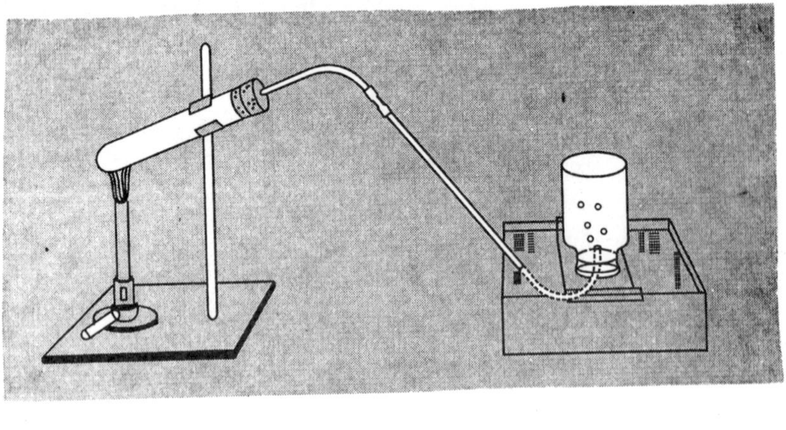

Fig. 16.3 : Collection of oxygen gas by displacement of water.

If we wish to collect some of the oxygen gas, in order to study some of its properties, we may use the apparatus shown in Figure . As oxygen gas is not very soluble in water, we may collect it by displacement of water as shown. As the gas is evolved, it displaces the water in the bottle, and we may collect the bottle full of the gas.

Potassium nitrate when heated yields oxygen, but the reaction differs from the decomposition of potassium chlorate in that only one-third of the oxygen in the potassium nitrate in released. This is evident from the equation

$$\underset{\text{potassium nitrate}}{2\,KNO_3} \rightarrow \underset{\text{potassium nitrate}}{2KNO_2} + O_2$$

Again it should be pointed out that not all oxygen-containing compounds give up their oxygen when heated; K_2SO_4 and SiO_2, for example, are stable toward heat.

(c) *The reaction of sodium peroxide with water*. Sodium peroxide, a white noncrystalline powder, reacts vigorously with water to give oxygen and sodium hydroxide:

$$\underset{\text{sodium peroxide}}{2\,Na_2O_2} + 2\,H_2O \rightarrow \underset{\text{sodium hydroxide}}{4NaOH} + O_2$$

The properties of oxygen

Oxygen is a colourless and odourless diatomic gas. It liquefies at -183°C and freezes at-219°C. Chemically it is very reactive, forming compounds with all other elements except the noble gases and, apart from the halogens and some unreactive metals, these can be made to combine directly with oxygen under the right conditions. The action of a silent electrical discharge upon oxygen gives ozone in concentrations up to 10 per cent.

In addition to its importance in normal respiration and combustion processes, oxygen is used in oxyacetylene welding and cutting, in the manufacture of many metals-particularly steel-and in hospitals, high altitude flying and in mountaineering. The combustion of fuels, e.g. kerosene in liquid oxygen, provides the tremendous thrust required to put satellites into space.

Chemical Properties

Oxygen is a chemically active element that combines either directly or indirectly with nearly all the other elements. The product formed when an element combines with oxygen is called an oxide. In combining with oxygen, the element increased in oxidation number and loses electrons, thereby undergoing oxidation. The oxygen receives the electrons and hence is the oxidizing agent, decreases in oxidation number from zero to-2, and is itself reduced.

1. Action with Metals. The metals down to and including copper in the activity series combine directly with oxygen in the air to form oxides. In most cases, the action is slow at ordinary temperatures but many of the more active metals burn readily when heated in the presence of pure oxygen.

$$2Na + O_2 \rightarrow Na_2O_2 \text{ (sodium peroxide)}$$
$$2Ca + O_2 \rightarrow 2CaO \text{ (calcium oxide)}$$
$$2Mg + O_2 \rightarrow 2MgO \text{ (magnesium oxide)}$$
$$4AI + 3O_2 \rightarrow 2AI_2O_3 \text{ (aluminum oxide)}$$
$$3Fe + 2O_2 \rightarrow Fe_3O_4 \text{ (iron (II, III)oxide)}$$

The oxide formed by burning iron in pure oxygen (Fe3O4) is different from the one produced when iron rusts (Fe_2O_3).In Fe_3O_4, one atom of iron can be thought of as having an oxidation number of +2 and the other atoms of iron an oxidation number of +3 each. The two different oxidation numbers for iron in Fe_3O_4, which may also be represented as $FeO.Fe_2O_3$, are indicated in the name iron (II, III) oxide. However one cannot distinguish the Fe (II) atoms from the Fe (III) atoms due to the rapidity of electron exchange between them.

Many oxides of metals dissolve in water to form aqueous solutions of hydroxide bases and hence are often called basic oxides, or basic anhydrides. The solutions turn red litmus blue, a property of aqueous solutions of bases.

$$Na_2O + H_2O \rightarrow 2Na + 2OH\text{- (sodium hydroxide)}$$
$$CaO + H_2O \rightarrow Ca^2 + 2OH \text{ (calcium hydroxide)}$$

There are, however, many metal oxides, such as Fe_3O_4, which are not significantly soluble in water and thus cannot be considered to be basic oxides.

2. Action with Nonmetals. Many of the nonmetals combine directly with oxygen to form oxides. Although oxygen does not combine directly with the halogens (fluorine, chlorine, bromine, and iodine), oxides of these elements are known. Equations for some typical reactions of nonmetals with oxygen are.

$$2H_2 + O_2 \rightarrow 2H_2O \text{(water)}$$

$$C + O_2 \rightarrow CO_2 \text{(carbon dioxide)}$$

$$S + O_2 \rightarrow SO_2 \text{(sulphur dioxide)}$$

$$P_4 + 5O_2 \rightarrow P_4O_{10} \text{(phosphorus(V)oxide)}$$

Elements which exhibit more than one oxidation number may react with oxygen to form more than one oxide. For example, carbon forms CO and CO_2, sulphur forms SO_2 and SO_3, and phosphorus forms P_4O_6 and P_4O_{10}.

Most oxides of nonmetals dissolve in water to form aqueous solutions of acids and, hence, are often called acidic oxides or acidic anhydrides. The solutions turn blue litmus red, a characteristic of aqueous solutions of acids.

$$CO_2 + H_2O \rightarrow H_2CO_3 \text{ (carbonic acid)}$$

$$SO_2 + H_2O \rightarrow H_2SO_3 \text{ (sulphurous acid)}$$

$$P_4O_{10} + 6H_2O \rightarrow 4H_3PO_4 \text{ (phosphoric acid)}$$

3. Action with Compounds. If a compound is composed of elements which when free will oxygen, the compound ordinarily will also react with oxygen to form oxides of the constituent elements. For example, hydrogen sulfide contains the elements hydrogen and sulphur, both of which will react directly with oxygen when they are free. Thus hydrogen sulfide burns in oxygen, forming water and sulphur dioxide.

$$2H_2S + 3O_2 \rightarrow 2H_2O + 2SO_2$$

Additional examples of compounds that react with oxygen and the equations for the reactions follow.

$$CS_2 + 3O_2 \rightarrow CO_2 + 2SO_2$$
carbon disulfide

$$C_{12}H_{22}O_{11} + 12O_2 \rightarrow 12\,CO_2 + 11H_2O$$
sugar

$$CH_4 + 2O_2 \rightarrow CO_2 + 2H_2O$$

$$2ZnS + 3O_2 \rightarrow 2ZnO + 2SO_2$$
Zinc sulfide

An important point that can be made now is that an equation usually represents only the main reaction which takes place under a given set of condition. There may be by-products, but these are not included in writing the equation for the principal reaction. In the equation for burning sulphur in oxygen, for example, we write $S + O_2 \rightarrow SO_2$; however, when the reaction takes place, small amounts of white smoke (SO_3) are observed. When hydrogen sulfide is burned, the chief products are water and sulphur dioxide, but we may also get small amounts of elemental sulphur or of sulphur trioxide. In the burning of carbon disulfide, possible products, in addition to carbon dioxide and sulphur dioxide, are elemental carbon, carbon monoxide, elemental sulphur, and sulphur trioxide.

Certain oxides, in which all the valence electrons of the element with which the oxygen is combined are not already involved in bonding, will unite with oxygen.

Uses of Oxygen. Oxygen has a number of important uses. In certain diseases such as pneumonia, tuberculosis, gas poisoning, and other respiratory impairments, pure oxygen may be administered in an "oxygen tent". Oxygen is often mixed with various anesthetic agents such as ethylene or nitrous oxide.

The largest amount of commercial oxygen is used in oxyacetylene torches and the hydrogen-oxygen blowpipe for welding purposes. Since oxidation occurs so rapidly in an atmosphere of pure oxygen, high temperatures may be reached, sufficient for melting and cutting metals.

Airplanes carry tanks of oxygen which may be used when flying at high altitudes. Flights into the stratosphere where the concentration of oxygen is low necessitate the use of oxygen for the respiratory needs of the passengers. Oxygen is also used in pulmotors, basal metabolism machines, submarines, and diving bells.

Ozone, O_3-an allotrope of oxygen

Occurrence

Ozone is too reactive to remain for long in the atmosphere at sea level, but at a height of about 20 kilometers it is formed from atmospheric oxygen by the energy of sunlight. This ozone layer protects the earth's surface from an excessive concentration of ultraviolet radiation. Very small concentration of ozone are formed in the vicinity of electrical machinery.

The structure of ozone

The two oxygen-oxygen bond lengths in the ozone molecule are identical and intermediate in value between those for an oxygen-oxygen double bond and an oxygen-oxygen single bond. The molecule is angular as expected (minimisation of electron pair repulsion p.75) with a bond angle of about 117°. It is a resonance hybrid of two main forms:

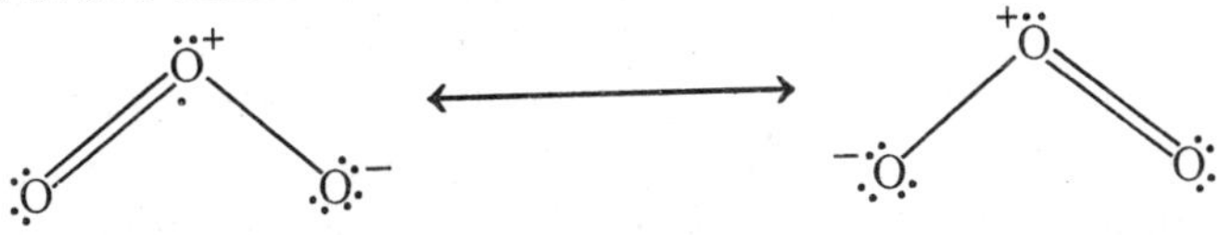

Preparation—When a slow dry stream is passed through a silent electrical discharge, upto 10% converssion to ozone occurs $3O_2 \rightarrow 2O_3$.

Ozone is prepared commercially by an electrical ozonizer which is more complicated in design but more efficient that the laboratory apparatus. An alternative method of commercial production of ozone is by electrolysis of very cold, concentrated sulphuric acid. Ozone is also formed, but in small quantities, by several methods which are not suitable for commercial production; for example, it is produced by the slow oxidation of phosphorus, by a jet of burning hydrogen, by lightning, and by most electrical discharges in air. Significant amounts of ozone are apparently formed as ultraviolet light from the sun falls upon the oxygen in the upper atmosphere. Ozone introduced in the atmosphere has been found to react with engine exhaust products to from substance which are important factors contributing to smog. A great deal of the deterioration of automobile tires is due to contact with ozone in the air. Considerable discussion has arisen recently in connection with the fear that exhaust gases from supersonic airliners may react with ozone in the upper atmosphere causing a decrease in the concentration to such an extent that the ozone layer would no longer constitute a protective layer against the harmful rays of the sun. It is not yet known whether this would happen, but if it does it could have a drastic effect on the equilibria involved in the life processes on the earth.

Properties

Ozone has a characteristic smell and in small concentrations it is harmless. However, if the concentration rises to above about 100 parts per million breathing becomes uncomfortable and it causes headaches. When pure it is a pale blue gas; liquid ozone is a darker blue and solid ozone is violetblack.

Ozone is thermodynamically unstable with respect to oxygen, since its decomposition into oxygen results in the liberation of heat (ΔH is negative) and an increase in entropy (ΔS is positive since the triatomic molecule is broken down into the simpler diatomic oxygen). These two effects reinforce each other, resulting in a large negative free energy change (ΔG) for its conversion into oxygen:

$$\Delta G = \Delta H\text{-}T\Delta S$$

It is not really surprising, therefore, that high concentrations of ozone can be dangerously explosive.

Ozone is a much more powerful oxidising agent than molecular oxygen: for instance, it will oxidise lead sulphide to lead sulphate, iodide ions to iodine and iron (II) ions to the iron (III) state:

$$PbS + 4O_3 \rightarrow PbSO_4 + 4O_2$$

$$2I^- + H_2O + O_3 \rightarrow 2OH^- + I_2 + O_2$$

$$2Fe^{2} + 2H^+ + O^3 \rightarrow 2Fe^{3+} + H_2O + O_2$$

The iodine liberated when ozone reacts with an excess of potassium iodide solution (acidified to remove the hydroxyl ions that are formed) can be titrated with a standard solution of sodium thiosulphate. This is a quantitative method of estimating the gas.

Organic compounds containing carbon-carbon double bonds react with ozone to give ozonides, which can be hydrolyzed to give aldehydes and ketones. This reaction has been used to determine the position of carbon-carbon double bonds in organic molecules, e.g.

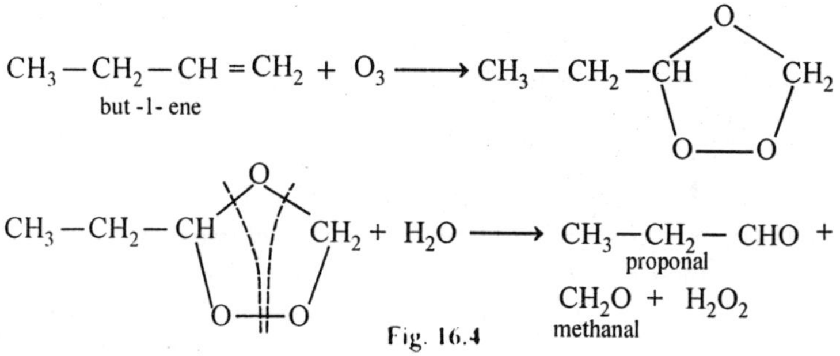

Fig. 16.4

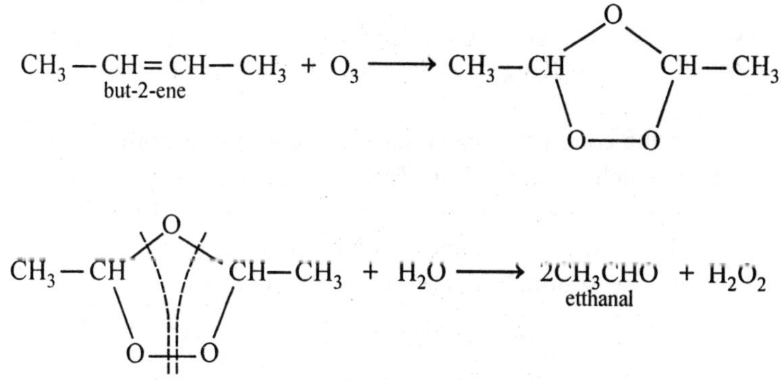

Ozone 'tails' mercury, partially converting it into the oxide so that it sticks to the walls of the vessel containing it. This serves as a conclusive test for the presence of ozone in the presence of excess oxygen.

Ozone has a considerable number of uses. In the treatment of air, ozone is used to eliminate bad odours by oxidizing the substances which cause the smell. A concentration of 0.01 parts per million is sufficient to deodorize the air in public buildings. A concentration of one part per million is noticeable and may be dangerous to health. There is still, however, some uncertainty regarding the toxic properties of ozone at low concentration. Another use of ozone is in the treatment of water. It removes undesirable odours and tastes, and it kills harmful bacteria. The action is one of oxidation, and ozone has the advantage that the only by-product of its action is ordinary oxygen. Ozone has been proposed as a rocket fuel.

Sulphur

Occurrence of sulphur—Sulphur was known to the ancients as brimstone which means "burning stone". It is often referred to in the Bible ancient priests thought of hell as a place not only of fire but of burning brimstone (choking odour of SO_2). The Greek philosophers considered sulphur to be an elementary substance, although their concept of as element was different from the view held today. It was extensively used in the treatment of disease and as a fumigant.[1] Deposits of sulphur are found in Sicily and in Louisiana and Texas, although proved reserves in these areas are being rapidly depleted. Recent discoveries in the Vera Cruz area of Mexico indicate large deposits that will be an important source of supply.

In the combined state, sulphur occurs in the form of metallic sulfides, such as *sphalerite*, ZnS; *galena*, PbS; *chalcocite*, Cu_2S; *chalcopyrite*, $CuFeS_2$; *iron pyrites*, FeS_2; cinnabar, HgS; and several other minerals. These compounds are economically important, since they are the sources of the metals Zn, Cu, Pb, Hg, and others. Some combined sulphur exists as sulfates of calcium (*gypsum*, $CaSO_4$ $2H_2O$) and barium (*barite*, $BaSO_4$). Many organic compounds contain sulphur; proteins in the body contain small amounts, as do certain compounds in onion, garlic, horseradish, and mustard. Sulphur appears to be one of the elements essential to vital life processes.

Production of Sulphur

Sulphur, the brimstone of the ancients, is widely distributed in nature, both as the free element in the form of metallic sulfides and sulphates. Iron pyrites, a sulfide of iron with the formula FeS_2

is one of the commonest of minerals, as is gypsum, or calcium sulfate. Many, if not most, of the commercially valuable metals such as zinc, lead. copper, and silver are, more frequently that not, found combined with sulphur. Great deposits of the element sulphur are found in certain parts of the world, particularly in Texas and Louisiana. This sulphur may have been formed through the action of water containing petroleum as a reducing agent on dissolved calcium and magnesium sulfates. The more important deposits of sulphur are often found associated with regions of the earth in which volcanic activity is notable. A large proportion of the world's sulphur is produced by the Frasch process. This is the process used in the southern United States. The sulphur is found as the element several hundred feet below the surface. and under layers of clay and limestone. The nature of the deposits and the poisonous gases present make it impossible for men to work in shafts such as are used in ordinary mining operations. In the Frasch process the sulphur is melted in the ground and pumped to the surface as a liquid. This procedure is possible because the sulphur melts at about 113°C. only a few degrees above the normal boiling point of water.

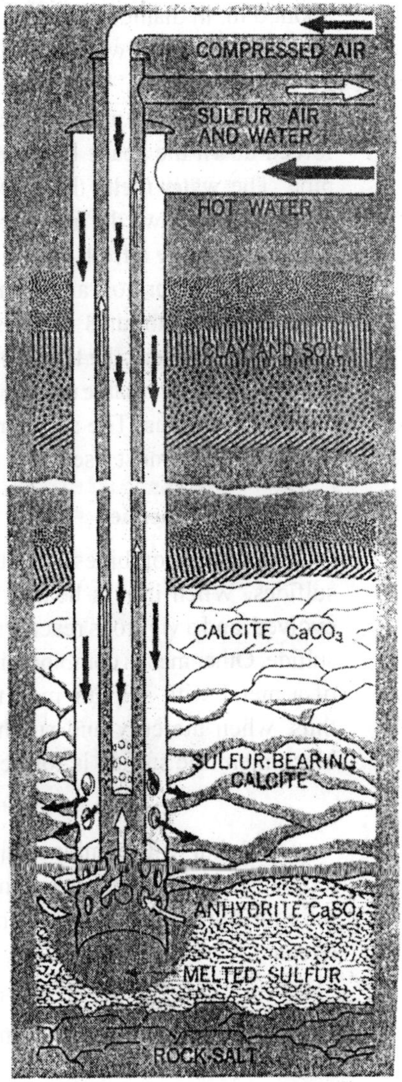

Fig. 16.5 : Diagrammatic representation of the Frasch process.

Three concentric pipes are sunk into the ground, the outermost about 6 in. in diameter and the innermost 1 in. It will be recalled that the boiling point of water rises with increasing pressure. At a pressure of 100 lb per sq in,. the water may be heated to 170°C. This is well above the melting point of the sulphur. This superheated water is forced down the space between the outermost and the intermediate pipe. The water melts the sulphur in the ground. Then compressed air is pumped down the innermost pipe. This air mixes with the molten sulphur, forming a froth which is forced up and emerges in the space between the innermost and the intermediate pipes. A fourth pipe with a diameter of about 8 in. is sometimes used to carry additional superheated water. The liquid product is collected in large bins where it solidifies into veritable mountains of yellow sulphur, containing many thousands of tons. The sulphur is very pure and requires no further processing for most uses.

Chemical properties

Sulphur combines directly with most of the metals to form sulfides. When iron is heated with sulphur, a great deal of heat is evolved, and a vigorous reaction ensures, with the formation of ferrous sulfide. Other metals combine similarly. In some cases, the combination of a metal with sulphur may take place at ordinary temperatures; thus, when mercury and sulphur are ground together in a mortar, black mercuric sulfide is formed:

$$Hg + S \rightarrow HgS$$

Sulphur will combine directly with certain of the nonmetals, although the combination does not take place as readily as with metals. Sulphur burns with a blue flame in air or oxygen to form sulphur dioxide:

$$S + O_2 \rightarrow SO_2$$

Examples of other compounds formed by sulphur with nonmetals are carbon disulfide, CS_2, phosphorus sesquisulfide, P_4S_3, and sulphur chloride, S_2Cl_2

Sulphur exhibits oxidation states of -2, +4, and + 6. As a member of group VIA, it tends to acquire two electrons to complete a stable octet configuration in the sulfide ion, S^{--}. Sulphur also forms covalent bonds with various other nonmetals to assume oxidation numbers of

+ 4, as in sulphur dioxide and sulfites, and + 6, as in sulphuric acid and sulfates.

The allotropy of sulphur

Unlike oxygen which is a discrete molecule, two atoms being united by a double bond, sulphur atoms show a marked reluctance to double bond with themselves and the two main allotropes of sulphur contain S_8 molecules, in which single bonds unite sulphur atoms into a puckered octagonal ring (Fig.16.6). The high molecular weight of these S_8 structural units explains why sulphur, unlike oxygen, is a solid. The bond angles of 105° are consistent with the simple theory of electron pair repulsion.

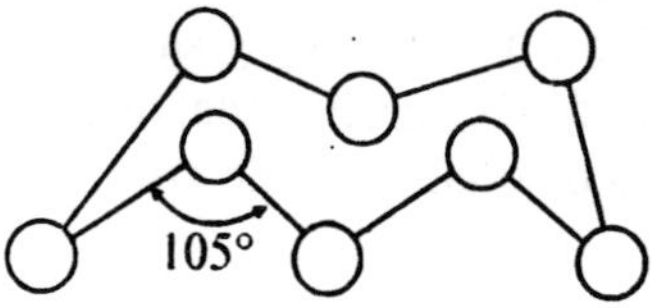

Fig. 16.6 : The structure of the S_8 molecule.

Rhombic sulphur

This is the form of sulphur normally encountered and consists of S_8 structural units packed together to give crystals whose shape is shown in Fig.16.7(a). Fairly large crystals can be obtained by allowing a solution of powdered roll sulphur in carbon disulphide to evaporate slowly; they are yellow, transparent and have a density of 2 g cm^{-3}.

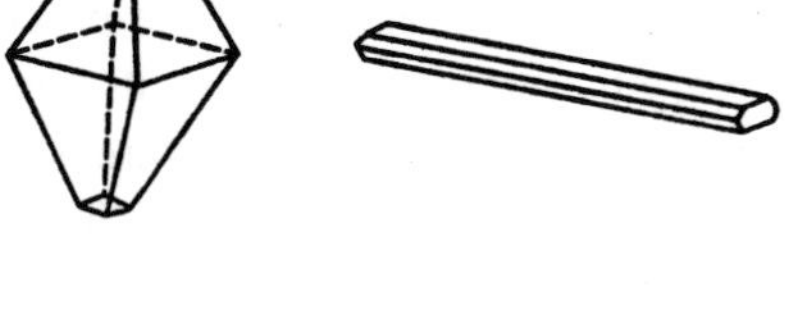

Fig. 16.7 : The crystal shapes of (a) rhombic sulpur. (b) monoclinic sulphur.

Monoclinic sulphur

This form of sulphur is formed when molten sulphur is allowed to crystallize above 95 6° C. Like rhombic sulphur it consists of S_8 structural units, but these are arranged differently in the crystal lattice (Fig.16.7(b)). The temperature of 95·6° C is the transition temperature for sulphur; below this temperature rhombic sulphur is the more stable allotrope and above it, monoclinic sulphur is the more stable of the two forms. This type of allotropy, in which a definite transition point exists where two forms become equally stable, is called enantiotropy.

$$\text{Rhombic sulphur} \overset{95.6^{\circ}C}{\rightleftharpoons} \text{Monoclinic sulphur}$$

Crystals of monoclinic sulphur are amber-yellow in colour and have a density of 1 96 g cm^{-3}. As they gradually change over into rhombic sulphur below 95 6° C, each crystal retains its overall shape but changes into a mass of small rhombic crystals.

Amorphous sulphur

A number of forms of sulphur which possess no regular crystalline form can be obtained when sulphur is liberated in chemical reactions, e.g. by the action of dilute hydrochloric acid on a solution of sodium thiosulphate:

$$S_2O_3^{2-} + 2H^+ \rightarrow H_2O + SO_2 + S$$

Plastic sulphur

This is obtained, as an amber-brown soft and elastic solid, by pouring nearly boiling sulphur into cold water. It consists of a completely random arrangement of chains of sulphur atoms, which, when stretched, align themselves parallel to each other. On standing, it slowly changes over into rhombic sulphur, as the chains of sulphur atoms break and reform the S_8 cyclic units.

Uses of sulphur- Sulphur is used in the processing of rubber. Natural rubber is very thermoplastic: that is, its consistency changes markedly is added to the rubber and the mixture is heated, the rubber becomes much tougher and less affected by heat. This process is called vulcanization. It is very essential to the manufacture of rubber goods, particularly for automobile tires which must remain strong at high temperatures.

Large amounts of sulphur are used, mixed with slaked lime, for spraying fruit trees, etc., to kill fungus growth. Sulphur is also used as the starting point in the manufacture of other sulphur compounds such as sulphur dioxide, sulphuric acid, hydrogen sulfide, sodium sulphide etc. Sulphur is a component black powder, together with carbon and potassium nitrate, constituent of coal gas and sewer gas. The offensive odour of rotten eggs is largely due to the presence of this substance.

Preparation of hydrogen sulfide-Sulphur does not combine appreciably with hydrogen at ordinary temperatures, but at an elevated temperature combination takes place readily to form hydrogen sulfide:

$$H_2 + S \rightleftharpoons H_2S$$

The reaction, however, is reversible, and since hydrogen sulfide itself is quite unstable toward heat, the yield is small.

In the laboratory it is most conveniently prepared by the action of a non-oxidizing acid on certain metallic sulfides:

$$FeS + (\text{dilute})\ H_2SO_4 \rightarrow FeSO_4 + H_2S\uparrow$$

$$ZnS + 2HCI \rightarrow ZnCI_2 + H_2S\uparrow$$

These equations are further examples of the general method of preparing an acid by treating a salt of the desired acid with a less volatile acid. Though volatile, hydrochloric acid may be used since it is less volatile than hydrogen sulfide from water solution.

Properties

Hydrogen sulfide is a colourless gas at ordinary temperatures, with a foul odour like that of rotten eggs. It may be condensed to a liquid at-59.6° and converted to a solid at -83°. It is not very soluble in water-at ordinary temperatures, one liter of water dissolves about three liters of the gas. It is very poisonous, since it paralyzes the nerve centres of the hear and lungs-more than one-half of one per cent in the air may be fatal to humans and animals. Fortunately, it is easily detected even in small quantities because of its characteristic odour.

Hydrogen sulfide reacts with water to form a solution with acidic properties:

$$H_2S + H_2O \rightleftharpoons H_3O^+ + HS^- \qquad K_i = 1.2 \times 10^{-7}$$

$$HS^- + H_2O \rightleftharpoons H_3O^+ + S^{--} \quad K_i = 1 \times 10^{-15}$$

or

$$H_2S + 2H_2O \rightleftharpoons 2H_3O^+ + S^{--}$$

The ionization is very slight; consequently, hydrosulphuric acid is classed as a weak acid. It exhibits typical weak acid reactions, reacting with bases to form salts; for example,

$$2NaOH + H_2S \rightleftharpoons Na_2S + 2H_2O$$

Hydrogen sulfide burns in the air with a blue flame, producing sulphur dioxide and water.

$$2H_2S + 3\,O_2 \rightarrow 2H_2O + 2\,SO_2$$

If combustion takes place in a limited supply of air (for example, in an open bottle) free sulphur is deposited:

$$2H_2S + O_2 \rightarrow 2H_2O + 2\,S$$

Hydrogen sulfide is a good reducing agent; note these reactions

$$3H_2S + \text{(dilute) } 2\,HNO_3 \rightarrow 3\,S + 2\,NO + 4\,H_2O$$

$$H_2S + I_2 \rightarrow S + 2\,HI$$

$$H_2SO_4 + H_2S \rightarrow S + SO_2 + 2\,H_2O$$

$$2\,H_2S + SO_2 \rightarrow 3\,S + 2\,H_2O$$

Certain metallic ions produce sulfides with hydrogen sulfide such as lead sulfide (black), copper sulfide (black), cadmium sulfide (yellow), antimony sulfide (orange), manganese sulfide (pink). The formation of a coloured sulfide may be used as a basis for a test for the presence of hydrogen sulfide. If a piece of filter paper saturated with lead nitrate solution is placed in a stream of hydrogen sulfide gas, a black deposit of lead sulfide is formed on the paper:

$$Pb^{++} + H_2S \rightarrow PbS \downarrow + 2\,H^+$$

This test useful in locating breaks in sewer lines, as well as in pipes carrying coal gas or natural gas, which usually contain a little hydrogen sulfide.

• *Uses of hydrogen sulfide*- The use of hydrogen sulfide is confined largely to the chemical laboratory, where it is employed in the separation and identification of metallic ions. Its use as an analytical reagent is very important. Let us consider for a moment the way in which it functions as a qualitative reagent.

The equilibrium

$$H^2S \rightleftharpoons H^+ + HS^- \rightleftharpoons 2\,H^+ + S^-$$

may be shifted by the addition of an acid or base. If acid is added, repression of ionization occurs, resulting in a decrease of the sulfide ion concentration. The addition of hydroxyl ions from a base causes a shift in equilibrium to the right because of the formation of un-ionized water by the combination of hydrogen and hydroxyl ions. This results in an increase of the sulfide ion concentration. It is evident, then, that the sulfide ion concentration can be controlled by proper adjustment of the acidity or basicity of the solution. Since the metallic sulfides have different solubility product constants, theoretically we could successively precipitate the metallic ions as sulfides by proper control of the acidity of the solution. Actually, in qualitative analysis the concentration of hydrogen ion is adjusted to 0.3 molar, in which solution the sulfides of arsenic, antimony, mercury, tin, lead, copper, bismuth, and cadmium as precipitated. Other metallic sulfides are not precipitated because the sulfide ion concentration in the solution is not sufficient to exceed the solubility product of these other sulfides. By making the solution basic the sulfide ion concentration is increased to the point where another group of sulfides precipitate. By means of hydrogen sulfide, the metallic ions may be readily separated into three groups: (1) those sulfides which are insoluble in dilute acid; (2) those sulfides which are insoluble in basic solution; and (3) those sulfides which are soluble in both acid and basic solutions.

Sulphur Dioxide

When sulphur burns in air or in oxygen, the product, as already pointed out, is sulphur dioxide.

$$S + O_2 \rightarrow SO_2$$

This is a typical oxidation in which the oxidation state of the sulphur goes from 0 to +4. Sulphur dioxide may also be prepared in other ways, two of which will be mentioned. If copper is treated with dilute sulphuric acid, no reaction takes place, but if the sulphuric acid is concentrated and hot, sulphur dioxide is liberated.

$$Cu + 2H_2SO_4 \text{ (conc, hot)} \rightarrow CuSO_4 + SO_2 + 2H_2O$$

Another convenient method for making sulphur dioxide is to treat sodium sulphite, Na_2SO_3, or sodium bisulfite, $NaHSO_3$, with dilute hydrochloric or sulphur acid.

$$NA_2SO_3 + 2H_2SO_4 \rightarrow 2NaHSO_4 + H_2O + SO_2$$

$$NaHSO_3 + HCI \rightarrow NaCI + H_2O + SO_2$$

Sulphur dioxide is also often formed as a by-product in smelting operations.

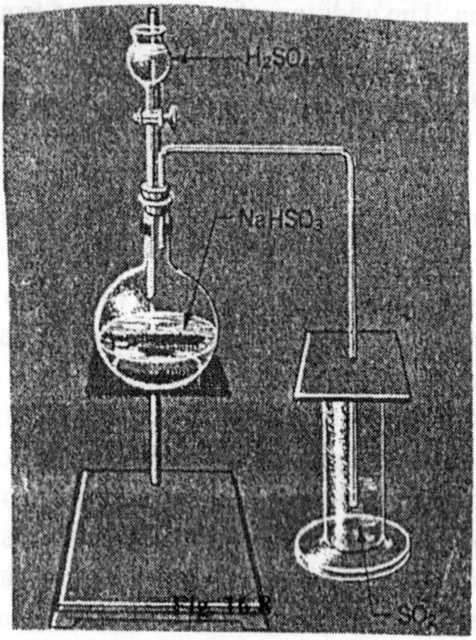

Fig 16.8

Sulphur dioxide is a colourless gas with a very strong, sharp, choking odour. It is highly poisonous, but under normal conditions accidental breathing of the gas rarely occurs. This is because the powerful odour of the gas gives ample warning of its presence.

The arrangement of electrons in sulphur dioxide present a peculiar problem. The sulphur atom has six electrons in its valence shell, and the oxygen also has six. Electron diffraction and other methods tell us that the two sulphur-oxygen bonds are approximately at right angles to each other and that the distance between sulphur and oxygen is the same in each case. Furthermore, whenever an atom or a molecule posesses an electron which is not paired with another electron, that substance is attracted to a magnet. Sulphur dioxide is not so attracted, and hence all electrons present must be paired. If both sulphur and oxygen are to have an octet of electrons as in argon, then the only possible arrangement would appear to be one in which two pairs of electrons are shared by one of the oxygens and only one by the other. This might be represented as follows:

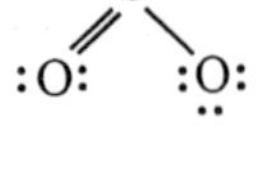

or

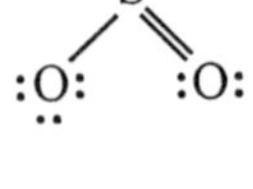

The difficulty with either one of these formulas is that in every known case of a double bond (two pairs of electrons) the distance between the atoms so bonded is less than it is for a single bond. The explanation currently accepted is that a combination of both formulas given above would be a more nearly satisfactory representation. Such a condition, in which no one electronic formula gives a structure which is consistent with known properties, is called resonance. We shall have occasion to refer to other examples in later chapters.

Under moderate pressure, or at somewhat lowered temperature, sulphur dioxide becomes a colourless liquid.

Many of the important chemical properties of sulphur dioxide are concerned with the solution of the gas in water. Pure sulphur dioxide is used chiefly in the manufacture of sulfuric acid, and in mechanical refrigerators. But its use for the latter purpose is being superseded by less poisonous substances. Sulphur dioxide gas is also applied as a disinfectant. Its use for this purpose, particularly for infected building, goes back many thousands of years, as is obvious from the quotation with which this chapter starts. But at the present time sulphur dioxide is being displaced by more effective substances.

The reader may wonder why some compounds are called dioxides and some peroxides, although in the formulas SO_2 and BaO_2 there is considerable superficial resemblance. The reason for this distinction is as follows: any compound containing two oxygen atoms would normally be called dioxide, but some dioxides have the property of forming hydrogen peroxide which has unique oxidizing properties. If SO_2 is treated with dilute sulphuric acid, no reaction occurs. But if BaO_2 is treated with dilute sulfuric acid, hydrogen peroxide is liberated.

$$BaO_2 + H_2SO_4 \rightarrow BaSO_4 + H_2O_2$$

The hydrogen peroxide so formed is easily identified by its oxidizing properties. An oxide which, when treated with dilute acid, forms hydrogen peroxide is called a peroxide. Other compounds, the formulas of which contain two atoms of oxygen, are generally called dioxides. There are differences in the mode of linkage between the atoms in peroxides and dioxides.

Sulphur dioxide dissolves to a considerable extent in water. The solution so formed consists in part of sulfurous acid.

$$SO_2 + H_2O \rightleftharpoons H_2SO_3$$

Sulfurous acid is a weak acid which ionizes to a slight degree into bisulfite ion, HSO_3^- and to an even smaller degree to sulphite ion, SO_3^{2-}.

$$H_2SO_3 \rightleftharpoons H^+ + HSO_3^-$$
$$HSO_3^- \rightleftharpoons H^+ + SO_3^{2-}$$

The solution smells very strongly of sulphur dioxide, all of which may be expelled by a few moments' boiling.

The formation of an acid by the oxide of an element as above, serves to classify the element. Many elements form oxides which, like that of sulphur, give acids on treatment with water. Such elements are called non-metals. Another example is phosphorus, an oxide of which, P_2O_5, yields phosphoric acid when it is treated with water.

$$P_2O_5 + 3H_2O \rightarrow 2H_3PO_4$$

In contrast to the nonmetals, certain other elements give oxides which, in water, form bases. Examples are sodium and calcium.

$$Na_2O + H_2O \rightarrow 2NaOH$$
$$CaO + H_2O \rightarrow Ca(OH)_2$$

Elements, the oxides of which form bases with water, are called metals. The distinction between metals and nonmetals is often used as a definition of these two classes of elements. As we shall see later, however, there are other properties which also serve to distinguish one class of elements from another.

Sulphur, as pointed out above, is capable of exhibiting several different oxidation states from-2 to +6. Because of this the sulphur in sulphurous acid, with an oxidation state of +4, is capable of being either oxidized or reduced. A typical case of oxidation occurs when sulphurous acid is shaken with air, especially with the addition of a drop of cupric chloride as catalyst. The sulphurous acid is oxidized to sulphuric acid, the oxidizing agent being, of course, the oxygen of the air.

$$2H_2SO_5 + O_2 \rightarrow 2H_2SO_4$$

The principal use of sulphurous acid is as a bleaching agent. Large quantities are used in the bleaching of delicate substances such as silk, paper, straw, or wool, which would be injured by more vigorous agent such as chlorine. As a preservative for fruits and vegetables, sulphur dioxide is effective but it is not harmless. Pure-food laws require that sulphur dioxide be almost entirely absent from the final product.

Sulphurous acid reacts with bases to form bisulfites and sulphites. With sodium hydroxide the reactions are as follows:

$$H_2SO_3 + NaOH \rightarrow \underset{\text{Sodium bisulfite}}{NaHSO_3} + H_2O$$

$$H_2SO_3 + 2NaOH \rightarrow \underset{\text{Sodium sulfite}}{Na_2SO_3} + 2H_2O$$

As previously stated, all sulphites and bisulfite give sulphur dioxide when they are treated with acid.

The "sulphite" process for the manufacture of paper pulp involves the compound calcium bisulfite, $Ca(HSO_3)_2$, which dissolves the lignin from wood, leaving relatively pure cellulose, from which paper is made.

Sulphuric acid, H_2SO_1

Preparation of sulphuric acid

Essentially the preparation of this acid involves the conversion of sulphur dioxide into sulphur trioxide which is then dissolved in water to form sulphuric acid.

(a) Sources of sulphur dioxide

The percentages of sulphuric acid manufactured from different sources of sulphur in Great Britain are as follows:

Imported sulphur-mostly Frasch sulphur from the USA, burnt to give sulphur dioxide (42 per cent).

Anhydrite (Ca^2 + SO_4^{2-})-mined in Great Britain, producing sulphur dioxide (16 percent).

Iron pyrites (FeS_2)-roasted in air to form sulphur dioxide (16 per cent).

Spent oxide (Fe_2S_3) from town gas purification-treated to give sulphur dioxide (12 per cent).

Hydrogen sulphide from oil refining-controlled burning in air producing sulphur dioxide (8 per cent).

Perhaps the most significant development during the last few years has been the isolation of sulphur from natural gas, particularly in France, which has now become the third largest producer of sulphur. Hydrogen sulphide from this source and from crude oil is extracted and burnt in air to form sulphur dioxide: this is then reacted with more hydrogen sulphide to form sulphur:

$$2H_2S + 3O_2 \rightarrow 2H_2O + 2SO_2$$
$$2H_2S + SO_2 \rightarrow 2H_2O + 3S$$

Sulphur dioxide is converted into sulphuric acid by 'the lead chamber process' and by 'the Contact process'. Since the former process only accounts for about 15 per cent of the sulphuric acid produced in Great Britain (no new chamber process plants are built nowadays) and in any event is adequately described in O-level texts, the latter process only is described here.

(b) The Contact process

Sulphur dioxide, intensively purified to remove impurities such as arsenic which would 'poison' the catalyst, and air in slight excess are passed over a vanadium (V) oxide catalyst at a temperature of about 450°C. The reaction is exothermic and the temperature rises to about 600°C; the gases are now cooled by passing through a heat exchanger (the heat extracted being used to heat up the initial sulphur dioxide/air mixture), and then absorbed in 98 per cent sulphuric acid to give 100 per cent sulphuric acid or, if the absorption is carried further, fuming sulphuric acid or oleum, $H_2S_2O_7$. Dilution of the fuming sulphuric acid with water gives sulphuric acid. Any unchanged sulphur dioxide (about 2 per cent of the initial amount) must be removed since it is poisonous; this is done by absorbing in calcium hydroxide solution.

$$2SO_2 + O_2 \text{ ? } 2SO_3 \qquad \Delta H^{\theta}_{298} = -94.6 kJ / mol\ of SO_3$$
$$SO_3 + H_2SO_4 \rightarrow H_2S_2O_7$$
$$H_2S_2O_7 + H_2O \rightarrow 2H_2SO_4$$

Since the catalytic oxidation of sulphur dioxide to sulphur trioxide is an exothermic reaction, a high yield of the trioxide will be favoured by a reasonably low temperature, but the temperature must not be too low or the rate of reaction will become too slow. A temperature of about 450°C is the optimum temperature and a conversion rate of about 95 to 98 per cent is achieved.

The reaction is also attended by a decrease in volume, hence in increase in pressure should increase the equilibrium yield of sulphur trioxide and also the rate of reaction. Since both yield and reaction rate are quite satisfactory at ordinary pressure it has been decided that the extra expense of making pressure equipment would not be justified.

Physical properties of sulphuric acid

Pure sulphuric acid is a viscous liquid which has a density of 1.85 g cm^{-3} and a freezing point of 10.5°C. In the absence of water it will not turn blue litmus red nor will it react with metals to form hydrogen. It decomposes on boiling to form sulphur trioxide and steam, and a constant boiling point mixture is formed, containing 98.3 per cent of acid.

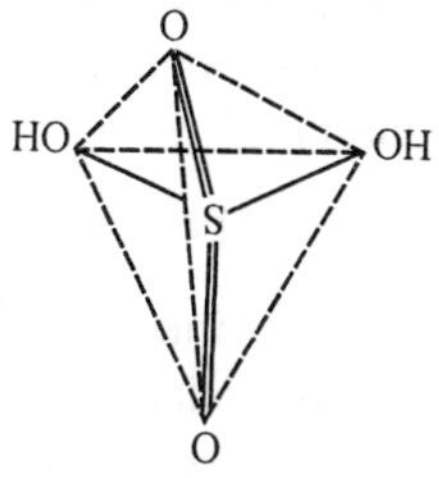

Fig. 16.9 : The structure of the sulphuric acid molecule

Structure of sulphuric acid

Pure sulphuric acid is covalent, its molecule having an approximately tetrahedral structure and containing 6-valent sulphur (Fig.16.9). Its high boiling point (270°C) and viscosity are presumably due to the presence of hydrogen bonding which link the molecules into larger aggregates:

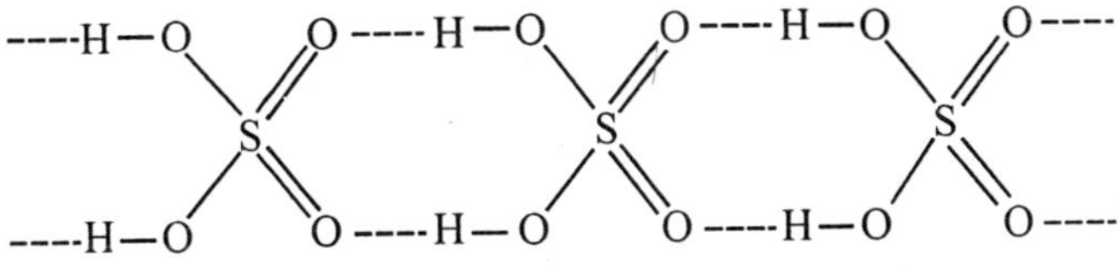

Chemical properties of sulphuric acid

These are most conveniently discussed under several heading.

(a) Acid properties

Concentrated sulphuric acid reacts violently with water to give a solution which exhibits the properties of a strong acid ,e.g. it displaces carbon dioxide from a carbonate, neutralizes a base to form a salt and water, and reacts with metals high up in the electrode potential series to form a salt and hydrogen.

The ionisation of sulphuric acid in water takes place in two stages, i.e. it is dibasic; the first ionisation is virtually complete, whereas the second takes place to the extent of about 10 per cent:

$$H_2SO_4 + H_2O \ ? \ H_3O^+ + HSO_4^- \text{ (virtually complete)}$$
$$HSO_4^- + H_2O \ ? \ H_3O^+ + SO_4^{2-} \text{ (about 10 complete)}$$

An aqueous solution of the acid therefore contains H_3O^+ ions and very many more HSO_4^- than SO_4^{2-} ions. In the presence of OH- ions, which are stronger proton accepts than the neutral water molecules, more extensive ionisation of the HSO_4^- ions takes place and SO_4^{2-} ions predominate:

$$HSO_4^- + OH \rightarrow H_2O + SO_4^{2-} \text{ (equilibrium to the right)}$$

This explains why the interaction of sodium hydroxide solution and dilute sulphuric acid produces sodium sulphate, and why special conditions are needed to crystallize out the hydrogen sulphate.

(b) Dehydrating properties

Concentrated sulphuric acid has such an affinity for water that it will remove it from mixtures and compounds with the evolution of much heat (the acid should always be diluted by pouring it into water while stirring, and not the other way round). Gases are dried in the laboratory by passing them through a wash bottle containing concentrated sulphuric acid, except of course the gases that chemically combine with the acid.

Many organic compounds lose the elements of water when treated with the concentrated acid; thus glucose gives carbon, formic acid produces carbon monoxide, and oxalic acid gives an equimolecular mixture of carbon monoxide and carbon dioxide:

$$C_6H_{12}O - 6H_2O \rightarrow 6C$$
$$HCOOH - H2O \rightarrow CO$$

$$\begin{array}{c} COOH \\ | \\ COOH \end{array} - H_2O \rightarrow CO + CO_2$$

The dehydration of organic compounds is of industrial importance; thus diethyl ether is made by dehydrating ethanol and the insecticide DDT is made by dehydrating a mixture of chlorobenzene and trichloroacetaldehyde (chloral):

$$2C_2H_5OH \xrightarrow{-H_2O} \underset{\text{diethyl ether}}{C_2H_5{-}O{-}C_2H_5}$$

$$2C_6H_5Cl + CCl_3CHO \xrightarrow{-H_2O} \underset{\text{DDT}}{ClC_6H_4{-}\begin{array}{c} CCl_3 \\ | \\ C \\ | \\ H \end{array}{-}C_6H_4Cl}$$

(c) Oxidizing properties

Hot concentrated sulphuric acid functions as an oxidizing agent, although less effectively than concentrated nitric acid. A number of reduction products of sulphuric acid can be formed, their relative proportions in any one particular reaction depending upon such factors as concentration of the acid, strength of the reducing agent present, and temperature. These possible reduction products are shown schematically below (underlined).

$$H_2SO_4 \xrightarrow{-[O]} (H_2SO_3) \rightarrow H_2O + \underline{SO_2}$$

$$H_2SO_4 \xrightarrow{-4[O]} \underline{H_2S}$$

which may be followed by oxidation of the hydrogen sulphide to sulphur:

$$3H_2S + H_2SO_4 \rightarrow 4H_2O + 4\underline{S}$$

Non-metals such as carbon and sulphur reduce sulphuric acid when heated to sulphur dioxide, they themselves being oxidized to their dioxides:

$$C + 2H_2SO_4 \rightarrow 2H_2O + CO_2 + 2SO_2$$

$$S + 2H_2SO_4 \rightarrow 2H_2O + 3SO_2$$

With metals, it is possible to obtain sulphur dioxide, hydrogen sulphide and sulphur in addition to the metallic sulphate; further reaction between the metallic sulphate and hydrogen sulphide can give rise to significant amounts of the metallic sulphide. These reactions are, therefore, quite complex; however, in general it can be said that sulphur dioxide is the main reduction product when metals low down in the electrode potential series are used, e.g. copper, but that increasing amounts of hydrogen sulphide are formed with more reactive metals, e.g.zinc. The following equations, therefore, do no more than indicate the main reactions that take place:

$$Cu + 2H_2SO_4 \rightarrow CuSO_4 + 2H_2O + 2SO_2$$

$$\underset{\text{98\% acid}}{Zn + 2H_2SO_4} \rightarrow ZnSO_4 + 2H_2O + 2SO_2$$

$$\underset{\text{90\% acid}}{4Zn + 5H_2SO_4} \rightarrow 4ZnSO_4 + 4H_2O + H_2S$$

Compounds oxidized by concentrated sulphuric acid include hydrogen bromide and hydrogen iodide:

$$2HBr + H_2SO_4 \rightarrow 2H_2O + SO_2 + Br_2$$

$$8HI + H_2SO_4 \rightarrow 4H_2O + H_2S + 4I_2$$

These reactions explain why the free halogen is always liberated, together with the halogen hydride, when bromides and iodides are treated with concentrated sulphuric acid.

(d) Displacement reactions

When concentrated sulphuric acid is added to a nitrate there is little sign of any reaction in the cold; nevertheless an equilibrium reaction is set up:

$$NO_3^- + H_2SO_4 \rightleftharpoons HSO_4^- + HNO_3$$

On warming the equilibrium moves to the right, since nitric acid, which has a lower boiling point than sulphuric acid, is removed from the system as a vapour.

Other displacement reactions involving sulphuric acid are similar and take place because the substance that is displaced has a lower boiling point than sulphuric acid, e.g.

$$Cl^- + H_2SO_4 \rightleftharpoons HCl + HSO_4^-$$

Hydrogen chloride is readily displaced from this system since it is a gas.

In some instances the substance displaced is unstable and breaks down to give volatile products which are lost from the system, thus enabling the reaction to proceed, e.g.

$$SO_3^{2-} + 2H_2SO_4 \rightleftharpoons 2HSO_4^- + H_2SO_3 \rightleftharpoons H_2O + SO_2$$

In the above reaction the loss of sulphur dioxide as a gas allows the reaction to proceed to completion.

These reactions, which depend on the high boiling point of concentrated sulphuric acid, explain why this acid is useful in analysis for detecting the presence of certain acid radicals.

(e) Reactions with organic compounds

Alkenes undergo addition reactions with concentrated sulphuric acid; if the addition products are diluted with water and heated, alcohols are produced. This type of reaction is of industrial importance , e.g.

$$\underset{\text{but-1-ene}}{CH_3{-}CH_2{-}CH{=}CH_2} + H_2SO_4 \rightarrow CH_3{-}CH_2{-}\underset{\displaystyle OSO_3H}{\underset{|}{CH}}{-}CH_3$$

$$CH_3{-}CH_2{-}\underset{\displaystyle OSO_3H}{\underset{|}{CH}}{-}CH_3 + H_2O \rightarrow \underset{\text{butan-2-ol}}{CH_3{-}CH_2{-}\underset{\displaystyle OH}{\underset{|}{CH}}{-}CH_3} + H_2SO_4$$

Sulphuric acid also reacts with long chain hydrocarbons with the elimination of water, e.g.

$$C_{18}H_{30} + H_2SO_4 \rightarrow \underset{\text{a sulphonic acid}}{C_{18}H_{29}SO_3H} + H_2O$$

The sodium salts of sulphonic acids are used in detergent preparations, e.g. sodium benzene dodecyl sulphonate, $C_{18}H_{29}SO_3^- Na^+$, is the main surface active constituent of Tide, Daz, Surf and Omo. The importance of dehydration reactions involving sulphuric acid and organic compounds has been mentioned earlier, e.g. the preparation of diethyl ether and DDT.

Uses of sulphuric acid

Sulphuric acid is used on an enormous scale; in Great Britain alone over $2\frac{1}{2}$ milllion tonnes of the concentrated acid are manufactured. Just over one third of the total output is used in the manufacture of fertilizers, e.g. 'superphosphate' and ammonium sulphate. Other uses (in order of consumption of the acid) include.

Paint manufacture—titanium (IV) oxide and lithopone.
Synthetic fibres—rayon and plastics.
Acid production—hydrochloric and hydrofluoric acids.
Metallurgy—pickling steel and sulphates for electrolysis.
Detergents—mostly alkyl-aryl sulphonates.
Petroleum refining—mostly extraction of alkenes.
Organic chemicals—dyestuffs and explosives and drugs.

When used in the manufacture of dyestuffs and explosives its purpose is to enhance the nitrating property of nitric acid.

17

P-Block Elements : (Halogens)—Group VIIA (17)

Introduction

The ground-state configuration of the halogens, ns^2np^5, is one electron short of that of the text rare gas. Accordingly, these elements acquire one electron to form ions derivatives of the type X^-, or they can form a simple covalent bond, X—R. Covalent bonds formed to the lighter elements in this family are generally very polar; if the molecule formed by the halogens is unsymmetrical, the compounds are among the most polar known. In addition, the halogens which are heavier than fluorine can expand their coordination number in the same way as the other representative elements in Periods 3—7. Thus, iodine uses all seven outer electrons to form covalent bonds in the compound IF_7 just as sulfur is capable of using all of its outer electrons in forming SF_6. Not only is it possible to achieve high coordination numbers among the heavier halogens, but compounds in which the halogen exhibits a high positive oxidation state, such as in $HClO_4$, are formed. Fluorine, which has the highest electronegativity among all of the elements, shows no tendency to form positively charged species.

The first halogen displays properties markedly different from those of the other memebrs of the family. Thus, chlorine, bromine, and iodine passes smoothly varying physical properties, but those of fluorine do not follow the same trends. For example, the ionization potential, electronegativity, heats of dissociation and of hydration, and electrode potentials are all higher for fluorine than expected from the trends established by the other halogens. Also, the ionic and covalent radii, melting and boiling points, and the electron affinity of fluorine are lower than expected. The differences between fluorine and the heavier members of the family appear to be more pronounced than the corresponding differences in other families.

About twenty isotopes of astatine are known but all are relatively short-lived. The most stable isotope has a half-life of 8.3 hours; the chemistry of astatine has been obtained from tracer studies. In general, astatine exhibits behaviour expected from the halogen heavier than iodine(*I*).

Some physical data of elements

	Atomic Number	Electronic Configuration	Atomic Radius (Å½	Ionic Radius (A) X	M.p. °C	B.p. C°
N	7	2.5 $1s^2 2s^2 2p^3$	0.74		-210	-1958
P	15	2.8.5 ...$2s^2 2p^6 3s^2 3p^3$	1.10		44.1 (white)	280.5 (white)
As	33	2.8.18.5 ...$3s^2 3p^6 3d^{10} 4s^2 4p^3$	1.21	0.69		610 (sublimation)
Sb	51	2.8.18.8.5 ...$4s^2 4p^6 4d^{10} 5s^2 5p^3$	1.41	0.90	630.5	1380
Bi	83	2.8.18.32.18.5 ...$5s^2 5p^6 5d^{10} 6s^2 6p^3$	1.52	1.20	271	1450

Fluorine

Occurence and preparation of fluorine

The most important naturally occurring compound of fluorine is fluorspar (Calcium fluoride), $Ca^{2+}(F^-)_2$; cryolite, $(Na^+)_3AlF_6^{3-}$.

The only possible method of preparing fluorine is by electrolysis, and even then the chemicals used must be free from traces of water which react rapidly with the free halogen. The electrolyte used is hydrogen fluoride to which some potassium fluoride has been added to increase the electrical conductivity. One type of cell which operates at about 250°C employs an equimolar mixture HF—KF; carbon is used as the anode and a steel vessel serves as the cathode. The overall process involves discharge of fluorine at the anode and hydrogen at the cathode; hydrogen fluoride is added to the cell periodically to maintain the correct composition of the mixture.

Fluorine discharged at the anode: $2F^- \rightarrow F_2 + 2e$

Hydrogen discharged at the cathode: $2H^+ + 2e \rightarrow H_2$

The fluorine is contaminated with hydrogen fluoride which is removed by passing it over sodium fluoride pellets:

$$Na^+F^- + HF \rightarrow Na^-HF_{2^-}$$

sodium hydrogen fluoride

Properties of fluorine

Fluorine is a pale yellow gas with an irritating smell; it is extremely poisonous. It liquefies at -187°C and freezes at -223°C.

It is the most chemically reactive non-metal, combining directly with all other non-metals except nitrogen, oxygen and the noble gases (except krypton and xenon). Most metals catch fire in fluorine and even gold and platinum are attakced if heated in a stream of the gas. Elements generally attain their highest valency state in combination with fluorine, e.g. the highest fluoride of sulphur is SF_6 but the highest chloride is Scl_4, and silver forms a 2-valent fluoride but only a 1-valent chloride.

Fluorine is the most electronegative of elements (a Pauling electro-negativity value of 4.0 compared with 3.0 for chlorine) and will displace all other halogens from their ionic halides, e.g.

$$2Na^+Cl^- + F_2 \rightarrow 2Na^+F^- + Cl_2$$

or $$2Cl^- + F_2 \rightarrow 2F^- + Cl_2$$

It displaces oxygen from silicon dioxide and water (some hydrogen peroxide and ozone are formed as well); with cold dilute solutions of alkalis it gives oxygen difluoride, F_2O, and with warm concentrated alkalis oxygen is liberated:

$$SiO_2 + 2F_2 \rightarrow SiF_4 + O_2$$

$$2H_2O + 2F_2 \rightarrow 4H^+F^- + O_2$$

(cold dilute) $2OH^- + 2F_2 \rightarrow F_2O + 2F^- + H_2O$

(warm conc.) $4OH^- + 2F_2 \rightarrow O_2 + 4F^- + 2H_2O$

Uses

Of late a large number of uses have been found for fluorine. For example.

1. It is of considerable use in the preparation of *fluorocarbons* which are non-inflammable and chemically inert and are used as solvents, lubricants and insulators.
2. It is used as a refrigerant in many of the cooling processes.
3. A new series or organic compounds of considerable practical use have also been isolated under the name of "*Freons*" used in refrigeration and air-conditioning and a series of transparent fluorine plastics are of considerable practical use.
4. It finds considerable use as DDFT which, similar to DDT, is extremely efficient as a fungicide and fumigant.
5. In nuclear physics and high voltage electricity, fluorine as SF_6 finds very great use *e.g.* in the separation of the isotopes of uranium.

Chlorine Occurrence

Chlorine is found in combination with many of the metals as chlorides, by far the most abundant of which is NaCl. Sea water contains nearly 3 percent NaCl. Evaporation of inland bodies of water has produced large and extensive deposits of it. The Great Salt Lake in Utah contains nearly 20 per cent. Large deposits of rock salt, which is impure NaCl, are found in New York, Michigan, Louisiana, Kansas, Oklahoma, and Texas.

In addition to sodium chloride, sea water contains smaller per-cent-ages of magnesium chloride, potassium chloride, and calcium chloride. Compounds of chlorine are found in the body—sodium chloride in the blood and hydrochloric acid in the gastric juices.

Preparation

(a) *Electrolysis of brine.* Chlorine is prepared commercially by the electrolysis of an aqueous sodium chloride solution. The processes taking place during electrolysis may be illustrated by means of a diagram (Figure 17.1). Sodium chloride is ionized in the solid and when the compound is dissolved in water, the sodium and chlorine ions are dispersed in the solution. Water itself ionizes slightly into hydrogen ions and hydroxyl ions. Thus, in the solution, four ions are present, two cations and two anions. When the electric current is passed through the solution, chlorine ions and hydroxyl ions migrate toward the anode (positive pole), while the sodium ions and hydrogen

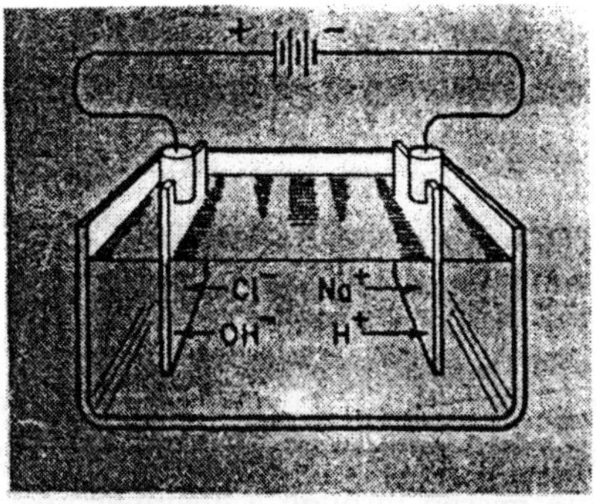

Figure 17.1 : Electrolysis of an aqueous sodium chloride solution yields hydrogen at the cathode and chlorine at the anode.

ions move toward the cathode (negative pole). Since hydrogen ions accept electrons more readily than sodium ions, the former ions are discharged at the cathode and form hydrogen atoms. Two hydrogen atoms then combine to form a molecule of hydrogen which escapes from the solution as a gas. At the anode, chlorine ions are discharged in preference to hydroxyl ions, since chlorine ions give up electrons more readily than hydroxyl ions. As a result, free chlorine is formed at the anode. The solution contains an excess of sodium and hydroxyl ions which may be recovered in the form of solid sodium hydroxide by evaporation of the water present. The equation for the electrolysis process is

$$2NaCl + 2H_2O \xrightarrow[\text{current}]{\text{electric}} 2NaOH + H_2 + Cl_2$$

It the actual commercial process it is necessary to keep the chlorine separated from the aqueous sodium hydroxide solution, since chlorine reacts with the latter to form sodium hypochlorite. This may be accomplished by separating the anode and cathode compartments by means of a porous diaphragm.

(b) Oxidation of chlorides. Free chlorine may be obtained by the removal of an electron from the chloride ion:

$$Cl^- \rightarrow Cl + e(\text{oxidation})$$

The removal of electrons from a substance is termed *oxidation*; substances capable of removing electrons are termed *oxidizing agents*.

A few substances which may oxidize chlorides will be discussed here.

Hydrogen chloride reacts slowly with oxygen at elevated temperatures and is the basis

$$4HCl + O_2 \rightarrow 2Cl_2 + 2H_2O$$

for the Deacon process, used extensively at one time but now replaced by cheaper methods of production. In this process hydrogen chloride and oxygen gases are passed through a chamber heated to about 400°C which contains pumice impregnated with a copper salt to act as a catalyst.

In the laboratory, manganese dioxide is usually employed as the oxidizing agent. Chlorine is collected by displacement of air (Figure 17.2):

$$MnO_2 + 4HCl \rightarrow MnCl_2 + 2H_2O + Cl_2$$

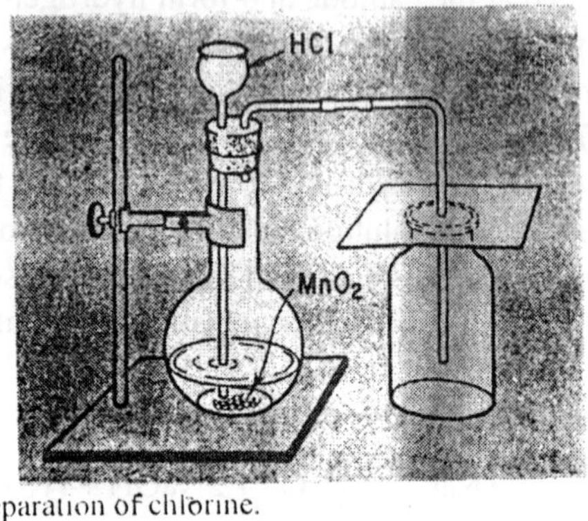

Fig. 17.2 : Laboratory preparation of chlorine.

Instead of hydrochloric acid, sodium chloride and sulfuric acid may be used with the manganese dioxide. This change is essentially the same as the one above since we may assume that the sulfuric acid first acts with the salt to form hydrogen chloride, which is then oxidized by the manganese dioxide:

$$2NaCl + 2H_2SO_4 + MnO_2 \rightarrow MnSO_4 + Na_2SO_4 + Cl_2 + 2H_2O$$

Other oxidizing agents may be employed with HCl:

$$2KMnO_4 + 16HCl \rightarrow 2KCl + 2MnCl_2 + 5Cl_2 + 8H_2O$$
$$K_2Cr_2O_7 + 14HCl \rightarrow 2KCl + 2CrCl_3 + 3Cl_2 + 7H_2O$$
$$HNO_3 + 6HCl \rightarrow NOCl + Cl_2 + 2H_2O$$
$$KClO_3 + 6HCl \rightarrow KCl + 3Cl_2 + 3H_2O$$
$$PbO_2 + 4HCl \rightarrow PbCl_2 + Cl_2 + 2H_2O$$

Physical properties

Chlorine is a greenish-yellow gas at ordinary temperatures and has a sharp, disagreeable odour. It is very irritating to the mucous lining of the lungs, and its inhalation in large quantities may be fatal; for this reason, chlorine has been used as a war gas. The gas is moderately soluble in water; 2.26 volumes dissolve in 1 volume of water at 20°. The gas condenses to a liquid at a temperature of -34.5° at a pressure of one atmosphere and freezes to a solid at -101.6°. Chlorine is easily liquefied under pressure and is stored as a liquid in steel cylinders.

Chemical properties

Because of its great activity, chlorine reacts with a great number of elements and compounds, some of the more important of which will be listed here.

(a) Metals. Chlorine combines directly with practically all of the metallic elements to form chlorides. The more active metals unite vigorously. Powdered antimony when sprinkled into a jar containing chlorine bursts into flame with the formation of a white cloud of antimony chloride. A few typical unions with metals are:

$$2Na + Cl_2 \rightarrow 2NaCl$$
$$Zn + Cl_2 \rightarrow ZnCl_2$$
$$2Sb + 3Cl_2 \rightarrow 2SbCl_3$$

(b) Nonmetals. Most of the nonmetals combine directly with chlorine. Phosphorus burns in chlorine forming phosphorus trichloride, PCl_3, a colourless liquid, and with an excess of chlorine forms PCl_5, a yellow solid:

$$2P + 3Cl_2 \rightarrow 2PCl_3$$

$$2P + 5Cl_2 \rightarrow 2PCl_5$$

Sulfur and chlorine combine at an elevated temperature to form sulfur monochloride, S_2Cl_2, a yellow liquid:

$$2S + Cl_2 \rightarrow S_2Cl_2$$

Chlorine and hydrogen combine slowly in the dark, but unite explosively in strong light with the formation of hydrogen chloride:

$$H_2 + Cl_2 \rightarrow 2HCl$$

(c) Compounds of hydrogen. Chlorine readily removes hydrogen from hydrocarbons (compounds of carbon and hydrogen) to form hydrogen chloride and carbon in each case. If a lighted jet of methane gas (this gas is the main constituent of natural gas) is introduced into an atmosphere of chlorine, combustion continues with the formation of a black sooty deposit of carbon:

$$CH_4 + 2Cl_2 \rightarrow C + 4HCl$$

A piece of filter paper which has been saturated with turpentine, $C_{10}H_{16}$, inserted into a jar of chlorine rapidly takes fire with the formation of a deposit of carbon on the sides of the vessel:

$$C_{10}H_{16} + 8Cl_2 + 10C + 16HCl$$

Other hydrocarbons act similarly with free chlorine.

(d) Water. Chlorine with water forms a solution of hydrochloric and hypochlorous acids:

$$Cl_2 + H_2O \rightleftharpoons HCl + HClO$$

Hypochlorous acid is quite unstable and breaks down on exposure to sunlight:

$$2HClO \rightarrow 2HCl + O_2$$

As a result, the equilibrium above is displaced to the right and eventually the reaction becomes complete to the right when the hypochlorous acid has completely decomposed.

(e) Bases. Since a solution of two acids is produced when chlorine is dissolved in water, we might suppose that the addition of a base to such a solution would result in the neutralization of the

acids to form salts. This is the case with the formation of a chloride salt and a hypochlorite salt:

$$Cl_2 + 2NaOH \rightarrow NaCl + NaClO + H_2O$$

Uses

Chlorine is used extensively as a bleaching aent. If a piece of coloured calico is dipped into a water solution of chlorine, bleaching of the colour takes rapidly. The hypochlorous acid which is present in the solution seems to be the active bleaching agent, since dry chlorine is not very effective. Chlorine is too corrosive to be used with fibers of animal origin (wool, silk, etc.). Much of the chlorine produced commerically is used for bleaching wood pulp, which is used in the production of paper and rayon.

Pathogenic organisms are destroyed in water containing as little as one or two parts of chlorine per million; hence chlorine finds extensive use in the sterilization of drinking water.

Chlorine to be used for bleaching purposes is often transported in the form of bleaching powder, which is produced from lime and chlorine:

$$CaO + Cl_2 \rightarrow CaOCl_2$$

The chlorine is readily liberated from this compound by treatment with dilute acid:

$$CaOCl_2 + H_2SO_4 \rightarrow CaSO_4 + H_2O + Cl_2$$

Chlorine was the first of the war gases to be used in World War I. The gas was liberated when the wind was in the proper direction, and since the gas is nearly two and one-half times as heavy as air, it remained close to the ground where it was most effective. Most of the poisonous gases developed later in that war were chlorine compounds. Among the more effective gases were *mustard gas* $(CH_2ClCH_2)_2S$, *phosgene*, $COCl_2$, and *chlorpicrin*, CCl_3NO_2.

In addition to the preparation of chlorides of many of the elements, chlorine also finds use in the preparation of dyes and drugs.

A particularly effective new insecticide is a chlorine compound $(C_6H_4Cl)_2C_2HCl_3$, named dichloro-diphenyl-trichloro-ethane (DDT).

Bromine

History and Occurrence

Balard in 1826 isolated a heavy dark brown liquid from sea salt. The liquid proved to be an element and was given the name "bromine."

from the Greek word *bromos*, meaning stench. Bromides are found in natural salt brines and in deposits from such brines. The Stassfurt salt deposits in Germany, and brines from wells near Midland, Michigan, furnish small quantities of the element. Bromides are also found in sea water from the element is extracted on a commercial scale.

Preparation

Since chlorine is more active than bromine, the latter may be liberated from its salts by treatment with chlorine water:

$$2NaBr + Cl_2 \rightarrow 2NaCl + Br_2$$

Similar to chlorine, it may also be prepared by electrolysis of bromides or by the oxidation of hydrobromic acid.

It is frequently prepared in the laboratory by treating a bromide salt with an oxidizing agent and sulfuric acid; for example,

$$2NaBr + +MnO_2 + 2H_2SO_3 \rightarrow Na_2SO_4 + MnSO_4 + Br_2 + 2H_2O$$

Commercially, bromine is obtained from sea water by displacement of the bromide present with free chlorine according to equation (1) above. Sea water is pumped into large towers, where it is circulated in contact with chlorine. The liberated bromine is swept upward in the towers by an air stream and collected in a solution of sodium carbonate:

$$3Br_2 + 3Na_2CO_3 \rightarrow 5NaBr + NaBrO_3 + 3CO_2$$

Bromine is readily liberated from the solution by treatment with sulfuric acid:

$$5NaBr + NaBrO_3 + 3H_2SO_4 \rightarrow 3Na_2SO_4 + 3H_2O + 3Br_2$$

Iodide

Occurrence : Iodine was discovered in 1812 by courtois who extracted crystals of voilet solid from the ashes of sea weed. Two years later, Gay-Lussac proved the substance to be an element and named it "iodine" from the Greek, meaning violet.

The principal source of iodine is sodium iodate, $NaIO_3$, which occurs in small amounts in Chile saltpeter, $NaNO_3$. Free iodine is liberated from the former compound by reduction with $NaHSO_3$.

Iodine in the form of sodium iodide is found in very small amounts in sea water. Certain seaweeds called *kelp* also contain small quantities and have been employed as a commercial source of the element.

Preparation

Iodine is prepared in the laboratory by the methods usually employed for the preparation of chlorine and bromine:

$$2NaI + MnO_2 + 2H_2SO_4 \rightarrow Na_2SO_4 + MnSO_4 + I_2 + 2H_2O$$

Iodides are more easily oxidized than either chlorides or bromides.

Since iodine is less active than chlorine or bromine, both of these elements displace iodine from iodides:

$$2NaI + Br_2 \rightarrow 2NaBr + I_2$$

$$2NaI + Cl_2 \rightarrow 2NaCl + I_2$$

Properties and Uses

Iodine is a dark gray or black solid at ordinary temperatures. When heated below its melting point of 113°, iodine vapourizes rapidly. The vapour may be condensed as crystals on a cold surface, without apparent formation of liquid. The property of changing directly from solid to gas and from gas to solid without apparently liquefying is termed *sublimation.*

Iodine is only slightly soluble in water, but is readily soluble in a solution which contains potassium iodide because of the combination:

$$I_2 + KI \rightarrow KI \cdot I_2 \text{ or } KI_3$$

It is readily soluble in alcohol, carbon tetrachloride, and ether, and, as is the case with bromine, these solvents are often used in extracting te element from water solutions. Starch forms an unstable blue compound with iodine and is used as a test for the presence of free iodine.

Although less active than the other halogens, iodine nevertheless combines with most of the metals and many of the nonmetals to form iodides.

Everyone is familiar with the use of iodine as an antiseptic. A solution of iodine in alcohol is known as "tincture" of iodine. Silver iodide, like silver bromide, is sensitive to light and finds use in the photographic industry.

The presence of iodine in the body seems to be essential for good health. Iodine appears to perform a vital function in the thyroid gland, in which it is found in the compound, iodothyrin. If the body is not furnished with iodine in some form, the vital gland does not function properly. Under normal circumstances, we derive sufficient iodine from foods such as butter, spinach, beans, and particularly sea foods.

Astatine

Astatine does not occur in nature, but over twenty artificial isotopes have been made. All of these radioactive. The most stable isotopes are ^{210}At (half life 8.3 hours), and ^{211}At (half life 7.5 hours). The latter was first made in 1940 by a nuclear reaction in which bismuth was bombarded with high energy α particles.

$$^{209}_{83}Bi + ^{4}_{2}He \rightarrow ^{211}_{85}At + 2^{1}_{0}n$$

Tracer methods were used to study the chemistry of ^{211}At, using minute quantities of about 10^{-14} mole. This isotope decays by orbital electron capture and by α-emission. Astatine appears to resemble iodine quite closely.

Type of Bonds Formed and Oxidation States

Most compounds formed by the halogens and metals are ionic. However, covalent halides are formed in a few cases where the metal ions are very small and have a high charge. (The structures of $BeCl_2$ and $AlCl_3$ are unusual.)

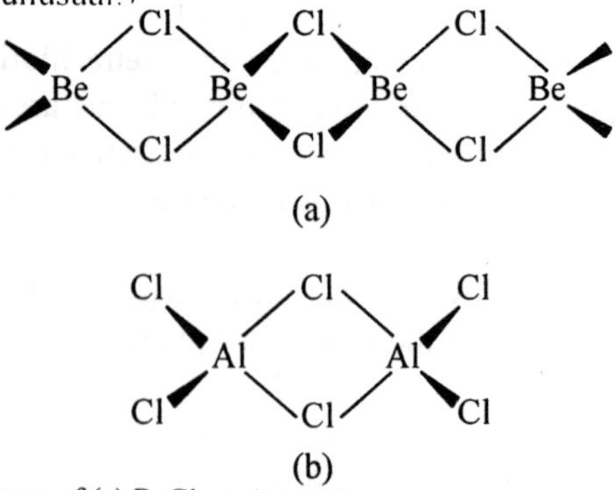

Fig. 17.3 : Structure of (a) $BeCl_2$ and (b) $AlCl_3$

The halogens all have very high electronegativity values. When they react with metals there will be a large electronegativity difference: hence they form ionic bonds. Halide ions are produced quite easily. This is shown by the large electron affinity values. Note that energy is evolved when a gaseous halogen atom gains an electron, and also because of their large positive standard electrode potentials for (The standard electrode potentials may be converted to an energy term using the relationship $\Delta G° = -nFE°$, where n is the number of electrons (2 in this case). and F is the Faraday constant 96486 Kj mol^{-1}.) The $E°$ values decrease down the group and thus the energy evolved on forming halide ions also decreases down the group. Many iodides are partly covalent. For example. CdI_2 forms a layer structure, and all the iodides have much lower melting points than the fluorides. When two halogen atoms form a molecule, they form a covalent bond. Most compounds between the halogens and non-metals are also covalent. Fluorine is always univalent, and since it is the most electronegative element it always has the oxidation number (–1). With Cl, Br and I, a covalence of one is the most common. The oxidation state may be either (–1) or (+1) depending on which atom in the moleucle has the greater electronegativity.

Halogen Iodides

Of all the halogen compounds, the most important practically are the hydrogen halides and the salts of the hydrohalic acids.

All the hydrogen halides are colourless gases possessing a pungent odour, and condensing readily into liquids. The hydrogen halides are very soluble in water. For instance, one volume of water will dissolve about 500 volumes of hydrogen chloride at 0° C; the solubility of hydrogen bromide and hydrogen iodide is about the same. Hydrogen fluoride is soluble in water in all proportions.

The aqueous solutions of the hydrogen are typical acids, only hydrofluoric acid being comparatively little ionized. The rest of them are among the most active acids.

All the hydrogen halides fume in the air, due to the formation of minute droplets of the corresponding acids with the water vapours in the air.

The stability of the hydrogen halides to heat falls abruptly from HF to HI due to the decrease in the electron affinity of the halogen atoms. While HF does not ionize into atoms to any perceptible degree even at very high temperatures, hydrogen iodide decomposes considerably into iodine and hydrogen at as low a termperature as 300°C.

The general method of preparing the hydrogen halides is by the action of concentrated sulphuric acid on the salts of the hydrohalic acids. Thus, for instance, hydrogen chloride is usually obtained by heating sodium chloride with concentrated sulphuric acid:

$$2NaCl + H_2SO_4 = Na_2SO_4 + 2HCl$$

Since glass vessels must not be heated too strongly, only one of the hydrogen atoms of sulphuric acid takes part in the reaction when it is carried out in the laboratory, resulting in the acid salt:

$$NaCl + H_2SO_4 = NaHSO_4 + HCl$$

Hydrogen halide constituent atoms are linked together by polar covalent bonds. The polarity of these molecules is a function of the electronegativity of the halogen atom, it is greatest for HF and least for HI.

All of the hydrogen halides decompose into their constituent elements when heated sufficiently. Of the four hydrogen halides, hydrogen iodide is the most easily decomposed, while hydrogen chloride and hydrogen fluoride show only slight dissociation at 1,000°C. The stability of these gases decreases, then, as the formula weight increases.

The hydrogen halides fume in moist air, and all are very soluble in water. With the exception of hydrogen fluoride, the hydrogen halides are strong electrolytes; they ionize completely in dilute aqueous solution. The ionization of hydrogen chloride is given by the equation $HCl \rightarrow H^+ + Cl^-$. If the concentration of these acids in water is greatly increased, the molecular form (HX) appears and escapes from solution.

The anhydrous hydrogen halides are rather inactive chemically and do not attack dry metals at ordinary temperatures. Neither do they conduct electricity in the liquid state; this fact indicates that they are molecular rather than ionic compounds.

Hydrogen fluoride is obtained from calcium fluoride, a natural compound of fluorine:

$$CaF_2 + H_2SO_4 = CaSO_4 + 2HF$$

As hydrogen fluoride attacks glass vigorously, the reaction is usually carried out in lead vessles, a protective layer of PbF_2 forming on their walls, and keeping the metal from further corrosion.

When dry it is known as anhydrous hydrofluoric acid and is available as a liquid which is usually stored in mild steel cylinders. Its boiling point (19.5°C) is considerably higher than that of hydrogen chloride (—84°C) and this is because the individual HF units are associated into larger aggregates by means of hydrogen bonding. Extreme care should be observed when handling this substance since its vapour is highly toxic and it causes painful burns if it comes in contact with the skin (a face shield and thick rubber gloves must be worn).

A solution of hydrogen fluoride in water behaves as a weak acid, in marked contrast with aqueous solutions of the other hydrogen halides which are strongly acidic. In order to trace the reason for this weakness (the dissociation constant K, for an aqueous solution of hydrogen fluoride at 25°C is 7×10^{-4}) it is necessary to consider a number of individual energy terms; when this is done, it turns out that the very high H—F bond energy is the principal reason why ionization into H_3O^+ and F^- ions in water is not very extensive:

$$HF + H_2O \rightleftharpoons H_3O^+ + F^- \text{ (equilibrium well over to the left)}$$

An aqueous solution of hydrofluoric acid attacks most metals with the formation of the metallic fluoride; these salts can also be obtained by the action of the acid on the metallic oxide, hydroxide or carbonate.

Moist hydrogen fluoride and aqueous solutions of the acid attack silica and glass, so the solution is stored in polythene containers.

$$SiO_2 + 4HF \rightarrow SiF_4 + 2H_2O$$

$$SiF_4 + 2HF \rightarrow (H^+)_2SiF_6^{2-}$$

fluorosilicic acid

Unlike the other hydrogen halides it forms acid salts, e.g. potassium hydrogen fluoride, $K^+HF_2^-$. Which can be made by passing hydrogen fluoride over potassium fluoride.

Gaseous HF is very toxic, and should be handled only in a good fume cupboard. Solutions of HF are called hydrofluoric acid, and this is very corrosive. Hydrofluoric acid is normally handled in metal apparatus made of copper or Monel, because hydrofluoric acid attacks glass with the formation of fluorosilicate ions $[SiF_6]^{2-}$

$$SiO_2 + 6HF \rightarrow [SiF_6]^{2-} + 2H^+ + 2H_2O$$

Surprisingly little corrosion occurs at concentractions above 80%. The main uses of HF are as follows:

1. Two thirds are used to make chlorofluorocarbons (Freons). These are sometimes called CFCs. They are used as refrigerating fluids and as the propellant in aerosols. The use of CFCs is being phased out because they damage the ozone layer in the upper atmosphere.

$$CCl_4 + 2HF \xrightarrow[+SbCl_5]{\text{anhydrous conditions}} \underset{\text{Freon}}{CCl_2F_2} + 2HCl$$

2. About 14% is used to make AlF_3 and synthetic cryolite used in the electrolytic extraction of aluminium.
3. About 2% is used for uranium processing (through intermediates UF_4 and UF_6).
4. About 4% is used in the anhydrous form as an alkylation catalyst in the petrochemical industry, for making long chain alkylbenzene compounds. These are then converted into alkylbenzene sulphonates and used as detergents.
5. Aqueous HF is used for pickling steel (about 4%) and for etching glass, making herbicides and many metal fluorides and also BF_3.

$$B_2O_3 + 6HF \xrightarrow{\text{conc.}H_2SO_4} 2BF_3 + 3H_2SO_4 \cdot H_2O$$

$$Al_2O_3 + 6HF \rightarrow 2AlF_3 + 3H_2O$$

HCl

1. At one time HCl was made exclusively by the 'salt cake' method. In this method, concentrated H_2SO_4 was added to rock salt (NaCl). The reaction was endothermic, and was

performed in two stages at different termperatures. The first of the reactions was carried out at about 150°C. The solid NaCl reacted with H_2SO_4 and became coated with insoluble $NaHSO_4$. This prevented further reaction and accounts for the name 'salt cake'. In the second stage, the mixture was heated to about 550°C when further reaction with H_2SO_4 occurred and Na_2SO_4 was formed. This by-product was sold, mostly for paper making (Kraft process).

$$NaCl + H_2SO_4 \xrightarrow{150°C} HCl_{(g)} + NaHSO_4$$

$$NaCl + NaHSO_4 \xrightarrow{550°C} HCl_{(g)} + Na_2SO_4$$

2. Large amounts of impure HCl have become available in recent years as a by-product from the heavy organic chemical industry. For example HCl is produced in the conversion of 1,2-dichloethane $CH_2Cl—CH_2Cl$ to vinyl chloride $CH_2=CHCl$, and in the manufacture of chlorinated ethanes and chlorinated fluorocarbons. This is now the largest source of HCl.
3. High purity HCl is made by direct combination of the elements. A gaseous mixture of H_2 and Cl_2 is explosive. However, the reaction proceeds quitely if the gases are burnt in a hydrogen—chlorine flame in a special combustion chamber. The process is strongly exothermic.
4. HCl is conveniently made in the laboratory by treating NH_4Cl with concentrated H_2SO_4. NH_4Cl costs more than NaCl (which was used in the 'salt cake' process). However, NH_4Cl is preferred because NH_4HSO_4 is soluble, and the reaction does not stop at the halfway stage.

$$2NH_4Cl + H_2SO_4 \rightarrow 2HCl + (NH_4)_2SO_4$$

Hydrogen chloride gas is very soluble in water. Aqueous solutions of HCl are sold as *hydrochloric acid.* A saturated solution at 20°C contains 42% HCl by weight, and 'concentrated' acid normally contains about 38% HCl by weight (approx. 12M). Pure hydrochloric acid is colourless, but technical grades are sometimes yellow because of contamination by Fe(III). The largest use is for 'pickling' metals, that is removing oxide layers from the surface. It is also used to make metal chlorides, in the manufacture of dyestuffs, and in the sugar industry.

Gaseous HCl is conveniently prepared in the laboratory from concentrated HCl and concentrated H_2SO_4.

Properties & Uses

Pure hydrochloric acid is a colourless liquid with the pungent odour characteristic of hydrogen chloride. A saturated solution of hydrogen chloride at 18° C contains 42 per cent HCl. Ordinary concentrated hydrochloric acid contains about 37 per cent HCl and has a specific gravity of 1.19. This acid fumes in the air due to the liberation of hydrogn chloride and for this reason is called *fuming* hydrochloric acid. The technical acid is yellow due to the presence of impurities (mainly $FeCl_3$) and contains around 27.5 per cent HCl. Synthetic hydrochloric acid contains 31 per cent HCl.

Hydrochloric acid is one of the most important acids in chemical practice. It reacts readily with many metals, liberating hydrogen and forming salts, known as metal chlorides or just chlorides. Hydrochloric acid is used for the preparation of metal chlorides, for the production of ammonia, glue, carbonic acid, a number of organic preparations, in the leather industry (to treat leather before chrome tanning), in the food industry, in tinning and soldering and in many other branches of industry. A small quantity of hydrochloric acid is contained in the gastric juices of man and animals and plays an important part in digestion.

The chlorides of most metals are readily soluble in water and therefore are widely used in laboratory practice when it is necessary to introduce any definite metal ion into a reaction.

HI is made by the reaction of concentrated phosphoric acid H_3PO_4 on metai iodides, in a similar reaction to the 'salt cake' process for HCl. Note that a non-oxidizing acid such as phosphoric acid must be used. Concentrated H_2SO_4 is a strong oxidizing agent and would oxidize HI to I_2.

The usual laboratory preparation involved reducing iodine with red phosphorus in water. HI, water is added to a mixture of phosphorus and iodine.

$$\overset{\text{red}}{2P} + 3I_2 \rightarrow 2PI_3 \xrightarrow{+6H_2O} 6HI + 2H_3PO_3$$

Properties

Hydrogen iodide is a colourless irritating gas, which fumes in moist air. It is remarkably soluble in water. A solution saturated with hydrogen iodide contains at 10° as much as 70 per cent of the acid by weight. Its constant boiling mixture has the composition of 57 per cent of acid.

Hydrogen iodide is easily decomposed on heating:

$$2HI \rightleftharpoons H_2 + I_2$$

The salts of the acid are known as *iodides.*

Hydrogen iodide is a strong reducing agent. It is easily oxidized by (a) oxygen, (b) hydrogen peroxide, (c) persulphate and peraids, (d) nitric acid, (e) chlorine and bromine, (f) nitrous acid, (g) ferric and cupric salts, (h) nitrates in dilute acid solution and even (i) sulphuric acid. In all these reactions the potential equation is

$$2HI + O \rightarrow H_2O + I_2$$

Some of thesr reactions are given along with the precipitation of iodine. A few typical reactions are given below. They may be performed with acidified solution of potassium iodide:

(a) $4HI + O_2 \rightarrow 2H_2O + 2I_2$

(b) $2HI + H_2O_2 \rightarrow 2H_2O + I_2$

(c) $2HNO_3 + 6Hi \rightarrow 4H_2O + 2NO + 3I_2$

(d) $K_2S_2O_8 + 2HI \rightarrow K_2SO_4 + H_2SO_4 + I_2$

(e) $\begin{bmatrix} H_2SO_4 + 6HI \rightarrow 4H_2O + 3I_2 + S \\ H_2SO_4 + 8HI \rightarrow H_2S + 4H_2O + 4I_2 \end{bmatrix}$

(f) $2CuSO_4 + 4HI \rightarrow Cu_2I_2 + I_2 + 2H_2SO_4$

(g) $2FeCL_3 + 2HI \rightarrow 2FeCl_2 + 2HCl + I_2$

Uses

HI is used as a reducing agent in organic chemistry.

Hydrobromic Acid, HBr

Preparation

(a) It may be preapared from the elements by passing them through a red hot tube filled with finely divided platnium.

(b) It cannot be prepared by the action of acids like H_2SO_4 on a bromide, because some of the HBr gets easily oxidized to give out bromine. However, a dilute solution may be obtained by this method.

$$KBr + H_2SO_4 \rightarrow KHSO_4 + HBr$$

$$H_2SO_4 + 2HBr \rightarrow 2H_2O + SO_2 + Br_2$$

In this reaction, phosphoric acid may be used in place of sulphuric acid, but it is explosive.

$$KBr + H_3PO_4 \rightarrow HBr + KH_2PO_4$$

(c) It is best prepared by the action of bromine (10 parts on a mixture of phosphorus (1 part) and water (20) parts):

$$2P + 5Br_2 + 8H_2O \rightarrow 10HBr + 2H_3PO_4$$

The reaction takes place with violence, and the vapours of hydrobromic acid thus formed are passed through a U-tube containing amorphous phosphorus to free it from bromine.

(d) It can also be prepared by passing a current of H_2S through bromine covered with a layer of water. The hydrobromic acid is formed which is contaminated with H_2S and sulphur, bromides, and, therefore, to purify, it is text passed through a solution of KBr in HBr containing a little phosphorus in suspension:

$$H_2S + Br_2 \rightarrow 2HBr + S$$

$$\text{or } 2H_2S + 3Br_2 \rightarrow 4HBr + S_2Br_2$$

(e) It can also be obtained by passing sulphur dioxide in bromine covered with water. The acid is then distilled over to free it from sulphuric acid formed in the reaction:

$$2H_2O + Br_2 + SO_2 \rightarrow 2HBr + H_2SO_4$$

Properties

The acid is very much similar to hydrogen chloride in smell, solubility and fuming nature. Its saturated solution contains 69 per cent of the acid by weight. It forms a *constant boiling mixture,* which boils at 126°C and contains 48 per cent of HBr.

Its salts with alkalies are known as *bromides.*

Hydrogen bromide is more easily oxidized than hydrogen chlorides. It is oxidized by even sulphuric acid.

$$2HBr + H_2SO_4 \rightarrow 2H_2O + SO_2 + Br$$

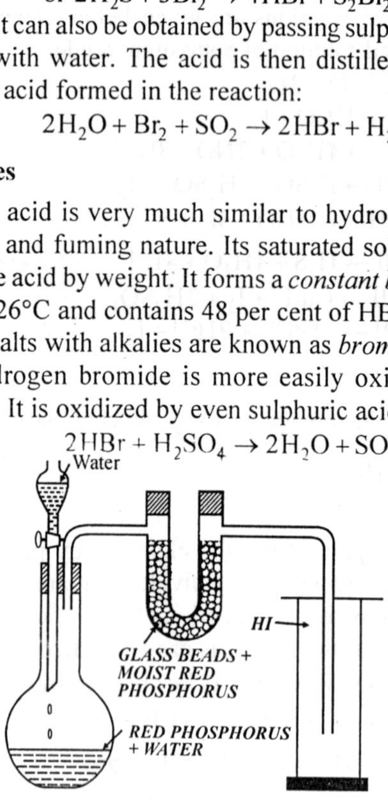

Uses

It is used in preparing bromic derivatives of unsaturated organic compounds. Potassium or sodium bromide which is the salt of HBr is used in medicine as a sleeping dose.

Unique Properties of HF

The unexpectedly high boiling point of HF arises because of the hydrogen bonds formed between the F atom of one molecules and the H atom of another molecule. This links the molecules together as $(HF)_n$, and they form zig-zag chains in both the liquid and the solid. Some hydrogen bonding also occurs in the gas, which consists of a mixture of cyclic $(HF)_6$ polymers, dimeric $(HF)_2$, and monomeric HF. HCl, HBr and HI are not hydrogen bonded in the gas and liquid, though HCl and HBr are weakly hydrogen bonded in the solid.

Hydrogen bonds are generally weak (5—35 Kj mol^{-1}) compared with normal covalent bonds (C—C 347 Kj mol^{-1}), but their effect is highly significant. The most electronegative elements fluorine and oxygen (and to a lesser extent chlorine) form the strongest hydrogen bonds. HF(g)). (The bond energy of the hydrogen bond in F—H...F is 29kJ mol^{-1} in $HF_{(g)}$).

In the gaseous state the hydrides are essentially covalent. However, in aqueous solutions they ionize. H^+ are not produced since the proton is transferred from HCl to H_2O, thus giving $[H_3O]^+$. HCl, HBr and Hi ionize almost completely and are therefore strong acids. HF only ionizes slightly and is therefore a weak acid.

$$HCl + H_2O \rightarrow [H_3O]^+ + Cl^-$$

Though HCl, HBr and HI completely ionize in water, the degree of ionization is much less in poorer ionizing solvents such as anhydrous acetic acid. HCl ionizes less than HI in glacial acetic acid as solvent. Thus in acetic acid, HI is the strongest acid, followed by HBr and HCl, and HF is the weakest.

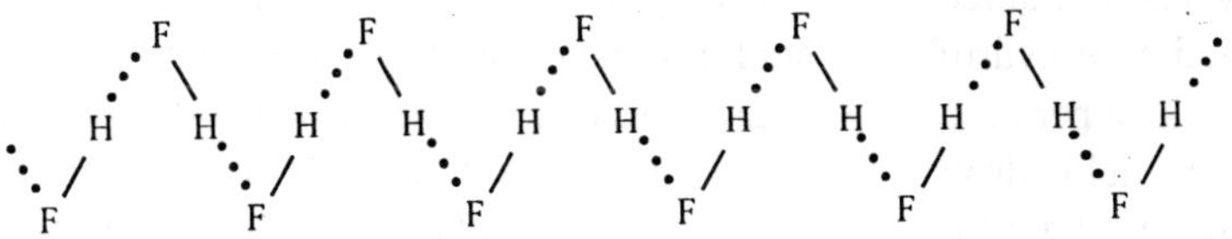

The Ionization of Hydrofluoric Acid

Hydrofluoric acid is weak (slightly ionized) whereas the other hydrohalic acids are strong (highly ionized). In order for ionization to occur, the bond in H—X must be broken, and then hydration of the resultant ions must take place. The H—F bond is about twice as difficult to break as is the H—I bond, and the strengths of the H—Br and H—Cl bonds lie between those of the other two. The greater stregth of the H—F bond as compared to that of H—I is considered to result from the smaller interatomic distance in HF (HF, 1.0 Å; HI, 1.7Å). The ionization of hydrogen fluoride in dilute solution may be written as

$$HF \rightleftharpoons H^+ + F^-$$

However, a second ion is formed in this process due to hydrogen bonding; it is the hydrogen difluoride ion.

$$F^- + HF \rightleftharpoons HF_2^-$$

In very concentrated solution of hydrofluoric acid, the degree of ionization rises sharply. This behaviour is opposite to that observed for all other weak electrolytes and results from an increase in the concentration of the dimer, H_2F_2, which is a stronger acid thatn the monomer, HF.

$$H_2F_2 \rightleftharpoons H^+ + HF_2^-$$

The Action of Hydrofluoric Acid on Glass

The most characteristic behaviour of hydrofluoric acid is its action on sand, which is silicon dioxide, and glass, which is a mixture of silicates (mostly calcium silicate). The reaction of hydrofluoric acid with sand (SiO_2) and with calcium silicate ($CaSiO_3$) is given by

$$SiO_2(s) + 4HF(g) \rightleftharpoons SiF_4\,(g) + 2H_2O\,(l) + 45.7\text{ kcal}$$

$$CasiO_3 + 6HF \rightarrow CaF_2 + SiF_4\uparrow + 3H_2O \quad (\Delta H^\circ = -45.7\text{ kcal})$$

The silicon escapes in these reactions as a volatile compound, silicon tetrafluoride. When one mole of SiF_4 is formed from the elements, the amount of energy released (361 kcal) is exceedingly large. This large amount of energy provides the driving force for the conversion of silicon-oxygen compounds to silicon tetrafluoride by the action of hydrofluoric acid. For example, in the reaction with SiO_2, 80 kcal per mole of SiO_2 are released, and the escape of the gaseous SIF_4 helps drive the reaction to completion. Glass objects such as electric light bulbs can be frosted, or etched, by hydrofluoric acid.

Thermometers, burets, and other glassware may be marked by use of this acid. In order that a restricted area may be etched, the glass is first covered with paraffin, and then, with a sharp-pointed instrument, part of the paraffin is removed from the glass. Extreme care should be taken in handling hydrofluoric acid for it is very corrosive to the skin and causes painful, slow-healing burns. Hydrofluoric acid is also a local anesthetic, and if in contact with the skin may not be noticed until much damage has been done.

The interhalogen compounds are volatile and covalent, being prepared by synthesis from the elements or by further halogenation, as in the fluorination of BrF_3 and IF_5 to form BrF_5 and IF_7 respectively. The less electronegative element is placed on the left in formulae and the naming is then straight-forward. In each case the smaller atom is present in larger atomic proportion: as the atomic volume increases, the capacity for covalency increases. The compounds are very reactive, some reaction being extremely violent. Chlorine trifluoride and bromine trifluoride react violently with organic material and set fire to asbestos: asbestos wool burns with incandescence. These compounds are important in the preparation of uranium hexafluoride in atomic energy work.

Some of the interhalogens ionize slightly (i.e. undergo auto-ionization) and are ionizing solvents so that non-protonic acid-base systems may be developed. Here, an acid gives the cation of the parent solvent and the base, the anion:

$$2ICl_3 \rightleftharpoons ICl_2^+ + ICl_4^-$$

$$2BrF_3 \rightleftharpoons BrF_2^+ + BrF_4^-$$

$$2IF_5 \rightleftharpoons \underset{\substack{\uparrow \\ \text{acids}}}{IF_4^+} + \underset{\substack{\uparrow \\ \text{bases}}}{IF_6^-}$$

The following bases and acids, have been prepared for the system

bases : $KBrF_4$, $AgBrF_4$, $Ba(BrF_4)_2$

acids: $BrF_2.SbF_6$, $(BrF_2)_2SnF_6$, $BrF_2.AuF_4$, etc.

Many non-protonic systems do not have high dielectric constants showing that this is not an essential characteristic of an ionizing medium.

Individual Members

(A) AB type. ***(i)*** **Iodine monochloride, ICI.**

When chlorine is passed over iodine, the monochloride is formed as a dark liquid : I_2+Cl_2→2ICl. The liquid solidifies on standing at the room temperature (16°—25°). From this product, two forms of ICl can be separated by fractional melting, the *metastable form* melting at 14°, and the *stable form* at 27.2°. The two can also be separated from the liquid cooled below 14°, by sowing into it a crystal of one or the other form.

The monochloride is soluble in water and is hydrolyzed by it.

$$ICl + H_2O \rightleftharpoons IOH + HCl$$

(ii) Iodine monobromide, IBr. It is obtained by the direct combination of halogens : I_2+Br_2 = 2IBr (m. pt. 36°, b. pt. 116°, black crystals).

(B) AB_3 type. ***(i)*** **Iodine trichloride, ICl_3.** The action of excess of chlorine on iodine or on the monochloride gives the trichloride.

$$ICI + Cl_2 \rightleftharpoons ICl_3$$

The reaction is, however, reversible. It is also obtained by the action of hydrochloric acid gas on heated iodine pentoxide :

$$I_2O_5 + 10HCl \rightarrow 2ICl_3 + 5H_2O + 2Cl_3$$

It is a lemon yellow crystalline solid.

When hydriodic acid reacts with chlorine, both the chlorides are formed the one as red liquid drops and the other as lemon yellow crystals:

$$4HI + 4Cl_2 \rightarrow 4HCl + I_2 + ICl + ICl_3$$

(ii) **Bromine trifluoride, BrF_3.**

It is formed from bromine (or HBr) and fluorine. It is a colourless fuming liquid (b. pt. 127°) which is being made use of a *fluorinating agent.*

(C) AB_5 type. *Iodine pentafluoride, IF_5.* Is formed from iodine and fluorine or from fluorine and heated iodine pentoxide, I_2O_5 (m. pt. 80., b. pt. 97°). The *pentafluoride,* BrF_5, is colourless liquid (b. pt. 40.4°) obtained by passing fluorine through bromine trifluoride at 90°. the *heptafluoride,* IF_7, is a colourless liquid, boiling at 4.5°.

(D) AB_7 type. Iodine heptafluoride, IF_7. It is formed by heating IF_5 with F2 at 270°—300°C. It is a colourless gas (m.p.= 5.5°C, subl = 4.5°C). It reacts vigorously with non-metals but less so with metals. It is slowly decomposed by water.

Structure of some interhalogen compounds.

(i) **AB_3 type molecules.** ClF_3 and BrF_3 both have **T-shaped structure** with bond angles equal to 87° 40' and 86° 12.6' respectively. Two bond lengths are the same while the third one is different (see Fig. 17.4). The central chlorine and bromine atoms, in these molecules are sp^3d hybridized and hence the arrangement of five electron pairs in trigonal bipyramidal. Two *equatorial* positions of the trigonal bipyramid are occupied by two lone pairs (*see* Fig. 17.5).

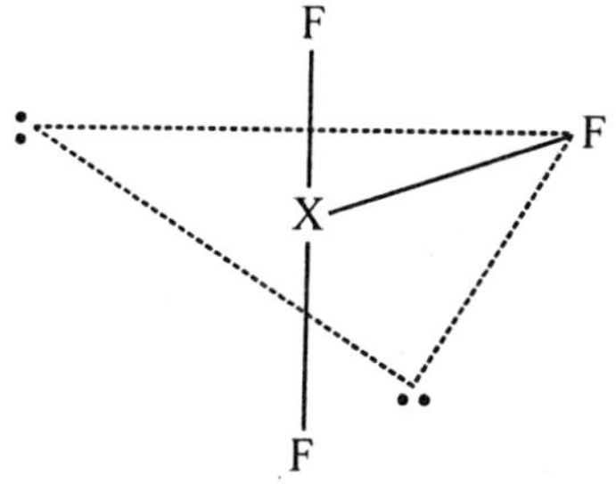

Fig. 17.4 : Trigonal bipyramidal arrangement of five electron pairs in XF_3 molecule (X = cl or Br)

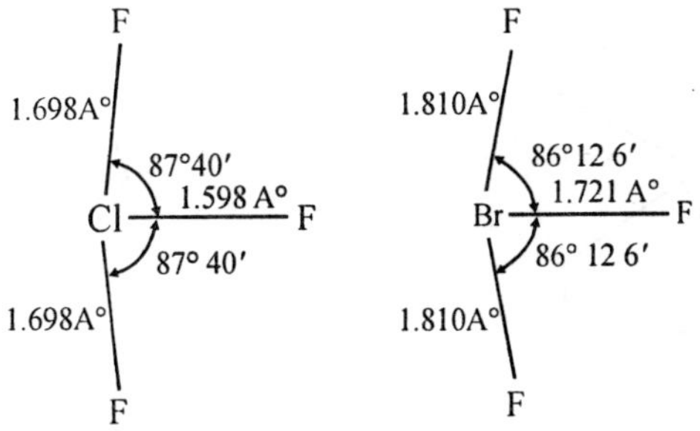

Fig. 17.5 : T-shaped structure of (a) ClF_3 and (b) BrF_3 molecules.

Electronic configuration of Cl and Br atoms (central atoms) in the excited state are : $3s^2\ 3p^2\ 3p^1\ 3p^1\ 3d^1$ and $4s^2\ 4p^2\ 4p^1\ 4p^1\ 4d^1$ respectively.

ICl_3 is also T-shaped but is dimerized as shown below:

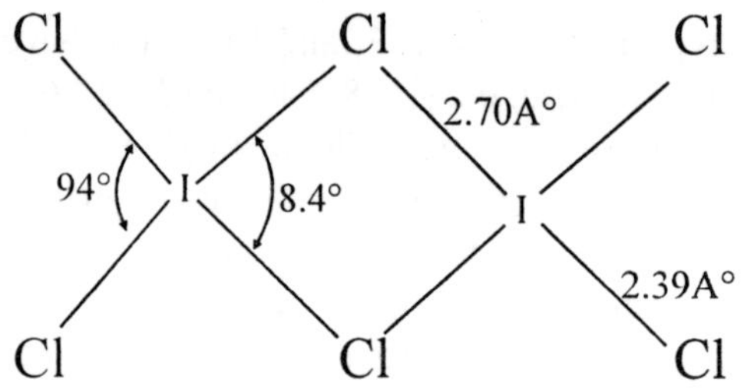

Fig. 17.6 : T-shaped structure of dimerized molecule of ICl_3.

(ii) **Ab_5 type molecules.** Let us consider IF5 molecule in which iodine atom is sp^3d^2 hybridized (Electron configuration of iodine atom in the doubly excited state is $5s^2\ 5p^1\ 5p^1\ 5p^1\ 5d^1\ 5d^1$). One of the octahedral positions is occupied by a lone pair (*see* Fig. 17.7). The presence of one lone pair distorts the shape of the molecule from regular octahedral to square pyramidal.

(iii) **AB_7 Type.** In IF_7 iodine atom is sp^3d^3 hybridized and the moleucle has petnagonal bipyramidal shape (*see* Fig. 17.8).

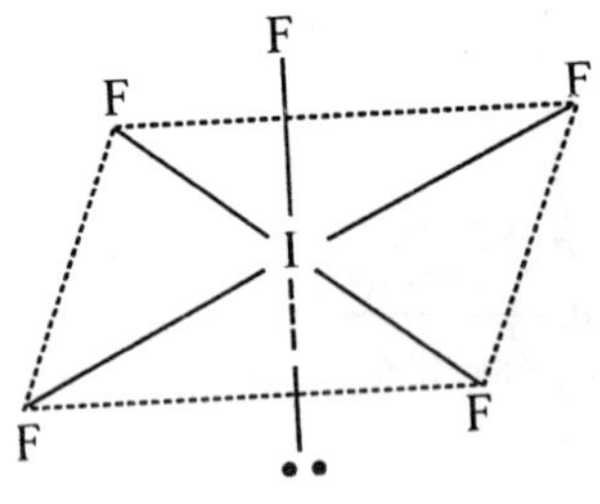

Fig. 17.7 : Squars pyramidal shape of IF_5 molecule

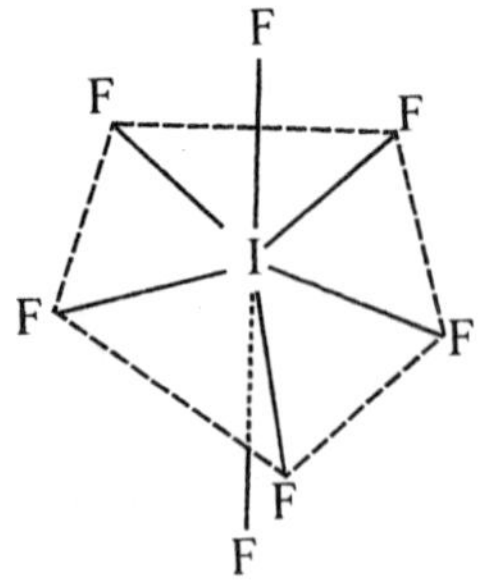

Fig. 17.8 : Pentagonal bipyramidal shape of IF_7 molecule.

Halogen Oxides

The compounds with oxygen probably shown greater differences between the different halogens than any other class of compound. Differences between F and the others arise for the usual reasons (small size, lack of *d* orbitals and high electronegativity). In addition oxygen is less electronegative than F, but more electronegative than Cl, Br and I. Thus binary compounds of F and O are fluorides of oxygen rather than oxides of fluorine. The other halogens are less electronegative than oxygen and thus form oxides. There is only a small difference in electronegativity between the halogens and oxygen, so the bonds are largely covalent. Rather surprisingly I_2O_4 and I_4O_9 are stable and ionic.

Most of the halogen oxides are unstable, and tend to explode when subjected to shock, or sometimes even when exposed to light. The iodine oxides are the most stable, then the chlorine oxides, but the bromine oxides all decompose below room temperature. The higher oxidation states are more stable than the lower states. Compounds ClO_2.Cl_2O, I_2O_5 and OF_2 are the most important.

Fluorine

OF_2 Oxygen difluoride

OF_2 is a pale yellow gas, formed by passing F_2 into dilute (2%) NaOH.

$$2F_2 + 2NaOH \rightarrow 2NaF + H_2O + OF_2$$

OF_2 is a strong oxidizing agent, and has been used as a rocket fuel. It reacts vigorously with metals, S, P and the halogens, giving fluorides and oxides. It dissolves in water and gives a netural solution, so it is not an acid anhydride. With NaOH it gives fluoride ions and dioxygen.

O_2F_2 Dioxygen difluoride

O_2F_2 is an unstable orange—yellow solid, and is a violent oxidizing and fluorinating agent. It is formed by passing an electric discharge through a mixture of F_2 and O_2 at very low pressure and at liquid air temperature. It decomposes at —95°C. Its structure is similar to that of H_2O_2, except that the O—O bond length of 1.22Å is much shorter than the O—O distance of 1.48Å H_2O_2. The O—F bond lengths are 1.58Å, which is much longer than in OF_2.O_4F_2 is made in a similar way, and apparently contains a chain of four oxygen atoms, O_5F_2 and O_6F_2 have been reported.

Bromine. In contrast to chlorine and iodine, the oxides of bromine have not been extensively studied. The data in Table 14.5 indicate that these compounds are thermally less stable than chlorine oxides; nothing is known of their structures.

Bromine monoxides, Br_2O, is a dark brown liquid which can be prepared by passing bromine over mercuric oxides. The substance decomposes at temperatures above —40°C to bromine and oxygen; hence it is poorly characterized in terms of its physical properties. Bromine monoxide dissolves in CCl_4 to give a stable green solution in which it exhibits a moleuclar weight corresponding to Br_2O. This solution reacts with aqueous base to form hypobromite ion, but when pure Br_2O reacts with aqueous base a mixture of hypobromite and bromate ions is formed. Bromine monoxides is an oxidizing agent capable of converting I^- to iodine.

Bromine dioxide, BrO_2, is formed as a yellow solid when a low-pressure mixture of oxygen and bromine is subjected to an electrical discharge at liquid-air temperatures. The substance is stable below —40° but decomposes into the elements at higher temperatures. It can be pyrolyzed in a vacuum to Br_2O and a white solid which may be Br_2O_7 (the latter point is not certain). Bromine dioxide, like the analogous ClO_2, dissolved in water to give bromine-contaiing anions in different oxidation states, namely, Br^-, OBr^-, BrO_2^-, and BrO_3^-.

When bromine is treated with ozone at about 0°C a white solid is formed which has been formulated as Br_3O_8 or BrO_3. The product is stable at —80°; a transition between two modifications has been reported at —35°C. It is soluble in water, giving a colourless solution which is acidic and oxidizing.

Iodine (X). Of the three oxides reported for iodine, only I_2O_7 appears to be related to the analogous compounds of the other halogens. The compound I_2O_4, which appears to be the dimer of the dioxide IO_2, has been formulated as iodyl iodate [(IO)(IO_3)]. It has been suggested that the oxide I_4O_9 is best formulated as $I(IO_3)_3$, that is as the iodate of trivalent iodine. Both of these descriptions emphasize the basic nature of iodine.

The compound I_2O_4 is formed by warming iodic acid (HIO_3) with concentrated sulfuric acid for several days. The product is insoluble in cold water but dissolves in hot water yield iodic acid and iodine

$$5I_2O_4 + 4H_2O \rightarrow 8HIO_3 + I_2;$$

in aqueous base it yields a mixture of iodate and iodide ions. It is an oxidizing agent, giving a mixture of chlorine and ICl when treated with hydrochloric acid. Above 85° I_2O_4 decomposes into I_2O_5 and iodine. Spectroscopic data suggest that I_2O_4 consists of a series of I—O chains crosslinked by IO_3 units (*37*). The formulation of I_2O_4 as iodyl iodate is supported by the fact that iodyl sulfate $[(IO)_2(SO_4)_2$, formed as a yellow solid when H_2SO_4 acts upon HIO_3 (or I_2O_7), can be hydrolyzed to I_2O_4.

The compound I_4O_9 is a very hygrosopic, pale yelow solid prepared by treating iodine with a mixture of ozone and oxygen at room temperature or by warming iodic acid with orthophosphoric acid. The compound decomposes above 75° to give I_2O_5, I_2, and O_2. Water reacts with I_4O_9 to give iodine and iodic acid.

Iodine pentoxide, which is the anhydride of iodic acid, can be prepared as a white infusible solid by pyrolyzing I_2O_4 or I_4O_9, oxidizing I_2 with nitric acid, or dehydrating HIO_3 at about 200°C. It behaves as an oxidizing agent, the reduction product being I_2. Iodine pentoxide has found use as a reagent for the quantitative determination of carbon mon-oxide. The I_2 which is formed when carbon monoxide reacts with anhy-drous I_2O_5 can be determined using conventional iodometric techniques.

The valence states of the chlorine atom in the oxides of chlorine are not certain. In chlorine monoxide, ClOCl, each chlorine atom is probably tetrahedrally hybridized and three of the hybrid orbitals are occupied by lone pairs. The orbital structures of the molecules of the chlorine oxides (omitting Cl_2O_6) can be written:

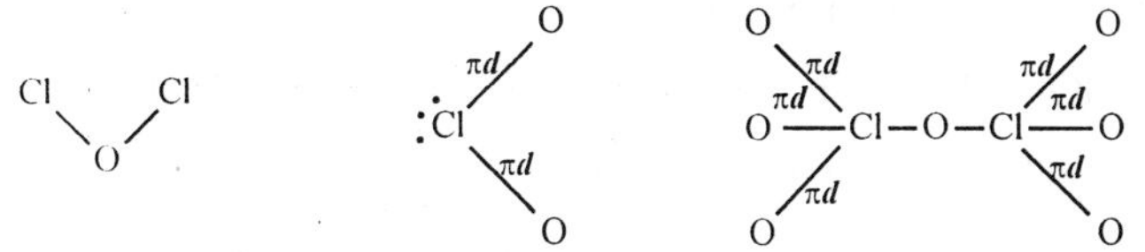

Chlorine resembles nitrogen in forming oxides having molecular structures in which unpaired electrones are present. Chlorine also resembles nitrogen in forming at least on atomic group, ClO_2, in which the same number of chlorine and oxygen nuclei may enter into stable combination with varying numbers of electrons. ClO_2 exists as the neutral chlorine dioxide, as the cation ClO_{2+}, and as the anion ClO_{2-}

Chlorine monoxide, Cl_2O, is made by passing chlorine over cooled precipitated mercuric oxide:

$$2Cl_2 + 2HgO = HgO, HgCl_2 + Cl_2O,$$

or by distilling a fairly concentrated solution of hypochlorous acid at low pressure. Chlorine monoxide explodes readily but not violently on heating. It may be distilled in the absence of organic matter. It attacks the eye and mucous membranes. It is a strong oxidizing agent, and converts most metals into a mixture of oxides and chlorides. It is very soluble in water, with which is reacts to give hypochlorous acid of which is the anhydride, and in carbon tetrachloride. It slowly attacks mercury. The composition of chlorine monoxide gas is given by its explosion on heating.

$$2Cl_2O = 2Cl_2 + O_2,$$

and the subsequent absorption of the chlorine by potassium hydroxide.

Chlorine dioxide, ClO_2, is made by the action of chlorine on silver chlorate:

$$2AgClO_3 + Cl_2 = 2AgCl + 2ClO_2 + O_2,$$

or by the reduction of potassium chlorate either by disproportionation on treatment with sulphuric acid:

$$3KClO_3 + 3H_2SO_4 = HClO_4 + 3KHSO_4 + H_2O + 2ClO_2,$$

or by the action of oxalic acid:

$$2KClO_3 + 2C_2H_2O_4 = K_2C_2O_4 + 2H_2O + 2CO_2 + 2ClO_2,$$

or of sulphur dioxide:

$$2NaClO_3 + SO_2 + H_2SO_4 = 2NaHSO_4 + 2ClO_2.$$

Chlorine dioxide explodes violently on the smallest provocation. In solution in carbon tetrachloride. however, in the dark and in the absence of traces of chlorine monoxide, it is stable for a long time. Chlorine dioxide is extremely soluble in water. The hydrate ClO_2, $6H_2O$ is crystalline, yellow, and stable up to 18°. In the dark the aqueous solution is stable, and gaseous chlorine dioxide can be expelled from it by boiling. Chlorine dioxide is decomposed by light, when pure or when in aqueous solution, to a mixture of chlorine, oxygen, chlorine, trioxide, and chlorine heptoxide.

Chlorine dioxide dissolves in potassium hydroxide solution according to the equation:

$$2ClO_2 + 2KOH = KClO_2 + KClO_3 + H_2O.$$

Chlorine dioxide reacts with hydrogen peroxide to yield chlorous acid, $HClO_2$, and with potassium permanganate solution to yield potassium chlorate, $KClO_3$, and a precipitate of manganese dioxide.

Chlorine dioxide is paramagnetic. The vapour density is normal, and no association occurs in the liquid state, or in solution in water or in carbon tetrachloride.

Chlorine hexoxide, Cl_2O_6, is made by the action of chlorine dioxide on ozone. When the mixed gases are cooled with ice, brown drops of a solution of chlorine dioxide in chlorine hexoxide are formed. The chlorine dioxide is removed by distillation at low pressure.

Chlorine hexoxide decomposes even at 0° to a mixture of other oxides of chlorine, and ultimately to chlorine and oxygen. If it is mixed with water vapour and cooled, it yields a mixture of chloric and perchloric acid:

$$Cl_2O_6 + H_2O = HClO_3 + HClO_4,$$

but an explosion occurs if it is added to liquid water.

Liquid chlorine hexoxide is diamagnetic. The molecular weight of chlorine hexoxide dissolved in carbon tetrachloride is 10% less than that corresponding to the formula Cl_2O_6.

Chlorine heptoxide, Cl_2O_7, is made:

(i) by allowing a mixture of phosphorus pentoxide and anhydrous or highly concentrated perchloric acid to stand at —10° for 24 hours; the product is distilled in vacum;

(ii) by heating potassium perchlorate with chlorosulphonic acid $HOClSO_2$, under reduced pressure; a good yield is obtained.

but the chlorine heptoixde is contaminated with sulphur compounds.

Chlorine heptoxide explodes when struck, or in contact with a flame. It has no action in the cold with sulphur, phosphorus, wood, or paper, but it explodes on contact with iodine. With water it gives perchloric acid of which it is the anhydride. The Rama spectrum of chlorine heptoxide indicates that the molecule consists of two ClO_3 groups linked by an oxygen atom.

Very unstable but less explosive than the dioxide, chlorine hexoxide has been prepared by the action of chlorine dioxide on ozonized oxygen. It is a dark red liquid (m.p. +3.5°C) which dissociates on heating and breaks down eventually to the elements. It reacts violently with water but with cooled water vapour, perchloric acid monohydrate (oxonium perchlorate) and chloric acid result:

$$Cl_2O_6 + 2H_2O \rightarrow (H_3O^+)(ClO_{4^-}) + HClO_3$$

odxonium percholrate
or perchloric acid
monhydrate
$HClO_4.H_2O$

It is a mixed anhydride; aqueous alkali forms a mixture of chlorate and perchlorate,

$$Cl_2O_6 + 2OH^- \rightarrow ClO_{3^-} + ClO_{4^-} + H_2O$$

Chlorine heptoxide, Cl_2O_7

The dehydration of perchloric acid yields chlorine heptoxide. The anhydrous or highly concentrated acid is cautiously added to phosphorus pentoxide, cooled in a freezing-mixture. Chlorine heptoxide may be distilled off under reduced pressure,

$$4HClO_4 + P_4O_{10} \rightarrow 2Cl_2O_7 + 4HPO_2$$

metaphos-
phoric acid

Chlorine heptoxide is a colourless, oily liquid (m.p. —91.5°C, b.p. 80°C). Although the most stable oxide of chlorine, it is easily detonated be heat or shock. It slowly decomposes on standing and with water slowly forms perchloric acid,

$$Cl_2O_7 + H_2O \rightarrow 2HClO_4$$

It is perchloric anhydride.

The Oxyacids of the Halogens and Their Salts

There is no definite evidence for the existence of oxyfluorine acids or oxyfluorine salts. The known oxyacids for three of the other halogens, including their formulas and the oxidation states of the halogen in them, are given in Table 17.1.

Table 17.1 : Oxyacids of the Halogens

Oxidation States	*Oxyacids of Chlorine*	*Oxyacids of Bromine*	*Oxyacids of Iodine*
+1	HClO	HBrO	HIO
+3	$HClO_2$	$HBrO_2$	
+5	$HClO_3$	$HBrO_3$	HIO_3
+7	$HClO_4$	$HBrO_4$	HIO_4
		H_5IO_6	

The oxygen compounds of chlorine, bromine and iodine differ sufficiently to make generalizations concerning them impossible. However, trends in properties of the oxychlorine acids and salts are well established (Table 17.2), and these trends hold, in a general way, for the analogous bromine and iodine compounds. The greater the oxygen content in members of a series of oxygen compounds

Table 17.2 : Trends in Properties of the Oxyacids of Chlorine and their Sodium Salts

Oxidation State	*Acid*	*Thermal Stability and Acid Strength*	*Oxidizing Power*	*Salt*	*Thermal Stability*	*Oxidizing Power and Anion Base Strength*
+1	HClO	NaClO				
+3	$HClO_2$	$NaClO_2$				
+5	$HClO_3$	$NaClO_3$				
+7	$HClO_4$	$NaClO_4$				

(oxyacids on their salts) of a halogen, the greater is their thermal stability. The acids decompose at much lower temperatures than do their salts. These rules apply generally to the oxygen compounds of sulfur, nitrogen, and phosphorus, as well as to those of the halogens.

The salts undergo disproportionation, or auto-oxidation-reduction, upon heating. This, too, is characteristic of the oxyralts of sulfur and phosphorus (but ot of nitrogen). Examples are

$$\underset{\text{Sodium hypochlorite}}{3NaClO} \rightarrow \underset{\text{Sodium chlorate}}{NaClO_3} + 2NaCl$$

$$\underset{\text{Sodium sulfite}}{4Na_2SO_3} \rightarrow \underset{\text{Sodium sulfate}}{3Na_2SO_4} + Na_2S$$

$$\underset{\text{Disodium hydrogen phosphite}}{4Na_2HPO_3} + \underset{\text{Sodium phosphate phosphate}}{2Na_3PO_4} + PH_3 + \underset{\text{Disodium hydrogen}}{Na_2HPO_4}$$

The same sort of reaction is characteristic of the elements.

$$2NaOH + Cl_2 \rightarrow NaCl + NaClO + H_2O$$

$$6NaOH + 3S \rightarrow 2Na_2S + Na_sSO_3 + 3H_2O$$

$$2Na_2S_x \;\; Na_2S_2O_3$$

$$3NaOH + P_4 + 3H_2O \rightarrow 3NaH_2PO_2 + PH_3$$

The activity of the oxyacids of the halogens as oxidizing agents is related inversely to the number of oxygen atoms in the moleucle, i.e., the activity is greatest for the HXO acids and least for the HXO_4 acids. Note that the oxidizing strength is great for the acids of lowest thermal stability.

In a series of oxyacids of any element, the acidic strength, as measured in terms of extent of ionization in aqueous solution, is greater the larger the number of oxygen atoms in the molecule. Phosphoric acid is stronger than phosphorous acid, sulfuric is stronger than sulfurous, perchloric is stronger than chloric, and so on.

The electron (Lewis) formulas for the oxyacids of chlorine are

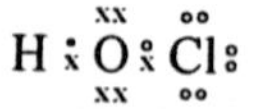

Hypochlorous acid

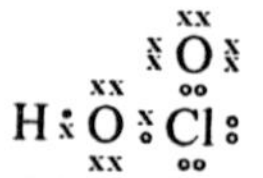

Cholorous acid

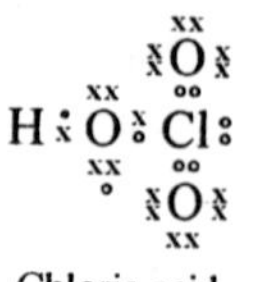

Chloric acid

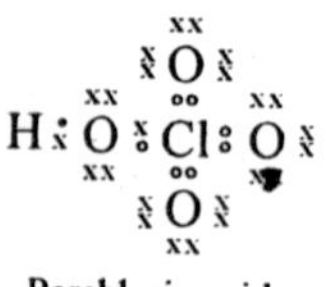

Perchloric acid

In all four acids the hydrogen is bonded through an oxygen atom. The hydrogn is attached to the oxygen by a polar covalent bond and the oxygen atoms are attached to the chlorine atom by bonds that are essentially covalent in character. The oxidation state of the chlorine in these acids is dependent upon the number of its electrons that is shares with oxygen atoms. Chlorine shares on of its electrons with oxygen in HClO, three in $HClO_2$, five in $HClO_3$, and seven in $HClO_4$. Thus, the corresponding oxidation states exhibited by chlorine are +1, +3, +5 and +7, respectively. The oxidation states are positive because oxygen is more electronegative than chlorine. As the positive oxidation state of chlorine increases, the bonding electrons are shifted away from the hydrogen atom, and the tendency for the molecule to lose a proton to water and thereby form hydronium ions in solution increases. This accounts for the coorrelation of the strength of the oxyacids as acids with the number of oxygen atoms in the moleucle. It follows that the tendency of an oxohalogen anion to accept protons increases as the oxygen content of the anion decreases. This tendency of a negative ion to accept protons is a measure of its strength as a base.

The oxyacids of chlorine are more acidic than the corresponding oxyacids of bromine; and those of bromine are more acidic than the corresponding ones of iodine. These facts can be accounted for by the decrease in electronegativity with increasing size of the halogen and the resultant decreases in the tendency for the oxyacid to lose a proton to water.

Hypohalous Acids. When fluorine is passed into liquid water, the reaction products include HF, O_2, and OF_2. However, if fluorine gas at low pressure is passed over water at 0°C, the gaseous products also include hypofluorous acid, HOF. This substance, the only hypohalous acid isolated and characterized at this point in time, is a white solid melting at —117°. It appears to be marginally stable at room temperature, decomposing to oxygen and HF

$$2HOF \rightarrow 2HF + O_2.$$

Hypofluorous acid reacts with water, forming H_2O_2, not oxygen

$$HOF + H_2O \rightarrow HF + H_2O_2.$$

Chlorine, bromine, and iodine are slightly soulble in water (Table 14.1), but not all of the halogen exists in the molecular form X_2, in solution. A disproportionation reaction occurs for a portion of the dissolved halogen molecules

$$X_2 + H_2O \rightleftharpoons H^+ + X + HOX;$$

the extent of this reaction decreases with increasing atomic number. The fraction of X_2 present in a saturated solution which undergoes disproportionation is Cl_2, 0.33; Br_2, 0.007; and I_2, .005. The hypohlaous acids are all very weak, the acidity decreasing with increasing atomic number. This trend parallels the increasing metallic character of the halogens in the same series. Indeed, it has been estimated that hypoiodous acids has a basic ionization constant

$$HOI \rightleftharpoons HO^- + I^+$$

several orders of magnitude larger than its ionization constant as an acid (Table 14.6).

The dissolution of halogens in aqueous basic solution leads to a more complex series of species. First a very rapid reaction occurs to form hypohalite ion

$$X_2 + 2OH^- \rightarrow X^- + OX^- + H_2O,$$

but the hypohalite ion undergoes further disproportionation, giving halate ion

$$3OX^- \rightleftharpoons 2X^- + XO_3^-$$

The rate of the latter reaction increases with increasing atomic number of the halogen. Thus, ClO^- disproportionates very slowly at room temperature but rapidly at temperatures above 75°; hypobromite disproportionates rapidly at room temperatures, but the reaction is sufficiently slow at 0°C that large quantities of BrO^- can be kept under these condition. The rate of disproportionation of IO^- is sufficiently rapid at all temperatures that solution of I_2 in aqueous base gives quantitiatvie yields of iodate, IO_3^-

$$3I_2 + 6OH^- \rightarrow 5I^- + IO_3^- + 3H_2O.$$

Thus, it is possible to obtain aqueous solution containing XO^- and XO_3^- by dissolving the corresponding halogen in aqueous base under different conditions.

The most suitable method for the preparation of hypohalite anions involves the reaction of halogen with an aqueous suspension of Hgo.

$$2X_2 + 2HgO + H_2O \rightleftharpoons HgO.HgX_2 + 2HOX$$

The halide reduction product is removed from solution and HgO is sufficiently insoluble to keep the solution from becoming too basic, which would encourage the disproportionation reaction. The hypohalous acids and the corresponding hypohalites are oxidizing agents. Little is known constitution of the free acids because attempts to prepare pure samples of HOX (X = Cl, Br, I) lead to decomposition products. For example, if a 25% aqueous solution of HClO is distilled at low pressure, pure Cl_2O is evolved. The only hypohalite salts that have been isolated are those of the alkali metals and the alkaline-earth metals. On the basis of the electron-pair-repulsion argument we might expect the hypohalous acids to be bent with an XOH bond angle near 109°.

$$H:\ddot{\underset{..}{O}}::\ddot{\underset{..}{X}}:$$

The most important salt of the hypohalous acids is sodium hypochlorite, Na^+OCl^-, which is made by electrolyzing a cold solution of sodium chloride, and allowing the products of electrolysis (sodium hydroxide solution and chlorine) to mix:

$$2OH^- + Cl_2 \rightarrow Cl^- + OCl^- + H_2O$$

It is sold under a variety of trade names like Parazone, Domestos and Milton as a dilute aqueous solution (which also contains sodium chloride) for use as a mild disinfectant and bleaching agent.

Sodium hypochlorite is alkaline in solution because hydroysis takes place, i.e. the OCl^- ion is a strong base:

$$OCl^- + H_2O = HOCl + OH^-$$

Both hypochlorous acid and the hypochlorite ion are strong oxidizing agents, the former more powerful than the latter. Thus in acid solution iodide ions are oxidised to iodine, and iron (II) ions to iron (III) ions:

$$2I^- + H^+ + HOCl \rightarrow I_2 + H_2O + Cl^-$$

$$2Fe^{2+} + H^+ + HOCl \rightarrow 2Fe^{3+} + H_2O + Cl^-$$

Hypochlorous acid and hypochlorites decompose in the presence of sunlight into oxygen and chloride ions:

$$2OCl^- \rightarrow 2Cl^- + O_2$$

On warming to about 75°C, disproportionation into the chloride and chlorate takes place:

$$3OCl^- \rightarrow 2Cl^- + ClO_3^-$$

Bleaching powder, which is a mixture of several substances, contains calcium hypochlorite. It is made by passing chlorine over moist slaked lime; as its name implies, it is used for bleaching fabrics.

Hypochlorous acid, HClO, is produced by the action of chlorine on water:

Equation

The action proceeds to completion in the presence of mercuric oxide which removes both H^+ ions and Cl^- ions from the solution, thus:

$$HgO + 2H^+ + 2Cl^- = HgCl_2(\text{covalent}) + H_2O$$

The mercuric chloride is precipitated as the basic chloride, HgO, $HgCl_2$, which together with excess of mercuric oxide is filtered off, and the filtrate is distilled.

The aqueous solution of hypochlorous acid is pale yellow. If a 25% aqueous solution hypochlorous acid is distilled at a low pressure, pure chlorine monoxide is evolved:

$$2HClO = Cl_2O + H_2O;$$

if a more dilute solution is distilled the distillate is an aqueous solution of the acid. The volatility of hypochlorous acid from a dilute solution may be due to the evolution of chlorine monoxide which combines with water on cooling, or it may be due to the volatility of the undissociated acid in steam. The evolution of chlorine monoxide is unlikely, as the equilibrium constant,

$$K = \frac{[Cl_2O]}{[HClO]^2}$$

for a 5% solutionof hypochlorous acid is only 9.6×10^{-4}.

Hypochlorous acid is a very weak acid. Its dissociation constant at 25° is 3.2×10^{-8}. Its salts are much hydrolyzed in solution, and are decomposed by carbon dioxide. The aqueous solution dissolves magnesium with the liberation of hydrogen, but less electropositive metals such as iron and aluminium liberate a mixture of hydrogen and chlorine, and cobalt, nickel, and copper liberate a mixture of chlorine and oxygen.

Hypochlorous acid oxidizes P, As, Sb, S, Se, and their hydrides, and it liberates chlorine from hydrochloric acid. It reacts with hydrogen peroxide to form hydrochloric acid, water, and oxygen. It liberates iodine from potassium iodide, and converts chromium salts to chromates, manganese salts to hydrated manganese dioxide, and lead salts to lead peroxide.

The hypochlorite ion is less easily reduced than the parent acid, but it slowly decomposes according to the equation:

$$2ClO^- = 2Cl^- + O_2.$$

Catalysts and rough surfaces assist the evolution of oxygen; among the materials used for this purpose are glass, platinum black, and the oxides of cobalt and nickel. If a hypochlorite solution is lightly acidified the liberated hypochlorous acid attacks the hypochlorite ion thus:

$$2HClO + ClO^- = ClO_3^- + 2H^+ + 2Cl^-.$$

Hypobromous acid, HBrO, can be obtained only in solution, and is always contaminated with its decomposition products, bromine and bromic acid. The solution is made (i) by the hydrolysis of bromine:

$$Cl_2 + H_2O \rightleftharpoons H^+ + Cl^- + HClO.$$

using precipitated mercuric oxide to remove the H^+ and the Br^- ions; this process is less efficient than the corresponding process for the preparation of hypochlorous acid, (ii) by the distillation of a mixture of bromine water and silver sulphate. The distillate is a dilute solution of hypobromous acid contaminated with bromine which can be removed by passing a stream of nitrogen through the solution.

Hypobromous acid is decomposed by rough surfaces with the liberation of oxygen; bromine (and not hydrobromic acid) is set free at the same time:

$$5HOBr = HBrO_3 + 2H_2O + 2Br_2.$$

This reaction differs from the corresponding reaction for hypochlorous acid, because the reaction:

$$HBrO + Br^- + H^+ = Br_2 + H_2O,$$

takes place more readily than the corresponding reaction involving chlorine. In slightly acid solution hypobromous acid decomposes to form bromic acid, again with the liberation of bromine:

$$2HBrO + BrO^- = BrO_3^- + 2H^+ + 2Br^-,$$

$$HBrO + Br^- + H^+ = Br_2 + H_2O.$$

Perchloric acid, $HClO_4$, a colourless mobile liquid, m.p. — 112°, b.p. 19° under 11mm pressure, is made by treating potassium perchlorate with concentrated (90-92%) sulphuric acid and distilling under reduced pressure (10-20 mm). If perchloric acid is heated under atmospheric pressure it breaks up, often explosively, to steam, oxygen, and oxides of chlorine. Anhydrous perchloric acid oxidizes organic substances, frequently with violence. The Raman spectrum of anhydrous perchloric acid indicates that the four oxygen atoms in the molecule are arranged tetrahedrally round the chlorine atom. Certain lines in the spectrum are identical with those of the spectrum of chlorine heptoxide. The spectrum is different from that of the ion ClO_4^-, and hence there is no dimer $[H_2ClO_4^+][ClO_4^-]$. When in contact with other acids, however, perchloric acid acts as an electron donor. It combines with hydrogen fluoride to form the rate cation $[H_2F^+]$.

Anhydrous perchloric acid is hygroscopic; it dissolves in water with a hissing sound to form hydrates with 1, 2, 2½, 3, 3½, molecules of water of crystallization. An aqueous solution of perchloric acid may also be made by evaporating a solution of ammonium perchlorate in a mixture of concentrated hydrochloric acid and concentrated nitric acid.

The monohydrate $HClO_4$, H_2O should be formulated $[H_3O^+][ClO_4^-]$. It is crystalline with the same lattice as ammonium perchlorate. It melts at 50° to a liquid ten times more viscous than the anhydrous acid. It can be heated nearly to its b.p. 110° without decomposition. Aqueous solutions of perchloric acid contain the ions $[H_3O^+]$ and $[ClO_4^-]$. Perchloric acid is the strongest of all acids. The perchlorate ion is not an oxidizing agent except to very strong reducing agents such as titanous chloride, chromous chloride, sodium dithionite and alkaline ferrous hydroxide. It is not attacked by SO_2, H_2S, HCl, HI, or HNO_2. The metals iron and zinc dissolves in dilute aqueous solution of perchloric acid to yield hydrogen; if the solution is concentrated, however,the volume of hydrogen liberated is less than stoichiometric quantity. The perrchlorate ion is tetrahedral, and is very resistant to deformation.

Halous acid

The only halous acid brown for certain is halrous acid HIO_2, only exist in solid.

Chlorine atom exists in the oxidation state (+III). $HClO_2$ is made by treating barium chlorite with H_2SO_4, and filtering off the $BaSO_4$.

$$Ba(Cl_2)_2 + H_2SO_4 \rightarrow 2HClO_2 + BaSO_4$$

Salts of $HClO_2$ are called chlorites, and are made either from ClO_2 and sodium hydroxide. or ClO_2 and sodium peroxide.

$$\underset{\text{chlorite}}{2ClO_2 + 2NaOH} \rightarrow \underset{\text{chlorate}}{NaClO_2} + NaClO_3 + H_2O$$

$$2ClO_2 + Na_2O_2 \rightarrow 2NaClO_2 + O_2$$

Chlorites are used as bleaches. They are stable in alkaline solution even when boiled. but in acid solution they disproportionate, particularly when heated.

(+III) (+IV) (-I) oxidation states

$$5HClO_2 \rightarrow 4ClO_2 + NaCl$$

Iodic acid, HIO_3,a colourless solid, is made by dissolving iodine pentoxide in the smallest possible volume of warm water. On cooling, rhombic crystals of iodic acid are deposited. When iodic is heated it decomposes; at 110°h HI $_3O_8$is formed and at 195°this compound decomposes to iodine pentoxide and steam.

Iodic acid is very soluble in water; the solution first reddens and then bleaches litmus. The solid acid inflames when heated with sulphur, phosphorus, or organic matter, and its solution oxidizes sulphur dioxide, hydrogen sulphide, and hydriodic acid with the liberation of iodine.

X-ray analysis shows that the molecule of iodic acid has a pyramidal configuration. In the crystal each iodine atom is surrounded by six oxygen atoms situated at the apices of a distorted octahedron. The distances I-O for the oxygen atoms which are part of the trigonal pyramid are 1.55, and for those which are not are 2.45, 2.70, and 2.95. Each IO_3 group is linked to two other IO_3 groups by hydrogen bonds, and hence the lattice may be regarded as a chain structure reminiscent of the chain of bicarbonate ions in potassium bicarbonate.

The iodates. The alkali metal and alkaline earth iodates can be made by the action of iodine on the mental hydroxide:

$$3I_2 + 6OH^- = IO_3^- 5I^- + 3H_2O.$$

Barium iodate is sparingly soluble, and hence is easily made by this method.

Potassium iodate, KIO_3, can be made :

(i) *by the action of iodine on potassium hydroxide,*

(ii) by double decomposition between barium iodate and potassium sulphate,

(iii) by the direct action of oxygen on potassium iodide at 100 atm pressure,

(iv) by the action of iodine on potassium chlorate in slightly acid solution:

$$5KClO_3 + 3I_2 + 3H_2O = 3(KIL_3, HIO_3) + 2Kcl + 3HCL.$$

The acid salt KIO_2, HIO_3 is sparingly soluble and crystallizes from the solution.

The normal iodates of most metals are soluble ,but those of Ag,TI(ous),Cu(ic), Ba,and Pb are sparingly soluble.Lithium forms a normal iodate only, but the other alkali metals form acid iodates, MIO_3, HIO_3. Alkali metal iodates do not detonate when brought into contact with carbon or sulphur, but the mixtures explode when heated. The iodates are oxidizing agents in aqueous solution. The metal iodates decompose when heated; the mode of decomposition is determined by the electro-negativity of the metal. The iodates of the most electropositive metals and silver on heating to 500^{0}give the metal iodide and liberate oxygen; the iodates of less electro-positive metals decompose to the metal oxide, liberating iodine. Sodium iodate has intermediate properties. When heated it liberates iodine and leaves a residue of sodium iodide and sodium oxide. Ammonium iodate decomposes completely when heated,thus:

$$2NH_4 = N_2 + I_2 + O_2 + 4H_2O.$$

The iodate ion,IO_3-, has a pyrimidal structure. Experimental evidence has been obtained showing that in the crystal of sodium iodate the group $IO_3$is a discrete ion, and is not part of a perovskite structure.

Bromic acid, $HBrO_3$, is known only in solution. A solution of the acid is obtained:

(i) by shaking silver bromate with bromine water , and filtering off the precipited silver bromide:

$$3H_2O + 5AgBrO_3 + 3Br_2 = 6HBrO_3 + 5agBr,$$

(ii) by treating barium bromate with dilute sulphuric acid, and removing the excess acid with baryta water, and filtering off the barium sulphate.

The aqueous solution of bromic acid may be concentrated up to 5% by evaporation on a water bath, and concentrated up to 50% by evaporation in a vacuum desiccator. An aqueous solution containing 50% of bromic acid corresponds to the composition $HBrO_37H_2O$. Attempts to concentrate the solution further cause the decomposition of the acid:

$$4HBrO_3 = 2Br_2 + 5O_2 + 2H_2O,$$

The aqueous solution of bromic acid is a strong oxidizing agent. It oxidizes sulphur dioxide to sulphuric acid, hydrogen sulphide to sulphur, and hydrogen bromide to bromine. In all these reactions bromic acid is reduced to bromine.

The bromates. Potassium bromate is made by dissolving a slight excess of bromine in hot concentrated potassium hydroxide solution:

$$3Br_2 + 6KOH = KBrO_3 + 5KBr + 3H_2O.$$

The potassium bromate is less soluble than potassium bromide, and it separates on cooling. Two variants of this reaction are described by the equations:

$$6KOCl + Br_2 = 2KBrO_3 + 4KCl + Cl_2, \text{ and}$$
$$3Cl_2 + KBr + 6KOH = KBrO_3 + 6KCl + 3H_2O.$$

The bromates of Na, K, Tl(ous) are anhydrous; the bromates of Li, Ca,Sr,Ba,Zn, and Pb have one molecule of water of crystallization;those of Mg and Zn have six molecules of water of crystallization. The bromates of Tl (ous), Ag,Ba, and Pb are sparingly soluble in water (of the order of 0.5gm/100gm at 20^0), and those of the other metals mentioned above are fairly soluble (of the order of 50gm/100gm at 20^0). The bromates are decomposed by heat, but the course of the reaction depends on the nature of the metal. Bromates of K, Hg(ous) ,and Ag give the metal bromide and oxygen; bromates of Mg,Zn, and Al give the metal oxide and bromine and oxygen; bromates of Pb and Cu give the metal bromide and oxide. Ammonium bromate and lead bromate are explosive.

Chloric acid, $HClO_3$, is made in aqueous solution by treating barium chlorate (see below) with dilute sulphuric acid, filtering and evaporating. The solution may be heated to its boiling point, and concentrated until its composition reaches $HClO_3 7H_2O$, without decomposing the chloric acid. If a dilute solution is allowed to stand, or if the concentration is carried too far, chloric acid decomposes to the equation:

$$3HClO_3 = HClO_4 + 2O_2 + Cl_2 + H_2O.$$

The concentrated aqueous solution of chloric acid is a strong oxidizing agent. It attacks organic substances, and dissolves certain metals to form the metal chlorties.

The Chlorates. Sodium chlorate is made by the electrolysis of a solution of sodium chloride in a cell which permits the mixing of the chlorine liberated at the anide with the OH^- formed at the cathode. The reaction occurs:

$$Cl_2 + OH^- = HOCl + Cl,$$

and at the anode the hypochlorite ion is oxidized to the chlorate ion:

$$HOCl- + 8H_2O - 16e = 3ClO3- + 2Cl- + 16H+.$$

Some oxygen is also liberated at the anode by the discharge of hydroxyl ions:

$$4OH^- - 4e = 2H_2O + O_2.$$

The electrolyzed solution is evaporated to precipitate the uncharged sodium chloride which is filtered off. Sodium chlorate is crystallized from the filtrate.

The metal chlorate are soluble. Barium chlorate is made by evaporating a mixed solution of sodium chlorate and barium chloride. The less soluble sodium chloride crystallizes first, and when the hot filtered solution is evaporated, the hydrate, $Ba(ClO_3)_2, H_2O$, crystallizes. Potassium chlorate is made by adding potassium chloride to sodium chlorate solution; potassium chlorate, which is slightly soluble, crystallizes.Potassium chlorate, when heated just above its m.p.370^0,is converted to a mixture of potassium chloride and potassium perchlorate:

$$4KClO_3 = KCl + 3KClO_4$$

If manganese dioxide is present it promotes an alternative mode if decomposition:

$$2KClO_3 = 2KCl + 3O_2,$$

Which takes place below the m.p. of potassium chlorate. Potassium chlorate forms explosive mixtures with sulphur, phosphorus, and charcoal. With concentrated hydrochloric acid it yield a mixture of chlorine and chlorine dioxide:

$$8KClO_3 + 24HCl = 8KCl + 9Cl_2 + 6ClO_2 + 12H_2O.$$

and with concentrated sulphuric acid a mixture of perchloric acid and chlorine dioxide(p.937).

Perhalic acids. Perchloric and periodic acids are well characterized compounds, perbrombic acid and its salts have been isolated recently. Anhydrous perchloric acid can be prepared by distilling a mixture of potassium perchlorate and concentrated sulfuric acid under reduced pressure. Anhydrous perchloric acid is hygroscopic; it dissolves in water to form a series of hydrates. $HClO_4.nH_2O$(n=1,2,2.5,3,3.5).The anhydrous acid is a vigrous oxidizing agent toward organic compounds, the reaction often occurring with explosive violence. It is a strong acid in aqueous media and even acts as a proton donor in such acidic media as H_2SO_4,HNO_3, and acetic acid. Perchlorate ion is an oxidizing agent, but only towards very strong reducing agents. Perchlorate can be prepared by the electrolytic oxidation of chlorates,by the solution of metals more active than hydrogen in aqueous perchloric acid.

Perchlorate ion is tetrahedral with bond distances shorter than those in the chlorate ion, suggesting even more double-bond character in the Cl-Obonds.There is no information on the structure of anhydrous perchloric acid. The monohydrate, $HClO_4.H_2O$,appears to be better formulated as hydronium perchlorate, $H_3O^+ClO_4^-$. This substance melts at 50°without decomposition to form a very viscous liquid which itself can be heated to 110°without decomposition.

Although numerous earlier attempts to synthesize perbromates were unsuccessful and reasons for their non-existence discussed, aqueous solutions containing this species were prepared by electrochemically oxidizing solutions containing BrO_3^-.Chemical oxidation of BrO_3^- can be effected in aqueous solution with XeF_2.Perbromic acid has been identified by its mass spectrum.

Three iodine-containing acids or their salts are known in which iodine exhibits an oxidation number of 7:HIO_4,H_3IO_5, and H_5IO_6.These acids can be imagined as hydrates of the hypothetical oxide I_2O_7.In these terms the acids HIO_4,H_3IO_5, and H_5IO_6 are the 1-,3-, and 5-hydrates, respectively. The best known of these acids is H_5IO_6, called paraperiodic acid. A convenient laboratory preparation of this substance involves treatment of the barium salt with concentrated HNO_3.Paraperiodic acid melts with decomposition at 140°to give iodic acid, water, and oxygen. Under vacuum at 80° it loses water to form HIO_4.which is known as metaperiodic acid. Salts of mesoperiodic acid,H_3IO_5.are also known, although the pure acid has never been reported.

Salts of the periodic acids can be formed by several indirect methods. For example, oxidation of $NaIO_3$in aqueous alkali with chlorine or peroxydisulfate ion gives $Na_3H_2IO_6$, which is the acid salt of H_5IO_6. This salt can converted into the meta salt ($NaIO_4$) by treatment with concentrated nitric acid.

The Periodic acids are strong acids and are stronger oxidizing agents than iodic acid. They are often used as reagents for analytical purposes because they generally react smoothy and rapidly. Ozone can sometimes be detected when periodic acids react, but peroxides are never formed.

The metaperiodate ion IO_4^-, is isostructural with the perchlorate ion as would be expected. The results of a variety of physical measurements on aqueous periodate solutions indicate that the species IO_4^-. $H_3IO_6^{2-}$, H_2IO6^{3-}, and $H_2I_2O^4_{10}{}^-$ are (43,44). Paraperiodic acid usually acts as a dibasic acid ($K_1 = 2.3 \times 10^{-2}$, $K_2 = 10^{-6}$, $K_3 = 2.5 \times 10^{-13}$) commonly forming acid salts with the stiochiometry $M_2H_3IO_6$. The structure of the ammonium salt shows the presence of IO_6^{5-} octahedra linked by hydrogen bonds. Thus it would appear that paraperiodic acid is best formulated as $(HO)_5IO$.

Perchloric acid,$HCIO_4$, a colourless mobile liquid, m.p.–112°,b.p.19°under II mm pressure , is made by treating potassium perchlorate with concentrated (90-92%) sulphuric acid and distilling under reduced pressure (10-20mm).If perchloric acid is heated under atmospheric pressure it breaks up, often explosively, to steam, oxygen, and oxides of chlorine. Anhydrous perchloric acid oxidizes organic substances, frequently with violence. The Raman spectrum of anhydrous perchloric acid indicates that the four oxygen atoms in the molecule are arranged tyetrahedrally round the chlorine atom.Certain lines in spectrum are identical with those of the spectrum of chlorine heptoxide. The spectrum is different from that of the ion CIO_4^-, and hence there is no dimer $[H_2XIO_4^+][CIO_4^-]$. When in contact with other acids, however, perchloric acis acts as an electron donor. It combines with hydrogen fluoride to form the rare cation $[H_2F^+]$.

Anhydrous perchloric acid is hydroscopic; it dissolves in water with a hissing sound to form hydrates with 1,2,2½,,3,3½, molecules of water crystallization.An aqueous solution of perchlorate in a mixture of concentrated hydrochloric acid and concentrated nitric acid.

The monohydrate $HCIO_4,H_2O$ should be formulated $[H_3O^+]$ $[CIO_4^-]$. It is crystalline with the same lattice as ammonium perchlorate. It melts at 50°to a liquid ten times more viscous than the anhydrous acid. It can be heated nearly to its b.p.110° without decomposition. Acqueous solutions of perchloric acid contain the ions $[H_3O^+]$ and $[CIO_4^-]$. Perchloric acid is the strongest of all acids. The perchlorate ion is not an oxidizing agent expcept to very strong reducing agents such as titanous chloride, chromous chloride, sodium dithionite and alkaline ferrous hydroxide. It is not attacked by SO_2,H_2S,HCI,HI.or HNO_2.The metals iron and zinc dissolve in dilute, aqueous solutions of perchloric acid to yield hydrogen; if the solution is concentrated, however, the volume of hydrogen liberated is less than the stoichiometric quantity. The perchlorate ion is tetrahedral , and is very resistant to deformation.

The perchlorates of the akali metals and the alkaline earth metals are stable to heat; lithium perchlorate may be heated to 300°without decomposition, and potassium perchlorate to 400°. Ammonium perchlorate *deflagrates* with a yellow flame if heated to 200°:

$$2NH_4ClO_4 = N_2 + Cl_2 + 2O_2 + 4H_2O.$$

The anhydrous perchlorates of the alkali metals and some amines are dimorphic. Below the following temperatures the crystals are rhombic, and above these temperatures the forms are more symmetrical: Na 308 , K 300°, Rb279°, Cs219°, $NH_4$240°. The perchlorates of the alkali metals and the alkaline earth metals fall into two classes, those which are hydrated and very soluble and those which are anhydrous sparingly soluble.

18

Transition Elements (d-Block Elements) (Scandium)–Group IIIB (3)

d-Block Elements and their Characteristics

***d*-Block Elements (Transition Elements).** In these elements, either in their atomic state or in any of their common oxidation state, *the last electron enters the d-orbital of the pen-ultimate* (inner to ultimate) *shell, i.e. the last electron goes to (n-1) d-orbital.* In other words, in these elements *(n-1) d*-orbital is being progressively filled. Hence the name *d*-**Block Elements.** With a few exceptions, in the atoms of these elements the *ns*-orbital is completely filled (s^2 configuration). Consequently, the valence shell configuration of most of those elements varies from $(n-1)d^1\ ns^2$ (Group III B) to $(n-1)\ d^{10}\ ns^2$ (Group II B) configuration. The elements of groups III B, IVB, V B, VI B, VII B, I B and II B belong to this block.

These elements are located in the middle of the periodic table (*i.e.* between *s* and *p*-block elements) and consist of metals only. *d*-block elements are also called **Transition Elements,** since their properties are intermediate between those of *s* and *p*-block elements. These are classified into *four series viz.* 3*d*, 4*d*, 5*d* and 6*d* series corresponding to the filling of 3*d*, 4*d*, 5*d*, and 6*d* orbitals of 3rd (M-shell), 4th (N-shell), 5th (O-shell) and 6th (P-shell) shells respectively.

(a) 3d-series (1st transition series). This series has *ten* elements, beginning with Sc (Z = 21) and ending at Zn (Z = 30). The elements of this series lie in 4th period.

(b) 4d-series (2nd transition series). This series also has *ten* elements, beginning with Y (Z = 39) and ending at Cd (Z = 48). The elements of this series lie in 5th period.

(c) 5d-series (3rd transition series). This series also has *ten* elements, viz. La (Z = 57), Hf (Z = 72) to Hg (Z = 80). The elements of this series lie in 6th period.

(d) 6d-series (4th transition series). At present this series has three elements viz, Ac (Z = 89), Ku (Z = 104) and Ha (Z = 105). The elements of this series lie in 7th period which is an incomplete period.

General Properties of Transition Elements

(i) Metallic Character. All the transition elements are metals, since the number of electrons in the outermost shell is very small, being equal to 2. They are hard, malleable and ductile and possess high tensile strength. They exhibit all the three types of structures: face centered cubic *(fcc).* hexagonal close packed *(hcp),* and body centered cubic *(bcc).* Metals of VIII and IB Groups are softer and more ductile than other metals.

It appears that covalent and metallic bonding both exist in the atoms of transition metals. *The presence of unfilled d-orbitals favours covalent bonding.*

These metals are good conductors of heat and electricity.

(ii) Melting and Boiling Points. These are very heavy metals and have high melting and boiling points. The last member of each series, *i.e.,* Zn, Cd, and Hg have comparatively lower values. This is because these metals have *completely filled d-orbitals with no unpaired electrons (n)* as is evident from their valence-shell electronic configurations given below:

$$\text{Zn} \longrightarrow 3d^{10}\,4s^2 \text{ witn } n = 0$$
$$\text{Cd} \longrightarrow 4d^{10}\,5s^2 \text{ with } n = 0$$
$$\text{and} \quad \text{Hg} \longrightarrow 5d^{10}\,6s^2 \text{ with } n = 0$$

(iii) Atomic Volume and Densities. The atomic volumes of the metals are much lower than those of alkali and alkaline earth metals. As the inner orbitals get filled, the increased nuclear charge pulls the electronic cloud inward. The atomic volume, therefore, decreases. The decrease in atomic volume increases the density. Accordingly, the densities of transition elements are very high as compared to those of the metals of Groups I A and II A.

Electronic configurations of d-block elements

Classification of series	*Element*	*Symbol*	*Atomic Number (Z)*	*Observed valence shell configuration*
	Scandium	Sc	21	$3d^1\ 4s^2$
	Titanium	Ti	22	$3d^2\ 4s^2$
	Vanadium	V	23	$3d^3\ 4s^2$
Elements of *3d*-series (1st)	Chromium	Cr	24	$3d^5\ 4s^1$
transition series – Period 4).	Manganese	Mn	25	$3d^5\ 4s^2$
	Iron	Fe	26	$3d^6\ 4s^2$
	Cobalt	Co	27	$3d^7\ 4s^2$
	Nickel	Ni	28	$3d^8\ 4s^2$
	Copper	Cu	29	$3d^{10}\ 4s^1$
	Zinc	Zn	30	$3d^{10}\ 4s^2$
	Yttrium	Y	39	$4d^1 5s^2$
	Zirconium	Zr	40	$4d^1 5s^2$
	Niobium	Nb	41	$4d^4 5s^1$
Elements of *4d*-series (2nd)	Molybdenum	M	42	$4d^5 5s^1$
transition series – Period 5)	Technetium	Tc	43	$4d^5 5s^2$
	Ruthenium	Ru	44	$4d^7 5s^1$
	Rhodium	Rh	45	$4d^8 5s^1$
	Palladium	Pd	46	$4d^{10} 5s^0$
	Silver	Ag	47	$4d^{10} 5s1$
	Cadmium	Cd	48	$4d^{10} 5s^2$
	Lanthanum	La	57	$4f^0 5d^1 6s^2$
	Hafnium	Hf	72	$4f^{14} 5d^2 6s^2$
	Tantalum	Ta	73	$4f^{14} 5d^3 6s^2$
Elements of 5*d*-series (3rd)	Tungsten	W	74	$4f^{14} 5d^4 6s^2$
transition series – Period 6).	(Wolfram)			
	Rhenium	Re	75	$4f^{14} 5d^5 6s^2$
	Osmium	Os	76	$4f^{14} 5d^6 6s^2$
	Iridium	Ir	77	$4f^{14} 5d^7 6s^2$
	Platinum	Pt	78	$4f^{14} 5d^9 6s^1$
	Gold	Au	79	$4f^{14} 5d^{10} 6s^1$
	Mercury	Hg	80	$4f^{14} 5d^{10} 6s^2$
Elements of *6d*-series (4th)	Actinium	Ac	89	*6d17s2*
transition series –				
Incomplete Period 7).	Kurchatovium	Ku	104	$6d^2 7s^2$
	Hahnium	Ha	105	$6d^3 7s^2$

(iv) Ionization Potential. The ionization potential values of most of the *d*-block elements lie between those of *s* and *p*-block elements. *Thus d-block elements are less electropositive than the s-block elements and more so than p-block elements.* Consequently, these elements *do not form ionic compounds so readily as s-block elements* (alkali and alkaline earth metals). In general, compounds with metals in lower valency state are ionic and others covalent. The tendency to form ionic compounds decreases as the size of the atom increase.

(v) Low Reactivity. These elements have low reactivity and this tendency is most pronounced in Au and Pt which are, therefore, called *noble metals.* Their low reactivity is due to the high values of their melting and boiling points, heats of sublimation and ionization potential.

(vi) Colour. The transition metals usually form coloured compounds. If an ion is coloured, it means that it absorbs light of a particular colour whose wavelength is in the visible region *viz* that of $\lambda = 4000 - 7000$ Å and reflects (*i.e.* transmits) the light which has the colour complementary to that of the absorbed light. This complementary colour which is actually the colour of the reflected light is the colour of the substance.

Ion	*Configuration*	*n*	*Colour*
Sc^{3+}	$3d_o$	0	Colourless
Ti^{4+}	$3d_o$	0	Colourless
Cu^{+}	$3d^{10}$	0	Colourless
Zn^{2+}	$3d^{10}$	0	Colourless
Ti^{3+}	$3d^{1}$	1	Purple
V^{3+}	$3d^{2}$	2	Green
Cr^{3+}	$3d^{3}$	3	Green
Fe^{2+}	$3d^{6}$	4	Pale green or faintly green
Fe^{3+}	$3d^{5}$	5	Yellow

The colour is due to the promotion of unpaired *d*-electrons from one energy level to the other in *d*-level. In general the atoms, ions or molecules which have unpaired electrons in *d*-orbital (*i.e.* partially filled *d*-orbital) are coloured because it is only in case of such orbitals that the promotion of unpaired *d*-electrons to higher energy level within the same sub-shell in possible. On the other hand, the atoms, ions or molecules having all the electrons paired in *d*-orbitals (*i.e.* completely filled *d*-orbitals) or no electrons (*i.e.* empty *d*-orbitals) are colourless, since in case of such *d*-orbitals there are no vacant *d*-orbitals to accept the electrons promoted to them. Thus, if n is supposed to be the number of unpaired electrons, then the atoms, ions or molecules having $n = 0$ *are colourless* while those having $n = 1.2,...$ *are coloured.* For example,

(vii) Variable Valency. Transition elements show variable valency. This is because in their case the core of kernel left after their atom loses valency electron is unstable and tends to lose one or more electrons further. For example, in case of iron we can represent the two valencies as below.

Fe (26)= 2, 8, 14, 2

Fe^{2+} (24)= 2, 8, 14 (Core)

The core being usable loses one more electron to give Fe^{3+} ion

Fe^{3-} (23) 2, 8, 13

(viii) Catalytic Properties. Most of the transition metals, their alloys and compounds are used as catalysts. Some common examples are: Pt, Ni, Fe, Cr, V_2O_5 etc. Catalytic power of these metals is due to either the use of their *d-orbitals.* Thus, the transition elements having all the electrons paired are diamagnetic. Fe and Cr are ferromagnetic and can be magnetized.

(x) Tendency to Form Complexes. The cations of transition elements have a great tendency to form complexes with several ligands. This tendency is due to the following reasons.

(a) The cations are relatively very small in size and hence have high positive charge density which makes it easy for the cations to accept the lone pairs of electrons from the ligands.

(b) The cations save vacant (n – 1) *d*-orbitals which are of appropriate energy to accept lone pair of electrons from the ligands for bonding with them.

(xi) Formation of Interstitial Compounds. Small non-metallic atoms (*e.g.* H, B, C, N etc.) are able to fit into the interstitial sites of transition metal lattices to form the compounds which are called **interstitial compounds.** These have variable composition and cannot, therefore, be expressed by simple formulae. *i.e.*, these are *non-stoichiometric.*

Scandium

Element	Symbol	Atomic weight	Atomic number
Scandium....	Sc	44.96	21
Yttrium....	Y	88.92	39
Lanthanum....	La	138.92	57
Actinium....	Ac	227	89

General Features of the Scandium Subgroup. The scandium subgroup includes besides those indicated above, the elements with atomic numbers from 58 to 71, known as the rare-earth elements or lanthanides.

All the elements of the scandium subgroup, including the lanthanides, contain two electrons in the outermost layer of their atoms. In their second last layer scandium, yttrium, lanthanum and actinium, as well as the lanthanides, gadolinium and lutetium, contain nine electrons, all the rest of the lanthanides containing eight.

The maximum valency of the scandium subgroup elements, as a rule, equals three. However, cerium also forms a number of derivatives in which it is tetravalent. Oxides of tetravalent praseodymium and terbium are also known, but they are rather unstable.

In nature these elements occur usually in intimate association with one another and with the elements of the fourth group, zirconium, hafnium and thorium. One of the main sources for their preparation is the mineral *monazite,* which is a mixture of phosphates of cerium, lanthanum, etc.

Table 18.1 : *Properties of the Metals in sub-Group III T*

	Sc	Y	La
Atomic number	21	39	57
m.p., °C	1400	1500	920
b.p., °C	2500	3000	4200
Specific gravity	30	46	615
Ionic radius, M^{3+}	081	093	1.06
Atomic radius (12 co-ordination)	166	1 82	1 87

Scandium

The valency system of the scandium atom in the ground state has the electron configuration

3d	4s
↑	↓↑

The three electrons are readily expelled from the atom no evidence for the existence of the state Sc^{11} has been found. In all the scandium compounds except the neutral complex ScA_3, and the complex anions $[ScF_4^-]$, $[ScF_5^{2-}$, and $[ScF6^{3-}]$, the scandium atom appears to be exerting ionic bonds, or at least bonds which possess much ionic character.

Scandium metal is obtained by the electrolysis of fused scandium chloride. It is dimorphic and crystallizes with the close-packed hexagonal structure or the close-packed cubic structure. It is easily oxidized. When heated in nitrogen it yields the interstitial nitride ScN, which is decomposed by, water with the liberation of ammonia.

Table 18.2 : *Examples of Compounds formed by the Elements in sub-Group III T*

Valency system of atom $\overbrace{d\ d\ d\ d\ d}^{n-1}\ \overbrace{s\ p\ p\ p}^{n}$

The value of *n* are : Sc 4, Y 5, La 6.

Normal prehybridisation electron configuration : $\overbrace{d\ d\ d\ d\ d}^{n-1}\ \overbrace{s\ p\ p\ p}^{n}$ ↑ ↑ ↑

Oxidation state	Nature of unit	Hybridization	Sc	Yt	La
	Interstitial compounds		ScN		LaN
	Simple ionic compounds		$Sc(NO_3)_3$, $Sc_2(SO_4)_3$		$La(ClO_4)_3$
III	Giant molecule compounds		Sc_2O_3, ScOOH, ScF_3		La_2O_3, LaO.OH
	Neutral complexes	Octahedral	ScA_3		
	Complex anions	Tetrahedral	$K[ScF_4]$		
		Trigonal bipyramidal	$K_2[ScF_5]$		
		Octahedral	$K_3[ScF_6]$		
	π-complexes		$(\pi\text{-}C_5H_5)_3Sc$		

Scandium oxide, Sc_2O_3, a white powder, has the Mn_2O_3 structure. It is more basic than aluminium oxide, but less basic calcium oxide. It absorbs carbon dioxide from the air., Scandium hydroxide is thrown down as a gelatinous precipitate on the addition of an alkali to a solution of scandium chloride. The lattice of scandium hydroxide is similar to that of γ-AlO.OH.

Scandium fluoride is insoluble in water, but dissolves in the presence of excess HF or NH_4F to give $[ScF_6]^{3-}$. The crystals have the ReO_3 structure. The other halides of scandium are soluble in water, and scandium chloride is soluble in alcohol. Scandium sulphate and nitrate are soluble; the carbonate, oxalate, and phosphate are insoluble. Scandium nitrate and scandium carbonate easily decompose to scandium oxide on heating; scandium sulphate on heating also yields the oxide but less readily.

Scandium (like zirconium and thorium) forms a thiosulphate which is easily hydrolyzed to an insoluble basic salt $Sc(OH)S_2O_3$; the thiosulphates of yttrium and lanthanum are soluble and not easily hydrolyzed, and hence these metals may be separated from scandium.

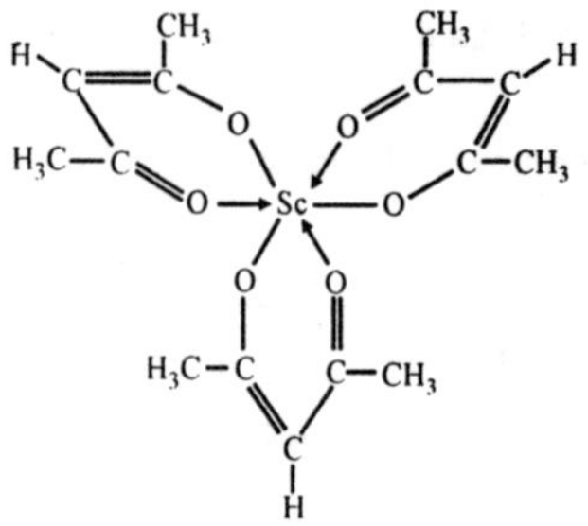

Fig.

The scandium atom, by accepting additional electrons in the valency system assume the tetrahedral, trigonal-bipyramidal, or octahedral valence states. The orbital diagrams of certain complexes of the scandium atom are given below, and the octahedral formula of ScA_3 is shown opposite.

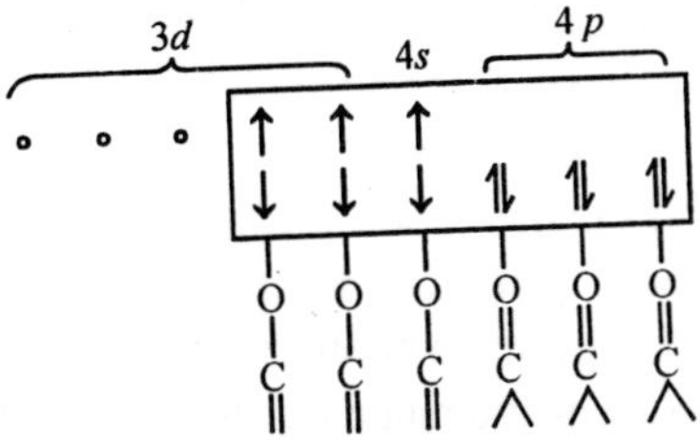

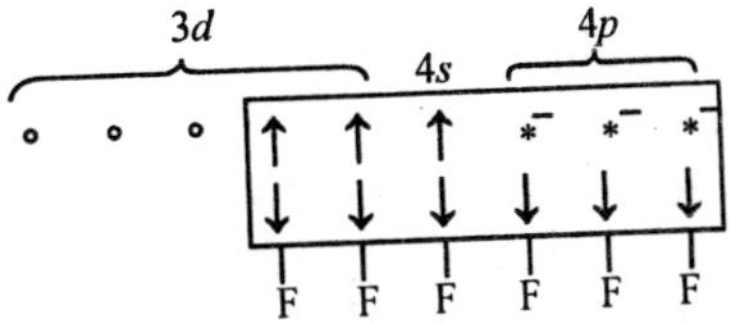

Fig. 18.1 : Structural formula and orbital orbital diagram for scandium triacetyl-$[ScF_6]^{3-}$, octahedral.

Fig. 18.2 : Orbital diagram for the anion $[ScF_6]^{3-}$, octahedral.

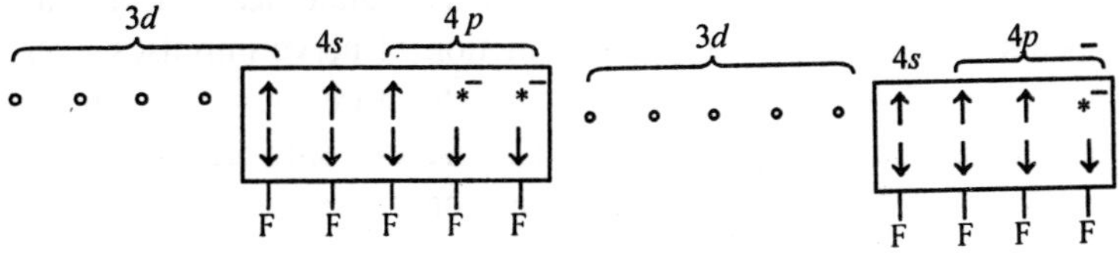

Fig. 18.3 : (a) Orbital diagram for the anion $[ScF_5]^{2-}$, trigonal-bipyramidal.
(b) Orbital diagram for the anion $[ScF_4]$ tetrahedral.

COMPLEXES

Despite the charge of +3, the metal ions in this group do not have a strong tendency to form complexes. This is because of their fairly large size. Sc^{3+} is the smallest ion in the group and forms complexes more readily than the other elements. These include $[Sc(OH)_6]^{3-}$ and $[ScF_6]^{3-}$, both of which have a coordination number of 6 and are octahedral in shape.

$$ScF_3 + 3NH_4F \rightarrow 3NH_4^+ + [ScF_6]^{3-}$$

By far the most common donor atom in complexes is O, and complexes occur particularly with chelating ligands. Thus complexes are formed with strong complexing agents such as oxalic acid, citric acid, acetylacetone and EDTA. The complexes of the larger metals Y and La often have higher coordination numbers than 6, and coordination number 8 is the most common. Thus [Sc(acetylacetone)$_3$] has a coordination number of 6 and is octahedral. Y(acetylacetone)$_3$(H_2O)] has a coordination number of 7 and the structure is a capped trigonal prism. In [La(acetylacetone)$_3$. (H_2O)$_2$] the coordination number is 8 and the structure is a distorted square antiprism. In [La. EDTA (H_2O)$_4$]. 3H_2O the EDTA forms four bonds from O atoms and two from N atoms, and the O atoms in the four water molecules all form bonds to La, giving a coordination number of 10. The ions NO_3^- and SO_4^{2-} also act as bidentate ligands and form complexes with high coordination numbers. In $(Sc(NO_3)_5]^{2-}$ four NO_3^- groups are bidentate and one is unidentate giving a coordination number of 9. In $[Y(NO_3)_5]^{2-}$ all five NO_3^- groups are bidentate and the coordination number is 10. In $La_2(SO_4)_3 . 9H_2O$ half of the La atoms are 12-coordinate.

Lanthanum salts have been used as a biological tracer. La^{3+} appears to replace Ca^{2+} in the conduction of nerve impulses along the axons of nerve cells, and also in structure promoting in cell membranes. La^{3+} is readily detected by electron spin resonance. The similarity in biological role may arise because the ions are similar in size (La^{3+} = 1.032 Å, Ca^{2+} = 1.00 Å).

19

d-Block Elemenets (Titanium and Zrconium)—Group IVB (4)

Elements	Symbol	Atomic wt	Atomic no.
Titanium	Ti	47.90	22
Zirconium	Za	91.22	40
Hafnium	Hf	178.2	72

General Features. The titanium sub group includes the fourth group elements in the periodic table, titanium, zirconium and hafnium situated in the even series of the long periods. Formaly this subgroup also included. The element of thorium which is now, included in the actinide group.

Titanium; at. wt. 47.90. Titanium is very abundant in nature; it constitutes 0.61 per cent of the earth's crust by weight.

The most important titanium minerals are the *titanomagnetites* $FeTiO_3$. *n* Fe_3O_4, *ilmenite* $FeTiO_3$, *sphere* or *titanate* $CaTiSiO_5$ and *rutile* TiO_2. The most important of them as a raw material for the production of titanium are the titanomagnetites.

The largest deposits of titanium ores in the U.S.S.R. are in the Urals.

Metallic titanium has a specific gravity of 4.54 and melts at 1,725°C. It is prepared in the free state from its dioxide by aluminothermy. Reduction of ferrotitanic ores results in an alloy of titanium and iron known as ferrotitanium and used in metallurgy for steel production. The addition of about 0.1 per cent titanium to steel greatly improves its quality. The action of titanium is based partly on its ability to combine with the nitrogen contained in molten steel and thus to prevent the separation of the latter as bubbles when the steel solidifies; as a result, castings made of this steel are homogenous and contain no cavities.

Under ordinary conditions titanium is not very active but at a high temperature it combines readily with the halogens, oxygen, sulphur, nitrogen and other elements.

However, the significance of titanium as a very valuable structural material became known comparatively recently, after the development of commercial methods of preparing the metal in its pure form.

Pure titanium is obtained by reducing its chloride with metallic sodium or magnesium in an atmosphere of hydrogen, or by decomposing titanium iodide on a red-hot surface.

Titanium can be obtained also by the reduction of its dioxide TiO_2 with metallic magnesium. However, the titanium produced by this method is not so pure.

Pure titanium is readily amenable to mechanical treatment. It forges well and can be rolled into sheets and bands and even into foil. Titanium is just a little heavier than aluminium, but it is three times as strong. This promise great opportunities for its application in aircraft engineering. The resistance of titanium to sea water makes it a good plating for ships not requiring anti-corrosive coatings. Titanium can be employed as a material for responsible parts in chemical engineering and in turbine manufacture. The use of titanium or titanium alloy parts in internal combustion engines decreases the weight of the latter by about 20 per cent.

The extensive use of titanium in engineering led to a rapid rise in its output. It will suffice to mention that while only 75 tons of titanium were produced in the U.S.A. in 1950, this figure had risen to 1,000 tons by 1952 and was about 20,000 tons in 1955.

The U.S.S.R possesses large reserves of titanium ores, enabling extensive production of metallic titanium.

At a high temperature titanium combines readily with the halogens, oxygen, sulphur, nitrogen and other elements. Titanium combines with carbon to form a carbide. Carbides of titanium and tungsten with admixtures of cobalt give alloys almost as hard as diamond.

Titanium dioxide TiO_2 can be prepared by calcining titanium in air or in an atmosphere of oxygen, as a white refractory substance insoluble in water and dilute acids. Titanium dioxide is amphoteric in nature, but both its basic and its acidic properties are manifested very weakly.

Titanium dioxide is used for the manufacture of a white oil paint of very high covering power (*titanium white*) and also for the manufacture of refractory glass, glazes, enamels and heat-resistant laboratory glassware.

Occurrence and Extraction of Titanium

Titanium is the seventh most abundant metal in the earth's crust and occurs principally as rutile (titanium (IV) oxide), TiO_2, and ilmenite, $FeO.TiO_2$.

The metal is obtained by first converting the dioxide into titanium (IV) chloride (a covalent liquid) by mixing it with carbon and heating in a stream of chlorine :

$$TiO_2 + C + 2Cl_2 \longrightarrow TiCl_4 + CO_2$$

After fractional distillation, the tetrachloride is reduced to metallic titanium heating either with magnesium or sodium under a blanket of argon (traces of air make the metal brittle) :

$$TioCl_4 + 2Mg \longrightarrow Ti + 2MgCl_2$$

Properties of Titanium

Titanium has a high melting point and a density of 4.43 g cm^{-3} (about 60 per cent that of steel and about as strong). It is unattacked by air and water at ordinary temperatures, but at high temperatures it burns in oxygen and nitrogen and will decompose steam on really strong heating.

On heating in chlorine it forms the tetrachloride; it is also attacked by hot concentrated hydrochloric acid and cold concentrated sulphuric acid to give Ti^{3+} salts and hydrogen.

Technological difficulties delayed production of the metal until about 1950 since traces of air and carbon render the hot metal brittle. Its great strength, comparatively low density and its resistance to corrosion make it an excellent material for the construction of supersonic aircraft, space vehicles and nuclear reactors.

The metal titanium combines the properties of low density with high strength and corrosion resistance.

Uses

(*i*) Alloys of titanium are used as deoxidizers and in the construction of tanks, automobiles and ships. **Ferro-titanium** which is an alloy of titanium with iron is obtained in the electric furnace by reducing ilmenite with carbon or aluminium.

$$FeTiO_3 + 3C \longrightarrow Fe + Ti + 3CO$$

$$FeTiO_3 + 2Al \longrightarrow Fe + Ti + Al_2O_3$$

Titanium dioxide is used in the manufacturing of paints and pigments. Mixed with barium sulphate it is sold under the trade name "*Titanox*".

Titanium tetrachloride is used in making smoke screens while the dichloride and oxalate are used in dyeing and mordanting.

Titanium II Compound

All the compounds, containing a titanium atom in oxidation state II are solids, and are very readily oxidized by air.

Titanium monoxide, TiO, consists of yellow crystals with the sodium chloride structure (the colour may be due to contamination with iron). It is made by heating titanic oxide and titanium to 1550°. It dissolves in hydrochloric and sulphuric acid, liberating hydrogen.

Titanium dichloride, $TiCl_2$, is made by heating titanous chloride $TiCl_3$, to redness; disproportionation occurs and the volatile titanic chloride. $TiCl_4$, distills off. Titanium dichloride is a black powder with the CdI_2 crystal structure. $TiCl_2$ itself disproportionates at 475° to titanium and titanic chloride. Titanium dichloride is violently oxidized by air, and it dissolves in acids with the evolution of hydrogen. A black **titanium di-iodide** is made by reducing titanic iodide with silver to mercury at a red heat. It is dimorphic and one of the forms has the CdI_2 structure. $TiBr_2$ is known.

Titanium III Compound

The titanium atom in oxidation state III is present in certain compounds in which the bonding probably has ionic character, such as titanous oxide, Ti_2O_3, titanous sulphide Ti_2S_3, and titanous chloride. $TiCl_3$, and in certain anionic and cationic octahedral complexes. All the titanous compounds are coloured; absorption of radiation occurs when the *d* electron makes the "forbidden" transition from a t_{2g} to an e_g orbital.

Complex Ions Containing Ti(III)

Titanous chloride forms a hydrate, $TiCl_3, 6H_2O$, which undergoes the same colour changes as $CrCl_3, 6H_2O$; the compound $[Ti, 6H_2O]Cl_3$ present in green. A titanium hexammine chloride is known, $[Ti, 6NH_3]Cl_3$, for which the orbital diagram is shown in Fig. 19.1.

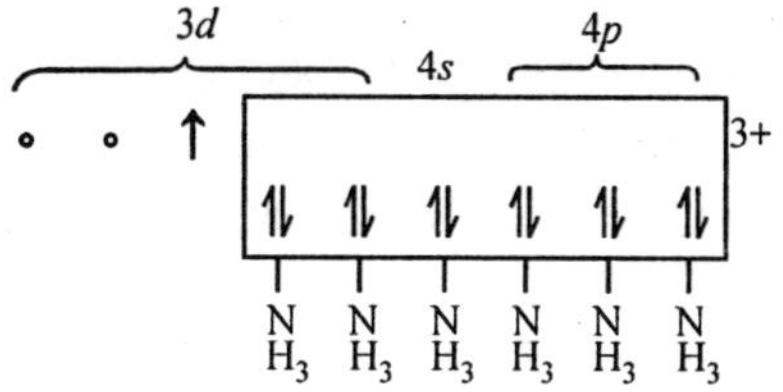

Fig. 19.1 : Orbital diagram for cation (Ti. $6NH_3]^{3+}$.

and a potassium di-oxalato complex, $K[Tiox_2, 2H_2O]$, for which th orbital diagram is shown in Fig. 19.2 :

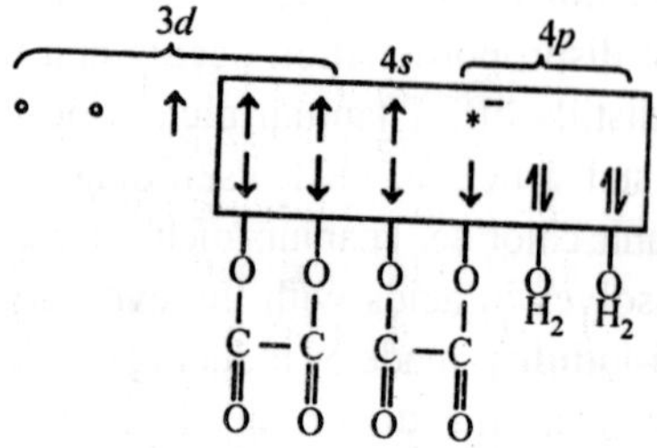

Fig. 19.2 : Orbital diagram for the anion $[Ti(C_2O_4)_2(H_2O)_2]^-$.

$(NH_4)_3[TiF_6]$ is isomorphic with the corresponding fluoro anions formed by Cr^{III} V^{III}, and Fe^{III}.

Titanium (IV) Compound

Among the simple compounds in this Section are TiO_2, and the titanic halides $Tihal_4$. Ti^{IV} forms no simple oxalate, nitrate, or sulphate, but it is present in several anionic complexes such as $K_2[TiF_6)$, and cationic complexes such as $[TiA_3]$ $[FECl_4]$. There are also certain peroxy compounds of Ti^{IV}.

Zirconium

Zirconium (Zirkonium); *Zirconium* is rather abundant in the earth's crust, but it is highly dispersed and occurs rarely in considerable accumulations. In the U.S.S.R. deposits of zirconium ores have been found in the Donates Basin.

In the elemental state zirconium is a hard lustrous metal; its specific gravity is 6.5 and melting point 1,860°C. It is used in metallurgy as an admixture to pig iron, raising the quality of the cast metal. Steels containing zirconium are suitable for the manufacture of armour, armour-piercing shells, etc.

For a long time extensive application of Zirconium was impeded by the difficulty of its preparation. However, lately the extraction of zirconium from its ores has greatly increased owing to the fact that the pure metal possesses a number of very valuable properties.

Its high melting point, sufficient strength and great resistance to corrosion, in combination with its almost complete inability to capture thermal neutrons, make pure zirconium a good structural material for atomic reactors.

Admixture of zirconium in copper greatly enhance the strength of the latter without hardly decreasing its conductivity. A magnesium alloy containing 4 or 5 per cent zinc and 0.6 to 0.7 per cent zirconium is twice as strong as pure magnesium and does not lose its strength even at 200°C. In 1953, 56.5 per cent of the magnesium alloys manufactured in Great Britain contained zirconium as one of their admixtures. The quality of aluminium alloys can also be considerably improved by adding zirconium to them.

Zirconium dioxide ZrO_2 is excellent refractory owing to its high melting point (about 2,700°C), its very low coefficient of expansion and its stability against chemical action. It is used for the manufacture of various refractory wares, such as crucibles. In the glass industry zirconium dioxide is used to manufacture refractory glasses, and in the ceramic industry for the production of enamels and glazes.

Zirconium carbide ZrC is employed, owing to its great hardness, as a grinding material and as a substitute for diamonds in cutting glass. Wide application of zirconium is impeded as yet by the difficulty of extracting it from its ores.

20

d-Block Elements (Vandium)—Group VB (5)

Table 20.1 :

Elements	Symbol	Atomic wt	Atomic no.
Vanadium	V	50.95	23
Niobium	Nb	92.91	41
Tantalum	Ta	180.95	73

General Features of the Vanadium Subgroup. The vanadium subgroup includes the three elements, *vanadium, niobium* and *tantalum,* of the fifth group of the Periodic Table, located in the even series of the long periods. Formerly this subgroup included also protactinium, which is now usually included in the actinide group. Having two or even only one electron in the outermost layer of their atoms, the elements of the vanadium subgroup differ from the elements of the main subgroup (nitrogen, phosphorus, etc.) in the predominance of metallic properties and in the absence of hydrogen compounds. But the highest valency derivatives of the elements of both subgroups resemble each other in many respects.

The most typical compounds of vanadium and its analogues are those in which they are pentavalent. Their highest oxides are of the nature of anhydrides, forming respectively vanadic, niobic and tantalic acids each of which has a number of corresponding salts. The lower oxides possess basic properties only.

In the free state vanadium, niobium and tantalum are greyish-white metals, very indifferent to all kinds of chemical actions and with high melting points. The most important practically is vanadium.

Vanadium

Vanadium compounds are rather abundant in nature but they are very dispersed and do not form considerable accumulations. For this reason vanadium is considered a rare element, although its total content in the earth's crust is estimated at 0.02 per cent and is a little higher than that of copper.

The richest deposits of vanadium ores are in South America, in Peru. The world output of vanadium (not counting the U.S.S.R.) amounts to several thousand tons per year and falls mainly on Peru, the U.S.A., South-Western Africa and Northern Rhodesia.

The main sources of vanadium in the U.S.S.R. are iron and poly-metallic ores, which contain small quantities of vanadium. Usually either an alloy of vanadium and iron, known as *ferrovanadium,* or salts of vanadic acid, are produced from the ores.

Pure vanadium is a very hard, light metal, its specific gravity being 5.8 and its melting point 1,735°C. It does not oxidize in the air and is indifferent to hydrochloric and sulphuric acids.

Oxidation States

The maximum oxidation state for this group is (+ V). All three elements show the full range of oxidation states from (– I) to (+ V). For vanadium the (+ II) and (+ III) states are reducing, (+ IV) is stable, and (+ V) slightly oxidizing. For Nb and Ta the (+ V) state is by far most stable and the best known, although lower oxidation states are known.

V (+ V) is reduced by zinc and acid to V^{2+}, Nb (+ V) is reduced to Nb^{3+} but TA (+ V) is not reduced. This illustrates the increasing stability of the (+ V) state on descending the group. At the same time the lower oxidation states become less stable. *This is the opposite trend to that in the main group.*

General Properties

V, Nb and Ta are silvery coloured metals with high melting points. V has the highest melting point in the first row transition elements. This is associated with the maximum participation of *d* electrons in metallic bonding. The melting points of Nb and Ta are high but the maximum melting point in the second and third row transition elements occurs in the next group (Group 6) with Mo and W.

The pure metals V, Nb and Ta are moderately soft and ductile, but traces of impurities make them harder and brittle. They are extremely resistant to corrosion due to the formation of a surface film of oxide. At root temperature they are not affected by air, water or acids, other than HF with which they form complexes. V also dissolves in oxidizing acids such as hot concentrated H_2SO_4, HNO_3 and aqua regia. V is unaffected by alkali, showing that it is completely basic, but Nb and Ta dissolve in fused alkali.

Table 20.1 : Some physical properties

	Covalent radius (Å)	Ionic radius (Å) M^{2+}	M^{3+}	Melting point (°C)	Boiling point (°C)	Density ($g\,cm^{-3}$)	Pauling's electronegativity
V	1.22	0.79	0.640	1915	3350	6.11	1.6
Nb	1.34	–	0.72	2468	4758	8.57	1.6
Ta	1.34	–	0.72	2980	5534	16.65	1.5

At high temperatures all three metals react with many non-metals. The products are often interstitial compounds which are non stoichiometric.

V forms many different positive ions, but Nb and Ta form virtually none. Thus though Nb and Ta are metals, their compounds in the (+ V) state are mostly covalent, volatile and readily hydrolyzed—properities associated with non-metals.

The tendency to form simple ionic compounds decreases as the oxidation state increases. Even though V^{2+} and V^{3+} are reducing, they exist both in the solid and in solution (as hexahydrate ions). They have an extensive aqueous chemistry. The oxidation state (+ IV) is dominated by the VO^{2+} ion. This is very stable and exists in a wide range of compounds both as solids and in solution (as the hydrated ion). Some covalent (+ IV) compounds such as VCl_4 also exist. The (+ V) state may be covalent as in VF_5, or form VO_2^+ or VO_4^{3-} hydrated ions. The chemistry of Nb and Ta is largely confined to the (+ V) state.

The basic properties of the oxides M_2O_5 increase down the group. V_2O_5 is amphoteric but mainly acidic. It dissolves slightly in water, giving a pale yellow acidic solution. It dissolves readily in NaOH, forming colourless solutions which contain a wide range of vanadate ions. The ions formed depend on the pH: various isopolyvanadates at intermediate pH and orthovanadate VO_4^{3-} at high pH. The aqueous chemistry of the polymerized vanadates is quite complex. V_2O_5 also dissolves slightly in concentrated H_2SO_4, forming the pale yellow VO_2^+ ion. Nb_2O_5 and Ta_2O_5 and are rather unreactive but are amphoteric. They have only very weak acidic properties. Niobates and tantalates are only formed by fusing with NaOH. They are decomposed by weak acids or CO_2, and only partially imitate the behaviour of the isopolyvanadates.

Whilst V^{2+} and V^{3+} are well known, Nb(II), Ta(III) and Nb(III) and Ta(III) are not ionic but exist as cluster compounds M_6X_{12}, in which groups of metal atoms are bonded together.

Colour

Colour in transition metal compounds very commonly arises from *d-d* electronic transitions. It can also arise from defects in the solid state and from charge transfer spectra. (Charge transfer spectra are discussed under SnI_4 in Chapter 13.) The oxidation states below (+ V) are coloured because they have an incomplete *d* shell of electrons and give *d-d* spectra. However, the (+ V) state has a d^0 configuration and so colourless compounds would be expected. NbF_5, TaF_5 and $TaCl_5$ are white, but V_2O_5 is red or orange. $NbCl_5$ is yellow, $NbBr_5$ is orange and NbI_5 is brass coloured. The colours arise because of charge transfer.

Uses of Vanadium

Vanadium has a slightly higher boiling point than titanium and a density of 6.07 g cm^{-3}. It resembles titanium is being strong and resistant to corrosion at ordinary temperatures, and in combining with oxygen and nitrogen at elevated temperatures. On heating with chlorine it forms the tetrachloride, VCl_4.

Cold dilute acids do not attack the metal, but it dissolves slowly in hot dilute nitric acid and hot concentrated sulphuric and hydrochloric acids.

Vanadium is used mainly as an alloying ingredient in steel, e.g. for the construction of high-speed tools and exhaust valves.

Compounds of Vanadium

Vanadium has oxidation states of + 5, + 4, + 3 and + 2. These are all readily demonstrated by shaking a solution of ammonium vanadate in dilute sulphuric acid with zinc amalgam. The colour of the solution changes from pale yellow to blue, to green, and finally to lavendar (Table 20.3) :

Table 20.2 : The different oxidation states of vanadium in aqueous solution

Ion	VO_3^- Metavanadate	$VO2^+$ Vanadium (IV)	V^{3+} Vanadium (III)	V^{2-} Vanadium (II)
Oxidation state	+ 5	+ 4	+ 3	+ 2
Colour	pale yellow	blue	green	lavender

Notice that the V^{4+} cation does not exist as such in aqueous solution since it is highly charged and has a small ionic radius. Ions such as this are able to exert a profound polarizing effect on neighbouring water molecules and in extreme cases, detach O^{2-} ions, i.e. VO^{2-} can be formally regarded as V^{4+} (O^{2-}).

Vanadium (V) compounds

Vanadium (V) oxide, V_2O_5, is obtained by the action of heat on ammonium metavanadate, NH_4VO_3, :

$$2NH_4VO_3 \longrightarrow V_2O_5 + 2NH_3 + H_2O$$

It is an orange coloured solid and its main use is as a catalyst in the Contact process.

When vanadium (V) oxide is dissolved in strongly alkaline solution the orthovanadate anion, VO^{3-}_4, is obtained, i.e. sodium orthovanadate, Na_3VO_4, can be obtained by treating vanadium (V) oxide with a concentrated solution of sodium hydroxide :

$$V_2O_5 + 6OH- \longrightarrow 2VO_4^{3-} + 3\,H_2O$$

Polyvanadate anions (condensed anions containing vanadium and oxygen) can be obtained by lowering the pH of the solution, but the nature of many of the species present still remains obscure. However, experimental evidence is available which indicates the formation of the anion $V_3O_9^{3-}$:

$$3VO_4^{3-} + 6H^+ \,.....\, V_3O_9^{3-} + 3H_2O$$

Ammonium metavanadate is precipitated by àdding ammonium chloride to a solution of sodium orthovanadate. Ammonium chloride provides NH^{4+} ions which can combine with OH– ions to form weakly dissociated ammonium hydroxide, i.e. the pH will be lowered. Polyvanadate ions will be formed which have the empirical formula VO_3^- : thus the formation of ammonium vanadium (V) oxide forms the VO_2^+ cation, and can thus act as a basic oxide :

$$V_2O_5 + 2\,H^+ \longrightarrow 2VO_2^+ + H_2O$$

If the solution is made less acidic a number of complex vanadates, the nature of which is still rather obscure, are formed.

Vanadium (IV) Compound

Vanadium dioxide, VO_2, exists as dark blue crystals with the rutile structure. It is made by heating V_2O_5 with V_2O_3 (or with oxalic acid or sulphur dioxide). It dissolves easily in acids to give vanadium oxo-compounds such as $VOCl_2$, and in alkalis to give vanadites derived from the acid $H_2V_4O_9$. Vanadium dioxide by heating in air, or with nitric acid, to V_2O_5.

Vanadium tetrafluoride, VF_4, a hygroscopic brown powder, is made by the addition of anhydrous hydrogen fluoride to vanadium tetrachloride at a low temperature. At 325° in an atmosphere of nitrogen it disproportionates of VF_3 and VF_5. With alkali fluorides in a alkaline solution it forms salts $K_2[VOF_4]$.

Vanadium tetrachloride,† VCl_4, is a dark-red dense liquid, m.p. –28°, b.p. 154°. It is made by passing air-free chlorine over ferrovanadium at a red heat. VCl_4 distils over and the less volatile $FeCl_3$ remains behind. VCl_4 is soluble in water which hydrolyses it to blue solution of $VoCl_2$:

$$VCl_4 + H_2O = VOCl_2 + 2HCl.$$

It is soluble in carbon disulphide, carbon tetrachloride, and other organic solvents. VCl_4 is stable above 600°. At ordinary temperatures, and more rapidly at 150°, it disproportionates thus :

$$2VCl_4 = 2VCl_3 + Cl_2,$$

but at high temperatures vanadium trichloride disproportionates :

$$2VCl_3 = VCl_4 + VCl_2.$$

The vapour density corresponds to VCl_4, and electron diffraction experiments show that the molecule is tetrahedra. In CCl_4 solution at –20° vanadium tetrachloride is dimeric.

Vanadium monoxodihalides. VOF_2, is a non-sublimable yellow powder, insoluble in any solvent. It is made by heating $VOBr_2$ to red heat in hydrogen fluoride. $VOCl_2$ exists as green deliquescent crystals decomposed by water. It is made :

(i) by passing mixture of $VOCl_3$ vapour and hydrogen through a red hot tube,

(ii) by the hydrolysis of VCl_4,

(iii) by the action of concentrated hydrochloric acid on vanadium pentoxide :

$VOBr_2$ is pale yellow powder which sublimes with decomposition at 330°. $VO(CN)_2$ is known.

Vanadium (III) Compound

This and the following section discuss compounds and ions which are formed in principle by the reduction of V^V compounds, and therefore contain V^{IV} and V^{III}. A few of them have discrete molecular structures, and others are giant molecules with structures fairly obviously related to the valencies of the atoms It is therefore possible to consider them under the headings V^{IV} and V^{III}.

Vanadic oxide, V_2O_3, a black, almost infusible powder with the corundum lattice, is made:

(i) by reducing vanadium pentoxide with hydrogen or carbon monoxide at a high temperature,

(ii) by the electrolysis of a weakly alkaline phosphate melt of vanadium pentoxide diluted with sodium chloride; this method is claimed to give a pure product.

Vanadic oxide is slowly oxidised by cold air to the blue dioxide, but when heated in air it inflames and gives the pentoxide. Vanadic oxide dissolves in nitric acid or hydrochloric acid to yield salts. It is wholly basic.

Vanadic sulphide, V_2S_3, is formed by the action of hydrogen sulphide on vanadic oxide. It is the most stable of the vanadium sulphides. At 400° it combines with sulphur to yield the pentasulphide.

Vanadic hydroxide, $V(OH)_3$, is obtained as a green flocculent precipitate by adding ammonia or alkali to a solution of vanadium trichloride. It rapidly absorbs oxygen from the air. In the absence of air it dissolves in acids to form vanadic salts. It is wholly basic.

Vanadium trihalides. Vanadium trifluoride differs from the other compounds in its insolubility in water and in organic solvents. The other vanadium trihalides are coloured, hygroscopic, very soluble solids, all of which form hydrates such as $[V,6H_2O]Cl_3$.

Vanadium oxomonohalides. There is no fluorine compound VOF.

VOCl consists of brown crystals which sublime only at high temperatures. It is made by hydrolyzing vanadium trichloride with water, by heating VCl_3 in a stream of carbon dioxide to 700°, or by heating $VOCl_2$ in hydrogen to red heat. VOCl dissolves with difficulty in water, but readily in nitric acid. Violet octahedral crystals of **VOBr** are made by heating $VOBr_2$ in vacuo at 360°. The crystals decompose at 480° in vacuo to give VBr_3 and V_2O_3. VOBr is slightly soluble in water, and generally less soluble in organic solvents.

Vanadium II Compounds

All the bivalent vanadium compounds are *coloured* and *paramagnetic* corresponding to three unpaired electrons (V2+ → $3d^34s^0$). The compounds having V^{2+} ion (vanadous ion) are *strong reducing agents*. The violet coloured aqueous solution of V^{2+} ion changes to green due to the oxidation of V^{2+} to V^{3+}, with the liberation of H_2 (reduction of water)

$$\underset{\substack{\text{Violet} \\ (V = +2)}}{V^{2+}} + H^+ \rightarrow \underset{\substack{\text{Green} \\ (V = +3)}}{V^{3+}} + \frac{1}{2}H_2$$

(1) Vanadium dichloride, VCl_2. *Vanadium dichloride,* VCl_2 is prepared by passing a mixture of hydrogen and vanadium tetrachloride through a glass tube heated to redness; it furnishes pale green crystals.

(2) Hypovanadous sulphate, $VSO_4 .7H_2O$. *Hypovanadous sulphate,* $VSO_4.7H_2O$, has been obtained by reducing a sulphuric acid solution of vanadium pentoxide by means of an electric current in the absence of air. The solution, when evaporated *in vacuo* over sulphuric acid, furnishes violet monoclinic crystals, isomorphous with ferrous and magnesium sulphates. Just like ferrous ammonium sulphate, a double sulphate *ammonium hypovanadous sulphate* $(NH_4)_2SO_4$. VSO_4 $6H_2O$, is also known.

21

d-Block Elements (chromium, molybdenum and tungsten)—Group VIB (6)

Element	symbol	Atomic weight	Atomic number
Chromium	Cr	52.01	24
Molybdenum	Mo	95.95	42
Tungsten	W	183.92	74

The metallic elements of the sixth group of the Periodic Table are *chromium, molybdenum* and *tungsten,* located in the even series of the long periods and constituting the secondary subgroup of the sixth group, otherwise known as the *chromium subgroup.* This subgroup formerly included the element uranium, which is now regarded as one of the actinides. As the chemical properties of uranium resemble those of the other elements of the chromium subgroup, we shall consider it together with these elements.

The outermost electron layer of the atoms of the chromium sub -group elements contains one or two electrons, which accounts for the metallic nature of these elements and their difference from the elements of the main subgroup. At the same time, their maximum positive valency equals 6, as, besides the outer electrons, a corresponding number of electrons from the incomplete second last layer can also take part in the formation of bonds.

Chromium and its analogues do not form hydrides. Their most typical compounds are their derivatives of highest valency, which in many respects resemble the corresponding compounds of sulphur.

The most important element of the subgroup under consideration is chromium, which has found diverse applications both in the free state and as compounds. The analogues of chromium, molybdenum and tungsten, are classed as rare elements; like chromium, they are of great practical importance.

Occurrence, Metallurgy, and Properties of Chromium

Chromium does not occur free in nature. Its most important ore is chromite, $FeCr_2O_4$, sometimes called **chrome iron ore.** Practically all of the chromium used in the United States is imported. Because chromium is essential to the production of special alloy steels and the supply is often short, it is classed as a strategic material by the United States Government.

Reduction of chromite by carbon in an electric furnace yields an alloy of iron and chromium, called **ferrochrome,** which is used in making chromium steels.

$$FeCr_2O_4 + 4c \rightarrow Fe + 2\ Cr + 4CO$$

Pure chromium is prepared by a thermite reaction, reducing chromium (III) oxide with aluminum.

$$Cr2O3(s) + 2Al(s) \rightarrow Al2O3(s) + 2Cr(s) + 128{,}1000\ cal$$
$$(\Delta H^{\circ}_{298} = -128{,}100\ cal)$$

Chromium can also be produced by the electrolytic reduction of its compounds from aqueous solution, a process which is used in chromium plating.

Chromium is a very hard, silvery-white crystalline metal. It assumes a passive state by becoming coated with a thin layer of oxide, which protects it against further corrosive attack. This property, along with its metallic luster, accounts for its extensive use in the plating of iron and copper objects, such as plumbing fixtures and automobile trim. Passivity is produced by the action of concentrated nitric acid or chromic acid, or exposure of the metal to air. When not in the passive state chromium is a transition metal which shows several oxidation states, the principal ones being +2, +3, and +6. The compounds of chromium are highly coloured, each colour exhibited being dependent to some extent upon a particular oxidation state and upon the structure of the compound.

Compounds of Chromium

Chromium has oxidation states of + 6 (oxidising), + 3 (the most stable) and + 2 (reducing). As in the case of vanadium, these are readily demonstrated by shaking an acidified solution of potassium dichromate with zinc amalgam: the colour changes are orange to green and then to blue (Table 21.1):

Ion	$Cr2O_7^{2-}$ Dichromate	Cr^{3+} Chromium (III)	Cr^{2+} Chromium (II)
Oxidation + 6	+ 3	+ 2	
Colour	orange	green	blue

Chromium (VI) compounds

Chromium (VI) oxide, CrO_3

This is prepared by adding concentrated sulphuric acid to a cold concentrated solution of potassium dichromate; red crystals of chromium (VI) oxide separate from the solution.

Chromium (VI) oxide is very soluble in water, forming 'chromic acid' which is used for cleaning laboratory glassware. Although chromic acid has never been isolated, it seems likely that a solution of chromium (VI) oxide in water contains H_2CrO_4 (which is the parent acid of chromates) and $H_2Cr_2O_7$ (from which dichromates are derived), i.e. chromium (VI) oxide is an acidic oxide.

Chromium (VI) oxide is powerfully oxidizing, thus it immediately inflames ethanol and on heating it forms chromium (III) oxide with the liberation of oxygen:

$$4\,CrO_3 \rightarrow 2Cr_2O_3 + 3O_2$$

Chromates and dichromates

Sodium chromate, Na_2CrO_4, and sodium dichromate, $Na_2Cr_2O_7$, are prepared industrially from chromite.

The chromate ion, CrO_4^{2-}, can only exist in solution under alkaline conditions, e.g. potassium and sodium chromates are stable in aqueous solution, since salt hydrolysis ensures that the solution is alkaline. These solutions are yellow in colour (the colour of the chromate ion); insoluble chromates are often yellow if the cation is colourless. e.g. $PbCrO_4$, although silver chromate, Ag_2CrO_4, is red. Chromates are generally isomorphous with the corresponding sulphates.

In the laboratory sodium chromate may be made by the oxidation of a chromium (III) salt with sodium peroxide in aqueous solution:

$$2Cr^{3+} + 16OH^- \rightarrow 2CrO_4^{2-} + 8H_2O + 6e \text{ (oxidation)}$$

$$\underset{\text{peroxide ion}}{3O_2^{2-}} + 6H_2O + 6e \rightarrow 12OH^- \text{ (reduction)}$$

or $$2Cr^{3+} + 4OH^- + 3O_2^{2-} \rightarrow 2CrO_4^{2-} + 2H_2O$$

The sodium peroxide provides sufficient OH^- ions in aqueous solution so it is not necessary to add further alkali.

The addition of acid to an aqueous solution of a chromate results in the elimination of water and the formation of a condensed anion—the dichromate anion $Cr_2O_7^{2-}$. The colour of the solution changes from yellow (the colour of the chromate ion) to orange (the colour of the dichromate ion). The reaction is easily reversed by adding alkali, i.e. OH^- ions:

$$\underset{\text{yellow}}{2CrO_4^{2-}} + 2H^+ \rightarrow \underset{\text{orange}}{Cr_2O_7^{2-}} + H_2O$$

Sodium and potassium dichromates are common laboratory reagents The former, which is more soluble in water than potassium dichromate, is used in the presence of acid as an oxidizing agent, thus a solution of it acidified with sulphuric acid is used for oxidizing alcohols to aldehydes or ketones, and aldehydes to fatty acids. Pieces of filter paper impregnated with an acidified solution of sodium dichromate can be used to test for gaseous reducing agents. e.g. sulphur dioxide and hydrogen sulphide:

$$Cr_2O_7^{2-} + 8H^+ + 3H_2S \rightarrow 2Cr^{3+} + 7H_2O + 3S$$

The orange spot is turned green owing to the formation of chromium (III) ions.

Sodium dichromate cannot be used as a primary standard in volumetric analysis, since the solid is too deliquescent. However, potassium dichromate is an ideal primary standard since it does not hydrate, and as its solubility in water increases rapidly with increasing temperature, it is readily obtained pure by recrystallization. It is made by mixing hot concentrated solutions of sodium dichromate and potassium chloride, filtering off the precipitated sodium chloride, and cooling the resulting solution, when the less soluble potassium dichromate crystallizes from solution.

Potassium dichromate is used in volumetric analysis to estimate reducing agents, e.g. Fe^{2-} ions in acid solution. Unlike a solution of potassium permanganate it can be used in the presence of hydrochloric acid, i.e. dichromate ions do not oxidize Cl^- ions to chlorine, but the stronger oxidizing permanganate ions produce chlorine under these conditions.

When a solution of potassium dichromate is added to an acidified solution of an iron (II) salt the orange dichromate ions are reduced to the green chromium (III) ions. The indicator used is barium diphenylamine sulphonate which is less readily oxidized than iron (II) ions, but which is oxidised to give a blue colour once all the iron (II) ions have been converted to the iron (III) state. Since the iron (III) ions would affect the indicator and thus give rise to an inaccurate end-point, they are removed by adding some phosphoric acid (before titration) with which they form a fairly stable complex. The relevant redox equations are:

$$Cr_2O_7^{2-} + 14H^+ 6e \rightarrow 2Cr^{3+} + 7H_2O \text{ (reduction)}$$

$$6Fe^{2+} \rightarrow 6Fe^{3+} + 6e \text{ (oxidation)}$$

Industrially sodium dichromate is used in the tanning of leather, as a source of other chromium compounds, e.g. chromium (VI) oxide, in the preparation of the electrolyte used in chrome plating, as a source of pure chromium, and as a general oxidizing agent.

Chromyl chloride, CrO_2Cl_2

When concentrated sulphuric acid is added to a mixture of solid potassium dichromate and sodium chloride, chromyl chloride is evolved as a dark red vapour on heating gently. It condenses to a dark red covalent liquid which is immediately hydrolyzed by solutions of alkalis to give the chromate:

$$CrO_2Cl_2 + 4OH^- \rightarrow CrO_4^{2-} + 2Cl^- + 2H_2O$$

Since bromides and iodides do not give analogous compounds this is a specific test for chloride ions.

Potassium chlorochromate, $K(CrO_3Cl)$

This is an orange-coloured ionic solid and can be prepared by boiling a solution of potassium dichromate with concentrated hydrochloric acid, Potassium chlorochromate separates as a solid on cooling.

$$Cr_2O_7^{2-} + 2HCl^- \rightarrow \underset{\text{chlorochromate}}{2CrO_3Cl^-} + H_2O$$

It reacts with solutions of alkalis to give the chromate and chloride ions:

$$CrO_3Cl^- + 2OH^- \rightarrow CrO_4^{2-} + Cl^- + H_2O$$

Similarities between chromium (VI) and sulphur (VI) compounds

Apart from the fact that chromium (VI) compounds are powerful oxidizing agents and highly coloured, there is a certain resemblance to sulphur (VI) compounds. For instance sulphur trioxide. SO_3, gives rise to sulphates and pyrosulphates containing respectively the SO_4^{2-} and $S_2O_7^{2-}$ ions, cf. CrO_4^{2-} and $Cr_2O_7^{2-}$. Other similarities include the formation of sulphuryl chloride, SO_2Cl_2, and chromyl chloride. CrO_2Cl_2, both of which are hydrolyzed by alkali to give the SO_4^{2-} and $CrO_4$2– ions respectively. Similarly the chlorosulphate, SO_3Cl^-, and chlorochromate, CrO_3Cl^-, ions are hydrolyzed by alkali to the sulphate and chromate ions.

Except for chromium (VI) oxide, whose structure is uncertain, sulphur (VI) and chromium (VI) compounds have similar structures (see pp. 316, 318, 317 for the structures of SO_4^2, SO_2Cl_2 and $HOSO_2Cl$—the acid from which the SO_3Cl– ion is derived).

Chromium (IV) compounds

This appears to be an unstable oxidation state of chromium.

Chromium tetrafluoride, CrF_4, is a dark coloured amorphous solid which gives a blue vapour. It is made by the action of fluorine at 350° on the metal, CrF_3, or $CrCl_3$. It is hydrolyzed by water to give Cr^{III} and Cr^{VI}.

Chromium tetrachloride, $CrCl_4$, is a gas made by heating a mixture of chromium trichloride and chlorine to 600-700°. It is condensed by rapid cooling with solid CO_2, but it begins to lose chlorine at – 80°. $CrBr_4$ is known.

Chromium dioxide, CrO_2, a black powder with the rutile crystal structure, is obtained by heating $Cr(OH)_3$ in oxygen at 350°. It is ferromagnetic and shows metallic conductance. The salts $Ba_2Cr^{IV}O_4$ and $Sr_2Cr^{IV}O_4$ contain discrete ions CrO_4^{4-}. They are stable and have magnetic moments of 2.8 B.M.

Chromium (V) compounds

There are few representatives of chromium compounds in which the chromium atom is in the Cr^V oxidation state, and none of them is well characterized.

Chromium pentafluoride, CrF_5, made by combination of the elements, is a red, fairly volatile solid, m.p. 30°. It is at once hydrolyzed by water to Cr^{III} and Cr^{VI}.

$KCrOF_4$ and **$AgCrOF_4$** are light purple powders obtained by the action of bromine trifluoride on the metal dichromate; in place of potassium dichromate an equi-molecular mixture of CrO_3 an KCl may be used.

$AgCrOF_4$ reacts with moist air to give silver chromate:

$$AgCrOF_4 + 2H_2O + O = AgCrO_4 + 4HF.$$

The orbital diagram of the anion $[CrOF_4]^-$ is :

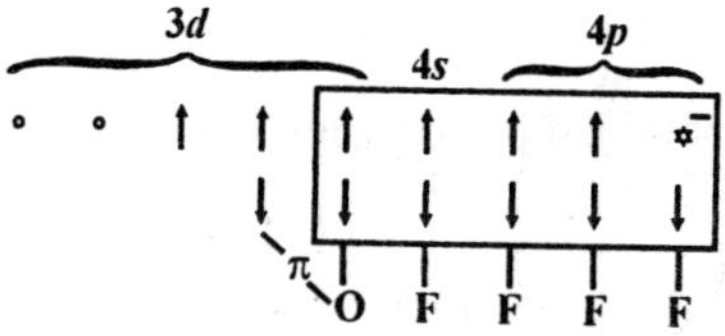

Fig. 21.1 : Orbital diagram for the anion $[CrOF_4]^-$.

The yellow pyridine and quinoline salts of the ion $[CrOCl_4]^-$ are known. Garnet red salts of the formula $M_2(CrOCl_5)$ where M is K, Rb, Cs, or NH_4, are made by reducing chromium trioxide at 0° with very concentrated hydrochloric acid. The orbital diagram for the complex anion is shown in Fig 21.2 :

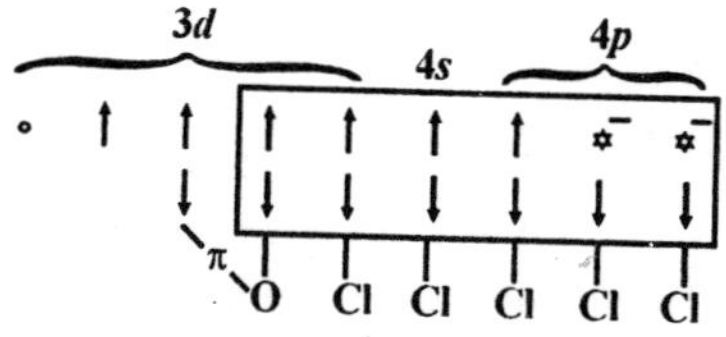

Fig. 21.2 : Orbital diagram for the anion $[CrOCl_5]^{2-}$.

$Na_3Cr^VO_4$ and **Li3CrvO4** contain discrete tetrahedral CrO_4^{3-} ions with magnetic moments equivalent to one unpaired electron.

Chromium (III) oxide is a dark green solid having a similar ionic structure to that of aluminium oxide. It is amphoteric, dissolving in acids to give chromium (III) ions and in concentrated solutions of alkalis to give chromites:

$$Cr_2O_3 + 6H^+ \rightarrow 2Cr^{3+} + 3H_2O$$

$$Cr_2O_3 + 6OH^- + 3H_2O \rightarrow 2Cr(OH)_6^{3-}$$

Chromium (III) compounds

Chromium (III) oxide, Cr_2O_3

Chromium (III) oxide may be obtained by heating chromium in a stream of oxygen, or by heating ammonium dichromate:

$$(NH_4)_2Cr_2O_7 \rightarrow Cr_2O_3 + 4H_2O + N_2$$

Industrially it is obtained by reducing sodium dichromate with carbon (p. 360).

Chromium (III) hydroxide, $Cr(OH)_3$

Chromium (III) hydroxide, usually represented by the formula $Cr(OH)_3$, although doubtless of more complex structure, is obtained as a green precipitate when alkali is added to a solution containing chromium (III) ions:

$$Cr^{3+} + 3OH^- \rightarrow Cr(OH)_3$$

Like chromium (III) oxide it dissolves in acids and concentrated alkalis, i.e. it is amphoteric.

Chromium (III) chloride, $CrCl_3$

Anhydrous chromium (III) chloride is a reddish-violet solid obtained when chlorine is passed over strongly heated chromium. The substance is insoluble in water unless either a trace of chromium (II) chloride, $CrCl_2$, is present, or a reducing agent such as tin (II) chloride is added to reduce some chromium (III) to chromium (II). Since the function of chromium (II) ions is catalytic, it would appear that a high energy of activation barrier opposes the dissolution of pure solid chromium (III) chloride.

When crystallized from aqueous solution a dark green hexahydrate is obtained. When a solution of this hydrated chromium (III) chloride is treated with a solution of silver nitrate, only one third of the chlorine precipitates out as silver chloride, i.e. only one third of the chlorine in this compound is ionic; this experimental behaviour is indicative of the formula:

$$[Cr\,(H_2O)_4Cl_2]^+\,Cl-.\,2H_2O$$

It is possible to form two isomers of the above compound. In one all the chlorine can be precipitated as silver chloride, while in the other only two thirds of the chlorine is converted to the insoluble chloride. These two compounds must have the respective formulae:

$$[Cr\,(H_2O)_6]^{3+}\,(Cl^-)_3 \qquad [Cr(H_2O)_5Cl]^{2+}\,(Cl^-)_2.H_2O$$

grey-blue light green

Notice that the successive replacement of neutral water molecules, in the octahedral chromium (III) complex by chloride ions, reduces the charge carried by the complex one unit at a time.

Chromium (III) sulphate, $Cr_2(SO_4)_3.xH_2O$

A number of different hydrates of chromium (III) sulphate can be obtained, with colours ranging from violet to green. One such hydrate gives no precipitate of barium sulphate when treated with a solution containing barium ions, i.e. the sulphate ions are part of the chromium (III) complex; while another gives the theoretical yield of barium sulphate, showing that the sulphate ions are not part of a complex.

Chromium (III) sulphate forms an alum of formula $K^+Cr^{3+}(SO_4^{2-})_2.\ 12\ H_2O$ when crystallized from aqueous solution in the presence of an equivalent amount of potassium sulphate. A convenient way of making chrome (III) alum involves the reduction of a solution of potassium dichromate, acidified with sulphuric acid, using sulphur dioxide:

$$K^2Cr_2O_7 + H_2SO_4 + 3SO_2 \rightarrow K_2SO_4 + Cr_2(SO_4)_3 + H_2O$$

On crystallization, purple octahedral crystals of chrome (III) alum separate from the solution.

Complexes containing chromium (III)

Chromium (III) forms a great number of complexes in which it is surrounded octahedrally by neutral ligands such as water and ammonia (Fig. 21.3) or by charged ions such as Cl^-. Some co-ordinating substances occupy two of the octahedral positions, i.e. they are bidendate, e.g. diaminoethane (usually known as ethylenediamine) $H_2N—CH_2—CH_2—NH_2$ (Fig. 21.4) Notice that the ethylenediamine complex can exist in two forms, neither of which has either a centre of symmetry or a plane of symmetry: one form is the mirror image of the other, i.e. they are optical isomers. Preparation of optically active complexes results in the formation of racemic mixtures, many of which have been resolved.

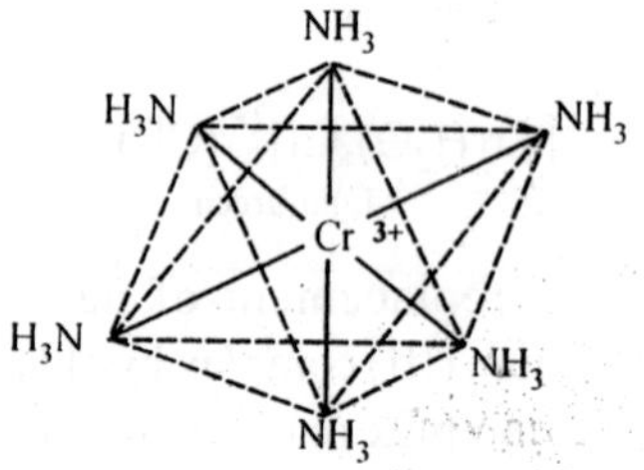

Fig. 21.3 : *The structure of the hexammaine chromium* (III) ion

Complexes can be formed which contain more than one type of ligand bonded to the central chromium (III) ion, e.g. the hydrates of chromium (III) chloride discussed previously (p. 364).

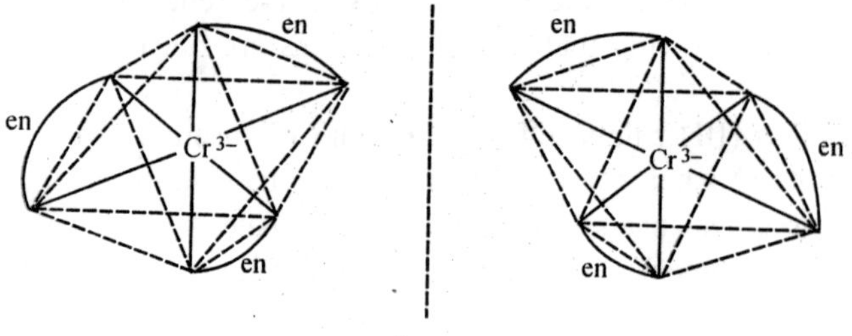

Fig. 21.4 : The optical isomers of tris (ethylenediamino) chromium (III) ion.

There are two forms of the complex $[Cr(NH_3)_4(Cl)_2]^+$ which bears one unit of positive charge, since two Cl^- ions neutralize two of the three charges on Cr^{3+}. One form contains adjacent chlorine atoms, while the other form contains chlorine atoms lying along the diagonals of a square; they are named 'cis' and 'trans' respectively and are geometrical isomers. Since both the cis- and trans-isomers have a plane of symmetry, they are not optically active (Fig. 21.5).

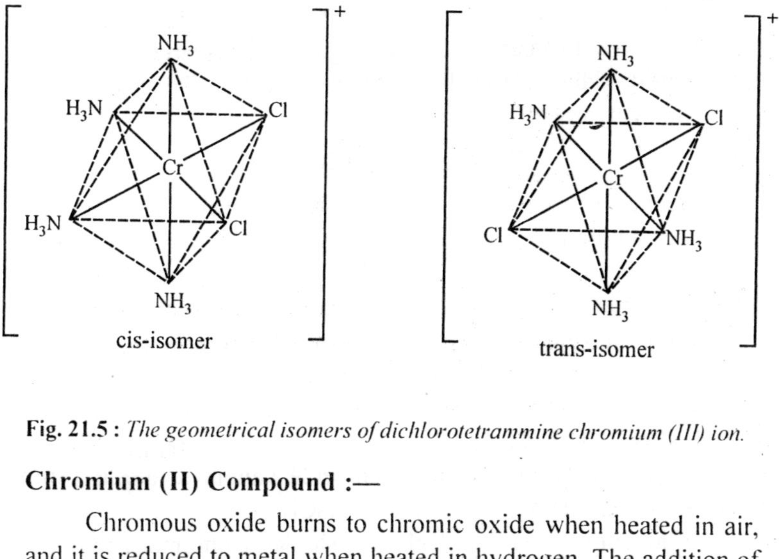

Fig. 21.5 : *The geometrical isomers of dichlorotetrammine chromium (III) ion.*

Chromium (II) Compound :—

Chromous oxide burns to chromic oxide when heated in air, and it is reduced to metal when heated in hydrogen. The addition of alkali to chromous chloride solution precipitates a yellow hydroxide, probably $Cr(OH)_2$, which reacts with water liberating hydrogen.

Chromous halides. Chromous fluoride, CrF_2, is made by the action of hydrogen fluoride on red hot chromium, or on chromous chloride in the cold. It is a non-volatile solid, m.p. 1100°, slightly soluble in water and insoluble in alcohol. The colour has been described as green or white. Crystalline CrF_2 has a rutile structure which is distorted on account of the Jahn-Teller effect.

Chromous chloride, $CrCl_2$, in the anhydrous condition exists as readily fusible, white, deliquescent crystals. It is made by the action of hydrogen chloride on chromium at a white heat, or by the reduction of chromic chloride with hydrogen at a red heat. Chromous chloride is very soluble and forms the hydrates:

I 50° II

$$CrCl_2, 6H_2O \rightarrow CrCl_2,4H_2O \rightarrow CrCl_2,4H_2O \rightarrow CrCl_2,3H_2O \rightarrow CrCl_2,2H_2O$$

The hydrate $CrCl_2,4H_2O$ is readily prepared by passing hydrogen chloride through a solution of chromous acetate in concentrated hydrochloric acid cooled to 0°. All air in the apparatus must previously have been displaced by hydrogen. The tetrahydrate is precipitated. Chromous chloride is obtained in solution by the action of dilute hydrochloric acid on pure electrolytic chromium.

Chromous chloride is a powerful reducing agent. Its solution is difficult to handle as it rapidly absorbs oxygen from the air; it is a more effective absorbent for oxygen in gas analysis than either quinone dithionite or sodium pyrogallate. Chromous chloride liberates hydrogen from dilute hydrochloric acid:

$$2HCl + 2CrCl_2 = 2CrCl_3 + H_2,$$

and it reduces aqueous solutions of potassium permanganate, Fehling's solution, ammoniacal silver oxide, iodine, and indigo. It reduces mercuric salts to mercurous, and solutions of gold, platinum, or tin salts to the metal. Chromous chloride reacts with nitric oxide in neutral solution to form ammonia, and in acid solution to form hydroxylamine. It reacts with ether to form the ethylate $CrCl_2,OC_2H_5$; the ethylate is not a reducing agent, and forms the neutral chromic complex $[(NH_3)_3CrCl_2OC_2H_5]^0$.

The density of chromous chloride vapour at about 1500° corresponds to the dimeric formula, Cr_2Cl_4.

Chromous bromide, $CrBr_2$, is similar to chromous chloride.

Chromous iodide, CrI_2, is dark red, m.p. 790-795°. It is made by combination of the elements at a high temperature.

Chromous sulphate, $CrSO_4$. The anhydrous salt is unknown. Chromous sulphate is present in the blue solution obtained by adding much granulated zinc to an acid solution of chromic sulphate, but the blue colour disappears as soon as the reaction moderates. A solution of chromous sulphate is conveniently made by electrolytic reduction of chromic sulphate solution using a platinum cathode. Addition of alcohol to the solution precipitates the pentahydrate, $CrSO_4,5H_2O$, which is stable in dry air. A blue heptahydrate, $CrSO_4.7H_2O$, can also be obtained, isomorphous with ferrous sulphate heptahydrate. Double salts are known such as $(NH_4)_2SO_4,CrSO_4,6H_2O$. The ammoniacal solution of chromous sulphate absorbs acetylene; the aqueous solution absorbs oxygen and nitric oxide.

MOLYBDENUM

Occurrence

The chief ores of molybdenum are *molybdenite*, MoS_2; *wulf-fenite*, $PbMoO_4$; and *molybdite*, also known as *molybdic ochre*, $Fe_2O_3.3MoO_3\,7\frac{1}{2}H_2O$.

Extraction

(*i*) **From Molybdenite.** The mineral is roasted, as long as sulphur dioxide is given off.

$$2MoS_2 + 7O_2 \rightarrow MoO_3 + 4SO_2$$

The residue, which is yellow when hot and white when cold, is dissolved in dilute ammonium hydroxide and the solution is evaporated until ammonium molybdate crystallises

$$MoO_3 + 3NH_4OH \rightarrow (NH_4)_2MoO_4 + H_2O$$

(*ii*) **From Wulfenite.** The ore is powdered and concentrated by the froth flotation process and then is fused with Na_2CO_3 whereby Na_2MoO_4 is obtained. The molten mass is extracted with hot water. The other heavy impurities like PbO etc. being insoluble separate out on filtration. The solution is mixed with NH_4OH and $(NH_4)_2CO_3$ to precipitate the remaining lead as $PbCO_3$. It is filtered and the filtrate after acidifying with HNO_3 is evaporated to dryness and then treated with $NH_4OH.H_2MoO_4$ is converted into $(NH_4)_2MoO_4$ which on heating gives MoO_3.

$$\underset{\text{Wulfenite}}{PbMoO_4} + Na2CO_3 \rightarrow Na_2MoO_4 + PbO + CO_2$$

$$Na_2MoO_4 + 2HNO_3 \rightarrow 2NaNO_3 + H_2MoO_4$$

$$H_2MoO_4 + 2NH_4OH \rightarrow (NH_4)_2MoO_4 + 2H_2O$$

heat

$$(NH_4)_2MoO_4\ 2\ MoO_3 + 2NH_3 + H_2O$$

Metallurgy

The metal is obtained by passing dry hydrogen over ammonium molybdate or molybdenum trioxide at red heat. The trioxide may be heated with "misch metal" (a mixture of rare earth metals as cerium, lanthanum, neodymium and praseodymium in the form of an alloy, or with aluminium *'Goldschmidt aluminothermic process'* to yield the metal

$$MoO_3 + 2Al \rightarrow Mo + Al_3O_3.$$

Properties

Molybdenum is a grey metallic powder, which remains untarnished in air at ordinary temperatures, but which when heated, passes gradually into the trioxide. It is insoluble in hydrochloric, hydrofluoric and dilute sulphuric acids, but soluble in nitric and strong sulphuric acids, in aqua regia, in chlorine water, and can be fused with potassium hydroxide or potassium nitrate. It has a sp. gr. 8.6 and m. pt. 2250°, having the highest melting point of all metals except osmium. The pure metal absorbs carbon above 1500°.

Uses

Molybdenum finds a wide range of applications, both as a metal and in its various compounds. It is used in artificial lighting devices, *e.g.*, as a support of tungsten filament in incandescent bulbs. Its wire is very effective as a winding for electric resistance furnaces, as it is cheaper than platinum, and is still capable of producing a higher temperature. Along with tungsten, it has also been used as a thermocouple for high temperature measurements. Molybdenum is also useful in dentistry and in the manufacture of standard weights. Molybdenum alloys, and specially the molybdenum steels, are of technical importance.

Molybdenum IV compounds

The compounds in this section include the oxide MoO_2, the sulphide MoS_2, the tetrahalides, and many salts containing molybdenum in the anion. The molybdenum atom in the anions often contains a lone electron pair in one of the *d* orbitals.

Molybdenum dioxide, MoO_2, a reddish-brown powder, is made:

(i) by the regulated oxidation of molybdenum,

(ii) by the action of steam on molybdenum at 800°,

(iii) by the reduction of MoO_3 with hydrogen at 470°,

Molybdenum dioxide is insoluble in alkaline or acid solutions. It unites with chlorine at 600° to give MoO_2Cl_2, and it is oxidized by silver nitrate (which is reduced to silver) to molybdenum trioxide. At 500° it is reduced by hydrogen to the metal. Molybdenum dioxide has a distorted rutile crystal structure; two metal atoms in adjacent octahedra are drawn together. It is a fairly good conductor of electricity.

Molybdenum disulphide, molybdenite, MoS_2, can be made by the action of hydrogen sulphide on molybdenum trioxide, or by fusing molybdenum trioxide with sulphur and potassium carbonate. It looks like graphite, and like graphite has a layer lattice; each molybdenum atom is at the centre of a trigonal prism, and six sulphur atoms are at the apices. It is diamagnetic. At about 200° the crystals have metallic conductivity. On strong heating molybdenum sulphide is decomposed into its elements.

The only member of the **molybdenum tetrahalides** which is well-known is the chloride $MoCl_4$, which is made:

(i) by heating molybdenum dioxide with a solution of chlorine in carbon tetrachloride at 250°,

(ii) by heating molybdenum trichloride to a red heat in a stream of carbon dioxide, when it disproportionates giving Mo_6Cl_{12} (non-volatile) and volatile $MoCl_4$.

Molybdenum tetrachloride is an easily volatile brown powder. It is soluble in alcohol without decomposition, but it reacts with water to give the hydrolysis products of $MoCl_3$ and $MoCl_5$. When heated it disproportionates to $MoCl_3$ and $MoCl_5$. MoF_4 is a green non-volatile solid.

Mo^{IV} complexes. A neutral complex is $MoCl_4(OA_SPh_3)_4$. Anionic complexes are exemplified by the dark green Na_2MoF_6 and Na_2MoCl_6. The salt $K_4[MO(CN)_8]2H_2O$ is made as above, or by the action of a large excess of potassium cyanide on $MoCl_3,3H_2O$, or on $MoOCl_3$. The salt is yellow. It is little affected by heat, alkalis, acids, or oxidizing agents. It is diamagnetic. In the crystal the anion is dodecahedral with the eight CN groups at the apices and the molybdenum atom at the centre. The orbital diagram is shown in Fig. 21.6:

The d_{xy} orbital has much the lowest energy of all the *d* orbitals in a dodecahedral ligand field, and its occupation by an electron pair stabilises the anion.

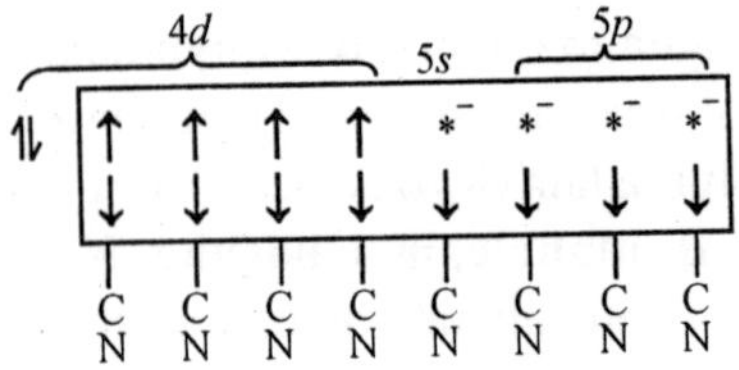

Fig. 21.6 : Orbital diagram for the anion $[Mo(CN)_8]^{4-}$.

Other salts are $K_4[MoO_2(CN)_6]$,xAq. violet, and $K_3[Mo(OH)_3(CN)_4$,H_2O,xAq. blue. The orbital diagram of the anion of the blue salt is shown in Fig. 21.7:

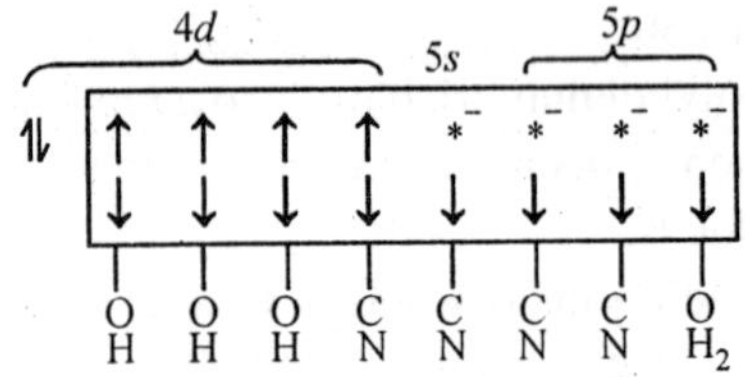

Fig. 21.7 : Orbital diagram for the anion $[Mo(OH)_3(CN)_4(H_2O)]^{3-}$.

Molybdenum V compounds

The existence of the oxide Mo_2O_5 is doubtful, and the dark violet powder which is assigned this formula is probably a mixture of MoO_2 and MoO_3. Molybdenum pentasulphide, Mo_2S_5 is made by reducing a solution of ammonium molybdate in 20% sulphuric acid with zinc until wine red. The solution is saturated with hydrogen sulphide, and a dark brown precipitate of $Mo_2S_5,3H_2O$ is obtained. The precipitate is insoluble in water, but soluble in strongly acid or alkaline solutions. Nearly complete dehydration can be effected by warming the precipitate in a stream of carbon dioxide.

Molybdenum pentahalides. Molybdenum pentafluoride, MoF_5, an olive green solid, is obtained by the action of fluorine on the metal at 400°, or on molybdenum hexacarbonyl at –75°. If the solid is distilled at + 170° in vacuo a light yellow liquid is condensed which solidifies at + 67° and has the composition MoF_5. It is isostructural with NbF_5.

Molybdenum pentachloride, $MoCl_5$, exists as dark green hygroscopic crystals, m.p. 194°, b.p. 268°. It is made by heating molybdenum powder in chlorine. The vapour is dark red. When heated in air it is oxidized to the white oxochloride, MoO_2Cl_2. It reacts violently with water to give the green oxochloride, $MoOCl_3$, and with alcohol to liberate hydrogen chloride. It forms the complex $MoCl_5,2(C_2H_5)_2O$ with ether, but it is reduced by pyridine to a complex $MoCl_4$,(pyr)x. The density of $MoCl_5$ vapour is normal, and electron diffraction experiments show that in the vapour the molecule is a trigonal bipyramid. The crystal structure is probably analogous to that of Nb_2Cl_{10}. $MoCl_5$ is paramagnetic, so there can be little coupling between the two Mo atoms.

Complexes. The neutral complex $MoCl_3,2(C_2H_5)_2O$ has been mentioned A complex anion is present in the green salt $K_2[MoOCl_5]$ which is made by the reduction of potassium molybdate at a platinised platinum cathode. The salt is soluble in water to give a reddish-brown solution: the colour fades is insoluble MoO(OH), is precipitated by hydrolysis. The magnetic moment of the anion corresponds to the presence of one unpaired electron.

Molybdenum VI compounds

Molybdenum trioxide, MoO_3, is made :

(i) by burning metallic molybdenum in oxygen,
(ii) by heating molybdic acid or ammonium molybdate,
(iii) by oxidizing molybdenum dioxide with nitric acid or silver nitrate,
(iv) by heating molybdenite, MoS_2, in air.

Molybdenum trioxide is a white solid which on heating turns yellow and sublimes; it melts at 795° to a yellow liquid, b.p. 1155°. It is sparingly soluble in cold (0.107 gm/ 100gm at 18°), more soluble in hot water (2.06 gm/100 gm), but it does not give molybdic acid.

Molybdenum trioxide is reduced by hydrogen at 500° to give reddish-brown MoO_2 at 1200° it gives a grey powder of metallic molybdenum. It is also reduced to metal by carbon at a good heat. It reacts with dry gaseous hydrogen chloride at 150-200° to give the crystalline compound MoO_3,2HCl. It dissolves in excess of ammonia solution to give ammonium molybdate, $(NH_4)_2$,MoO_4, but on evaporation the solution deposits crystals of the paramolybdate $(NH_4)_6$ Mo_7O_{24},$4H_2O$. Molybdenum trioxide dissolves in alkaline hydroxide and carbonate to give X_2MoO_4, but it' fused with calcium carbonate at 1000° it gives Ca_3MoO_3. If a solution of molybdenum trioxide in hydrochloric acid is electrolyzed an aqueous solution of molybdenum trichloride is obtained.

In the crystal of molybdenum trioxide the atoms are arranged in MoO_6 octahedra. Each octahedron shares two corners and two edges with other octahedra, thus giving rise to a layer structure. The structure of MoO_3 differs from those of CrO_3 and WO_3. The MoO_3 structure is common in sulphides but rare in oxides.

MoS_3. The hydrated compound is precipitated by passing hydrogen sulphide into a slightly acid solution of a molybdate. It dissolves in alkali sulphide solution to give thiomolybdates such as K_2MoS_4, containing the discrete tetrahedral ion MoS_4^{2-}.

Molybdenum hexafluoride, MoF_6, a white crystalline compound, m.p. 17.5°, b.p. 35°, is the only hexahalide of molybdenum. It is made:

(i) by the action of fluorine diluted with nitrogen on molybdenum at 350°,
(ii) by the action of hydrogen fluoride on Mo_6Br_{12} at 800-860°,
(iii) by the action of bromine trifluoride on molybdenum or on Mo_6Br_{12},
(iv) by the action of fluorine on molybdenum hexacarbonyl above 50°,
(v) by the action of SF_4 on MoO_3.

Molybdenum hexafluoride is quite stable, and it is not attacked by dry air, chlorine, or sulphur dioxide, but it is immediately hydrolyzed by water. It reacts slowly with NaF to give MoF_6.2NaF. The results of experiments on the Raman and infrared spectra indicate that the molecule is a regular octahedron.

Molybdenum oxohalides. $MoOF_4$, a colourless deliquescent crystalline compound, m.p. 97° b.p. 180°, is obtained by the action of hydrogen fluoride on the product (perhaps $MoOCl_4$) resulting from the action of chlorine on partially reduced molybdenum dioxide. $MoOF_4$ dissolves in water to form a colourless solution which on evaporation leaves molybdenum trioxide.

MoO_2F_2, a colourless solid, is made by the action of hydrogen fluoride on MoO_2Cl_2.

MoO_2Cl_2, a white or pale yellow crystalline compound, is made:

(i) by passing dry chlorine over MoO_2 heated to 7000,

(i) by heating MoO_3 to 150-300 in dry gaseous hydrogen chloride.

It sublimes easily, and melts only under pressure. It is easily soluble in water, alcohol, ether, and acetone; in the last two solvents its molecular weight is normal. It is the starting point for the preparation of other molybdenum compounds. The configuration of the molecule in the vapour is tetrahedral.

Molybdenum (III) compounds

(1) **Molybdenum trichloride, $MoCl_3$** is the most important trihalide and is prepared by the reduction of $MoCl_5$ with H_2 it 250°C.

$$MoCl_5 + H_2 \xrightarrow{250°C} MoCl_3 + 2HCl$$

It is in soluble is H_2O.

When trichloride is heated in an inert atmosphere, it yields di- and tetrachlorides the tetra is volatile of the two and leaves behind the dichloride on distillation

$$2MoCl3 \rightarrow MoCl_3, + MoCl_4$$

The chloride, in fact, exists in its trimeric form, Mo_3Cl_6, or rather $[Mo_3Cl_4]Cl_2$, since it gives rise to $[Mo_3Cl_4]$ $(OH)_2.8H_2O$ and $[Mo_3Cl_4](NO_3)2$ when dissolved in alkalies and is re-acidified.

(2) **Oxyhalides.** The oxyhalides of Mo(+3) are of $MoOX.4H_2O$ type (X=F, Cl, Br). The oxychloride, MoOCl is precipitated by the addition of acetare to a solution of Mo(+3) salt in HCl. The oxybromide is also prepared similarly.

(3) **Molybdenum trisulphide, Mo_2,S_3.** It is obtained as a precipitate on adding $(NH_4)_2S$ to a solution containing a salt of Mo(+3).

(4) **Other compounds.** When a molybdate solution is reduced with zinc and sulphuric acid, the orange coloured solution of *molybdic sulphate,* $Mo_2(SO_4)_3$ is obtained. This when treated with sodium hydroxide yields brown black precipitate of *molybdic hydroxide,* $Mo(OH)_3$. The hydroxide dissolves in hydrochloric acid to give the *trichloride,* $MoCl_3$, which undergoes marked hydrolysis in the solution.

Molybdenum (II) compounds

The only known compounds of Mo(+2) are dihalides *viz* $MoCl_2$, $MoBr_2$, and MoI_2. These can be prepared by the direct union of the elements at suitable high temperatures.

These are coloured, non-volatile solids, insoluble in H_2O and exhibit little reducing power.

TUNGSTEN

Occurrence

Tungsten is a rare element, found combined in minerals which are usually associated with tin ores. The chief minerals are *wolframite* $(Fe, Mn)WO_4$; *Scheelite* $CaWO_4$ and *Hubnerite,* $MnWO_3$, all the three containing 70 to 80 per cent of WO_3.

Extraction

Wolframite is the usual source of tungstic acid. The mineral is fused with a mixture of dry sodium carbonate and sodium nitrate. The fused mass is treated with water when sodium tungstate formed on fusion goes into the solution. The solution when acidified with hydrochloric acid yields a precipitate of tungstic acid.

$$4FeWO_4 + O_2 + 4Na_2CO_3 \rightarrow 4Na2WO_4 + 2Fe_2O_3 + 4CO_2 \uparrow$$

soluble insoluble

$$2MnWO_4 + O_2 + 2Na_2CO_3 \rightarrow 2Na2WO_4 + 2MnO_2 + 2CO_2 \uparrow$$

Soluble Insoluble

$$Na_2WO_4 + 2HCl \rightarrow 2NCl + H_2WO_4$$

Metallurgy

The element tungsten may be obtained by (*i*) heating the tungstic acid in a current of hydrogen; (*ii*) heating the chloride, WCl_6, in a current of hydrogen; (*iii*) heating the acid or trioxide, WO_3 with carbon, zinc or aluminium *(alumino-thermic* process with less of aluminium is preferred); (*iv*) passing an electric current in vacuum through filaments of an amalgam of tungsten and mercury, cadmium and mercury being expelled and the tungsten left.

Properties

Physical properties of tungsten depend on the degree of purity and the methods of production. The density varies from 19 to 21. The metal is mere ductile than even gold. It has the lowest vapour pressure and the highest melting point of any metal. The metal is hard, brittle and crystalline in the powdered form. It remains unaffected by air and water at ordinary temperatures, but is readily oxidized at higher temperatures. The vapours of sulphur and phosphorus react vigorously with the metal. Though not attacked by the alkaline solutions, the metal can be fused with alkali nitrates, peroxides and carbonates, or acid sulphates. Sulphuric and hydrochloric acids have no action on the metal except on boiling and when used concentrated. Nitric or hydrofluoric acid has no action, but the metal dissolves readily in a mixture of the two.

Uses

The metal is largely used in electrical contrivances, lighting devices and in the formation of alloys and steels. Tungsten molybdenum thermocouples are useful in high temperature work. Tungsten filaments give brilliant light and hence used in lamps. As the metal resists both acids and alkalies, it is used in the manufacture of gauzes in centrifugal machines. The oxide of tungsten is used in ceramics to produce various shades of yellow in glass and porcelain. Sodium tungstate is used to decolourise acetic acid, in fire-proofing of cloth and also as a mordant for silk. Calcium tungstate is used for the preparation of X-ray fluorescent screens.

Tungsten (II) compounds

WCl_2, WBr_2, and WI_2 are infusible, non-volatile, coloured powders made by reducing higher halides. The chloride attacks water liberating hydrogen and in this and in other reactions it shows itself to be a powerful reducing agent. In this it differs from Mo_6Cl_{12}. WBr_2 and WI_2 are insoluble in water. The tungsten dihalides are certainly polymerized but their structures are unknown

The acid $(H_3O)_2[W_6Cl_8,Cl_6]6H_2O$ exists as yellow needles.

Tungsten (III) compounds

In contrast to the chemistry of chromium, tungsten forms no simple compounds in the state W^{III}, and few complexes. A complex anion is present in the salt M [WCl] where M represents NH_4, K, Rb, Cs, or Tl, or a complex cation such as $[Co(NH_3)_6]^{3+}$ or $[Ag(NH_3)_2]^+$. The alkali metal salts are anhydrous and greenish yellow, and they may be crystallized from water. The salts of complex cations are almost insoluble. It has been shown by X-ray analysis that the alkali metal salts are isomorphous, and that the anion consists of two octahedra with one face in common (compare $Fe_2(CO)_9$). The anion is diamagnetic, and hence there is coupling between the tungsten atoms. $[Cr2Cl_9]^{3-}$ and $[Tl_2Cl_9]^{3-}$ are isostructural with the tungsten anion, but there is no metal-metal coupling.

Tungsten (IV) compounds

Tungsten (again in contrast to chromium) forms a number of stable compounds in which it is present in the w^{IV} state, including an oxide, a sulphide, and complex anions.

Tungsten dioxide, a brown powder with the rutile crystal structure, m.p. about 1600° in nitrogen, is made by the reduction of WO_3. It is easily oxidized to WO_3, and reduced by hydrogen to the metal. Chlorine oxidises it to WO_2Cl_2. Tungsten disulphide WS_2 has much the same properties and the same crystal structure as MoS_2.

All four tetrahalides of tungsten are known, WF_4, a red-brown non-volatile substance, is obtained by the reduction of WF_6 with benzene at 110°. WCl_4 is made by the reduction of WCl_6 with hydrogen at a high temperature, or by the action of CCl_4 on WO_2 at 250°. WCl_4 and WBr_4 are infusible, non-volatile, hygroscopic black substances. The chloride is diamagnetic and isomorphous with $NbCl_4$.

The potassium hydroxychlorotungstate, $K_2[W(OH)Cl_5]$ is a dark green solid which dissolves in water to a red solution not changed by exposure to air. The complex cyanide, $K_4[W(CN)_8]$ is yellow, stable to air, and is not hydrolyzed in solution.

Tungsten (v) compounds

Among the simple compounds, pentahalides *viz* WCl_5 and WBr_5 are known.

WGl_5 is obtained as dark green needles from the thermal decomposition of WCl_6 and WBr_5 (black) can be prepared by the direct combination of the elements at red heat or by the reaction between WCl_6 and HBr at 230-250°C. Both the pentahalides are soluble in a number of organic solvents. They are easily hydrolyzed by water.

Tungsten (VI) compounds

The compounds of tungsten in this state are some of the most familiar and best-characterised.

Tungsten trioxide, WO_3, is a lemon yellow powder, orange when hot, melting at 1473° to a green liquid, b.p. above 1750°. It is made by heating tungsten in air.

Tungsten trioxide is completely volatile in chlorine at 500°, perhaps as a result of conversion to the oxochloride. It is reduced by hydrogen to blue W_4O_{11}, or to WO_2, or to the metal. It is converted by CCl_4 in a sealed tube at 200° to $WOCl_4$, and at 280° to WCl_6. It is reduced by carbon monoxide at 800° to WO_2. When heated with silver in vacuo a mixture of Ag_2WO_4 and the oxide W_3O_8 is produced. When heated with calcium carbonate at 1000° Ca_3WO_3 is formed. Tungsten trioxide is insoluble in water or acids, but it dissolves in alkaline solutions to give tungstates. The crystal structure of tungsten trioxide is the same as that of CrO_3.

The tungsten hexahalides. The preparations and properties of the tungsten hexahalides are summarized in Table 21.2. It should be noted that tungsten forms a hexachloride and a hexabromide. No other element forms a hexabromide, and only one other element, uranium, forms a hexachloride. Many other elements besides tungsten form hexafluorides, including: S, Se, Te, Mo, U, Re, Os, Ir, Pt, Np, Pu.

Table 21.2. : *The Hexahalides of Tungsten*

WF6	WCl6	WBr6	
Appearance liquid	Fuming volatile crystals	Violet or steel blue	Blue-black needles
m.p., °C	2.5	27	(Decomposes on gentle heating)
b.p., °C	19.5	347	
Preparation	(i) Action of fluorine on metal (ii) Act ion of HF on WCl_6 in the cold	(i) Action of dry chlorine on reduced W at low red heat	Pass Br2 vapour diluted with N_2 over warmed tungsten

		(ii) Action of Cl_2, SCl_2, or PCl_5, on WO_3, or WS_3	
Properties	Hydrolysed by water Gives coloured solutions with organic solvents Attacks all metals except Au and Pt	Hydrolysed by water to $WOCl_4$ Soluble in organic solvents Is reduced by H_2 to WCl_5 and WCl_2, by HI at 400° to WI_2, by HBr at 275° to WBr_5	Hydrolyzed by water Gives violet-red solutions with organic solvents
Structure	Vapour density is normal. Raman and infrared spectra show octahedral configuration	Vapour density is normal.† Electron diffraction shows octahedral configuration	

The tungsten oxohalides. WOF_4, made by passing a mixture of oxygen and fluorine over the heated metal, is a colourless, volatile solid. $WOCl_4$ and WO_2Cl_2 are well characterized crystalline compounds. $WOCl_4$, scarlet crystals, m.p. 209°, b.p. 233°, is formed by passing moist chlorine over tungsten, by the action of oxygen or moisture on WCl_6, or by treating WO_3 with PCl_5, CCl_4, or WCl_6 vapour. It reacts violently with water to give WO_3. Its vapour density is normal up to 440°. WO_2Cl_2 consists of yellow cr m.p. 265°. It is made by passing a stream of chlorine diluted with nitrogen over WO_2 at 540°. It is volatile, but less volatile than $WOCl_4$. On heating strongly it disproportionates, thus:

$$2WO_2Cl_2 = WO_3 + WOCl_4.$$

It is stable in moist air, and is only slowly hydrolyzed by water.

22

d-Block Elements (Manganese)—Group VIIB (7)

Element	Atomic Symbol	Atomic weight	number
Manganese	Mn	54.94	25
Technetium	Tc	99*	43
Rhenium	Re	186.31	75

* Stablest isotope

The secondary subgroup of the seventh group includes the elements of the even series of the long periods *manganese, technetium* and *rhenium*. The relation between these elements and those of the main subgroup of the seventh group, i.e., the halogens, is about the same as between the elements of the main and secondary subgroups of the sixth group. Having only two electrons in the outermost layer of their atoms, manganese and its analogues are incapable of gaining electrons, and, contrary to the halogens, do not form hydrides. However, the highest oxygen compounds of these elements have some resemblance to the corresponding compounds of the halogens, as seven electrons can participate in the formation of bonds with oxygen just as is the case with the halogens. Therefore, their highest positive valency equals 7.

Of the elements of the manganese subgroup the most important is manganese itself. Rhenium, discovered in 1925, is a very rare element, but, owing to a number of valuable properties, has already found certain practical applications. Technetium does not occur in nature. It was produced artificially in 1937 by bombarding molybdenum nuclei with deutrons, heavy hydrogen nuclei, accelerated by means of a cyclotron. Technetium was the first element obtained artificially, "technically," which accounts for its name.

Occurrence and Extraction of Manganese

Manganese

Typical ores of manganese are pyrolusite, MnO_2, and hausmannite, Mn_3O_4. The former is the more important.

Pure manganese is obtained by the reduction of Mn_3O_4 with aluminium, followed by distillation of the manganese in vacuo :

$$3Mn_3O_4 + 8Al \rightarrow 4Al_2O_3 + 9Mn$$

Reduction of MnO_2 with aluminium is dangerously explosive, and this oxide is first converted into Mn_3O_4 by heating strongly in air before reduction is attempted:

$$3MnO_2 \rightarrow Mn_3O_4 + O_2$$

Most manganese is used in the form of alloys with iron, and these are obtained by reducing mixed manganese and iron ores with coke in an electric furnace.

Properties and Uses of Manganese

Manganese has a comparatively low melting point (1247°C) and a density of 7.21 g cm^{-3}. It is a hard, brittle metal and, although not readily attacked by air at ordinary temperature, it is fairly reactive. e.g. it combines with a variety of non-metals on heating (these include oxygen, nitrogen, chlorine and sulphur) and will liberate hydrogen from warm water. Dilute acids such as hydrochloric acid and sulphuric acid readily dissolve the metal, forming manganese (II), Mn^{2+}. ions and liberating hydrogen.

Pure manganese has few uses since it is brittle and is attacked by water, but industrially it is very important, in the form of alloys with iron. Ferro-manganese (Mn, 80 per cent; Fe, 20 per cent) is used in the manufacture of alloy steels, the presence of manganese conferring hardness, toughness and resistance to wear. Spiegeleisen

(Mn, 25 per cent; Fe, 70 per cent; C, 5 per cent) is added to molten steel for the purpose of removing combined oxygen and sulphur, thereby improving its mechanical properties.

General Properties

Mn is more reactive than its neighbours in the periodic table. It reacts slowly with H_2O, liberating H_2, and it dissolves readily in dilute acids. The finely divided metal is pyrophoric in air, but the massive metal does not react unless heated. When strongly heated the massive metal reacts with many non-metals such as O_2, N_2, Cl_2 and F_2. forming Mn_3O_4, Mn_3N_2, $MnCl_2$ and a mixture of MnF_2 and MnF_3. The melting point of the metal is appreciably lower than for the earlier first row elements Ti, V and Cr. It is unusual in that it has four different solid structures (α or body-centred cubic, cubic close-packed, β and χ). The α form is stable at room temperature and has a body-centred cubic structure.

Tc and Re metals are less reactive than Mn. They do not react with H_2O, or non-oxidizing acids. They do not dissolve in HCl and HF, but they react with oxidizing acids such as concentrated HNO_3 and H_2SO_4, forming pertechnic acid $HTcO_4$ and perrhenic acid $HReO_4$. (This is not the usual reaction of acid + metal to give a salt + H_2. For example, with concentrated HNO_3, the NO_3^- ion is a stronger oxidizing agent than H_3O^+, and NO_2 is evolved.) Tc and Re undergo similar reactions with H_2O_2 and bromine water. The massive metals tarnish (oxidize) slowly in moist air, but the powdered metals are more reactive. Heating with O_2 gives Tc_2O_7 and Re_2O_7 which are both low melting (119.5°C and 300°C respectively) and volatile. Heating with F_2 gives TcF_5 and TcF_6, and ReF_6 and ReF_7.

Table 22.1 : Some physical properties

	Covalent radius (Å)	Ionic radius (Å) M^{2+}	M^{3+}	Melting point (°C)	Boiling point (°C)	Density (g cm−3)	Pauling's electro-negativity
Mn	1.22	0.67	0.645^h 0.58^l	1244	2060	7.43	1.5
Tc	1.34	–	–	2200	4567	11.5	1.9
Re	1.34	–	–	3180	5650	21.0	1.9

h = high spin value l = low spin radius.

Many ionic compounds of Mn are known including Mn^{2+}, Mn^{3+}, MnO_4^{2-} and MnO_4^-. In contrast Tc and Re have virtually no aqueous ionic chemistry apart from the oxoions TcO_4^- and ReO_4^-. The elements Tc and Re have a marked tendency to form metal-metal bonds in the lower oxidation states (+ II). (+111) and (+IV).

Trace amounts of manganese are essential for plant and animal growth. For this reason small amounts of $MnSO_4$ are often added to fertilizers.

The basic character of any element changes with the oxidation state. Low oxidation states are more basic and high oxidation states are more acidic. MnO and Mn_2O_3 are basic oxides and are ionic. MnO_2 is amphoteric and does not exist as Mn^{4+} ions. Mn(+V) is rather uncommon. Mn(+ VI) is represented by manganates such as $Na_2[MnO_4]$. This may be regarded as a salt of the unstable acidic oxide MnO_3. which does not exist in the free state. Mn(+VII) occurs as Mn_2O_7 which is strongly acidic. The corresponding acid, permanganic acid $HMnO_4$, is a very strong acid.

Almost all manganese compounds are coloured. Mn^{2+} is pale pink, and MnO_2 is black, both because of *d-d* transitions. The (+VII) oxidation state has a d^0 configuration and would be expected to be colourless. Whilst perrhenates ReO_4^- containing Re(+VII) are colourless, permanganates MnO_4^- containing Mn(+VII) are intensely coloured. The purple-black colour arises from charge transfer spectra.

Mn resembles iron in its physical and chemical properties. Mn is harder and more brittle than iron, but melts at a lower temperature (Mn = 1244°C, Fe = 1535 °C). All three metals Mn, Tc and Re are usually obtained as grey powders, but in the massive form they look like platinum. Re has the second highest melting point of all the metals (W = 3380°C; Re = 3180°C).

Manganese is quite electropositive and dissolves in cold dilute non-oxidizing acids. It is not very reactive towards non-metals at room temperature, but reacts readily on heating.

Mn is much more reactive than Re. Similar behaviour is observed in adjacent groups: Cr is more reactive than W, and Fe more reactive than Os. This decrease in electropositive character is opposite to the trend shown in the main groups of the periodic table. In addition, Re tends to attain a higher oxidation state than Mn when they both react with the same element. A comparison of some of the reactions of Mn and Re is given in Table 22.2.

Table 22.2 : Some reactions of manganese and rhenium

Reagent	Mn	Re
N_2	Mn_3N_2 formed at 1200°C	No reaction
C	Mn_3C	No reaction
H_2O	$Mn^{2+} + H_2$	No reaction
Dilute acid	Mn^{2+} +	H_2 No reaction
Strong acid	Mn^{2+} +	H_2 Dissolves slowly
Halogens	MnX_2 and MnF_3	ReF_6, $ReCl_5$, $ReBr_3$
S	MnS	ReS_2
O_2	Mn_3O_4	Re_2O_7

Group 7 (manganese group) elements have seven outer electrons, but the similarity to Group VIIA elements, the halogens, is very slight except in the highest oxidation state. Mn_2O_7 and Cl_2O_7 may be compared; MnO_4^- and ClO_4^- are isomorphous and have similar solubilities, and IO_4^- and ReO_4^- are quite similar. There are much closer similarities between Mn and its horizontal neighbours Cr and Fe. The chromates CrO_4, manganates MnO_4^{2-} and ferrates FeO_4^{2-} are similar. The solubilities of the lower oxides are also similar, which accounts for the occurrence of iron and manganese together.

Compounds of Manganese

Manganese shows important oxidation states of +7 (powerfully oxidizing), +6, +4, +3 and +2 (the most stable). Reduction of manganese (VII) to manganese (II) is readily achieved with a variety of reducing agents, e.g. shaking an acidified solution of potassium permanganate with zinc amalgam :

$$\underset{\substack{\text{permanganate} \\ \text{purple} \\ +7}}{2MnO_4^-} + 16^{H+} + 5Zn \rightarrow \underset{\substack{\text{pale pink} \\ +2}}{2Mn^{2+}} + 5Zn^{2+} + 8H_2O$$

Manganese (VII) compounds

Manganese (VII) oxide, Mn_2O_7

This oxide has been obtained by adding potassium permanganate to well-cooled concentrated sulphuric acid. It is a dark coloured covalent liquid and **no attempt should be made to prepare it, since**

it decomposes explosively into manganese **(IV) oxide and oxygen**

$$2Mn_2O_7 \rightarrow 4MnO_2 + 3O_2$$

Manganese (VII) oxide dissolves in water to give a purple solution which contains 'permanganic acid', and although the pure acid, $HMnO_4$ has never been isolated, the potassium salt is well known.

Potassium permanganate, $KMnO_4$

This is the best known compound containing manganese (VII), and can be prepared in the laboratory by fusing manganese (IV) oxide and potassium hydroxide in the presence of an oxidizing agent such as potassium chlorate. Potassium manganate (VI), K_2MnO_4, a green solid is obtained which is then crushed, dissolved in water and the resulting mixture boiled. Under these conditions potassium manganate (VI) begins to disproportionate, since it is only stable in strongly alkaline solution, and potassium permanganate and manganese (IV) oxide are formed. To assist this disproportionation, a stream of carbon dioxide is bubbled through the solution and this removes the hydroxyl ions as hydrogen carbonate ions and thus allows the reaction to go to completion (see second equation below):

$$3MnO_2 + 6OH^- + ClO_3^- \rightarrow 3MnO_4^{2-} + 3H2O + Cl-$$

manganate (VI) green

$$3MnO_4^{2-} + 2H_2O \rightarrow 2MnO_4^- + MnO_2 + 4OH^-$$

manganate (VI) permanganate

$$4OH^- + 4CO_2 \rightarrow 4HCO_3^-$$

After completion of the reaction, the mixture can be freed from solid manganese (IV) oxide by filtration through glass wool, and the resulting solution evaporated until crystallization of potassium permanganate begins.

Industrially, potassium permanganate is obtained from the potassium manganate (VI) by oxidation with chlorine; this reaction avoids the formation of unwanted manganese (IV) oxide:

$$2MnO_4^{2-} + Cl_2 \rightarrow 2MnO_4^- + 2Cl^-$$

Powerful oxidizing agents such as sodium bismuthate oxidise a solution of a manganese (II) salt to the permanganate ion, MnO_4^-. This reaction is carried out in the cold in the presence of dilute nitric acid, and is often used as a confirmatory test for manganese (II) ions, since the permanganate ion is intensely coloured even in very dilute solution:

$$2MnO^{2+} + 8BiO_2^- + 14H^+ \rightarrow 2MnO_4^- + 5Bi^{3+} + 7H_2O$$

Potassium permanganate is a powerful oxidizing agent, e.g. it will oxidise hydrochloric acid to chlorine and hydrogen peroxide to oxygen. Strips of filter paper impregnated with potassium permanganate solution and acidified with dilute sulphuric acid are decolourised in the presence of some gaseous reducing agents; this is a convenient test for sulphur dioxide and hydrogen sulphide, e.g.

$$2MnO_4^{2-} + 6H^+ + 5H_2S \rightarrow 2Mn^{2+} + 8H_2O + 5S$$

Under alkaline conditions potassium permanganate is reduced first to potassium manganate (VI) and then to manganese (IV) oxide, e.g. alkenes give diols (glycols) when treated with an alkaline solution of potassium permanganate, the purple solution turning green as potassium manganate (VI) is formed; a brown residue of manganese (IV) oxide is eventually formed.

Potassium permanganate is used as a volumetric reagent but, unlike potassium dichromate, it is not a primary standard since it cannot be obtained in a high degree of purity; samples of the substance contain some manganese (IV) oxide and aqueous solutions slowly deposit brown manganese (IV) oxide on standing. It is standardized against a standard solution of iron (II) ammonium sulphate acidified with dilute sulphuric acid. Potassium permanganate solution is run into the acidified iron (II) salt from a burette and the purple colour is discharged as virtually colourless manganese (II) ions are formed; at the end-point the solution turns pink, the colour of potassium permanganate in extremely dilute solution; thus it acts as its own indicator. The relevant partial equations are :

$$MnO_4^- + 8H^+ + 5e \rightarrow Mn^{2-} + 4H_2O \quad \text{(reduction)}$$

$$5Fe^{2-} \rightarrow 5Fe^{3-} + 5e \quad \text{(oxidation)}$$

A standard solution of potassium permanganate can be used for estimating iron (II) salts and oxalates, previously acidified with dilute sulphuric acid before titration, e.g.

$$2MnO_4^- + 16H^+ + 10e \rightarrow 2Mn^{2-} + 8H_2O \qquad \text{(reduction)}$$

$$5\begin{vmatrix} COO^- \\ COO^- \end{vmatrix} \rightarrow 10CO_2 + 10e \qquad \text{(oxidation)}$$

Similarities between manganese (VII) and chlorine (VII) compounds

Apart from the fact that manganese (VII) compounds are highly coloured, there is a certain amount of resemblance to chlorine (VII) compounds. For example, manganese (VII) oxide, Mn_2O_7, and chlorine heptoxide. Cl_2O_7, are explosive covalent liquids. The permanganate ion, MnO_4^-, and the perchlorate ion, ClO_4^-, are oxidizing agents, the former more **powerful than** the latter; corresponding permanganates and perchlorates are isomorphous.

Manganese (IV) Compound

The compounds of manganese in the Mn^{IV} oxidation state include the giant molecule compound manganese dioxide, the manganites such as K_2MnO_3, and the anionic complexes present in the salts $KMnF_5$, and K_2MnCl_6. The ion Mn^{4+} is present in manganese dioxide, but the ion is unstable in water. There are no simple covalent

Manganese dioxide, MnO_2, is a black solid with the rutile lattice. It is made by heating manganous nitrate; different detailed procedures are recommended by different authors. At about 500° manganese dioxide loses oxygen to give Mn_2O_3 and Mn_3O_4. Manganese dioxide is insoluble in water. It is a fairly good conductor of electricity; the specific conductance at 0° is 0.16 $ohm^{-1}\ cm^{-1}$.

Manganese dioxide dissolves in hydrochloric acid liberating chlorine:

$$MnO_2 + 4HCl = Mn^{4+} + 4Cl^- + 2H_2O,$$

$$Mn^{4+} + 2Cl^- = Mn^{2+} + Cl2,$$

and in sulphuric acid liberating oxygen; the violet colour which appears in the reaction with sulphuric acid is due to the manganous ion Mn^{2+}. Manganese dioxide reacts with hydrogen fluoride at 450-500° to yield manganous fluoride:

$$2MNO_2 + 4HF = 2MnF_2 + O_2 + 2H_2O.$$

An aqueous suspension of manganese dioxide reacts with sulphur dioxide in two ways according to the equations:

(a) $MnO_2 + SO_2 = MnSO_4$ (BaO_2 and PbO_2 undergo corresponding reactions),

(b) $2MnO_2 + SO_2 = Mn_2O_3 + SO_3$,

$MN_2O_3 + 3SO_2 = Mn_2(SO_3)_3$,

$Mn_2(SO_3)_3 = MnSO_3 + MnS_2O_6$.

The precipitate obtained by reducing potassium permanganate in alkaline solution has the composition H_2MnO_3. This hydrated manganese dioxide reddens blue litmus, and dissolves in concentrated alkalis to give manganites.

Manganese tetrafluoride, MnF4, can be made by direct combination of the elements under controlled conditions. It is a volatile, blue-grey solid, and is highly reactive.

The pink salt **$KMnF_5$**, is made by the action of bromine trifluoride on potassium permanganate; excess of bromine trifluoride is removed by evaporation in vacuo at 20°. There are several salts of the general formula $K_2[MnX_6]$, where X represents F, Cl, or CN.

Potassium hexafluoromanganite, $K_2[MnF_6]$, exists as golden yellow crystals which at high temperatures have the cubic lattice of $K_2[PtCl_6]$. It is made by treating potassium manganate, K_2MnO_4, with concentrated hydrofluoric acid, or by treating a mixture of equal molecular proportions of potassium permanganate and potassium chloride with bromine trifluoride. It is hydrolyzed by water to manganese dioxide. $K_2[MnCl_6]$ is obtained by the action of gaseous hydrogen chloride on potassium permanganate dissolved in glacial acetic acid, and $K_2Mn(CN)_6]$ by the action of potassium cyanide on potassium permanganate in saturated aqueous solution. Both salts are red and are decomposed by water.

Manganese (VI) Compound

The only compounds in which manganese is present in the oxidation state Mn^{VI} are the manganates.

Potassium manganate, K_2MnO_4, is prepared by heating manganese dioxide with potassium hydroxide and potassium nitrate (air is used as the oxidant in industrial processes). The product is a green mass which is extracted with water containing a little alkali. A solution of potassium manganate can be obtained by boiling a strong solution of potassium permanganate with potassium hydroxide:

$$4KMnO_4 + 4KOH = 4K_2MnO_4 + 2H_2O + O_2.$$

Potassium manganate can be isolated as dark green crystals, soluble in alkaline aqueous solution, but hydrolyzed by water:

$$3K_2MNO_4 + 2H_2O = 2K\ MnO_4 + MnO_2 + 4KOH.$$

If heated to 500° the crystals lose oxygen and yield K_2MnO_3. Potassium manganate is isomorphous with potassium sulphate, and $Na_2MnO_4,10H_2O$ is isomorphous with $Na_2SO_4,10H_2O$. The ion $[MnO_4]^{2-}$ is tetrahedral and contains one unpaired electron.

Manganese (III) Compound

Manganic Compounds Consisting of Giant Molecules Have Some Ionic Character

It is doubtful whether the ion Mn3+ exists; manganic salts are unstable and in solution either disproportionate or readily undergo hydrolysis.

Manganic oxide, Mn_2O_3, a black powder, is made by igniting manganese dioxide or a manganous salt in air at 500-900°. If heated in air above 950°, or in hydrogen above 230°, it loses oxygen to yield Mn_3O_4. It dissolves in dilute hydrofluoric acid or in hydrocyanic acid. Manganic oxide reacts with sulphur dioxide to yield manganous sulphite, manganous dithionate, and some manganous sulphate. Manganic oxide is the typical example of the *C*-M_2O_3 crystal structure. The lattice may be regarded as a fluorite lattice from which a quarter of the anions have been removed, and the positions of the cations have been slightly adjusted. **Mn_3O_4,** obtained as above, has a distorted spinel structure $Mn^{II}Mn^{III}O_4$.

Manganic hydroxide, MnOOH, is the product of the combined oxidation and hydrolysis of a manganous salt. It is made by passing chlorine into a suspension of manganous carbonate in water :

$$3MnCO_3 + Cl_2 + H_2O = 2MnOOH + Mncl2 + 3CO_2.$$

When treated with nitric acid it disproportionates thus:

$$2MnOOH + 2HNO_3 + Mn(NO_3)_2 + MnO_2 + 2H_2O.$$

Manganic fluoride, MnF_3, a red crystalline solid, is made by the action of fluorine on manganous iodide. On heating it dissociates into fluorine and manganous fluoride. Fluorine diluted with nitrogen reacts at 100-150° with potassium permanganate, or with any oxide of manganese stable at that temperature, to give mainly manganic fluoride with a little manganous fluoride. An aqueous solution of manganic fluoride is made by dissolving manganic oxide in hydrofluoric acid. The red dihydrate, $MnF_3,2H_2O$, may be crystallized from the solution. Manganic fluoride is hydrolyzed if the solution is diluted. Crystalline manganic fluoride furnishes a typical example of the Jahn—Teller effect.

Manganic chloride, $MnCl_3$, stable only below –40°, is obtained in etheral solution by saturating a suspension of manganese dioxide in ether with gaseous hydrogen chloride at –70°.

Manganic sulphate, $Mn_2(SO_4)_3$, a dark green powder, stable to 300°, is made by heating precipitated manganese dioxide with concentrated sulphuric acid at 138°. The solid product is drained on a porous plate, washed with concentrated nitric acid, and heated to 130° to expel the nitric acid:

$$4MnO_2 + 6H_2S0_4 = 2Mn_2(SO_4)_3 + 6H_2O + O_2.$$

Manganic sulphate dissolves in water to give a violet solution from which a precipitate of MnOOH separates by hydrolysis.

23

d-Block Elements (Iron, Cobalt, Nickel)—Group VIII (8,9,10)

Element	Symbol	Atomic weight	weight number
Iron	Fe	55.85	26
Cobalt	Co	58.94	27
Nickel	Ni	58.69	28

Introduction

Most of the elements of the eighth group have two electrons in the outermost electron layer of their atoms, so that their properties are predominantly metallic. Besides the outer electrons, they are capable of yielding electrons from their incomplete second last layer, manifesting a valency of +3, +4. However valency exceeding 4 is manifested very rarely.

A comparison of the physical and chemical properties of the elements of the eighth group shows that iron, cobalt and nickel, situated in the first long period, resemble each other very closely but differ pronouncedly from the elements of the two other triads. For that reason they are usually united into a separate **iron subgroup**. The other six elements of the eighth group are united under the generic name of **platinum metals.**

History and Importance of Iron

Iron was known at least as early as 4000 B.C. and probably was used to some extent during prehistoric times. Because free iron is not commonly found in nature, that first used in prehistoric times may well have been of meteoric origin. The early discovery and application of iron to the manufacture of tools and weapons resulted from the wide distribution of iron and the ease with which the iron compounds in the ores could be reduced by carbon. Charcoal was the form of carbon used for a long time in the reduction process, but the production and use of iron on a large scale did not begin until about 1620, when coal was introduced as the reducing agent.

Iron is the most widely used of all the metals. This is because its ores are abundant and widely distributed, the metal is easily and cheaply produced from the ores, and its properties can be varied over a wide range by the addition of other substances and by different methods of treatment,, such as tempering and annealing. More iron is used than all the other metals combined (14 times as much).

Occurrence of Iron

Iron ranks second in abundance among the metals (aluminum is first) and fourth among all elements in the earth's crust. The central core of the earth is largely iron, as has been indicated by studies of the earth's density and the rate of transmission of earthquake shock. Metallic meteors are usually about 90 per cent iron, and the remainder is principally nickel. Practically all rocks, minerals, and soils, as well as plants, contain some iron. Iron is present in the hemoglobin of the blood, which acts as a carrier of oxygen.

The important iron-bearing ores are **hematite,** Fe_2O_3, and **magnetite**, Fe_2O_4. The presence of Fe_2O_3 accounts largely for the red colouration of many rocks and soils., Hematite often occurs as the reddish-brown hydrate $2Fe_2O_3.3H_2O$, which is called **limonite.** Magnetite is a black. crystalline mineral, which is strongly magnetic, sometimes possessing polarity, in which case it is called **lodestone.** Black crystalline magnetite interlocked with crystals of silica in the form of hard rock is called **taconite, Siderite** is a white or brownish carbonate, $FeCO_3$. **Pyrite,** FeS_2, occurs as pale yellow crystals with a metallic luster which are easily mistaken for gold; hence the name "food's gold." Pyrite is quite abundant and useful as a source of sulfur for the manufacture of sulfuric acid; but because roasting does not remove all of the sulfur, pyrite is usually considered unsatisfactory as a source of iron, although low-grade pig iron can be made from it.

With the rather rapid depletion of domestic high-grade iron ore (over 50 per cent iron) deposits, the steel industry has turned to the conquest of the taconite ores (25-30 per cent iron). Taconite is the extremely hard rock that makes up some 95 per cent by volume of famous Mesabi Range's vast iron-bearing deposits in upper Minnesota. Once regarded as useless for steel production, the taconite ores are now used extensively as a result of the development of satisfactory techniques for mining and concentrating them.

Extraction of Iron

The extraction of Iron is carried out in a blast furnace which can vary in size and can be between 25 and 60 metres in height and up to 10 metres in diameter. It is constructed form steel with the inner regions lined with firebricks. A charge of iron ore, limestone and coke in the correct proportions is fed into the top of the furnace through a cone and hopper arrangement. Preheated air at a temperature of about 600°C is injected into the furnace through a number of pipes called tuyeres (pronounced 'tweers'); the tuyeres are fed from a 'bustle' pipe encircling the blast furnace. The blast furnace is provided with two tap holes which plugged with clay; molten iron is tapped from the lower one and molten slag from the other. The production of iron is a continuous process and, depending upon its size, a blast furnace can produce from 1000 to 1800 tonnes of iron every twenty-four hours.

The energy and reducing agent required for the smelting of iron is obtained by the combustion of coke, the temperature of the charge increasing steadily as it falls through the ascending combustion gases :

$$2C + O_2 \rightarrow 2CO$$

At a temperature of about 700°C the iron ore is reduced to spongy iron by the carbon monoxide :

$$Fe_2O_3 + 3CO \rightarrow 2Fe + CO_2$$

Since this reaction is reversible at high CO/CO_2 pressure ratio favours the reduction to iron.

The limestone decomposes at about 800°C and the calcium oxide reacts with sandy impurities to form a slag of calcium silicate. More carbon monoxide is produced by the reduction of carbon dioxide :

$$CaCO_3 \rightarrow CaO + CO_2$$

$$CaO + \underset{\text{impurity}}{SiO_2} \rightarrow \underset{\text{slag}}{CaSiO_3}$$

$$C + CO_2 \rightarrow 2CO$$

instance silica is reduced to silicon and this enters the iron as ferrosilicon :

$$SiO_2 + 2C \rightarrow Si + 2CO$$

A similar reaction occurs with any phosphates present and some Fe_3P is retained by the iron. Other impurities present in the iron include iron (II) sulphide and manganese, which form an alloy.

The molten metal is either run out into moulds of sand, when it is known as a pig-iron, or more generally conveyed directly in the liquid form to steelmaking plants. The slag is tapped from the furnace as a liquid and can be used in concrete or blown into a 'woolly' material and used for insulation.

The hot gases emerging from the top of the furnace contain appreciable amounts of carbon monoxide and are burnt in Cowper stoves to preheat the air for the blast.

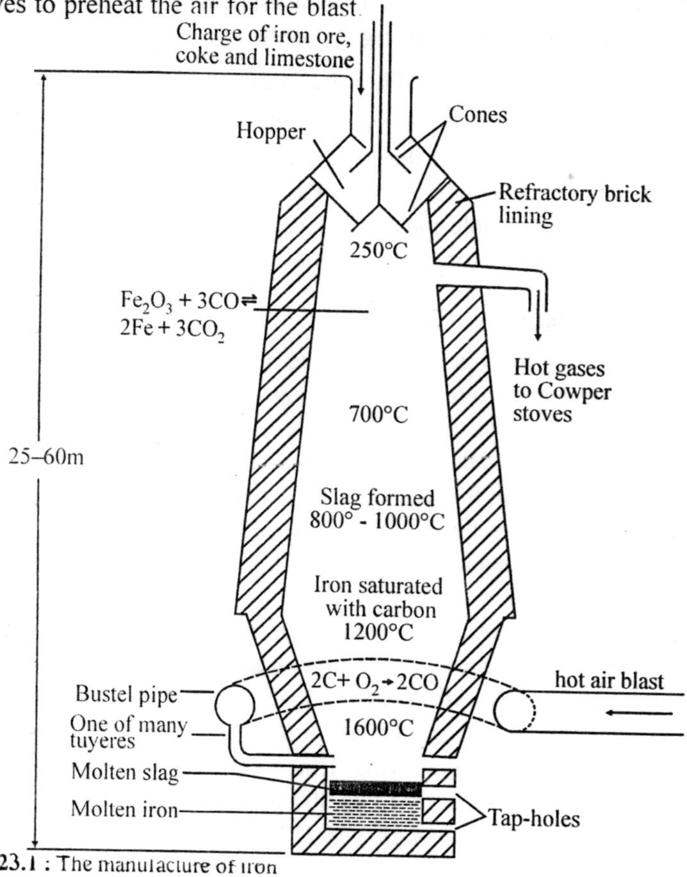

Fig. 23.1 : The manufacture of iron

Cast Iron

Iron castings are made by igniting a mixture of pig-iron, scrap iron and coke in cupola furnaces by a blast of hot air. The molten iron is poured into moulds to make articles such as manhole covers, guttering, machinery frames and drainpipes. Cast iron expands slightly on solidifying and therefore faithfully reproduces the shape of the mould. It is extremely hard, but unfortunately is very brittle and will fracture if struck by a sharp blow. The impurities in cast iron lower the melting point from 1528°C for pure iron to approximately 1200°C.

Wrought Iron

This is made by heating impure iron with haematite so that impurities are oxidized. Carbon is converted to carbon monoxide and silicon and manganese to a slag.

$$Fe_2O_3 + 3C \rightarrow 2Fe + 3CO$$

As the impurities are removed, the melting point of the iron rises to about 1500°C and the pasty mass is removed from the furnace as balls and worked under a hammer to squeeze out the slag. Wrought iron is tough, malleable and ductile; it can be worked by a blacksmith into chains, railway carriage couplings and ornamental gates, etc.

The Manufacture of Steel

The term **steel** is used to denote many widely different alloys of iron. It is made from pig iron by removing the impurities and later adding substances such as manganese, chromium, nickel, tungsten, molybdenum, and vanadium to produce alloys with properties which make them suitable for specific uses. Most steels also contain small but definite percentage of carbon (0.04-2.5%). Thus, a large part of the carbon contained in pig iron must be removed in the manufacture of steel.

Although there are many grades of steel, steels can be classified into three main categories : (a) **carbon steels,** which are primarily iron and carbon; (b) **stainless steel,** low-carbon steels containing over 12% chromium; and (c) **alloy steels,** specialty steels which contain large amounts of alloying elements to impart special properties for specific uses. Of the 141,000,000 tons of steel produced in the United States in a recent single year, 124,600,000 tons (88.4%) were carbon steels, 14,800,000 tons (10.5%) were stainless steels, and 1,60,000 tons (1.1%) were alloy steels. The 141,000,000 tons produced in the United States constituted 22.5% of the world production of 625,000,000 tons of steel.

Two processes are used chiefly in the production of steel; they are the **open-hearth process** and the **basic oxygen process,** the latter of which accounts for the bulk of the total production of steel in the United States. The open-hearth process was used for 90% of the steel produced in the United States immediately prior to the introduction of the basic oxygen process in this country in 1954. Steels for special purposes are made by the electric furnace method, which accounts for 14% of total steel production in this country.

The Bessemer process, formerly quite a significant factor in steel production, accounted for only about 5% of total production in 1950 and accounts for almost none at the present time.

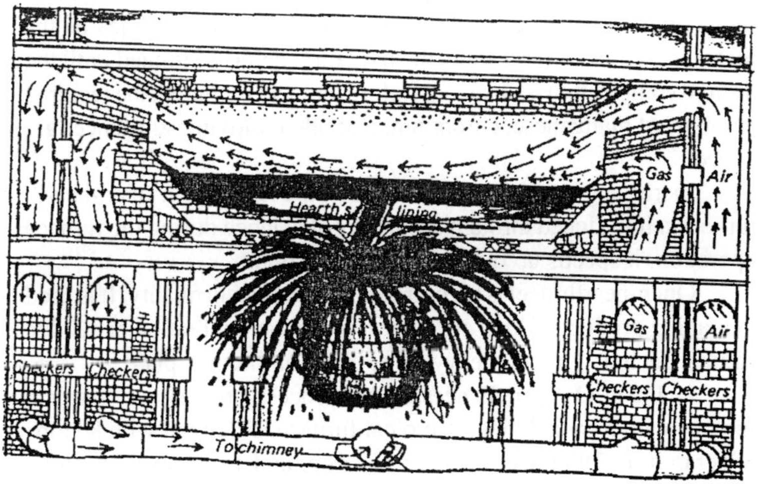

Fig. 23.2 : Diagram of an open-hearth furnace for making steel.

1. Open-Hearth Process. A diagram of the open-hearth furnace is given in Fig. 36-3. The charge, which typically is about two hundred tons, consists of pig iron, scrap iron, and iron ore. It is placed on a shallow hearth and heated from above by a gaseous fuel containing an excess of air. The mixture of fuel and air required for the oxidation of such impurities as carbon and sulfur in the charge is preheated by passage through a checkework of hot brick. The bricks are heated by the hot gases which are produced during the oxidation of the impurities in the charge. The directions of the entering gases and the gaseous products are reversed at intervals so that each chamber of bricks is alternately heated and used to preheat the incoming fuel and air mixture. The iron ore and the rust on the scrap iron, which are parts of the charge, aid in the oxidation of the impurities.

When the charge contains elements whose oxides are acidic (e.g., silicon, phosphorus, and sulfur) the lining of the hearth is made of basic materials, principally magnesium and calcium oxides. The acidic oxides react with the basic lining, forming a slag. Most of the open-hearth steel is made in basic-lined furnace. If the charge contain basic oxides an acid lining composed chiefly of sand or siliceous materials is employed.

About eight hours are required to oxidize most of the carbon in the charge of an open-hearth furnace. At intervals, samples of the metal are removed for testing, the oxidation being continued until the desired composition is obtained. Before the steel is poured, the required amounts of manganese, chromium, and other alloying substances are added.

Increasing use is being made of pure oxygen in open-hearth furnaces. Introduced through a water-cooled lance during the furnace operation, it speeds up oxidation of the carbon.

Open-hearth steel is high-grade and uniform in quality. It is used in making products which are subject to frequent vibration or sudden strains, such as girders, heavy rails, guns, and armor plate.

2. Basic Oxygen Process. This process utilizes a cylindrical furnace with a basic lining. A typical furnace is charged with 80 tons of scrap iron, 200 tons of molten pig iron, and 18 tons of limestone (to form slag), and operates under computer control. A jet of high-purity (99.5%) oxygen at 140-180 pounds pressure per square inch is directed vertically through a water-cooled lance, 50 feet long and 10 inches in diameter, to a point 4 to 8 feet below the surface of the white-hot, molten charge. The reaction produces a vigorous boiling action that reaches all parts of the charge. In the central reaction zone, temperatures reach a level close to the boiling point of iron. The entire steel-making cycle is completed in one hour to less—much faster than the average heat time for open-hearth furnaces. Electrostatic precipitators clean the gases resulting from the furnace reactions, making the furnaces virtually smokeless. The steel produced is of extremely high and uniform quality. The basic oxygen process, because of its speed, efficiency, and the high quality of its product, is rapidly increasing in importance relative to the open-hearth process, rising from 6% of total steel production in 1962 to well over half the total production in 1975.

3. Other Processes. Various types of electric furnaces are used in the United States of obtain high-grade steel. Three distinct advantages of the electric furnace are (a) the temperature and composition of the metal bath can be controlled with high precision, (b) high temperatures are easily attainable, making possible the reduction of special steels with high melting points, and (c) the furnace can accept an all-scrap charge (whereas basic oxygen furnaces are limited to a maximum of about 35% scrap, since excess scrap would have too great a cooling effect on the charge). However, the cost of power limits the use of the electric furnace. **Electric steel** is used as tool steel in the making of such articles as files, drills, knives, razor blades, watch springs, and axes.

A process under developments is the **direct reduction process,** which could eliminate the need for the blast furnace. In one form, this process utilizes a rotary kiln which takes a charge of iron ore, carbon, and flux. Combustible gases, heated by an electric furnace are used to heat the charge. The product goes from the kiln to an electric furnace for rapid reduction to pig iron or a low-grade steel.

The Structure of Steel

A complete discussion of the structure of steel would be very complex and the following is a short and somewhat oversimplified account.

Below about 910°C the structure of iron is body-centred cubic, i.e. each atom of iron is surrounded by eight nearest neighbours situated at the corners of a cube; above this temperature the structure of iron changes to that of a face-centred lattice, i.e. each atom of iron is surrounded by twelve nearest neighbours. The presence of carbon in liquid iron lowers the temperature at which one structure changes over to the other; furthermore, whereas the body-centred structure of iron can accommodate virtually no carbon in solid solution, the face-centred lattice can hold up to 1.7 per cent carbon in solid solution. When liquid iron containing about 0.3 per cent of carbon is cooled it crystallises in the face-centred arrangement but changes over into the body-centred structure at about 810°C. The carbon separates as a distinct phase in the form of cementite, Fe_3C and the final solid contains regions of iron and regions consisting of a sandwich structure of iron alternating with cementite, such a sandwich structure being known as pearlite. The two distinct regions have an important part of play in determining the properties of different steels. The regions of iron confer ductility and the regions of pearlite confer hardness and strength. In steel containing about 0.3 per cent of carbon the regions of iron predominate over the regions of pearlite and the result is a ductile steel; as the carbon content increases so too does the relative proportion of pearlite to iron with the result that a harder less ductile steel is produced.

Properties of Steel

The properties of steel can be altered by varying the amount of carbon added during manufacturing (Table 23.1). Mild steel contains between 0.1 and 0.4 per cent of carbon and is used for making such articles as motor car bodies, tinplate, nuts and bolts and piping. Hard steel containing between 0.5 and 1.5 per cent carbon, can be further hardened tempered by heat treatment and is used for tools. There is a vast range alloy steels; a few examples are given in the table below.

Table 23.1 : Some alloy steels

Name of Alloy	Percentage Composition (Approx.)	Use	Special Features
Stainless steel	73Fe. 18Cr. 89NI + C	Cutlery	Resists corrosion
Tungsten steel	94Fe.5W + C	High speed cutting tools	Hardness
Invar	64Fe.36Ni	Watches	Small coefficient of expansion
Manganese steel	86FE.13Mn + C	Rock drills	Toughness
Permalloy	78Ni.21Fe. + C	Electro-magnets	Strongly magnetised by electric current Loses magnetism when current in switched off

Hardening and tempering of steel

Hard steel can be further hardened by heating it to red heat (850°C) followed by plunging it into cold water. This process, known as 'quenching', makes the steel extremely hard but brittle. This phenomenon can be explained as follows : above 850°C the structure is face-centred cubic and the carbon is present in solid solution. Rapid cooling does not allow sufficient time for the carbon to separate as cementite, Fe_3C, and instead carbon is deposited; the presence of carbon atoms makes it impossible for the iron to adopt a true body-centred structure and a distorted modification which is hard but brittle is the result.

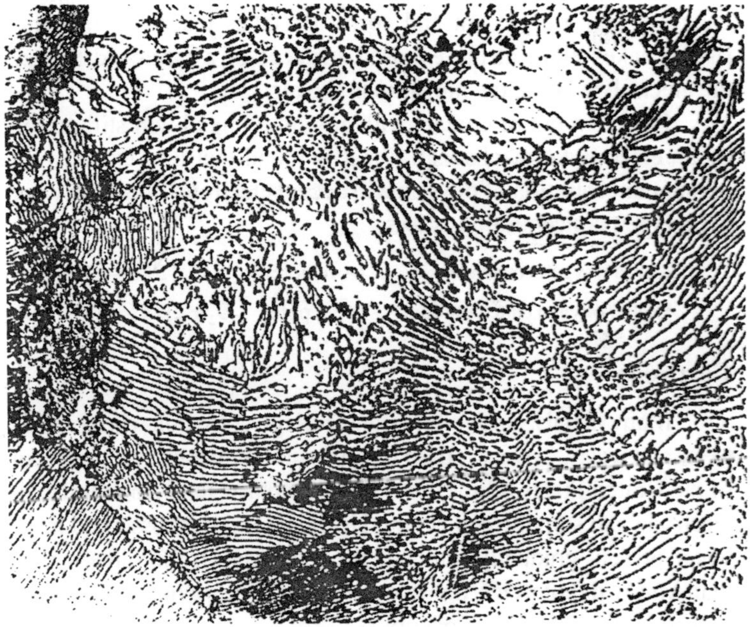

Fig. 23.3 : A photograph of an annealed specimen of 0.7% carbon steel showing pearlite structure under magnification

If the steel is now heated to a temperature between 230° and 300°C, and allowed to cool slowly, it retains most of its hardness but the brittleness disappears; this process is known as 'tempering' or 'annealing'. During tempering the surface of the steel takes on certain colours, being yellow at 230°C and blue at

300°C. These colours are due to the formation of a very thin oxide film on the surface of the steel, the thicken of the film determining the final colour. Watch springs blue, since tempering at 300°C removes all brittleness and produces an extremely tough and springy steel.

During 'tempering' microscopically small areas of cementite are formed and this helps to relieve the strain present in 'quenched steel' without significantly altering the hardness of the steel.

The Rusting of Iron and Steel

If an electrolyte is in contact with two metals which are themselves in contact, the three essential elements of an electrochemical cell are present and corrosion takes place. For instance, rusting takes place at the exposed surfaces of a scratched tinplate, the electrolyte being the moisture in the air.

Pure iron out of contact with any other metal will still corrode by an electrolytic mechanism, for it is found that iron exposed to a higher concentration of oxygen will accept electrons from iron where the oxygen concentration is less.

Lower oxygen concentration	(– ve pole)	$2Fe \rightarrow 2Fe^{2+} + 4e$
Higher oxygen concentration	(+ ve pole)	$O_2 + 2H_2O + 4e \rightarrow 4OH^-$

Corrosion

Many metals, particularly iron, undergo corrosion when exposed to air and water. Losses caused by corrosion of metals cost the people of the United States billions of dollars annually, tends to polarize the electrode and stop the cell action. However, oxygen dissolved in any water present gradually removes the hydrogen in the same manner that MnO_2 depolarizes the dry cell, and electrochemical reaction proceeds; i.e., corrosion occurs.

$$2H2 + O2 \rightarrow 2H_2O$$

Iron (II) ions combine with hydroxide ions of the electrolytic solution, forming $Fe(OH)_2$. Iron (II) hydroxide is then readily oxidized by air to $Fe_2O_3.xH_2O$, which is **iron rust.**

Many methods and devices have been employed to prevent or retard corrosion. Some of these methods are as follows:

Iron can be protected against corrosion by coating it with (a) an organic material such as paint, lacquer, grease, or asphalt; (b) another metal such as zinc, copper, nickel, chromium, tin, cadmium, or lead; (c) a ceramic enamel, like that used on sinks, bathtubs stoves, refrigerators, and washers; or (d) an adherent oxide, which may be formed by exposing iron to superheated steam, thereby giving it an adherent coating of magnetic oxide, Fe_3O_4.

A piece of iron can also be protected by treating with a compound that provides a suitable coordinating group, such as phosphate, which ties the surface iron up in a complex such that it cannot react to form $Fe_2O_3.xH_2O$. Alloys of iron with certain other elements are often corrosion resistant. Typical examples are **stainless steel** (Fe, Cr, and Ni) and **duriron** (Fe and Si).

Another method of preventing the corrosion of iron or steel involves an application of electrochemistry called **cathodic protection.** For example, corrosion of iron or steel water tanks can be retarded by suspending several stainless steel anodes in the tank; the tank serves as cathode. A small current is passed continuously through the system, with the natural salts in the water making it conducting. A slight cathodic evolution of a protective coat of hydrogen takes place on the wall of the tank. Cathodic protection against corrosion is also used on iron and steel which is in contact with soil, such as underground pipe lines. Other applications of cathodic protection do not involve the use of a current from an external source. In these cases, the iron becomes cathodic when connected by a wire to a more active metal such as zinc, aluminum, or magnesium. The difference in activity of the two metals causes a current to flow between them, producing corrosion on the more active metal and furnishing cathodic protection to the iron.

Properties of Iron

Pure iron is silvery-white, capable of taking a high polish, ductile, relatively soft, and high in tensile strength. These properties are greatly modified by the amount and nature of the impurities present, especially carbon. Pure iron is attracted by a magnet but does not retain magnetism. Iron dissolves in hydrochloric acid and dilute sulfuric acid, forming hydrogen and iron (II) ions, Fe^{2+}. Hot concentrated sulfuric acid forms Fe^{3+}, and SO_2 is evolved. Cold concentrated nitric acid induces **passivity** in iron. When in the passive condition iron does.

Table 23.2 : Alloy Steels

Name	Composition	Characteristic Properties	Uses
Manganese	10-18% Mn	Hard, tough, resistant to wear	Railroad rails, safes, armor plate, rock crushing machinery
Silicon	1-5% Si	Hard, strong, highly magnetic	Magnets
Duriron	12-15% Si	Resistant to corrosion, acids	Pipes, kettles, condensers, etc.
Invar	36% Ni	Low coefficient of expansion	Meter scales, measuring tapes, pendulum rods
Chrome-Vanadium	1-10% Cr 0-15% V	Strong, resistant to strains	Axles
Stainless steel	14-18% Cr 7-9% Ni	Resistant to corrosion	Cutler, instruments
Permalloy	78% Ni	High magnetic susceptibility	Ocean cables
High-speed steels	14-20% W or 6-12% Mo	Retain temper at high temperatures	High-speed cutting tools
Nickel	2-4% Ni	Hard and elastic, resistant to corrosion	Drive shafts, gears, cables

It has been shown that iron will not rust in dry air, nor in water which is free from dissolved oxygen. It follows then that both air and water are involved in the corrosion process. The presence of an electrolyte in the water accelerates corrosion, particularly when solution is acidic. Strained metals corrode more rapidly than unstrained ones. Heated portions of a metal corrode more rapidly than unheated ones. Finally, iron in contact with a less active metal such as tin, lead, or copper corrodes more rapidly than when either alone or in contact with a more active metal such as zinc.

An electrochemical theory helps to explain why the presence of a less active metal than iron accelerates the corrosion of iron and the presence of a more active metal slows down the corrosion. It appears that very small electrochemical cells are set up when corrosion takes place. When iron is in contact with water containing an electrolyte, the half-reaction given below tends to occur.

$$\text{Anodic oxidation : } Fe \rightarrow Fe^{2+} + 3e^-$$

Sometimes, one part of a piece of iron is more active than the rest and tends to go into solution more easily. This means that the more active part has a higher potential than the rest of the iron piece, and this part tends to dissolve in the electrolyte. On the less active parts of the iron (lower electrode potentia!), hydrogen tends to form according to the half-reaction

$$\text{Cathodic reduction}: 2H^+ 2e^- \rightarrow H_2$$

The accumulation of hydrogen on the surface of the less active part of the iron not react with dilute acids which otherwise would readily dissolve it. The passivity is easily destroyed by a scratch or by shock. Hot dilute nitric acid oxidizes iron to Fe^{3+} with the evolution of nitric oxide. When heated to redness, iron reduces steam to produce hydrogen and iron (II, III), oxide, Fe_3O_4, which is the same oxide that is produced when iron burns in oxygen. When exposed to air at ordinary temperatures in iron becomes oxidized, with a lose coating of partially hydrated iron (III) oxide, $Fe_2O_3.xH_2O$.

Iron forms two principal series of compounds in which the oxidation states are + 2 and + 3; compounds in which iron is + 6 are rare and less important. The + 2 and + 3 series are called iron (II), or ferrous, and iron (III), or ferric, respectively. The oxides and hydroxides of iron (II) and (III) are basic character with little or no acid-forming property.

Compounds of Iron

Iron shows oxidation states of + 3 (the most stable) and + 2 (reducing). There is also an unstable oxidation state of + 6 which is powerfully oxidizing. The presence of five singly occupied 3*d* orbitals in the Fe^{3+} ion and only four singly occupied 3*d* orbitals in the Fe^{2+} ion is often cited as an explanation of the easy oxidation of Fe^{2-} to Fe^{3+}.

Iron (VI) compounds

The ferrate (VI) ion, FeO_4^{2-}, contains iron in the + 6 oxidation state, and potassium ferrate (VI), K_2FeO_4, can be obtained by fusing a mixture of iron filings and potassium nitrate (oxidizing agent); when the cooled mass is poured into water a dark red solution is obtained, the colour being due to the FeO_4^{2-} ion. Barium ferrate (VI). $BaFeO_4$, which is a purple solid. can be precipitated by adding barium chloride solution.

The ferrate (VI) ion is tetrahedral and a very powerful oxidizing agent, e.g. it will oxidise chromium (III) ions to chromate ions; corresponding ferrates (VI) and sulphates are isomorphous.

Neither the oxide, FeO_3, nor the parent acid of the ferrates (VI), H_2FeO_4, is known.

Iron (III) compounds

Iron (III) oxide, Fe_2O_3

Iron (III) oxide occurs naturally as haematite; it can be prepared as a rust-coloured solid by heating iron (III) hydroxide or iron (II) sulphate :

$$2FeSO_4 \rightarrow Fe_2O_3 + SO_2 + SO_3$$

Note that iron (II) sulphate is oxidized to iron (III) oxide in the above reaction.

Naturally occurring iron (III) oxide has a similar structure to that of aluminium oxide, i.e. it is ionic.

Iron (III) hydroxide, $Fe_2O_3.xH_2O$

Iron (III) hydroxide, which is best represented as iron (III) oxide associated with an indefinite number of molecules of water, is precipitated as a rust-coloured gelatinous solid when aqueous solutions containing hydroxyl and iron (III) ion are mixed. It reacts readily with dilute acids to give iron (III) salts but since it also reacts with concentrated solutions of alkalis to give ferrates (III), e.g. $NaFeO_2$, it must be regarded as amphoteric.

Colloidal iron (III) hydroxide is obtained as a rust-coloured 'solution' when a few drops of iron (III) chloride solution are added to a large volume of boiling water.

Tri-iron tetroxide (magnetic oxide of iron), Fe_3O_4

This oxide occurs naturally as the ore magnetite which is magnetic. It is a black solid and is formed when steam is passed over heated iron. When reacted with acids it gives a mixture of iron (II) and iron (III) salts (ratio 1 : 2) and this is consistent with a structure containing Fe^{2-} and Fe^{3+} ions (ratio 1 : 2) with enough oxide ions to maintain electrical neutrality.

Uses of iron (III) oxides

Haematite, Fe_2O_3, and magnetic, Fe_3O_4, are major sources of iron and steel. In addition, iron (III) oxide is used as an abrasive (jeweller's rouge) and as a pigment (yellow, red and purple forms can be obtained by a variety of methods).

Iron (III) halides

Iron (III) fluoride, chloride and bromide can be made by heating iron in the presence of the halogen; pure iron (III) iodide cannot be isolated.

Iron (III) chloride, $FeCl_3$, and iron bromide, $FeBr_3$, dark red covalent solids which in the vapour state exist as dimerised molecules Fe_2Cl_6 and Fe_2Br_6 respectively; these structure are similar to that of aluminium chloride. Both compounds dissociate on heating to give the iron (II) halide and the halogen and both dissolve very readily in water to give the yellow $[Fe(H_2O)_6]^{3+}$ ion and the halide ion, cf. aluminum chloride; crystallization from solution gives the hexahydrate, e.g. $FeCl_3.6H_2O$.

Anhydrous iron (III) chloride like aluminium chloride is used as a Friedel-Crafts catalyst and this action depends upon the formation of the $FeCl_4^-$ complex ion.

It is not really surprising that iron (III) iodide cannot be obtained in the pure state, since a solution containing iron (III) readily oxidizes iodide ions to iodine.

$$2Fe^{3+} + 2I^- \rightarrow 2Fe^{2-} + I_2$$

The instability of FeI_3 with respect to FeI_2 and iodine means that is not a sufficiently powerful oxidizing agent to convert Fe^{II} to Fe^{III}; this behaviour is very much in line with the more extensive dissociation of iron (III) bromide into iron (II) bromide and bromine on heating as compared with the corresponding chloride.

Iron (III) sulphate, $Fe_2(SO_4)_3$

Iron (III) sulphate can be obtained by heating an aqueous acidified solution of iron (II) sulphate with hydrogen peroxide (oxidizing agent) :

$$2FeSO_4 + H_2SO_4 + H_2O_2 \rightarrow Fe_2(SO_4)_3 + 2H_2O$$

On crystallizing the solution, the hydrate $Fe_2(SO_4)_3.9H_2O$ separates out

Iron (III) sulphate forms alums, the best known being ammonium iron (III) sulphate, $NH_4 + Fe^{3+}(SO_4^{2-})_2.12H_2O$.

Qualitative tests for iron (III) ions

When aqueous solutions containing iron (III) ions and hydroxyl ions are mixed a rust-coloured gelatinous precipitate of iron (III) hydroxide is formed.

The addition of potassium hexacyanoferrate (II) solution to an aqueous solution containing iron (III) ions produces a dark blue precipitate (Prussian blue) :

$$\underset{\text{hexacyanoferrate (II)}}{K^+ + [Fe^{II}(CN)_6]^{4-}} + Fe^{3+} \rightarrow \underset{\text{Prussian blue}}{K^IFe^{III}[Fe^{II}(CN)_6]^{4-}}$$

The addition of a solution of potassium hexacyanoferrate (III). $(K+)_3[Fe^{III}(CN)_6]^{3-}$, to a solution containing iron (III) ions produces a brown or green colouration.

The addition of potassium thiocyanate solution to a solution containing iron (III) ions produces an intense red colouration, the colour being due to the $[Fe(CNS)]^{2+}$ ion :

$$Fe3+ \underset{\text{thiocyanate}}{CNS-} \rightarrow [Fe(CNS)]^{2+}$$

This is extremely sensitive test for iron (III) ions and the presence of trace amounts of these ions in an iron (II) solution produces a red colouration. The test must therefore be performed under controlled conditions, i.e. equal volumes of potassium thiocyanate solution are added to equal volumes of solutions containing iron (II) ions and suspected iron (III) ions respectively; a much more intense colouration is produced with the iron (III) solution.

The qualitative tests for iron (II) ions are given on p. 382.

Iron (II) compounds

Iron (II) oxide, FeO

Iron (II) oxide can be obtained as a black solid by heating iron (II) oxalate (the carbon monoxide provides a reducing atmosphere and prevents aerial oxidation to iron (III) oxide):

$$FeC_2O_4 \rightarrow FeO + CO + CO_2$$

The oxide as made by the above method is pyrophoric, becoming incandescent when sprinkled into the air. It has the sodium chloride structure, indicating that it is ionic, but the crystal lattice is somewhat deficient in iron (II) ions, and it is non-stoichiometric. If heated in an inert atmosphere to a high temperature and allowed to cool slowly it tends to disproportionate into iron and magnetic oxide of iron :

$$4FeO \rightarrow Fe + Fe_3O_4$$

Iron (II) hydroxide, $Fe(OH)_2$

Iron (II) hydroxide when pure is a white solid, but it is normally obtained as a green gelatinous precipitate by adding sodium hydroxide solution to an aqueous solution containing iron (II) ions :

$$Fe^{2+} + 2\,OH^- \rightarrow Fe(OH)_2$$

The green colour is due to the presence of some iron (III) hydroxide produced either by aerial oxidation or by there being some iron (III) ions in the iron (II) solution. Prolonged standing in air results in complete conversion into iron (III) hydroxide, essentially hydrated iron (III) oxide.

Iron (II) halides

Anhydrous iron (II) chloride, $FeCl_2$, can be made by passing a stream of dry hydrogen chloride over the heated metal, the anhydrous fluoride is made in a similar manner using a stream of dry hydrogen fluoride. Note that the action of either chlorine or fluorine on the heated metal produces the iron (III) halide.

Both these halides dissolve in water from which the hydrates $FeF_2.8H_2O$ and $FeCl_2.6H_2O$ can be crystallized; the hydrated salts can also be made by the action of dilute hydrofluoric and dilute hydrochloric acid on the metal.

Iron (II) fluoride and iron (II) chloride, whether anhydrous or hydrated, are predominantly ionic solids.

Iron (II) bromide is obtained by heating iron in the presence of bromine vapour; an excess of metal is used to prevent the formation of any iron (III) bromide. Iron (II) iodide is made by a similar method from iron and iodine but, since iron (III) iodide does not exist, it is not necessary in this case to use an excess of the metal.

Iron (II) sulphate, $FeSO_4$

The action of dilute sulphuric acid on iron produces a solution of iron (II) sulphate and hydrogen is liberated. When the solution is crystallized the green heptahydrate, $FeSO_4.7H_2O$, is deposited.

Since iron (II) salts readily undergo aerial oxidation to the iron (III) state, the preparation is best carried out by maintaining a reducing atmosphere above the solution.

Careful heating of the heptahydrate produces the white anhydrous salt, but further heating produces iron (III) oxide (not iron (II) oxide since this is readily oxidized), sulphur dioxide and sulphur trioxide :

$$FeSO_4.7H_2O \rightarrow FeSO_4 + 7H_2O$$
$$2FeSO_4 \rightarrow Fe_2O_3 + SO_2 + SO_3$$

A solution of iron (II) sulphate contains the pale green $[Fe(H_2O)_6]^{2+}$ ions; in the presence of nitric oxide one molecule of water is displaced from the complex and replaced by NO to give :

$$[Fe(H_2O)_5(NO)]^{2+}$$

This complex ion is dark brown, and the above reaction is the basis of the 'brown ring' test for ionic nitrates.

Iron (II) sulphate is one of the cheapest industrial chemicals and is used to make iron (III) oxide (used as a pigment), Prussian blue (a pigment) and inks.

Iron (II) ammonium sulphate, $FeSO_4(NH_4)_2SO_4.6H_2O$

This double salt is obtained by crystallizing a solution containing equivalent amounts of iron (II) sulphate and ammonium sulphate. Unlike hydrated iron (II) sulphate it neither effloresces nor oxidize exposure to the atmosphere. It is used as a primary volumetric standard, especially for standardizing potassium permanganate solutions.

Iron (II) oxalate, FeC_2O_4

This is a lemon-coloured solid and it is of significance that its aqueous solution is also yellow. Since $[Fe(H_2O)_6]^{2+}$ ions are pale green, the yellow colour must be due to the presence of an iron (II) oxalate complex which is stable in aqueous solution. The following equilibrium is suggested :

$$[Fe(H_2O)_6]^{2-} + 2C_2O_4^{2-} \rightarrow [Fe(C_2O_4)_2(H_2O)_2]^{2-} + 4H_2O$$

pale green yellow

(Equilibrium displaced largely to the right)

Notice that the oxalate ion is bidentate, i.e. each oxalate ion occupies two of the octahedral positions in the complex. Iron (II) oxalate is more correctly formulated as $Fe_{II}[Fe_{II}(C_2O_4)_2]$.

Iron (II) sulphide, FeS

This solid is obtained by heating a mixture of iron filings and sulphur :

$$Fe + S \rightarrow FeS$$

Samples of iron (II) sulphide having slightly variable compositions, but all deficient in iron, have been reported, i.e. the sulphide is non-stoichiometric. The formula of one specimen has been reported as $Fe_{0.858}S$.

The action of dilute acids on iron (II) sulphide produces hydrogen sulphide, and this is a convenient way of making the gas in the laboratory :

$$FeS + 2H^+ \rightarrow Fe^{2+} + H_2S$$

Iron (II) carbonate, $FeCO_3$

Iron (II) carbonate occurs naturally as the ore siderite; it precipitates as a white solid when solutions containing iron (II) ions and carbonate ions are mixed :

$$Fe^{2+} + CO_3^{2-} \rightarrow FeCO_3$$

Since it oxidises on standing in the air, eventually forming iron (III) hydroxide (essentially hydrated iron (III) oxide), it must be obtained using solutions made up from 'boiled out' distilled water; air must also be excluded.

It is of significance that iron (III) carbonate does not exist and this is because hydrate iron (III) ions are more acidic than hydrated iron (II) ions (the smaller iron (III) ions polarise co-ordination water molecules to a greater extent than the larger iron (II) ions which are also less heavily charged). A strong base such as CO_3^{2-} abstracts protons very readily from hydrated iron (III) ions, eventually forming (III) hydroxide (hydrated iron (III) oxide) :

$$[Fe(H_2O)_6]3+ \; 3CO_3^{2-} \rightarrow Fe(H2O)3(OH)3 + 3HCO_3^-$$

(equilibrium displaced largely to the right)

$$2Fe(H_2O)_3(OH)_3 \rightarrow Fe_2O_3 + 9H_2O$$

hydrated iron (III) oxide

Aluminium carbonate does not exist for a similar reason.

Qualitative tests for iron (II) ions

An aqueous solution containing iron (II) ions produces a green gelatinous precipitate of iron (II) hydroxide when a solution of an alkali is added :

$$Fe^{2+} + 2OH^- \rightarrow Fe(OH)_2$$

The addition to potassium hexacyanoferrate (II) solution to an aqueous solution containing iron (II) produces a white precipitate which rapidly turns blue.

The addition of potassium hexacyanoferrate (III) solution to an aqueous solution containing iron (II) iron produces a dark blue precipitate (Turn-bull's blue). Turnbull's has been shown to be identical with Prussian blue, and it is likely that iron (II) ions first reduce the hexacyanoferrate (III) to hexacyanoferrate (II) :

$$Fe^{2+} + [Fe^{III}(CN)_6]^{3-} \rightarrow Fe^{3-} + [Fe^{II}(CN)_6]^{4-}$$

followed by :

$$K^+ + [Fe^{II}(CN)_6]^{4-} + Fe^{3+} \rightarrow K^IFe^{III}[Fe^{II}(CN)_6]^{4-}$$

Turnbull's blue or Prussian blue

The addition of potassium thiocyanate solution to a solution containing iron (II) ions does not produce any colouration if iron (III) ions are completely absent. However, in view of the ready oxidation of Fe^{2+} to Fe^{3+} and the extreme sensitivity of the thiocyanate test for iron (III) ions, iron (II) solutions often give some colouration with the reagent.

Cobalt

Occurrence and Metallurgy of Cobalt

The common cobalt minerals are **cobaltite,** CoAsS, and **smaltite,** $CoAs_2$, with the richest deposits of these minerals being found in Ontario, Canada. However nearly all the world's production

+of cobalt is obtained as a by-product of the metallurgy of nickel, copper, iron, silver, and other metals. The metallurgy of cobalt involves its separation from nickel, copper, and iron; conversion to Co_3O_4; and the reduction of the oxide with aluminum or hydrogen.

$$3Co_3O_4(s) + 8Al(s) \rightarrow 9Co(s) + 4Al_2O_3(s) + 963{,}000 \text{ Cal}$$
$$(\Delta H^0_{2987} = -963{,}000 \text{ cal})$$

(Notice the very large amount of energy evolved here. Very few reactions involve this much energy.) Purification of the metal is affected by electrolytic deposition on a rotating, stainless steel cathode.

Properties and Uses of Cobalt

Cobalt is similar to iron in appearance, except that is has a faint tinge of pink. Like iron and nickel, it is magnetic. It is slowly soluble in warm dilute hydrochloric or sulfuric acid and more rapidly soluble in dilute nitric acid. Like iron and nickel, cobalt is rendered passive by contact with concentrated nitric acid. It is not oxidized on exposure to the air, but at red heat cobalt reduces hydrogen in steam, with evolution of the hydrogen. The halogens, except fluorine, convert it to cobalt (II) halides. When cobalt is heated with fluorine, **cobalt (III) fluoride,** CoF_3, is formed.

Cobalt is alloyed with iron and small percentage of other metals in making high-speed cutting tools and surgical instruments. Permanent magnets are made from the alloys **Alnico** (Al, Ni, Co, Fe), **Hiperco** (Co, Fe, Cr), and **Vicalloy** (Co, Fe, V).

Finely divided cobalt metal is used as a catalyst in the hydrogenation of carbon monoxide and carbon dioxide, with the formation of hydrocarbons, and for the oxidation of ammonia.

The Oxides of Cobalt

The oxides, of cobalt, like the oxides of iron, are solids, and their compositions depend on the presence of Co^{2+} and Co^{3+} ions in the holes between the close-packed oxygen ions. It is therefore convenient to discuss them together. Cobalt forms no oxo-ion.

Cobaltous oxide, CoO, is a pale green solid, m.p. 1935°. It has the sodium chloride lattice, the oxygen ions are close-packed and cobaltous ions are in all the octahedral holes. It is made by heating cobaltous nitrate or carbonate in vacuo to 1000°, or by heating cobaltous hydroxide. It is predominantly a basic oxide but it dissolves in concentrated sodium hydroxide solution to give a deep blue solution. This solution is readily oxidized by air to give a black precipitate of hydrated cobaltic oxide. Cobaltous oxide absorbs oxygen, and the product on heating yields the oxide Co_3O_4 with the spinel structure.

Co_3O_4 is a black solid which is made by the ignition of cobaltous nitrate below 1000°, by heating Co_2O_3, H_2O, or by heating cobaltous oxide in air. It is isomorphous with Fe_3O_4, and it must therefore have a corresponding distribution of Co^{3+} and Co^{2+} ions in the tetrahedral and octahedral holes in the lattice. At 900° Co_3O_4 loses oxygen and yields cobaltous oxide. It absorbs oxygen up to the composition Co_2O_3 without change in the spinel lattice. The cobaltic oxide thus produced must therefore correspond to γ-Fe_2O_3.

The Hydroxides of Cobalt

Cobaltous hydroxide, $Co(OH)_2$, exists in a blue form and in a pink. The blue form is precipitated by adding an alkali to a solution of a cobaltous salt at 0°; the pink by adding a solution of a cobaltous salt to an alkali. The blue form changes to the pink form on warming. The colours are not explained; they may be due to difference in particle size. Cobaltous hydroxide has the CdI_2 layer lattice. It is fairly readily oxidized by air to hydrated cobaltic oxide.

Cobaltic hydroxide. The composition of cobaltic hydroxide is Co_2O_3, $2H_2O$. It can be made by precipitating a cobaltic salt with alkali, or by oxidizing cobaltous oxide or Co_3O_4 in the presence of moisture.

Sulphides of Cobalt

Cobaltous sulphide, CoS, is a red solid, m.p. above 1100°. If hydrogen sulphide is passed into a solution of cobaltous chloride containing acetic acid a black crystalline precipitate of cobaltous sulphide is obtained which is soluble in cold dilute hydrochloric acid only with difficulty. If sodium sulphide or ammonium sulphide is used as the precipitant (air being excluded) the precipitate is amorphous and is easily soluble in cold dilute hydrochloric acid with the evolution of hydrogen sulphide. If air is present the precipitate contains oxygen. Cobaltous sulphide has the NiAs structure.

There are other sulphides of cobalt CoS_2, Co_3S_8, and Co_2S_3. The first three are obtained together when cobalt is heated with sulphur. CoS_2 has the pyrites structure, and hence contains the ion Co^{2+}. The structure of Co_3S_4 and Co_9S_8 are in some respects similar to the spinels; the sulphur atoms are close-packed and the cobalt atoms are in the lattice holes.

Co_2S_3 is precipitated if hydrogen sulphide is passed into a solution of a cobaltic ammine, but it is not stable, and on warming it changes to a mixture of the cobaltous sulphides CoS and CoS_2.

(i) **Cobaltous chloride, $CoCl_2.6H_2O$.** When cobalt powder comes in contact with chlorine gas, it takes fire, forming the blue anhydrous chloride. Its solution in water is violet if concentrated, but pink when dilute. It is also soluble in alcohol forming a blue solution.

Its crystals may be obtained by dissolving the oxide or carbonate in hydrochloric acid. Like ferrous chloride, it combines with ammonia to form a rosy compound, $CoCl_2.6NH_3$.

Dilute cobaltous chloride writes invisibly on paper, but the letters become distinctly blue on warming in air. It is due to the loss of water. On leaving in air, the colour again disappears. For this property, it is used as a *sympathetic ink.*

Cobalt chloride solution also behaves peculiarly when heated. The rose colour at 30° may change to blue at 50°. It has been suggested by Basset that the rosy solution contains hydrated cobalt ions, as $[Co(H_2O)_6[_{2+}$, $[Co(H_2O)_{10}]^{2+}$ etc. and the blue solution contains $CoCl_4^-$ or $[CoCl_3(H_2O)^-$. These cations and anions are formed in the following way:

$$CoCl_2 \rightleftharpoons Co^{2+} + 2Cl^-$$
$$Co^{2+}\ 6H_2O \rightleftharpoons [Co(H_2O)_6]^{2+}$$

This hydrated cation breaks up at a higher temperature and combines with the chloride anion:

$$Co^{2+} + 4Cl^- \rightarrow CoCl_4^{2-}$$

or $$2CoCl_2 \rightleftharpoons Co.CoCl_4 \rightleftharpoons Co^{2+} + CoCl_4^{2-}$$

and $$Co^{2+} + H_2O + 3Cl^- \longrightarrow \left[Co \begin{matrix} H_2O \\ Cl_3 \end{matrix} \right]$$

Similar changes in colour occur for identical reason, on the addition of hydrochloric acid or alcohol to a cobalt chloride solution.

(ii) **Cobalt nitrate, Co(NO3)2.6H2O.** It is obtained by dissolving the metal, oxide or carbonate in nitric acid. The anhydrous salt is made by the action of nitrogen pentoxide on the hydrated salt. The hexahydrate forms pink slightly deliquescent crystals, highly soluble in water. On heating it decomposes to give Co_3O_4. It is used as a reagent in blow-pipe analysis and in the preparation of cobalt pigments and cereamics.

(iii) **Cobalt sulphate, $CoSO_4.7H_2O$.** It is isomorphous with green and white vitriols and is formed when cobalt oxide or carbonate is treated with dilute sulphuric acid. It gives red crystals. When strongly heated, the crystals become first anhydrous and finally yield Co_3O_4. It readily forms red double salts with alkali metal sulphates, $M_2 + SO_4.CoSO_4.6H_2O$ where $M^+ = K^+, Rb^+, Cs^+, NK_4^+, TI^+$. They are all monoclinic and isomorphous with one another and also with corresponding compounds with $Fe^{2+}, Ni^{2+}, Zn^{2+}, Mg^{2+}$ etc. The double sulphates, $(NH_4)_2SO_4.CoSO_4.6H_2O$ is used in cobalt plating.

(iv) **Cobalt silicate or smalt, $K_2O.CoO.3SiO_2$.** It is one of the most important compounds of cobalt and is obtained by fusing cobalt oxide with sand and sodium carbonate, thus furnishing deep blue glassy mass. It is ground and is used as a blue pigment. Blue colours imparted to porcelain and glass are due to the formation of cobalt silicate, $CoSiO_3$.

(v) **Thenards blue, $CoO.Al_2O_3$.** The blue pigment is most likely cobalt aluminate and is obtained by mixing cobalt phosphate or some other cobalt compound with freshly precipitated aluminium hydroxide; the mass is then heated to redness and is then ground with water.

(vi) **Cobalt zincate or Rinman's Green, ZnO.CoO or $CoZnO_2$.** When zinc oxide is heated with cobalt nitrate, **a** green-coloured product is obtained which may be regarded as a solid solution of cobalt zincate, $CoZnO_2$, in zinc oxide. This solid solution is sold under the name of *Rinman's Green,* to be used as a pigment.

Cobalt Salt

Cobaltic salts can often be prepared by electrolyzing a solution of a cobaltous salt dissolved in a solution of the parent acid of the salt. Cobaltic fluoride hydrate, cobaltic acetate, cobaltic carbonate, cobaltic oxalate, and cobaltic sulphate can all be made by this process, or by slight modifications of it. All cobaltic salts are strong oxidizing agents, as they tend to revert to the corresponding cobaltous salts; this tendency is so great that in the absence of complexing agents the Co^{3+} ion liberates oxygen from water.

Cobaltic sulphate exists as blue crystals of the hydrate $Co_2(SO_4)3$, $18H_2O$. It is stable if dry, but if moist it changes to cobaltous sulphate with the liberation of oxygen. The cobaltic ion plays the part of a trivalent ion in the alums. Potassium cobaltic alum is diamagnetic.

Cobaltic fluoride, CoF_3. The only cobaltic halide which has certain existence is cobaltic fluoride, which is known in the anhydrous state, and as the hydrate $CoF_3,3\frac{1}{2}H_2O$. Anhydrous cobaltic fluoride, made by heating Co or $CoCl_2$ in fluorine at 350°, is a pale brown crystalline solid. When heated it begins to decompose at about 250° if heated in a stream of carbon dioxide at 350° it is completely decomposed to cobaltous fluoride. It liberates oxygen from water and chlorine from hydrochloric acid, and it fluorinates many elements such as silicon, phosphorus, bromine, iodine, and sulphur. It is insoluble in alcohol, ether, and benzene. Its crystals possess the ReO_3 structure, in which they resemble FeF_3 and AlF_3. The hydrate $CoF_3,3\frac{1}{2}H_2O$ can be made by electrolyzing a saturated solution of cobaltous fluoride in hydrofluoric acid; it is precipitated as a green powder.

Cobaltic Complexes

The range of complexes, neutral, anionic, and cationic formed by cobalt in the cobaltic state is one of the widest formed by any element. The range includes numerous cationic complexes, many of which are polynuclear and contain two, three, or four cobalt atoms in the cation, as for example in the sulphate:

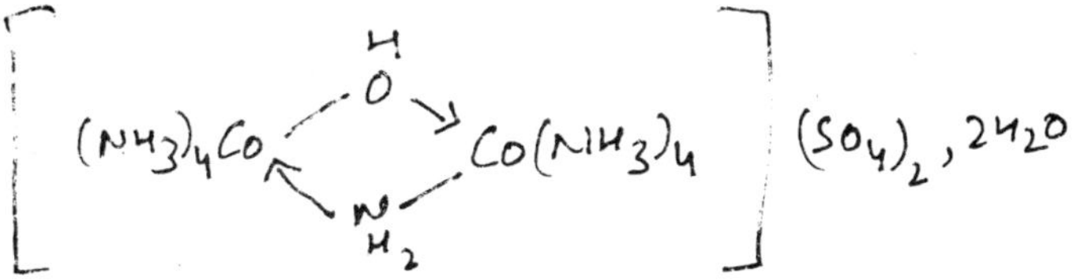

Cobalt does not appear to form polynuclear anionic complexes. The co-ordination number of the cobalt atom is always 6.

Nickel

Occurrence and Extraction of nickel

Ores of nickel are frequently associated with those of cobalt, typical ones being millerite, NiS, pentlandite (a mixed sulphide of iron and nïckel), and garnierite, $(Ni,Mg)SiO_3.xH_2O$.

The extraction of nickel involves a number of rather complex stages which ultimately result in the conversion of the ores into nickel (II) oxide, NiO. The oxide is then reduced to nickel by heating with carbon:

$$NiO + C \longrightarrow Ni + CO$$

Nickel of purity 99.99 per cent can be obtained from nickel (II) oxide by reduction with synthesis gas (a mixture of carbon monoxide and hydrogen) at a temperature of about 350°C. The impure nickel which is formed is then heated to about 60°C in a stream of carbon monoxide; volatile nickel carbonyl, $Ni(CO)_4$, distills leaving impurities behind and is decomposed to give very pure nickel by heating to about 200°C.

$$NiO + CO \rightarrow \underset{\text{impure}}{Ni} + CO2$$

$$\underset{\text{impure}}{Ni} + 4CO \xrightarrow{60°C} Ni(CO)_4$$

$$Ni(CO)_4 \xrightarrow{200°C} \underset{\text{Pure}}{Ni} + 4CO$$

This method of producing extremely pure nickel is known as the Mond process.

Properties and Uses of Nickel

The properties of nickel are very similar to those of cobalt; it melts at 1452°C and has a density of 8.90 g cm^{-3}. It is unreactive towards air or water at ordinary temperatures, but combines directly with oxygen on heating to give nickel (II) oxide, NiO. Dilute mineral acids only slowly attack the metal with the formation of nickel (II), Ni^{2+}. ions. However, concentrated nitric acid renders it passive. Since it is not attacked by fused sodium hydroxide, nickel crucibles are convenient receptacles in which to carry out alkali fusions.

Finely divided nickel, which can be made by reduction of nickel (II) oxide with hydrogen, readily absorbs hydrogen and is used as a catalyst in the hydrogenation of alkenes, e.g. unsaturated inedible oils which contain carbon-carbon double bonds can be hydrogenated to saturated solids, which are edible fats.

Other alloys containing nickel include monel metal (nickel and copper) which is used in the construction of chemical plant, and cupronicke! (also nickel and copper), which is used in 'silver' coinage.

Compounds of nickel

Although a few compounds are known which contain Ni^{III} (e.g. an impure oxide, $Ni_2O_3.2H_2O$) and Ni^{iv} (e.g. an impure oxide, NiO_2, and a complex fluoride, K_2NiF_6), the only important oxidation state of nickel is +2.

Nickel (O) Compound

$K_4[Ni(CN)_4]$ is one of the compounds which contain a nickel atom in the zero oxidation state. It is a yellow solid made by reducing the salt K2[Ni1(CN)$_3$] with potassium or calcium in liquid ammonia. It turns black in air, and decomposes water with the evolution of hydrogen.

NiCN and $K_2[Ni(CN)_3]$ are two compounds in which the nickel atom is present in oxidation state I. $K_2[Ni(CN)_3]$ is a dark oily liquid. It is made by reducing a cold saturated solution of $K_2[Ni(CN)_4]$ with sodium amalgam or some other powerful reducing agent. The product is precipitated with alcohol. It rapidly decomposes in air. It dissolves in water to give an intensely red coloured solution which rapidly decomposes in the cold to give metallic nickel and $K_2IINi(CN)_4]$; if the solution is boiled hydrogen is evolved and $K_2[Ni(CN)_4]$ is formed. The solution is more stable in alkaline conditions. The solution is oxidized by air, $AgNO_3$, $PbCl_2$, and $HgCl_2$ to divalent nickel. The solution absorbs carbon monoxide and nitric oxide.

$K_2[Ni(CN)_3]$ is diamagnetic. The anion is probably dimeric with metal to metal bonds.

NiCN is made by acidifying the red aqueous solution of $K_2[Ni(CN)_3]$; the insoluble orange NiCN is precipitated. The precipitate dissolves in potassium cyanide solution to form the red solution again. NiCN is oxidized in air to $Ni(CN)_2$ and NiO. It absorbs carbon monoxide to give Ni(CN)CO.

Nickel oxide, NiO, is a green solid, insoluble in water, but soluble in acids. It is made by igniting the hydroxide of the carbonate in the absence of air at 600°, or by heating the nitrate to 1000°. The product obtained by heating nickel in air is nickel oxide contaminated with the metal and with Ni_2O_3. Nickel oxide has the sodium chloride lattice.

Nickel trioxide, Ni_2O_3, has not been isolated, but it is assumed to be formed on charging an Edison accumulator, when the two reversible reactions proceed to the right:

$$Fe^{2+} + 2e \rightarrow Fe; 2Ni^{2+} \rightarrow 2Ni^{3+} + 2e.$$

If a nickel salt is treated with bromine and sodium hydroxide a black precipitate is obtained, which if dried and gently warmed at 138° has the composition Ni_2O_3,H_2O. It decomposes if the removal of the water is attempted.

Nickel hydroxide, $Ni(OH)_2$, is obtained as a green precipitate when sodium hydroxide is added to a solution of a nickel salt. It cannot be made by the hydration of nickel oxide. It is almost insoluble in water, but is soluble in acids, and it dissolves in aqueous ammonia, probably to form $[Ni (NH_3)_4,](OH)_2$. Nickel hydroxide has the CdI_2 structure in which there are no hydrogen bonds; it is isomorphous with cobaltous hydroxide.

Nickel sulphide, NiS, is formed as a black precipitate when ammonium sulphide is added to a neutral or slightly acid solution of a nickel salt. It is not precipitated in acid solution, but once precipitated it will not redissolve in any acid except aqua regia. It has the nickel arsenide crystal structure. NiS, has the pyrites structure.

Nickel (II) Compound

Among the compounds in which a nickel atom displays ionic character are nickel carbonate, $NiCO_3$, nickel acetate $Ni (OCOCH_3)_2$, nickel nitrate $Ni (NO_3)_2$, nickel sulphate $NiSO_4$, nickel perchlorate $Ni (ClO_4)_2$, and the nickel halides.

Nickel nitrate, $Ni(NO_3)_2$, is made in solution by dissolving nickel hydroxide or carbonate in nitric acid. It crystallizes as a hydrate which may have 9, 6, 4, or 2 molecules of water of crystallization. The emerald green hexahydrate is isomorphous with the corresponding cobaltous compound. Nickel nitrate is said to be obtainable in the anhydrous state if the dihydrate is heated with a mixture of 100% nitric acid and nitrogen pentoxide.

Nickel sulphate forms hydrates with 7, 6, 5, 4, 3, 2, and 1 molecules of water of crystallization. The heptahydrate is a vitriol.

isomorphous with those formed by Mg, Mn, Fe, Co, Zn. The hexahydrate is dimorphic. The Ni^{2+} ion plays the part of a bivalent ion in the double sulphates $M_2 + M^{2+} (SO_4)_2,6H_2O$. Anhydrous nickel sulphate is greenish-yellow. It is insoluble in cold water by which it is only very slowly hydrated.

Complexes of Nickel

Nickel (II) complexes are fairly numerous and the majority of them are octahedral, e.g. the green $Ni(H_2O)_6]^{2+}$ and the blue $[Ni(NH_3)_6]^{2-}$. A well-known planar nickel complex is $[Ni(CN)4]^2$ and another is nickel dimethyl-glyoxime, which precipitates as a red solid when dimethyl-glyoxime is added to a solution of a nickel salt made just alkaline by the addition of ammonia; the formation of the latter complex is used as a test for Ni^{2+} ions.

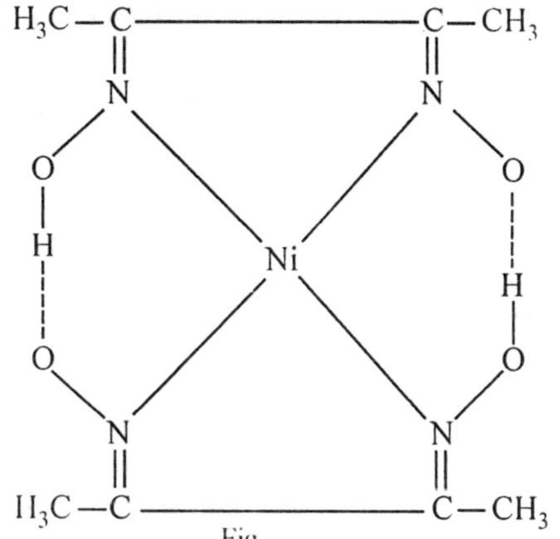

Fig.

If dimethyl-glyoxime is represented as DMG, the equation for its formation can be represented as:

$$2DMG + Ni^{2+} + 2OH^{-} \rightarrow Ni(DMG)_2 + 2H_2O$$

During the reaction two protons (one from each dimethyl-glyoxime molecule) are neutralized by the ammonia solution. The complex, which is thus uncharged, is held in the planar configuration by two hydrogen bonds (shown by dotted lines).

24

d-Block Elements (Platinum)— Group VIII

The elements of the second and third triads of the eight group in the Periodic Table, *ruthenium*, *rhodium*, *palladium*, *osmium*, *iridium* and *platinum*, are combined under the generic name of *platinum metals*.

Element	Symbol	Atomic weight	Atomic number
Ruthenium	Ru	101.1	44
Rhodium	Rh	102.91	45
Palladium .	Pd	106.7	46
Osmium .	Os	190.2	76
Iridium . .	Ir	192.2	77
Platinum .	Pt	195.23	78

General Features of the Platinum Metals : The platinum metals form a group of rather scarce metals, but they are so similar to one another in properties that it is very difficult to separate them. On the other hand, they resemble the elements of the iron group very little.

The platinum metals occur in nature almost exclusively in the native state, usually all together, but are never found in iron ores.

Being greatly disseminated through various rocks, the platinum metals became known to mankind comparatively recently. The existence of platinum was established earlier than that of the others, in 1750. Then, in the early XIX century, palladium, rhodium, osmium and iridium were discovered. The last platinum metal, ruthenium, was observed only in 1844 by Prof. Klaus of the Kazan University who named it in honour of Russia (Ruthenia—Russia).

The most important constant of the platinum metals are given in Table 24.1.

Table 24.1 : Chief Physical Constants of the Platinum Metals

Constant	Ruthe-nium Ru	Rhodium Rh	Palladium Pd	Somium Os	Iridium Ir	Platinum Pt
Atomic weight....	101.1	102.91	106.7	190.2	192.2	195.23
Nuclear charge....	44	45	46	76	77	78
Atomic radius, Å	1.32	1.34	1.37	1.34	1.35	1.38
Specific gravity....	12.2	12.44	11.9	22.46	22.5	21.45
Melting point. de-gress C....	2,450	1,960	1,552	2,700	2.443	1,769

The platinum metals belong to the less active chemical elements and are very stable against various kind of chemical action. Some of the fail to dissolve not only in acids but even in aqua regia.

Although the platinum metals may manifest various degrees of valency in their compounds, their typical compounds are those in which they are tetravalent.

The most important of the platinum metals from a practical standpoint, is platinum itself.

(vii) $AuCl_3$ solution gives a precipitate of purple of Cassius with $SnCl_2$ solution while $PtCl_4$ remains unaffected.

$$2AuCl_3 + 3SnCl_2 \rightarrow 2Au + 3SnCl_4$$
$$SnCl_4 + 4H_2O \rightarrow Sn(OH)_4 + 4HCl$$

Occurrence

Platinum mostly occurs in the native state, though in one mineral *sperrylite*, $PtAs_2$, it has been found combined with arsenic from Urals, California or Choco (South America) contain, besides platinum, other metals of the platinum group along with copper, iron and gold.

From Urals—76.4% Pt, 0.40 Au, 11.7 Fe, 4.3 Ir, 0.3 Rh, 1.4 Pd. 4.1 Cu, 0.5 Os-Ir and some sand.

Choco ores contain 86% platinum.

Extraction

(1) From native metal by Wet process. The platiniferrous sands and gravels obtained from the river beds or rocks are washed with a strong current of water on the *Wilfley's tables* so that the sand and other lighter materials are washed away and the heavier metallic particles ore left behind. The platinum concentrate thus obtained can be used to isolate platinum by the following steps :

(i) The gold is removed by amalgamation, then (ii) the rest of the ore is dissolved in *aqua regia*, the solution evaporated to dryness and heated to 125°; Palladium and rhodium are in the process reduced to insoluble chlorides, $PdCl_2$ and $RhCl_2$ and the platinic chloride remains as such which is then dissolved in water and acidified. (iii) To the solution of platinic chloride is then added ammonium chloride which forms an insoluble double salt, ammonium chloroplatinate, that get precipitated.

$$2NH_4Cl + PtCl_4 \rightarrow (NH_4)_2PtCl_6$$

The double of iridium, $(NH_4)_2IrCl_6$ is soluble in water and, therefore, remains in the solution. (iv) The double chloride is then ignited when it decomposes to form spongy platinum.

$$(NH_4)_2PtCl_6 \rightarrow 2NH_4Cl + 2Cl_2 + Pt$$

It is again fused for purification in lime crucible in presence of oxygen.

(2) From the Mond's process residue. The non-volatile residue left after the refining of nickel by Mond's process contains the platinum metals. Platinum metal can be obtained from this residue by the following steps :

(*i*) Concentration. The residue is smelted with lead oxide, carbon and a flux when a lead alloy of the rare metals is obtained and the impurities are removed as slag which is rejected. This lead alloy is cupelled to get lead free concentrate of the metals. Lead is removed as its oxide, PbO.

(*ii*) Treatment of concentrate with aqua-regia. The concentrate obtained as above is heated with hot aqua regia which dissolves Pt, Pd. Au and traces of Ir as their chloro complexes viz. $H_2[PtCl_6]$, $H_2[PdCl_6]$, $H[AuCl_4]$ and $H_2[IrCl_6]$ and Ru, Ir, Ag and Os remain in the residue as insoluble chlorides.

(*iii*) Separation of Gold. The solution containing the chloro complexes of Pt(+4), Pd(+4), Au(+3) and Ir(+4) is treated with $FeSO_4$ solution when $H[AuCl_4]$ is reduced to elementary gold and $H_2[PdCl_6]$ gets reduced to $H_2[PdCl_4]$, but the chloro-complexes : $H_2[PtCl_6]$ and $H_2[IrCl_6]$ remain uneffected.

$$\left[Fe^{2+} \rightarrow Fe^{3} + e^{-}\right] \times 3$$

$$AuCl_4^- + 3e^- \rightarrow 4Cl^- + Au^\circ \downarrow$$

$$3Fe^{2+} + AuCl_4^- \rightarrow 3Fe^{3+} + 4Cl^- + Au^\circ \downarrow$$

$$PdCl_6^{2-} + 2Fe^{2+} \rightarrow PdCl_4^{2-} + Cl^- + 2Fe^{3+}$$

Thus at the end of this step we have gold in the insoluble portion and $H_2[PtCl_6]$, $H_2[PdCl_4]$ and traces of $H_2[IrCl_6]$ in the soluble portion.

(iv) Separation of Pd and Ir. The soluble portion obtained as above and containing the chloro complexes of Pt(+4), Pd(+2) and Ir(+4) is treated with NH_4Cl which converts $H_2[PtCl_6]$ into *insoluble* $(NH_4)_2(PtCl_6)$ and $H_2[IrCl_6]$ into *soluble* $(NH_4)_2[IrCl_6]$.$H_2[PdCl_4]$ remains unaffected (soluble portion).

$$H_2\left[PtCl_6\right] + 2NH_4Cl \rightarrow \underset{\text{Insoluble}}{(NH_4)_2\left[PtCl_6\right]} + 2HCl$$

$$H_2\left[IrCl_6\right] + 2NH_4Cl \rightarrow \underset{\text{Soluble}}{(NH_4)_2\left[IrCl_6\right]} + 2HCi$$

$$H_2\left[PdCl_4\right] \rightarrow \text{Remains unaffected (soluble portion)}$$

The soluble portion which contains $(NH_4)_2[IrCl_6]$ and $H_2[PdCl_4]$ is treated with NH_4OH followed by HCl to precipitate yellow crystals of dichlorommine palladium (II), $[Pd(NH_3)_2Cl_2]$ which on ignition gives Pd.

$$H_2\left[PdCl_4\right] + 6NH4OH + 2HCl \rightarrow 4NH4Cl + \underset{\text{ppt}}{\left[Pd(NH_3)_2Cl_2\right]} + 6H_2O$$

The soluble portion which contains $(NH_4)_2[IrCl_6]$ obtained in step (*iv*) is dried and ignited to get spongy platinum.

$$(NH_4)_2[PtCl_6] \xrightarrow{\text{heat}} \underset{\text{Sprong platinum}}{Pt} + 2NH_4Cl + 2Cl_2$$

It is purified by again converting spongy Pt into $(NH_4)_2[PtCl_6]$ and repeating the above process.

The insoluble portion obtained after the treatment of the concentrate with hot aqua regia contains chlorides of Ru, Rh, Ir, Ag and Os [*See step (ii) given above*]. This portion can be used to get these metals.

Properties

Platinum is a white lustrous malleable metal with a specific gravity of 21.45, melting at 1,769°C. In the air platinum does not change even if heated very intensively. It is not attacked by the separate acid. Platinum with dissolve only in aqua regia, but much less readily than gold. Platinum is extracted from its ores with the aid of aqua regia.

Owing to its refractoriness and resistance to chemical action platinum is widely used in scientific and technical laboratories; in analytical work use is made of platinum crucibles, evaporating dishes, electrodes, etc.

Platinum is employed as a catalyst for accelerating many chemical process. Finely divided platinum is capable of adsorbing considerable quantities of hydrogen and oxygen (as many as 100 volumes per volume of platinum). In the adsorbed state the hydrogen and oxygen become very active; this forms the basis for the use of platinum as a carrier of hydrogen and oxygen.

In its compounds platinum is predominantly bi- and tetravalent. It is capable of forming complex compounds in both states; the tetravalent compounds of platinum are the more important.

When platinum is dissolved in aqua regia the result is chloroplatinic acid $H_2[PtCl_6]$, which separates out when the solution is evaporated as reddish-brown crystals of the composition $H_2[PtCl_6].6H_2O$. This is the usual commercial preparation of platinum. The potassium, and its formation is employed in chemical analysis for the detection of potassium.

Uses of Platinum

(*i*) **In scientific instruments and laboratory apparatus.** Due to its remarkable properties like high melting point and corrosion resistance, platinum is largely used to manufacture scientific instruments and laboratory apparatus such as crucibles, dishes, basins etc. Pt is also used for contacts in electrical apparatus and for making electrodes.

***(ii)* As a catalyst.** Pt is also used as a catalyst in a number of inorganic and organic reactions. Some important applications of platinum as catalyst are given below.

(a) Manufacture of H_2SO_4. SO_3 and O_2 combine to form SO_3 in the presence of platinised asbestos in the manufacture of H_2SO_4 by *contact process.*

(b) Manufacture of HNO_3. NH_3 is oxidized to NO in the presence of platinum gauge in the manufacture of HNO_3 by *Ostwald's process.*

$$4NH_3 + 5O_2 \xrightarrow[\text{gauge}]{\text{platitnum}} 4NO + 6H_2O$$

(c) Preparation of formaldehyde. Vapours of CH_3OH and O_2 are passed on platinum black to get HCHO.

$$2CH_3OH + O_2 \xrightarrow[\text{black}]{\text{platitnum}} 2HCHO + 2H_2O$$

(d) Synthesis of HBr and HI. Mixture of H_2 and halogens is passed over red hot finely divided platinum.

$$H_2 + Br_2 \rightleftharpoons 2HBr$$

(e) Decomposition of H_2O_2.

$$2H_2O_2 \xrightarrow[\text{black}]{\text{platitnum}} 2H_2O + O_2 \uparrow$$

(f) In hydrogenation and dehydrogenation industries. Since finely divided platinum has the valuable property of absorbing several hundred times its own volume of hydrogen gas, the metal is widely used for many hydrogenation and dehydrogenation reactions. Large quantities of the metal are also used in the petroleum industry, synthetic fibre industry and glass industry.

***(iii)* In jewellery.** Pt is also used in jewellery as a setting for diamonds. For this purpose the metal is alloyed with metals like IR, Ru or Cu. Pt—Ir alloy is also used in pen points, standard weights, coin etc.

***(iv)* In dentistry.** The metal is used for making artificial teeth, pins dental filling and support for dentures.

(v) Pt-salts are used in printing and toning to photography. Pt wire is also used in the qualitative analysis in chemistry. In the medical fields many instruments like X-ray tubes, radium applicators, lypodermic needles require platinum.

Platinum Compounds

Platinum chiefly forms two series of compounds—*platinous* and *platinic* corresponding to + 2 and + 4 oxidation states of platinum respectively. The compounds in + 1 (e.g. PtCl), + 3 ($PtCl_3$), + 5 (PtF_5) and + 6 (PtO_3) oxidation states are also known but these are rare.

Platinum (II) Compound

The platinous compounds include the hydroxide $Pt(OH)_2$, the oxide PtO, the sulphide PtS, and the dark-coloured insoluble dihalides, *Pthal*$_2$.

Platinous hydroxide, $Pt(OH)_2$, a black powder, is an unstable substance made by the action of hot potassium hydroxide solution on platinous chloride or $K_2[PtCl_4]$ in an inert atmosphere. It is reduced by hydrogen peroxide to metal, and is oxidized by ozone or potassium permanganate to the dioxide, PtO_2.

Platinous oxide, PtO, is a grey powder. It is made by heating the hydrate, but it cannot be obtained in pure condition because the hydrate cannot be completely dehydrated without loss of oxygen. It is soluble in acids but not in alkalis.

Platinous sulphide, PtS, is a green powder, which is made either by fusing sulphur and platinum, or by fusing platinous chloride with sulphur and sodium carbonate. Its crystal structure is similar to that of palladous oxide and palladous sulphide.

Platinum (III) Compound

The compounds are either simple, such as Pt_2O_3aq., Pt hal$_3$, $Pt(CN)_3$, or complex, such $[Pt(N(C_2H_5)_3Cl]\ Cl_2$, $[Pt(N(C_2H_5)_3)_2Cl_3]^0$, $K_3[PtCl_5]$. The molecular weight of none of these compounds is known, and there is no confirmation of the monomeric formulae written above. It is not possible to conclude definitely whether Pt^{III} is a stable state of the platinum atom, or whether compounds written as PtX_3 are not in fact $PtX_2.PtX_4$. It has been shown that $Pt(NH_3)_2Br_3$ and certain similar complexes consists of square Pt^{II} complex molecules and octahedral Pt^{IV} complex molecules linked by bromine bridges :

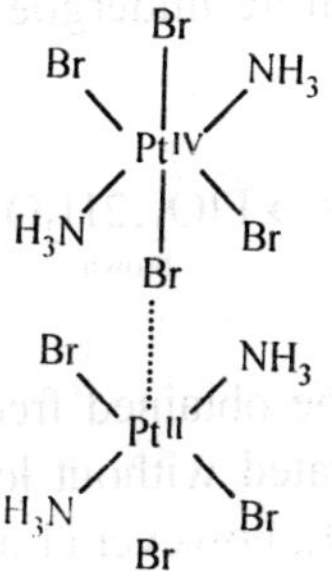

The hydrated oxide Pt_2O_3, aq. can be made as a dark brown precipitate by treating platinum trichloride with potassium hydroxide. It cannot be dehydrated. It is not oxidized by air. It dissolves in concentrated alkali but not in acids.

Complexes which may contain Pt^{III} are made by gently oxidising the complexes of Pt^{II}. The salt $[Pt(N(C_2H_5)_3)_4Cl]Cl_2$ is brilliant red. It is made by oxidizing the colourless $[Pt^{II}(N(C_2H_5)_3)_4]Cl_2, 2H_2O$ with hydrogen peroxide. The red salt, however, dissolves in hot water to a pale yellow solution which behaves as through it contains Pt^{II} and Pt^{IV}; if solutions of Pt^{IV}: if solutions of Pt^{II} and Pt^{iv} are mixed the red salt is precipitated.

Platinum (IV) Compound

Among the compounds of Pt^{IV} are the hydrated oxide, PtO_2, the platinum tetrahalides, and numerous complexes which are always octahedral. Pt^{IV} forms as many complexes as Pt^{II}, the most stable being those in which nitrogen atoms are linked to the platinum atom. The ammine complexes include all possible compounds from $[Pt(NH_3)_6]Cl_4$ to $K_2[PtCl_6]$, and the Cl^- ions can be replaced in turn by OH^- until $K_2[Pt(OH)_6]$ is reached. CNS^- and NO_2^- can also replace Cl^-. The ammonia molecules can be replaced by hydrazine, hydroxylamine, or ethylendeiamine.

PtO_2, aq. If an aqueous solution of platinum tetrachloride is boiled with sodium hydroxide, and acetic acid is added, a white precipitate is obtained which turns yellow on boiling. It has the composition PtO_2, $3H_2O$. It is amphoteric and dissolves in alkalis to give $Na_2[Pt(OH)_6]$. The trihydrate undergoes dehydration in the stages :

$$\underset{\text{pale yellow soluble in HCl or NaOH}}{PtO_2, 3H_2O} \xrightarrow{\text{over conc. } H_2SO_4} \underset{\text{brown}}{PtO_2, 2H_2O} \xrightarrow{100^\circ} \underset{\text{black insoluble}}{PtO_2, H_2O}$$

Platinum dioxide cannot be obtained free from water, as the monohydrate cannot be dehydrated without loss of some oxygen. The dissociation pressure of PtO_2, however (1 atm at 400°), is lower than that of PtO or Pt_2O_3.

Platinum (V) Compound

PtF_5 is produced by the action of fluorine on $PtCl_2$ at 350°; it rapidly disproportionates PtF_6 and PtF_4.

Platinum (VI) Compound

Platinum hexafluoride, PtF_6, is obtained when fluorine vapour interacts with a platinum filament near a liquid nitrogen cold finger. It exists as dark red crystals, m.p. 61.3°, b.p. 69.1°. The vapour resembles bromine vapour. The electron affinity of PtF_6 is great enough to ionise oxygen to give the salt $[O_2]^+[Pt^VF_6]^-$, and xenon to give $[Xe]^+[Pt^VF_6]^-$.

25

d-Block Elements (Copper, Silver, Gold)—Group IB (11)

Element	Symbol	Atomic weight	Atomic number
Copper....	Cu	63.54	29
Silver....	Ag	107.880	47
Gold	Au	197.0	79

General Features of the Copper Subgroup. The copper subgroup includes three elements, *copper*, *silver* and *gold*, which begin the odd series of the long periods in the Mendeleyev Table. Like the alkali metals, all three elements have one electron each in the outermost layer of their atoms; their second last electron layers, however, have a different structure, consisting of eighteen electrons, whereas in the atoms of the alkali metals, except lithium, this layer has only eight electrons.

Atomic Radii and Ionization Potentials of the Copper Subgroup and the Alkali Metals

Element	Atomic radius, Å	Ionization potential, V	Element	Atomic radius Å	Ionization potential, V
Copper....	1.27	7.72	Potassium....	2.36	4.32
Silver....	1.44	7.51	Rubidium....	2.53	4.10
Gold....	1.44	9.22	Caesium....	2.74	3.87

The difference in structure of the second last layer accounts for the rather pronounced difference in the properties of the elements of these subgroups, depending, apparently, on the relative sizes of their atoms. The atomic radii and ionization potentials of the elements of the copper subgroup are compared with those of the alkali metals located in the same periods.

The data given in this table show that in the atoms of copper, silver and gold the outer electron is much closer to the nucleus and must therefore be attracted by it more strongly than in the atoms of the corresponding alkali metals. Accordingly, the ionization potential, i.e., the energy that has to be expended to tear the electron away from the atom, is much higher in the copper subgroup elements than in the alkali metals. Indeed, the alkali metals part with their subgroup hold on to it rather tenaciously. That is why they are much more difficult to oxidized, and their ions on the other hand, more easy to reduce; they do not decompose water, their hydroxides are comparatively weak bases, etc. At the same time, the 18-electron layer, stable in other elements, is not quite so in these elements, and is capable of yielding some of its electrons.

Copper, silver and gold are known as the 'coinage metals' and are resistant to chemical attack (gold in particular is chemically very unreactive). All three metals exhibit an oxidation state of + 1, and in this state, their compounds are largely covalent or at least possess a considerable degree of covalent character although the principal oxidation states for copper, silver and gold are + 2, + 1 and + 3 respectively. In the higher oxidation states the Group 1B metals utilize their *d* electrons in addition to their single *s* electrons; thus their ions contain an incomplete *d* level and are typically transitional. Unlike the Group 1A metals, copper, silver and gold form numerous complexes in all their oxidation states.

Since there are no smooth gradiations in properties along the series copper, silver and gold, these three metals are discussed separately.

Copper

Occurrence. Copper ores are widely distributed over the earth's surface. Large deposits of native copper are found in the Lake Superior region of Michigan, on the island of Cyprus, and in Bolivia. However, the metal usually occurs in the combined state as sulphides, oxides, or carbonates. Most of the copper produced in this country comes from sulphide minerals, chief of which is *chalcopyrite*, $CuFeS_2$. Another important sulphide mineral is *chalcocite*, Cu_2S. In general, sulphide ores are of low grade, the copper content usually running from 1 to 10 per cent. Oxide and carbonate ores have been formed as a result of the action of dissolved oxygen and carbon dioxide in ground water. Cuprite, Cu_2O, a reddish mineral, and *malachite*, $CuCO_3$. $Cu(OH)_2$, a green mineral, are important so-called oxidized copper minerals. Principal copper-producing localities in the United States are Montana, Arizona, and Utah. South America and central Africa have large producing copper mines.

Copper in fairly high percentage has been found in the feathers of certain birds of brilliant plumage. It is interesting that the touraco bird of Africa has quite a high copper content in its feathers, which have the red colour of cuprite and the green colour of malachite. Traces of copper are found in several sea plants and marine animals. In oysters, it appears that a copper compound is substituted for an iron compound as an oxygen carrier in the haemoglobin of the blood. There is some evidence to indicate that copper in minute amounts may play a vital role in the health and well-being of man and animals. For example, a healthy cow's liver contains between 40 and 200 parts per million of copper.

Metallurgy

The ore is crushed and ground wet. Sulphides are concentrated by flotation. Low-grade ore is treated differently: after leaching out with acidified water, copper is precipitated by the addition of scrap iron. Other grades are smelted after concentration.

If necessary, sulphur in excess of matte requirements is removed by a preliminary roasting in air at 1200-1500°C. Sulphur dioxide and oxides of volatile impurities, such as arsenic, are evolved.

$$2CuFeS_2 + O_2 \rightarrow Cu_2S + 2FeS + SO_2$$

Hot calcine, or wet concentrate, if no preliminary roasting is done, is smelted with a flux of limestone or silica, or both, in a reverberatory furnace. Earthy material forms a slag, which floats and may be removed. Air for the combustion is preheated by hot waste gases. A fusion of copper(I) and iron(II) sulphide, called a matte, with a copper content of 40-45% is.

Air under pressure is passed through the molten matte for about 10 hours in converters lined with magnesite bricks. Iron(II) sulphide is oxidized preferentially, iron passing to the slag and sulphur being evolved as dioxide,

$$2FeS + 3O_2 \rightarrow 2FeO + 2SO_2$$
$$FeO + SiO_2 \rightarrow FeSiO_3$$

A silicate slag is formed by use of siliceous copper ore as flux. The reaction is exothermic and is maintained at about 1300°C by addition of cold scrap copper. Copper(I) sulphide is partly oxidized to the oxide, when copper is formed by autoreduction :

$$2Cu_2S + 3O_2 \rightarrow 2Cu_2O + 2SO_2$$
$$Cu_2S + 2Cu_2O \rightarrow 6Cu + SO_2$$

Molten copper is cast into water-cooled moulds. Expelled sulphur dioxide leaves the surface roughened, so the product, 98-99.5% pure, is called 'blister' copper.

High-quality blister copper is furnace refined. The process is called 'poling' because of the use of wooden poles. Air is blown through the molten copper, to which a silica flux has been added. Impurities are oxidized; sulphur forms sulphur dioxide which escapes with other volatile matter while other impurities pass into the slag. Partial oxidation of copper to copper(I) oxide also occurs. To counter this, the surface of the molten copper is sealed with coke of low sulphur content and the ends of green hardwood trees submerged below the surface. Hydrocarbon decomposition products of wood reduce any copper(I) oxide to copper. The metal is cast into moulds and quenched with water.

Where the precious metal content is high electrolysis follows a preliminary furnace refining. The electrolyte contains copper(II) sulphate and dilute sulphuric acid. Anodes of impure copper are hung interleaved with the cathodes of pure copper. Copper passes from anode to the electrolyte and so the cathodes while noble impurities drop into the anode sludge, which is periodically worked up for gold and silver. Deposited copper is melted, furnace refined and cast. The purity of the electrolyte must be maintained to avoid contamination of copper.

Properties

Copper is a soft metal with a melting point of 1083° and a boiling point of about 2300°. By reflected light appears red in colour and by transmitted light in thin sheets its colour is green. The metal is very ductile and malleable, and, while not as tough as iron, it may be used for many structural purposes. It does not respond to heat treatment as does steel, but the addition of small quantities of alloying element such as manganese and silicon yields a product which may be hardened by heating.

Copper lies low in the activity series and thus is quite resistant to corrosion. Dry air, water, and nonoxidizing acids do not attack the metal. However, long exposure to air and moisture gradually forms a tightly adhering coating of a green basic carbonate on the surface. Corrosion then cases, in direct contrast to the rusting of iron, which in time changes the metal over completely to iron oxide. Hydrochloric acid does not act on copper unless oxygen is present, in which case is gradually dissolves according to the equation :

$$2Cu + 4HCl + O_2 \rightarrow 2CuCl_2 + 2H_2O$$

Nitric acid acts readily on the metal to form copper nitrate, but hydrogen is *not* liberated; instead the other products are oxides of nitrogen and water. Hot concentrated sulphuric acid dissolves the metal to form copper sulphate, sulphur dioxide, and water :

$$Cu + 2H_2SO_4 \rightarrow CuSO_4 + SO_2 + 2H_2O$$

Copper is an excellent conductor of heat and is second only to silver as a conductor of electricity.

Uses. Copper is malleable and ductile, of excellent thermal and electrical conductance, with marked resistance to corrosion. The properties of copper are altered appreciably by impurities. Copper, of a purity not less than 99.9%, is used for electrical purposes : about 40% of world production going as wire to the electrical industries for use the windings of electric motors, power lines, submarine cables, telephone lines and, with cadmium added for strength, for overhead wires on electrified railways and for extra long spans of high-voltage transmission lines of the National Grid. Copper is electroplated for two main purposes. A comparatively thin coating is applied to articles before nickel- and chromium-plating while thickness of copper may be built-up, or electro-formed, by electrolysis for printing rollers, wave-guides and moulds for gramophone records.

Copper alloys are widely used and being the most important group of non-ferrous alloys, are considered separately in the next chapter. Besides the general engineering purposes to which copper and its alloys are put, further uses include heat exchange units and boilers where good thermal properties are paramount, and in the chemical, brewing and jam-making industries where corrosion resistance is important and building, plumbing and marine engineering. Alloys of copper are also used in coinage.

Oxidation States of Copper

Copper forms two principal series of compounds, which are based upon the oxidation states of + 1 and + 2, respectively. When the one electron in the outermost shell of the copper atom is involved in bond formation, the oxidation state of + 1 is exhibited. When an additional electron from the underlying shell is removed, the copper(II) ion results. When two electrons from the underlying shell are used in bonding, copper exhibits an oxidation state of + 3: copper(III) compounds are rare and relatively unimportant, however. Copper(I) forms a complete series of binary compounds such as the halides and the oxide, but the ternary oxygen salts are few in number of readily decomposed by water. In contrast, copper(II) forms a complete series of oxygen salts, while some of its binary compounds are unstable and break down spontaneously.

Copper(I) Compounds

Solid salts of copper(I) show a variety of colours, but their solutions are colourless.

1. Copper(I) Oxide. Cu_2O may be prepared as a reddish-brown precipitate by boiling copper(I) chloride with sodium hydroxide.

$$2CuCl + 2OH^- \rightarrow \underline{Cu_2O} + 2Cl^- + H_2O$$

When basic solutions of copper(II) salts are heated with reducing agents, copper(I) oxide precipitates.

$$2Cu^{2+} + 2OH^- + 2e^-\text{(reducing agent)} \rightarrow \underline{Cu_2O} + H_2O$$

This reaction is the basis for the test for the presence of reducing sugars in the urine in the diagnosis of diabetes. The reagent most often used for this purpose is Benedict's solution, which is a solution of copper(II) sulphate, sodium carbonate, and sodium citrate. The addition of a reducing sugar, such as glucose, causes the precipitation of the reddish-brown copper(I) oxide.

2. Copper(I) Hydroxide. The compound is formed as a yellow precipitate when a cold solution of copper(I) chloride in hydrochloric acid is treated with sodium hydroxide. The hydroxide is unstable and decomposes into the oxide and water upon heating.

Copper(I) chloride, CuCl

This is a white solid which is insoluble in water. It can be obtained by boiling a solution containing copper(II) chloride, excess of copper turnings and concentrated hydrochloric acid. Copper(I) is present in this solution as the $[CuCl_2]^-$ complex ion :

$$Cu + Cu^{2+} \rightleftharpoons 2Cu^+$$

$$+4Cl^-$$

$$\upharpoonleft\downharpoonright$$

$$2[CuCl_2]^-$$

On pouring the solution into air-free distilled water, copper(I) chloride precipitates as a white solid. It must be rapidly washed, dried and sealed in the absence of air, since a combination of air and moisture oxidizes it to copper(II) chloride.

Copper(I) chloride is essentially covalent and its structure is similar to that of diamond, i.e. each copper atom is surrounded tetrahedrally by four chlorine atoms and each chlorine atom is surrounded tetrahedrally by four copper atoms; in the vapour phase dimeric and trimeric species are present, i.e. $(CuCl)_2$ and $(CuCl)_3$. It is soluble in water in the presence of entities such as Cl^-, $S_2O_3^{2-}$ and NH_3, with which it forms complex ions, e.g. $[Cu(NH_3)_2]^+$.

Copper(I) chloride is used in conjunction with ammonium chloride as a catalyst in the dimerization of ethyne to but-1-ene-3-yne (vinylacetylene), mixture of copper(I) chloride and hydrochloric acid is used for converting benzene diazonium chloride into chlorobenzene (Sandmeyer reaction) :

$$C_6H_5N_2^+Cl \xrightarrow{CuCl\ HCl} C_6H_5Cl + N_2$$

An ammoniacal solution of copper(I) chloride absorbs carbon monoxide and precipitates copper(I) carbide when ethyne is bubbled through it; the latter reaction can be used for purifying ethyne since the carbide readily evolves ethyne when treated with dilute acid.

Copper(I) sulphate, Cu_2SO_4

Copper(I) sulphate is obtained as a white solid when copper(I) oxide is heated with anhydrous dimethyl sulphate :

$$Cu_2O + (CH_3)_2SO_4 \rightarrow Cu_2SO_4 + (CH_3)_2O$$

In the absence of moisture it is stable but, on dissolving, it disproportionates rapidly into copper(II) sulphate and copper :

$$Cu_2SO_4 \rightarrow CuSO_4 + Cu$$

Copper(II) compounds

This is the most common oxidation state of copper and in aqueous solution copper(II) salts are blue, the colour being due to the presence of $[Cu(H_2O)_6]^{2+}$ ions.

1. Copper(II) Oxide. CuO, is a black insoluble compound that may be obtained by heating the carbonate or nitrate, or by heating finely divided copper in oxygen.

$$CuCO_3 \rightarrow CuO + CO_2$$

$$100{,}500\text{cal} + 2Cu(NO_3)_2(s) \rightarrow 2CuO(s) + 4NO_2(g) + O_2(g)$$

$$(\Delta H^\circ_{298} = 100{,}500\text{cal})$$

$$2Cu(s) + O_2(g) \rightarrow 2CuO(s) + 37{,}600\text{cal}$$

$$(\Delta H^\circ_{298} = -37{,}600\text{cal})$$

Because of the ease with which the copper can be reduced either to oxidation state + 1 or to the metal, hot copper(II) oxide is a good oxidizing agent. As such, it is used in the analysis of organic compounds to determine the percentage of carbon that they contain. Copper(II) oxide oxidizes the carbon to carbon dioxide, which is absorbed by sodium hydroxide, and the gain in weight of the sodium hydroxide is taken as a measure of the quantity of carbon dioxide produced and absorbed. The copper oxide is continuously regenerated by passing a stream of air through the tube in which the organic matter and the copper oxide are rearing.

Copper(II) hydroxide, $Cu(OH)_2$

Copper(II) hydroxide is precipitated as a blue-green gelatinous solid when an aqueous solution of a copper(II) salt is made alkaline with sodium hydroxide solution. It can be filtered and dried at 100°C to the composition $Cu(OH)_2$, however, heating the infiltered suspension to about 80°C results in decomposition into copper(II) oxide and water.

Copper(II) hydroxide is readily soluble in an aqueous solution of ammonia with the formation of the intensely blue $[Cu(NH_3)_4]^{2+}$ ion :

$$Cu(OH)_2 \rightleftharpoons Cu^{2+} + 2OH^-$$
$$+$$
$$4NH_3$$
$$\updownarrow$$
$$[Cu(NH_3)_4]^{2+}$$

It reacts readily with dilute acids to give copper(II) salts but, since the freshly precipitated solid is also slightly soluble in sodium hydroxide solution, it must be considered to be somewhat amphoteric :

$$Cu(OH)_2 + 2H^+ \longrightarrow Cu^{2+} + 2H_2O$$
$$Cu(OH)_2 + 2OH^- \rightleftharpoons Cu(OH)_4^{2-}$$

Copper(II) sulphide, CuS

Copper(II) sulphide precipitates as a black solid when hydrogen sulphide is passed into an aqueous solution of a copper(II) salt, even in the presence of a large concentration of hydrogen ions, since it has a very low solubility product :

$$Cu^{2+} + H_2S \rightarrow CuS + 2H^+$$

Copper(II) Sulphate. This sulphate is the most important of the copper salts. The anhydrous salt is colourless, but when it is crystallized from aqueous solution the blue pentahydrate, $[Cu(H_2O)_4]SO_4{\cdot}H_2O$, known as **blue vitriol**, is formed. As indicated by the formula, four of the water molecules are coordinated to the copper(II) ion and the fifth is attached to both the sulphate and two of the coordinated water molecules by hydrogen bonds. The structure of the pentahydrate may be represented as shown.

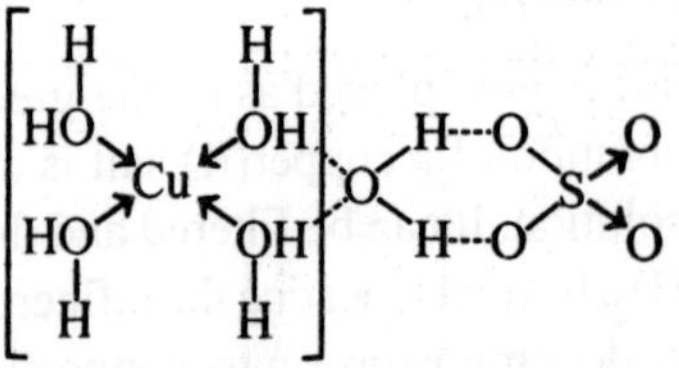

The dehydration of this salt proceeds in steps, yielding successively $CuSO_4 \cdot 3H_2O$ (with loss of the two nonhydrogen-bonded waters), $CuSO_4 \cdot H_2O$ [with loss of the other two water molecules coordinated to the Cu(II) ion], and finally $CuSO_4$ (with loss of the water molecule attached to the sulphate ion).

Copper(II) sulphate is produced commercially by oxidizing the sulphate either directly to the sulphate or to the oxide, which is then converted to the sulphate by reaction with sulphuric acid. In the laboratory the anhydrous sulphate may be prepared by oxidizing copper with not concentrated sulphuric acid.

$$Cu(s) + 2H_2SO_4(l) \rightarrow CuSO_4(s) + SO_2(g) + 2H_2O(l) + 2{,}840ca$$

$$(\Delta H^\circ_{298} = -2{,}840\text{cal})$$

Aqueous solutions of copper(II) sulphate are acidic by hydrolysis,

$$[Cu(H_2O)_4]^{2+} + H_2O \rightleftharpoons [Cu(H_2O)_3OH]^+ + H_3O^+$$

or simply,

$$Cu^{2+} + H_2O \rightleftharpoons CuOH^+ + H^+$$

Copper(II) sulphate is used in the electrolytic refining of copper, in electroplating, in the Danier cell, in the manufacture of pigments, as a mordant in the textile industry, and in the prevention of algae in reservoirs and swimming pools. Because anhydrous copper(II) sulphate is insoluble in alcohol and ether and readily takes up water, turning blue, this salt is used in detecting water in liquids. Of course, copper(II) sulphate will not remove or detect water in liquids with which it reacts.

Copper(II) sulphate has many industrial uses, including the preparation of Bordeaux mixture (a fungicide) and the preparation of other copper compounds. It is also used in electroplating solutions, in textile dyeing, and as a timber preservative.

Copper(II) nitrate, $Cu(NO_3)_2.3H_2O$

Copper(II) nitrate is a blue deliquescent solid which may be obtained by reacting either copper(II) oxide or copper(II) carbonate with dilute nitric acid and crystallizing the resulting solution. Other hydrates containing six and nine molecules of water of crystallization are known. On heating, it readily decomposes into copper(II) oxide, nitrogen dioxide and oxygen.

Anhydrous copper(II) nitrate can be obtained by reacting copper with a solution of nitrogen dioxide in ethyl acetate, followed by isolation of the solid product and decomposition of the solvate at 90°C. It sublimes on heating to about 200°C and this seems to suggest that anhydrous solid is appreciably covalent.

Copper(II) carbonate, $aCuCO_3.bCu(OH)_2$

The pure substance is unknown and any method of preparation, e.g. the addition of the either sodium carbonate or sodium hydrogen carbonate solutions to a solution of a copper(II) salt, results in the formation of a green basic carbonate. The ore malachite is $CuCO_3.Cu(OH)_2$; another basic carbonate has the composition $2CuCO_3.Cu(OH)_2$.

Basic copper(II) carbonate is insoluble in water but dissolves in ammonia solution, since the dark blue $[Cu(NH_3)_4]^{2+}$ ion is formed and the reaction (see below) move to the right :

$$CuCO_3 \rightleftharpoons Cu^{2+} + CO_3^{2-}$$

$$+$$

$$4NH_3$$

$$\updownarrows$$

$$[Cu(NH_3)_4]^{2+}$$

Copper (II) halides

Copper(II) fluoride is a white crystalline solid, probably of ionic structure, formed by passing hydrogen fluoride over heated copper(II) oxide. Both anhydrous copper(II) chloride and copper(II) bromide form flat, covalent macromolecular chains in which adjacent copper atoms are linked by two separate halogen atoms :

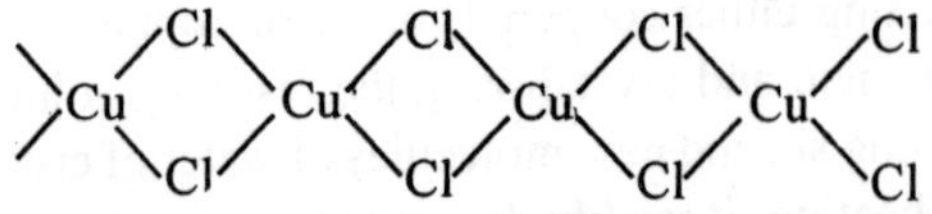

All three halides form hydrates. Addition of potassium iodide to a solution of copper(II) ions, results in the precipitation of white copper(I) iodide and the liberation of iodine :

$$2Cu^{2+} + 4I^- \rightarrow 2CuI \downarrow + I_2 \downarrow$$

Copper(II) chloride may be prepared by reacting copper(II) oxide or carbonate with hydrochloric acid and evaporation to crystallization. With care, blue crystals of the dihydrate, $CuCl_2.2H_2O$, which appear green when moist, separate. Further heating, preferably in a current of dry hydrogen chloride, yields dark brown anhydrous copper(II) chloride. Alternatively, copper may be heated in chlorine to give the anhydrous compound. It is deliquescent. The structure of they anhydrous compound has been given; that of the dihydrate is also covalent :

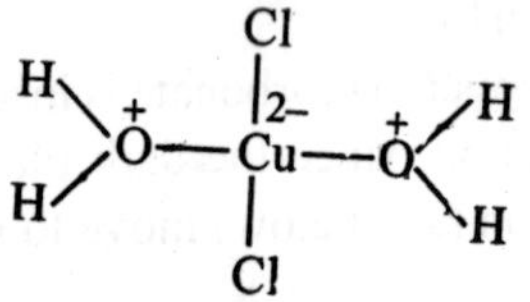

Copper(II) chloride is very soluble in water, 76.0 g dissolving in 100 g water at 25°C.

Concentrated solutions of copper(II) chloride are dark green-brown in colour, while dilute solution have the blue colour associated with the hydrated copper(II) ion. Addition of alkali-metal chlorides or passing in hydrogen chloride changes the colour to brownish-yellow. The phenomenon is linked with the formation of complexes and differing degrees of hydration and has probably not been fully elucidated. The blue colour is due to $Cu(OH_2)_4^{2+}$ and the dark colour to complex anions $CuCl_3^-$, mixed with unionized molecules :

$$CuCl_2 \xrightarrow[\text{solution}]{\text{aq.}} Cu(OH_2)_4^{2+} \text{ and } CuCl_4^{2-}$$

Various complexes may be isolated, including $Li(CuCl_3).2H_2O$ and $K(CuCl_3)$ in addition to the acids $HCuCl_3.3H_2O$ and $H_2CuCl_4.5H_2O$, all of which from dark red crystals. In addition, potassium chloride and copper(II) chloride will crystallize as the double salt, $2KCl.CuCl_2.2H_2O$.

At red heat, copper(II) chloride loses chlorine to form copper(I) chloride :

$$2CuCl_2 \rightarrow 2CuCl + Cl_2$$

Copper(II) chloride was used as a catalyst in the oxidation of hydrogen chloride to chlorine by air (Deacon's process).

Basic copper carbonate(II) $Cu_2(OH)_2CO_3$ is found in nature as the mineral malachite which has a beautiful emerald green colour. Can be synthesized by treating solutions of bivalent copper salts with soda :

$$2CuSO_4 + 2Na_2CO_3 + H_2O = \downarrow Cu_2(OH)_2 + 2Na_2SO_4 + CO_2$$

Is used for the preparation of cupric chloride, for the manufacture of blue and green mineral paints and in pyrotechnics.

Copper (cupric) acetate $Cu(CH_3COO)_2.H_2O$ is prepared by the action of acetic acid on metallic copper or cupric oxide. The commercial product is usually a mixture of basic salts of varying composition and colour (green and bluish green). It used under the name of neutral *verdigris* for the preparation of oil colours.

The double acetate-arsenite of copper (known as "Paris green") $Cu(CH_3COO)_2.Cu_3(AsO_3)_2$ is used to destroy agricultural pests.

Copper salts are employed for the manufacture of a large number of mineral paints of various colours : green, blue, brown, violet, and black.

All copper salts are poisonous, and that is why copper pots are tinned on the inside to prevent the formation of copper salts.

Complex compounds of copper. A very characteristic property of bivalent copper ions is their ability to combine with ammonia molecules to form what are known as complex ions.

If ammonia solution is added to a solution of copper sulphate, the result is a blue precipitate of the basic salt, which dissolves readily in an excess of ammonia, colouring the liquid an intense blue. Addition of an alkali to this solution brings down no $Cu(OH)_2$ is not reached even with a high concentration of hydroxyl-ion. Hence, it follows that the cupric ions react with the ammonia added to form new ions of some kind which give no insoluble compound with OH' ion. At the same time the SO_4'' ions remain unchanged, as can be proved experimentally by adding barium chloride to the ammonia solution, this immediately bringing down a precipitate of $BaSO_4$ (test of SO_4''-ion).

It has been established by investigation that the dark blue colouring of the ammonia solution is due to the presence in it of the complex ions $[Cu(NH_3)_4]^{\cdot\cdot}$, a result of the addition of four ammonia molecules to each cupric ion, if the water is evaporated, the $[Cu(NH_3)_4]^{\cdot\cdot}$ ions combine with the SO_4'' ions, and dark blue crystals separate out of the solution their composition being represented by the formula $[Cu(NH_3)_4]SO_4.H_2O$.

Thus, when copper sulphate(II) is treated with ammonia, the following reaction takes place :

$$Cu_2SO_4 + 4NH_3 = [Cu(NH_3)_4]SO_4$$

The ionic equivalent of this reaction is

$$Cu^{\cdot\cdot} + 4NH_3 = [Cu(NH_3)_4]^{\cdot\cdot}$$

Ions which, like $[Cu(NH_3)_4]^{\cdot\cdot}$, are formed by the addition of neutral molecules or other ions capable of existing independently in solution to the ion in question, are called complex ions. Salts containing such ions are referred to as complex salts. Complex acids and complex bases are also known, these compounds ionizing in solution into complex ions.

In writing the formulas of complex compounds the complex ion is usually enclosed in square brackets. This indicates that when the compound in question is dissolved in water, the complex ion remains in solution with breaking down into its composite parts.

Other salts of bivalent copper react with ammonia in a manner similar to copper sulphate(II). In all such cases the result is a dark blue solution containing the complex ion. $[Cu(NH_3)_4]^{\cdot\cdot}$.

Univalent copper combines with ammonia to give colourless complex ions of the composition $[Cu(NH_3)_2]^{\cdot}$.

Cupric hydroxide also dissolves in ammonia, forming a dark blue solution which contains $[Cu(NH_3)_4]^{\cdot\cdot}$-ion and hydroxyl-ion :

$$Cu(OH)_2 + 4NH_3 = [Cu(NH_3)_4]^{\cdot\cdot} + 2OH'$$

The resulting solution is capable of dissolving cellulose (cotton wool, filter paper, etc.). If the cellulose solution is forced through very fine apertures into dilute sulphuric acid, the cellulose separates out again in the form of fine shiny threads. That is how one of the types of artificial silk is made.

Cupric hydroxide dissolves also in very concentrated alkali solutions, forming bluish-violet solutions of cuprites, salts containing the complex ion $[Cu(OH)_4]''$.

$$Cu(OH)_2 + 2NaOH \rightleftharpoons Na_2[Cu(OH)_4]$$

or in the ionic form

$$Cu(OH)_2 + 2OH' \rightleftharpoons [Cu(OH)_4]''$$

Unlike the copper-ammonia complexes, the Cu'' ions do not add electrically neutral molecules in this case, but for negative OH' ions, forming complex anions instead of cations.

Silver

Occurrence. Silver sulphide, Ag_2S, is the principal source. It occurs as argentite or silver glance, and as complex sulphides with arsenic and antimony. These may occur alone or with base metal sulphides. About three-quarters of the silver produced comes as a by-product of the extraction of base metals chiefly lead and copper, but also zinc, nickel and tin. It is also found with gold. Silver chloride, AgCl, occurs naturally as the mineral ceragyrite. The American continent supplies most of the world's silver: since the 16th century Mexico has been the main supplier.

Metallurgy. Since most of the silver produced in this country is a by-product of the refining of copper and lead, the treatment of these latter metals for recovery of silver is of primary interest. In the electrolytic refining of copper, the silver finds its way into the anode sludge or mud along with gold, platinum, bismuth, and arsenic. The sludge, after treatment with dilute sulphuric acid removes most of the base metals, is dried and mixed with sodium carbonate, sodium nitrate, and silica. The mixture is highly heated to oxidize and flux away the remainder of the base metals. Silver remains along with any gold or platinum originally present.

Silver is recovered from lead by the Parkes process, which depends upon the greater solubility of silver in zinc. About one per cent of zinc is added to molten lead and the mixture thoroughly stirred. The silver concentrates in the layer of zinc which is skimmed off.

The zinc and silver are separated by distillation; zinc, being more volatile, passes into the distillate and may be used in treating subsequent batches of lead. The silver remaining usually contains gold and some lead and is refined by heating in a shallow hearth made of bone ash or other absorbent material. An oxidizing atmosphere is maintained which oxidizes and volatilizes some of the lead oxide; the remainder of the liquid lead oxide is absorbed by the bone ash, leaving a silver-gold alloy. The process is called *cupellation*. The silver and gold are separated or *parted* by mean of sulphuric acid or nitric acid which dissolves silver other metals present but has no effect on the gold. The silver which passes into solution may be further refined by electrolysis.

Silver ores, after being crushed finely, may be leached with a dilute sodium or potassium cyanide solution in the presence of air. Metallic silver or its compounds are soluble because of the formation of a complex cyanide :

$$4Ag + 8CN^- + O_2 + 2H_2O \rightarrow 4Ag(CN)_2^- + 4OH^-$$

The metal is precipitated from the cyanide solution by the addition of zinc shavings :

$$2Ag(CN)_2^- + Zn \rightarrow Zn(CN)_4^- + 2Ag$$

Properties. Silver is a white metal intermediate in hardness between copper and gold. It is ductile and malleable and the best conductor of heat and electricity known. Since silver lies low in the activity series, it is classed as an inactive metal. It is resistant to the action of air, water, alkalies, and the nonoxidizing acid. Sulphur and sulphur compounds attack silver, readily producing a black deposit of silver sulphide; this action is responsible for the tarnishing of the metal. Silver is also attacked by the halogen to form and adhering layer of halide salt which soon stops the action.

Use. Silver is noted for excellent qualities of electrical and thermal conductance, resistance to corrosion and reflectivity. About 90-95% of visible light may be reflected from highly polished silver but the surface is adversely affected by tranishing due to sulphur compounds. Both mechanical strength and hardness of silver are improved by alloying. In chemical engineering, silver linings are used in reaction vessels, where the high initial cost of offset by the recovery value when plant is scrapped. Sliver-lined moulds are used for casting very pure causting soda and caustic potash. Silver is used in electrical equipment as contact material in domestic switches, circuit-breakers and telephone relays. In electronic engineering, the 'skin effect', whereby current is carried by the out skin of a conductor, is utilized in radio-frequency circuits: here, silver-clad copper or silver electro deposited on brass, aluminium or steel gives the effective performance of pure silver.

Silver is used in electroplating. Articles of German silver (Cu 50, Zn 30, Ni 20 per cent by weight) are cathodes in an electrolyte of potassium dicyanoargentate(I), $K[Ag(CN)_2]$, with silver anodes. A fine, coherent, strongly adhering deposit results whereas with an electrolyte of silver nitrate, the relatively huge silver ion concentration gives a loose, spongy layer,

$$Ag(CN)_2^- \rightleftharpoons Ag^+ + 2CN^-$$

The concentration of silver ions is relatively minute but ions are replenished is required; the electrode processes are :

at the cathode, $Ag^+ + e \rightarrow Ag$

at the anode, $Ag - e \rightarrow Ag^+$

$$Ag(CN)_2^- \rightleftharpoons Ag^+ + 2CN^-$$

The concentration of the electrolyte is therefore maintained.

Pure silver is called 'fine silver' to distinguish it from an alloy called sterling silver or standard silver (Ag 92.5, Cu 7.5, percentages by weight). Except for a period in the 17th century, this alloy has been used for plate since 1300 and until 1921, for coinage. The alloy is of excellent appearance and durability, harder and more lustrous than silver itself but still liable to tranishing.

Compounds of silver

The important oxidation state of silver is + 1 although compounds of Ag(II) (e.g. AgF_2) and Ag(III) ($K^+[AgF_4]^-$ are known.

Compounds containing either silver(III) or silver(II) are powerful oxidising agents. The instability of Ag^{2+} with respect to Ag^+ in aqueous solution is indicated by the standard redox potentials for the systems Ag^{2+}/Ag^+ and Ag^+/Ag which are given below :

$$Ag^{2+} + e \rightarrow Ag^{+} \qquad E° = +1.98 \text{ v}$$

$$Ag^{+} + e \rightarrow Ag \qquad E?? = +0.80 \text{ v}$$

or $$Ag^{2+} + Ag \rightarrow 2Ag^{+} \; E^{\theta}_{total} = +1.18 \text{ v}$$

The positive e.m.f. of the above cell reaction implies that Ag^{2+} ions are thermodynamically capable of oxidizing silver to Ag^{+} ions. Notice that the disproportionation of Ag^{+} into Ag^{2+} and Ag cannot take place since the e.m.f. of this cell reaction is negative.

Silver oxide. Ag_2O, is a basic oxide with a covalent macromolecular structure. In the absence of a hydroxide of silver, it is precipitated as a dark brown, almost black, solid by addition of caustic alkali to silver nitrate solution,

$$2Ag^{+} + 2OH^{-} \rightarrow Ag_2O\downarrow + H_2O$$

After filtration, silver oxide may be washed and dried at about 80°C although it is extremely difficult to achieve complete dehydration.

A suspension of silver oxide in water is slightly alkaline, reacting blue to litmus. This basic tendency is shown by moist silver oxide when absorbing carbon dioxide from the air to form the carbonate.

$$Ag_2O + CO_2 \rightarrow Ag_2CO_3$$

Because of the insolubility of silver chloride, bromide and iodide, it is sometimes convenient to use most silver oxide for the mild hydrolysis of organic halides :

$$\underset{\text{iodoethane}}{CH_3.CH_2.I} + AgOH \rightarrow \underset{\text{ethanol}}{CH_3.CH_2.OH} + AgI$$

On the other hand, dry silver oxide yields an ether :

$$2CH_3.CH_2.I + Ag_2O \rightarrow \underset{\text{ethoxyethane (diethyl ether)}}{CH_3.CH_2.O.CH_2.CH_3} + 2AgI$$

Silver oxide is decomposed on heating of 160-2509°C:

$$2Ag_2O \rightarrow 4Ag + O_2$$

It is reduced by hydrogen below 100°C,

$$2Ag_2O + H_2 \rightarrow 2Ag + H_2O$$

and is reduced to the metal in reacting with hydrogen peroxide,

$$2Ag_2O + H_2O \rightarrow 2Ag + H_2O + O_2 \uparrow$$

Silver oxide may be precipitated from silver nitrate solution by aqueous ammonia, but the precipitate dissolves in excess to form the diammine-silver(I) cation, $Ag(NH_3)_2^+$. This solution may be used to detect reducing agents but should be freshly prepared, because on standing in air the very explosive fulminating silver, usually described as silver nitride, Ag_3N, is precipitated. In an alternative preparation, silver oxide is precipitated by the addition of dilute causting soda to silver nitrate solution and the precipitate of silver oxide just dissolved drop by drop addition of aqueous ammonia. The solution is called ammoniacal silver nitrate or ammoniacal silver oxide. By slow reduction, a silver mirror may be deposited on the walls of the containing vessel, while rapid reduction gives a coarse precipitate of silver. The reaction vessel must be clean to form a good mirror. Ethanal (acetaldehyde) is oxidized to acetic acid :

$$2Ag(NH_3)_2^+ + CH_3CHO + H_2O \rightarrow$$

$$2Ag + CH_3COO^- + 3NH_4^+ + NH_3$$

Silver acetylide, Ag_2C_2, is a white solid, explosive when dry and easily detonated by a blow or by heating to 120-140°C. With dilute nitric or sulphuric acid, pure ethyne (acetylene) is evolved :

$$Ag_2C_2 + 2H^+ \rightarrow 2Ag^+ + \begin{matrix} CH \\ ||| \\ CH \end{matrix}$$

Alternatively, warm potassium cyanide forms the complex dicyanoargentate(I), liberating ethyne:

$$Ag_2C_2 + 4CN^- + 2H_2O \rightarrow 2Ag(CN)_2 + 2OH^- + C_2H_2$$

Silver carbonate, Ag_2CO_3, is conveniently prepared as a yellow precipitate formed on mixing solutions of silver nitrate and sodium carbonate. It may also be precipitated by passing carbon dioxide into ammoniacal silver nitrate. Silver carbonate darkens in light and on heating decomposes to give as final products, silver, oxygen and carbon dioxide :

Silver cyanide, AgCN, is formed as a white precipitate on mixing silver nitrate solution and potassium cyanide and is soluble in an excess of the latter to form the complex dicyanoargentat(I) anion etc.,

$$Ag^+ + CN^- \rightarrow AgCN\downarrow$$

$$AgCN + CN^- \rightarrow Ag(CN)_2^- \text{ (and } Ag(CN)_3^{2-})$$

Silver cyanide has a linear macromolecular structure of repeating AgCN units.

The silver halides

Silver fluoride differs from the other halides and is unusual in forming hydrates. The anhydrous salt, formed by union of the elements, is deep yellow in colour, ionic (with the rock salt lattice) and forming two colourless hydrates, $AgF.4H_2O$ stable below 188.7°C and $AgF.2H_2O$ stable up to 39.5°C. Silver fluoride is very soluble in water (179 g saturating 100 g water at 25°C) and very deliquescent.

The other silver halides (AgCl, AgBr, AgI) are very insoluble in water and made by mixing solutions containing the appropriate ions. Curdy and partly colloidal precipitates are formed. Silver chloride is white, 3.0×10^{-3} g dissolving in 100 g water at 25°C, and ionic in structure. Silver bromide is cream, 5.5×10^{-4} g dissolving in 100 g water at 25°C and also ionic. Silver iodide is pale yellow, and has a complex covalent macromolecular structure with 5.6×10^{-6} g dissolving in 100 g water at 25°C.

Although silver chloride is precipitated as white curds, on heating it fuses at 449°C and solidifies on cooling to a translucent horny mass. Horn-silver is a naturally occurring form of silver chloride. Silver chloride is sensitive to light, and rapidly turns violet after precipitation in sunlight. It is used in printing-out papers in photography. Although very insoluble in water, silver, chloride will dissolve in reagents which form silver complexes : it is readily soluble in aqueous ammónia to give the diamminesliver(I) cation,

$$AgCl+2H_3 \rightleftharpoons [Ag(NH_3)_2]^+ + Cl^-$$

while potassium cyanide solution forms the dicyanoargentate(I) anion, $[Ag(CN)_2]^-$,

$$AgCl + 2CN^- \rightleftharpoons [Ag(CN)_2]^- + Cl^-$$

and aqueous sodium thiosulphate, a series of thiosulphatoargentate(I) anions,

$$[Ag(S_2O_3)]^-, [Ag(S_2O_3)_2]^{3-}, [Ag(S_2O_3)_2]^{5-}$$

the formation of which is used to remove unchanged silver halide after development in photography. Silver chloride is about one hundred times more soluble in normal hydrochloric acid than in water, due to the formation of chloroargentate(I) complexes, $AgCl_2^-$ and $AgCl_3^{2-}$. They are unimportant.

Much silver chloride (and thiocyanate) may be accumulated as silver residues in the laboratory after volumetric work. Because of the value placed on silver, recovery becomes worth while. Silver chloride may be fused with sodium carbonate, or fusion mixture, when silver results from the breakdown of carbonate and oxide,

$$2Na_2CO_3 + 4AgCl \rightarrow 4Ag + O_2 + 2CO_2 + 4NaCl$$

Silver chloride is reduced by zinc or aluminium in dilute acid, or by heating in hydrogen:

$$Zn + 2AgCl \rightarrow Zn^{2+} + 2Cl^- + 2Ag$$

$$2AgCl + H_2 \rightarrow 2Ag + 2HCl$$

Silver bromide is similar to silver chloride, but is even less soluble in water. It dissolves in the complex forming reagents, but is only sparingly soluble in aqueous ammonia. Silver bromide is more photosensitive than either silver chloride or silver iodide. Silver bromide is used in various photographic emulsions. Silver iodide is the most insoluble halide, virtually insoluble in ammonia and only moderately soluble in sodium thiosulphate solution.

Silver sulphide, Ag_2S

The familiar black stain that forms on the surface of silver articles which have been exposed to an atmosphere containing hydrogen sulphide is silver sulphide. It precipitates as a black solid when hydrogen sulphide is bubbled through a solution of a silver salt :

$$2Ag^+ + H_2S \rightarrow Ag_2S + 2H^+$$

Since it is exceedingly insoluble in water, it can be precipitated from solutions containing only a minute concentration of silver ions. e.g. from solutions in which the silver ions are bound in the form of stable complexes.

Silver nitrate, $AgNO_3$

This is the most important compound of silver and is prepared industrially by the action of nitric acid on silver. It is very soluble in water, giving a colourless solutions which contains hydrated Ag^+ ions. The solid, which is also colourless, readily decomposes on heating into silver, oxygen and nitrogen dioxide :

$$2AgNO_3 \rightarrow 2Ag + 2NO_2 + O_2$$

An ammoniacal solution of silver nitrate contains in $[Ag(NH_3)_2]^-$ complex ion, and this solution reacts with ethyne, forming a precipitate of silver carbide, Ag_2C_2, which explodes if struck when dry :

$$C_2H_2 + 2[Ag(NH_3)_2]^+ \rightarrow Ag_2C_2 + 2NH_4^+ + 2NH_3$$

The action of dilute nitric acid on sliver carbide regenerates pure ethyne and this reaction may be used to purify this hydrocarbon. e.g. to separate it from alkanes and alkenes.

Silver nitrate is used industrially for the preparation of other silver compounds, particularly silver halides which are used in photography. In the laboratory, a standard solution of silver nitrate is used for the volumetric estimation of chloride, bromide and iodide ions in neutral solution; one method uses a 5 per cent solution of potassium chromate as indicator. The procedure is as follows :

A measured amount of the neutral halide solution, e.g. sodium chloride, is taken and about 5 cm^3 of 5 per cent potassium chromate solution are added to act as indicator. Standard silver nitrate is added from a burette, when white silver chloride precipitates; at the end-point the slight excess of silver ions precipitates silver chromate (pink). Silver chromate is more soluble than silver chloride and thus does not precipitate until all the chloride ions have precipitated as silver chloride. Since silver chromate does not precipitate in acid solution the halide solution must be neutral (alkaline solutions would lead to the precipitation of silver oxide).

$$Ag^+ + Cl^- \rightarrow \underset{\text{white}}{AgCl}$$

$$\underset{\text{slight excess}}{2Ag^+} + CrO_4^{2-} \rightarrow \underset{\text{pink}}{Ag_2SO_4}$$

Silver sulphate, Ag_2SO_4

Silver sulphate precipitates as a white solid when a solution of a sulphate is added to a fairly concentrated solution of silver nitrate, i.e. it is moderately insoluble :

$$2Ag^+ + SO_4^{2-} \rightarrow Ag_2SO_4$$

It is of no real importance.

Complexes containing silver(I)

Typical complexes containing silver(I) include $[Ag(NH_3)_2]^+$, $[Ag(S_2O_3)_2]^{3-}$ and $[Ag(CN)_2]^-$; the thiosulphate complex is of importance in the 'fixing' process in photography, while the cyanide complex is used during the extraction of silver. The stability of these three complexes decreases in the order.

$$[Ag(CN)_2]^- > [Ag(S_2O_3)_2]^{3-} > [Ag(MH_3)_2]^+$$

The cyanide and thiosulphate complexes can be shown to be more stable than the diammine complex by relatively simple experiments. Thus the addition of chloride ions to a solution of a silver salt results in the formation of a white precipitate of silver chloride which readily dissolves in dilute ammonia solution. The addition of bromide ions to this solution results in the precipitation of silver bromide, i.e. although the equilibrium shown below is well over to the right, there are sufficient 'free' silver ions present to exceed the solubility product of silver bromide.

$$Ag^+ + 2NH_3 \rightleftharpoons [Ag(NH_3)_2]^+$$

The addition of solutions of either sodium cyanide or sodium thiosulphate results in the precipitate of silver bromide dissolving, i.e. the concentration of 'free' silver ions is reduced to such a low value that the solubility product of silver bromide is no longer exceeded. The stability constants (equilibrium constants) for solutions of the three complexes are given below :

$$Ag^+ + 2NH_3 \rightleftharpoons [Ag(NH_3)_2]^+ \qquad K = 1.7 \times 10^7 \text{ dm}^6 \text{ mol}^{-2}$$

$$Ag^+ + 2S_2O_3 \rightleftharpoons [Ag(S_2O_3)_2]^{3-} \qquad K = 1.7 \times 10^{13} \text{ dm}^6 \text{ mol}^{-2}$$

$$Ag^+ + 2CN^- \rightleftharpoons [Ag(CN)_2]^- \qquad K = 5.5 \times 10^{18} \text{ dm}^6 \text{ mol}^{-2}$$

Since the passage of hydrogen sulphide through an aqueous solution of the cyanide complex results in the precipitation of silver sulphide, the solubility product of silver sulphide must be exceedingly small, i.e., the minute concentration of 'free' silver ions in this solution is sufficient to cause silver sulphide to precipitate.

Photography. Silver salts, especially the chloride and the bromide, owing to their ability to decompose under the action of light with the liberation of metallic silver, are widely used in photography for the preparation of sensitized plates, films and papers.

Photographic plates and films are prepared as follow. A certain amount of gelatine is added to a warm solution of silver nitrate, which is then mixed with a solution of KBr. Silver bromide and potassium nitrate form according to the equation :

$$AgNO_3 + KBr = AgBr + KNO_3$$

Owing to the protective action of the gelatine, the silver bromide does not precipitate, but remains suspended in a very finely divided state. The resulting turbid liquid is referred to as "silver bromide emulsion" (though it would be more correct to call it a suspension). As AgBr is not sensitive enough when very finely divided, the emulsion is left standing for a time in a warm place until it becomes *ripe*, i.e., until the degree of dispersity of the silver bromide decreases and larger particles are formed; this increases the sensitivity of the AgBr by hundreds of times.

After ripening, the emulsion is cooled, and the resulting gelatine jelly is crushed and washed with water to remove the KNO_3. Then it is melted and applied to glass plates or celluloid films. All these processes are carried out in red light which does not act on silver bromide.

If a photographic plate is exposed to light is darkens very slowly. However, this process can be greatly accelerated by the action on the plate of various substances called *developers*. It is remarkable that is a developer is capable of decomposing (reducing) silver bromide only after the latter has been exposed to light for at least a very short time (fractions of a second).

To obtain an image of any subject on a photographic plate, the latter is placed in camera and is exposed for a very short time to the rays admitted by the lens (rather, by a system of converging lenses) of the camera, which throw the image of the subject onto the plate. If the plate is then immersed in a developer solution (this being done in a dark room by red light) the exposed parts of the plate darken rapidly, and the image of the subject photographed appears on the plate. This process is called *developing*. After the image has been developed, it is fixed by immersing the plate in a solution of sodium thiosulphate (hypo) $Na_2S_2O_2$. The undecomposed silver bromide dissolves in this solution, the silver and thiosulphate ions combining into complex ions, probably of the composition $[Ag(S_2O_3)]'$:

This gives the *negative*, a reversed image, on which the light parts of the subject appear dark and vice versa.

The normal image or *positive* can be printed from the negative on sensitized paper, prepared in the same manner as the plate. For this purpose the negative is placed on the paper and exposed to light for a short time, then the paper is developed and fixed.

Some types of paper have silver chloride in their sensitized layer instead of silver bromide. Printing on such paper does not require subsequent developing, the paper being held in the light under the negative until a visible image appears on it, after which the prints is fixed.

Gold

History and occurrence. Without question, gold was one of the first metals known to man. History records the use of the metal for jewellery and ornamental purposes in very early times. The fact that the metal occurs free, is pleasing in appearance, and is easily worked would account for its early use. The symbol, Au, is derived from the Latin, *aurum*. Gold occupies a unique place in chemical history since the principal pursuit of the alchemists was to find a method for transmutation of the baser metals into gold.

Gold nearly always occurs in the negative state either in placer deposits in river sands or in veins disseminated in quartz. Native gold is occasionally found in the form of nuggets; one such nugget found in Australia weighed more than one hundred pounds and was worth $ 50,000. The only combined source of gold of any consequence is the telluride mineral, $AgAuTe_2$, small deposits of which are found in Colourado. Some gold is usually associated with copper and lead ores. While gold is widely distributed over the earth, relative few deposits are worth working. A very small amount of gold is present in sea water-less than 0.00005 gram per ton of water. More than 50 per cent of the world's supply of gold comes from mines in South Africa.

Metallurgy. *Placer mining* is based upon the high specific gravity of gold, which occurs usually as small particles mixed with sand and gravel. The sand may be washed in pans or long troughs called sluices, and since the particles of gold are heavier, they settle to the bottom and are retained behind cleats or riffles, while the lighter sand particles are washed away.

Gold ore obtained from hard rock vein deposits is crushed to a powdery condition, after which a water suspension of the ore is passed over copper plates covered with a coating of mercury. Mercury amalgamates with the gold and at intervals the copper plates are scraped. The amalgam (Hg-Au) is heated and the mercury distilled away from the gold, which remains as a residue. The recovered mercury is used over and over again. The tailings from this treatment may be leached with cyanide solution, which removes more of the gold.

If a gold ore is "free milling" (gold not intimately associated with other metal sulphides) and if the gold is in the ore in a quite finely divided state, it may be treated with sodium cyanide solution after having first been ground to a porous granular mass. Like silver, gold is quite soluble in NaCN or KCN solution in the presence of air :

$$4Au + 8CN^- + O_2 + 2H_2O \rightarrow 4Au(CN)_2 + 4OH^-$$

The gold may be recovered from the cyanide solution by adding zinc dust or allowing the solution to flow over zinc shavings; the zinc replaces and precipitates the gold.

The recovery of gold and silver as by-products of the smelting of copper and lead ores has already been discussed.

An alloy of gold and silver may be treated with nitric acid and the silver dissolved away from the gold, provided the content of the of gold in the alloy is not greater than 25 per cent. An alloy of greater gold content is not soluble in the acid. More silver may be added to lower the gold content below 25 per cent or the alloy may be refined by electrolysis. The impure gold is made the anode in a cell with gold chloride and hydrochloric acid as the electrolyte and a sheet of pure gold as the cathode. Passage of the current causes gold to pass into solution at the anode and pure gold is deposited on the cathode.

Properties of gold

Gold has a melting point of 1063°C and a density of 19.3 g cm^{-3}. It is extremely malleable and ductile, e.g. it can be beaten into sheets no thicker than 0.00001 mm and pulled into wire of extremely small diameter; its thermal and electrical conductivities are very high. The metal is one of the most unreactive elements; it is not attacked by air, water or steam, and the common mineral acids leave the metal untouched. A mixture of concentrated hydrochloric acid and concentrated nitric acid (aqua regia), however, attacks the metal with the formation of the $[AuCl_4]^-$ complex ion which contains gold(III) :

$$Au + 6HNO_3 \rightleftharpoons Au^{3+} + 3NO_3^- + 3H_2O + 3NO_2$$

$$+$$

$$4Cl^- \text{(from the conc. } HCl) \rightleftharpoons [AuCl_4]^-$$

It is thought that the gold is first oxidized by nitric acid to Au^{3+} which is then removed by complexing with the chloride ions.

Uses. Gold is the bases of many currencies. The metal is very malleable and ductile; a 10 mm cube may be beaten out to the size of a football field and 1 g drawn into 2000 metres of fine wire. It is not tarnished in air and is used in jewellery and ornamentation, usually alloyed with copper, or sometimes silver and zinc, to increase hardness. The purity is reckoned in terms of 24 units or carats; pure gold is 24 carat, and 22 carat, called British Standard gold, has two parts in twenty four of copper. Electrodeposited gold and alloys with silver and platinum are used in electrical and electronic engineering where a constant contact resistance is desired. Gold is sometimes used as a lining to reaction vessels for extreme conditions but platinum, which has better mechanical properties, is preferred.

Gold (I) Compounds

Aurous alkyls and aryls. Gold aryls can be made in ethereal solution by the action of the appropriate aryl magnesium bromide on gold carbonyl chloride, AuClCO, but they are very unstable. Aurous alkyls are no known.

Aurous oxide, Au_2O, is precipitated as a violet powder when dilute potassium hydroxide solution is added to aurous chloride, or to $K[AuBr_2]$. If aurous oxide is heated to 225°C it yields a mixture of gold and auric oxide. It reacts with ammonia to form $NAu_3(NH_3)$.

Aurous sulphide, Au_2S, an insoluble dark brown solid, is made by passing hydrogen sulphide through a solution of $K[AuBr_2]$. It is unaffected by water or dilute acids, but it dissolves in aqua regia and in potassium cyanide solution.

Aurous cyanide, AuCN, exists as yellow, very insoluble, crystals. It is formed when auric cyanide decomposes, or by heating $K[Au(CN)_2]$ with hydrochloric acid to 50°. The arrangement of atoms in crystalline aurous cyanide is similar to that in argentous cyanide. Aurous cyanide is more stable than either cuprous cyanide or argentous cyanide, but if heated to a red heat is decomposes leaving a deposit of gold. It has a strong tendency to complex formation.

The aurous halides. Aurous fluoride does not exist; it would tend to be ionic, and the aurous ion is itself unstable.

Aurous chloride, AuCl, an insoluble yellow solid, is made by heating auric chloride to 175°. At higher temperatures aurous chloride decomposes to gold and chlorine. With water it disproportionates yielding gold and auric chloride. There is some evidence that aurous chloride is dimeric.

Aurous bromide, AuBr, is made by warming auric bromide at 115°. When heated or moistened it behaves like aurous chloride.

Aurous iodide, AuI, is less soluble than either the chloride or the bromide, and hence is easier to prepare, and more stable when made. It is formed when auric iodide decomposes at room temperature, when hydrogen iodide reacts with auric oxide, when iodine reacts with gold, and when potassium iodide reacts with auric chloride.

Gold (III) Compounds

The auric compounds are more stable than the aurous compounds. The simple ion Au^{3+} does not exist; auric compounds are covalent. In the auric compounds the co-ordination number of the gold atom is always four, and the arrangement of other atoms round the gold atoms is planar. Many auric compounds contain a gold atom with valencies :

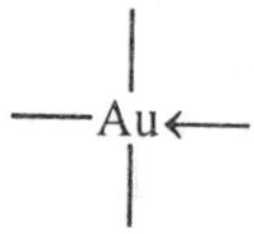

Because of the strong tendency of the auric gold atom to become 4 co-ordinate, the auric compounds fall into two classes, the giant molecule compounds and complexes.

Auric hydroxide, AuOOH, is obtained by treating with nitric acid the precipitate $Mg[AuO_2]$ formed by the action of magnesia on an aqueous solution of auric chloride. If the hydroxide is gently heated at 140°, auric oxide is obtained.

Auric oxide, Au_2O_3, is a brown solid. It decomposes at 155° to aurous oxide and oxygen. Auric oxide is soluble in potassium hydroxide to yield potassium aurate, $K[Au(OH)_4]$, H_2O, and a aqueous ammonia to give fulminating gold, a dark powder of uncertain composition and constitution.

The auric halides. *Auric fluoride*, AuF_3, can be made by dissolving gold in BrF_2 to give crystalline $[BrF_2^+][AuF_4^-]$, and heating this to 300°. AuF_3 is decomposed on heating to 500°, or on treatment with water. Its structure is uncertain. The other auric halides have dimeric planar structure :

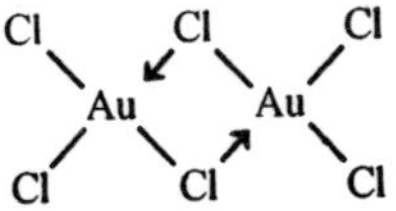

Auric chloride, $(AuCl_3)_2$, a brown solid, is made by heating at 120° crystals of aurichloric acid $H[AuCl_4]$, $4H_2O$. It can be also be obtained by the action of chlorine water on gold. It decomposes to aurous chloride and chlorine at 175°. It is soluble in water, alcohol, and ether. Molecular weight determinations have shown that the vapour is dimeric.

Auric bromide, $AuBr_3$, is a dark brown substance which very easily loses bromine to give aurous bromide. It is made by heating finely divided gold with bromine, or by gently warming $H[AuBr_4]$. It is dimeric in solution in bromine.

Auric iodide, AuI_3, is a dark green nearly insoluble solid which decomposes very easily. It is made by the action of auric chloride on potassium iodide solution:

$$AuCl_3 + 4KI = K[AuI_4] + 3KCl$$
$$3K[AuI_4] + AuCl_3 = 4AuI_3 + 3KCl.$$

Auric alkyls. These compounds furnish a good example of the tendency of Au^{III} to become tetravalent. There are three types of auric alkyls, but instead of the simple constitutions :

AuR_3 AuR_2Cl $AurCl_2$

the posses the complex constitutions:

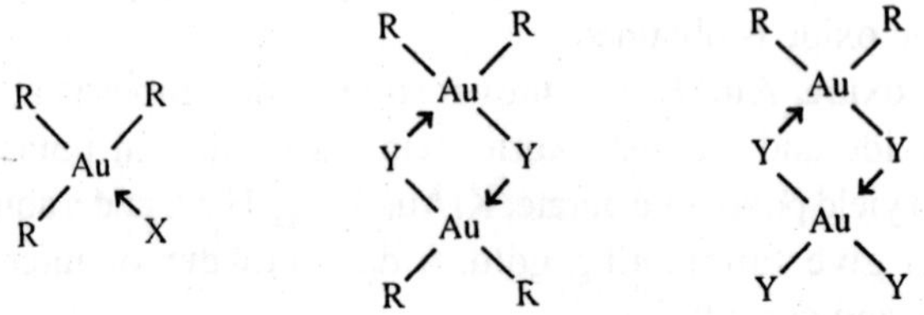

where R is a alkyl radical, X represents pyridine, or tolylamine $N(CH_2C_6H_5)\ H_2$, and Y represents Cl, Br, or CN.

The etherates and ammines of the auric trialkyls are very unstable; the diethyl etherate of the trimethyl derivative decomposes at – 35°.

26

d-Block Elements (Zinc, Cadmium and Mercury)—Group IIB (12)

26.1 : Some physical data of Group IIB elements

	Atomic Number	Electronic Configuration	Atomic Radius (Å)	Ionic Radius (Å) M^{2+}	M.p. °C	B.p. °C	Density ($g\ cm^{-3}$)	Standard Redox Potential (volts) $M^{2+} + 2e \rightarrow M$
Zn	30	2.8.18.2 .....$3s^2 3p^6 3d^{10} 4s^2$	1.25	0.74	419	907	7.1	− 0.76
Cd	48	2.8.18.18.2 .....$4s^2 4p^6 4d^{10} 5s^2$	1.41	0.97	321	768	8.6	− 0.40
Hg	80	2.8.18.32.18.2 .....$5s^2 5p^6 5d^{10} 6s^2$	1.44	1.10	− 39	357	13.6	+ 0.85

The element of this subgroup zinc, cadmium and mercury, are characterized by the presence of two electrons in the outer shell of their atoms and eighteen in the penultimate shell.

	Group IIA		Group IIB
Calcium	2.8.8.2	Zinc	2.8.18.2
Strontium	2.8.18.8.2	Cadmium	2.8.18.18.2
Barium	2.8.18.18.8.2	Mercury	2.8.18.32.18.2

Although there is little in common between comparable members of Groups $IIA_{(II)}$ and $IIB_{(II)}$, zinc does resemble magnesium in some respects (magnesium immediately precedes calcium in Group IIA$^{(II)}$). e.g. zinc and magnesium form a number of isomorphous compounds, their chlorides are partially covalent, deliquescent, and acidic in solution owing to solutions containing zinc and magnesium ions results in the precipitation of basic carbonates.

There is a marked difference in physical properties between Group IIB metals and those of Group IB (copper, silver and gold); for example, the melting points, boiling points and thermal and electrical conductivities of zinc, cadmium and mercury are very much lower than those of copper, silver and gold. These differences are surprising in view of the fact that both groups of metals are assigned completely filled *d* levels; clearly the *d* electrons of the Group IIB metals cannot participate in inter-atomic bonding to any great extent, neither are they readily available for thermal and electrical conductivity although the precise reason why this should be so is not fully understood. The fact that mercury is a liquid is attributed to the inert nature of the two 6s electrons (the inert pair effect), and bismuth; once again, this effect is not fully understood but it is a feature of the chemistry of the heavier elements.

Zinc and cadmium have negative electrode potentials (unlike copper and silver) and, except when very pure, displace hydrogen from dilute non-oxidizing acids (the pure metals have high hydrogen overpotentials; mercury, on the other hand, has a positive electrode potential and is chemically rather inert, like silver.

It is instructive to compare the standard electrode potentials of zinc and copper and to enquire into the reasons for the large difference between them :

$$Zn(s) \rightarrow Zn^{2+}(aq) + 2e \quad E^{\phi} = -0.76 \text{ v}$$
$$Cu(s) \rightarrow Cu^{2+}(aq) + 2e \quad E^{\phi} = +0.34 \text{ v}$$

Zinc, with a negative standard electrode potential, has a greater tendency to form hydrated ions than copper, with a positive standard electrode potential. The process of a metal passing into solution as its hydrated ions can be considered to take place in three hypothetical stages:

$$Zn(s) \rightarrow Zn(g) \qquad S = +130.5 \text{ kJ/mol}$$
$$Zn(g) \rightarrow Zn^{2+}(g) + 2e \qquad I = +2652 \text{ kJ/mol}$$
$$Zn^{2+}(g) + \text{water} \rightarrow Zn^{2+}(aq) \qquad \Delta H = -2046 \text{ kJ/mol}$$

The overall process is endothermic to the extent of 736.5 kJ/mol, i.e.

$$Zn(s) + water \rightarrow Zn^{2+}(aq) + 2e \quad \Delta H_{total} = +736.5 \text{ kJ/mol}$$

For copper the corresponding changes are :

$$Cu(s) \rightarrow Cu(g) \qquad S = +339.3 \text{ kJ/mol}$$
$$Cu(g) \rightarrow Cu^{2+}(g) + 2e \qquad I = +2715 \text{ kJ/mol}$$
$$Cu^{2+}(g) + water \rightarrow Cu^{2+}(aq) \quad \Delta H = -2102 \text{ kJ/mol}$$

The overall process is endothermic to the extent of 952.3 kJ/mol, i.e.

$$Cu(s) + water \rightarrow Cu^{2+}(aq) + 2e \quad \Delta H_{total} = +952.3 \text{ kJ/mol}$$

The process of zinc metal passing into solution as its hydrated ions is less endothermic to the extent of 215.8 kJ/mol than the similar process for copper metal. Of course this amount of heat is evolved when 1 mole of zinc displaces 1 mole of copper from copper(II) sulphate solution according to the equation :

$$Zn + Cu^{2+}(aq) \rightarrow Zn^{2+}(aq) + Cu \quad \Delta H = -215.8 \text{ kJ/mol}$$

It can be seen that the larger ionisation energy(I) of the copper atom is almost counter balanced by the larger hydration energy energy (ΔH) of the copper(II) ion, i.e. the sum of $I + \Delta H$ for both zinc and copper is almost identical. The difference between the sublimation energies (S) for the two metals is almost 209 kJ/mol; thus the principal reason why zinc has a negative electrode potential as compared with that of copper is because zinc has a lower sublimation energy, i.e. zinc has lower melting and boiling points than copper. Notice that rubidium is a more reactive metal than silver for a different reason.

Zinc cadmium and mercury are non-transition metals, i.e. they do not exhibit oxidation states in which d electrons are involved; but, like transition metals, they do form a wide range of complexes, particularly with ligands such as ammonium, cyanide ions and halide ions. In general the mercury complexes are the most stable and the zinc complexes the least stable. although this trend is reversed when the bonded atom is oxygen, e.g. zinc oxide is amphoteric and forms the zincate anion, $[Zn(OH)_4]^{2-}$, whereas neither cadmium oxide nor mercury(II) oxide is amphoteric. All three metals exhibit a principal oxidation state of + 2 and, in addition, mercury forms the unique mercury(I) ion, Hg_2^{2+}, in which two mercury atoms are covalently bonded. Mercury, in particular, displays a marked tendency to form covalent bonds, which accounts for the large number of organometallic compounds formed by this element.

Zinc and cadmium are sufficiently similar elements to be treated comparatively; there are few points of similarity between cadmium and mercury, and the latter is treated separately.

Zinc

Occurrence

Zinc probably never occurs in a native state, but is widely distributed in compounds, the chief mineral being the sulphide, ZnS, or *sphalerite*. The element has been found in alloys dating back to ancient times.

When zinc occurs as the sulphide, it is first converted to the oxide by roasting in air.

$$2ZnS + 3O_2 \rightarrow 2ZnO + 2SO_2$$

Some zinc, however, occurs as the oxide, so that the roasting step may be eliminated. A considerable quantity of zinc is sold as zinc oxide, so that no other process may be necessary. When the metal itself is required, the zinc oxide must be reduced. There are two way in which this reduction is achieved industrially. The reducing agent may be carbon, in which case a mixture of the oxide and coke is strongly heated.

$$ZnO + C \rightarrow Zn + CO$$

The escaping carbon monoxide is burned, and the zinc volatilizes into receivers where it is collected as a liquid.

The other process for zinc reduction utilizes electrolysis from a solution of zinc sulphate in water. It is rather surprising that a metal standing fairly high in the electromotive series can be deposited by electrolysis from a water solution, and the process requires considerable care. The zinc oxide is dissolved in sulphuric acid. The solution is then fed to large tanks containing alternate rows of aluminium cathodes and lead anodes. During electrolysis the zinc deposits on the cathodes from which it is stripped off from time to time. While the zinc is being deposited, the concentration of sulphuric acid in the cells rises.

$$2Zn^{2+} + 2H_2O(\text{electrolysis}) \rightarrow 2Zn + 4H^+ + O_2$$

The product at the anode is oxygen, which is allowed to escape. The sulphuric acid thus generated is used for dissolving more zinc oxide, so that little acid has to be added after the process is in operation.

Properties of zinc. Zinc is a crystalline, silver-white metal which is brittle at ordinary temperatures but may be rolled into sheets at a temperature of 120° to 150°. Above 200° it again become brittle. On contact with air an adherent film of basic carbonate forms on the surface of the metal which prevents further corrosion.

Zinc is fairly high in the activity series and therefore quite an active metal. It reacts readily with acids to produce hydrogen and displaces less active metals from their salts. The action of acids on impure zinc is much more rapid than on very pure zinc, since in the latter case bubbles of hydrogen gas collect on the surface of the metal and slow down the action. If another metal is present as an impurity, the hydrogen is liberated from the surface of the contaminating metal rather than from the zinc. An electric couple of facilitate the action is probably set up between the two metals.

Uses of zinc. Principal uses and distribution of zinc are shown in Figure 26.1. The chief use in the manufacture of *galvanized iron*. A thoroughly cleaned iron sheet is dipped into molten zinc, and a thin layer of zinc forms on the surface of the iron. Galvanized iron is much more resistant to corrosion than plain iron, since the zinc forms a protective basic carbonate coating. In another process called *sherardizing*, zinc dust and zinc oxide are sprinkled on the surface of an iron sheet and heated. A tightly adhering layer is thus formed. In *electroglavanizing* of iron objects, the iron article to be coated is made the cathode in a cell and drawn slowly through the bath containing zinc ions. Although this method is more expensive than the others, it has the advantage of forming a layer of an desired thickness, and the layer is of uniform thickness and adheres more tightly to the iron. Because of its resistance to corrosion, galvanized iron is used for roofing, gutters, and cornices.

Galvanized iron has the advantage over tin plate (sheet iron coated with tin) of resisting corrosion longer after the surface is broken and the iron exposed. Since zinc is more active than iron, it corrodes first, thus protecting the iron. Once the surface of tin plate is broken, the iron corrodes first, since it is more active than tin. Important alloys of zinc are brass (Cu, Zn), *bearing metal* (Zn, Cu, Sn, Sb) and *German silver* (Cu, Zn, Ni).

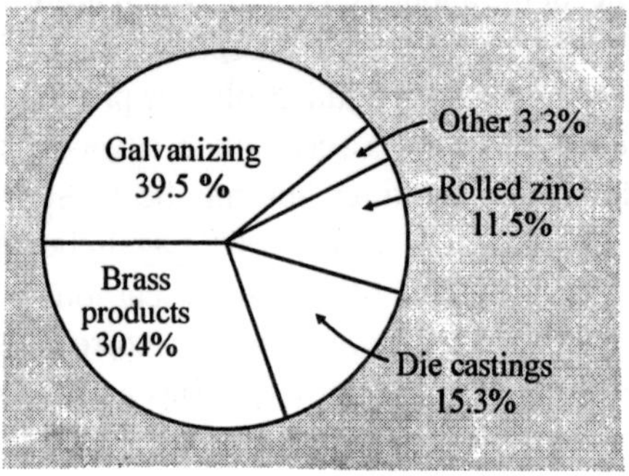

Fig. 26.2 : Uses of metallic zinc

Compound of Zinc

Zinc sulphate $ZnSO_4$ crystallizes out of aqueous solution as a crystal hydrate of the composition $ZnSO_4.7H_2O$, in which form it is known as *white vitriol*. Is used in dyeing and cotton printing, for the preparation of metallic zinc by electrolysis and is medicine as an astringent. It is also an initial material for the preparation of other zinc compounds.

Zinc chloride $ZnCl_2$. Solutions of this salt are used for impregnating railways ties to protect them from rotting. Zinc chloride is used also in soldering. It cleanses the surface to be soldered from oxides and prevents oxidation of the metal at the moment of soldering.

Some zinc chloride is employed in the manufacture of vegetable parchment paper. The unglued paper is treated with a concentrated solution of zinc chloride which partially decomposes the surface layers of the cellulose, filling the pores of the paper with the decomposition products. Then the paper is thoroughly washed and treated with a solution of glycerine to make it soft and flexible.

Zinc oxide, ZnO, the widest used zinc compound, is a white powder at ordinary temperatures which charges to a yellow colour on heating. When cooled, it again becomes white. Zinc oxide is obtained by burning zinc in air. by heating the basic carbonate, or by roasting the sulphide. Principal use of ZnO is in rubber manufacture as a filler, particularly in automobile tires. As a body for paints, it has the advantage over white lead of not darkening on exposure to an atmosphere containing hydrogen sulphide. Its covering power, however, is inferior to that of white lead. Thus, it is used primarily for interior paints. Because of its mild alkaline and antiseptic properties, zinc oxide is also used in ointments and medicinally as an antiseptic powder.

Zinc hydroxide, $Zn(OH)_2$, forms as a white precipitate when sodium or potassium hydroxide is added to a solution of a zinc salt :

$$Zn^{++} + 2OH^- \rightarrow Zn(OH)_2 \downarrow$$

The hydroxide is amphoteric and dissolves in both acids and bases. An excess of sodium hydroxide will redissolve the precipitated zinc hydroxide with the formation of zincate ions :

$$Zn(OH)_2 + 2OH^- \rightarrow Zn(OH)_4^{--} \text{ or } ZnO_2^{--} \cdot 2H_2O$$

The hydroxide is also soluble in ammonium hydroxide because of the formation of an ammonia complex :

$$Zn(OH)_2 + 4NH_3 \rightarrow Zn(NH_3)_4^{++} + 2OH^-$$

Zinc sulphide, ZnS, a white salt, is precipitated in neutral or alkaline solution with hydrogen sulphide :

$$Zn^{++} + H_2O + 2OH^- \rightarrow ZnS \downarrow + 2H_2O$$

Zinc sulphide of an ingredient of *lithopone*. a mixture of barium sulphate and zinc sulphide, used as a paint body and produced by the reaction :

$$BaS + ZnSO_4 \rightarrow BaSO_4 \downarrow + ZnS \downarrow$$

Zinc chloride, $ZnCl_2$, hydrolyzes readily in water to give an acid solution. The salt is highly poisonous and because of its antiseptic action is used as a wood preservative. It finds limited use as a cleaner for metals before soldering and mixed with zinc oxide produces a certain type of cement.

Zinc Nitrate

Zinc nitrate can be obtained by the methods similar to those used to prepare sulphates. Zinc nitrate is very deliquescent and since it is soluble in ethanol it presumably possess considerable covalent character. Zinc nitrate on heating decompose into the metallic oxide, No_2 and oxygen.

$$2Zn(NO_3)_2 \rightarrow 2ZnO + 4No_2 + O_2$$

Zinc carbonate ($ZnCo_3$)

Zinc carbonate precipitate as white solids when sodium hydrogen carbonate solu. is added to an aqueous solution of the metallic salt.

Cadmium

Occurrence and metallurgy. Because of its close similarity to zinc, cadmium is a constituent of zinc ores in the proportion of about 1 part in 200. The mineral CdS (called *greenokite*) is also known but is relatively unimportant as a source of cadmium. Cadmium is obtained as a by-product in the smelting of zinc ores. After zinc sulphide ore is roasted, both cadmium and zinc are present as oxides. The cadmium oxide is more easily reduced than zinc oxide, and cadmium is more volatile than zinc; hence the greater portion of cadmium metal is found in the first portion of the distillate. By fractional distillation, highly pure cadmium is recovered.

Properties and uses. Cadmium, a white metal malleable than zinc, is very similar to zinc in chemical properties, appearing just below zinc in the activity series. Its compounds, like those of zinc, are all divalent. The addition of a strong base to a solution of a soluble salt produces a white precipitate of cadmium hydroxide. Unlike zinc hydroxide, cadmium hydroxide is not amphoteric; it will not dissolve in an excess of the base. However, the hydroxide is soluble in ammonium hydroxide because, like zinc, cadmium ions form a complex with ammonia, $Cd(NH_3)_4^{++}$.

Cadmium is becoming of increasing economic importance. Its principal uses are in low-melting alloys for fuses and automatic fire sprinklers, and in production of bearing metal. It offers promise as a plating material in place of zinc and tin. Alloyed with copper, it produces a product of durability, hardness, and tensile strength which is used for telegraphic and power transmission lines. An alloy of cadmium is used in making jewellery. Cadmium rods find use in atomic energy piles because the metal is a good neutron absorber.

Compounds of Cadmium

Cadmium oxides, CdO and Cd_2O. Cadmium forms two oxides—*cadmous*, Cd_2O, and *cadmium*, CdO. Cadmous oxide is obtained by fusing cadmium metal in cadmium chloride and treating the product with water; at first cadmous hydroxide, CdOH, is precipitated which on gently heating yields cadmous oxide :

$$CdCl_2 + Cd \rightarrow 2CdCl$$
$$CdCl + H_2O \rightarrow CdOH + HCl$$
$$2CdOH \rightarrow Cd_2O + H_2O$$

Cadmous oxide is a yellow power.

When the metal burns in air, brown fumes of cadmium oxide CdO, are obtained. This oxide has a rich brown colour.

Cadmium hydroxide, $Cd(OH)_2$. When caustic soda solution is added to a cadmium salt solution, a white precipitate of cadmium hydroxide, $Cd(OH)_2$ is obtained. This precipitate is soluble in acids and in ammonia, but not is excess of causting alkali. With ammonia, complex hydroxide, $Cd(NH_3)_2(OH)_2$ is obtained. Cadmium hydroxide absorbs carbon dioxide from air.

Cadmium sulphide, CdS. In nature, it occurs as *greenokite*. It is precipitated in neutral, alkaline or feebly acidic media by passing a current of hydrogen sulphide through a cadmium salt solution. The colour of sulphide is usually yellow, but may also be orange : depending on the concentration of the solution, temperature and other precipitation conditions. If the acid concentration is more than 1.3 normal, the sulphide is not precipitated. The orange-red precipitate obtained by passing hydrogen sulphide through a cadmium chloride solution has probably to constitution of cadmium thiochloride.

$$Cl—Cd—S—Cd—Cl$$

Cadmium sulphide is used as a pigment under the name *cadmium yellow*. It also forms **cadimopone**, similar to lithopone.

Cadmium sulphate, $3CdSO_4.8H_2O$. It is obtained by dissolving cadmium oxide in dilute sulphuric acid and crystallizing the salt from the solution by concentration. It yields monoclinic crystals. soluble in water (100 gms. of water dissolve 76.3 gms. of the anhydrous salt at 50°). It forms double salt of the type K_2SO_4. $CdSO_4.6H_2O$, isomorphous with the corresponding double sulphates of zinc, magnesium and several other bivalent metals.

In the Weston Standard Cadmium Cell, the electrodes are of cadmium amalgam and mercury respectively and the liquid is saturated solution of cadmium and mercurous sulphates. The e.m.f. of the cell is 1.019 volts at 20°. The cell has a negligible temperature coefficient, since solubility of cadmium sulphate does not vary much with temperature.

Cadmium chloride, $2CdCl_2.5H_2O$. The salt is obtained by dissolving the oxide in hydrochloric acid. It differs from the zinc salt in that it is not hydrolyzed by water and is not deliquescent; it is efflorescent instead.

Like mercuric chloride, it is a poor electrolyte, i.e. it is feebly ionized in water. This is due to the formation of an autocomplex viz. $Cd[CdCl_4[.CdCl_2 \rightarrow Cd^{2+} + 2Cl^-; Cd^{2+} + CdCl_2 + 2Cl^{2-} \rightarrow Cd[CdCl_4]$.

Cadmium iodide, CdI_2, is only feeble ionized in solution. It probably forms a complex ion in which cadmium exists in the negative ion :

$$2CdI_2 \rightleftharpoons CdCdI_4 \rightleftharpoons Cd^{2+} + CdI_4{}^{2-}$$

On account of the formation of this soluble complex, insoluble cadmium salts as CdS can be dissolved in a solution of precipitate of $Cd(NH_3)_4I_4$ is formed.

Cadmium iodide is soluble in alcohol and is used in photography.

Cadmium carbonate $CdCO_3$. When an excess of ammonium carbonate is added to a solution of cadmium chloride, a precipitate is obtained which can be dissolved by adding a suitable quantity of ammonia solution. This solution when concentrated on the water bath yields minute crystals of normal cadmium carbonate. When, however, alkali carbonate solution is added to a cadmium salt solution, the precipitate obtained is of an-*identifinite* composition.

Cadmium cyanide $Cd(CN)_2$. Cadmium salts with potassium cyanide give a precipitate of cadmium cyanide, which is soluble in excess of the reagent, forming $Cd(CN)_2.2KCN$ or $K_2Cd(CN)_4$. Hydrogen sulphide passed through the solution of this complex yields a precipitate of cadmium sulphide (cf. copper sulphide which is not precipitated under similar conditions).

Complex compounds. Cadmium like other transition metals also forms a number of complexes like $[Cd(NH_3)_4](OH)_2$, $[Cd(NH_3)_4]_4$, $K_2[Cd(CN)_4]$, $Cd[CdI_4]$ etc. as already mentioned above

Mercury

Occurrence. The principal ore of mercury is cinnabar, mercury(II) sulphide, HgS. The mercury content by weight is usually less than 1% but Spanish ore may contain 5-7%. Mercury is produced from ore mined in Spain, Italy, Yugoslavia, Peru and U.S.A. The principal countries in which mercury is extracted are Spain, the U.S.A. and Yugoslavia.

Production. Beyond hand-picking, crushing and mechanical shifting there is no preliminary treatment of the ore. Mercury(II) sulphide is roasted in air and mercury vapour escapes with sulphur dioxide ånd other gases,

$$HgS + O_2 \rightarrow Hg + SO_2$$

The concentration of mercury in the gaseous product is exceedingly small and present special problems. Further, mercury vapour is very toxic. The most efficient furnaces are the rotary type used in the U.S.A. Ore passes down the rotating kiln, 15-16 m long and 1-1.5 m diameter, inclined at a gradient of about 1 in 12, or less. Flames from an oil burner play directly on to the ore reaching the bottom, and hot gases pass up the kiln, decomposing the mineral. Mercury vapour passes through water sprays or electrostatic dust precipitators and so the condensing system of many connected vertical pipes of glazed stoneware, alloy steel or painted sheet iron. Condensed mercury drips through the open bottoms of the pipes into water a trough. Filtration through canvas separates mercury from solid [soot] which is worked up of further yields of mercury.

Properties and Uses of Mercury

Mercury is a silvery-white metal and is the only metal which is liquid at room temperature. Due to its low freezing point (-38.87°C), it high boiling point (356.6°C), its uniform coefficient of expansion, and the fact that it does not wet glass, mercury is an excellent thermometric substance. Because of its chemical inactivity mobility, high density, and electrical conductivity, it is used extensively in barometers, vacuum pumps, and liquid seals, and for electrical contacts. With the exceptions of iron and platinum, all metals readily dissolve in, or are wet by, it is less active than the alkali metal alone. Amalgams of tin, silver and gold are used in dentistry.

Mercury vapour lamps and fluorescent lamps make use of the fact that mercury vapour is conductor of electricity and, while conducting, it emits light.

Mercury vapour lamps fluorescent lamps make use of the fact that mercury vapour is a conductor of electricity and, while conducting, it emits light.

Mercury does not change chemically in air at ordinary temperatures but is slowly oxidized when heated, forming **mercury(II) oxide**, HgO, which decomposes at still higher temperatures. Being below hydrogen in the activity series, mercury is not attacked by hydrochloric acid but does dissolve in an excess of oxidizing acids such as cold nitric or hot concentrated sulphuric acid. With excess acid, mercury(II) salts are formed.

$$3Hg + 8H^+ + 2NO_3^- \rightarrow 3Hg^{2+} + 2NO\uparrow + 4H_2O$$

$$Hg + 2H_2SO_4 \rightarrow Hg^{2+} + SO_4^{2-} + SO_2\uparrow + 2H_2O$$

When the mercury is in excess these oxidizing acids convert the metal to the corresponding mercury(I), Hg_2^{2+}, salt rather than to the mercury(II), Hg^{2+}, salts.

The halogens attack mercury forming the halides, and sulphur combines with it to produce mercury(II) sulphide. Mercury is displaced from solutions of its ions by all metal except silver, gold, and the platinum metals (Ru, Rh, Pd, Os, Ir and Pt). A copper wire soon becomes amalgamated when immersed in a solution of a mercury compound.

Mercury forms two series of compounds; in these series the metal shows oxidation states of + 1 and + 2. The mercury atom uses both its valence electrons in bonding in all of its compounds. For example, in mercury(II) chloride, $HgCl_2$, the chloride atoms are covalently bonded to the mercury atom in a linear molecule, as shown by the electronic formula

$$\underset{\times\times}{\overset{\times\times}{{}^{\times}_{\times}Cl}}{}^{\bullet}_{\bullet}Hg{}^{\bullet}_{\bullet}\underset{\times\times}{\overset{\times\times}{Cl}}{}^{\times}_{\times}$$

With the strongly electronegative sulphate and nitrate groups, mercury(II) forms ionic compounds in which dipositive mercury(II) ions, Hg^{2+}, are present. In the series of compounds in which mercury atoms serves to form an electron-pair bond between the two mercury atoms. The Hg_2Cl_2 molecule has a linear covalent structure.

$$:\ddot{\underset{..}{Cl}}:Hg:Hg:\ddot{\underset{..}{Cl}}:$$

The equal sharing of the electron pair between the two mercury atoms does not contribute to the oxidation state of either atom, but the unequal sharing of an electron pair by each mercury atom with an electronegative chlorine atom gives each mercury atom an oxidation state of + 1. When the second electron of each mercury atom is lost completely by electron transfer, as in the formation of $Hg_2(NO_3)_2$, the double mercury(I) ion, Hg + Hg is formed.

The Hg_2^{2+} ion is the most common ionic species with a **metal-metal bond**. Evidence from various sources points to the dimeric nature of this ion. X-ray studies of crystalline mercury(I) chloride show that mercury atoms occur in pairs, and magnetic studies of solutions containing univalent mercury show that it is not paramagnetic, meaning that there are no unpaired electrons as there would be if such solutions were to contain Hg^+ ions with an odd number of electrons per ion. On the contrary, $Hg : Hg^{2+}$ contains as even number of electrons so that all the electrons present are assumed to exist in pairs. Thus, it is correct to write the formulas of mercury(I) compounds as Hg_2X_2, instead of HgX, where X is an acid radical.

Compounds of mercury

Mercury exhibits oxidation states of + 2 and + 1; in the latter, two mercury atoms are united by a single covalent bond as in the $[Hg—Hg]^{2+}$ ion. Mercury compounds are more covalent than those of zinc and cadmium and, indeed, covalency is the rule rather than the exception. Mercury(II) compounds form stable complexes with a variety of ligands; indeed, they are generally many orders of magnitude more stable than the corresponding complexes of zinc and cadmium.

Mercury(II) oxide, HgO, is a basic and dimorphous oxide. A red form results from heating mercury in air or oxygen just below the boiling point (357°C). It reverts to mercury on stronger heating, the colour darkening.

$$2Hg + O_2 \rightleftharpoons 2HgO$$

The red form is also produced by thermal decomposition of mercury(II) nitrate, obtained by the action of nitric acid on mercury,

$$2Hg(NO_3)_2 \rightarrow 2HgO + 4NO_2\uparrow + O_2\uparrow$$

or on the commercial scale by gentle ignition of mercury and mercury and mercury(II) nitrate,

$$Hg(NO_3)_2 + Hg \rightarrow 2HgO + 2NO_2\uparrow$$

Mixing boiling solutions of mercury(II) chloride and sodium carbonate and continued heating of the precipitated carbonate in the liquid also produces the red variety,

$$HgCl_2 + CO_3^{2-} \rightarrow HgO\downarrow + CO_2\uparrow + 2Cl^-$$

The yellow modification is precipitated when mercury(II) chloric solution is poured into an excess of aqueous caustic soda, there being no mercury(II) hydroxide.

$$HgCl_2 + 2OH^- \rightarrow HgO\downarrow + H_2O + 2Cl^-$$

The yellow form is slightly metastable with respect to the red : the difference being due to particle size. A basic chloride results from the addition of alkali to mercury(II) chloride solution.

Mercury(II) oxide does not react with alkali, but forms salts with acids. It will dissolve in alkali metal halide solution on warming.

Mercury(II) oxide is thermally unstable and readily decomposes into mercury and oxygen. This decomposition is of historical interest, since it was the method used by priestly to obtain oxygen :

$$2HgO \rightarrow 2Hg + O_2$$

Mercury(II) sulphide, HgS

Mercury(II) sulphide occurs as the red ore cinnabar. It is obtained as a black solid by passing hydrogen through an aqueous solution of a mercury (II) salt; its solubility product is so low that it precipitates in the presence of a high concentration of hydrogen ions, which considerably reduces the already low concentration of sulphide ions provided by the dissolved hydrogen sulphide :

$$Hg^{2+} + H_2S \rightarrow HgS + 2H^+$$

Both forms of the compound have different crystal structures and the black form, which is metastable with respect to the other, changes over into the red variety on heating.

Mercury(II) sulphate, $HgSO_4$

Mercury(II) sulphate may be obtained by the action of hot concentrated sulphuric acid on mercury; it is a white hygroscopic solid. On heating it decomposes into mercury(I) sulphate, mercury, sulphur dioxide and oxygen :

$$3HgSO_4 \rightarrow Hg_2SO_4 + Hg + 2SO_2 + 2O_2$$

Mercury(II) nitrate, $Hg(NO_3)_2$

Mercury(II) nitrate is obtained by the prolonged treatment of mercury with hot concentrated nitric acid; it can be crystallized as white deliquescent crystals. On heating it decomposes into mercury, nitrogen dioxide and oxygen :

$$Hg(NO_3)_2 \rightarrow Hg + 2NO_2 + O_2$$

Mercury(II) carbonate, $HgCO_3.HgO$

The pure substance is unknown and any method of preparation results in the formation of a basic carbonate, i.e. the action of sodium carbonate or sodium hydrogen carbonate solutions on an aqueous solution of a mercury(II) salt. Typical basic carbonates are $HgCO_3.2HgO$ and $HgCO_3.3HgO$.

The mercury(II) halides

Anhydrous mercury(II) fluoride is a colourless, crystalline salt of ionic structure which forms a yellow dihydrate, $HgF_2.2H_2O$, also indicative of ionic bonding. It is hydrolyzed by excess water to basic salts and mercury(II) oxide. The other halides are covalent, little hydrolyzed, sublime readily and are readily soluble in organic solvents. Their structures are covalent; that of mercury(II) iodide being macromolecular. They have low solubilities in water, mercury(II) chloride being much more soluble in hot water than cold (5.6 g saturating 100 g water at 10°C and 54 g at 100°C), mercury(II) bromide being much less soluble than the chloride, and the iodide, insoluble. Electrical conductance measurements show that mercury(II) chloride and bromide are little ionized in solution, the degree being less than 0.01 in centinormal solution :

$$HgCl_2 \rightleftharpoons HgCl^+ + Cl^-$$

Mercury(II) chloride, also called corrosive sublimate, is very poisonous. Very dilute solutions may be used as an antiseptic because it kills bacteria. Mercury(II) chloride may be prepared by the commercial methods : heating mercury in chlorine or heating a mixture of mercury(II) sulphate and sodium chloride, with a little manganese(IV) oxide to oxidize dimercury(II) compounds to the mercury(II) state, when mercury(II) chloride sublimes on to a suitably cooled surface :

$$Hg + Cl_2 \rightarrow HgCl_2$$
$$HgSO_4 + 2NaCl \rightarrow HgCl_2 + Na_2SO_4$$

The sublimate may be purified by dissolving in hot water and cooling, when colourless needle-like crystals separate. Mercury(II) chloride is very soluble in ethanol, ether and ethyl acetate. It is not affected by hot concentrated nitric acid and volatilizes away when treated with boiling concentrated sulphuric acid.

Mercury(II) chloride is readily reduced to dimercury(II) chloride and mercury. In solution, tin(II) chloride precipitates white dimercury(II) chloride, rapidly turning dark grey as mercury is formed :

$$2HgCl_2 + SnCl_2 \rightarrow Hg_2Cl_2 \downarrow + SnCl_4$$
$$Hg_2Cl_2 \downarrow + SnCl_2 \rightarrow 2Hg \downarrow + SnCl_4$$

On warming, excess formic acid brings about a similar reduction :

$$2HgCl_2 + H.COOH \rightarrow Hg_2Cl_2 \downarrow + CO_2 \uparrow + 2HCl$$
$$Hg_2Cl_2 + H.COOH \rightarrow 2Hg \downarrow + CO_2 \uparrow + 2HCl$$

Mercury(II) chloride forms complexes with ammonia which are referred to by the descriptions, fusible white precipitate and infusible white precipitate, the latter volatilizing before fusion. Fusible white precipitate is the diammine, obtained by passing dry ammonia into a solution of mercury(II) chloride in ethyl acetate.

$$HgCl_2 + NH_2 \rightarrow HgCl_2.2NH_2$$

Mercury(II) iodide, HgI_2

This compund precipitates as a yellow solid which then turns rapidly red when the correct amount of potassium iodide solution is added to a solution of mercury(II) chloride :

$$Hg^{2+} + 2I^- \rightarrow HgI_2$$

On heating to about 127°C the red form changes into a yellow variety which has a different crystal structure (polymorphism). On cooling again the change is reversed and this change in crystal structure may be accelerated by scratching the material with a glass rod.

$$\underset{\text{red}}{HgI_2} \underset{\text{below } 127°C}{\overset{\text{above } 127°C}{\rightleftharpoons}} \underset{\text{yellow}}{HgI_2}$$

Mercury(II) iodide readily dissolves in an excess of potassium iodide solution with formation of the $[HgI_4]^{2-}$ complex ion :

$$HgI_2 + 2I^- \rightarrow [HgI_4]^{2-}$$

Salts containing this complex anion include Ag_2HgI_2 (yellow) and Cu_2HgI_4 (red). Which are both insoluble.

Nessler's reagent is a solution of K_2HgI_4 to which some potassium hydroxide has been added. It reacts with traces of ammonia to give a yellow colouration; larger quantities of the gas produce a brown precipitate.

The treatment of mercury with nitric acid and alcohol yields *mercuric fulminate*, $Hg(ONC)_2$, a highly unstable and explosive compound. It is used in percussion caps as a detonator to initiate the explosion of TNT or gunpowder.

Mercuric thiocyanate, $Hg(CNS)_2$, may be prepared from a soluble mercury salt and a thiocyanate. It burns with a copious ash which is often referred to as "Pharaoh's serpents."

oxygen bonds would be formed if these compounds were attacked by either air or water.

Mercury(I) compounds

An interesting feature of mercury is its ability to form the Hg_2^{2+} ion in which the two mercury atoms are united by a single covalent bond, i.e., the simple Hg^+ ion does not exist.

Evidence in favour or the Hg_2^{2+} ion as opposed to the Hg^+ ion is considerable and includes the following experimental results :

(a) Mercury(I) compounds are not paramagnetic but if the formula of the mercury(I) ions was Hg^+ this would contain an unpaired 6*s* election. In Hg_2^{2+} both 6*s* electrons originating from each mercury atom are paired and this explains its lack of paramagnetism.

(b) When an excess of mercury is shaken with an aqueous solution of a mercury(II) salt, an equilibrium is established and mercury, mercury(II) ions, and mercury(I) ions are present.

Attempts to prepare mercury(I) hydroxide, mercury(I) oxide and mercury(I) sulphide have failed and, in order to understand the reason for this, it is necessary to consider the equilibrium :

$$Hg_2^{2+} \rightleftharpoons Hg + Hg^{2+}$$

The above disproportionation reaction has an equilibrium constant, K", of about 6.0 × 10^{-3} at 25°C, i.e.

$$K'' = \frac{[Hg^{2+}]}{[Hg_2^{2+}]} = 6.0 \times 10^{-3}$$

The low value for the equilibrium constant implies that under normal conditions there is little tendency for the mercury(I) ion to disproportionate into the mercury(II) ion and mercury. However, the addition of an anion, e.g. the sulphide ion, which forms a more insoluble compound with the mercury(II) ion that with the mercury(I) ion, causes the disproportionation reaction to proceed further to the right :

$$Hg_2^{2+} \rightleftharpoons Hg + Hg^{2+}$$

$$+$$

$$S^{2-}$$

$$\upharpoonleft\downharpoonright$$

$$HgS$$

(equilibrium displaced to the right by the removal of Hg^{2+} ions as the very insoluble mercury (II) sulphide)

The addition of hydrogen sulphide to an aqueous solution of a mercury(I) salt thus results in the precipitation of mercury(II) sulphide and free unstable, and mercury :

$$Hg_2^{2+} + 2OH^- \rightarrow Hg + HgS + 2H^-$$

Similarly, the addition of sodium hydroxide solution to a solution of a mercury(I) salt forms mercury(II) oxide, mercury(II) hydroxide is very unstable, and mercury :

$$Hg_2^{2+} + 2OH^- \rightarrow Hg + HgO + H_2O$$

Mercury(I) compounds form very few complexes principally because mercury(II) compounds form very stable ones with many ligands; thus once again the disproportionation reaction is drive well over to the right :

$$Hg_2^{2+} \rightleftharpoons Hg + Hg^{2+}$$

$$+$$

$$\text{Complexing ligands}$$

$$\upharpoonleft\downharpoonright$$

$$Hg^{2+}\ \text{complex}$$

(equilibrium displaced to the right by the removal of Hg^{2+} ions as complex ions)

Two of the best known compounds of mercury(I) are mercury(I) chloride and mercury(I) nitrate.

Mercury(I) chloride, Hg_2Cl_2

Mercury(I) chloride precipitates as a white solid when solutions containing mercury(I) and chloride ions are mixed :

$$Hg_2^{2+} + 2Cl^- \rightarrow Hg_2Cl_2$$

It may also be obtained by subliming a mixture of mercury(II) chloride and mercury :

$$HgCl_2 + Hg \rightarrow Hg_2Cl_2$$

Molecular weight determinations on mercury(I) chloride vapour at high temperatures indicate the formula HgCl, but the low molecular weight is in reality due to complete dissociation into mercury(II) chloride and mercury vapour :

$$Hg_2Cl_2 \rightleftharpoons HgCl_2 + Hg$$

The action of ammonia solution on mercury(I) chloride gives a black deposit of mercury and a compound containing mercury(II), i.e. disproportionation occurs :

$$Hg_2Cl_2 + 2NH_3 \rightarrow Hg + HgNH_2Cl + NH_4Cl$$

The compound $HgNH_2Cl$ contains the mercury-nitrogen covalent bond, the formation of which assists in moving the disproportionation reaction from left to right. This reaction is a convenient test for the mercury(I) ion.

Mercury(I) chloride (sometimes called calomel) is employed in the calomel electrode which is used as a reference standard in redox potential measurements. It consists of a molar solution of potassium chloride in contact with mercury and mercury(I) chloride, i.e. the potassium chloride solution is saturated with mercury(I) chloride; electrical contact is established by means of a platinum wire dipping into the pool of mercury. The electrode potential of this arrangement, measured against a standard hydrogen electrode at 25°C, is + 0.285 v. It is much more convenient to use than the hydrogen electrode.

Mercury(I) nitrate, $Hg_2(NO_3)_2$

Mercury(I) nitrate is formed by the action of cold dilute nitric acid on an excess of mercury. It can be crystallized as a white solid hydrate. $Hg_2(NO_3)_2.2H_2O$. Although readily soluble in water it is also hydrolyzed by it and a white basic nitrate. $Hg(OH)NO_3$, slowly deposits. The solution can be cleared by the addition of dilute nitric acid which dissolves the basic salt.

Mercury(I) nitrate readily decomposes on heating into mercury(II) oxide and nitrogen dioxide; on further heating the mercury(II) oxide decomposes into mercury and oxygen :

$$Hg_2(NO_3)_2 \rightarrow 2HgO + 2NO_2$$
$$2HgO \rightarrow 2Hg + O_2$$

27

F-Block elements
The Lanthanides, the actinides

THE LANTHANIDES

Introduction

Much of the revival of interest in the chemistry of a lanthanide or rare earth elements has arisen from the discovery of the actinides and the use of the earlier series as a model for the interpretation and prediction of the properties of members of the later one. There are, nevertheless, other features of general interest about this series of elements. Some of these concern the effects of a slow but steady decrease in atomic and ionic sizes with increase in atomic number (the so-called lanthanide contraction): the metals in their normal tripositive oxidation state resemble one another much more closely than members of any other series of elements, and the separation of some of the series, however, exhibit oxidation states other than, three, and the occurrence of these less common oxidation states can be partly correlated with electronic configurations, such correlation often being expressed in the statement that extra stability with respect to oxidation or reduction is associated with the configurations f^0, f^7 and f^{14}.

Yttrium, having an ionic radius of 0.88 Å (about equal to those of dysprosium and holmium) is often classed with the lanthanides, with which occurs naturally; scandium forms a smaller ion, but like yttrium it has only one oxidation state, and it, too, should be mentioned here. The differences between scandium and the lanthanides are differences of degree; not surprisingly, scandium (ionic radius 0.68Ã) is more prone to complex formation, its hydroxide is a weaker base, its chloride more volatile and its nitrate more easily decomposed by heat, than the corresponding compounds of yttrium, lanthanum and the lanthanides. Although it forms compounds of formula ScO, $ScCl_{1.5}$, $ScBr_{1.5}$ and $ScI_{2.2}$, none of these is a genuine derivative of an oxidation state lower than three; ScO and $ScI_{2.2}$ are metallic conductors, and $ScCl_{1.5}$ and $ScBr_{1.5}$ appear to contain metal clusters. There are at present to features of special interest in the chemistry of scandium or (with the exception of a stable volatile complex $Cs^+[Y(CF_3COCHCODF_3)_4]^-$) of yttrium, and in the rest of this chapter as we shall restrict attention to lanthanum and the following fourteen elements.

Some Physical Data of the First Inner Transition Series Elements

Elements	Symbol	Atomic Number	Electronic Configuration	Atomic Radius (Å)	Ionic Radius (Å) M3+	M.p. °C (aprox.)	Density (g cm−3)
Lanthanum	La	57	2.8.18.18.8(1).2 .....$4f^0 5s^2 5p^6 5d^1 6s^2$	1.87	1.25	920	6.19
Cerium	Ce	58	2.8.18.18.(1).8.2 .....$4f^2 5s^2 5p^6 5d^0 6s^2$	1.81		804	6.78
Praseodymium	Pr	59	2.8.18.18(3).8.2 .....$4f^3 5s^2 5p^6 5d^0 6s^2$	1.82		935	6.78
Neodymium	Nd	60	2.8.18.18(4).8.2 .....$4f^4 5s^2 5p^6 5d^0 6s^2$	1.82	1.17	1024	7.00
Prometheum	Pm	61	2.8.18.18(5).8.2 .....$4f^5 5s^2 5p^6 5d^0 6s^2$				
Samarium	Sm	62	2.8.18.18(6).8.2 ... $4f^6 5s^2 5p^6 5d^0 6s^2$		1.14	1052	6.93
Europium	Eu	63	2.8.18.18(7).8.2 .....$4f^7 5s^2 5p^6 5d^0 6s^2$	2.04	1.12	1150	5.24
Gadolinium	Gd	64	2.8.18.18(7).8(1).2 ....$4f^7 5s^2 5p^6 5d^1 6s^2$	1.79	1.11	1327	7.95
Terbium	Tb	65	2.8.18.18(9).8.2 ....$4f^9 5s^2 5p^6 5d^0 6s^2$	1.77	1.08	1450	8.33
Dysprosium	Dy	66	2.8 18.18(10).8.2 ....$4f^{10} 5s^2 5p^6 5d^0 6s^2$	1.77	1.07	1500	8.56

Holmium	HO	67	2.8.18.18(11).8.2 .....$4f^{11}5s^2 5p^6 5d^0 6s^2$	1.76	1.06	1500	8.76
Erbium	Er	68	2.8.18.18(12).8.2 .....$4f^{12}5s^2 5p^6 5d^0 6s^2$	1.75	1.04	1527	9.16
Thulium	Tm	69	2.8.18.18(13).8.2 .....$4f^{13}5s^2 5p^6 5d^0 6s^2$	1.74	1.02	1600	9.35
Ytterbium	Yb	70	2.8.18.18(14).8.2 .....$4f^{14}5s^2 5p^6 5d^0 6s^2$	1.92	1.00	824	7.01
Lutetium	Lu	71	2.8.18.18(14).8(1).2 .....$4f^{14}5s^2 5p^6 5d^1 6s^2$	1.74	0.99	1700	9.74

Electronic Structure

The electronic structures of the lanthanide metals are shown in Table. Lanthanum (the *d*-block element preceding this series) has the electronic structure : xenon core $5d^1 6s^2$. It might be expected that the 14 elements from cerium to lutetium would be formed by adding 1,2,3,...14 electrons into the $4f$ level. However, it is energetically favourable to over the single $5d$ electron into the $4f$ level in most of the elements, but not in the cases of Ce, Gd and Lu. The reason why Gd has a $5d^1$ arrangement is that this leaves a half filled $4f$ level, which gives increased stability. Lu has a $5d^1$ arrangement because the f shell is already full. The lanthanides are characterized by the uniform (+ III) oxidation state shown by all the metals. They typically form compounds which are ionic and trivalent. The electronic structures of the ions are $Ce^{3+} f^1$, $Pr^{3+} f^2$, $Nd^{3+} f^3$, ..$Lu^{3+} f^{14}$.

The $4f$ electrons in the antepenultimate shell are very effectively shielded from their environment outside the atom by the $5s$ and $5p$ electrons. Consequently the $4f$ electrons do not take part in bonding. They are neither removed to produce ions nor do they take any significant part in crystal field stabilization of complexes. Crystal field stabilization is very important with the *d*-block elements. The octahedral splitting of f orbitals Δ_0 is only about 1 kj mol^{-1}. Whether the f orbitals are filled or empty has little effect on the normal chemical properties. However, it does affect their spectra and their magnetic properties.

Discovery and Occurrence of the Lanthanides

As early as 1794 the Swedish chemist Gadolin announced the discovery of an oxide which the named yttria and it was not until about fifty years later that this so called substance was broken down into three portions named yttria, erbia and terbia. Over the years further separations were achieved, including the isolation of an oxide called lutetia. Of course, the names of the lanthanides as we now know them are obtained by changing the ending a in the oxide into -um, e.g. yttria changes to yttrium.

The major source of the lanthanides is monozite sand which is composed chiefly of the phosphates of thorium, cerium, neodymium and lanthanide ions, and the only lanthanide that does not occur naturally is promethium, which is made artificially by nuclear reactions. In view of the very similar properties of this group of elements, it is only comparatively recently that methods of separation have been worked out which have enabled the very pure elements to be made commercially available.

Separation and extraction of the lanthanides

All the lanthanides form + 3 ions, M^{3+}, whose ionic decrease progressively with increasing atomic number from lanthanum (La^{3+}, 1.25Å) to lutetium (Lu^{3+}, 0.99Å); this is the so-called lanthanide contraction. As the ionic radii contract along the lanthanide series, the ability to form complex ions increases and this is the basis of their separation on an ion exchange column.

A solution containing + 3 lanthanide ions is placed at the top of a cation exchange column and the lanthanide ions release an equivalent amount of hydrogen ions from the column :

$$M^{3-} + 3H^+R^- \rightarrow M^{3+}(R^-)_3 + 3H^+$$

cation exchange

A citrate buffer solution (which complexes with the lanthanide ions) is slowly run down the column and the cations partition themselves between the column itself and the moving citrate solution. Since the smaller ions show a greater preference for complexing with the citrate solution, these ions are the first to emerge from the column. By the correct choice of conditions the lutetium ion, Lu^{3+}, emerges first from the column, followed by the cations ytterbium, thulium, erbium, etc., in order of increasing ionic radius.

Lanthanum, cerium, praseodymium, neodymium and gadolinium may be obtained by reduction of their trichlorides with calcium at about 1000°C. e.g.

$$2PrCl_3 + 3Ca \rightarrow 3CaCl_2 + 2Pr$$

Terbium, dysprosium, holmium. erbium and thulium are obtained by a similar process except that their trifluorides are used, since their trichlorides are too volatile, e.g.

$$2HoF_3 + 3Ca \rightarrow 3CaF_2 + 2Ho$$

Europium, samarium and ytterbium are obtainable by chemical reduction of their trioxides.

Lanthanide Contraction

Covalent and ionic radii normally increase on descending a groups in the periodic table due to the presence of extra filled shells of electrons. On moving from left to right across a period, the covalent and ionic radii decrease. This is because the extra orbital electrons incompletely shield the extra nuclear charge. Thus all the electrons are pulled in closer. The shielding effect of electrons decreases in the order $s > p > d > f$. The contraction in size from one element to another is fairly small. However, the additive effect over the 14 lanthanide elements form Ce to Lu is about 0.2Å, and this is known as the lanthanide contraction.

The hardness, melting points and boiling points of the elements all increase from Ce to Lu. This is because the attraction between the atoms increases as the size decreases.

The properties of an ion depend on its size and its charge. The Ln^{3+} lanthanide ions charge by only a small amount from one element to the next (Table 27.1), and their charge is the same, and so their chemical properties are very similar. Since Lu^{3+} is the smallest ion it is the most heavily hydrated. Though the lanthanides do not form complexes very extensively, since Lu^{3+} is the smallest ion the complexes formed by Lu^{3+} are the strongest, Lu^{3+} and Ce^{3+} are the largest ions so $La(OH)_3$ and $Ce(OH)_3$ are the strongest bases.

The lanthanide contraction reduces the radii of the last four element in the series below that for Y in the preceding transition series. Since the size of the heavier lanthanide ions particularly Dy^{3+} and Ho^{3+}, are similar to that of Y^{3+} it follows that their chemical properties are also very similar. As a result the separation of these elements is very difficult.

Table 27.1 : Ionic radii of Sc^{3+}, Y^{3+} and La^{3+} and the Ln^{3+} ions (Å)

Sc														
0 745														
Y														
0.900														
La	Ce	Pr	And	Pm	Sm	Eu	Gd	Tb	Dy	Ho	Er	Tm	Yb	Lu
0.032	1.02	0.99	0.983	0.97	0.958	0.947	0.938	0.923	0.912	0.901	0.890	0.880	0.868	0.8691

Ionic radii depend on how many electrons are removed. For simplicity, the radii of ions with the same charge are compared in Table 29.9. A similar change in size across the series is observed if covalent radii are compared (Table 27.2).

Because of this contraction in size across the lanthanide series, the elements which follow in the third transition series are considerably smaller than would otherwise be expected. The normal size increase Sc $\rightarrow$ Y $\rightarrow$ La disappears after the lanthanides. Thus pairs of elements such as Zr/Hf, Nb/Ta and Mo/W are almost identical in size. The close similarity of properties in such a pair makes chemical separation very difficult. The sizes of the third row of transition elements are very similar to those of the second row of transition elements (see Table 27.1). Thus the second and third rows of transition elements resemble each other more closely than do the first and second rows.

Table 27.1 Covalent radii of the transition elements (Å)

Sc	Ti	V	Cr	Mn	Fe	Co	Ni
1.44	1.32	1.22	1.17	1.17	1.17	1.16	1.15
Y	Zr	Nb	Mo	Tc	Ru	Rh	Pd
1.62	1.45	1.34	1.29	-	1.24	1.25	1.28
La	Hf	Ta					
W	Re	Os	Ir	Pt			
1.69	1.44	1.34	1.30	1.28	1.26	1.26	1.29

* 14 Lanthanide elements

Oxidation State (II)

The majority of the lanthanides form at least one halide of formula MX_2, but if M is lanthanum, cerium, praseodymium or gadolinium the halide is a metallic conductor and hence should be formulated $M^{3+}(X^-)_2(e)$. Up to the present time only neodymium, samarium, europium, dysprosium, thulium and ytterbium have been found to form saline dihalides, and of these elements only samarium, europium and ytterbium have an extensive chemistry in this oxidation state. The position of the equilibrium

$$MXn(s) \rightleftharpoons MXn{-}1(s) +$$

is most likely to lie on the right-hand side when X I, and the iodides are the most easily obtained compounds of these metals in the dipositive state. When M Eu, thermal decomposition of the trichloride, -bromide or -iodide gives a dihalide, but for M = Sm or Yb only the triiodide decomposes, the trichloride and tribromide requiring reduction with hydrogen at a high temperature.

There is, however, another route to the dihalides by the reverse of the disproportionation reaction which is their normal mode of decomposition:

$$2MX_3(s) + M(s) \rightarrow 3MX_2(s)$$

This is the method which has been used at about 500° for the preparation of $NdCl_2$ and NdI_2, $DyCl_2$, and TmI_2, the only compounds of neodymium (II), dysprosium (II) and thulium (II) yet known. For all three of these elements this oxidation state is very powerfully reducing and cannot exist in aqueous media. The structures of $NdCl_2$ and TmI_2 are known, the compounds being isomorphous with $PbCl_2$ and CdI_2 respectively.

Samarium, europium and ytterbium can be obtained in the dipositive state in aqueous solution by electrolytic reduction, reduction with an alkali metal amalgam or (in the case of europium only) metallic iron or zinc. The Eu^{3+}/Eu^{2+} half-cell has a standard potential of –0.43 V, very close to that of the Cr3+/Cr2+ system; semiquantitative estimates of E^0 for Sm_{3+}/Sm^{2+} and Yb_{3+}/Yb_{2+} give values of –1.5 and –1.1 V respectively. It follows from these values that solutions of europium (II) can be used for the preparation of many europium (II) compounds, but that solutions of samarium (II) and ytterbium (II) are so unstable with respect to decomposition of water that pure samarium (II) and ytterbium (II) compounds are unlikely to be obtained in aqueous media. Europium (II) has recently attracted some attention as a one-electron reducing agent.

In addition to dihalides, lower oxides of samarium, europium and ytterbium can be obtained, reduction of M_2O_3 by the metal itself or by lanthanum being the method employed. In their structural chemistry, Sm^{2+}, Eu^{2+} and Yb^{2+} (ionic radii 1.11, 1.10 and 0.93 Å respectively) closely resemble Ca^{2+}, Sr^{2+} and Ba^{2+}: the monoxides all have the NaCl structure; SmF_2 and EuF_2 have the CaF_2, structure; $SmBr_2$ and $EuBr_2$ are isomorphous with $SrBr_2$; and YbI_2 is isomorphous with CaI_2 (and hence with CdI_2).

It would be interesting to discuss the variation in stability of the dipositive oxidation state of the lanthanides from a thermochemical viewpoint, but unfortunately experimental values for the third ionization potentials of these elements are available for only a few of them; lattice energies for the trihalides are also inaccessible. Nevertheless, it seems likely from the very smooth variation in interatomic distances in, say, fluorides and oxides that there is a smooth variation in the difference between the lattice energies of MX_3 and MX_2, and that the third ionization potentials therefore play a decisive part in determining the stabilities of the trihalides with respect to decomposition into dihalides and halogens. Without going into detail here, it can be said that the likely variation in ionization potentials among the lanthanides can be discussed in the same way as for ordinary transition elements, (i.e. in terms of the effects of increasing nuclear charge and loss of exchange energy on ionization, in this way it can be shown that the loss of exchange energy on removal of an *f* electron from the configuration f^n rises to a maximum at $n = 7$, drops to zero at $n = 8$ and then rises to another maximum. If we are considering the change $M^{2+} \rightarrow M^{3+}$, $n = 7$ corresponds to M = Eu. Thus there should be a sharp drop in the third ionization potential between europium and gadolinium, making $Gd^{3+} + e$ much more stable with respect to Gd^{2+} than $Eu^{3+} + e$ is with respect to Eu^{2+}. Whilst this argument is reasonably convincing so far as preparation of a lower halide by reduction or decomposition is concerned, it is not, of course, a satisfactory treatment of the preparation of a dihalide from a trihalide and a metal, which poses a much more complicated problem. To consider this matter in detail is beyond the scope of this book; but it should be said that, with the aid of some estimated thermochemical data, a discussion of the problem that correctly predicts which elements can form dihalides has recently been given.

Oxidation State (IV)

Only cerium has a tetrapositive oxidation state capable of existence in the presence of water (and then only because the oxidation of water by cerium (IV), though thermodynamically possible, is very slow); the other elements to exhibit this oxidation state do so only in fluorides and (in some cases) oxides. It will therefore be convenient to deal with cerium first.

The only binary compounds of cerium (IV) yet known are CeO_2 (made by heating Ce $(OH)_3$ in air or by oxidation of a solution of a cerium (III) salt with hypochlorite) and CeF_4 (made from CeF_3 and F_2). Several solid complexes are known, however, among them $(NH_4)_2[Ce(NO_3)_6]$ and $(pyH_2)_2[CeCl_6]$. The solution chemistry of cerium (IV) is rather like that of zirconium (IV) insofar as hydrolytic cationic polymerization and formation of an acid-insoluble phosphate are concerned, but cerium (IV) is also a powerful oxidizing agent. In M perchloric acid, E^0 for the reaction

$$Ce^{4+} + e \rightarrow Ce^{3+}$$

is + 1.70 V. Values of E^0 in M HNO_3, H_2SO_4 or HCl are lower, being + 1.61, 1.44 and 1.28 V respectively, from which it is deduced that solutions in at least three of the acids contain the cerium in the form of a complex ion. Cerium (IV) sulphate is a familiar reagent in titrimetric analysis, in which it is often used in place of permanganate or dichromate; as a one-electron oxidant it is less prone to undergo side-reactions than the more familiar reagents.

Praseodymium forms PrO_2 (made by heating Pr_2O_3 in oxygen at 500° and 100 atm pressure) and a number of fluoro compounds, including impure PrF_4, Na_2PrF_6 and K_2PrF_6, all made in the dry way. An oxide of approximate composition Pr_6O_{11} is obtained by heating the nitrate, carbonate or hydroxide of praseodymium (III) in air; several other phases intermediate between $PrO_{1.5}$ and $PrO_{2.0}$ also exist in different temperature ranges. Praseodymium (IV) compounds liberate oxygen on treatment with acids, and there are no oxy-salts of the metal in this oxidation state. Terbium (IV) is represented by TbO_2 and TbF_4 (and some ill-defined complexes of the latter) and there is also an intermediate oxide of approximate composition Tb_4O_2; again an oxy-salts are known. For neodymium and dysprosium, fluorination of mixtures of composition 3CsCl: MF_3 yields mixtures believed to contain Cs_3NdF_7 and Cs_3DyF_7 respectively; these are the only derivatives yet made of the metals in a higher oxidation state.

The oxides CeO_2, PrO_2 and TbO_2 all have the fluorite structure; CeF_4 and TbF_4 are isomorphous with UF_4; and $(NH_4)_2[Ce(NO_3)_6]$, contains a twelve-coordinated metal atom, the six nitrate ions all acting as bidentate ligands.

The discussion in the preceding section concerning the variation in ionization potential with atomic number can be extended to cover the fourth ionization potential. The abrupt fall in the ionization potential for the configuration *f*8 then means that terbium should have a much lower fourth ionization potential than gadolinium (i.e. be more easily oxidized). Thus we see the exchange energy factor provides a good explanation of the relative stabilities of europium (II) and terbium (IV). It does not, however, suffice to account for all the other 'anomalous' oxidation states, and until more thermodynamic data are available for all the lanthanides a full thermochemical treatment will be impossible. In the meantime, the qualitative idea of the stability of the half-filled shell remains a useful half-truth.

Oxidation State (+III) Compounds

The anhydrous halides MX_3 can be made by heating the metal and halogen, or by heating the oxide with the appropriate ammonium halide.

$$Ln_2O_3 + 6NH_4Cl \xrightarrow{300°c} 2LnCl_3 + 3H_2O$$

The fluorides are very insoluble, and can be precipitated from solutions of Ln^{3+} by addition of Na^+F^- or HF. This is used as a test for the lanthanides in qualitative analysis. However, with excess F^-, the smaller lanthanide ions may form soluble complexes $[LnF(H_2O)_n]^{2+}$. The chlorides are deliquescent and soluble, and crystallize with six or seven molecules of water of crystallization. If the hydrated halides are heated, they form oxohalides instead of dehydrating to anhydrous halides.

$$LnCl_3.6H_2O \xrightarrow{heat} LnOCl + 5H_2O + 2HCl$$

Heating $CeX_3.(H_2O)_n$ results in CeO_2. The bromides and iodides are similar to the chlorides.

At elevated temperatures, the lanthanides react with B, giving LnB_4 and LnB_6.

On arc-melting the metals with C in an inert atmosphere they form carbides of stoichiometry LnC_2 and $Ln_4(C_2)_3$. The carbides can also be made by reducing Ln_2O_3 with C in an electric furnace. LnC_2 are more reactive than CaC_2. They react with water, giving ethyne and also some hydrogen, C_2H_4 and C_2H_6. They also show metallic conductivity. They do not contain Ln(+II) and are best described as acetylides of Ln^{3+} and C_2^{2-} with the extra electron in a conduction band.

$$2LnC_2 + 6H_2O \rightarrow 2Ln(OH)_3 + 2C_2H_2 + H_2$$

$$C_2H_2 + H_2 \rightarrow C_2H_4 \xrightarrow{+H_2} C_2H_6$$

At elevated temperatures the metals also react with N. P. As. Sb and Bi, giving LnN etc. The latter is hydrolyzed by water in a similar way to AlN:

$$LnN + 3H_2O \rightarrow Ln(OH)_3 + NH_3$$

A wide variety of oxosalts are known, including nitrates, carbonates, oxalates, sulphates, phosphates and also salts of strongly oxidizing ions such as perchlorates.

Reactions of the Lanthanides

In chemical reactivity they resemble calcium; thus they readily tarnish in air and burn to give oxides (all give the trioxide, except for cerium which forms CeO_2). Other non-metals which combine directly with them include nitrogen, sulphur, the halogens and hydrogen. The hydrides are non-stoichiometric but their composition approaches that required for the general formula MH_3; like calcium hydride, these hydrides liberate hydrogen from water and are probably essentially ionic in structure. Other similarities to calcium include the liberation of hydrogen from water and the vigorous evolution of this same gas from dilute non-oxidizing acids, e.g.

$$2La + 6H_2O \rightarrow 2La\,(OH)_3 + 3H_2$$

Lanthanum oxide, La_2O_3, is a strong base and combines vigorously with water to form the hydroxide, $La\,(OH)_3$, cf. the reaction of calcium oxide with water. The oxides and hydroxides of the other lanthanides become progressively weaker as bases (another manifestation of the lanthanide contraction). Lanthanide compounds are generally predominantly ionic and usually contain the lanthanide metal in its +3 oxidation state.

Colour and Absorption Spectra Of Ln^{3+} Ions

Some of the Ln^{3+} ions are coloured in the solid and in aqueous solution.

The following points may be noted from the Table.

(i) The Ln^{3+} cations containing x electrons in 4*f*-level have the same colour as the Ln^{3+} cations containing (14–x) electrons in the 4*f*-orbital, *e.g.*, Pr^{3+} and Tm^{3+} ions having 2 and (14–2) *i.e.*, 12 electrons in the 4*f*-orbital respectively have the same colour namely green. Thus *4f-orbitals are the source of the colour of Ln^{3+} ions.*

(ii) The colours of the cations *viz.*, La^{3+} to Gd^{3+} repeat themselves at least qualitatively from Lu^{3+} back to Gd^{3+}. Thus it can be concluded that the colours depend on the number of unpaired electrons in the 4*f*-orbital, but the situation seems to be more complex, *e.g.*, although $Eu^{3+}(4f6)$ and $Sm^{2+}(4f6)$ ions have the same number of unpaired electrons in the 4*f*-level (equal to six), they have different colours (Eu^{3+}– colourless, Sm^{2+} — reddish).

(iii) Ln^{3+} ions having *empty, half-filled* or *completely filled 4f-orbital (i.e.,* $4f^0$, $4f^7$ and $4f^{14}$ configuration) are colourless, *e.g.*, La^{3+} {$4f^0$, $Gd^{3+}(4f7)$ and $Lu^{3+}(4f^{14})$ ions are colourless.

Since the colour is due to the absorption of light by the cations in the visible region of the spectrum (as in case of *d*-block elements), it seems that the bands in the absorption spectra are due to the transition of the electrons from one level to the other within the 4f-level. Such type of transition of the electrons is called **f → f transition.**

The absorption spectra of Ln^{3+} ions (except Ce^{3+} and Yb^{3+} ions) contain *very sharp bands* when compared to those of *d*-block elements. Many of these bands are *line-like* and become even narrower as the temperature is lowered. These bands appear due to *f—f transitions.*

Magnetic Properties

The electronic configuration of the lanthanides is 2, 8, 18, $4s^2$ $4p^6 4d^{10} 4f^{0-14}$, $5s^2 5p^6 5d^{0-1}$, $6s^2$. These elements generally form trivalent cations (Ln^{3+}) by losing two electrons from 6*s*-orbital and once electron from 5*d* or 4*f*-orbital. Thus Ln^{3+} cations have partly filled 4*f*-orbital which is the source of their magnetic properties. Since in these cations 4*f*-orbital is quite deeply seated, it is *well shielded* from the ligands surrounding the Ln^{3+} ion by the overlying 5*s* and 5*p*-subshells. The surrounding ligands are, therefore, not able to interact with the 4*f*-electrons to quench *(i.e.,* nullify) the orbital magnetic moment. Consequently for determining the effective magnetic moment, μ_{eff} for Ln^{3+} ions both spin moment and orbital moment of the electron are considered. The orbital magnetic moment which is ignored in case of *d*-block elements, cannot be ignored here. Thus the spin-only formula, *viz.,*

$$\mu_{eff} = \sqrt{n(n+2)}\,BM$$

where n = number of unpaired electrons which is used to calculate μ_{eff} for *d*-block elements cannot be used in case of Ln^{3+} ions. The formula used for Ln^{3+} ions is :

$$\mu_{eff} = g\sqrt{J(J+1)}\,BM$$

where g = Lande splitting factor, J = total angular momentum which is a vector sum of L (orbital angular momentum) and S (spin angular momentum), i.e., J (L + S). Only Sm^{3+} ions do not obey the relationship represented by equation (1).

The calculated [from equation (1)] and experimental values of μ_{eff} (in BM) for the sulphates of Ln^{3+} ions show that La^{3+} and Lu^{3+} ions, which are $4f^0$ and $4f^{14}$ type respectively are *diamagnetic* while the remaining Ln^{3+} ions are *paramagnetic.*

28

The Actinides

The actinides are the elements following actinium in the seventh Period of the periodic table. They are radioactive, and therefore of the greatest importance in the study of nuclear chemistry and of the principles which underlie the conversion of mass into energy. For this reason, actinide compounds have been the subject of intensive research, and although many of them have been isolated only in milligram quantities or less, a large body of accurate data concerning their properties has been accumulated. Experimental work on properties of the actinides and their compounds is somewhat different from normal work in chemistry and physics because only thorium and uranium occur naturally in quantities which permit work on a macrochemical scale; the other elements are made artificially by nuclear bombardment, and the quantities obtainable are such that microchemical methods must be used, or are so very small that tracer techniques must be employed. Since the radioactivity of the actinides renders them injurious to health, experiments must normally be conducted by remote control. Even when larger quantities of material are available much of the experimental work is done using microchemical methods, because the small quantity of the elements used minimizes the danger to health and the interference of the radioactivity with the course of the experiments.

The object of work on the actinides has been largely concerned with the preparation and isolation of pure substances to facilitate experiments on nuclear physics. The orbital structures of the actinide atoms are not easy to determine because the energy differences between the orbitals near the periphery are not large, and hence the correlation between the properties of the actinide compounds and their orbital structures cannot be made with certainty. In this chapter no attempt is made to give a complete account of all the actinides. A general account is given first followed by a more detailed treatment of thorium and uranium.

The electron configuration of the atoms of the actinides in the ground state is set out in Table 28.1.

Table 28.1 : *Electron Configuration of the Three Outer Shell of the Atoms of the Actinides in the Ground State*

Symbol of element	Name of element	Atomic Z	Quantum number			Quantum shell n = 5		shell n = 6		n = 7	Oxidation state
		s	p	d	f	s	p	d	f	s	
Ac	Actinium	89	2	6	10		2	6	1	2	III
Th	Thorium	90	2	6	10		2	6	2	2	IV
Pa	Protactinium	91	2	6	10	2	2	6	1	2	IV, V
U	Uranium	92	2	6	10	3	2	6	1	2	III, IV, V, VI
Np	Neptunium	93	2	6	10	4	2	6	1	2	III, IV. V, VI
Pu	Plutonium	94	2	6	10	6	2	6	2		III, IV V
Am	Americium	95	2	6	10	7	2	6	2		III, IV). V, VI
Cm	Curium	96	2	6	10	7	2	6	1	2	III, (IV)
Bk	Berkelium	97	2	6	10	8	2	6	1	2	III, IV
Cf	Californium	98	2	6	10	10	2	6		2	III
E	Einsteinium	99	2	6	10	11	2	6		2	
Fm	Fermium	100	2	6	10	12	2	6		2	
Md	Mendelinium	101	2	6	10	13	2	6		2	
No	Nobelium	102	2	6	10	14	2	6		2	
Lw	Laurencium	103	2	6	10	14	2	6	1	2	

There is agreement that for all the actinides the quantum shells n = 1, 2, 3, and 4, and the orbitals 5s, 5p, 5d, 6s, 6p, and 7s are completely occupied. There has been much discussion of the distribution of electrons between the 5f and the 6d orbitals. The electron configuration in Table 28.1 is that recommended by Katz and Seaborg.and is in accord with the view that actinium in Period 7 corresponds to lanthanum in Period 6, and that the actinides form a set of 14 elements which are to be placed below the lanthanides in the periodic table. According to this view the two sets of inner transition elements correspond as shown in Table 28.2.

Table 28.2 : *Relation of Actinides to Lanthanides*

Lanthanides	57	58	59	60	61	62	63	64	65
		66	67	68	69	70	71		
	La	Ce	Pr	Nd	Pm	Sm	Eu	Gd	Tb
Dy	Ho	Er	Tm	Yb	Lu				
Actinides	89	90	91	92	93	94	95	96	97
98	99	100	101	102	103				
	Ac	Th	Pa	U	Np	Pu	Am	Cm	Bk
Cf	E	Fm	Md	No	Lw				

Table 28.2 indicates that all the actinides have been identified. If it is possible to bring about nuclear changes which produce elements of atomic number higher than 103 it is predicted that the elements will be transition elements which will take their places in the periodic table under hafnium, tantalum, and tungsten. Classification by chemical experiment will be difficult, however, as the half life periods of the new elements will be short.

Occurrence and Preparation of the Actinides

The atomic numbers, names, probable electronic structures of the ground state gaseous atoms and ionic radii in the tripositive and tetrapositive states of actinium. Thorium and uranium occur naturally in relatively large amounts; actinium and protactinium are present in very small amounts as decay products; and neptunium and plutonium are also found in traces as a result of the slight natural conversion of ^{238}U under the influence of slow neutrons Neptunium and plutonium are, however, almost entirely made by neutron irradiation of uranium in a pile.

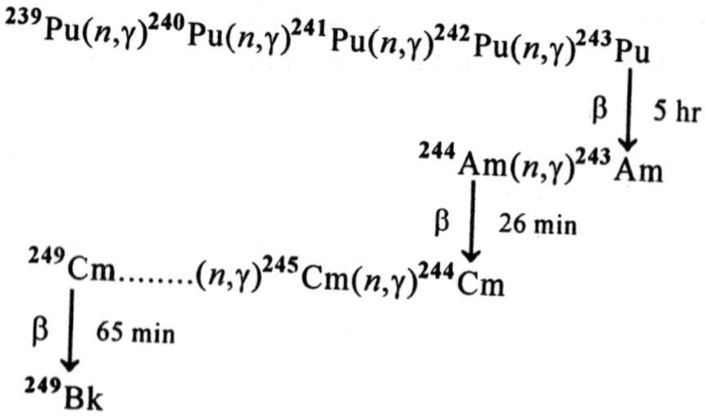

The very high flux of neutrons in a pile leads to the production of small amounts of higher members of the series by a process of multiple neutron capture. Thus, for example, if a 239_{Pu} target is used the following series of reactions will occur :

The process of synthesis involves successive neutron captures until the accumulation of mass renders the nucleus unstable, when β-decay occurs. The berkelium isotope shown above will itself capture neutrons, and the isotopes ^{253}Cf and ^{254}Cf result in turn. As a consequence of multiple neutron capture minute amounts of the higher transuranic elements accumulate with plutonium in the uranium rods during their irradiation in the reactor. It is also interesting to note that einsteinium and fermium were first detected as by-products of a thermonuclear explosion, the result of multiple neutron capture in the very high concentration of neutrons produced.

The transuranic elements may also be synthesized with the aid of high-voltage machines such as the cyclotron; for example,

$$^{238}U(\alpha,n)^{241}Pu \xrightarrow[1\ yr]{\beta} {}^{241}Am \xrightarrow[500\ yr]{\alpha} {}^{247}Np$$

$$^{239}Pu(\alpha,n)^{242}Cm \xrightarrow[5\ mth]{\alpha} {}^{238}Pu$$

Numerous other reactions of this sort have been carried out. The following illustrate the production of some of the higher members of the series: $^{239}Pu(\alpha,2n)\ ^{241}Cm$; $^{243}Pu(\alpha,2n)\ ^{246}Bk$; $^{242}Cm(\alpha,2n)\ ^{244}Cf$; ^{249}Cf; $Cd(d,3n)\ ^{248}Es$; $249Cf(\alpha,2n)\ ^{251}Fm$; $253Es(\alpha,2n)\ ^{256}Mv$.

An important advance in the technique for producing the very heavy elements entails the use of heavy ion bombardments, especially with ions derived by stripping off all electrons from ^{11}B, ^{12}C,

^{14}N and ^{16}O.[4] The bombardment of a target by such ions after acceleration in a cyclotron or similar accelerator results in the gain in a single reaction of considerably more nuclear charge and mass than in a-particle bombardment; examples of its application are

$$^{238}U(^{16}O,4n)^{250}Fm \quad ^{238}U(^{14}N,p+4n)^{247}Cf$$

$$^{252}Cf(^{11}B,6n)^{257}LW$$

Table 28.3 : *Electronic configurations and radii of actinium and later elements*

Atomic number	Name	Symbol	Probable electronic configuration*	Radii† M^{3+}	M^{4+}
89	Actinium	Ac	$6d^17s2$	1.08	
90	Thorium	Th	$6d^27s2$		0.99
91	Protactinium	Pa	$5f^26d^17s^2$		0.96
92	Uranium	U	$5f3^26d^17s^2$	1.00	0.93
93	Neptunium	Np	$5f57s2$	0.99	0.92
94	Plutonium	Pu	$5f^67s^2$	0.97	0.90
95	Americium	Am	$5f^77s^2$	096	0.89
96	Curium	Cm	$5f^76d^17s^2$	0.95	
97	Berkelium	Bk	$5f^86d^17s^2$	0.93	
98	Californium	Cf	$5f^{10}7s^2$	0.92	
99	Einsteinium	Es	$5f^{11}7s^2$		
100	Fermium	Fm	$5f^{12}7s^2$		
101	Mendelevium	Mv	$5f^{13}7s^2$		
102	Nobelium	No	$5f^{14}7s^2$		
103	Lawrencium	Lw	$5f^{14}6d^17s^2$		
104	Rutherfordium	i‡ Rf	$5f^{14}6d^27s^2$		

There are several important factors which limit the range and amounts of the heaviest elements that can be produced by the methods described. If, for example, ^{252}Cf were required as a target it would take 5-10 years to produce about one milligram by multiple neutron capture from one kilogram of ^{239}Pu at a neutron flux to 3×10^{14} neutrons cm^{-2} sec^{-1}; and the yield of product from such a target would be smaller by many orders of magnitude. Fortunately, however, the decay properties and yield of the product can be predicted, and with the aid of ion-exchange resins the separation of minute amount of product is comparatively easy. Many of the heavier isotopes are also short-lived in the sense of radioactive decay of the normal type (usually α-particle emission) or may undergo spontaneous fission; the longest-lived isotope of fermium, for example, is ^{254}Fm, with half-lives of 3.3 hours and 220 days for α-decay and fission respectively. Thus (to mention the most stable isotopes) ^{241}Am and ^{244}Cm are available in gram quantities, ^{249}BK, ^{249}Cf and ^{251}Cf in milligrams and ^{253}Es in micrograms; for element beyond einsteinium the highest amounts yet obtained are a few atoms, and these elements can be studied only in solution.

The present limit to the elements known and studied chemically is rutherfordium, three isotopes of which, ^{257}Rf, ^{259}Rf and ^{261}Rf, are obtained by bombarding ^{249}Cf wit ^{12}C and ^{13}C ions and ^{248}Cm with ^{18}O ions respectively; all decay by α-emission with half-lives of a few seconds. It seems likely that few more elements may be prepared, but still shorter half-lives may make chemical characterization impossible.

General Properties

Some properties of the elements are given in Table 28.5. The elements are all silvery metals. Their melting points are moderately high, but are considerably lower than those for the transition elements. The size of the ions decreases regularly along the series, because the extra charge on the nucleus is poorly screened by the *f* electrons. This results in an 'actinide contraction' similar to the lanthanide contraction. Comparison of the M^{3+} ionic radii with whose for lanthanides shows that the actinide and lanthanide ions are very similar in size. Hence their chemical properties are alike. However, the actinides have much higher densities and a much greater tendency to form complexes.

The actinides are reactive metals like lanthanum and the lanthanides. They react with hot water, and tarnish in air, forming an oxide coating. In the case of Th this coating is protective, but this is not so with the others. The metals react readily with HCl, but reaction with other acids is slower than expected. Concentrated HNO_3 passivates Th, U and Pu. The metals are basic and do not react with NaOH. They react with oxygen, the halogens and with hydrogen. The hydrides are non-stoichiometric and have ideal formulae MH_2 or MH_3.

Table 28.4 : Availability of various isotopes

Tonnes	Kilograms	100 grams	Milligrams	Micrograms
232Th	237Np	231Pa	244Pu	257Fm
238U	239Pu	238Pu	249Bk	
		242Pu	242Cm	
		241Am	252Cf	
		243Am	253Es	
		244Cm	254Es	

The metals are usually obtained by electrolysis of fused salts, or by reducing the halides with Ca at high temperatures.

Table 28.5 : Some properties of the actinides

	m.p. (°C)	b.p. (°C)	Density (gcm^{-3})	Radius M^{3+} (Å)	Radius M^{4+} (Å)
Ac	817	2470	—	1.12	–
Th	1750	4850	11.8	(1.08)	0.94
Pa	1552	4227	15.4	1.04	0.90
U	1130	3930	19.1	1.025	0.89
Np	640	5235	20.5	1.01	0.87
Pu	640	(3230)	19.9	1.00	0.86
Am	1170	2600	13.7	0.975	0.85
Cm	1340		13.5	0.97	0.85
Bk	986		14.8	0.96	0.83
Cf	(900)			0.95	0.82
Es	(860)				

Thorium[1,2,6]

Dissolution of thorium and other oxides in hydrochloric acid, followed by solvent extraction with tributyl phosphate, effects a separation of the thorium. Alternatively, extraction from 3M nitric acid solution by methyl isobutyl ketone in the presence of excess of aluminium nitrate may be used. Naturally occurring thorium consists mainly of ^{232}Th, of half-life 10^{10} years; the chemistry of the element is therefore studied almost entirely by conventional methods.

The metal is obtained by reduction of the tetrafluoride with calcium in the presence of zinc chloride as a flux, or by electrolysis of a fused mixture of ThF_4, KCl and NaCl. It may be obtained very pure by thermal decomposition of the vapour of the tetraiodide on a hot tungsten wire.

The reactivity of metallic thorium is comparable to that of the lanthanide metals, but the chemistry is almost entirely that of thorium (IV). $E°$ for the Th^{IV}/Th system is –1.9 V (compared with –1.7 and –1.5 V for Hf^{IV}/Hf and Zr^{IV}/Zr^{IV} respectively), and there is no evidence for a lower oxidation state in aqueous media. A golden-yellow iodide, ThI_3, may be obtained by heating the tetraiodide with the metal, and a sulphide, ThS, is obtained by an analogous method; the conductivity of these substances, however, indicates that they are really derivatives of thorium (IV) with electrons in conduction bands.

Thorium dioxide is converted by carbon tetrachloride at 600° into the anhydrous tetrachloride, which is soluble in water in the presence of hydrochloric acid, and from solutions of which the hydroxide and hydrated fluoride may be precipitated. Ignition of the hydroxide yields ThO_2; the anhydrous fluoride is best made by the action of hydrogen fluoride on the oxide at 600°. The nitrate and sulphate are readily soluble, but the iodate, oxalate and phosphate are insoluble even in dilute mineral acid. A peroxy derivative $Th(O_2)_2SO_4.3H_2O$ of unknown structure is also insoluble in dilute acid. The remarkable borohydride $Th(BH_4)_4$ is a white solid (m.p. 204°) which is stable in dry air and can be sublimed *in vacuo* at 150°; it is obtained by the interaction of aluminium borohydride and thorium tetrafluoride:

$$4Al(BH_4)_3 + 3ThF4 \rightarrow 3Th(BH_4)_4 + 4AlF_3$$

As might be expected from the high charge, the thorium (IV) ion is extensively hydrolyzed in aqueous solution, the main products being $Th(OH)^{3+}$ and $Th(OH)_2^{2+}$ (and polymers thereof). A cation chain of composition $Th(OH)_2^{2+}$ is found in solid $Th(OH)_2CrO_4.H_2O$ and $Th(OH)_2SO_4$; in it there 'are two parallel rows of hydroxyl groups with thorium atoms on alternate sides, so that each thorium atom is in contact with four hydroxyl groups and each hydroxyl group with two thorium atoms, giving a structure which may be represented as follows:

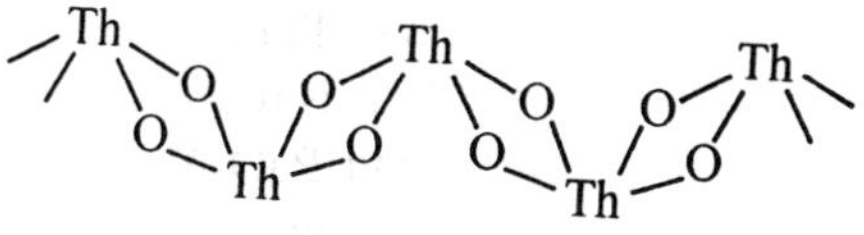

Complex formation with anions (e.g. F^-, Cl^-, NO_3^-, $C_2O_4^{2-}$) also takes place in aqueous media.

Protactinium

Traces of Pa (about 0.1 ppm of ^{231}Pa) are found in the uranium ore pitchblende UO_2. The Pa is formed as a decay product of ^{235}U in the actinium decay series. It also occurs in the neptunium and uranium natural decay series. It is difficult to isolate Pa from minerals. It is usually prepared artificially either from Th by the reaction given above, or obtained as ^{231}Pa from processing spent uranium fuel elements removed from nuclear reactors.

The study of protactinium is a story of scientific cooperation which is rarely encountered. The chemistry of this element was almost unknown until 1960 when A.G. Maddock and a team at the UK Atomic Energy Authority extracted 130 g of the element from over 50 tonnes of waste from the extraction of U. They sent samples to major laboratories throughout the world. This rapidly produced most of the information on this element.

The chemistry of Pa is particularly difficult to study because the compounds hydrolyze readily. In addition the ions polymerize forming colloidal precipitates in water and most acids. These precipitates are adsorbed on to the vessel walls or onto other precipitates. Colloidal precipitates are not formed when Pa is handled in fluoride solutions, because it forms fluoride complexes.

Pa can be extracted from solutions in HCl or HNO_3 by tributyl phosphate. It is then precipitated as PaF_4. and reduced to the metal with Ba at 1400°C. Pa has been obtained in 100g quantities. The metal is shiny, malleable, and tarnishes in air.

The most stable oxidation state is (+V). Pa_2O_5 is obtained as a white solid by igniting Pa compounds in air. The oxide is weakly acidic as it is attacked by fused alkali. Heating in a vacuum, or reduction with hydrogen at 1500°C, gives a black non-stoichiometric phase $PaO_{2.3}$ and eventually PaO_2. PaF_5 can be made by the action of F_2 on PaF_4, or BrF_3 on Pa_2O_5. It is reactive, and can be sublimed. $PaCl_5$ and oxohalides $PaOX_3$ and PaO_2X are also known. Pa(V) complexes with oxalate, citrate, tartrate, sulphate and phosphate ions are known, and some unusual fluoride complexes have been studied. In $Rb[PaF_6]$ the Pa atom is eight-coordinate. The complex $K_2[PaF_7]$ contains nine-coordinate PaF_9 groups. These form two F bridges to neighbouring groups on either side, giving a chain. In $Na_3[PaF_8]$ the $[PaF_8]^{3-}$ ion is a slightly distorted cube. In the (+IV) state, PaO_2 and the halides PaX_4 are all known, and also oxohalides $PaOX_2$. Reduction of aqueous Pa(V) solutions with zinc amalgam or Cr^{2+}

gives Pa(IV), but this is readily oxidized by air. Spectra of $PaCl_4$ in HCl or $HClO_4$ are similar to those for $Ce^{3+}(4f^1)$. This suggests that Pa(IV) has a $5f^1$ configuration, and is an actinide. Pa(III) has been detected polarographically.

Uranium

Although uranium is regarded as the third member of the actinides, in many of its properties it closely resembles molybdenum and tungsten. A comparison of the types of compounds formed by uranium with those formed by the elements in sub-Group VI T can be made. The chief differences are (i) the absence of divalent compounds of uranium, (ii) the absence of complexes of trivalent and pentavalent uranium, and (iii) the existence of the ion UO_2^{2+}, a type of ion which is peculiar to the actinides.

Uranium metal is highly reactive. At slightly elevated temperatures it reacts with all the common non-metals, including hydrogen and nitrogen, but it reacts with carbon only at about 2000°. With steam it gives a mixture of UO_2 and UH_3. it reacts with acids but not with alkalis.

Uranium hydride, UH_3, a black, brittle, metallic solid which conducts electricity, and inflames spontaneously in air, is made by heating uranium in hydrogen at 250-300°. If the hydride is heated above 300° hydrogen is released, leaving pure very finely divided uranium. The nature of uranium hydride is obscure. It is not interstitial, as considerable expansion occurs during its formation. Each uranium atom is surrounded by six other uranium atoms at a distance of 3.32 and by 12 hydrogen atoms. Each H atom occupies a large approximately tetrahedral hole in the metal crystal lattice.

"Interstitial" compounds. UC, UN, and UO all have the sodium chloride structure. UC is made from the elements at 2100°. At 2400°, with excess carbon, UC_2 is formed with a crystal structure similar to that of calcium carbide. UC_2 ignites in air at 400°, and decomposes cold water with the formation of hydrocarbons. U_2C_3 is also known. A second nitride, UN_2, with the fluorite structure, is stable under nitrogen at 100 atm pressure.

U^{III}. All the compounds of uranium in oxidation state III appear to be giant molecules. They include $U(BH_4)_3$, U_2S_3, and the three halides UF_3, UCl_3, and UBr_3. $U(BH_4)_3$, a brown-red non-volatile solid, which detonates violently, is made by heating $U(BH_4)_4$ at 100°. Grey crystals of U_2S_3 are obtained when UBr_3 is heated in hydrogen sulphide for 10 hours. The uranium trihalides are made by reducing the tetrahalides with hydrogen. All are red-brown solids. The fluoride is insoluble in water, and is resistant to acids except nitric acid, but the chloride and the bromide are hygroscopic and dissolve in water easily. The aqucous solution of the chloride evolves hydrogen and precipitates $U(OH)_4$.

Evidence of the inclusion of the 5*f* orbitals of uranium in the valence system of the atom is furnished by electron spin resonance experiments. These experiments show that if Nd^{3+} and U^{3+} are dissolved in single crystals of CaF_2, and SrF_2 both ions have the f^3 configuration, but a hyperfine structure due to interaction with the fluorine nuclei is present in the case of only. It is concluded that the 5*f* electrons in U^{3+} have a small overlap with orbitals of the fluorine atoms, but there is no corresponding overlap between the 4*f* electrons in Nd^{3+} and the fluorine atoms.

U^{IV}. The compounds of uranium in oxidation state IV include the oxide UO_2, the sulphide US_2, the borohydride $U(BH_4)_4$, the halides $Uhal_4$, a few salts of oxyacids, and a number of anionic complexes.

$U(BH_4)_4$, a dark green solid, is made by treating UF_4 with aluminium borohydride. It is stable in the cold, but it reacts with air, water, alcohol, and hydrochloric acid. It can be methylated with boron trimethyl, giving ultimately $U(BH_3CH_3)_4$. If heated to 100° it yields diborane and uranium triborohydride :

$$2U(BH_4)_4 = 2U(BH_4)_3 + B_2H_6 + H_2.$$

UO_2, a brown or black powder, m.p. 2176°, is made by reducing UO_3 or U_3O_8 with hydrogen. It dissolves in nitric acid to give uranyl nitrate, but it is insoluble in other acids. It has the fluorite structure.

US_2, a grey crystalline powder, is made from the elements. It is slowly attacked by water, and rapidly by acids.

Uranium tetrahalides. All the tetrahalides are known: UF_4, a green solid, m.p. about 1000°; UCl_4, hygroscopic dark green crystals, m.p. 567°, b.p. 618°; UBr_4, hygroscopic dark brown crystals; UI_4, deliquescent black crystals, m.p. about 500°. All except the bromide are made by direct combination of the elements; UBr_4 is made by the action of hydrogen bromide on UCl_4, or by the action of bromine on a mixture of U_3O_8 and carbon. All the tetrahalides are reduced by hydrogen to the trihalides. UF_4 is scarcely soluble in water or in acids, but it is hydrolyzed by boiling alkalis to $U(OH)_4$. The other halides are soluble in water and are hydrolyzed in solution.

Salts. Uranium in oxidation state IV forms a chlorate, a perchlorate, and a sulphate which crystallizes as the hydrates $U(SO_4)_2,4H_2O$ and $U(SO_4)_2,8H_2O$. The only nitrate is the basic salt $UO(N0_3)_2$. The salts are green, soluble in water, and easily hydrolyzed. There is little evidence for the existence of the simple ion U^{4+}.

Complexes. The complexes of U^{IV} include $K[UF_5]$, $NH_4[UOF_3]$, $M_2[UCl_6]$ where M any alkali metal, $M_2[UBr_6]$ where M = Na or K, and sulphato complexes such as $(NH_4)_4[U(SO_4)_4]$ and oxalato complexes such as $K_4[Uox_4],5H_2O$.

U^V. Uranium in the oxidation state V is present in UF_5, UCl_5, and probably in the oxide U_2O5, and in the ion UO_2^+.

U^{VI}. Uranium in oxidation state VI forms two simple covalent compounds UF_6 UCl_6, and a large number of stable compounds based on the group UO_2.

Uranium hexafluoride, UF_6, exists as very hygroscopic pale yellow crystals which sublime at 56.5°. It melts under pressure at 64°. It is made by the action of fluorine on uranium in the presence of a little chlorine, or by the action of fluorine on UCl_5. Uranium hexafluoride is not attacked by air, oxygen, chlorine, or iodine, but it is reduced by hydrogen to UF_4. It is rapidly hydrolyzed by water to UO_2F_2 or UO_3. acts with aromatic hydrocarbons forming HF and liberating carbon. It is a powerful fluorinating agent. The molecule is octahedral. The outer shells of the ion U^{VI} are isostructural with those of the Xe atom; KUF_7 and K_2UF_8 are known.

Uranium hexachloride, UCl_6, is one of the only two hexachlorides known, the other g WCl_6. It is made by disproportionation of UCl_5. It decomposes very easily.

The Group UO_2

The group UO_2 is so firmly knit that it has the properties of an electropositive metal. The classification that has been applied in preceding chapters of this book to the derivatives of metals (ionic compounds, giant molecule compounds, simple covalent compounds, neutral, anionic, and cationic complexes) can be applied to the derivatives of the group UO_2. UO_2 forms no simple covalent compounds. The crystal structures of certain giant molecule compounds of UO_2 have been ascertained, and these contain the linear group O—U—O. Whether the UO_2 group in all its compounds retains this arrangement is not known. The UO_2 derivatives are described below in the same order as if the group were a metal.

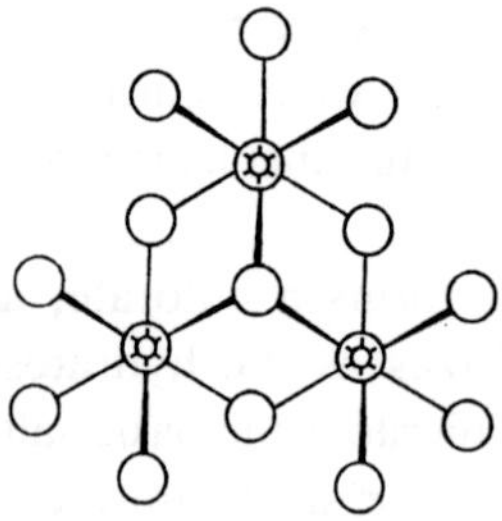

Fig. 28.1 : Part of one layer o the UO_2 structure. The three uranium atoms shown (small circles) lie in the plane of the paper.

It has been suggested that 5*f* bonding is present in the ions MO_2^{2+} and MO_2^{+} which are characteristic of certain actinides, but are not formed by any lanthanides. The oxygen compounds UO_3, U_3O_8, and UO_2F_2 have closely related crystal structures. The unit common to all these structures consists of a modified cadmium iodide structure with uranium atoms in the positions of the cadmium atoms. (Fig. 28.1). The layers, however, are not close-packed on one another, but are linked through additional oxygen atoms, A, placed between the uranium atoms on axes at right angles to the planes of the layers. The management in UO_3 is shown in Fig. 28.2. In UO_2F_2 the oxygen atoms in the layers replaced by fluorine atoms, and each oxygen atom A is bound to one uranium m only (Fig. 28.3). Two A oxygen atoms are interposed between each pair of layers, but there appears to be no chemical bond between them. The same arrangement is found in M_2UO_4, where M is Li, Na, K, $\frac{1}{2}$Ca, $\frac{1}{2}$Sr. In these compounds the ordination number of the uranium atom is 8, but its oxidation number remains as two electrons are transferred to each uranium atom from the six oxygen atoms which surround it in the layer.

Ionic compounds. *Uranyl nitrate*, $UO_2(NO_3)_2$, a yellow powder, is made by dissolving UO_3 in nitric acid. It forms hydrates with 2, 3, or 6 molecules of water. It is very soluble in water and is slightly hydrolyzed in solution. It dissolves in ether and acetone.

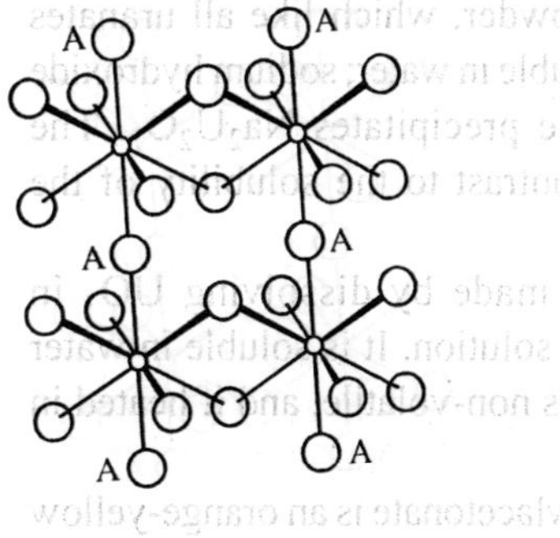

Fig. 28.2 : Diagrammatic perspective of four uranium atoms showing UO_2 structure in UO_3. The planes of the layers are at right angles to the plane of the paper.

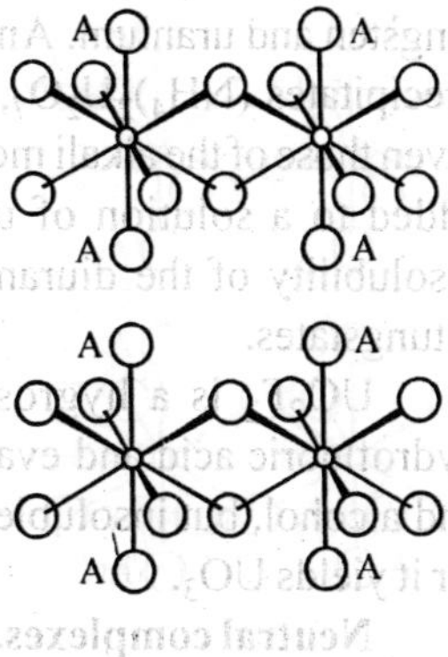

Fig. 28.3 : Diagrammatic perspective of four uranium atoms in the UO_4 ions and in UO_2F_2. The orientation is the same as that in Fig. 28.2.

The hexahydrate is insoluble in ether. Among the other uranyl salts are the perchlorate, the sulphate, and the oxalate.

Uranyl chloride, bromide, and Iodide are hygroscopic solids, soluble in water from which they crystallize as hydrates. Uranyl fluoride is probably a giant molecule compound (see below).

Giant molecule compounds. The crystal structures of UO_3, U_3O_8, and UO_2F_2 have been determined and have already been described.

UO_3, a hygroscopic powder, is made by heating uranyl nitrate, or ammonium uranate, or $UO_2(OH)_2$. On heating it loses oxygen and forms U_3O_8. It forms uranyl salts with acids, and uranates with metallic oxides.

U_3O_8, a green solid, is made by igniting UO_2, UO_3, or $(NH_4)_2U_2O_7$ in air. It is slightly volatile and can be sublimed at about 1300°; at higher temperatures some UO_2 is formed. It is scarcely attacked by hydrochloric acid, or by dilute sulphuric acid, but it is converted by nitric acid to $UO_2(NO_3)_2$, by hydrofluoric acid to UO_2F_2 and UF_4, and by concentrated sulphuric acid to $UO_2(SO_4)$ and $U(SO_4)_2$.

The uranates may be derived from the simple acid H_2UO_4, or they may be derived from condensed acids such as $H_2U_2O_7$. Salts of condensed acids up to $H_2U_6O_{19}$ are known. Uranium, however, does not form heteropoly acids; this is a clear mark of distinction between tungsten and uranium. Ammonia added to a solution of uranyl nitrate precipitates $(NH_4)_2U_2O_7$, a yellow powder, which like all uranates (even those of the alkali metals) is insoluble in water; sodium hydroxide added to a solution of uranyl acetate precipitates $Na_2U_2O_7$. The insolubility of the diuranates is in contrast to the solubility of the ditungstates.

UO_2F_2 is a hygroscopic solid made by dissolving UO_3 in hydrofluoric acid and evaporating the solution. It is soluble in water and alcohol, but insoluble in ether. It is non-volatile, and if heated in air it yields UO_3.

Neutral complexes. Uranyl acetylacetonate is an orange-yellow solid with the composition UO_2A_2,H_2O. If formed in the presence of quinoline or pyridine the composition is UO_2A_2B, but if formed in the presence of NH_3, $N(CH_3)H_2$, or aniline, the composition is UO_2A_2,HA,B, where B represents a molecule of the base.

Complex anions are numerous and of many types. Examples are present in the salts $K[UO_2F_3]$, $K_3[UO_2F_5]$, $K_4[UO_2F_6]$, $K_2[UO_2Cl_4]$, $K[UO_2(NO_3)_3]$, $K_2[UO_2(CN)_4]$ the only uranium cyanide derivative, and $K_2[UO_2(SO_4)_2(H_2O)_2]$. Many of these complex salts crystallize as hydrates.

Complex cations of the present in the salt $[UO_2(NH_3)_2]Cl_2$ are known. The ammonia molecules may be replaced by molecules of either, aniline, pyridine, or quinoline, and the chloride ion by the bromide or the nitrate ion.

Peroxyuranates. The compound $K_4[UO_8].10H_2O$, is made by treating uranyl nitrate with potassium peroxide solution. If uranyl nitrate solution is treated with hydrogen peroxide a compound of composition $UO_4,2H_2O$ is precipitated, which is stable up to 100°. This substance, however, does not react with bases and is not the parent acid of the peroxyuranates.

The mechanisms of some of these redox reactions have been studied with interesting results. When UO_2^{2+} is reduced by Cr^{2+} the bright-green ion $[(H_2O)_5CrOUO(H_2O)_n]^{4+}$ appears to be formed as an intermediate on the way to Cr^{3+} and U^{4+}. In the oxidation of U^{4+} by PbO_2, H_2O_2 or MnO_2, the oxygen in the UO_2^{2+} produced has been shown by ^{18}O tracer studies to come from the oxidant rather than from the water (with which UO_2^{2+} exchanges ^{18}O only very slowly). Further, although the U^{4+}/U^{3+} and UO_2^{2+}/UO_2^{+} systems are reversible, UO_2^{+}/U^{4+} and UO_2^{2+}/U^{4+} are not; electron transfer here is a fast process, but U—O bond fission or formation a relatively slow one.

We have already mentioned the uranyl complex sulphate and carbonate anions; stable complexes are also formed with acetate and, especially, fluoride in aqueous media. As in the case of other actinides, little is known about complexing of either UO_2^{2+} or U^{4+} with nitrogen-, phosphorus- or sulphur-containing ligands, but this may well be because few studies in non-aqueous media have yet been described. Within the framework of the class 'a' and class 'b' division of cations, however, those derived from uranium (and from the other actinides) are clearly in class 'a'.

Neptunium, Plutonium and Americium

These three elements show a close relationship, with the stability of high oxidation states decreasing as the atomic number increases. This factor is so important in the separation of the elements that it is desirable to begin this section by giving the potential diagrams for all three elements before describing individual compounds. Mention is made later of compounds containing neptunium (VII) and plutonium (VII) for which accurate data are not yet available.

We shall discuss disproportionation later; for the present it will suffice to point out that all the MO_2^{2+} ions are much less stable to reduction than UO_2^{2+} (though there is only a small difference between NpO_2^{2+} and PuO_2^{2+}), that Pu^{4+} is much more easily reduced than Np^{4+}, and that all higher oxidation states of americium are powerful oxidizing agents. Solvent-extraction and precipitation may then be used to effect separations from solutions of different oxidation potentials. The nitrates $MO_2(NO_3)_2$ may be extracted from nitrate solutions by diethyl ether or methyl isobutyl ketone; or the elements in the tetrapositive state may be extracted from 6M nitric acid solution with tributyl phosphate in kerosene. Only the tetra- and tripositive states can be precipitated (or coprecipitated) as fluorides; and coprecipitation with zirconium phosphate (or, rather surprisingly, bismuth phosphate) in acidic solution is almost specific for the tetrapositive states.

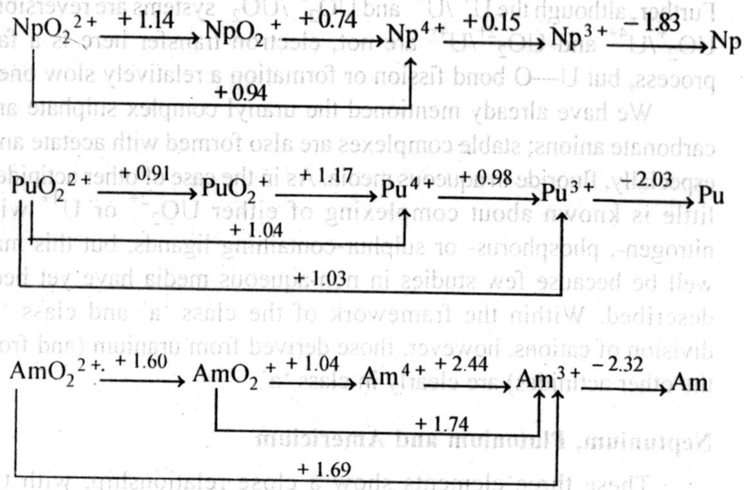

Fig. 28.4

In the preparation of ^{239}Pu (of half-life 2.4×10^4 years) in the pile by the sequence of reactions

the short half-life of ^{239}Np means that it is never present in large amounts, and its concentration decays rapidly after irradiation ceases; however, ^{237}Np (of half-life 2.2×10^6 years) is formed in small amounts by two successive (n,γ) reactions of ^{235}U and subsequent β-decay of ^{237}U. Dissolution of the irradiated uranium rods in nitric acid thus gives a solution containing UO_2^{2+}, NpO_2^{2+} and PuO_2^{2+} (as nitrates) and fission products. Solvent-extraction at this stage removes the actinides into the organic phase, from which neptunium and plutonium may be re-extracted by washing with an aqueous solution of a mild reducing agent, usually sulphur dioxide. Separation of neptunium and plutonium may be effected by bromate oxidation (which converts neptunium to NpO_2^+, but does not oxidize plutonium beyond Pu^{4+}) or hydrazine reduction (which converts Pu^{4+} to Pu^{3+}, but does not affect Np^{4+}), followed by precipitation of Pu^{4+} or Np^{4+}. Several other separation methods have been used, but all depend upon similar principles.

There are two isotopes of americium of half-life long enough for conventional studies, ^{241}Am (458 years) and ^{243}Am (7600 years). Both are obtained from ^{239}Pu by successive neutron captures to give ^{241}Pu and ^{243}Pu, which then undergo β-decay to give the americium isotopes. Separation of americium is readily achieved by oxidation of plutonium to PuO_2^{2+} and precipitation of AmF_3 or coprecipitation with LaF_3.

With these elements the effects of α-emission on the study of their chemistry become appreciable; there is, for example, considerable radiation-induced decomposition of water into hydrogen atoms and hydroxyl radicals, whose formation may lead to reduction of high oxidation states and production of hydrogen peroxide respectively. Thermal effects and radiation damage to solids become serious as half-lives decrease. Furthermore, the health hazards associated with working with strong α-emitters are very considerable unless rigid safety precautions are adopted; all of the work described in this and the following sections therefore has to be carried out in isolated systems.

The metals neptunium, plutonium and americium are all obtained by reduction of their fluorides with lithium or other electropositive metals at 1000—1200°. In general reactivity they resemble uranium, reacting readily with acids and with hydrogen. The stable oxides of all three elements are the dioxides, which are obtained by heating the nitrates or hydroxides of any oxidation states in air. Hydrated trioxides of neptunium and plutonium have been obtained by the action of ozone on the tetrahydroxides, but anhydrous trioxides are not known; Np_3O_8 is obtained by thermal decomposition of $(NH_4)_2Np_2O_7$, but Pu_3O_8 is less stable and has not been isolated.

A similar variation is found in the stabilities of halides. Neptunium and plutonium both form volatile hexafluorides closely resembling UF_6, but their preparation requires the use of fluorine at higher temperatures and the products are less stable with respect to decomposition into tetrafluoride and fluorine. Neptunium forms a tetrachloride and a tetrabromide, but neither compound can be isolated in the case of plutonium or americium, though chloro complexes of plutonium (IV), such as Cs_2PuCI_6, can be made from metal chlorides, plutonium trichloride and chlorine at 50°. The only halides, binary or complex, of americium (IV) are the tetrafluoride, obtained from the trifluoride and fluorine at 500°, and its complexes. A further interesting illustration of the decreasing stability of fluorine compounds of elements after uranium in high oxidation states is the action of NpF_6 and PuF_6 on sodium fluoride; elemental fluorine is liberated and complexes of neptunium (V) and plutonium (V) are formed. In anhydrous hydrogen fluoride, however, CsNpF6 disproportionates, forming Cs_2NpF_6 and NpF_6. In aqueous solutions, on the other hand, neptunium (V), present in NpO_2^+, is stable at pH 0 with respect to disproportionation, though the margin of stability is not large, K for the reaction

$$2NpO_2^+ + 4H^+ \rightarrow NpO_2^{2+} + Np^{4+} + 2H_2O$$

being 10^7. At higher acidities, NpO^{2+} becomes unstable, and in 8M acid K is about 200.

In the case of plutonium, successive potentials for the reduction of PuO_2^{2+} are so similar that PuO_2^{2+}, PuO_2^+, Pu^{4+} and Pu^{3+} are all present in M $HClO_4$ solution. Hydrolysis and complex formation also occur, and since kinetic factors are often involved too (the PuO_2^+/Pu^{4+} system, for example, is not reversible), the interpretation of equilibria and kinetics in plutonium-containing solutions is a matter of great difficulty.

The Later Actinide Elements

Much less is Known about the later elements curium, einsteinium, fermium, mendelevium, nobelium and lawrencium. This is due in part to their limited availability, and partly to their instability.

Table 28.3 : Halides of neptunium. plutonium and americium

+ III	NPF_3 purple-black	PuF_3	purple AmF_3 pink
	$NpCl_3$ white	$PuCl_3$ emerald green	$AmCl_3$ pink
	$NpBr_3$ green	$PuBr_3$ green	$AmBr_3$ white
	NpI_3 brown	PuI_3 brown	AmI_3 yellow
+ IV	NpF_4 green		
	$NpCl_4$ red-brown		
	$NpBr_4$ red-brown		
+ VI	NpF_6 brown	PuF_6 red-brown	

Interest in these elements is largely concerned with showing that the second half of the actinides resemble the lanthanides quite closely. In spite of similarities, the actinides can be separated from the lanthanides quite easily, as the actinides form complexes more readily. For example, with concentrated HCl the actinides form chloro complexes. If both groups of ions are adsorbed on a cation exchange column, the actinides can be eluted with concentrated HCl. The actinide ions are separated from each other by ion exchange using citrate solutions to elute them. The order in which the actinides are eluted shows a close similarity to the order in which the lanthanides are eluted.

The (+III) oxidation state is the most stable for all but one of the elements with atomic numbers 96-103, i.e. Cm. Bk, Cf, Es, Fm. Md, (No) and Lw. The exception is No, which is most stable as No(+II). This is stable because it has a favourable f^{14} electronic configuration.

There is evidence of (+II) states for the elements 98-102. i.e. californium to nobelium. Except for nobelium this state is reducing or strongly reducing in nature. Higher oxidation states of Cm(+ IV) are found in the solid but not in solution with compounds such as CmF_4 and $Rb[CmF_6]$. Bk(+IV) compounds are oxidizing. They exist in both the solid and in solution, and BkO_2 and $Cs_2[BkCl_6]$ have been isolated. Lawrencium exists only in the (+III) state and resists both oxidation and reduction, again illustrating the stability of an f^{14} electronic arrangement.

The first three elements curium, berkelium and californium have been obtained in milligram quantities. Their chemistry has been studied by normal small scale methods, and compounds have been isolated. The remaining elements have been studied by radioactive tracer methods, because they are only available in such minute amounts. In this technique the traces of heavy elements are precipitated, or form complexes, in the presence of major quantities of a carrier element which has similar chemical properties. Thus mendelevium is studied using the lanthanide element europium, and the mendelevium is detected and followed by its radioactivity.

□□□